Trigonometry

TENTH EDITION

Trigonometry

TENTH EDITION

Margaret L. Lial
American River College

John Hornsby
University of New Orleans

David I. Schneider
University of Maryland

Callie J. Daniels
St. Charles Community College

PEARSON

Boston Columbus Indianapolis New York San Francisco Upper Saddle River
Amsterdam Cape Town Dubai London Madrid Milan Munich Paris Montréal Toronto
Delhi Mexico City São Paulo Sydney Hong Kong Seoul Singapore Taipei Tokyo

Editor-in-Chief: Anne Kelly
Executive Content Editor: Christine O'Brien
Editorial Assistant: Judith Garber
Senior Managing Editor: Karen Wernholm
Senior Production Project Manager: Kathleen A. Manley
Digital Assets Manager: Marianne Groth
Associate Media Producer: Stephanie Green
Software Development: Kristina Evans, MathXL; Mary Durnwald, TestGen
Marketing Manager: Peggy Sue Lucas
Marketing Assistant: Justine Goulart
Senior Author Support/Technology Specialist: Joe Vetere
Rights and Permissions Advisor: Michael Joyce
Image Manager: Rachel Youdelman
Procurement Manager: Evelyn Beaton
Procurement Specialist: Debbie Rossi
Media Procurement Specialist: Ginny Michaud
Associate Director of Design: Andrea Nix
Senior Designer: Beth Paquin
Text Design, Production Coordination, Composition: Cenveo Publisher Services/
Nesbitt Graphics, Inc.
Illustrations: Cenveo Publisher Services/Nesbitt Graphics, Inc., Laserwords, Network Graphics
Cover Design: Jenny Willingham
Cover Image: Andriy Dykun/Fotolia

Library of Congress Cataloging-in-Publication Data

Trigonometry/Margaret L. Lial ... [et al.]. — 10th ed.
p. cm.
Previous ed. by Margaret L. Lial, John Hornsby, and David Schneider.
Includes index.
ISBN-13: 978-0-321-67177-6
ISBN-10: 0-321-67177-5
1. Trigonometry. I. Lial, Margaret L. Trigonometry. II. Title.
QA531.L5 2013
516.24—dc23 2011013314

3 4 5 6 7 8 9 10—CRK—16 15 14

PEARSON

www.PearsonSchool.com/Advanced

ISBN-10: 0-13-283258-5 (High School Binding)
ISBN-13: 978-0-13-283258-8 (High School Binding)

To my friend Joe Long, with thanks from all of your fans for the music and the memories—you are our favorite Season

E.J.H.

To my parents, James and Patricia Harmon

C.J.D.

Contents

Preface

WELCOME TO THE 10TH EDITION

As authors, we have called upon our classroom experiences, use of MyMathLab, suggestions from users and reviewers, and many years of writing to provide tools that will support learning and teaching. This new edition of *Trigonometry* continues our effort to provide a sound pedagogical approach through logical development of the subject matter. This approach forms the basis for all of the Lial team's instructional materials available from Pearson Education, in both print and technology forms.

Our goal is to produce a textbook that will be an integral component of the student's experience in learning trigonometry. With this in mind, we have provided a textbook that students can read more easily, which is often a difficult task, given the nature of mathematical language. We have also improved page layouts for better flow, provided additional side comments, and updated many figures.

We realize that today's classroom experience is evolving and that technology-based teaching and learning aids have become essential to address the ever-changing needs of teachers and students. As a result, we've worked to provide support for all classroom types—traditional, hybrid, and online. In the 10th edition, text and online materials are more tightly integrated than ever before. This enhances flexibility and ease of use for teachers and increases success for students. See pages xvii–xix for descriptions of these materials.

NEW TO THE 10TH EDITION

- In **Chapter 1** we begin our effort in using more side comments in Examples and providing better pairing of even and odd exercises in our exercise sets. In Section 1.4, we have provided a better visual in the figure (**Figure 34**) accompanying the explanation of the ranges of the sine and cosine functions (ratios) in conjunction with right triangles.

- In **Chapter 2** we have added new exercises for evaluating trigonometric expressions with function values of special angles (Section 2.2, Exercises 45–52). We have updated the discussion on using inverse trigonometric functions to find angle measures using a calculator (Section 2.3 throughout), and we have updated examples of solving right triangles to prepare students for the more challenging exercises in the Exercise sets (Sections 2.4 and 2.5).

- In **Chapter 3** we have added a figure (**Figure 2**) clarifying the concept of radian measure. In Section 3.2 there are new exercises involving application of the formula for the area of a circle (Exercises 61–64). Section 3.3 now includes an expanded explanation of what values a calculator returns for inverse trigonometric functions. We have added a new figure in Example 4 explaining how to use the concept of inverse functions. There is a new subhead "Expressing Function Values as Lengths of Line Segments" along with a new example (Example 6) and exercises (Exercises 87 and 88).

- In **Chapter 4** we have an updated figure and discussion relating the sine function to the unit circle (Section 4.1, **Figure 4**, and discussion). We have expanded the discussion of sketching graphs of translated trigonometric functions and have updated the guidelines for these sketches (Section 4.2). We have included new examples of connecting graphs with equations (Section 4.1, Example 6 and Section 4.3, Example 6), and have included new exercises of this type in the exercise sets and chapter review exercises.

- In **Chapter 5** we have rewritten the solution of Example 4 in Section 5.3 to illustrate how this standard type of problem can be solved by either using the Pythagorean identities or using angles in standard position. In Exercises 73–84 of Section 5.6, we now discuss the rarely-studied exact function values of 18° and 72° angles.
- In **Chapter 6** we have updated the discussion on finding inverse function values using a calculator (Section 6.1). We have included updated examples and many more exercises in which trigonometric equations are solved for *all solutions* in both degrees and radians. Several new figures are included relating solutions of trigonometric equations with angle measures and arc lengths on the unit circle (Sections 6.2 and 6.3). For equations involving inverse trigonometric functions that are solved for a specified variable, restrictions are now given so that each equation provides a one-to-one correspondence, and a new figure is given to provide conceptual understanding (Section 6.4).
- In **Chapter 7** we have updated and improved many of the illustrations in the examples and exercises. In Section 7.4 the new **Figures 35** and **37** illustrate operations with vectors geometrically, and we have also included the justification for the geometric interpretation of the dot product. Section 7.5 now includes a new example (Example 5) illustrating vectors applied to a navigation problem. This type of problem is always troublesome for students.
- In **Chapter 8** we have rewritten the introduction to the set of complex numbers and have included a new diagram illustrating the relationships among its subsets. Polar graphs now include underlying grids for easier placement of polar coordinates.
- For visual learners, numbered **Figure** and **Example** references within the text are set using the same typeface as the figure and bold print for the example. This makes it easier for the students to identify and connect them. We also have increased our use of a "drop down" style, when appropriate, to distinguish between simplifying expressions and solving equations, and we have added many more explanatory side comments. Interactive figures with accompanying exercises and explorations are now available and assignable in MyMathLab.
- Enhancing the already well-respected exercises, hundreds are new or modified, and many present updated real-life data. In addition, the MyMathLab course has expanded coverage of all exercise types appearing in the exercise sets, as well as the mid-chapter Quizzes and Summary Exercises.

FEATURES OF THIS TEXT

SUPPORT FOR LEARNING CONCEPTS

We provide a variety of features to support students' learning of the essential topics of trigonometry. Explanations that are written in understandable terms, figures and graphs that illustrate examples and concepts, graphing technology that supports and enhances algebraic manipulations, and real-life applications that enrich the topics with meaning all provide opportunities for students to deepen their understanding of mathematics. These features help students make mathematical connections and expand their own knowledge base.

- **Examples** Numbered examples that illustrate the techniques for working exercises are found in every section. We use traditional explanations, side comments, and pointers to describe the steps taken—and to warn students about common pitfalls. Some examples provide additional graphing calculator solutions, although these can be omitted if desired.

- **Now Try Exercises** Following each numbered example, the student is directed to try a corresponding odd-numbered exercise (or exercises). This feature allows for quick feedback to determine whether the student has understood the principles illustrated in the example.

- **Real-Life Applications** We have included hundreds of real-life applications, many with data updated from the previous edition. They come from fields such as astronomy, meteorology, environmental studies, construction, biology and life sciences, music, and physics.

- **Function Boxes** Beginning in Chapter 4, special function boxes (for example, see page 135) offer a comprehensive, visual introduction to each type of function and also serve as an excellent resource for reference and review. Each function box includes a table of values, traditional and calculator-generated graphs, the domain, the range, and other special information about the function. These boxes are now assignable in MyMathLab.

- **Figures and Photos** Today's students are more visually oriented than ever before, and we have updated the figures in this edition to a greater extent than in our previous few editions. Interactive figures with accompanying exercises and explorations are now available and assignable in MyMathLab.

- **Use of Graphing Technology** We have integrated the use of graphing calculators where appropriate, although ***this technology is completely optional and can be omitted without loss of continuity.*** We continue to stress that graphing calculators support understanding but that students must first master the underlying mathematical concepts. Exercises that require their use are marked with an icon.

- **Cautions and Notes** Text that is marked **CAUTION** warns students of common errors, and NOTE comments point out explanations that should receive particular attention.

- **Looking Ahead to Calculus** These margin notes offer glimpses of how the topics currently being studied are used in calculus.

SUPPORT FOR PRACTICING CONCEPTS

This text offers a wide variety of exercises to help students master trigonometry. The extensive exercise sets provide ample opportunity for practice, and the exercise problems increase in difficulty so that students at every level of understanding are challenged. The variety of exercise types promotes understanding of the concepts and reduces the need for rote memorization.

- **Exercise Sets** We have revised many drill and application exercises for better pairing of corresponding even and odd exercises, and answers to the odd exercises are provided in the Student Edition. In addition to these, we include writing exercises, optional graphing calculator problems, and multiple-choice, matching, true/false, and completion exercises. Those marked *Concept Check* focus on conceptual thinking. *Connecting Graphs with Equations* exercises challenge students to write equations that correspond to given graphs. Finally, MyMathLab offers Pencast solutions for selected Connecting Graphs with Equations problems.

- **Relating Concepts Exercises** Appearing in selected exercise sets, these groups of exercises are designed so that students who work them in numerical order will follow a line of reasoning that leads to an understanding of how various topics and concepts are related. All answers to these exercises appear in the student answer section, and these exercises are now assignable in MyMathLab.

- **Complete Solutions to Selected Exercises** Exercise numbers marked indicate that a full worked-out solution appears at the back of the text. These are often exercises that extend the skills and concepts presented in the numbered examples.

SUPPORT FOR REVIEW AND TEST PREP

Ample opportunities for review are found within the chapters and at the ends of chapters. Quizzes that are interspersed within chapters provide a quick assessment of students' understanding of the material presented up to that point in the chapter. Chapter "Test Preps" provide comprehensive study aids to help students prepare for tests.

- **Quizzes** Students can periodically check their progress with in-chapter quizzes that appear in all chapters. All answers, with corresponding section references, appear in the student answer section. These quizzes are now assignable in MyMathLab.

- **Summary Exercises** These sets of in-chapter exercises give students the all-important opportunity to work *mixed* review exercises, requiring them to synthesize concepts and select appropriate solution methods. The summary exercises are now assignable in MyMathLab.

- **End-of-Chapter Test Prep** Following the final numbered section in each chapter, the Test Prep provides a list of **Key Terms,** a list of **New Symbols** (if applicable), and a two-column **Quick Review** that includes a section-by-section summary of concepts and examples. This feature concludes with a comprehensive set of **Review Exercises** and a **Chapter Test.** The Test Prep, Review Exercises, and Chapter Test are assignable in MyMathLab.

- **Glossary** A comprehensive glossary of important terms drawn from the entire book follows Appendix D.

Student Supplements

Teacher Supplements

Supplements available in MyMathLab require an access code, which is available for purchase. See page xviii for more information.

Available for Purchase:

Student's Solutions Manual

- Provides detailed solutions to all odd-numbered text exercises

Video Lectures with Optional Captioning

- Available in MyMathLab®
- Feature Quick Reviews and Example Solutions:
  Quick Reviews cover key definitions and procedures from each section.
  Example Solutions walk students through the detailed solution process for every example in the textbook.
- Ideal for distance learning or supplemental instruction at home
- Include optional text captioning

Additional Skill and Drill Manual

- Provides additional practice and test preparation for students

MyNotes

- Available in MyMathLab
- Provide structure for student reading and understanding of the textbook
- Include textbook examples along with ample space for students to write solutions and notes
- Include key concepts along with prompts for students to read, write, and reflect on what they have just learned
- Customizable so that teachers can add their own examples or remove examples that are not covered in their courses

MyClassroomExamples

- Available in MyMathLab
- Offer structure for classroom lecture
- Include Classroom Examples along with ample space for students to write solutions and notes
- Include key concepts along with fill in the blank opportunities to keep students engaged
- Customizable so that teachers can add their own examples or remove Classroom Examples that are not covered in their courses

Most of the teacher supplements and resources for this text are available electronically to qualified adopters on the Instructor Resource Center (IRC). Upon adoption or to preview, please go to www.PearsonSchool.com/Access_Request and select Instructor Resource Center. You will be required to complete a brief one-time registration subject to verification of educator status. Upon verification, access information and instructions will be sent to you via email. Once logged into the IRC, enter your text ISBN in the **Search our Catalog** box to locate your resources.

Annotated Instructor's Edition

- Provides answers in the margins to almost all text exercises, as well as helpful Teaching Tips and Classroom Examples
- Includes sample homework assignments indicated by problem numbers underlined in blue within each end-of-section exercise set
- Sample homework problems assignable in MyMathLab

Online Instructor's Solutions Manual

- Provides complete solutions to all text exercises
- Available in MyMathLab or downloadable from the Instructor Resource Center

Online Instructor's Testing Manual

- Includes diagnostic pretests, chapter tests, final exams, and additional test items, grouped by section, with answers provided
- Available in MyMathLab or downloadable from the Instructor Resource Center

TestGen® CD-ROM

- Enables teachers to build, edit, print, and administer tests
- Features a computerized bank of questions developed to cover all text objectives

Online PowerPoint Presentation, Active Learning Questions, and Classroom Example PowerPoints

- Written and designed specifically for this text
- Include figures and examples from the text
- Provide active learning questions for use with classroom response systems, including multiple-choice questions to review lecture material (available in MyMathLab only)
- Provide Classroom Example PowerPoints that include full worked-out solutions to all Classroom Examples
- Available in MyMathLab or downloadable from the Instructor Resource Center

MEDIA RESOURCES

MathXL® for School (optional, for purchase only)—access code required, www.mathxlforschool.com

Included with purchase of Student Edition: Access to MathXL for School for 1 year (access after 1 year is available for purchase)

MathXL for School is a powerful online homework, tutorial, and assessment supplement that aligns to Pearson Education's textbooks in mathematics or statistics. With MathXL for School, teachers can:

- Create, edit, and assign auto-graded online homework and tests correlated at the objective level to the textbook
- Utilize automatic grading to rapidly assess student understanding
- Track both student and group performance in an online gradebook
- Prepare students for high-stakes testing, including aligning assignments to state and Common Core State Standards, where available
- Deliver quality, effective instruction regardless of experience level

With MathXL for School, students can:

- Do their homework and receive immediate feedback
- Get self-paced assistance on problems in a variety of ways (guided solutions, step-by-step examples, video clips, animations)
- Have a large number of practice problems to choose from, helping them master a topic
- Receive personalized study plans and homework based on test results

For more information and to purchase student access codes after the first year, visit our Web site at www.mathxlforschool com, or contact your Pearson School account executive.

MyMathLab® Online Course (optional, for purchase only)—access code required, www.mymathlab.com

MyMathLab is a text-specific, easily customizable, online course that integrates interactive multimedia instruction with textbook content. MyMathLab gives you the tools you need to deliver all or a portion of your course online.

MyMathLab features include:

- Interactive eText, including highlighting and note taking tools, and links to videos and exercises
- Rich and flexible course management, communication, and teacher support tools
- Online homework and assessment, and personalized study plans
- Complete multimedia library to enhance learning
- All teacher resources in one convenient location

For more information, visit www.mymathlab.com or contact your Pearson School account executive.

MyMathLab® Ready to Go Course (access code required, available for purchase)

These new Ready to Go courses provide students with all the same great MyMathLab features that they are used to, but the courses make it easier for teachers to get started. Each course includes preassigned homework and quizzes to make creating a course even simpler. Ask your Pearson representative about the details for this particular course, or request a copy of the course.

ACKNOWLEDGMENTS

We wish to thank the following individuals who provided valuable input into this edition of the text.

Mark Burtch – Austin Community College – Rio Grande
Tilak De Alwis – Southeastern Louisiana University
Richard G. Goldthwait – Youngstown State University
Rene Lumampao – Austin Community College – Rio Grande
Lucia Riderer – Citrus College
Mayada Shahrokhi – Lonestar College – CyFair
Magdalena Toda – Texas Tech University

Our sincere thanks to those individuals at Pearson Education who have supported us throughout this revision: Greg Tobin, Anne Kelly, Kathy Manley, and Christine O'Brien. Terry McGinnis continues to provide behind-the-scenes guidance for both content and production. We have come to rely on her expertise during all phases of the revision process. Marilyn Dwyer of Nesbitt Graphics, Inc., with the assistance of Carol Merrigan, provided excellent production work. Special thanks go out to Abby Tanenbaum for updating data in applications, and to Chris Heeren and Paul Lorczak for their excellent accuracy-checking. We thank Lucie Haskins, who provided an accurate index, and Becky Troutman, who revised our Index of Applications. A special thank you goes out to Dr. Mohammed Alaimia, KFUPM, Saudi Arabia, who provided assistance in revising many of the exercises in Chapter 6.

As an author team, we are committed to providing the best possible trigonometry course to help instructors teach and students succeed. As we continue to work toward this goal, we welcome any comments or suggestions you might send, via e-mail, to math@pearson.com.

Margaret L. Lial
John Hornsby
David I. Schneider
Callie J. Daniels

1 Trigonometric Functions

A sequence of *similar triangles,* a topic covered in this introductory chapter, can be used to approximate the spiral of the *chambered nautilus*.

1.1 Angles

- Basic Terminology
- Degree Measure
- Standard Position
- Coterminal Angles

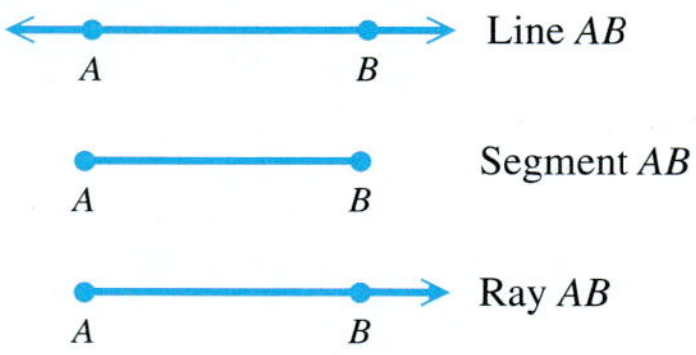

Figure 1

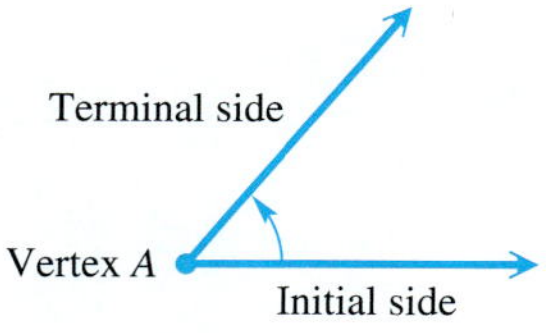

Figure 2

Basic Terminology

Two distinct points A and B determine a line called **line *AB*.** The portion of the line between A and B, including points A and B themselves, is **line segment *AB*,** or simply **segment *AB*.** The portion of line AB that starts at A and continues through B, and on past B, is the **ray *AB*.** Point A is the **endpoint of the ray.** See **Figure 1.**

In trigonometry, an **angle** consists of two rays in a plane with a common endpoint, or two line segments with a common endpoint. These two rays (or segments) are the **sides** of the angle, and the common endpoint is the **vertex** of the angle. Associated with an angle is its measure, generated by a rotation about the vertex. See **Figure 2.** This measure is determined by rotating a ray starting at one side of the angle, the **initial side,** to the position of the other side, the **terminal side.** ***A counterclockwise rotation generates a positive measure, and a clockwise rotation generates a negative measure.*** The rotation can consist of more than one complete revolution.

Figure 3 shows two angles, one **positive** and one **negative.**

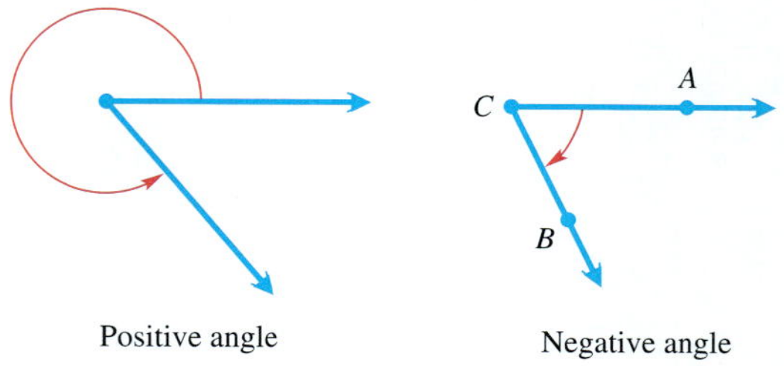

Figure 3

An angle can be named by using the name of its vertex. For example, the angle on the right in **Figure 3** can be named angle C. Alternatively, an angle can be named using three letters, with the vertex letter in the middle. Thus, the angle on the right also could be named angle ACB or angle BCA.

Degree Measure

The most common unit for measuring angles is the **degree.** Degree measure was developed by the Babylonians 4000 yr ago. To use degree measure, we assign 360 degrees to a complete rotation of a ray.* In **Figure 4,** notice that the terminal side of the angle corresponds to its initial side when it makes a complete rotation.

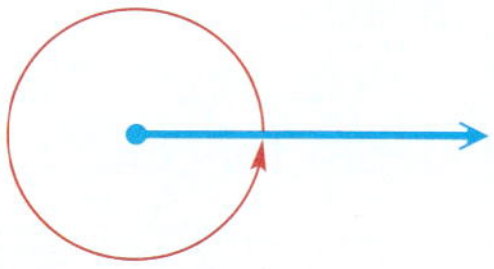

A complete rotation of a ray gives an angle whose measure is 360°. $\frac{1}{360}$ of a complete rotation gives an angle whose measure is 1°.

Figure 4

$$\text{One degree, written } 1^\circ\text{, represents } \frac{1}{360} \text{ of a rotation.}$$

Therefore, 90° represents $\frac{90}{360} = \frac{1}{4}$ of a complete rotation, and 180° represents $\frac{180}{360} = \frac{1}{2}$ of a complete rotation.

An angle measuring between 0° and 90° is an **acute angle.** An angle measuring exactly 90° is a **right angle.** The symbol ⌝ is often used at the vertex of a right angle to denote the 90° measure. An angle measuring more than 90° but less than 180° is an **obtuse angle,** and an angle of exactly 180° is a **straight angle.**

*The Babylonians were the first to subdivide the circumference of a circle into 360 parts. There are various theories about why the number 360 was chosen. One is that it is approximately the number of days in a year, and it has many divisors, which makes it convenient to work with.

In **Figure 5,** we use the **Greek letter θ (theta)*** to name each angle.

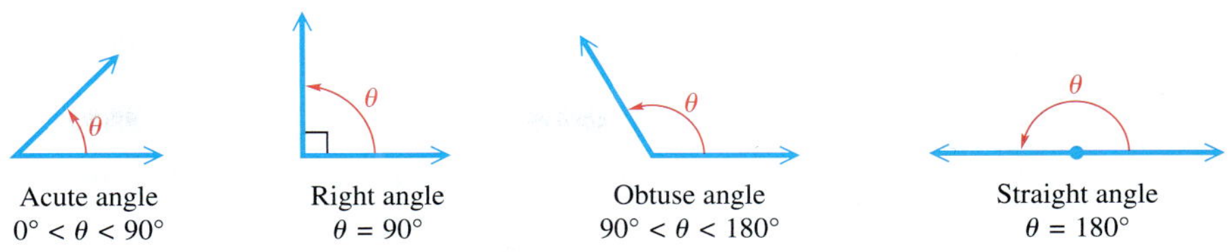

Figure 5

If the sum of the measures of two positive angles is 90°, the angles are **complementary** and the angles are **complements** of each other. Two positive angles with measures whose sum is 180° are **supplementary,** and the angles are **supplements.**

EXAMPLE 1 Finding the Complement and the Supplement of an Angle

For an angle measuring 40°, find the measure of **(a)** its complement and **(b)** its supplement.

SOLUTION

(a) To find the measure of its complement, subtract the measure of the angle from 90°.

$$90° - 40° = 50° \quad \text{Complement of } 40°$$

(b) To find the measure of its supplement, subtract the measure of the angle from 180°.

$$180° - 40° = 140° \quad \text{Supplement of } 40°$$

✔ *Now Try Exercise 1.*

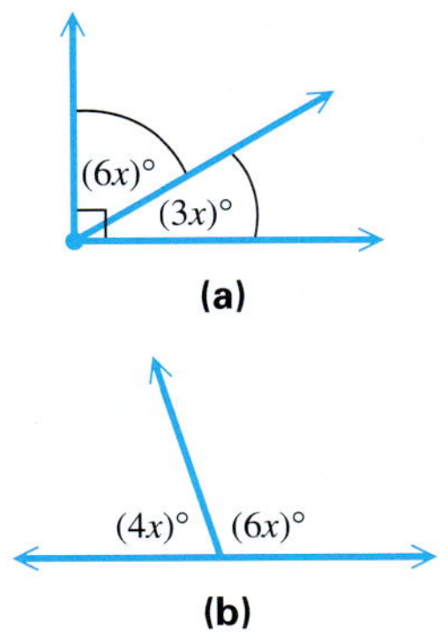

Figure 6

EXAMPLE 2 Finding Measures of Complementary and Supplementary Angles

Find the measure of each marked angle in **Figure 6.**

SOLUTION

(a) Since the two angles in **Figure 6(a)** form a right angle, they are complementary angles.

$$6x + 3x = 90 \quad \text{Complementary angles sum to } 90°.$$

$$9x = 90 \quad \text{Combine like terms.}$$

Don't stop here. $x = 10$ Divide by 9. (Appendix A)

Be sure to determine the measure of each angle by substituting 10 for x. The two angles have measures of $6(10) = 60°$ and $3(10) = 30°$.

(b) The angles in **Figure 6(b)** are supplementary, so their sum must be 180°.

$$4x + 6x = 180 \quad \text{Supplementary angles sum to } 180°.$$

$$10x = 180 \quad \text{Combine like terms.}$$

$$x = 18 \quad \text{Divide by 10.}$$

These angle measures are $4(18) = 72°$ and $6(18) = 108°$.

✔ *Now Try Exercises 13 and 15.*

*In addition to θ (theta), other Greek letters such as α (alpha) and β (beta) are often used.

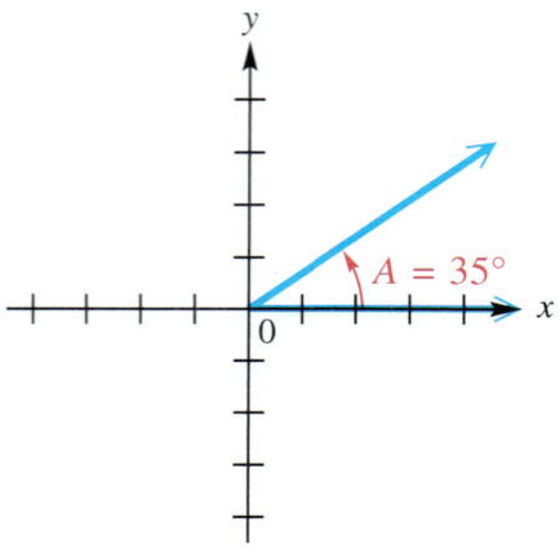

Figure 7

The measure of angle A in **Figure 7** is 35°. This measure is often expressed by saying that $m(\text{angle } A)$ is 35°, where $m(\text{angle } A)$ is read **"the measure of angle A."** It is convenient, however, to abbreviate the symbolism $m(\text{angle } A) = 35°$ as $A = 35°$.

Traditionally, portions of a degree have been measured with minutes and seconds. One **minute,** written **1′**, is $\frac{1}{60}$ of a degree.

$$1' = \left(\frac{1}{60}\right)^\circ \quad \text{or} \quad 60' = 1^\circ$$

One **second, 1″**, is $\frac{1}{60}$ of a minute.

$$1'' = \left(\frac{1}{60}\right)' = \left(\frac{1}{3600}\right)^\circ \quad \text{or} \quad 60'' = 1'$$

The measure $12^\circ\, 42'\, 38''$ represents 12 degrees, 42 minutes, 38 seconds.

EXAMPLE 3 Calculating with Degrees, Minutes, and Seconds

Perform each calculation.

(a) $51^\circ\, 29' + 32^\circ\, 46'$ **(b)** $90^\circ - 73^\circ\, 12'$

SOLUTION

(a)
$$\begin{array}{r} 51^\circ\, 29' \\ +\ 32^\circ\, 46' \\ \hline 83^\circ\, 75' \end{array}$$
Add degrees and minutes separately.

The sum $83^\circ\, 75'$ can be rewritten as follows.

$$83^\circ\, 75' = 83^\circ + 1^\circ\, 15' \qquad 75' = 60' + 15' = 1^\circ\, 15'$$
$$= 84^\circ\, 15' \qquad \text{Add.}$$

(b)
$$\begin{array}{r} 89^\circ\, 60' \\ -\ 73^\circ\, 12' \\ \hline 16^\circ\, 48' \end{array}$$
Write 90° as 89° 60′.

✔ *Now Try Exercises 37 and 41.*

Because calculators are so prevalent, angles are commonly measured in decimal degrees. For example, 12.4238° represents

$$12.4238^\circ = 12\frac{4238}{10{,}000}^\circ.$$

EXAMPLE 4 Converting between Decimal Degrees and Degrees, Minutes, and Seconds

(a) Convert $74^\circ\, 08'\, 14''$ to decimal degrees to the nearest thousandth.

(b) Convert 34.817° to degrees, minutes, and seconds to the nearest second.

SOLUTION

(a) $$74^\circ\, 08'\, 14'' = 74^\circ + \left(\frac{8}{60}\right)^\circ + \left(\frac{14}{3600}\right)^\circ \qquad 1' = \left(\frac{1}{60}\right)^\circ \text{ and } 1'' = \left(\frac{1}{3600}\right)^\circ$$

$$\approx 74^\circ + 0.1333^\circ - 0.0039^\circ$$

$$\approx 74.137^\circ \qquad \text{Add and round to the nearest thousandth.}$$

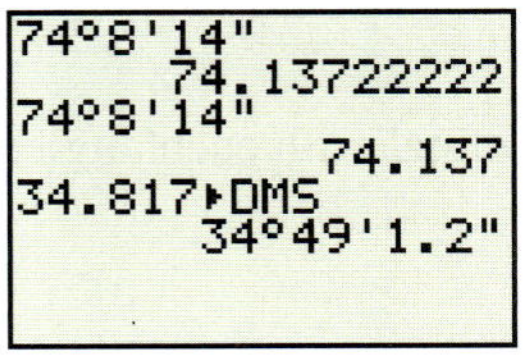

A graphing calculator performs the conversions in **Example 4** as shown above. The ▶DMS option is found in the ANGLE Menu of the TI-83/84 Plus calculator.

(b)

$34.817° = 34° + 0.817°$	Write as a sum.
$= 34° + 0.817(60')$	$1° = 60'$
$= 34° + 49.02'$	Multiply.
$= 34° + 49' + 0.02'$	Write as a sum.
$= 34° + 49' + 0.02(60'')$	$1' = 60''$
$= 34° + 49' + 1.2''$	Write as a sum.
$\approx 34°\,49'\,01''$	Approximate to the nearest second.

✔ ***Now Try Exercises 53 and 63.***

Standard Position An angle is in **standard position** if its vertex is at the origin and its initial side lies on the positive x-axis. The angles in **Figures 8(a) and 8(b)** are in standard position. An angle in standard position is said to lie in the quadrant in which its terminal side lies. An acute angle is in quadrant I (**Figure 8(a)**) and an obtuse angle is in quadrant II (**Figure 8(b)**). **Figure 8(c)** shows ranges of angle measures for each quadrant when $0° < \theta < 360°$.

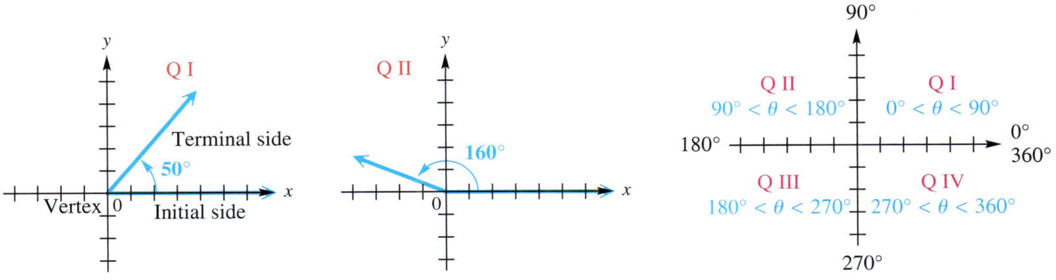

(a) **(b)** **(c)**

Figure 8

Quadrantal Angles

Angles in standard position whose terminal sides lie on the x-axis or y-axis, such as angles with measures 90°, 180°, 270°, and so on, are **quadrantal angles.**

Coterminal Angles A complete rotation of a ray results in an angle measuring 360°. By continuing the rotation, angles of measure larger than 360° can be produced. The angles in **Figure 9** with measures 60° and 420° have the same initial side and the same terminal side, but different amounts of rotation. Such angles are **coterminal angles.** ***Their measures differ by a multiple of* 360°.** As shown in **Figure 10,** angles with measures 110° and 830° are coterminal.

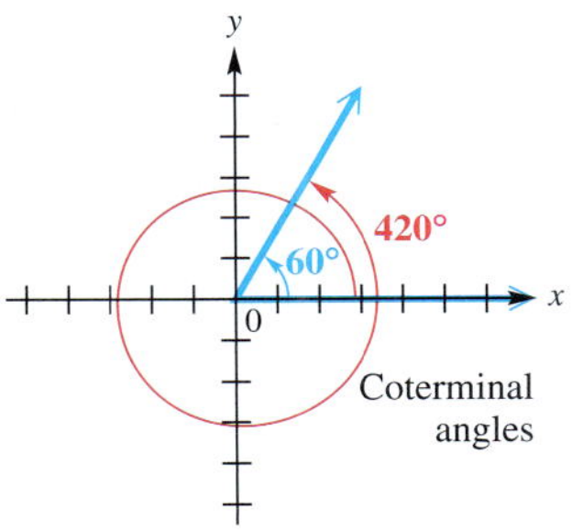

Figure 9

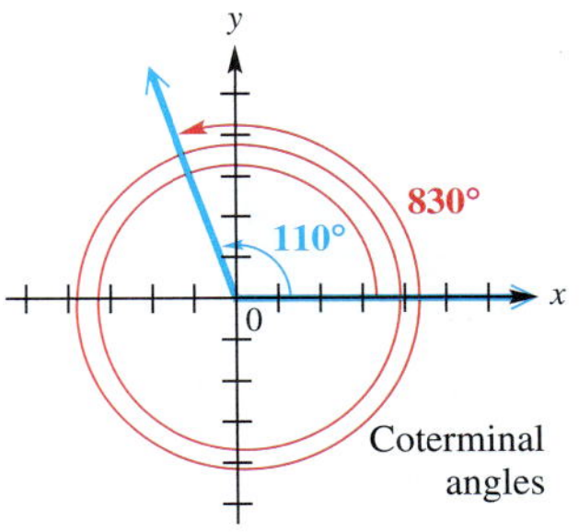

Figure 10

EXAMPLE 5 Finding Measures of Coterminal Angles

Find the angles of least positive measure that are coterminal with each angle.

(a) 908° **(b)** −75° **(c)** −800°

SOLUTION

(a) Subtract 360° as many times as needed to obtain an angle with measure greater than 0° but less than 360°. Since

$$908^\circ - 2 \cdot 360^\circ = 188^\circ,$$

an angle of 188° is coterminal with an angle of 908°. See **Figure 11.**

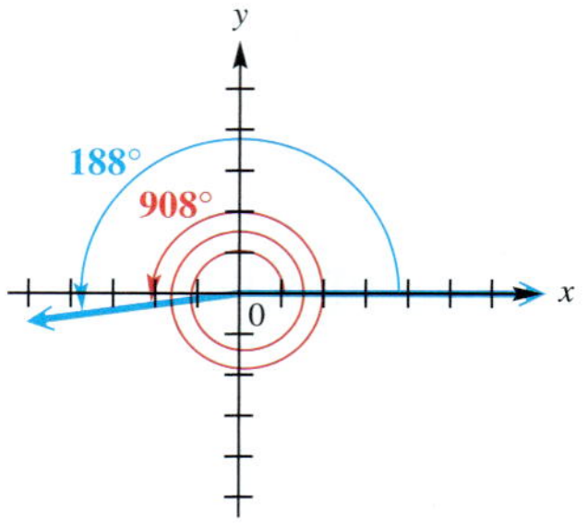

Figure 11

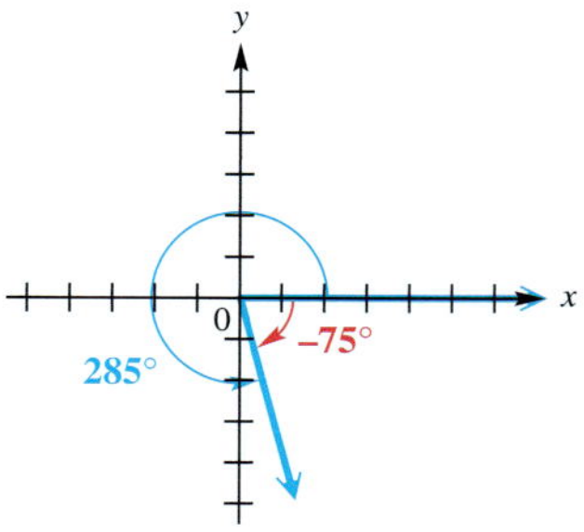

Figure 12

(b) See **Figure 12.** Use a rotation of

$$360^\circ - (-75^\circ) = 285^\circ.$$

(c) The least integer multiple of 360° greater than 800° is

$$360^\circ \cdot 3 = 1080^\circ.$$

Add 1080° to −800° to obtain

$$1080^\circ + (-800^\circ) = 280^\circ.$$

✔ ***Now Try Exercises 77, 87, and 91.***

Sometimes it is necessary to find an expression that will generate all angles coterminal with a given angle. For example, we can obtain any angle coterminal with 60° by adding an integer multiple of 360° to 60°. Let n represent any integer. Then the following expression represents all such coterminal angles.

$$60^\circ + \boldsymbol{n \cdot 360^\circ} \quad \text{Angles coterminal with } 60^\circ$$

The table below shows a few possibilities.

Value of n	Angle Coterminal with 60°
2	$60^\circ + 2 \cdot 360^\circ = 780^\circ$
1	$60^\circ + 1 \cdot 360^\circ = 420^\circ$
0	$60^\circ + 0 \cdot 360^\circ = 60^\circ$ (the angle itself)
−1	$60^\circ + (-1) \cdot 360^\circ = -300^\circ$

The table in the margin shows some examples of coterminal quadrantal angles.

Examples of Coterminal Quadrantal Angles

Quadrantal Angle θ	Coterminal with θ
0°	±360°, ±720°
90°	−630°, −270°, 450°
180°	−180°, 540°, 900°
270°	−450°, −90°, 630°

EXAMPLE 6 Analyzing the Revolutions of a Disk Drive

A constant angular velocity disk drive spins a disk at a constant speed. Suppose a disk makes 480 revolutions per min. Through how many degrees will a point on the edge of the disk move in 2 sec?

SOLUTION The disk revolves 480 times in 1 min, or $\frac{480}{60}$ times $= 8$ times per sec (since $60 \text{ sec} = 1 \text{ min}$). In 2 sec, the disk will revolve $2 \cdot 8 = 16$ times. Each revolution is 360°, so in 2 sec a point on the edge of the disk will revolve

$$16 \cdot 360° = 5760°.$$

A unit analysis expression can also be used.

$$\frac{480 \text{ rev}}{1 \text{ min}} \times \frac{1 \text{ min}}{60 \text{ sec}} \times \frac{360°}{1 \text{ rev}} \times 2 \text{ sec} = 5760° \quad \text{Divide out common units.}$$

✔ *Now Try Exercise 131.*

1.1 Exercises

Find (***a***) *the complement and* (***b***) *the supplement of an angle with the given measure.* ***See Examples 1 and 3.***

1. 30° **2.** 60° **3.** 45° **4.** 18°

5. 54° **6.** 89° **7.** 1° **8.** 10°

9. 14° 20′ **10.** 39° 50′ **11.** 20° 10′ 30″ **12.** 50° 40′ 50″

Find the measure of each unknown angle in Exercises 13–22. ***See Example 2.***

13.
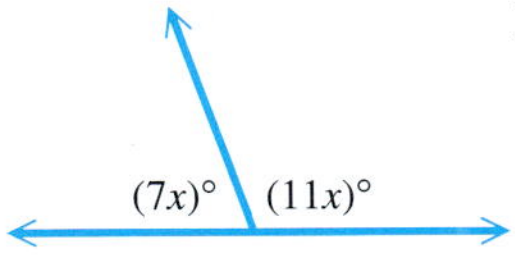

14.
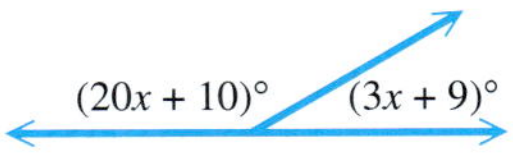

15.
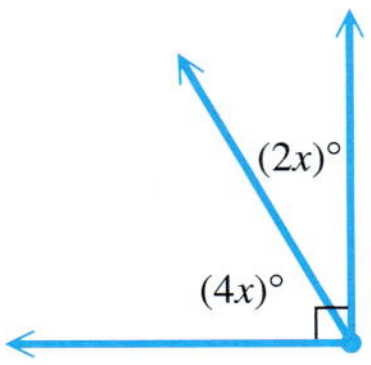

16.
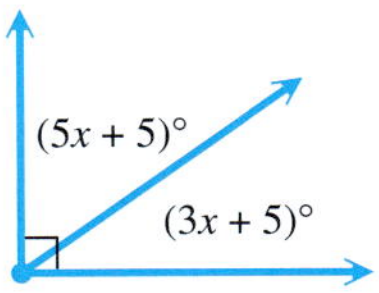

17.
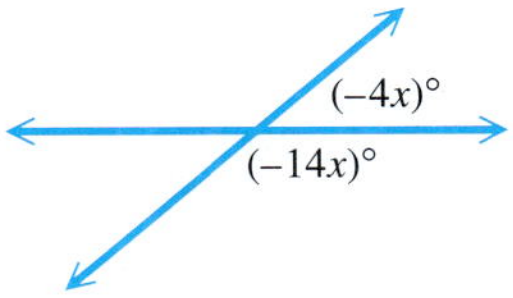

18.
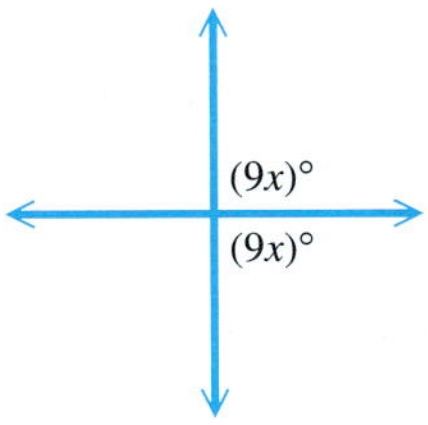

19. supplementary angles with measures $10x + 7$ and $7x + 3$ degrees

20. supplementary angles with measures $6x - 4$ and $8x - 12$ degrees

21. complementary angles with measures $9x + 6$ and $3x$ degrees

22. complementary angles with measures $3x - 5$ and $6x - 40$ degrees

23. *Concept Check* What is the measure of an angle that is its own complement?

24. *Concept Check* What is the measure of an angle that is its own supplement?

Find the measure of the smaller angle formed by the hands of a clock at the following times.

25.

26.

27. 3:15 **28.** 9:45 **29.** 8:20 **30.** 6:10

Concept Check *Answer each question.*

31. If an angle measures $x°$, how can we represent its complement?

32. If an angle measures $x°$, how can we represent its supplement?

33. If a positive angle has measure $x°$ between 0° and 60°, how can we represent the first negative angle coterminal with it?

34. If a negative angle has measure $x°$ between 0° and −60°, how can we represent the first positive angle coterminal with it?

Perform each calculation. **See Example 3.**

35. $62° 18' + 21° 41'$ **36.** $75° 15' + 83° 32'$ **37.** $97° 42' + 81° 37'$

38. $110° 25' + 32° 55'$ **39.** $71° 18' - 47° 29'$ **40.** $47° 23' - 73° 48'$

41. $90° - 51° 28'$ **42.** $90° - 17° 13'$

43. $180° - 119° 26'$ **44.** $180° - 124° 51'$

45. $26° 20' + 18° 17' - 14° 10'$ **46.** $55° 30' + 12° 44' - 8° 15'$

47. $90° - 72° 58' 11''$ **48.** $90° - 36° 18' 47''$

Convert each angle measure to decimal degrees. If applicable, round to the nearest thousandth of a degree. **See Example 4(a).**

49. $35° 30'$ **50.** $82° 30'$ **51.** $112° 15'$

52. $133° 45'$ **53.** $-60° 12'$ **54.** $-70° 48'$

55. $20° 54' 00''$ **56.** $38° 42' 00''$ **57.** $91° 35' 54''$

58. $34° 51' 35''$ **59.** $274° 18' 59''$ **60.** $165° 51' 09''$

Convert each angle measure to degrees, minutes, and seconds. Round answers to the nearest second, if applicable. **See Example 4(b).**

61. 39.25° **62.** 46.75° **63.** 126.76° **64.** 174.255°

65. −18.515° **66.** −25.485° **67.** 31.4296° **68.** 59.0854°

69. 89.9004° **70.** 102.3771° **71.** 178.5994° **72.** 122.6853°

Find the angle of least positive measure (not equal to the given measure) that is coterminal with each angle. **See Example 5.**

73. 32° **74.** 86° **75.** $26° 30'$ **76.** $58° 40'$

77. −40° **78.** −98° **79.** −125° **80.** −203°

81. 361° **82.** 541° **83.** −361° **84.** −541°

85. 539° **86.** 699° **87.** 850° **88.** 1000°

89. 5280° **90.** 8440° **91.** −5280° **92.** −8440°

Give two positive and two negative angles that are coterminal with the given quadrantal angle.

93. 90° **94.** 180° **95.** 0° **96.** 270°

Give an expression that generates all angles coterminal with each angle. Let n represent any integer.

97. 30° **98.** 45° **99.** 135° **100.** 225°

101. −90° **102.** −180° **103.** 0° **104.** 360°

105. Explain why the answers to **Exercises 103 and 104** give the same set of angles.

106. *Concept Check* Which two of the following are not coterminal with $r°$?

A. $360° + r°$ **B.** $r° - 360°$ **C.** $360° - r°$ **D.** $r° + 180°$

Concept Check *Sketch each angle in standard position. Draw an arrow representing the correct amount of rotation. Find the measure of two other angles, one positive and one negative, that are coterminal with the given angle. Give the quadrant of each angle, if applicable.*

107. 75° **108.** 89° **109.** 174° **110.** 234°

111. 300° **112.** 512° **113.** −61° **114.** −159°

115. 90° **116.** 180° **117.** −90° **118.** −180°

Concept Check *Locate each point in a coordinate system. Draw a ray from the origin through the given point. Indicate with an arrow the angle in standard position having least positive measure. Then find the distance r from the origin to the point, using the distance formula of* ***Appendix B.***

119. $(-3, -3)$ **120.** $(4, -4)$ **121.** $(-3, -5)$ **122.** $(-5, 2)$

123. $(\sqrt{2}, -\sqrt{2})$ **124.** $(-2\sqrt{2}, 2\sqrt{2})$ **125.** $(-1, \sqrt{3})$ **126.** $(\sqrt{3}, 1)$

127. $(-2, 2\sqrt{3})$ **128.** $(4\sqrt{3}, -4)$ **129.** $(0, -4)$ **130.** $(0, 2)$

Solve each problem. ***See Example 6.***

131. *Revolutions of a Turntable* A turntable in a shop makes 45 revolutions per min. How many revolutions does it make per second?

132. *Revolutions of a Windmill* A windmill makes 90 revolutions per min. How many revolutions does it make per second?

133. *Rotating Tire* A tire is rotating 600 times per min. Through how many degrees does a point on the edge of the tire move in $\frac{1}{2}$ sec?

134. *Rotating Airplane Propeller* An airplane propeller rotates 1000 times per min. Find the number of degrees that a point on the edge of the propeller will rotate in 1 sec.

135. *Rotating Pulley* A pulley rotates through 75° in 1 min. How many rotations does the pulley make in an hour?

136. ***Surveying*** One student in a surveying class measures an angle as 74.25°, while another student measures the same angle as 74° 20′. Find the difference between these measurements, both to the nearest minute and to the nearest hundredth of a degree.

137. ***Viewing Field of a Telescope*** As a consequence of Earth's rotation, celestial objects such as the moon and the stars appear to move across the sky, rising in the east and setting in the west. As a result, if a telescope on Earth remains stationary while viewing a celestial object, the object will slowly move outside the viewing field of the telescope. For this reason, a motor is often attached to telescopes so that the telescope rotates at the same rate as Earth. Determine how long it should take the motor to turn the telescope through an angle of 1 min in a direction perpendicular to Earth's axis.

138. ***Angle Measure of a Star on the American Flag*** Determine the measure of the angle in each point of the five-pointed star appearing on the American flag. (*Hint:* Inscribe the star in a circle, and use the following theorem from geometry: *An angle whose vertex lies on the circumference of a circle is equal to half the central angle that cuts off the same arc.* See the figure.)

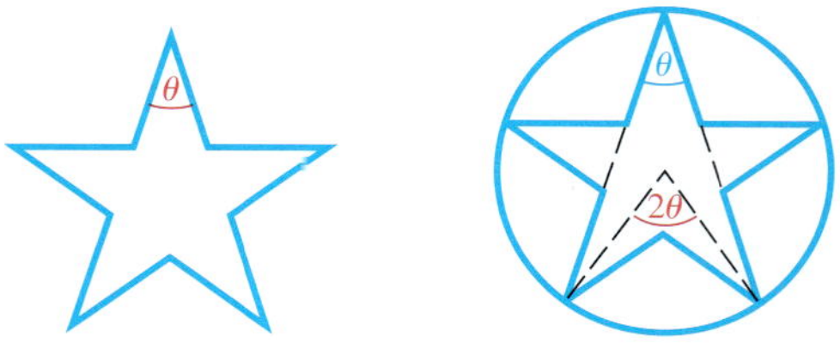

1.2 Angle Relationships and Similar Triangles

- Geometric Properties
- Triangles

Geometric Properties

In **Figure 13,** we extended the sides of angle NMP to form another angle, RMQ. The pair of angles NMP and RMQ are **vertical angles.** Another pair of vertical angles, NMQ and PMR, are also formed. Vertical angles have the following important property.

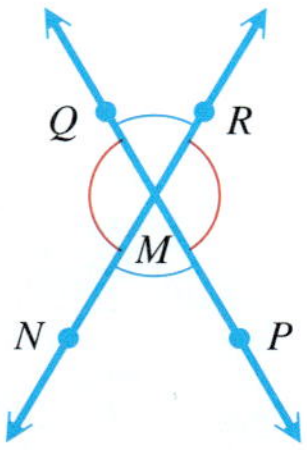

Vertical angles

Figure 13

Vertical Angles

Vertical angles have equal measures.

Parallel lines are lines that lie in the same plane and do not intersect. **Figure 14** shows parallel lines m and n. When a line q intersects two parallel lines, q is called a **transversal.** In **Figure 14,** the transversal intersecting the parallel lines forms eight angles, indicated by numbers.

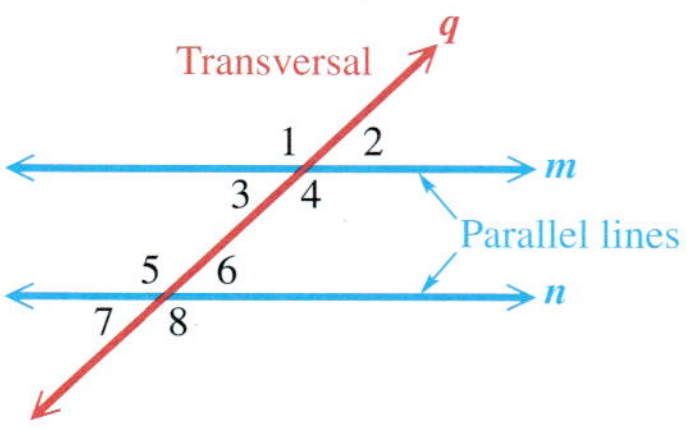

Figure 14

We learn in geometry that the degree measures of angles 1 through 8 in **Figure 14** possess some special properties. The following chart gives the names of these angles and rules about their measures.

Name	Sketch	Rule
Alternate interior angles	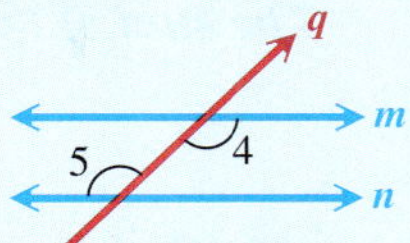 (also 3 and 6)	Angle measures are equal.
Alternate exterior angles	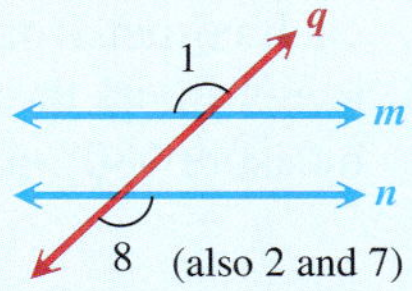 (also 2 and 7)	Angle measures are equal.
Interior angles on same side of transversal	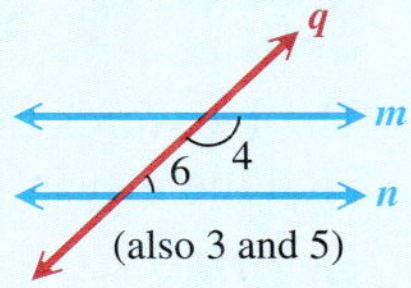 (also 3 and 5)	Angle measures add to 180°.
Corresponding angles	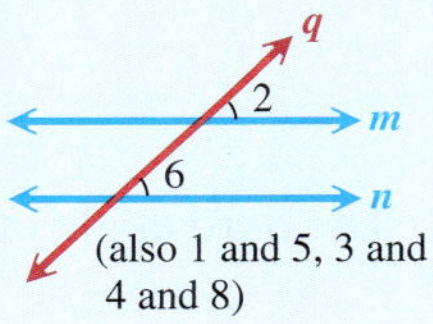 (also 1 and 5, 3 and 7, 4 and 8)	Angle measures are equal.

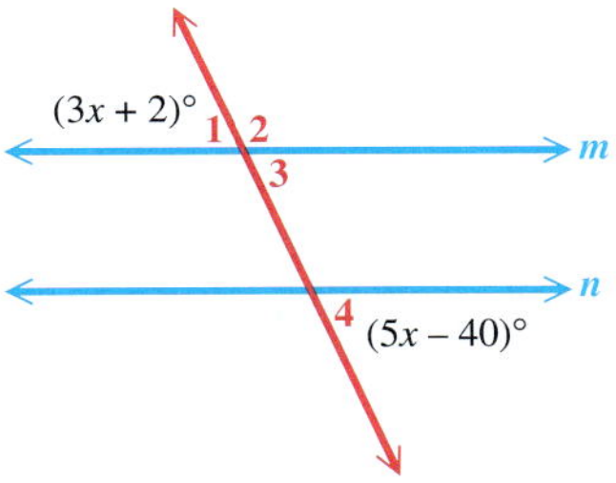

Figure 15

EXAMPLE 1 Finding Angle Measures

Find the measures of angles 1, 2, 3, and 4 in **Figure 15,** given that lines m and n are parallel.

SOLUTION Angles 1 and 4 are alternate exterior angles, so they are equal.

$$3x + 2 = 5x - 40 \quad \text{Alternate exterior angles have equal measures.}$$

$$42 = 2x \quad \text{Subtract } 3x \text{ and add 40. (Appendix A)}$$

$$21 = x \quad \text{Divide by 2.}$$

Angle 1 has measure

$$3x + 2 = 3 \cdot 21 + 2 \quad \text{Substitute 21 for } x.$$

$$= 65°, \quad \text{Multiply, and then add.}$$

and angle 4 has measure

$$5x - 40 = 5 \cdot 21 - 40 \quad \text{Substitute 21 for } x.$$

$$= 65°. \quad \text{Multiply, and then subtract.}$$

Angle 2 is the supplement of a 65° angle, so it has measure

$$180° - 65° = 115°.$$

Angle 3 is a vertical angle to angle 1, so its measure is 65°. (There are other ways to determine these measures.)

✔ ***Now Try Exercises 3 and 11.***

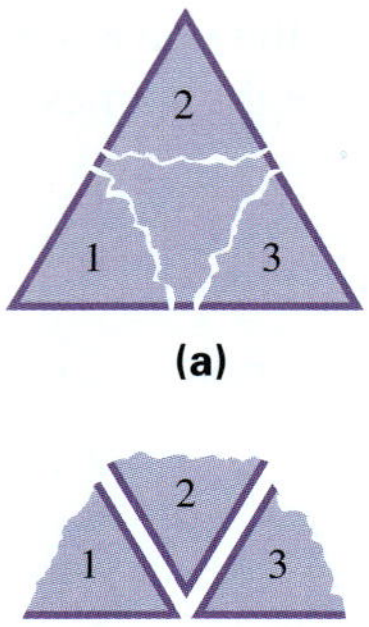

(a)

(b)

Figure 16

Triangles An important property of triangles, first proved by Greek geometers, deals with the sum of the measures of the angles of any triangle.

Angle Sum of a Triangle

***The sum of the measures of the angles of any triangle is* 180°.**

Although it is not an actual proof, we give a rather convincing argument for the truth of this statement, using any size triangle cut from a piece of paper. Tear each corner from the triangle, as suggested in **Figure 16(a).** You should be able to rearrange the pieces so that the three angles form a straight angle, which has measure 180°, as shown in **Figure 16(b).** (See also **Exercise 39.**)

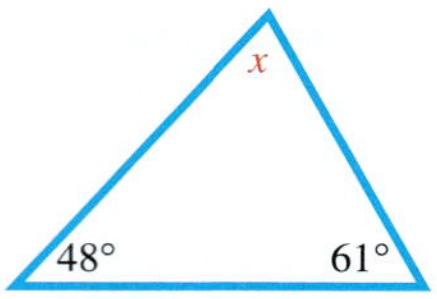

Figure 17

EXAMPLE 2 Applying the Angle Sum of a Triangle Property

The measures of two of the angles of a triangle are 48° and 61°. See **Figure 17.** Find the measure of the third angle, x.

SOLUTION

$$48^\circ + 61^\circ + x = 180^\circ \quad \text{The sum of the angles is } 180^\circ.$$

$$109^\circ + x = 180^\circ \quad \text{Add.}$$

$$x = 71^\circ \quad \text{Subtract } 109^\circ.$$

The third angle of the triangle measures 71°.

✔ *Now Try Exercises 5 and 15.*

We classify triangles according to angles and sides, as shown below.

Types of Triangles

	All acute	One right angle	One obtuse angle
Angles		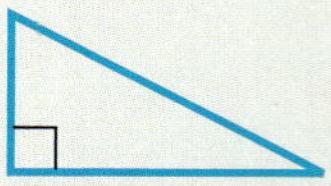	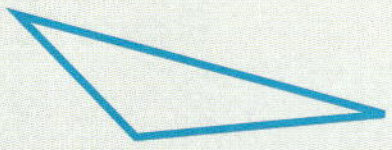
	Acute triangle	**Right triangle**	**Obtuse triangle**
	All sides equal	Two sides equal	No sides equal
Sides	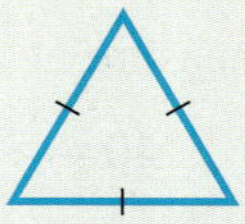	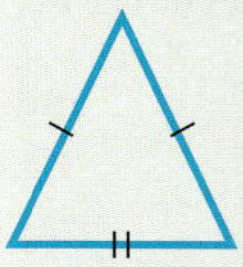	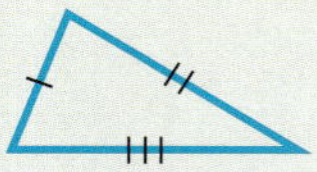
	Equilateral triangle	**Isosceles triangle**	**Scalene triangle**

Similar triangles are triangles of exactly the same shape but not necessarily the same size. **Figure 18** shows three pairs of similar triangles. The two triangles in **Figure 18(c)** have not only the same shape but also the same size. Triangles that are both the same size and the same shape are called **congruent triangles.**

If two triangles are congruent, then it is possible to pick one of them up and place it on top of the other so that they coincide. ***If two triangles are congruent, then they must be similar. However, two similar triangles need not be congruent.***

(a) (b) (c)

Figure 18

The triangular supports for a child's swing set are congruent (and thus similar) triangles, machine-produced with exactly the same dimensions each time. These supports are just one example of similar triangles. The supports of a long bridge, all the same shape but increasing in size toward the center of the bridge, are examples of similar (but not congruent) figures. See the photo.

Suppose a correspondence between two triangles ABC and DEF is set up as shown in **Figure 19.**

Angle A corresponds to angle D.
Angle B corresponds to angle E.
Angle C corresponds to angle F.
Side AB corresponds to side DE.
Side BC corresponds to side EF.
Side AC corresponds to side DF.

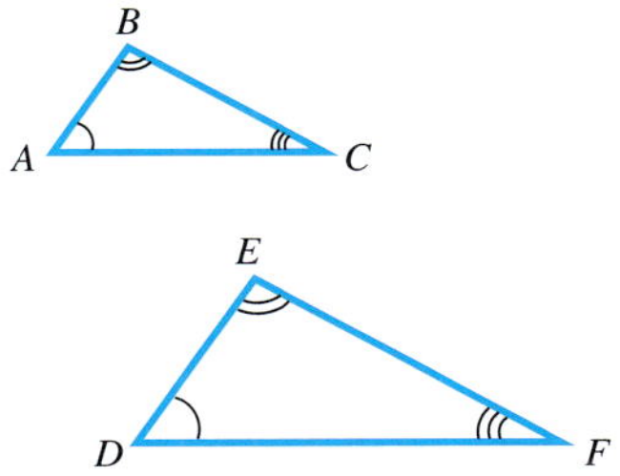

Figure 19

The small arcs found at the angles in **Figure 19** denote the corresponding angles in the triangles.

Conditions for Similar Triangles

For triangle ABC to be similar to triangle DEF, the following conditions must hold.

1. Corresponding angles must have the same measure.

2. Corresponding sides must be proportional. (That is, the ratios of their corresponding sides must be equal.)

EXAMPLE 3 Finding Angle Measures in Similar Triangles

In **Figure 20,** triangles ABC and NMP are similar. Find the measures of angles B and C.

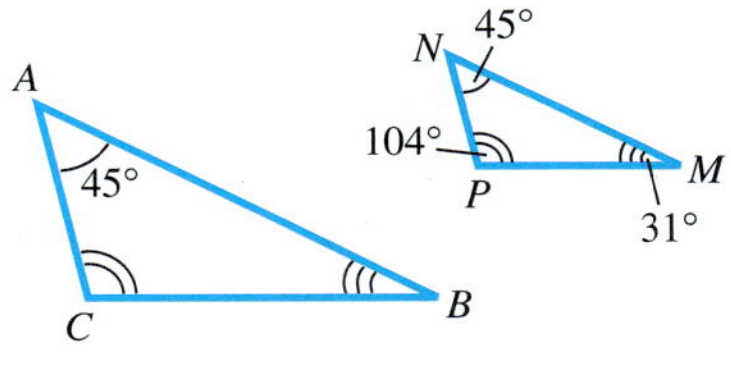

Figure 20

SOLUTION Since the triangles are similar, corresponding angles have the same measure. Since C corresponds to P and P measures 104°, angle C also measures 104°. Since angles B and M correspond, B measures 31°.

✔ *Now Try Exercise 47.*

EXAMPLE 4 Finding Side Lengths in Similar Triangles

Given that triangle ABC and triangle DFE in **Figure 21** are similar, find the lengths of the unknown sides of triangle DFE.

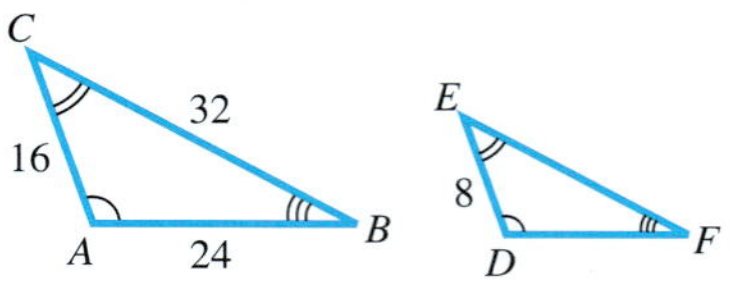

Figure 21

SOLUTION Similar triangles have corresponding sides in proportion. Use this fact to find the unknown side lengths in triangle DFE.

Side DF of triangle DFE corresponds to side AB of triangle ABC, and sides DE and AC correspond. This leads to the following proportion.

$$\frac{8}{16} = \frac{DF}{24}$$

Recall this property of proportions from algebra.

$$\textbf{If} \quad \frac{a}{b} = \frac{c}{d}, \quad \textbf{then} \quad ad = bc.$$

We use this property to solve the equation for DF.

$$\frac{8}{16} = \frac{DF}{24}$$

$$8 \cdot 24 = 16 \cdot DF \quad \text{Property of proportions}$$

$$192 = 16 \cdot DF \quad \text{Multiply.}$$

$$12 = DF \quad \text{Divide by 16.}$$

Side DF has length 12.

Side EF corresponds to CB. This leads to another proportion.

$$\frac{8}{16} = \frac{EF}{32}$$

$$8 \cdot 32 = 16 \cdot EF \quad \text{Property of proportions}$$

$$16 = EF \quad \text{Solve for } EF.$$

Side EF has length 16.

✔ *Now Try Exercise 53.*

EXAMPLE 5 Finding the Height of a Flagpole

Workers at the Morganza Spillway Station need to measure the height of the station flagpole. They find that at the instant when the shadow of the station is 18 m long, the shadow of the flagpole is 99 ft long. The station is 10 m high. Find the height of the flagpole.

SOLUTION **Figure 22** shows the information given in the problem. The two triangles are similar, so corresponding sides are in proportion.

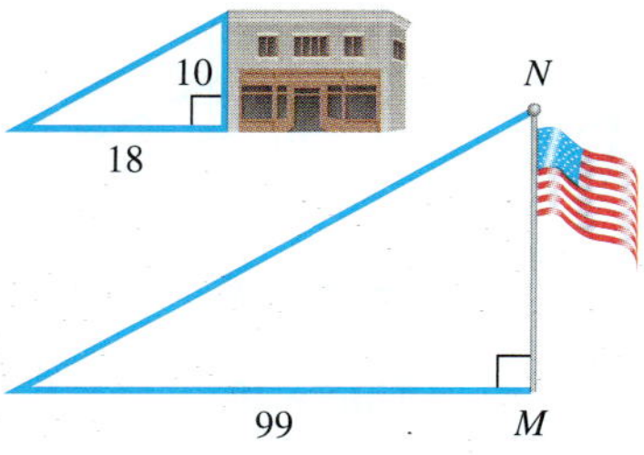

Figure 22

$$\frac{MN}{10} = \frac{99}{18} \quad \text{Corresponding sides are proportional.}$$

$$\frac{MN}{10} = \frac{11}{2} \quad \text{Write in lowest terms.}$$

$$MN \cdot 2 = 10 \cdot 11 \quad \text{Property of proportions}$$

$$MN = 55 \quad \text{Solve for } MN.$$

The flagpole is 55 ft high.

✔ *Now Try Exercise 57.*

1.2 Exercises

1. *Concept Check* Use the given figure to find the measures of the numbered angles, given that lines *m* and *n* are parallel.

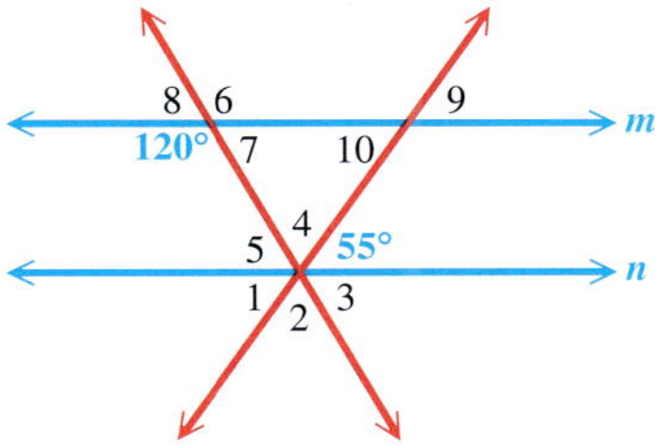

2. In the figure here, if the measure of one of the angles is known, explain how the remaining measures can be found.

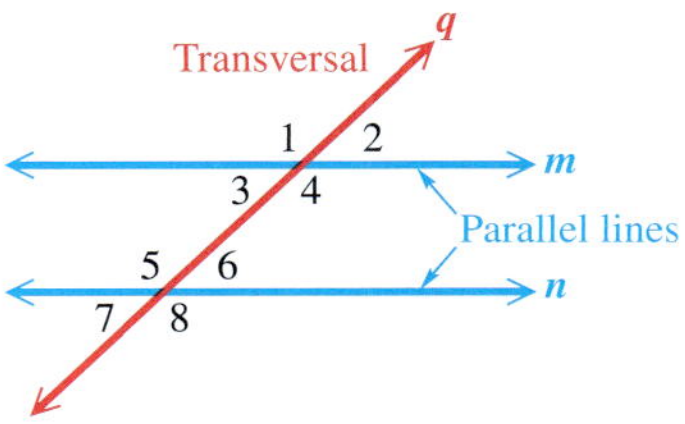

Find the measure of each marked angle. In Exercises 11–14, m and n are parallel. ***See Examples 1 and 2.***

3.

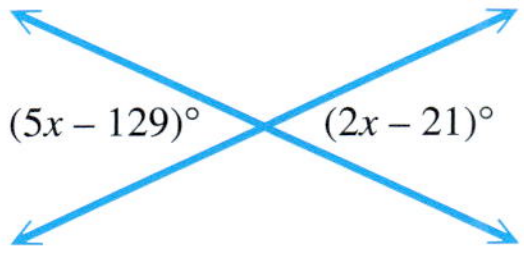

4.

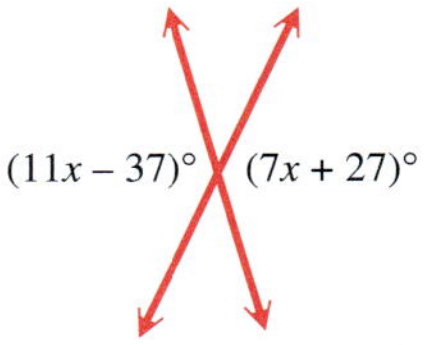

5.

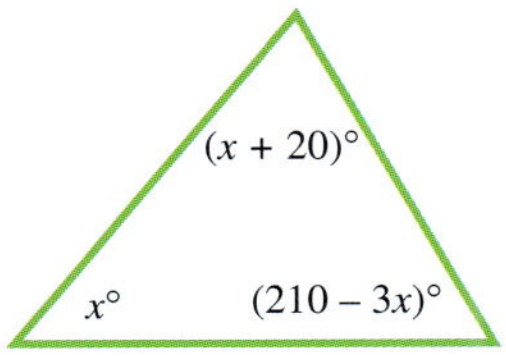

6.

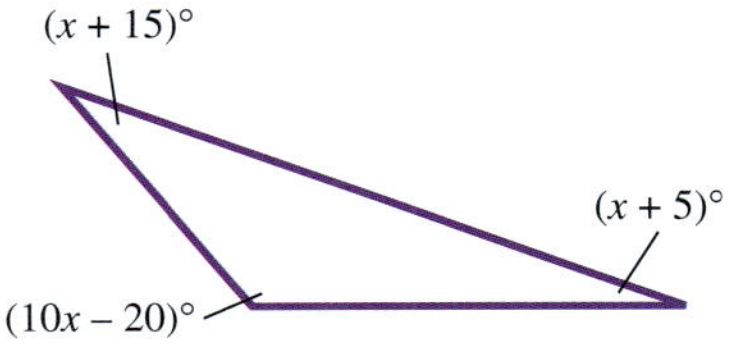

7.

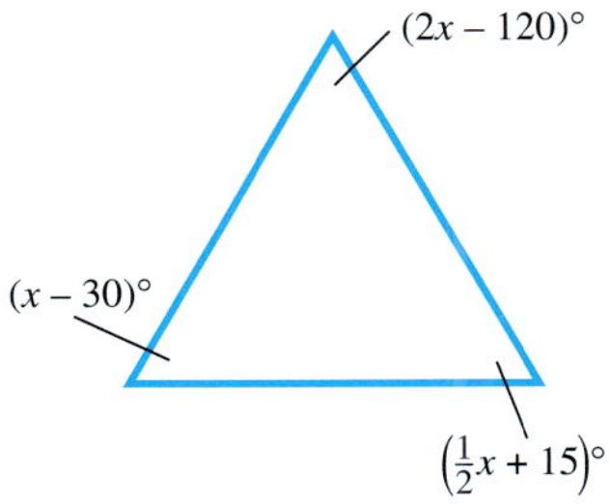

8.

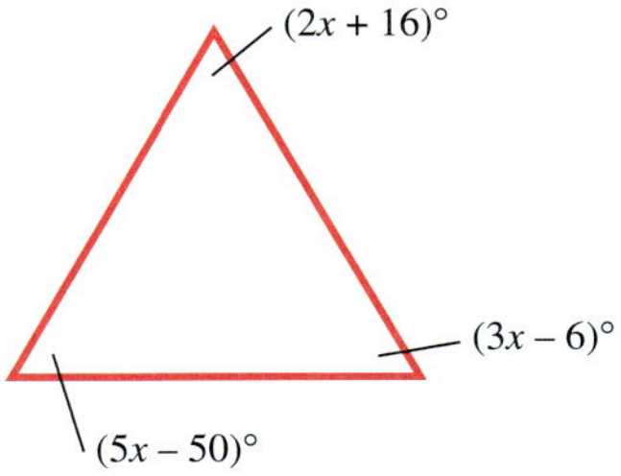

9.

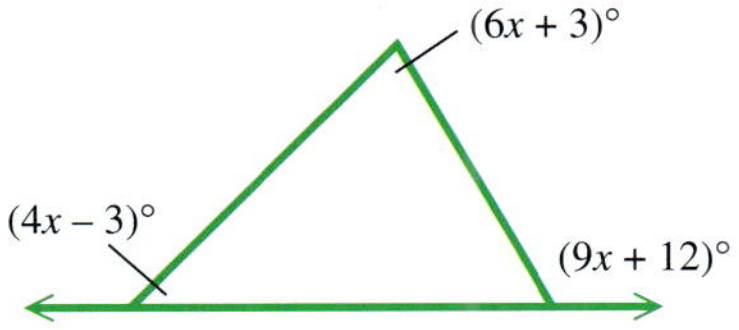

10.

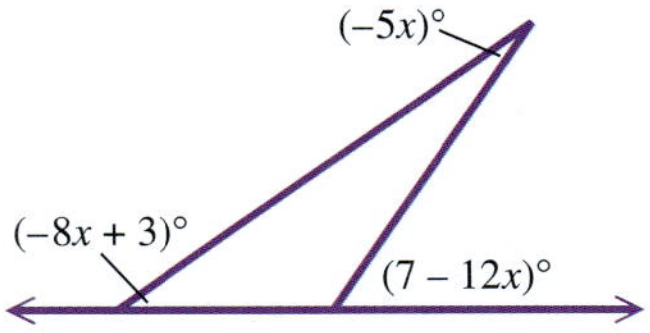

11.

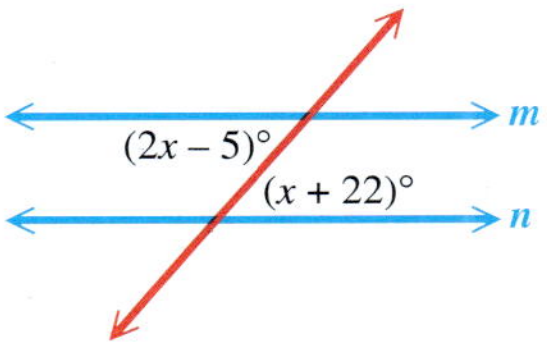

12.

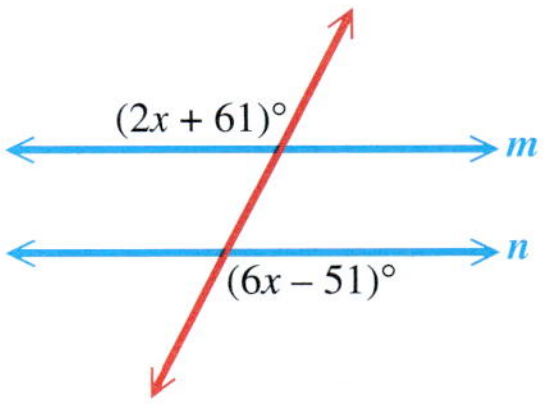

13.

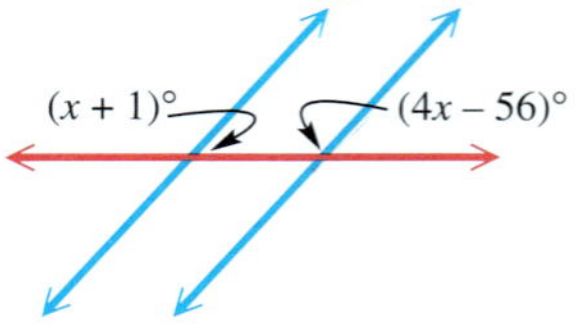

14.

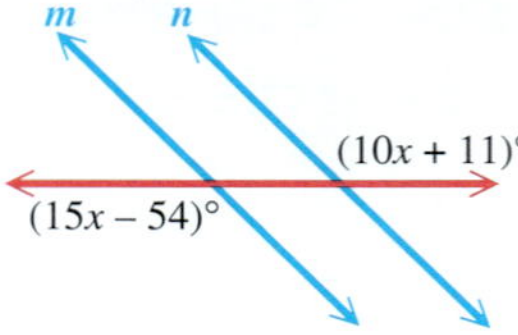

The measures of two angles of a triangle are given. Find the measure of the third angle. ***See Example 2.***

15. 37°, 52°

16. 29°, 104°

17. 147° 12′, 30° 19′

18. 136° 50′, 41° 38′

19. 74.2°, 80.4°

20. 29.6°, 49.7°

21. 51° 20′ 14″, 106° 10′ 12″

22. 17° 41′ 13″, 96° 12′ 10″

23. Can a triangle have angles of measures 85° and 100°? Explain.

24. Can a triangle have two obtuse angles? Explain.

Concept Check Classify each triangle in Exercises 25–36 as acute, right, *or* obtuse. *Also classify each as* equilateral, isosceles, *or* scalene. ***See the discussion following Example 2.***

25.

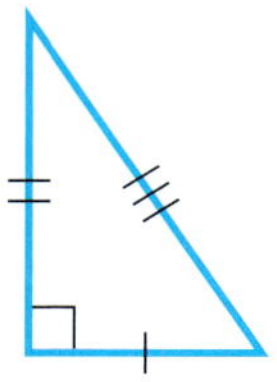

26.

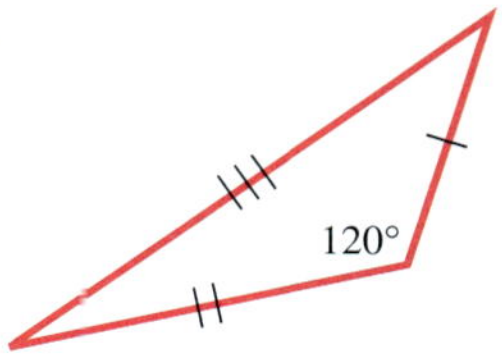

27.

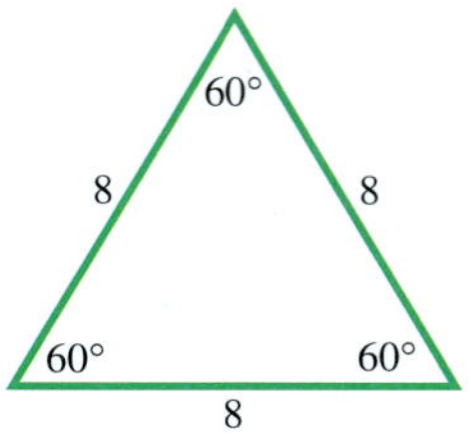

28.

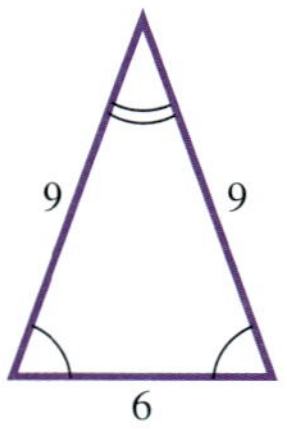

29.

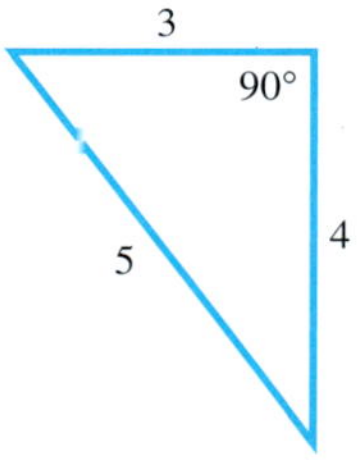

30.

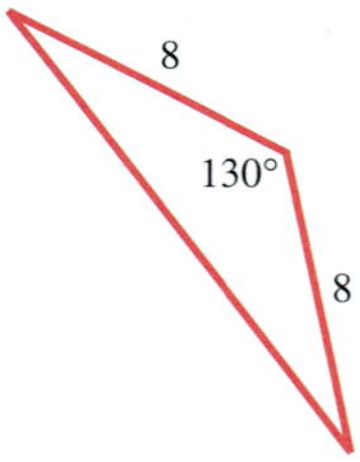

31.

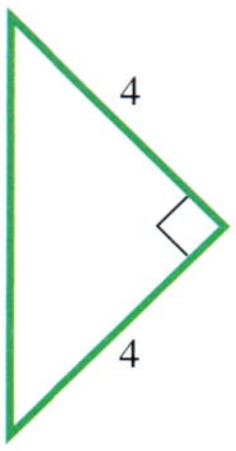

32.

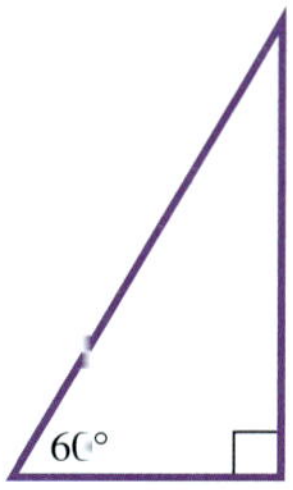

33.

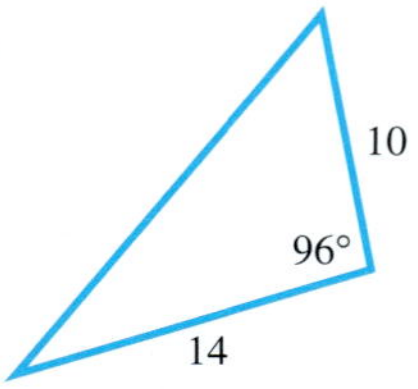

34.

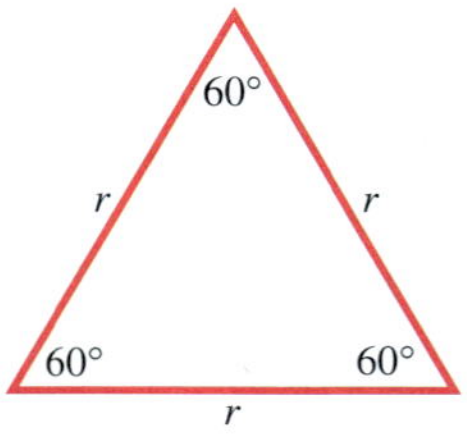

35.

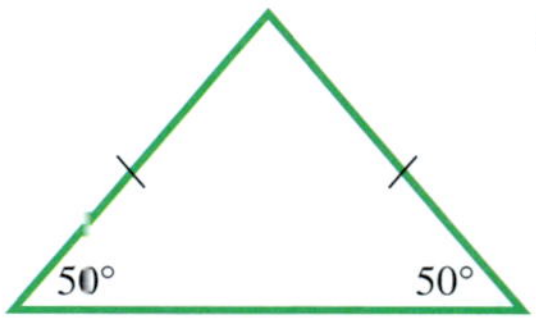

36.

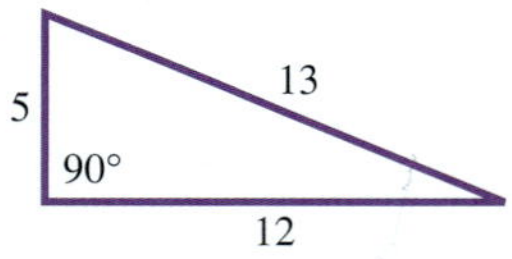

37. Write a definition of *isosceles right triangle.*

38. Explain why the sum of the lengths of any two sides of a triangle must be greater than the length of the third side.

39. Use this figure to discuss why the measures of the angles of a triangle must add up to the same sum as the measure of a straight angle.

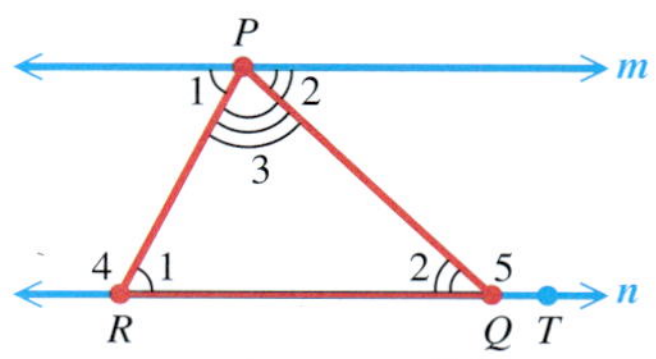

m and *n* are parallel.

40. *Carpentry Technique* The following technique is used by carpenters to draw a 60° angle with a straightedge and a pair of compasses. Explain why this technique works. (*Source:* Hamilton, J. E. and M. S. Hamilton, *Math to Build On,* Construction Trades Press.)

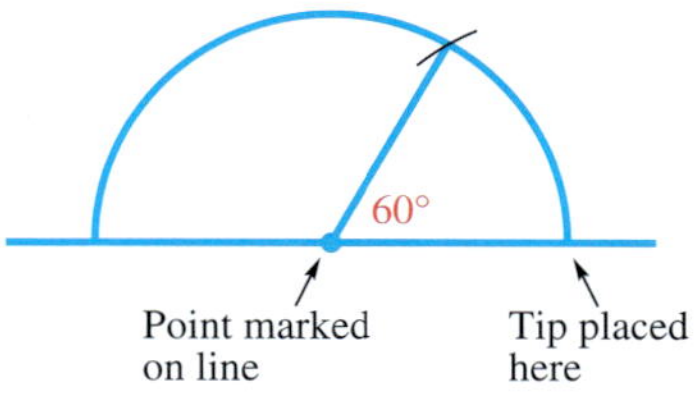

"Draw a straight line segment, and mark a point near the midpoint. Now place the tip on the marked point, and draw a semicircle. Without changing the setting of the pair of compasses, place the tip at the right intersection of the line and the semicircle, and then mark a small arc across the semicircle. Finally, draw a line segment from the marked point on the original segment to the point where the arc crosses the semicircle. This will form a 60° angle with the original segment."

Concept Check *Name the corresponding angles and the corresponding sides of each pair of similar triangles.* ***See the discussion preceding Example 3.***

41.

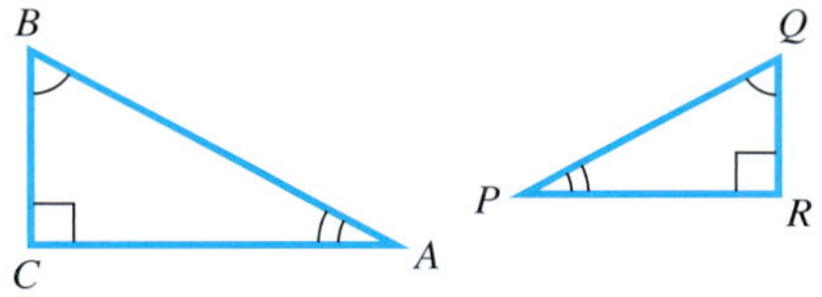

42.

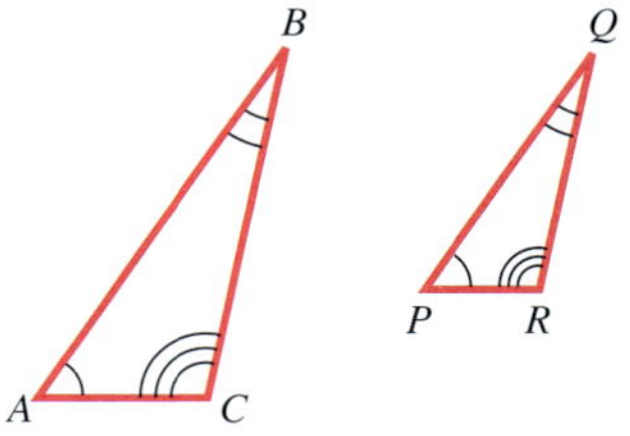

43. (*EA* is parallel to *CD*.)

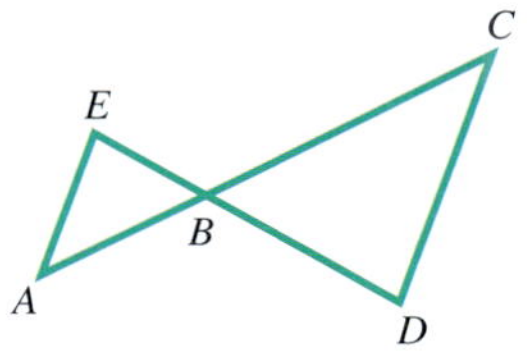

44. (*HK* is parallel to *EF*.)

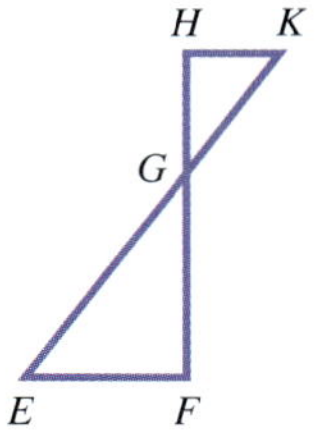

Find all unknown angle measures in each pair of similar triangles. ***See Example 3.***

45.

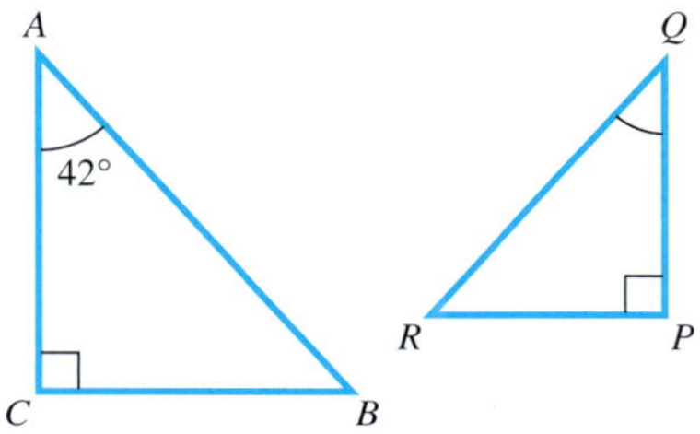

46.

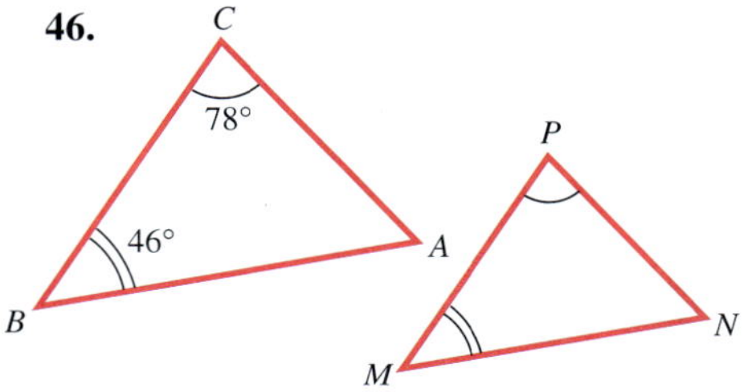

47.

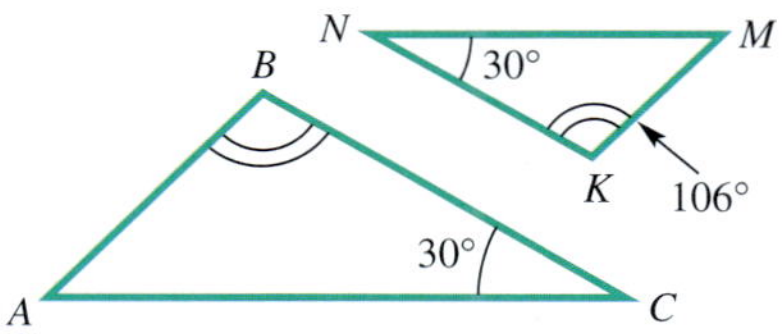

48.

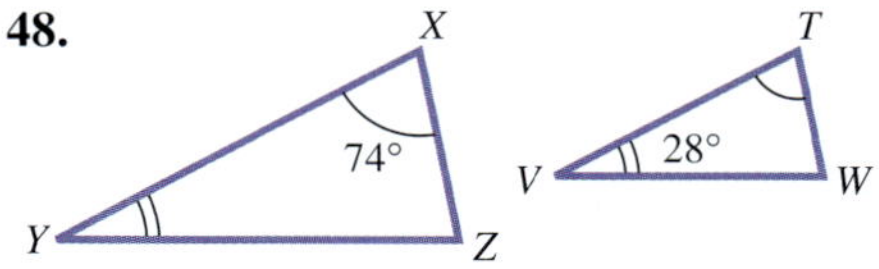

49.

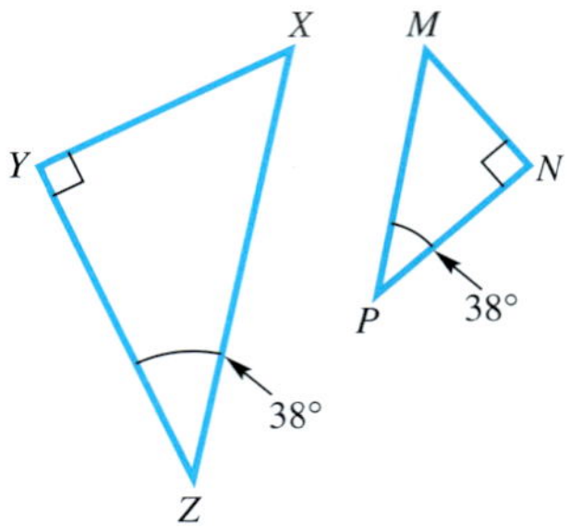

50.

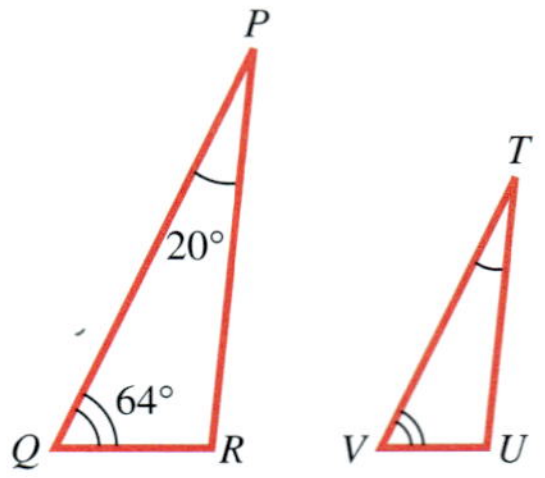

Find the unknown side lengths labeled with a variable in each pair of similar triangles. **See Example 4.**

51.

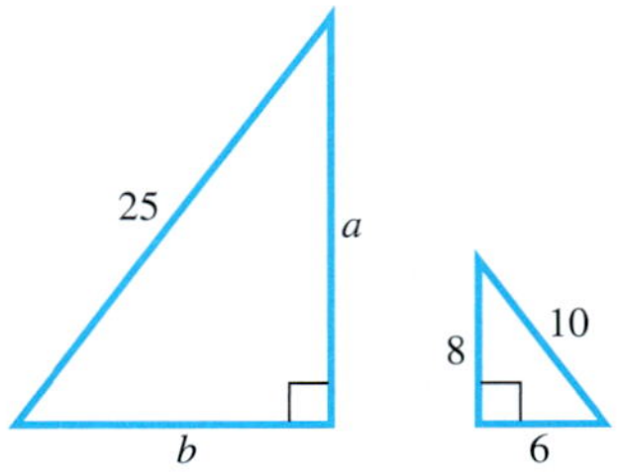

52.

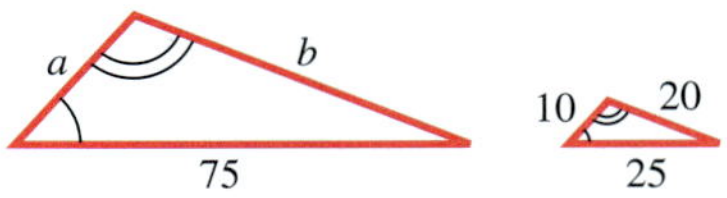

53.

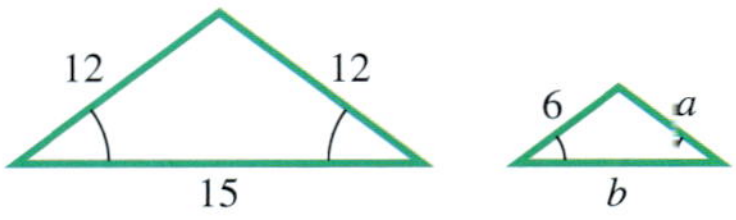

54.

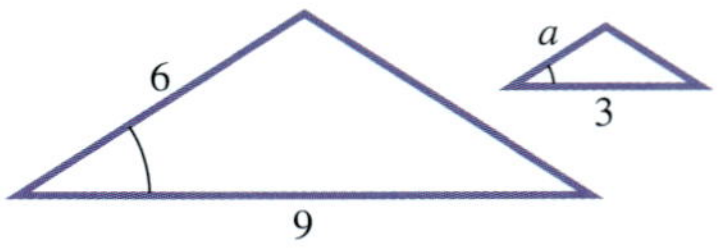

55.

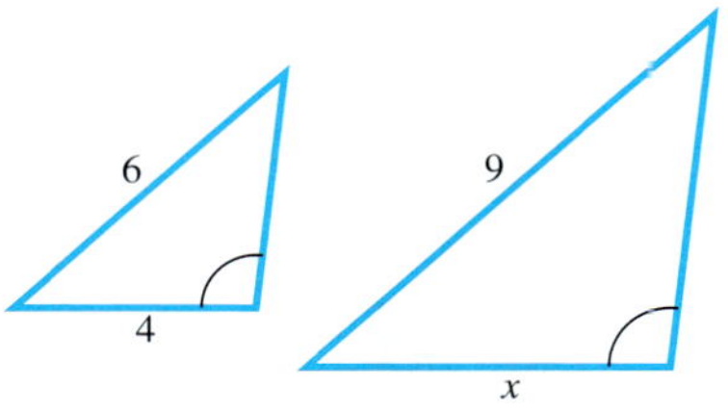

56.

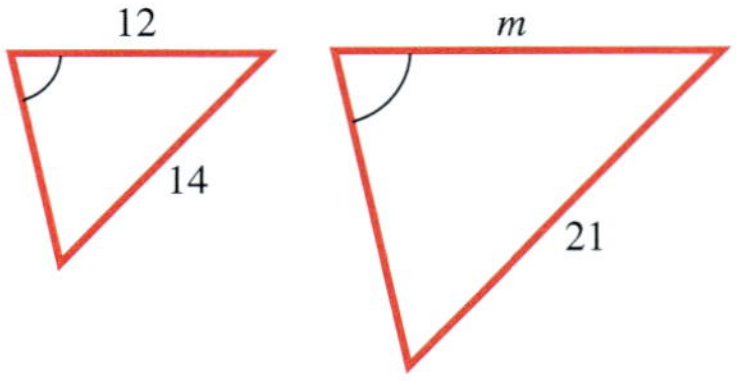

Solve each problem. **See Example 5.**

57. ***Height of a Tree*** A tree casts a shadow 45 m long. At the same time, the shadow cast by a vertical 2-m stick is 3 m long. Find the height of the tree.

58. ***Height of a Lookout Tower*** A forest fire lookout tower casts a shadow 180 ft long at the same time that the shadow of a 9-ft truck is 15 ft long. Find the height of the tower.

59. ***Lengths of Sides of a Triangle*** On a photograph of a triangular piece of land, the lengths of the three sides are 4 cm, 5 cm, and 7 cm, respectively. The shortest side of the actual piece of land is 400 m long. Find the lengths of the other two sides.

60. ***Height of a Lighthouse*** The Biloxi lighthouse in the figure casts a shadow 28 m long at 7 p.m. At the same time, the shadow of the lighthouse keeper, who is 1.75 m tall, is 3.5 m long. How tall is the lighthouse?

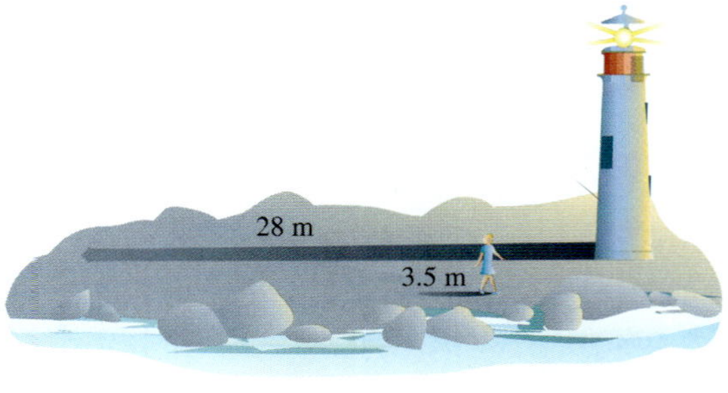

Not to scale

61. *Height of a Building* A house is 15 ft tall. Its shadow is 40 ft long at the same time that the shadow of a nearby building is 300 ft long. Find the height of the building.

62. *Height of a Carving of Lincoln* Assume that Lincoln was $6\frac{1}{3}$ ft tall and his head $\frac{3}{4}$ ft long. Knowing that the carved head of Lincoln at Mt. Rushmore is 60 ft tall, find how tall his entire body would be if it were carved into the mountain.

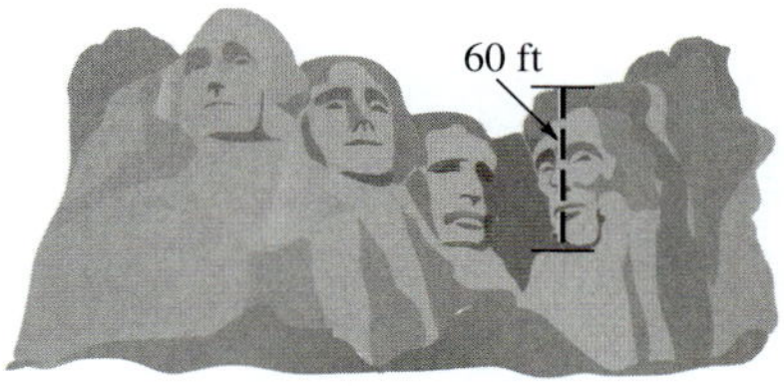

In each diagram, there are two similar triangles. Find the unknown measurement. (Hint: In the sketch for Exercise 63, the side of length 100 *in the small triangle corresponds to the side of length* 100 + 120 = 220 *in the large triangle.)*

63.

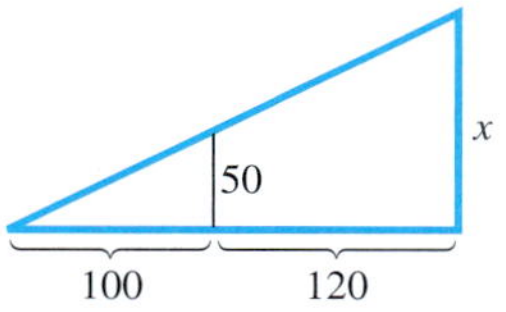

64.

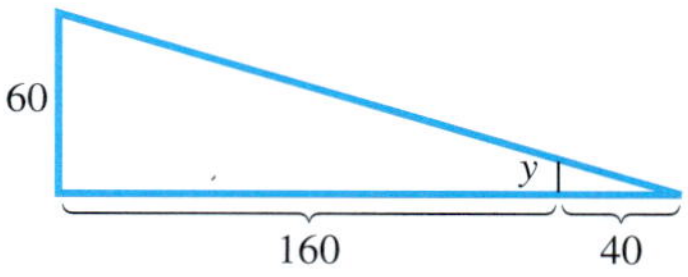

65.

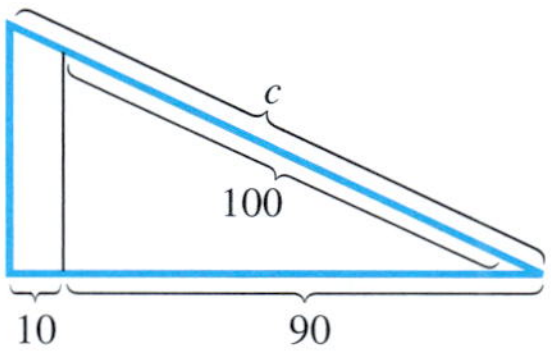

66.

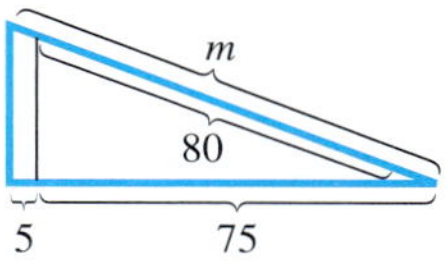

Solve each problem.

67. *Solar Eclipse on Earth* The sun has a diameter of about 865,000 mi with a maximum distance from Earth's surface of about 94,500,000 mi. The moon has a smaller diameter of 2159 mi. For a total solar eclipse to occur, the moon must pass between Earth and the sun. The moon must also be close enough to Earth for the moon's **umbra** (shadow) to reach the surface of Earth. (*Source:* Karttunen, H., P. Kröger, H. Oja, M. Putannen, and K. Donners, Editors, *Fundamental Astronomy,* Fourth Edition, Springer-Verlag.)

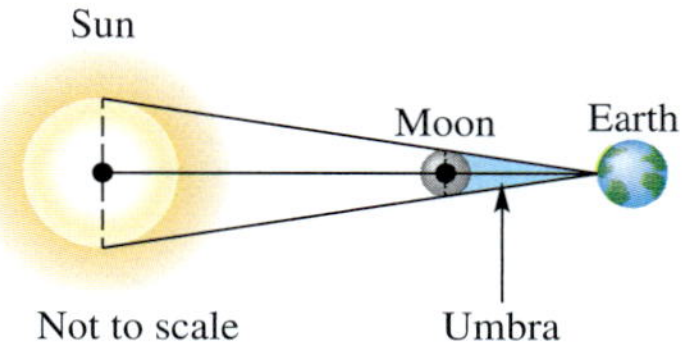

(a) Calculate the maximum distance that the moon can be from Earth and still have a total solar eclipse occur. (*Hint:* Use similar triangles.)

(b) The closest approach of the moon to Earth's surface was 225,745 mi and the farthest was 251,978 mi. (*Source: World Almanac and Book of Facts.*) Can a total solar eclipse occur every time the moon is between Earth and the sun?

68. *Solar Eclipse on Neptune* **(Refer to Exercise 67.)** The sun's distance from Neptune is approximately 2,800,000,000 mi (2.8 billion mi). The largest moon of Neptune is Triton, with a diameter of approximately 1680 mi. (*Source: World Almanac and Book of Facts.*)

(a) Calculate the maximum distance that Triton can be from Neptune for a total eclipse of the sun to occur on Neptune. (*Hint:* Use similar triangles.)

(b) Triton is approximately 220,000 mi from Neptune. Is it possible for Triton to cause a total eclipse on Neptune?

69. *Solar Eclipse on Mars* **(Refer to Exercise 67.)** The sun's distance from the surface of Mars is approximately 142,000,000 mi. One of Mars' two moons, Phobos, has a maximum diameter of 17.4 mi. (*Source: World Almanac and Book of Facts.*)

(a) Calculate the maximum distance that the moon Phobos can be from Mars for a total eclipse of the sun to occur on Mars.

(b) Phobos is approximately 5800 mi from Mars. Is it possible for Phobos to cause a total eclipse on Mars?

70. *Solar Eclipse on Jupiter* (**Refer to Exercise 67.**) The sun's distance from the surface of Jupiter is approximately 484,000,000 mi. One of Jupiter's moons, Ganymede, has a diameter of 3270 mi. (*Source: World Almanac and Book of Facts.*)

(a) Calculate the maximum distance that the moon Ganymede can be from Jupiter for a total eclipse of the sun to occur on Jupiter.

(b) Ganymede is approximately 665,000 mi from Jupiter. Is it possible for Ganymede to cause a total eclipse on Jupiter?

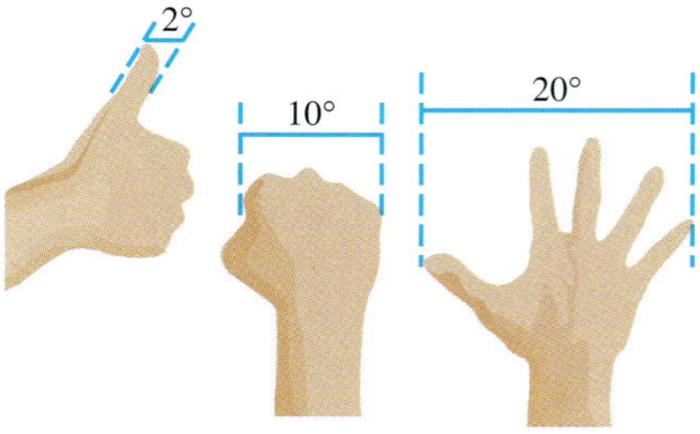

71. *Sizes and Distances in the Sky* Astronomers use degrees, minutes, and seconds to measure sizes and distances in the sky along an arc from the horizon to the zenith point directly overhead. An adult observer on Earth can judge distances in the sky using his or her hand at arm's length. An outstretched hand will be about 20 arc degrees wide from the tip of the thumb to the tip of the little finger. A clenched fist at arm's length measures about 10 arc degrees, and a thumb corresponds to about 2 arc degrees. (*Source:* Levy, D. H., *Skywatching,* The Nature Company.)

(a) The apparent size of the moon is about 31 arc minutes. What part of your thumb would cover the moon?

(b) If an outstretched hand plus a fist cover the distance between two bright stars, about how far apart in arc degrees are the stars?

72. *Estimates of Heights* There is a relatively simple way to make a reasonable estimate of a vertical height. Hold a 1-ft ruler vertically at arm's length as you approach the object to be measured. Stop when one end of the ruler lines up with the top of the object and the other end with its base. Now pace off the distance to the object, taking normal strides. The number of paces will be the approximate height of the object in feet.

Furnish the reasons in parts (a)–(d), which refer to the figure. (Assume that the length of one pace is *EF*.) Then answer the question in part (e).

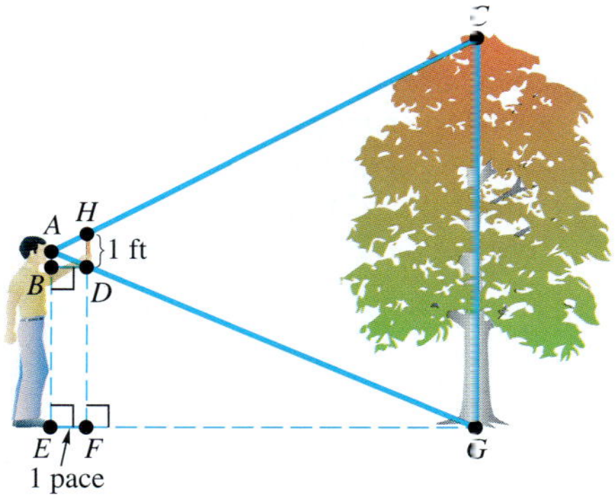

Reasons

(a) ***Step 1*** $CG = \frac{CG}{1} = \frac{AG}{AD}$ ________________

(b) ***Step 2*** $\frac{AG}{AD} = \frac{EG}{BD}$ ________________

(c) ***Step 3*** $\frac{EG}{EF} = \frac{EG}{BD} = \frac{EG}{1}$ ________________

(d) ***Step 4*** CG ft $= EG$ paces ________________

(e) What is the height of the tree in feet?

Chapter 1 Quiz (Sections 1.1–1.2)

1. For an angle measuring 19°, give the measure of **(a)** its complement and **(b)** its supplement.

Find the measure of each unknown angle.

2.

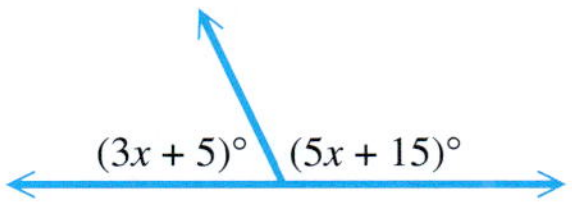

3.

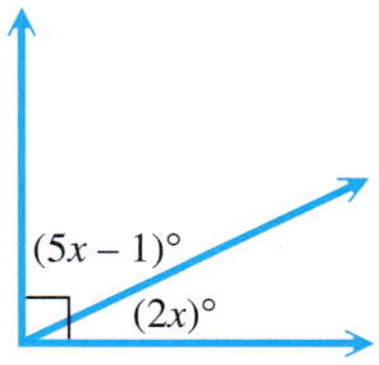

4.

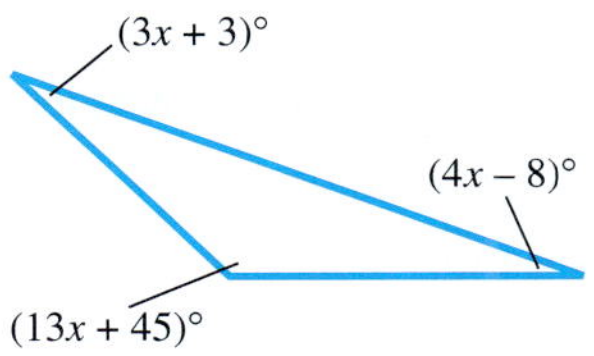

5.

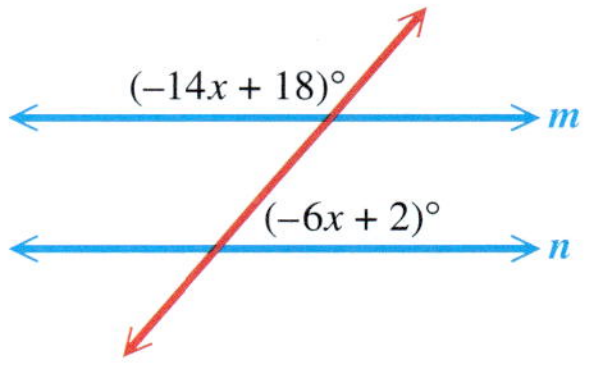

m and *n* are parallel.

6. Perform each indicated conversion.

(a) 77° 12′ 09″ to decimal degrees **(b)** 22.0250° to degrees, minutes, seconds

7. Find the angle of least positive measure (not equal to the given angle) coterminal with each angle.

(a) 410° **(b)** −60° **(c)** 890° **(d)** 57°

8. ***Rotating Flywheel*** A flywheel rotates 300 times per min. Through how many degrees does a point on the edge of the flywheel move in 1 sec?

9. ***Length of a Shadow*** If a vertical antenna 45 ft tall casts a shadow 15 ft long, how long would the shadow of a 30-ft pole be at the same time and place?

10. Find the values of *x* and *y*.

(a)

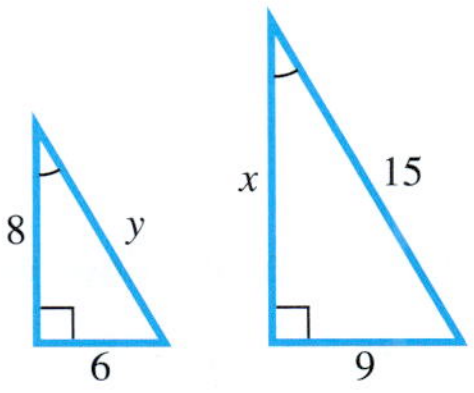

(b)

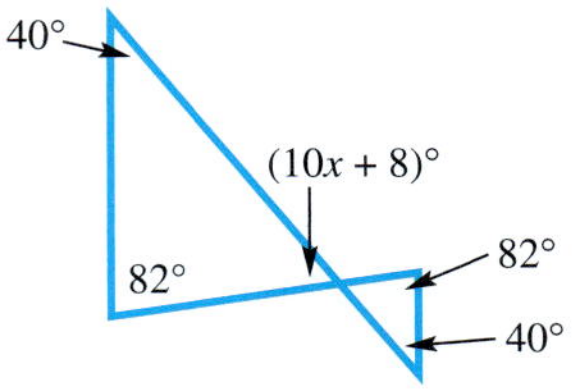

1.3 Trigonometric Functions

- Trigonometric Functions
- Quadrantal Angles

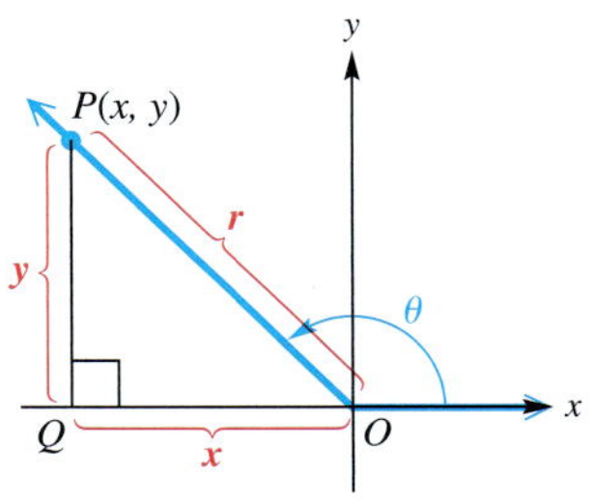

Figure 23

Trigonometric Functions To define the six **trigonometric functions,** we start with an angle θ in standard position and choose any point P having coordinates (x, y) on the terminal side of angle θ. (The point P must not be the vertex of the angle.) See **Figure 23.** A perpendicular from P to the x-axis at point Q determines a right triangle, having vertices at O, P, and Q. We find the distance r from $P(x, y)$ to the origin, $(0, 0)$, using the distance formula.

$$r = \sqrt{(x-0)^2 + (y-0)^2} \quad \text{(Appendix B)}$$

$$r = \sqrt{x^2 + y^2}$$

Notice that $r > 0$ since this is the undirected distance.

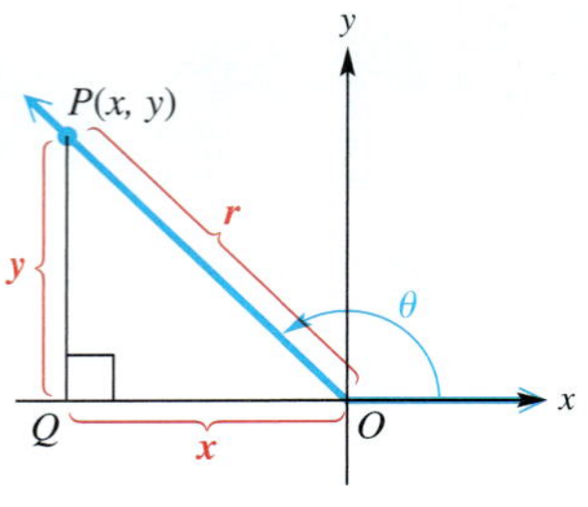

Figure 23 (repeated)

The six trigonometric functions of angle θ are **sine, cosine, tangent, cotangent, secant,** and **cosecant,** abbreviated **sin, cos, tan, cot, sec,** and **csc.**

Trigonometric Functions

Let (x, y) be a point other than the origin on the terminal side of an angle θ in standard position. The distance from the point to the origin is $r = \sqrt{x^2 + y^2}$. The six trigonometric functions of θ are defined as follows.

$$\sin\theta = \frac{y}{r} \qquad \cos\theta = \frac{x}{r} \qquad \tan\theta = \frac{y}{x} \quad (x \neq 0)$$

$$\csc\theta = \frac{r}{y} \quad (y \neq 0) \qquad \sec\theta = \frac{r}{x} \quad (x \neq 0) \qquad \cot\theta = \frac{x}{y} \quad (y \neq 0)$$

EXAMPLE 1 Finding Function Values of an Angle

The terminal side of an angle θ in standard position passes through the point $(8, 15)$. Find the values of the six trigonometric functions of angle θ.

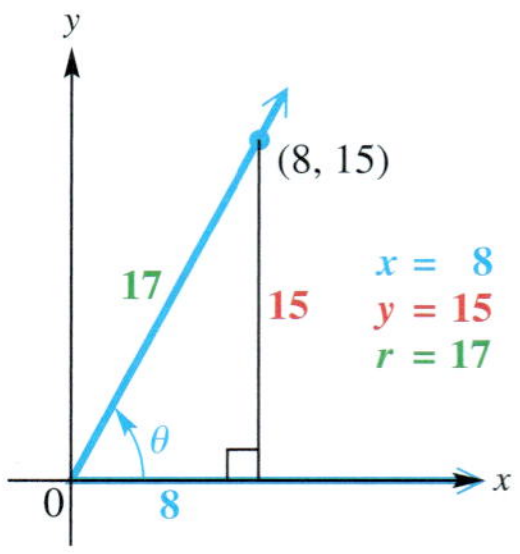

Figure 24

SOLUTION **Figure 24** shows angle θ and the triangle formed by dropping a perpendicular from the point $(8, 15)$ to the x-axis. The point $(8, 15)$ is 8 units to the right of the y-axis and 15 units above the x-axis, so $x = 8$ and $y = 15$. Now use $r = \sqrt{x^2 + y^2}$.

$$r = \sqrt{8^2 + 15^2} = \sqrt{64 + 225} = \sqrt{289} = 17$$

We can now find the values of the six trigonometric functions of angle θ.

$$\sin\theta = \frac{y}{r} = \frac{15}{17} \qquad \cos\theta = \frac{x}{r} = \frac{8}{17} \qquad \tan\theta = \frac{y}{x} = \frac{15}{8}$$

$$\csc\theta = \frac{r}{y} = \frac{17}{15} \qquad \sec\theta = \frac{r}{x} = \frac{17}{8} \qquad \cot\theta = \frac{x}{y} = \frac{8}{15}$$

✔ *Now Try Exercise 5.*

EXAMPLE 2 Finding Function Values of an Angle

The terminal side of an angle θ in standard position passes through the point $(-3, -4)$. Find the values of the six trigonometric functions of angle θ.

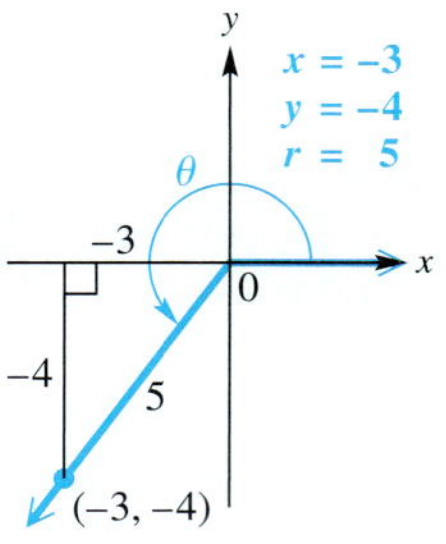

Figure 25

SOLUTION As shown in **Figure 25**, $x = -3$ and $y = -4$.

$$r = \sqrt{(-3)^2 + (-4)^2} \qquad r = \sqrt{x^2 + y^2}$$

$$r = \sqrt{25} \qquad \text{Simplify the radicand.}$$

$$r = 5 \qquad r > 0$$

Now use the definitions of the trigonometric functions.

$$\sin\theta = \frac{-4}{5} = -\frac{4}{5} \qquad \cos\theta = \frac{-3}{5} = -\frac{3}{5} \qquad \tan\theta = \frac{-4}{-3} = \frac{4}{3}$$

$$\csc\theta = \frac{5}{-4} = -\frac{5}{4} \qquad \sec\theta = \frac{5}{-3} = -\frac{5}{3} \qquad \cot\theta = \frac{-3}{-4} = \frac{3}{4}$$

✔ *Now Try Exercise 19.*

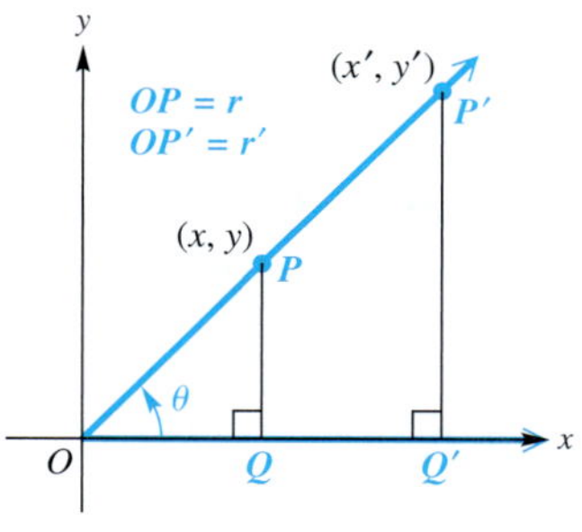

Figure 26

We can find the six trigonometric functions using *any* point other than the origin on the terminal side of an angle. To see why any point can be used, refer to **Figure 26,** which shows an angle θ and two distinct points on its terminal side. Point P has coordinates (x, y), and point $\boldsymbol{P'}$ (read **"*P*-prime"**) has coordinates (x', y'). Let r be the length of the hypotenuse of triangle OPQ, and let r' be the length of the hypotenuse of triangle $OP'Q'$. Since corresponding sides of similar triangles are proportional,

$$\frac{y}{r} = \frac{y'}{r'}, \quad \text{(Section 1.2)}$$

so $\sin \theta = \frac{y}{r}$ is the same no matter which point is used to find it. A similar result holds for the other five trigonometric functions.

We can also find the trigonometric function values of an angle if we know the equation of the line coinciding with the terminal ray. Recall from algebra that the graph of the equation

$$Ax + By = 0 \quad \text{(Appendix B)}$$

is a line that passes through the origin. If we restrict x to have only nonpositive or only nonnegative values, we obtain as the graph a ray with endpoint at the origin. For example, the graph of $x + 2y = 0,\ x \geq 0$, shown in **Figure 27,** is a ray that can serve as the terminal side of an angle θ in standard position. By choosing a point on the ray, we can find the trigonometric function values of the angle.

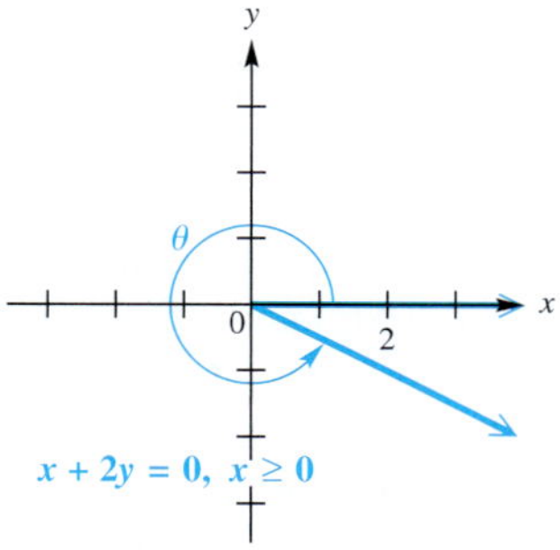

Figure 27

EXAMPLE 3 Finding Function Values of an Angle

Find the six trigonometric function values of the angle θ in standard position, if the terminal side of θ is defined by $x + 2y = 0,\ x \geq 0$.

SOLUTION The angle is shown in **Figure 28**. We can use *any* point except $(0, 0)$ on the terminal side of θ to find the trigonometric function values. We choose $x = 2$ and find the corresponding y-value.

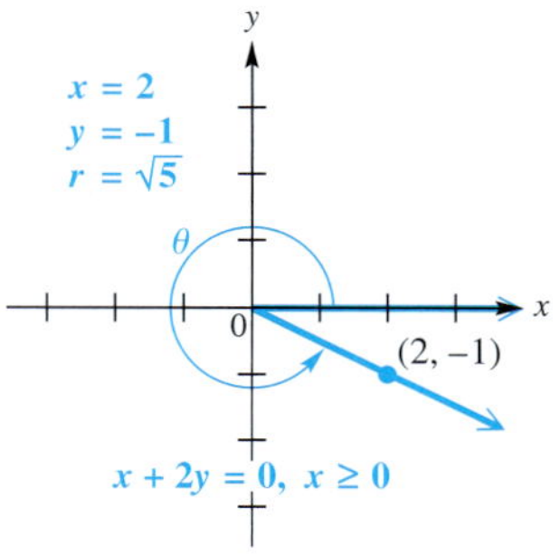

Figure 28

$$\begin{aligned} x + 2y &= 0, \quad x \geq 0 & \\ 2 + 2y &= 0 & \text{Let } x = 2. \\ 2y &= -2 & \text{Subtract 2. (Appendix A)} \\ y &= -1 & \text{Divide by 2.} \end{aligned}$$

The point $(2, -1)$ lies on the terminal side, and the corresponding value of r is $r = \sqrt{2^2 + (-1)^2} = \sqrt{5}$. Now we use the definitions of the trigonometric functions.

$$\sin \theta = \frac{y}{r} = \frac{-1}{\sqrt{5}} = \frac{-1}{\sqrt{5}} \cdot \frac{\sqrt{5}}{\sqrt{5}} = -\frac{\sqrt{5}}{5}$$

$$\cos \theta = \frac{x}{r} = \frac{2}{\sqrt{5}} = \frac{2}{\sqrt{5}} \cdot \frac{\sqrt{5}}{\sqrt{5}} = \frac{2\sqrt{5}}{5}$$

Multiply by $\frac{\sqrt{5}}{\sqrt{5}}$, which equals 1, to rationalize the denominators.

$$\tan \theta = \frac{y}{x} = \frac{-1}{2} = -\frac{1}{2}$$

$$\csc \theta = \frac{r}{y} = \frac{\sqrt{5}}{-1} = -\sqrt{5} \qquad \sec \theta = \frac{r}{x} = \frac{\sqrt{5}}{2} \qquad \cot \theta = \frac{x}{y} = \frac{2}{-1} = -2$$

✔ ***Now Try Exercise 45.***

Recall that when the equation of a line is written in slope-intercept form

$$y = mx + b,$$

the coefficient m of x is the slope of the line. In **Example 3,** the equation $x + 2y = 0$ can be written as $y = -\frac{1}{2}x$, so the slope is $-\frac{1}{2}$. Notice that $\tan \theta = -\frac{1}{2}$.

In general, it is true that $m = \tan \theta$.

NOTE The trigonometric function values we found in **Examples 1–3** are *exact.* If we were to use a calculator to approximate these values, the decimal results would not be acceptable if exact values were required.

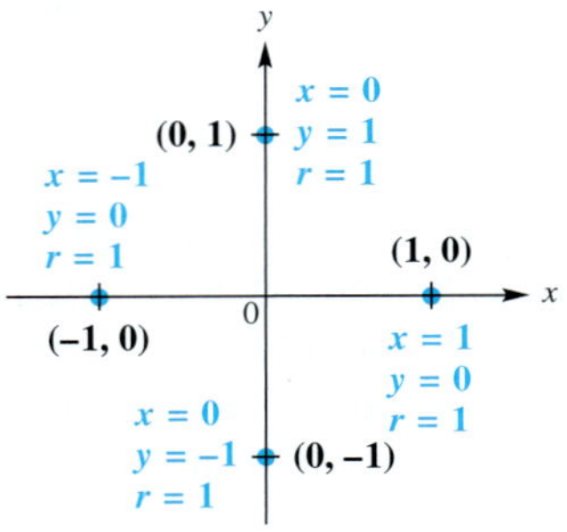

Figure 29

Quadrantal Angles If the terminal side of an angle in standard position lies along the y-axis, any point on this terminal side has x-coordinate 0. Similarly, an angle with terminal side on the x-axis has y-coordinate 0 for any point on the terminal side. Since the values of x and y appear in the denominators of some trigonometric functions, and since a fraction is undefined if its denominator is 0, some trigonometric function values of quadrantal angles (i.e., those with terminal side on an axis) are undefined.

When determining trigonometric function values of quadrantal angles, **Figure 29** can help find the ratios. Because *any* point on the terminal side can be used, it is convenient to choose the point one unit from the origin, with $r = 1$. (In **Chapter 3** we extend this idea to the *unit circle.*)

To find the function values of a quadrantal angle, determine the position of the terminal side, choose the one of these four points that lies on this terminal side, and then use the definitions involving x, y, and r.

EXAMPLE 4 Finding Function Values of Quadrantal Angles

Find the values of the six trigonometric functions for each angle.

(a) an angle of 90°

(b) an angle θ in standard position with terminal side through $(-3, 0)$

SOLUTION

(a) **Figure 30** shows that the terminal side passes through $(0, 1)$. So $x = 0$, $y = 1$, and $r = 1$. Thus, we have the following.

$$\sin 90° = \frac{1}{1} = 1 \qquad \cos 90° = \frac{0}{1} = 0 \qquad \tan 90° = \frac{1}{0} \ \text{(undefined)}$$

$$\csc 90° = \frac{1}{1} = 1 \qquad \sec 90° = \frac{1}{0} \ \text{(undefined)} \qquad \cot 90° = \frac{0}{1} = 0$$

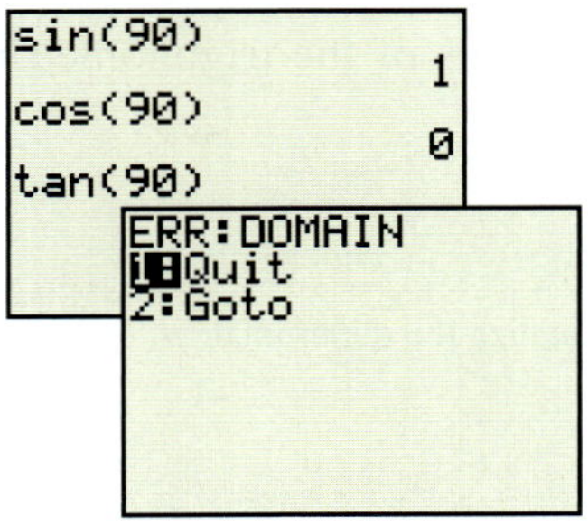

A calculator in degree mode returns the correct values for sin 90° and cos 90°. The second screen shows an ERROR message for tan 90°, because 90° is not in the domain of the tangent function.

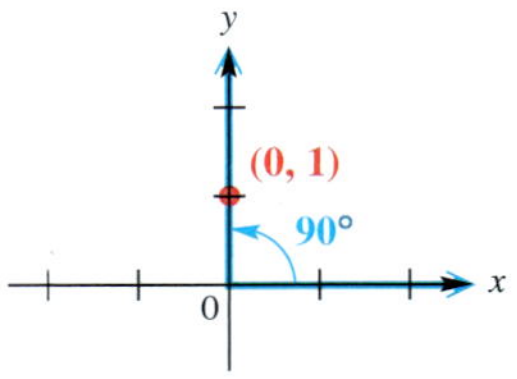

Figure 30

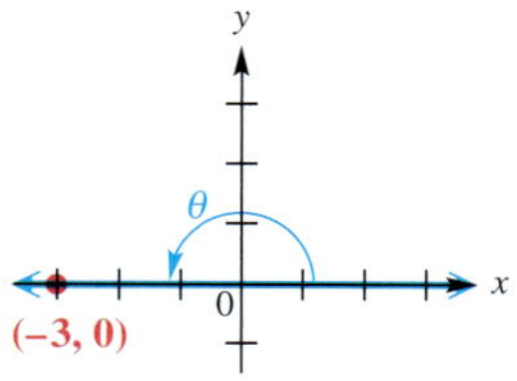

Figure 31

(b) **Figure 31** shows the angle. Here, $x = -3$, $y = 0$, and $r = 3$, so the trigonometric functions have the following values.

$$\sin\theta = \frac{0}{3} = 0 \qquad \cos\theta = \frac{-3}{3} = -1 \qquad \tan\theta = \frac{0}{-3} = 0$$

$$\csc\theta = \frac{3}{0}\ \text{(undefined)} \qquad \sec\theta = \frac{3}{-3} = -1 \qquad \cot\theta = \frac{-3}{0}\ \text{(undefined)}$$

Verify that these values can also be found by using the point $(-1, 0)$.

Now Try Exercises 13, 59, 61, 65, and 67.

The conditions under which the trigonometric function values of quadrantal angles are undefined are summarized here.

Conditions for Undefined Function Values

Identify the terminal side of a quadrantal angle.

- If the terminal side of the quadrantal angle lies along the y-axis, then the tangent and secant functions are undefined.
- If the terminal side of the quadrantal angle lies along the x-axis, then the cotangent and cosecant functions are undefined.

The function values of some commonly used quadrantal angles, 0°, 90°, 180°, 270°, and 360°, are summarized in the table. They can be determined when needed by using **Figure 29** and the method of **Example 4(a).**

For other quadrantal angles such as −90°, −270°, and 450°, first determine the coterminal angle that lies between 0° and 360°, and then refer to the table entries for that particular angle. For example, the function values of a −90° angle would correspond to those of a 270° angle.

Function Values of Quadrantal Angles

θ	$\sin\theta$	$\cos\theta$	$\tan\theta$	$\cot\theta$	$\sec\theta$	$\csc\theta$
0°	0	1	0	Undefined	1	Undefined
90°	1	0	Undefined	0	Undefined	1
180°	0	−1	0	Undefined	−1	Undefined
270°	−1	0	Undefined	0	Undefined	−1
360°	0	1	0	Undefined	1	Undefined

The values given in this table can be found with a calculator that has trigonometric function keys. ***Make sure the calculator is set in degree mode.***

TI-83 Plus

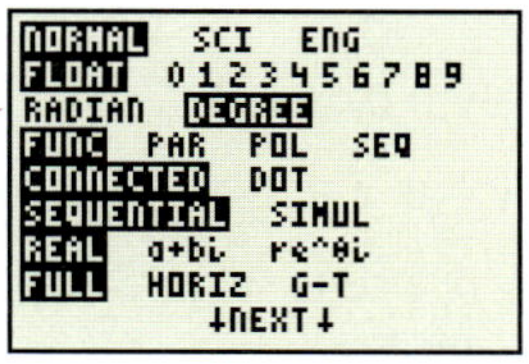

TI-84 Plus

Figure 32

CAUTION ***One of the most common errors involving calculators in trigonometry occurs when the calculator is set for radian measure, rather than degree measure.*** (Radian measure of angles is discussed in **Chapter 3.**) Be sure you know how to set your calculator in degree mode. See **Figure 32**, which illustrates degree mode for TI-83/84 Plus calculators.

1.3 Exercises

Concept Check *Sketch an angle θ in standard position such that θ has the least positive measure, and the given point is on the terminal side of θ. Then find the values of the six trigonometric functions for each angle. Rationalize denominators when applicable.* ***See Examples 1, 2, and 4.***

1. $(5, -12)$ **2.** $(-12, -5)$ **3.** $(-3, 4)$ **4.** $(-4, -3)$

5. $(-8, 15)$ **6.** $(15, -8)$ **7.** $(7, -24)$ **8.** $(-24, -7)$

9. $(0, 2)$ **10.** $(0, 5)$ **11.** $(-4, 0)$ **12.** $(-5, 0)$

13. $(0, -4)$ **14.** $(0, -3)$ **15.** $(1, \sqrt{3})$ **16.** $(-1, \sqrt{3})$

17. $(\sqrt{2}, \sqrt{2})$ **18.** $(-\sqrt{2}, -\sqrt{2})$ **19.** $(-2\sqrt{3}, -2)$ **20.** $(-2\sqrt{3}, 2)$

21. For any nonquadrantal angle θ, $\sin \theta$ and $\csc \theta$ will have the same sign. Explain why.

22. *Concept Check* How is the value of r interpreted geometrically in the definitions of the sine, cosine, secant, and cosecant functions?

23. *Concept Check* If $\cot \theta$ is undefined, what is the value of $\tan \theta$?

24. *Concept Check* If the terminal side of an angle θ is in quadrant III, what is the sign of each of the trigonometric function values of θ?

Concept Check *Suppose that the point (x, y) is in the indicated quadrant. Decide whether the given ratio is positive or negative. Recall that $r = \sqrt{x^2 + y^2}$. (Hint: Drawing a sketch may help.)*

25. II, $\frac{x}{r}$ **26.** III, $\frac{y}{r}$ **27.** IV, $\frac{y}{x}$ **28.** IV, $\frac{x}{y}$

29. II, $\frac{y}{r}$ **30.** III, $\frac{x}{r}$ **31.** IV, $\frac{x}{r}$ **32.** IV, $\frac{y}{r}$

33. II, $\frac{x}{y}$ **34.** II, $\frac{y}{x}$ **35.** III, $\frac{y}{x}$ **36.** III, $\frac{x}{y}$

37. III, $\frac{r}{x}$ **38.** III, $\frac{r}{y}$ **39.** I, $\frac{x}{y}$ **40.** I, $\frac{y}{x}$

41. I, $\frac{y}{r}$ **42.** I, $\frac{x}{r}$ **43.** I, $\frac{r}{x}$ **44.** I, $\frac{r}{y}$

In Exercises 45–56, an equation of the terminal side of an angle θ in standard position is given with a restriction on x. Sketch the least positive such angle θ, and find the values of the six trigonometric functions of θ. ***See Example 3.***

45. $2x + y = 0, x \geq 0$ **46.** $3x + 5y = 0, x \geq 0$

47. $-6x - y = 0, x \leq 0$ **48.** $-5x - 3y = 0, x \leq 0$

49. $-4x + 7y = 0, x \leq 0$ **50.** $6x - 5y = 0, x \geq 0$

51. $x + y = 0, x \geq 0$ **52.** $x - y = 0, x \geq 0$

53. $-\sqrt{3}x + y = 0, x \leq 0$ **54.** $\sqrt{3}x + y = 0, x \leq 0$

55. $x = 0, y \geq 0$ **56.** $y = 0, x \leq 0$

To work Exercises 57–77, begin by reproducing the graph in **Figure 29.** *Keep in mind that for each of the four points labeled in the figure, $r = 1$. For each quadrantal angle, identify the appropriate values of x, y, and r to find the indicated function value. If it is undefined, say so.* ***See Example 4.***

57. $\cos 90°$ **58.** $\sin 90°$ **59.** $\tan 180°$

60. $\cot 90°$ **61.** $\sec 180°$ **62.** $\csc 270°$

63. $\sin(-270°)$ **64.** $\cos(-90°)$ **65.** $\cot 540°$

66. $\tan 450°$ **67.** $\csc(-450°)$ **68.** $\sec(-540°)$

69. $\sin 1800°$ **70.** $\cos 1800°$ **71.** $\csc 1800°$

72. $\cot 1800°$ **73.** $\sec 1800°$ **74.** $\tan 1800°$

75. $\cos(-900°)$ **76.** $\sin(-900°)$ **77.** $\tan(-900°)$

78. Explain how the answer to **Exercise 77** can be given once the answers to **Exercises 75 and 76** have been determined.

Use the trigonometric function values of quadrantal angles given in this section to evaluate each expression. An expression such as $\cot^2 90°$ *means* $(\cot 90°)^2$, *which is equal to* $0^2 = 0$.

79. $\cos 90° + 3 \sin 270°$ **80.** $\tan 0° - 6 \sin 90°$

81. $3 \sec 180° - 5 \tan 360°$ **82.** $4 \csc 270° + 3 \cos 180°$

83. $\tan 360° + 4 \sin 180° + 5 \cos^2 180°$ **84.** $2 \sec 0° + 4 \cot^2 90° + \cos 360°$

85. $\sin^2 180° + \cos^2 180°$ **86.** $\sin^2 360° + \cos^2 360°$

87. $\sec^2 180° - 3 \sin^2 360° + \cos 180°$ **88.** $5 \sin^2 90° + 2 \cos^2 270° - \tan 360°$

89. $-2 \sin^4 0° + 3 \tan^2 0°$ **90.** $-3 \sin^4 90° + 4 \cos^3 180°$

91. $\sin^2(-90°) + \cos^2(-90°)$ **92.** $\cos^2(-180°) + \sin^2(-180°)$

If n is an integer, $n \cdot 180°$ *represents an integer multiple of* $180°$, $(2n + 1) \cdot 90°$ *represents an odd integer multiple of* $90°$, *and so on. Decide whether each expression is equal to* 0, 1, *or* -1 *or is* undefined.

93. $\cos[(2n + 1) \cdot 90°]$ **94.** $\sin[n \cdot 180°]$

95. $\tan[n \cdot 180°]$ **96.** $\tan[(2n + 1) \cdot 90°]$

97. $\sin[270° + n \cdot 360°]$ **98.** $\cot[n \cdot 180°]$

99. $\cot[(2n + 1) \cdot 90°]$ **100.** $\cos[n \cdot 360°]$

101. $\sec[(2n + 1) \cdot 90°]$ **102.** $\csc[n \cdot 180°]$

Concept Check *In later chapters we will study trigonometric functions of angles other than quadrantal angles, such as* $15°$, $30°$, $60°$, $75°$, *and so on. To prepare for some important concepts, provide conjectures in Exercises 103–106. Be sure that your calculator is in degree mode.*

103. The angles $15°$ and $75°$ are complementary. With your calculator determine $\sin 15°$ and $\cos 75°$. Make a conjecture about the sines and cosines of complementary angles, and test your hypothesis with other pairs of complementary angles. (*Note:* This relationship will be discussed in detail in **Section 2.1.**)

104. The angles $25°$ and $65°$ are complementary. With your calculator determine $\tan 25°$ and $\cot 65°$. Make a conjecture about the tangents and cotangents of complementary angles, and test your hypothesis with other pairs of complementary angles. (*Note:* This relationship will be discussed in detail in **Section 2.1.**)

105. With your calculator determine $\sin 10°$ and $\sin(-10°)$. Make a conjecture about the sines of an angle and its negative, and test your hypothesis with other angles. (*Note:* This relationship will be discussed in detail in **Section 5.1.**)

106. With your calculator determine $\cos 20°$ and $\cos(-20°)$. Make a conjecture about the cosines of an angle and its negative, and test your hypothesis with other angles. (*Note:* This relationship will be discussed in detail in **Section 5.1.**)

In Exercises 107–112, set your TI graphing calculator in parametric and degree modes. Set the window and functions (see the third screen) as shown here, and graph. A circle of radius 1 *will appear on the screen. Trace to move a short distance around the circle. In the screen, the point on the circle corresponds to an angle* T = 25°. *Since* $r = 1$, $\cos 25°$ *is* X = 0.90630779, *and* $\sin 25°$ *is* Y = 0.42261826.

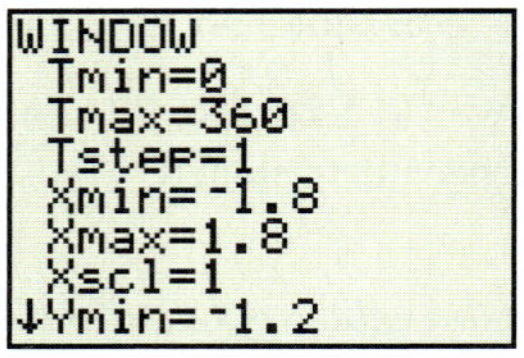

This screen is a continuation of the previous one.

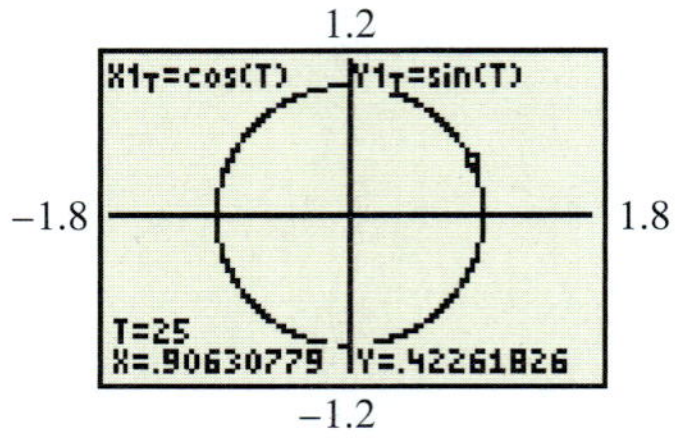

107. Use the right- and left-arrow keys to move to the point corresponding to 20° (T = 20). What are $\cos 20°$ and $\sin 20°$?

108. For what angle T, $0° \le \text{T} \le 90°$, is $\cos \text{T} \approx 0.766$?

109. For what angle T, $0° \le \text{T} \le 90°$, is $\sin \text{T} \approx 0.574$?

110. For what angle T, $0° \le \text{T} \le 90°$, does cos T equal sin T?

111. As T increases from 0° to 90°, does the cosine increase or decrease? What about the sine?

112. As T increases from 90° to 180°, does the cosine increase or decrease? What about the sine?

1.4 Using the Definitions of the Trigonometric Functions

- Reciprocal Identities
- Signs and Ranges of Function Values
- Pythagorean Identities
- Quotient Identities

Identities are equations that are true for all values of the variables for which all expressions are defined. Identities are studied in more detail in **Chapter 5.**

$$(x + y)^2 = x^2 + 2xy + y^2 \qquad 2(x + 3) = 2x + 6 \qquad \text{Identities (Appendix A)}$$

Reciprocal Identities

Recall the definition of a reciprocal: the **reciprocal** of the nonzero number x is $\frac{1}{x}$. For example, the reciprocal of 2 is $\frac{1}{2}$, and the reciprocal of $\frac{8}{11}$ is $\frac{11}{8}$. There is no reciprocal for 0. Scientific calculators have a reciprocal key, usually labeled $\boxed{1/x}$ or $\boxed{x^{-1}}$. Using this key gives the reciprocal of any nonzero number entered in the display.

The definitions of the trigonometric functions in the previous section were written so that functions in the same column were reciprocals of each other. Since $\sin \theta = \frac{y}{r}$ and $\csc \theta = \frac{r}{y}$,

$$\sin \theta = \frac{1}{\csc \theta} \quad \text{and} \quad \csc \theta = \frac{1}{\sin \theta}, \quad \text{provided } \sin \theta \neq 0.$$

Also, $\cos \theta$ and $\sec \theta$ are reciprocals, as are $\tan \theta$ and $\cot \theta$. The **reciprocal identities** hold for any angle θ that does not lead to a 0 denominator.

Reciprocal Identities

For all angles θ for which both functions are defined, the following identities hold.

$$\sin\theta = \frac{1}{\csc\theta} \qquad \cos\theta = \frac{1}{\sec\theta} \qquad \tan\theta = \frac{1}{\cot\theta}$$

$$\csc\theta = \frac{1}{\sin\theta} \qquad \sec\theta = \frac{1}{\cos\theta} \qquad \cot\theta = \frac{1}{\tan\theta}$$

The screens in **Figures 33(a) and (b)** show how to find csc 90°, sec 180°, and csc(−270°), using the appropriate reciprocal identities and the reciprocal key of a graphing calculator in degree mode. Attempting to find sec 90° by entering $\frac{1}{\cos 90°}$ produces an ERROR message, indicating that the reciprocal is undefined. See **Figure 33(c).** Compare these results with the ones found in the table of quadrantal angle function values in **Section 1.3.**

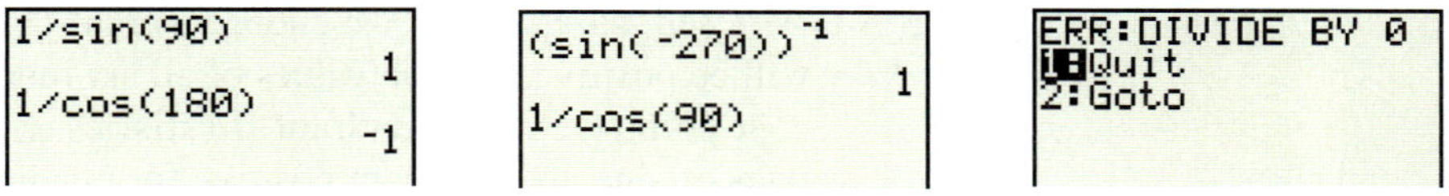

(a) (b) (c)

Figure 33

CAUTION ***Be sure not to use the inverse trigonometric function keys to find reciprocal function values.*** For example,

$$\sin^{-1}(90°) \neq \frac{1}{\sin(90°)}.$$

Inverse trigonometric functions are covered in **Section 2.3.**

The reciprocal identities can be written in different forms. For example,

$$\sin\theta = \frac{1}{\csc\theta} \quad \text{can be written} \quad \csc\theta = \frac{1}{\sin\theta}, \quad \text{or} \quad (\sin\theta)(\csc\theta) = 1.$$

EXAMPLE 1 Using the Reciprocal Identities

Find each function value.

(a) $\cos\theta$, given that $\sec\theta = \frac{5}{3}$ **(b)** $\sin\theta$, given that $\csc\theta = -\frac{\sqrt{12}}{2}$

SOLUTION

(a) Since $\cos\theta$ is the reciprocal of $\sec\theta$,

$$\cos\theta = \frac{1}{\sec\theta} = \frac{1}{\frac{5}{3}} = 1 \div \frac{5}{3} = 1 \cdot \frac{3}{5} = \frac{3}{5}. \quad \text{Simplify the complex fraction.}$$

(b) $\sin\theta = \dfrac{1}{-\frac{\sqrt{12}}{2}}$ $\quad$ $\sin\theta = \frac{1}{\csc\theta}$ and $\csc\theta = -\frac{\sqrt{12}}{2}$

$= -\dfrac{2}{\sqrt{12}}$ $\quad$ Simplify the complex fraction as in part (a).

$= -\dfrac{2}{2\sqrt{3}}$ $\quad$ $\sqrt{12} = \sqrt{4 \cdot 3} = 2\sqrt{3}$

$= -\dfrac{1}{\sqrt{3}}$ $\quad$ Divide out the common factor 2.

$= -\dfrac{1}{\sqrt{3}} \cdot \dfrac{\sqrt{3}}{\sqrt{3}}$ $\quad$ Rationalize the denominator.

$= -\dfrac{\sqrt{3}}{3}$ $\quad$ Multiply.

✔ *Now Try Exercises 1 and 9.*

Signs and Ranges of Function Values In the definitions of the trigonometric functions, r is the distance from the origin to the point (x, y). This distance is undirected, so $r > 0$. If we choose a point (x, y) in quadrant I, then both x and y will be positive, and the values of all six functions will be positive.

A point (x, y) in quadrant II satisfies $x < 0$ and $y > 0$. This makes the values of sine and cosecant positive for quadrant II angles, while the other four functions take on negative values. Similar results can be obtained for the other quadrants.

This important information is summarized here.

Signs of Function Values

θ in Quadrant	$\sin\theta$	$\cos\theta$	$\tan\theta$	$\cot\theta$	$\sec\theta$	$\csc\theta$
I	+	+	+	+	+	+
II	+	−	−	−	−	+
III	−	−	+	+	−	−
IV	−	+	−	−	+	−

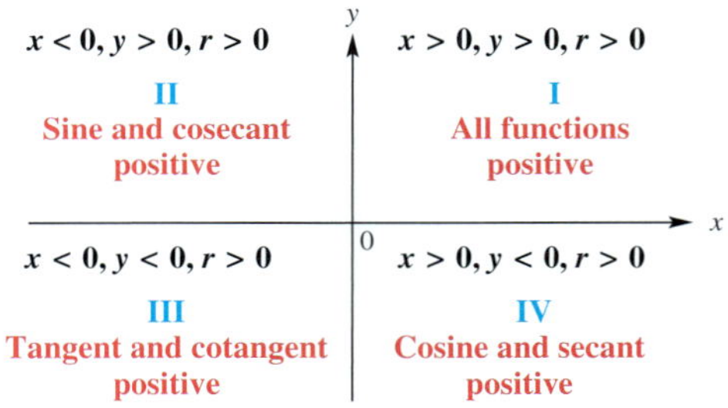

EXAMPLE 2 Determining Signs of Functions of Nonquadrantal Angles

Determine the signs of the trigonometric functions of an angle in standard position with the given measure.

(a) 87° $\qquad$ **(b)** 300° $\qquad$ **(c)** −200°

SOLUTION

(a) An angle of 87° is in the first quadrant, with x, y, and r all positive, so all of its trigonometric function values are positive.

(b) A 300° angle is in quadrant IV, so the cosine and secant are positive, while the sine, cosecant, tangent, and cotangent are negative.

(c) A −200° angle is in quadrant II. The sine and cosecant are positive, and all other function values are negative.

✔ *Now Try Exercises 19, 21, and 25.*

NOTE Because numbers that are reciprocals always have the same sign, the sign of a function value automatically determines the sign of the reciprocal function value.

EXAMPLE 3 Identifying the Quadrant of an Angle

Identify the quadrant (or possible quadrants) of an angle θ that satisfies the given conditions.

(a) $\sin \theta > 0$, $\tan \theta < 0$ **(b)** $\cos \theta < 0$, $\sec \theta < 0$

SOLUTION

(a) Since $\sin \theta > 0$ in quadrants I and II and $\tan \theta < 0$ in quadrants II and IV, both conditions are met only in quadrant II.

(b) The cosine and secant functions are both negative in quadrants II and III, so in this case θ could be in either of these two quadrants.

✔ *Now Try Exercises 35 and 41.*

Figure 34(a) shows an angle θ as it increases in measure from near 0° toward 90°. In each case, the value of r is the same. As the measure of the angle increases, y increases but never exceeds r, so $y \leq r$. Dividing both sides by the positive number r gives $\frac{y}{r} \leq 1$.

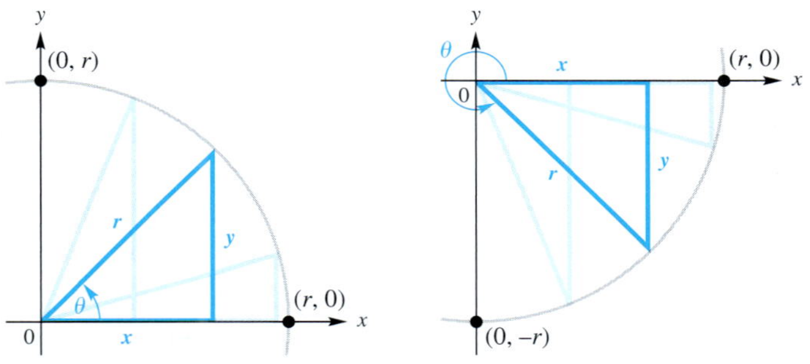

(a) (b)

Figure 34

In a similar way, angles in quadrant IV as in **Figure 34(b)** suggest that

$$-1 \leq \frac{y}{r},$$

so $$-1 \leq \frac{y}{r} \leq 1$$

and $$\mathbf{-1 \leq \sin \theta \leq 1.}$$ $\frac{y}{r} = \sin \theta$ for any angle θ. (Section 1.3)

Similarly, $$\mathbf{-1 \leq \cos \theta \leq 1.}$$

The tangent of an angle is defined as $\frac{y}{x}$. It is possible that $x < y$, $x = y$, or $x > y$. Thus, $\frac{y}{x}$ can take any value, so **tan θ can be any real number, as can cot θ.**

The functions $\sec \theta$ and $\csc \theta$ are reciprocals of the functions $\cos \theta$ and $\sin \theta$, respectively, making

$$\mathbf{\sec \theta \leq -1} \quad \text{or} \quad \mathbf{\sec \theta \geq 1} \quad \text{and} \quad \mathbf{\csc \theta \leq -1} \quad \text{or} \quad \mathbf{\csc \theta \geq 1.}$$

In summary, the ranges of the trigonometric functions are as follows.

Ranges of Trigonometric Functions

Trigonometric Function of θ	Range (Set-Builder Notation)	Range (Interval Notation)
$\sin\theta$, $\cos\theta$	$\{y \mid \lvert y\rvert \le 1\}$	$[-1, 1]$
$\tan\theta$, $\cot\theta$	$\{y \mid y \text{ is a real number}\}$	$(-\infty, \infty)$
$\sec\theta$, $\csc\theta$	$\{y \mid \lvert y\rvert \ge 1\}$	$(-\infty, -1] \cup [1, \infty)$

EXAMPLE 4 Deciding Whether a Value Is in the Range of a Trigonometric Function

Decide whether each statement is *possible* or *impossible*.

(a) $\sin\theta = 2.5$ **(b)** $\tan\theta = 110.47$ **(c)** $\sec\theta = 0.6$

SOLUTION

(a) For any value of θ, we know that $-1 \le \sin\theta \le 1$. Since $2.5 > 1$, it is impossible to find a value of θ that satisfies $\sin\theta = 2.5$.

(b) The tangent function can take on any real number value. Thus, $\tan\theta = 110.47$ is possible.

(c) Since $\lvert\sec\theta\rvert \ge 1$ for all θ for which the secant is defined, the statement $\sec\theta = 0.6$ is impossible.

✔ *Now Try Exercises 45, 49, and 51.*

The six trigonometric functions are defined in terms of x, y, and r, where the Pythagorean theorem shows that $r^2 = x^2 + y^2$ and $r > 0$. With these relationships, knowing the value of only one function and the quadrant in which the angle lies makes it possible to find the values of the other trigonometric functions.

EXAMPLE 5 Finding All Function Values Given One Value and the Quadrant

Suppose that angle θ is in quadrant II and $\sin\theta = \frac{2}{3}$. Find the values of the other five trigonometric functions.

SOLUTION Choose any point on the terminal side of angle θ. For simplicity, since $\sin\theta = \frac{y}{r}$, choose the point with $r = 3$.

$$\sin\theta = \frac{2}{3} \quad \text{Given value}$$

$$\frac{y}{r} = \frac{2}{3} \quad \text{Substitute } \tfrac{y}{r} \text{ for } \sin\theta.$$

Since $\frac{y}{r} = \frac{2}{3}$ and $r = 3$, then $y = 2$. To find x, use the equation $x^2 + y^2 = r^2$.

$$x^2 + y^2 = r^2$$

$$x^2 + 2^2 = 3^2 \quad \text{Substitute.}$$

$$x^2 + 4 = 9 \quad \text{Apply exponents.}$$

$$x^2 = 5 \quad \text{Subtract 4. (Appendix A)}$$

Remember *both* roots.

$$x = \sqrt{5} \quad \text{or} \quad x = -\sqrt{5} \quad \text{Square root property (Appendix A)}$$

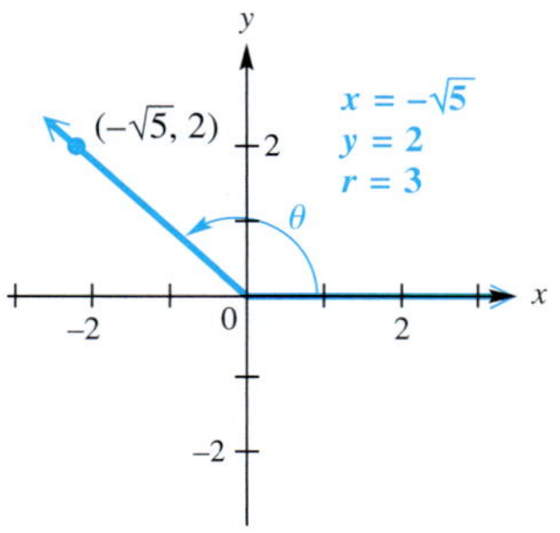

Figure 35

Since θ is in quadrant II, x must be negative. Choose $x = -\sqrt{5}$ so that the point $(-\sqrt{5}, 2)$ is on the terminal side of θ. See **Figure 35.** Now we can find the values of the remaining trigonometric functions.

$$\cos\theta = \frac{x}{r} = \frac{-\sqrt{5}}{3} = -\frac{\sqrt{5}}{3}$$

$$\sec\theta = \frac{r}{x} = \frac{3}{-\sqrt{5}} = -\frac{3}{\sqrt{5}} \cdot \frac{\sqrt{5}}{\sqrt{5}} = -\frac{3\sqrt{5}}{5}$$

$$\tan\theta = \frac{y}{x} = \frac{2}{-\sqrt{5}} = -\frac{2}{\sqrt{5}} \cdot \frac{\sqrt{5}}{\sqrt{5}} = -\frac{2\sqrt{5}}{5}$$

These have rationalized denominators.

$$\cot\theta = \frac{x}{y} = \frac{-\sqrt{5}}{2} = -\frac{\sqrt{5}}{2}$$

$$\csc\theta = \frac{r}{y} = \frac{3}{2}$$

✔ ***Now Try Exercise 71.***

Pythagorean Identities We derive three new identities from the relationship $x^2 + y^2 = r^2$.

$$\frac{x^2}{r^2} + \frac{y^2}{r^2} = \frac{r^2}{r^2} \qquad \text{Divide by } r^2.$$

$$\left(\frac{x}{r}\right)^2 + \left(\frac{y}{r}\right)^2 = 1 \qquad \text{Power rule for exponents; } \frac{a^m}{b^m} = \left(\frac{a}{b}\right)^m$$

$$(\cos\theta)^2 + (\sin\theta)^2 = 1 \qquad \cos\theta = \frac{x}{r},\ \sin\theta = \frac{y}{r} \text{ (Section 1.3)}$$

$$\mathbf{\sin^2\theta + \cos^2\theta = 1} \qquad \text{Apply exponents; commutative property}$$

Starting again with $x^2 + y^2 = r^2$ and dividing through by x^2 gives the following.

$$\frac{x^2}{x^2} + \frac{y^2}{x^2} = \frac{r^2}{x^2} \qquad \text{Divide by } x^2.$$

$$1 + \left(\frac{y}{x}\right)^2 = \left(\frac{r}{x}\right)^2 \qquad \text{Power rule for exponents}$$

$$1 + (\tan\theta)^2 = (\sec\theta)^2 \qquad \tan\theta = \frac{y}{x},\ \sec\theta = \frac{r}{x} \text{ (Section 1.3)}$$

$$\mathbf{\tan^2\theta + 1 = \sec^2\theta} \qquad \text{Apply exponents; commutative property}$$

Similarly, dividing through by y^2 leads to another identity.

$$\mathbf{1 + \cot^2\theta = \csc^2\theta}$$

These three identities are the **Pythagorean identities** since the original equation that led to them, $x^2 + y^2 = r^2$, comes from the Pythagorean theorem.

Pythagorean Identities

For all angles θ for which the function values are defined, the following identities hold.

$$\mathbf{\sin^2\theta + \cos^2\theta = 1} \qquad \mathbf{\tan^2\theta + 1 = \sec^2\theta} \qquad \mathbf{1 + \cot^2\theta = \csc^2\theta}$$

As before, we have given only one form of each identity. However, algebraic transformations produce equivalent identities. For example, by subtracting $\sin^2\theta$ from both sides of $\sin^2\theta + \cos^2\theta = 1$, we obtain an equivalent identity.

$$\cos^2\theta = 1 - \sin^2\theta \quad \text{Alternative form}$$

It is important to be able to transform these identities quickly and also to recognize their equivalent forms.

LOOKING AHEAD TO CALCULUS

The reciprocal, Pythagorean, and quotient identities are used in calculus to find derivatives and integrals of trigonometric functions. A standard technique of integration called **trigonometric substitution** relies on the Pythagorean identities.

Quotient Identities

Consider the quotient of $\sin\theta$ and $\cos\theta$, for $\cos\theta \neq 0$.

$$\frac{\sin\theta}{\cos\theta} = \frac{\frac{y}{r}}{\frac{x}{r}} = \frac{y}{r} \div \frac{x}{r} = \frac{y}{r} \cdot \frac{r}{x} = \frac{y}{x} = \tan\theta$$

Similarly, $\frac{\cos\theta}{\sin\theta} = \cot\theta$, for $\sin\theta \neq 0$. Thus, we have the **quotient identities.**

Quotient Identities

For all angles θ for which the denominators are not zero, the following identities hold.

$$\frac{\sin\theta}{\cos\theta} = \tan\theta \qquad \frac{\cos\theta}{\sin\theta} = \cot\theta$$

EXAMPLE 6 Using Identities to Find Function Values

Find $\sin\theta$ and $\tan\theta$, given that $\cos\theta = -\frac{\sqrt{3}}{4}$ and $\sin\theta > 0$.

SOLUTION Start with $\sin^2\theta + \cos^2\theta = 1$.

$$\sin^2\theta + \left(-\frac{\sqrt{3}}{4}\right)^2 = 1 \quad \text{Replace } \cos\theta \text{ with } -\tfrac{\sqrt{3}}{4}.$$

$$\sin^2\theta + \frac{3}{16} = 1 \quad \text{Square } -\tfrac{\sqrt{3}}{4}.$$

$$\sin^2\theta = \frac{13}{16} \quad \text{Subtract } \tfrac{3}{16}.$$

$$\sin\theta = \pm\frac{\sqrt{13}}{4} \quad \text{Take square roots.}$$

Choose the correct sign here.

$$\sin\theta = \frac{\sqrt{13}}{4} \quad \text{Choose the positive square root since } \sin\theta \text{ is positive.}$$

To find $\tan\theta$, use the quotient identity $\tan\theta = \frac{\sin\theta}{\cos\theta}$.

$$\tan\theta = \frac{\sin\theta}{\cos\theta} = \frac{\frac{\sqrt{13}}{4}}{-\frac{\sqrt{3}}{4}} = \frac{\sqrt{13}}{4}\left(-\frac{4}{\sqrt{3}}\right) = -\frac{\sqrt{13}}{\sqrt{3}}$$

$$= -\frac{\sqrt{13}}{\sqrt{3}} \cdot \frac{\sqrt{3}}{\sqrt{3}} = -\frac{\sqrt{39}}{3} \quad \text{Rationalize the denominator.}$$

✔ *Now Try Exercise 75.*

CAUTION ***In exercises like those of Examples 5 and 6, be careful to choose the correct sign when square roots are taken.*** You may wish to refer back to the diagrams preceding **Example 2.** They summarize the signs of the functions in the four quadrants.

EXAMPLE 7 Using Identities to Find Function Values

Find $\sin\theta$ and $\cos\theta$, given that $\tan\theta = \frac{4}{3}$ and θ is in quadrant III.

SOLUTION Since θ is in quadrant III, $\sin\theta$ and $\cos\theta$ will both be negative. It is tempting to say that since $\tan\theta = \frac{\sin\theta}{\cos\theta}$ and $\tan\theta = \frac{4}{3}$, then $\sin\theta = -4$ and $\cos\theta = -3$. This is *incorrect,* however, since both $\sin\theta$ and $\cos\theta$ must be in the interval $[-1, 1]$.

We use the Pythagorean identity $\tan^2\theta + 1 = \sec^2\theta$ to find $\sec\theta$, and then the reciprocal identity $\cos\theta = \frac{1}{\sec\theta}$ to find $\cos\theta$.

$$\tan^2\theta + 1 = \sec^2\theta \qquad \text{Pythagorean identity}$$

$$\left(\frac{4}{3}\right)^2 + 1 = \sec^2\theta \qquad \tan\theta = \frac{4}{3}$$

$$\frac{16}{9} + 1 = \sec^2\theta \qquad \text{Square } \frac{4}{3}.$$

$$\frac{25}{9} = \sec^2\theta \qquad \text{Add.}$$

Be careful to choose the correct sign here.

$$-\frac{5}{3} = \sec\theta \qquad \text{Choose the negative square root since } \sec\theta \text{ is negative when } \theta \text{ is in quadrant III.}$$

$$-\frac{3}{5} = \cos\theta \qquad \text{Secant and cosine are reciprocals.}$$

Since $\sin^2\theta = 1 - \cos^2\theta$,

$$\sin^2\theta = 1 - \left(-\frac{3}{5}\right)^2 \qquad \cos\theta = -\frac{3}{5}$$

$$\sin^2\theta = 1 - \frac{9}{25} \qquad \text{Square } -\frac{3}{5}.$$

$$\sin^2\theta = \frac{16}{25} \qquad \text{Subtract.}$$

Again, be careful.

$$\sin\theta = -\frac{4}{5}. \qquad \text{Choose the negative square root.}$$

✔ *Now Try Exercise 73.*

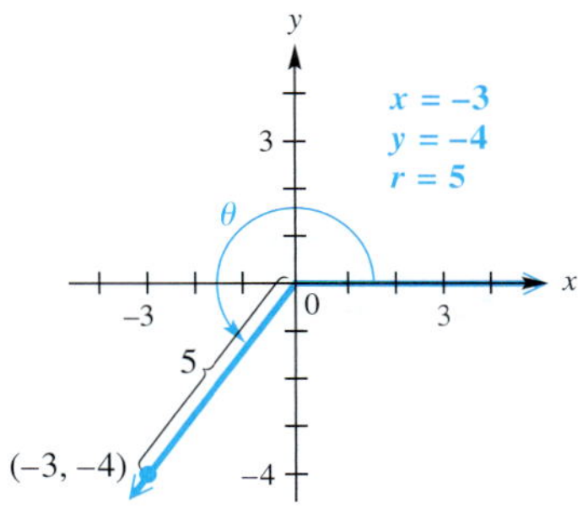

Figure 36

NOTE **Example 7** can also be worked by sketching θ in standard position in quadrant III, finding r to be 5, and then using the definitions of $\sin\theta$ and $\cos\theta$ in terms of x, y, and r. See **Figure 36.**

When using this method, be sure to choose the correct signs for x and y as determined by the quadrant in which the terminal side of θ lies. This is analogous to choosing the correct signs after applying the Pythagorean identities.

1.4 Exercises

Use the appropriate reciprocal identity to find each function value. Rationalize denominators when applicable. ***See Example 1.***

1. $\sec\theta$, given that $\cos\theta = \frac{2}{3}$

2. $\sec\theta$, given that $\cos\theta = \frac{5}{8}$

3. $\csc\theta$, given that $\sin\theta = -\frac{3}{7}$

4. $\csc\theta$, given that $\sin\theta = -\frac{8}{43}$

5. $\cot\theta$, given that $\tan\theta = 5$

6. $\cot\theta$, given that $\tan\theta = 18$

7. $\cos\theta$, given that $\sec\theta = -\frac{5}{2}$

8. $\cos\theta$, given that $\sec\theta = -\frac{11}{7}$

9. $\sin\theta$, given that $\csc\theta = \frac{\sqrt{8}}{2}$

10. $\sin\theta$, given that $\csc\theta = \frac{\sqrt{24}}{3}$

11. $\tan\theta$, given that $\cot\theta = -2.5$

12. $\tan\theta$, given that $\cot\theta = -0.01$

13. $\sin\theta$, given that $\csc\theta = 1.42716321$

14. $\cos\theta$, given that $\sec\theta = 9.80425133$

15. Can a given angle θ satisfy both $\sin\theta > 0$ and $\csc\theta < 0$? Explain.

16. Explain what is wrong with the following item that appears on a trigonometry test:

"*Find* $\sec\theta$, *given that* $\cos\theta = \frac{3}{2}$."

17. *Concept Check* What is **wrong** with the following statement: $\tan 90° = \frac{1}{\cot 90°}$?

18. *Concept Check* One form of a particular reciprocal identity is $\tan\theta = \frac{1}{\cot\theta}$. Give two other, equivalent forms of this identity.

Determine the signs of the trigonometric functions of an angle in standard position with the given measure. ***See Example 2.***

19. $74°$ **20.** $84°$ **21.** $218°$ **22.** $195°$

23. $178°$ **24.** $125°$ **25.** $-80°$ **26.** $-15°$

27. $855°$ **28.** $1005°$ **29.** $-345°$ **30.** $-640°$

Identify the quadrant (or possible quadrants) of an angle θ that satisfies the given conditions. ***See Example 3.***

31. $\sin\theta > 0, \csc\theta > 0$ **32.** $\cos\theta > 0, \sec\theta > 0$ **33.** $\cos\theta > 0, \sin\theta > 0$

34. $\sin\theta > 0, \tan\theta > 0$ **35.** $\tan\theta < 0, \cos\theta < 0$ **36.** $\cos\theta < 0, \sin\theta < 0$

37. $\sec\theta > 0, \csc\theta > 0$ **38.** $\csc\theta > 0, \cot\theta > 0$ **39.** $\sec\theta < 0, \csc\theta < 0$

40. $\cot\theta < 0, \sec\theta < 0$ **41.** $\sin\theta < 0, \csc\theta < 0$ **42.** $\tan\theta < 0, \cot\theta < 0$

43. Explain why the answers to **Exercises 33 and 37** are the same.

44. Explain why there is no angle θ that satisfies $\tan\theta > 0, \cot\theta < 0$.

Decide whether each statement is possible *or* impossible *for some angle θ.* ***See Example 4.***

45. $\sin\theta = 2$ **46.** $\sin\theta = 3$ **47.** $\cos\theta = -0.96$

48. $\cos\theta = -0.56$ **49.** $\tan\theta = 0.93$ **50.** $\cot\theta = 0.93$

51. $\sec\theta = -0.3$ **52.** $\sec\theta = -0.9$ **53.** $\csc\theta = 100$

54. $\csc\theta = -100$ **55.** $\cot\theta = -4$ **56.** $\cot\theta = -6$

Concept Check *Determine whether each statement is* possible *or* impossible *for some angle θ.*

57. $\sin\theta = \frac{1}{2}, \csc\theta = 2$ **58.** $\tan\theta = 2, \cot\theta = -2$ **59.** $\cos\theta = -2, \sec\theta = \frac{1}{2}$

60. Explain why there is no angle θ that satisfies $\cos\theta = \frac{1}{2}$ and $\sec\theta = -2$.

Use identities to solve each of the following. ***See Examples 5–7.***

61. Find $\cos\theta$, given that $\sin\theta = \frac{3}{5}$ and θ is in quadrant II.

62. Find $\sin\theta$, given that $\cos\theta = \frac{4}{5}$ and θ is in quadrant IV.

63. Find $\csc\theta$, given that $\cot\theta = -\frac{1}{2}$ and θ is in quadrant IV.

64. Find $\sec\theta$, given that $\tan\theta = \frac{\sqrt{7}}{3}$ and θ is in quadrant III.

65. Find $\tan\theta$, given that $\sin\theta = \frac{1}{2}$ and θ is in quadrant II.

66. Find $\cot\theta$, given that $\csc\theta = -2$ and θ is in quadrant III.

67. Find $\cot\theta$, given that $\csc\theta = -3.5891420$ and θ is in quadrant III.

68. Find $\tan\theta$, given that $\sin\theta = 0.49268329$ and θ is in quadrant II.

Find the five remaining trigonometric function values for each angle θ. ***See Examples 5–7.***

69. $\tan\theta = -\frac{15}{8}$, and θ is in quadrant II

70. $\cos\theta = -\frac{3}{5}$, and θ is in quadrant III

71. $\sin\theta = \frac{\sqrt{5}}{7}$, and θ is in quadrant I

72. $\tan\theta = \sqrt{3}$, and θ is in quadrant III

73. $\cot\theta = \frac{\sqrt{3}}{8}$, and θ is in quadrant I

74. $\csc\theta = 2$, and θ is in quadrant II

75. $\sin\theta = \frac{\sqrt{2}}{6}$, and $\cos\theta < 0$

76. $\cos\theta = \frac{\sqrt{5}}{8}$, and $\tan\theta < 0$

77. $\sec\theta = -4$, and $\sin\theta > 0$

78. $\csc\theta = -3$, and $\cos\theta > 0$

79. $\sin\theta = 0.164215$, and θ is in quadrant II

80. $\cot\theta = -1.49586$, and θ is in quadrant IV

Work each problem.

81. Derive the identity $1 + \cot^2\theta = \csc^2\theta$ by dividing $x^2 + y^2 = r^2$ by y^2.

82. Using a method similar to the one given in this section showing that $\frac{\sin\theta}{\cos\theta} = \tan\theta$, show that $\frac{\cos\theta}{\sin\theta} = \cot\theta$.

83. *Concept Check* *True* or *false*: For all angles θ, $\sin\theta + \cos\theta = 1$. If the statement is false, give an example showing why.

84. *Concept Check* *True* or *false*: Since $\cot\theta = \frac{\cos\theta}{\sin\theta}$, if $\cot\theta = \frac{1}{2}$ with θ in quadrant I, then $\cos\theta = 1$ and $\sin\theta = 2$. If the statement is false, give an explanation showing why.

Concept Check *Suppose that* $90° < \theta < 180°$. *Find the sign of each function value.*

85. $\sin 2\theta$

86. $\csc 2\theta$

87. $\tan\dfrac{\theta}{2}$

88. $\cot\dfrac{\theta}{2}$

89. $\cot(\theta + 180°)$

90. $\tan(\theta + 180°)$

91. $\cos(-\theta)$

92. $\sec(-\theta)$

Concept Check *Suppose that* $-90° < \theta < 90°$. *Find the sign of each function value.*

93. $\cos\dfrac{\theta}{2}$

94. $\sec\dfrac{\theta}{2}$

95. $\sec(\theta + 180°)$

96. $\cos(\theta + 180°)$

97. $\sec(-\theta)$

98. $\cos(-\theta)$

99. $\cos(\theta - 180°)$

100. $\sec(\theta - 180°)$

Concept Check *Find a value of each variable.*

101. $\tan(3\theta - 4°) = \dfrac{1}{\cot(5\theta - 8°)}$

102. $\cos(6\theta + 5°) = \dfrac{1}{\sec(4\theta + 15°)}$

103. $\sin(4\theta + 2°)\csc(3\theta + 5°) = 1$

104. $\sec(2\theta + 6°)\cos(5\theta + 3°) = 1$

105. *Concept Check* The screen below was obtained with the calculator in degree mode. How can we use it to justify that an angle of 14,879° is a quadrant II angle?

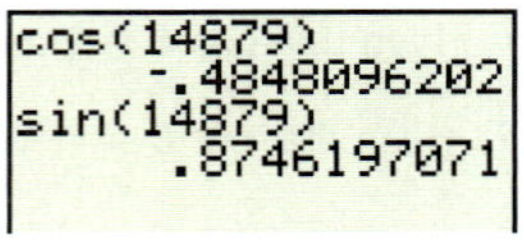

106. *Concept Check* The screen below was obtained with the calculator in degree mode. In which quadrant does a 1294° angle lie?

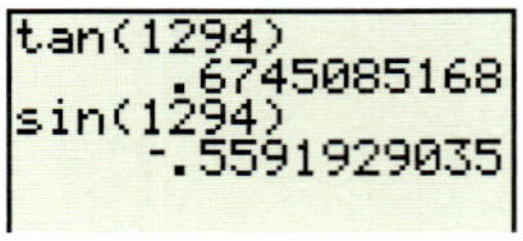

Chapter 1 Test Prep

Key Terms

1.1 line, line segment (or segment), ray, endpoint of a ray, angle, side of an angle, vertex of an angle, initial side, terminal side, positive angle, negative angle, degree, acute angle, right angle, obtuse angle, straight angle, complementary angles (complements), supplementary angles (supplements), minute, second, angle in standard position, quadrantal angle, coterminal angles

1.2 vertical angles, parallel lines, transversal, similar triangles, congruent triangles

1.3 sine (sin), cosine (cos), tangent (tan), cotangent (cot), secant (sec), cosecant (csc), degree mode

1.4 reciprocal

New Symbols

⌝	right angle symbol (for a right triangle)
θ	Greek letter theta
°	degree
′	minute
″	second

Quick Review

Concepts	Examples

1.1 Angles

Types of Angles

Two angles with a sum of 90° are complementary angles, and two angles with a sum of 180° are supplementary angles.

$70°$ and $90° - 70° = 20°$ are complementary.

$70°$ and $180° - 70° = 110°$ are supplementary.

$$1 \text{ degree} = 60 \text{ minutes} \quad (1° = 60')$$

$$1 \text{ minute} = 60 \text{ seconds} \quad (1' = 60'')$$

$$15°\,30'\,45'' = 15° + \frac{30°}{60} + \frac{45°}{3600}$$

$$= 15.5125° \qquad \text{Decimal degrees}$$

Coterminal angles have measures that differ by a multiple of 360°. Their terminal sides coincide when in standard position.

The acute angle θ in the figure is in standard position. If θ measures 46°, find the measure of a negative coterminal angle.

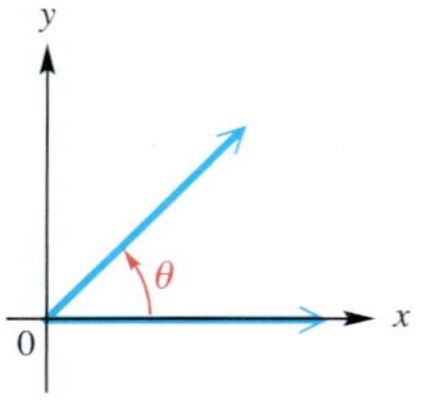

$$46° - 360° = -314°$$

Concepts	Examples

1.2 Angle Relationships and Similar Triangles

Vertical angles have equal measures.

The sum of the measures of the angles of any triangle is 180°.

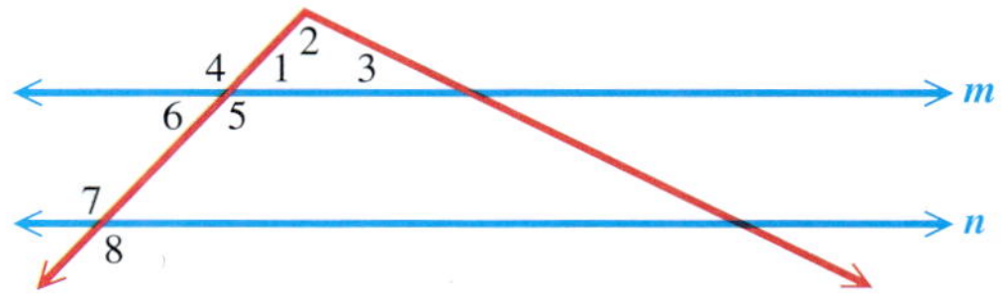

m and n are parallel lines.

Vertical angles 4 and 5 are equal.

The sum of angles 1, 2, and 3 is 180°.

When a transversal intersects parallel lines, the following angles formed have equal measure: alternate interior angles, alternate exterior angles, and corresponding angles. Interior angles on the same side of the transversal are supplementary.

Refer to the diagram above. Angles 5 and 7 are alternate interior angles, so they are equal. Angles 4 and 8 are alternate exterior angles, so they are equal. Angles 4 and 7 are corresponding angles, so they are equal. Angles 6 and 7 are interior angles on the same side of the transversal, so they are supplementary.

Similar triangles have corresponding angles with the same measures and have corresponding sides proportional.

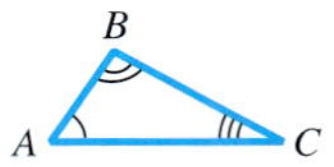

Corresponding angles as marked in triangles ABC and DEF are equal.

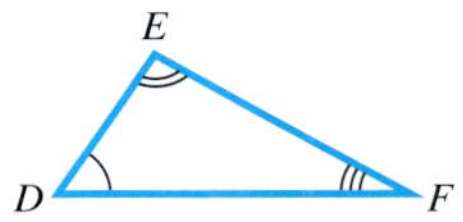

Also, $\frac{AB}{DE} = \frac{BC}{EF} = \frac{AC}{DF}$.

Congruent triangles are the same size and the same shape.

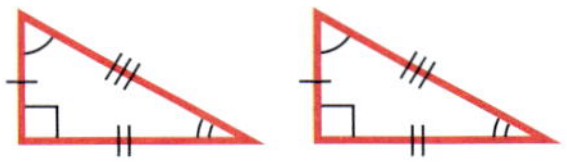

Corresponding angles are equal, and corresponding sides are equal.

1.3 Trigonometric Functions

Definitions of the Trigonometric Functions

Let (x, y) be a point other than the origin on the terminal side of an angle θ in standard position. Let $r = \sqrt{x^2 + y^2}$ represent the distance from the origin to (x, y). Then

$$\sin\theta = \frac{y}{r} \qquad \cos\theta = \frac{x}{r} \qquad \tan\theta = \frac{y}{x}\ (x \neq 0)$$

$$\csc\theta = \frac{r}{y}\ (y \neq 0) \quad \sec\theta = \frac{r}{x}\ (x \neq 0) \quad \cot\theta = \frac{x}{y}\ (y \neq 0).$$

See the summary table of trigonometric function values for quadrantal angles in **Section 1.3.**

If the point $(-2, 3)$ is on the terminal side of angle θ in standard position, then $x = -2$, $y = 3$, and

$$r = \sqrt{(-2)^2 + 3^2} = \sqrt{4 + 9} = \sqrt{13}.$$

Then

$$\sin\theta = \frac{3\sqrt{13}}{13}, \quad \cos\theta = -\frac{2\sqrt{13}}{13}, \quad \tan\theta = -\frac{3}{2},$$

$$\csc\theta = \frac{\sqrt{13}}{3}, \quad \sec\theta = -\frac{\sqrt{13}}{2}, \quad \cot\theta = -\frac{2}{3}.$$

1.4 Using the Definitions of the Trigonometric Functions

Reciprocal Identities

$$\sin\theta = \frac{1}{\csc\theta} \qquad \cos\theta = \frac{1}{\sec\theta} \qquad \tan\theta = \frac{1}{\cot\theta}$$

$$\csc\theta = \frac{1}{\sin\theta} \qquad \sec\theta = \frac{1}{\cos\theta} \qquad \cot\theta = \frac{1}{\tan\theta}$$

If $\cot\theta = -\frac{2}{3}$, find $\tan\theta$.

$$\tan\theta = \frac{1}{\cot\theta} = \frac{1}{-\frac{2}{3}} = -\frac{3}{2}$$

(continued)

Concepts	Examples
Pythagorean Identities $\sin^2\theta + \cos^2\theta = 1 \quad \tan^2\theta + 1 = \sec^2\theta \quad 1 + \cot^2\theta = \csc^2\theta$	Use the function values for the example from **Section 1.3** to illustrate the Pythagorean identities. $\sin^2\theta + \cos^2\theta = \left(\frac{3\sqrt{13}}{13}\right)^2 + \left(-\frac{2\sqrt{13}}{13}\right)^2 = \frac{9}{13} + \frac{4}{13} = 1$ $\tan^2\theta + 1 = \left(-\frac{3}{2}\right)^2 + 1 = \frac{13}{4} = \left(-\frac{\sqrt{13}}{2}\right)^2 = \sec^2\theta$ $1 + \cot^2\theta = 1 + \left(-\frac{2}{3}\right)^2 = \frac{13}{9} = \left(\frac{\sqrt{13}}{3}\right)^2 = \csc^2\theta$
Quotient Identities $\frac{\sin\theta}{\cos\theta} = \tan\theta \quad \frac{\cos\theta}{\sin\theta} = \cot\theta$	Use the function values for the example from **Section 1.3** to illustrate $\frac{\sin\theta}{\cos\theta} = \tan\theta$. $\frac{\sin\theta}{\cos\theta} = \frac{\frac{3\sqrt{13}}{13}}{-\frac{2\sqrt{13}}{13}} = \frac{3\sqrt{13}}{13}\left(-\frac{13}{2\sqrt{13}}\right) = -\frac{3}{2} = \tan\theta$
Signs of the Trigonometric Functions 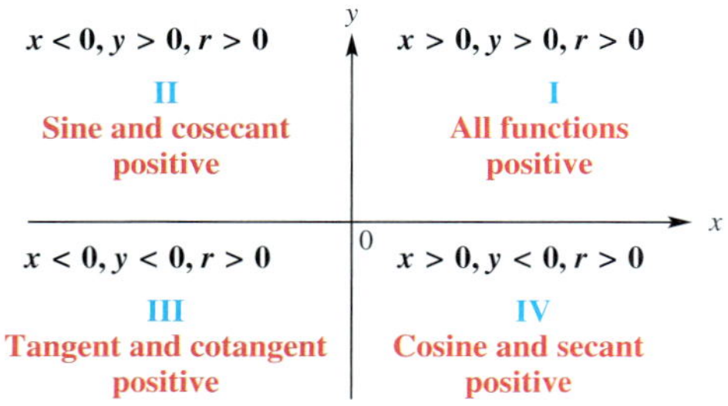	Identify the quadrant(s) of any angle θ that satisfies $\sin\theta < 0$, $\tan\theta > 0$. Since $\sin\theta < 0$ in quadrants III and IV, and $\tan\theta > 0$ in quadrants I and III, both conditions are met only in quadrant III.

Chapter 1 Review Exercises

1. Give the measures of the complement and the supplement of an angle measuring 35°.

Find the angle of least positive measure that is coterminal with each angle.

2. $-51°$ **3.** $-174°$ **4.** $792°$

5. Find the measure of each marked angle.

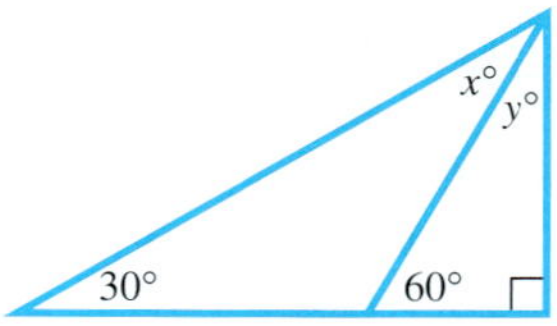

Work each problem.

6. *Rotating Pulley* A pulley is rotating 320 times per min. Through how many degrees does a point on the edge of the pulley move in $\frac{2}{3}$ sec?

7. *Rotating Propeller* The propeller of a speedboat rotates 650 times per min. Through how many degrees does a point on the edge of the propeller rotate in 2.4 sec?

Convert decimal degrees to degrees, minutes, seconds, and convert degrees, minutes, seconds to decimal degrees. Round to the nearest second or the nearest thousandth of a degree, as appropriate. Use a calculator as necessary.

8. $47° 25' 11''$ **9.** $119° 08' 03''$ **10.** $-61.5034°$ **11.** $275.1005°$

Find the measure of each marked angle.

12.

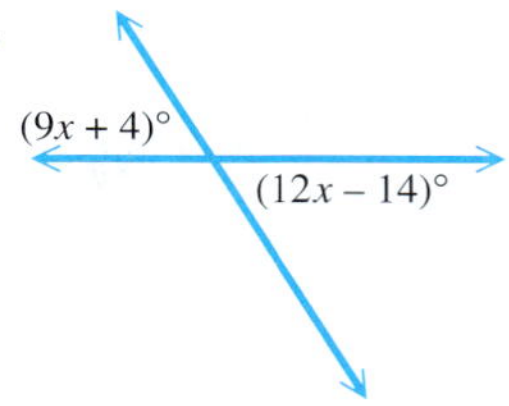

13.

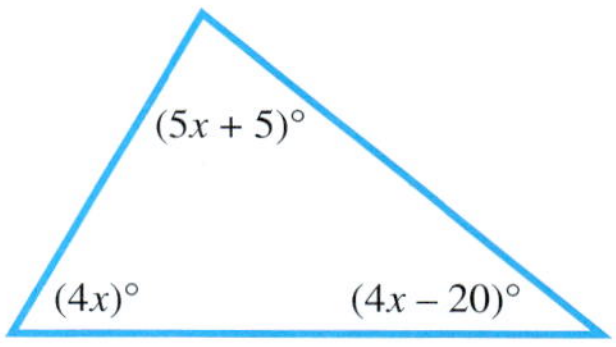

14. Express θ in terms of α and β.

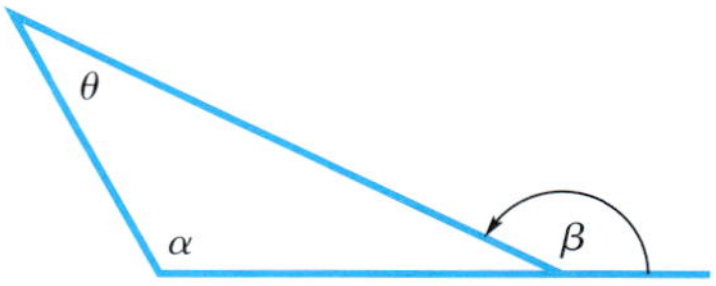

15. ***Length of a Road*** The flight path CP of a satellite carrying a camera with its lens at C is shown in the figure. Length PC represents the distance from the lens to the film PQ, and BA represents a straight road on the ground. Use the measurements given in the figure to find the length of the road. (*Source:* Kastner, B., *Space Mathematics,* NASA.)

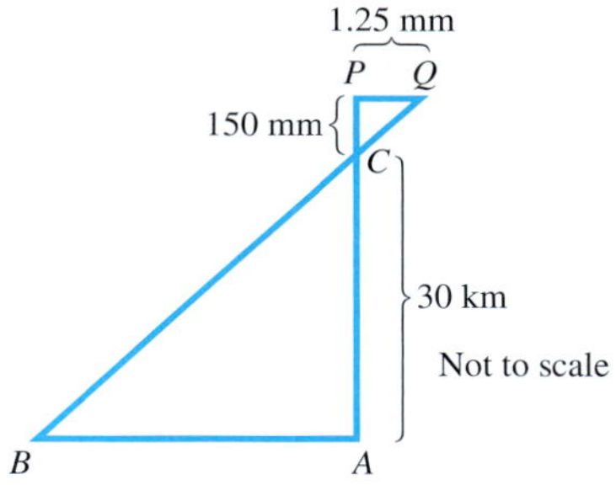

Find all unknown angle measures in each pair of similar triangles.

16.

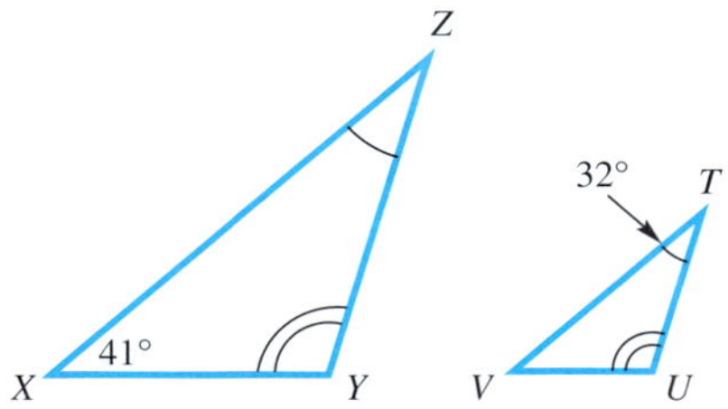

17.

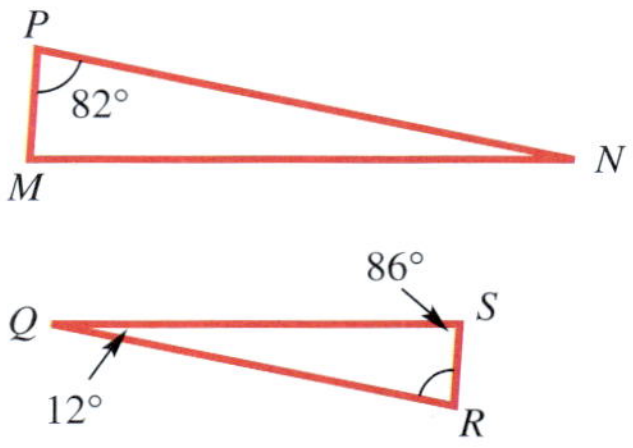

Find the unknown side lengths in each pair of similar triangles.

18.

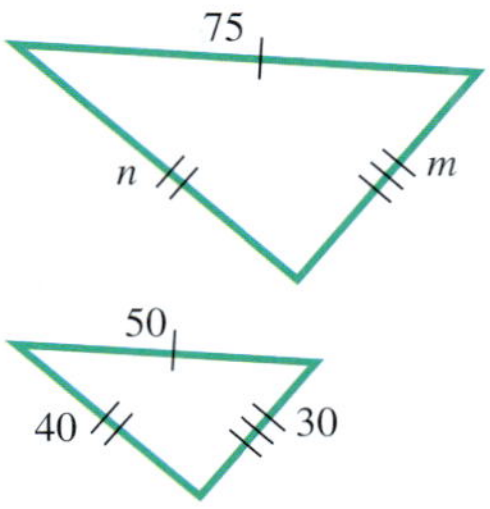

19.

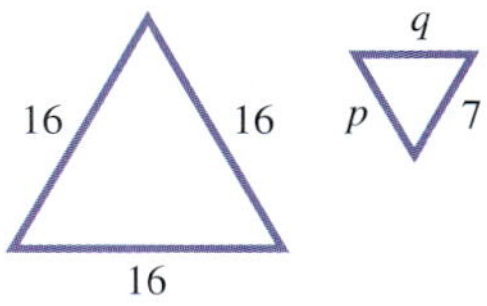

Find the unknown measurement. There are two similar triangles in each figure.

20.

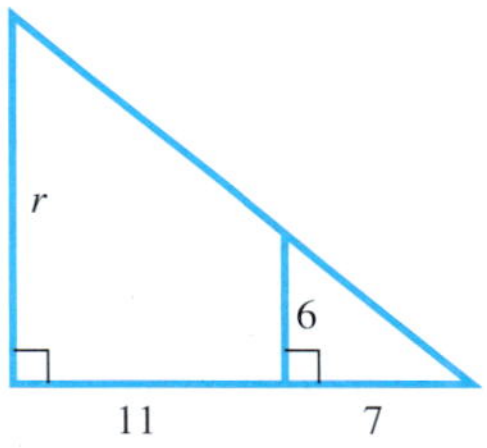

21.

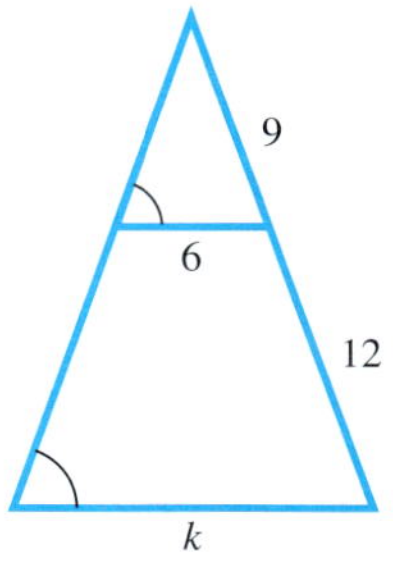

22. *Concept Check* Complete the following statement: If two triangles are similar, then their corresponding sides are ______________ and the measures of their corresponding angles are ______________.

23. *Length of a Shadow* If a tree 20 ft tall casts a shadow 8 ft long, how long would the shadow of a 30-ft tree be at the same time and place?

Find the six trigonometric function values for each angle. If a value is undefined, say so.

24.

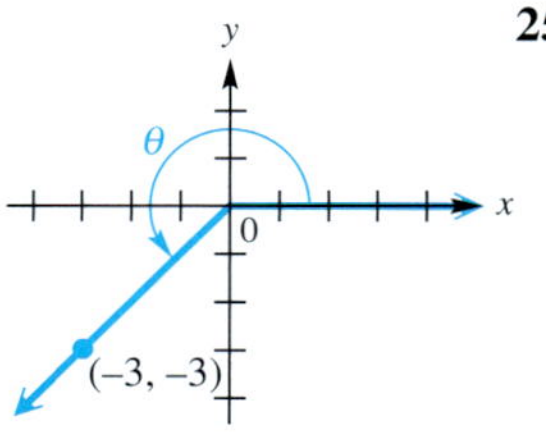

25.

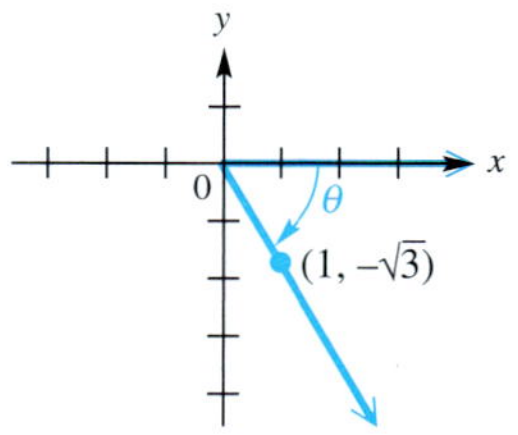

26.

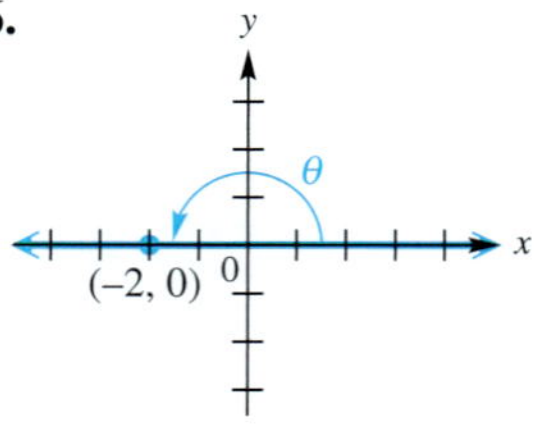

Find the values of the six trigonometric functions for an angle in standard position having each given point on its terminal side.

27. $(3, -4)$

28. $(9, -2)$

29. $(-8, 15)$

30. $(1, -5)$

31. $\left(6\sqrt{3}, -6\right)$

32. $\left(-2\sqrt{2}, 2\sqrt{2}\right)$

33. *Concept Check* If the terminal side of a quadrantal angle lies along the y-axis, which of its trigonometric functions are undefined?

34. Find the values of all six trigonometric functions for an angle in standard position having its terminal side defined by the equation $5x - 3y = 0$, $x \geq 0$.

In Exercises 35 and 36, consider an angle θ in standard position whose terminal side has the equation $y = -5x$, with $x \leq 0$.

35. Sketch θ and use an arrow to show the rotation if $0° \leq \theta < 360°$.

36. Find the exact values of $\sin\theta$, $\cos\theta$, $\tan\theta$, $\cot\theta$, $\sec\theta$, and $\csc\theta$.

Complete the table with the appropriate function values of the given quadrantal angles. If the value is undefined, say so.

	θ	$\sin\theta$	$\cos\theta$	$\tan\theta$	$\cot\theta$	$\sec\theta$	$\csc\theta$
37.	180°						
38.	−90°						

39. Decide whether each statement is *possible* or *impossible* for some angle θ.

(a) $\sec\theta = -\dfrac{2}{3}$ **(b)** $\tan\theta = 1.4$ **(c)** $\cos\theta = 5$

Find all six trigonometric function values for each angle θ. Rationalize denominators when applicable.

40. $\sin\theta = \frac{\sqrt{3}}{5}$, and $\cos\theta < 0$

41. $\cos\theta = -\frac{5}{8}$, and θ is in quadrant III

42. $\tan\theta = 2$, and θ is in quadrant III

43. $\sec\theta = -\sqrt{5}$, and θ is in quadrant II

44. $\sin\theta = -\frac{2}{5}$, and θ is in quadrant III

45. $\sec\theta = \frac{5}{4}$, and θ is in quadrant IV

46. *Concept Check* If, for some particular angle θ, $\sin\theta < 0$ and $\cos\theta > 0$, in what quadrant must θ lie? What is the sign of $\tan\theta$?

Solve each problem.

47. ***Swimmer in Distress*** A lifeguard located 20 yd from the water spots a swimmer in distress. The swimmer is 30 yd from shore and 100 yd east of the lifeguard. Suppose the lifeguard runs and then swims to the swimmer in a direct line, as shown in the figure. How far east from his original position will he enter the water? (*Hint:* Find the value of x in the sketch.)

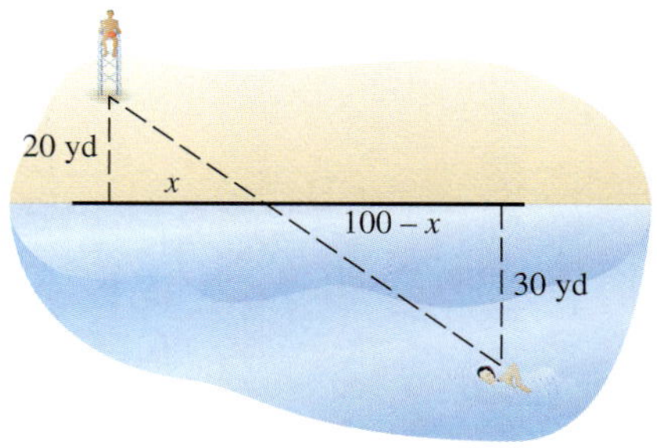

48. ***Angle through Which the Celestial North Pole Moves*** At present, the north star Polaris is located very near the celestial north pole. However, because Earth is inclined 23.5°, the moon's gravitational pull on Earth is uneven. As a result, Earth slowly precesses (moves in) like a spinning top, and the direction of the celestial north pole traces out a circular path once every 26,000 yr. See the figure. For example, in approximately A.D. 14,000 the star Vega—not the star Polaris—will be located at the celestial north pole. As viewed from the center C of this circular path, calculate the angle (to the nearest second) through which the celestial north pole moves each year. (*Source:* Zeilik, M., S. Gregory, and E. Smith, *Introductory Astronomy and Astrophysics,* Second Edition, Saunders College Publishers.)

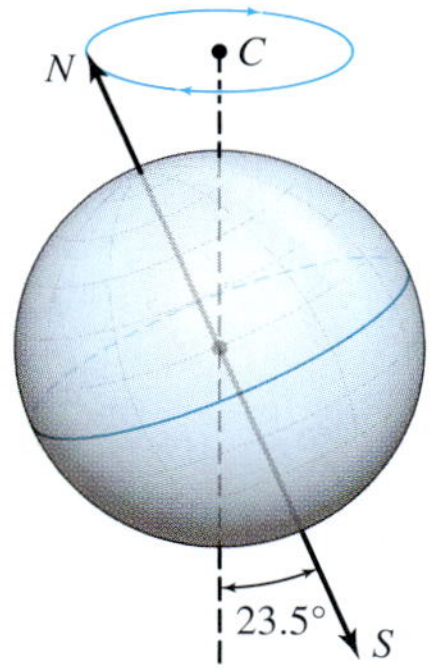

49. ***Depth of a Crater on the Moon*** The depths of unknown craters on the moon can be approximated by comparing the lengths of their shadows to the shadows of nearby craters with known depths. The crater Aristillus is 11,000 ft deep, and its shadow was measured as 1.5 mm on a photograph. Its companion crater, Autolycus, had a shadow of 1.3 mm on the same photograph. Use similar triangles to determine the depth of the crater Autolycus. (*Source:* Webb, T., *Celestial Objects for Common Telescopes,* Dover Publications.)

50. ***Height of a Lunar Peak*** The lunar mountain peak Huygens has a height of 21,000 ft. The shadow of Huygens on a photograph was 2.8 mm, while the nearby mountain Bradley had a shadow of 1.8 mm on the same photograph. Calculate the height of Bradley. (*Source:* Webb, T., *Celestial Objects for Common Telescopes,* Dover Publications.)

Chapter 1 Test

1. For an angle measuring 67°, give the measure of

(a) its complement **(b)** its supplement.

Find the measure of each unknown angle.

2.

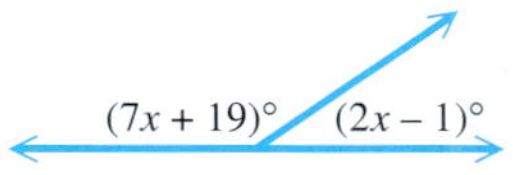

3.

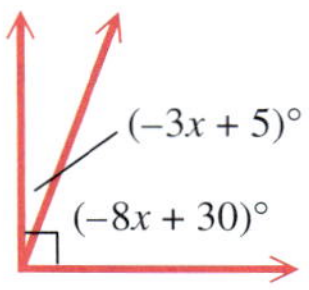

4.

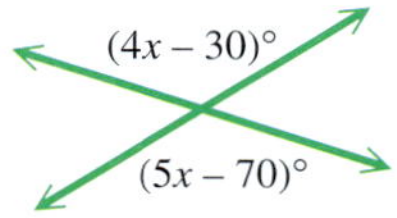

5.

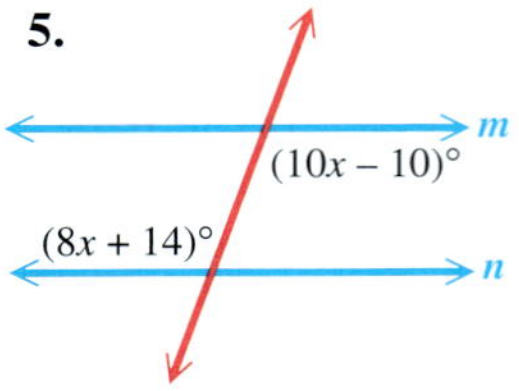

m and *n* are parallel.

6.

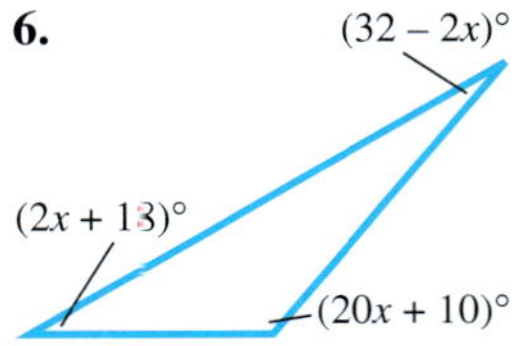

7.

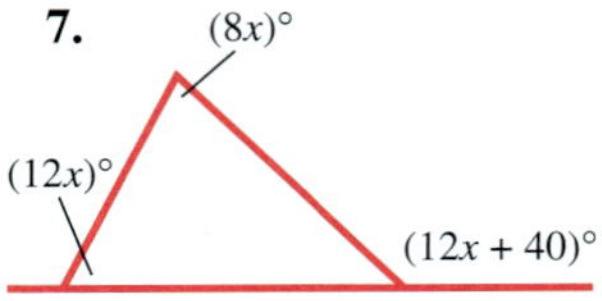

Perform each conversion.

8. $74° \, 18' \, 36''$ to decimal degrees

9. $45.2025°$ to degrees, minutes, seconds

10. Find the least positive measure of an angle that is coterminal with an angle of the given measure.

(a) $390°$ **(b)** $-80°$ **(c)** $810°$

11. *Rotating Tire* A tire rotates 450 times per min. Through how many degrees does a point on the edge of the tire move in 1 sec?

12. *Length of a Shadow* If a vertical pole 30 ft tall casts a shadow 8 ft long, how long would the shadow of a 40-ft pole be at the same time and place?

13. Find the unknown side lengths x and y in this pair of similar triangles.

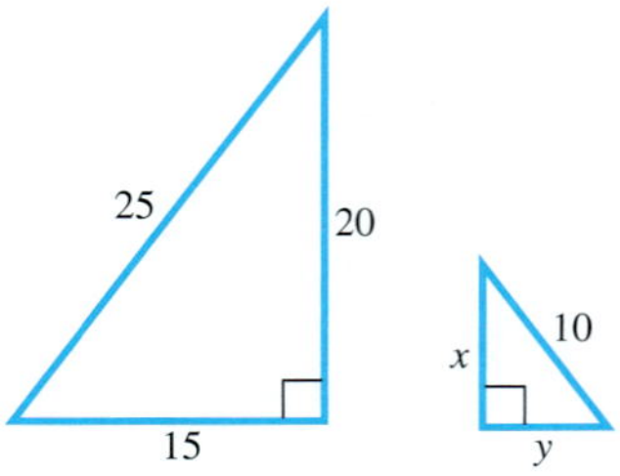

Draw a sketch of an angle in standard position having the given point on its terminal side. Indicate the angle of least positive measure θ*, and give the values of* $\sin \theta$, $\cos \theta$, $\tan \theta$, $\cot \theta$, $\sec \theta$, *and* $\csc \theta$. *If any of these are undefined, say so.*

14. $(2, -7)$

15. $(0, -2)$

16. Draw a sketch of an angle in standard position having the equation $3x - 4y = 0$, $x \leq 0$, as its terminal side. Indicate the angle of least positive measure θ, and give the values of $\sin \theta$, $\cos \theta$, $\tan \theta$, $\cot \theta$, $\sec \theta$, and $\csc \theta$.

17. Complete the table with the appropriate function values of the given quadrantal angles. If the value is undefined, say so.

θ	$\sin \theta$	$\cos \theta$	$\tan \theta$	$\cot \theta$	$\sec \theta$	$\csc \theta$
$90°$						
$-360°$						
$630°$						

18. If the terminal side of a quadrantal angle lies along the negative x-axis, which two of its trigonometric function values are undefined?

19. Identify the possible quadrant(s) in which θ must lie under the given conditions.

(a) $\cos \theta > 0, \tan \theta > 0$ **(b)** $\sin \theta < 0, \csc \theta < 0$ **(c)** $\cot \theta > 0, \cos \theta < 0$

20. Decide whether each statement is *possible* or *impossible* for some angle θ.

(a) $\sin \theta = 1.5$ **(b)** $\sec \theta = 4$ **(c)** $\tan \theta = 10{,}000$

21. Find the value of $\sec \theta$ if $\cos \theta = -\frac{7}{12}$.

22. Find the five remaining trigonometric function values of θ if $\sin \theta = \frac{3}{7}$ and θ is in quadrant II.

2 Acute Angles and Right Triangles

Trigonometry is used in safe roadway design to provide sufficient visibility around curves as well as a smooth-flowing, comfortable ride.

2.1 Trigonometric Functions of Acute Angles

- Right-Triangle-Based Definitions of the Trigonometric Functions
- Cofunctions
- Trigonometric Function Values of Special Angles

Right-Triangle-Based Definitions of the Trigonometric Functions

We used angles in standard position to define the trigonometric functions in **Section 1.3.** There is another way to approach them:

As ratios of the lengths of the sides of right triangles.

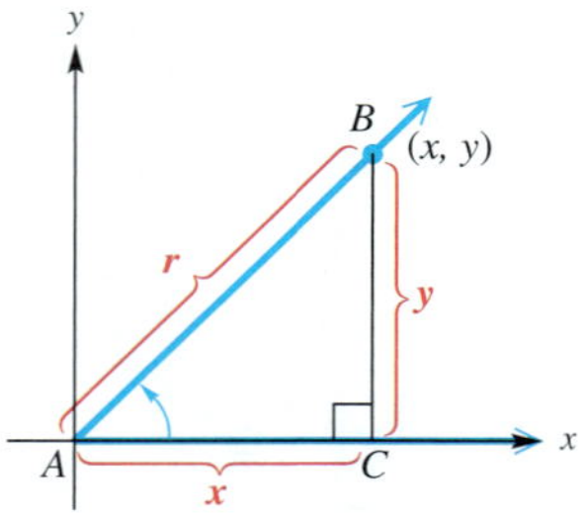

Figure 1

Figure 1 shows an acute angle A in standard position. The definitions of the trigonometric function values of angle A require x, y, and r. As drawn in **Figure 1,** x and y are the lengths of the two legs of the right triangle ABC, and r is the length of the hypotenuse.

The side of length y is called the **side opposite** angle A, and the side of length x is called the **side adjacent** to angle A. We use the lengths of these sides to replace x and y in the definitions of the trigonometric functions, and the length of the hypotenuse to replace r, to get the following right-triangle-based definitions.

Right-Triangle-Based Definitions of Trigonometric Functions

Let A represent any acute angle in standard position.

$$\sin A = \frac{y}{r} = \frac{\text{side opposite } A}{\text{hypotenuse}} \qquad \csc A = \frac{r}{y} = \frac{\text{hypotenuse}}{\text{side opposite } A}$$

$$\cos A = \frac{x}{r} = \frac{\text{side adjacent to } A}{\text{hypotenuse}} \qquad \sec A = \frac{r}{x} = \frac{\text{hypotenuse}}{\text{side adjacent to } A}$$

$$\tan A = \frac{y}{x} = \frac{\text{side opposite } A}{\text{side adjacent to } A} \qquad \cot A = \frac{x}{y} = \frac{\text{side adjacent to } A}{\text{side opposite } A}$$

NOTE We will sometimes shorten wording like "side opposite A" to just "side opposite" when the meaning is obvious.

EXAMPLE 1 Finding Trigonometric Function Values of an Acute Angle

Find the sine, cosine, and tangent values for angles A and B in the right triangle in **Figure 2.**

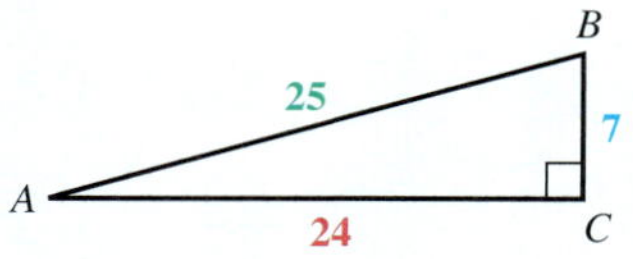

Figure 2

SOLUTION The length of the side opposite angle A is 7, the length of the side adjacent to angle A is 24, and the length of the hypotenuse is 25.

$$\sin A = \frac{\text{side opposite}}{\text{hypotenuse}} = \frac{7}{25} \qquad \cos A = \frac{\text{side adjacent}}{\text{hypotenuse}} = \frac{24}{25} \qquad \tan A = \frac{\text{side opposite}}{\text{side adjacent}} = \frac{7}{24}$$

The length of the side opposite angle B is 24, and the length of the side adjacent to B is 7.

$$\sin B = \frac{24}{25} \qquad \cos B = \frac{7}{25} \qquad \tan B = \frac{24}{7}$$

Use the relationships given in the box.

✔ *Now Try Exercise 1.*

NOTE Because the cosecant, secant, and cotangent ratios are the reciprocals of the sine, cosine, and tangent values, respectively, in **Example 1,**

$$\csc A = \frac{25}{7}, \quad \sec A = \frac{25}{24}, \quad \cot A = \frac{24}{7}, \quad \csc B = \frac{25}{24},$$

$$\sec B = \frac{25}{7}, \quad \text{and} \quad \cot B = \frac{7}{24}.$$

Cofunctions In **Example 1,** notice that $\sin A = \cos B$ and $\cos A = \sin B$. Such relationships are always true for the two acute angles of a right triangle.

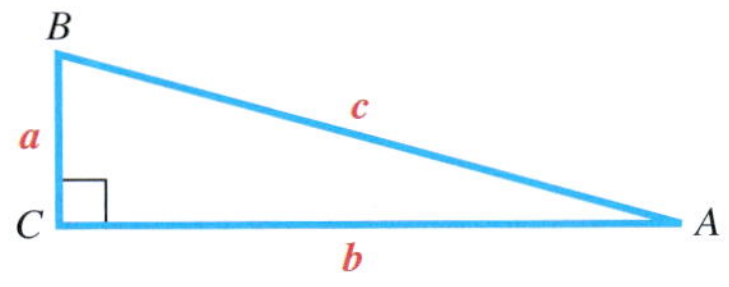

Whenever we use *A*, *B*, and *C* to name angles in a right triangle, *C* will be the right angle.

Figure 3

Figure 3 shows a right triangle with acute angles *A* and *B* and a right angle at *C*. The length of the side opposite angle *A* is *a*, and the length of the side opposite angle *B* is *b*. The length of the hypotenuse is *c*.

By the preceding definitions, $\sin A = \frac{a}{c}$. Also, $\cos B = \frac{a}{c}$. Thus,

$$\mathbf{\sin A = \frac{a}{c} = \cos B.}$$

Similarly, $\quad \mathbf{\tan A = \frac{a}{b} = \cot B} \quad$ and $\quad \mathbf{\sec A = \frac{c}{b} = \csc B.}$

Since the sum of the three angles in any triangle is 180° and angle *C* equals 90°, angles *A* and *B* must have a sum of $180° - 90° = 90°$. As mentioned in **Section 1.1,** angles with a sum of 90° are complementary angles. Since angles *A* and *B* are complementary and $\sin A = \cos B$, the functions sine and cosine are **cofunctions.** Tangent and cotangent are also cofunctions, as are secant and cosecant. And since the angles *A* and *B* are complementary, $A + B = 90°$, or $B = 90° - A$, giving the following.

$$\sin A = \cos B = \cos(90° - A)$$

Similar **cofunction identities** are true for the other trigonometric functions.

Cofunction Identities

For any acute angle *A*, cofunction values of complementary angles are equal.

$$\mathbf{\sin A = \cos(90° - A) \quad \sec A = \csc(90° - A) \quad \tan A = \cot(90° - A)}$$

$$\mathbf{\cos A = \sin(90° - A) \quad \csc A = \sec(90° - A) \quad \cot A = \tan(90° - A)}$$

EXAMPLE 2 Writing Functions in Terms of Cofunctions

Write each function in terms of its cofunction.

(a) cos 52° **(b)** tan 71° **(c)** sec 24°

SOLUTION

(a)

Cofunctions

$$\cos 52° = \sin(90° - 52°) = \sin 38° \qquad \cos A = \sin(90° - A)$$

Complementary angles

(b) $\tan 71° = \cot(90° - 71°) = \cot 19°$ **(c)** $\sec 24° = \csc 66°$

✔ ***Now Try Exercises 25 and 27.***

EXAMPLE 3 Solving Equations Using Cofunction Identities

Find one solution for each equation. Assume all angles involved are acute angles.

(a) $\cos(\theta + 4^\circ) = \sin(3\theta + 2^\circ)$ **(b)** $\tan(2\theta - 18^\circ) = \cot(\theta + 18^\circ)$

SOLUTION

(a) Since sine and cosine are cofunctions, $\cos(\theta + 4^\circ) = \sin(3\theta + 2^\circ)$ is true if the sum of the angles is 90°.

$$(\theta + 4^\circ) + (3\theta + 2^\circ) = 90^\circ \quad \text{Complementary angles (Section 1.1)}$$

$$4\theta + 6^\circ = 90^\circ \quad \text{Combine like terms.}$$

$$4\theta = 84^\circ \quad \text{Subtract } 6^\circ \text{ from each side. (Appendix A)}$$

$$\theta = 21^\circ \quad \text{Divide by 4.}$$

(b) Tangent and cotangent are cofunctions.

$$(2\theta - 18^\circ) + (\theta + 18^\circ) = 90^\circ \quad \text{Complementary angles}$$

$$3\theta = 90^\circ \quad \text{Combine like terms.}$$

$$\theta = 30^\circ \quad \text{Divide by 3.}$$

✔ *Now Try Exercises 31 and 33.*

Figure 4 shows three right triangles. From left to right, the length of each hypotenuse is the same, but angle A increases in measure. As angle A increases in measure from 0° to 90°, the length of the side opposite angle A also increases.

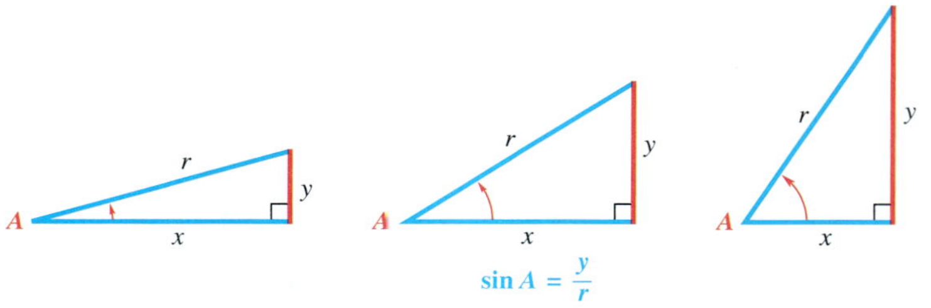

As A increases, y increases. Since r is fixed, $\sin A$ increases.

Figure 4

Since $$\sin A = \frac{\text{side opposite}}{\text{hypotenuse}} = \frac{y}{r},$$

as angle A increases, the numerator of this fraction also increases, while the denominator is fixed. Therefore, $\sin A$ *increases* as A increases from 0° to 90°.

As angle A increases from 0° to 90°, the length of the side adjacent to A decreases. Since r is fixed, the ratio $\frac{x}{r}$ decreases. This ratio gives $\cos A$, showing that the values of cosine *decrease* as the angle measure changes from 0° to 90°. Finally, increasing A from 0° to 90° causes y to increase and x to decrease, making the values of $\frac{y}{x} = \tan A$ increase.

A similar discussion shows that as A increases from 0° to 90°, the values of $\sec A$ increase, while the values of $\cot A$ and $\csc A$ decrease.

EXAMPLE 4 Comparing Function Values of Acute Angles

Determine whether each statement is *true* or *false*.

(a) $\sin 21° > \sin 18°$ **(b)** $\sec 56° \leq \sec 49°$

SOLUTION

(a) In the interval from $0°$ to $90°$, as the angle increases, so does the sine of the angle, which makes $\sin 21° > \sin 18°$ a true statement.

(b) For fixed r, increasing an angle from $0°$ to $90°$ causes x to decrease. Therefore, $\sec \theta = \frac{r}{x}$ increases. The given statement, $\sec 56° \leq \sec 49°$, is false.

✔ *Now Try Exercises 41 and 47.*

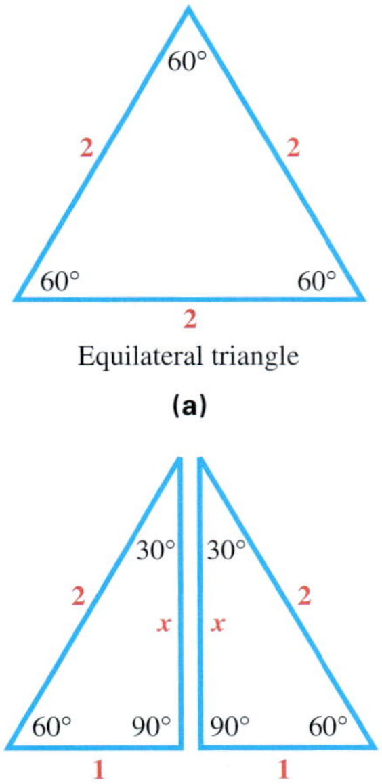

Equilateral triangle

(a)

30°–60° right triangle

(b)

Figure 5

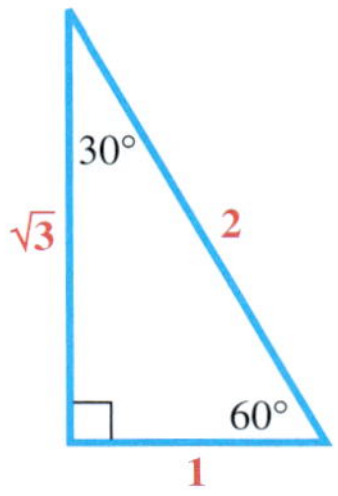

Figure 6

Trigonometric Function Values of Special Angles Certain special angles, such as $30°$, $45°$, and $60°$, occur so often in trigonometry and in more advanced mathematics that they deserve special study. We start with an equilateral triangle, a triangle with all sides of equal length. Each angle of such a triangle measures $60°$. Although the results we will obtain are independent of the length, for convenience we choose the length of each side to be 2 units. See **Figure 5(a).**

Bisecting one angle of this equilateral triangle leads to two right triangles, each of which has angles of $30°$, $60°$, and $90°$, as shown in **Figure 5(b).** An angle bisector of an equilateral triangle also bisects the opposite side; therefore, the shorter leg has length 1. Let x represent the length of the longer leg.

$$2^2 = 1^2 + x^2 \quad \text{Pythagorean theorem (Appendix B)}$$

$$4 = 1 + x^2 \quad \text{Apply the exponents.}$$

$$3 = x^2 \quad \text{Subtract 1 from each side.}$$

$$\sqrt{3} = x \quad \text{Square root property (Appendix A); choose the positive root.}$$

Figure 6 summarizes our results using a $30°$–$60°$ right triangle. As shown in the figure, the side opposite the $30°$ angle has length 1; that is, for the $30°$ angle,

$$\text{hypotenuse} = 2, \quad \text{side opposite} = 1, \quad \text{side adjacent} = \sqrt{3}.$$

Now we use the definitions of the trigonometric functions.

$$\sin 30° = \frac{\text{side opposite}}{\text{hypotenuse}} = \frac{1}{2}$$

$$\cos 30° = \frac{\text{side adjacent}}{\text{hypotenuse}} = \frac{\sqrt{3}}{2}$$

$$\tan 30° = \frac{\text{side opposite}}{\text{side adjacent}} = \frac{1}{\sqrt{3}} = \frac{1}{\sqrt{3}} \cdot \frac{\sqrt{3}}{\sqrt{3}} = \frac{\sqrt{3}}{3}$$

$$\csc 30° = \frac{2}{1} = 2$$

Rationalize the denominator.

$$\sec 30° = \frac{2}{\sqrt{3}} = \frac{2}{\sqrt{3}} \cdot \frac{\sqrt{3}}{\sqrt{3}} = \frac{2\sqrt{3}}{3}$$

$$\cot 30° = \frac{\sqrt{3}}{1} = \sqrt{3}$$

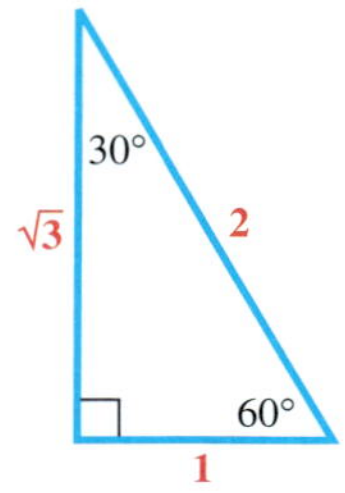

Figure 6 (repeated)

EXAMPLE 5 Finding Trigonometric Function Values for 60°

Find the six trigonometric function values for a 60° angle.

SOLUTION Refer to **Figure 6** to find the following ratios.

$$\sin 60^\circ = \frac{\sqrt{3}}{2} \qquad \cos 60^\circ = \frac{1}{2} \qquad \tan 60^\circ = \frac{\sqrt{3}}{1} = \sqrt{3}$$

$$\csc 60^\circ = \frac{2}{\sqrt{3}} = \frac{2\sqrt{3}}{3} \qquad \sec 60^\circ = \frac{2}{1} = 2 \qquad \cot 60^\circ = \frac{1}{\sqrt{3}} = \frac{\sqrt{3}}{3}$$

✔ ***Now Try Exercises 49, 51, and 53.***

NOTE The results in **Example 5** can also be found using the fact that cofunction values of complementary angles are equal.

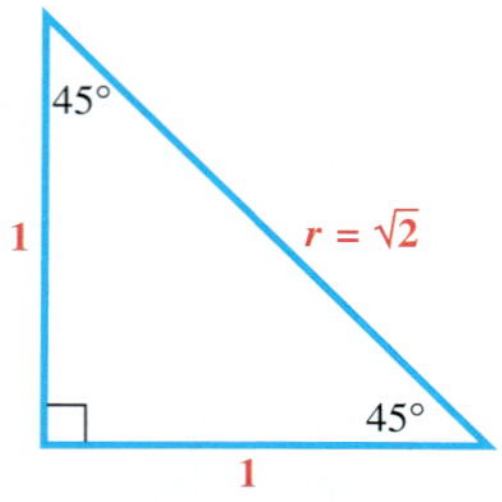

45°–45° right triangle

Figure 7

We find the values of the trigonometric functions for 45° by starting with a 45°–45° right triangle, as shown in **Figure 7.** This triangle is isosceles. For simplicity, we choose the lengths of the equal sides to be 1 unit. (As before, the results are independent of the length of the equal sides.) If r represents the length of the hypotenuse, then we can find its value using the Pythagorean theorem.

$$1^2 + 1^2 = r^2 \quad \text{Pythagorean theorem}$$
$$2 = r^2 \quad \text{Simplify.}$$
$$\sqrt{2} = r \quad \text{Choose the positive root.}$$

Now we use the measures indicated on the 45°–45° right triangle in **Figure 7.**

$$\sin 45^\circ = \frac{1}{\sqrt{2}} = \frac{\sqrt{2}}{2} \qquad \cos 45^\circ = \frac{1}{\sqrt{2}} = \frac{\sqrt{2}}{2} \qquad \tan 45^\circ = \frac{1}{1} = 1$$

$$\csc 45^\circ = \frac{\sqrt{2}}{1} = \sqrt{2} \qquad \sec 45^\circ = \frac{\sqrt{2}}{1} = \sqrt{2} \qquad \cot 45^\circ = \frac{1}{1} = 1$$

Function values for 30°, 45°, and 60° are summarized in the table that follows.

Function Values of Special Angles

θ	$\sin \theta$	$\cos \theta$	$\tan \theta$	$\cot \theta$	$\sec \theta$	$\csc \theta$
30°	$\frac{1}{2}$	$\frac{\sqrt{3}}{2}$	$\frac{\sqrt{3}}{3}$	$\sqrt{3}$	$\frac{2\sqrt{3}}{3}$	2
45°	$\frac{\sqrt{2}}{2}$	$\frac{\sqrt{2}}{2}$	1	1	$\sqrt{2}$	$\sqrt{2}$
60°	$\frac{\sqrt{3}}{2}$	$\frac{1}{2}$	$\sqrt{3}$	$\frac{\sqrt{3}}{3}$	2	$\frac{2\sqrt{3}}{3}$

TI-83

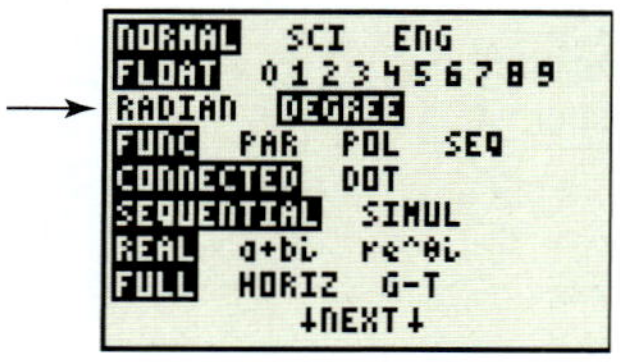

TI-84

Figure 8

NOTE You will be able to reproduce this table quickly if you learn the values of sin 30°, sin 45°, and sin 60°. Then you can complete the rest of the table using the reciprocal, cofunction, and quotient identities.

Since a calculator finds trigonometric function values at the touch of a key, why do we spend so much time finding values for special angles? We do this because a calculator gives only *approximate* values in most cases instead of *exact* values. A scientific calculator gives the following approximation for tan 30°.

$$\tan 30° \approx 0.57735027 \quad \approx \text{ means "is approximately equal to."}$$

Earlier, however, we found the exact value.

$$\tan 30° = \frac{\sqrt{3}}{3} \quad \text{Exact value}$$

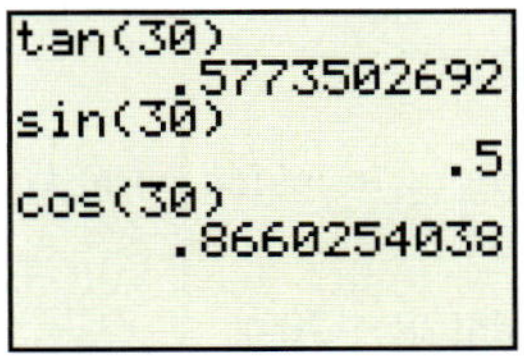

Figure 9

Figure 8 on the previous page shows the mode display screens for TI graphing calculators. **Figure 9** shows the output when evaluating the tangent, sine, and cosine of 30°. (The calculator must be in degree mode for the angle measure to be entered in degrees.) ■

2.1 Exercises

Find exact values or expressions for sin *A*, cos *A*, *and* tan *A*. ***See Example 1.***

1.

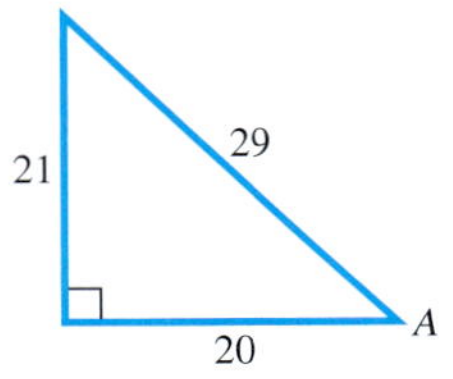

2.

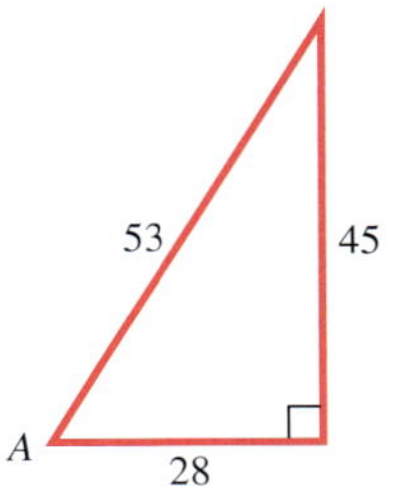

3.

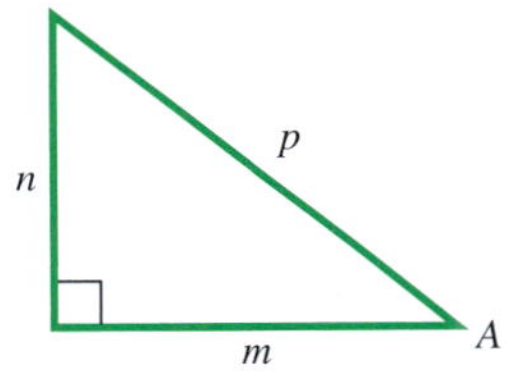

4.

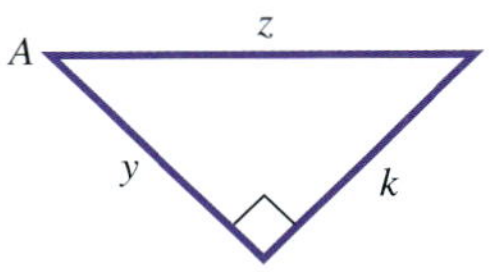

Concept Check *For each trigonometric function in Column I, choose its value from Column II.*

I		II		
5. sin 30°	**6.** cos 45°	**A.** $\sqrt{3}$	**B.** 1	**C.** $\frac{1}{2}$
7. tan 45°	**8.** sec 60°	**D.** $\frac{\sqrt{3}}{2}$	**E.** $\frac{2\sqrt{3}}{3}$	**F.** $\frac{\sqrt{3}}{3}$
9. csc 60°	**10.** cot 30°	**G.** 2	**H.** $\frac{\sqrt{2}}{2}$	**I.** $\sqrt{2}$

Suppose ABC is a right triangle with sides of lengths a, b, and c and right angle at C. (See **Figure 3.**) *Find the unknown side length using the Pythagorean theorem (**Appendix B**), and then find the values of the six trigonometric functions for angle B. Rationalize denominators when applicable.*

11. $a = 5,\ b = 12$ **12.** $a = 3,\ b = 4$ **13.** $a = 6,\ c = 7$

14. $b = 7,\ c = 12$ **15.** $a = 3,\ c = 10$ **16.** $b = 8,\ c = 11$

17. $a = 1,\ c = 2$ **18.** $a = \sqrt{2},\ c = 2$ **19.** $b = 2,\ c = 5$

20. ***Concept Check*** Give a summary of the six cofunction relationships.

Write each function in terms of its cofunction. Assume that all angles in which an unknown appears are acute angles. **See Example 2.**

21. $\cos 30°$ **22.** $\sin 45°$ **23.** $\csc 60°$

24. $\cot 73°$ **25.** $\sec 39°$ **26.** $\tan 25.4°$

27. $\sin 38.7°$ **28.** $\cos(\theta + 20°)$ **29.** $\sec(\theta + 15°)$

30. With a calculator, evaluate $\sin(90° - \theta)$ and $\cos \theta$ for various values of θ. (Include values greater than 90° and less than 0°.) What do you find?

Find one solution for each equation. Assume that all angles in which an unknown appears are acute angles. **See Example 3.**

31. $\tan \alpha = \cot(\alpha + 10°)$ **32.** $\cos \theta = \sin(2\theta - 30°)$

33. $\sin(2\theta + 10°) = \cos(3\theta - 20°)$ **34.** $\sec(\beta + 10°) = \csc(2\beta + 20°)$

35. $\tan(3B + 4°) = \cot(5B - 10°)$ **36.** $\cot(5\theta + 2°) = \tan(2\theta + 4°)$

37. $\sin(\theta - 20°) = \cos(2\theta + 5°)$ **38.** $\cos(2\theta + 50°) = \sin(2\theta - 20°)$

39. $\sec(3\beta + 10°) = \csc(\beta + 8°)$ **40.** $\csc(\beta + 40°) = \sec(\beta - 20°)$

Determine whether each statement is true *or* false. **See Example 4.**

41. $\sin 50° > \sin 40°$ **42.** $\tan 28° \leq \tan 40°$

43. $\sin 46° < \cos 46°$ (*Hint*: $\cos 46° = \sin 44°$) **44.** $\cos 28° < \sin 28°$ (*Hint*: $\sin 28° = \cos 62°$)

45. $\tan 41° < \cot 41°$ **46.** $\cot 30° < \tan 40°$

47. $\sec 60° > \sec 30°$ **48.** $\csc 20° < \csc 30°$

For each expression, give the exact value. **See Example 5.**

49. $\tan 30°$ **50.** $\cot 30°$ **51.** $\sin 30°$ **52.** $\cos 30°$

53. $\sec 30°$ **54.** $\csc 30°$ **55.** $\csc 45°$ **56.** $\sec 45°$

57. $\cos 45°$ **58.** $\cot 45°$ **59.** $\tan 45°$ **60.** $\sin 45°$

61. $\sin 60°$ **62.** $\cos 60°$ **63.** $\tan 60°$ **64.** $\csc 60°$

Relating Concepts

For individual or collaborative investigation *(Exercises 65–68)*

The figure shows a 45° central angle in a circle with radius 4 units. To find the coordinates of point P on the circle, **work Exercises 65–68 in order.**

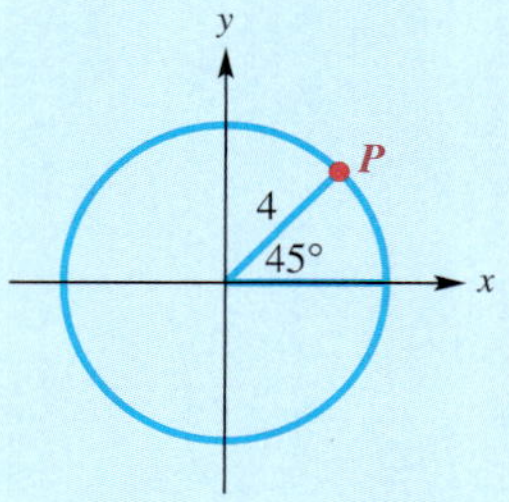

65. Sketch a line segment from P perpendicular to the x-axis.

66. Use the trigonometric ratios for a 45° angle to label the sides of the right triangle you sketched in **Exercise 65.**

67. Which sides of the right triangle give the coordinates of point P? What are the coordinates of P?

68. The figure at the right shows a 60° central angle in a circle of radius 2 units. Follow the same procedure as in **Exercises 65–67** to find the coordinates of P in the figure.

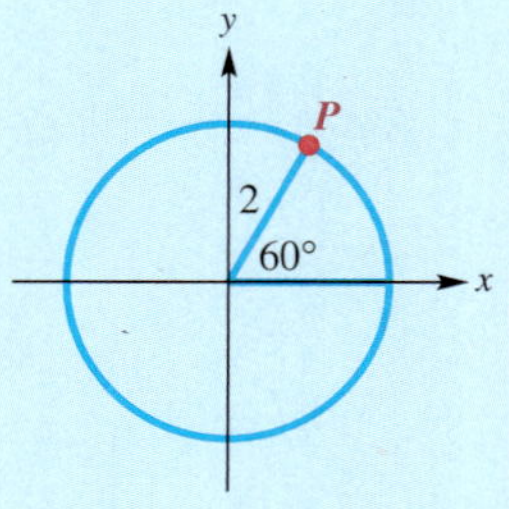

69. *Concept Check* Refer to the table. What trigonometric functions are y_1 and y_2?

$x°$	y_1	y_2
0	0	0
15	0.25882	0.26795
30	0.5	0.57735
45	0.70711	1
60	0.86603	1.7321
75	0.96593	3.7321
90	1	undefined

70. *Concept Check* Refer to the table. What trigonometric functions are y_1 and y_2?

$x°$	y_1	y_2
0	1	undefined
15	0.96593	3.8637
30	0.86603	2
45	0.70711	1.4142
60	0.5	1.1547
75	0.25882	1.0353
90	0	1

71. *Concept Check* What value of A between 0° and 90° will produce the output shown on the graphing calculator screen?

72. A student was asked to give the exact value of sin 45°. Using a calculator, he gave the answer 0.7071067812. The teacher did not give him credit. What was the teacher's reason for this?

73. With a graphing calculator, find the coordinates of the point of intersection of $y = x$ and $y = \sqrt{1 - x^2}$. These coordinates are the cosine and sine of what angle between 0° and 90°?

Concept Check *Work each problem.*

74. Find the equation of the line that passes through the origin and makes a 60° angle with the x-axis.

75. Find the equation of the line that passes through the origin and makes a 30° angle with the x-axis.

76. What angle does the line $y = \frac{\sqrt{3}}{3}x$ make with the positive x-axis?

77. What angle does the line $y = \sqrt{3}x$ make with the positive x-axis?

78. Construct a square with each side of length k.

(a) Draw a diagonal of the square. What is the measure of each angle formed by a side of the square and this diagonal?
(b) What is the length of the diagonal?
(c) From the results of parts (a) and (b), complete the following statement: In a 45°–45° right triangle, the hypotenuse has a length that is ________ times as long as either leg.

79. Construct an equilateral triangle with each side having length $2k$.

(a) What is the measure of each angle?
(b) Label one angle A. Drop a perpendicular from A to the side opposite A. Two 30° angles are formed at A, and two right triangles are formed. What is the length of the sides opposite the 30° angles?
(c) What is the length of the perpendicular in part (b)?
(d) From the results of parts (a)–(c), complete the following statement: In a 30°–60° right triangle, the hypotenuse is always ________ times as long as the shorter leg, and the longer leg has a length that is ________ times as long as that of the shorter leg. Also, the shorter leg is opposite the ________ angle, and the longer leg is opposite the ________ angle.

Find the exact value of each part labeled with a variable in each figure.

80.

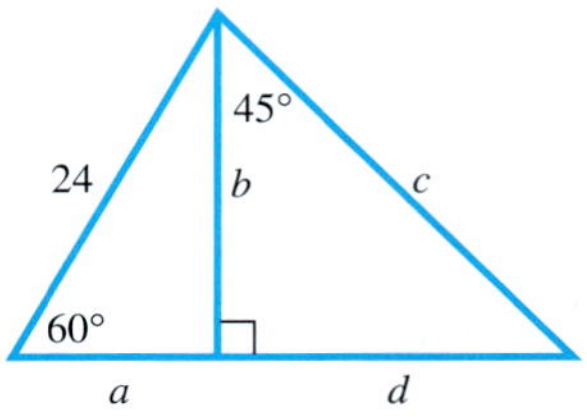

81.

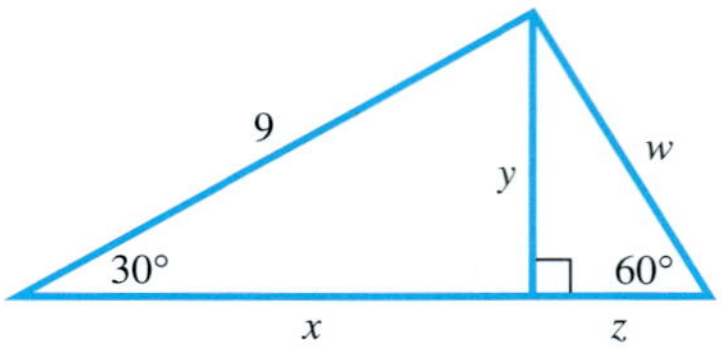

82.

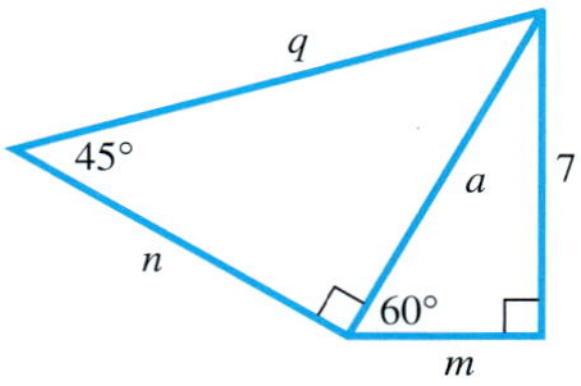

83.

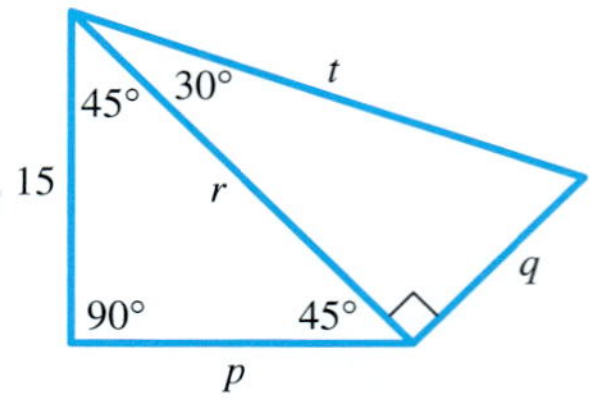

Find a formula for the area of each figure in terms of s.

84.

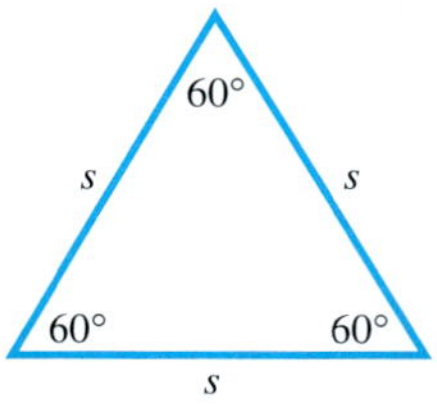

85.

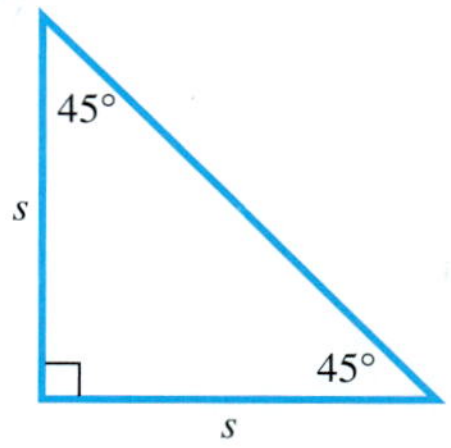

86. *Concept Check* Suppose you know the length of one side and one acute angle of a right triangle. Is it possible to determine the measures of all the sides and angles of the triangle?

2.2 Trigonometric Functions of Non-Acute Angles

- Reference Angles
- Special Angles as Reference Angles
- Finding Angle Measures with Special Angles

Reference Angles Associated with every nonquadrantal angle in standard position is a positive acute angle called its *reference angle.* A **reference angle** for an angle θ, written θ', is the positive acute angle made by the terminal side of angle θ and the x-axis.

Figure 10 shows several angles θ (each less than one complete counterclockwise revolution) in quadrants II, III, and IV, respectively, with the reference angle θ' also shown. In quadrant I, θ and θ' are the same. If an angle θ is negative or has measure greater than 360°, its reference angle is found by first finding its coterminal angle that is between 0° and 360°, and then using the diagrams in **Figure 10.**

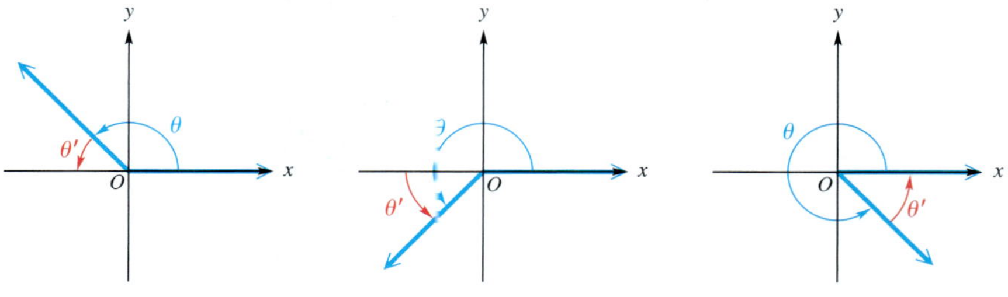

θ in quadrant II $\qquad$ θ in quadrant III $\qquad$ θ in quadrant IV

Figure 10

CAUTION A common error is to find the reference angle by using the terminal side of θ and the y-axis. ***The reference angle is always found with reference to the x-axis.***

EXAMPLE 1 Finding Reference Angles

Find the reference angle for each angle.

(a) 218° **(b)** 1387°

SOLUTION

(a) As shown in **Figure 11(a),** the positive acute angle made by the terminal side of this angle and the x-axis is

$$218^\circ - 180^\circ = 38^\circ.$$

For $\theta = 218^\circ$, the reference angle $\theta' = 38^\circ$.

(b) First find a coterminal angle between 0° and 360°. Divide 1387° by 360° to get a quotient of about 3.9. Begin by subtracting 360° three times (because of the whole number 3 in 3.9).

$$1387^\circ - 3 \cdot 360^\circ = 1387^\circ - 1080^\circ \quad \text{Multiply. (Section 1.1)}$$
$$= 307^\circ \quad \text{Subtract.}$$

The reference angle for 307° (and thus for 1387°) is $360^\circ - 307^\circ = 53^\circ$. See **Figure 11(b).**

Now Try Exercises 1 and 5.

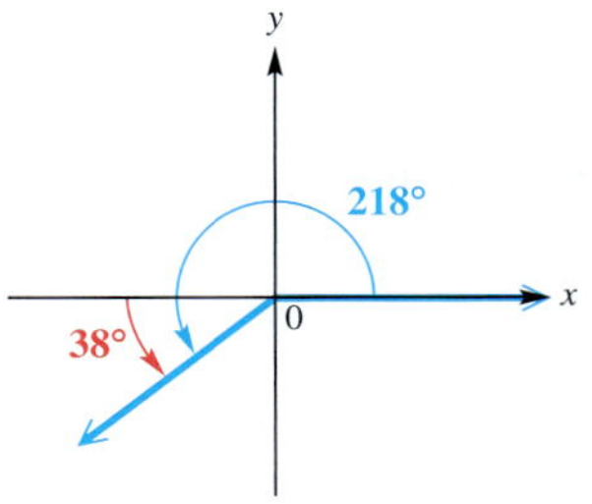

218° – 180° = 38°

(a)

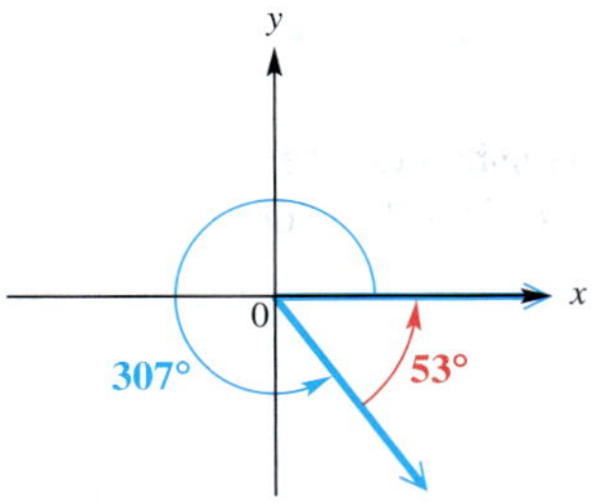

360° – 307° = 53°

(b)

Figure 11

The preceding example suggests the following table for finding the reference angle θ' for any angle θ between 0° and 360°.

Reference Angle θ' for θ, where $0^\circ < \theta < 360^\circ$*

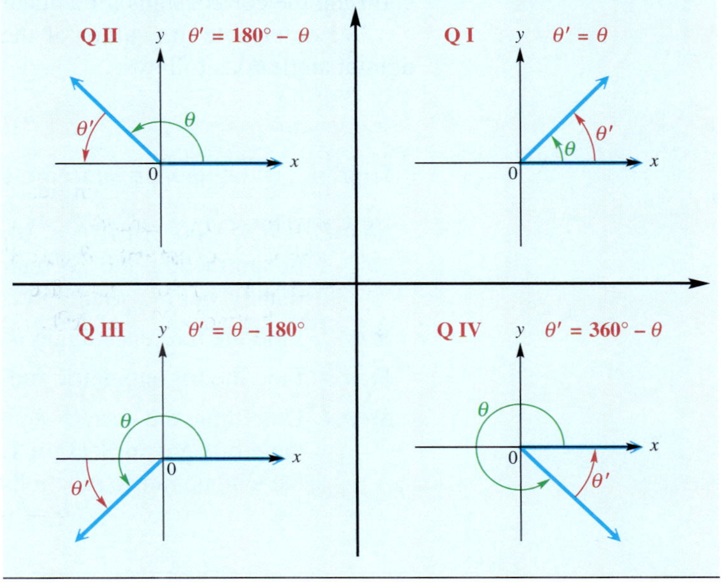

*The authors would like to thank Bethany Vaughn and Theresa Matick, of Vincennes Lincoln High School, for their suggestions concerning this table.

Special Angles as Reference Angles We can now find exact trigonometric function values of angles with reference angles of 30°, 45°, or 60°.

EXAMPLE 2 Finding Trigonometric Function Values of a Quadrant III Angle

Find the values of the six trigonometric functions for 210°.

SOLUTION An angle of 210° is shown in **Figure 12.** The reference angle is

$$210° - 180° = 30°.$$

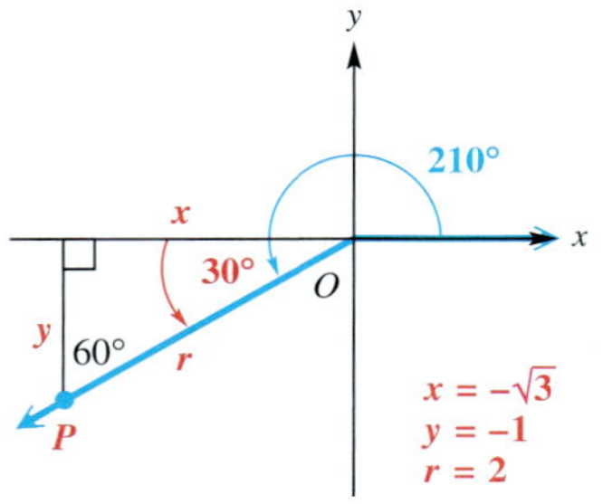

Figure 12

To find the trigonometric function values of 210°, choose point P on the terminal side of the angle so that the distance from the origin O to P is 2. By the results from 30°–60° right triangles, the coordinates of point P become $(-\sqrt{3}, -1)$, with $x = -\sqrt{3}$, $y = -1$, and $r = 2$. Then, by the definitions of the trigonometric functions in **Section 2.1,** we obtain the following.

$$\sin 210° = \frac{-1}{2} = -\frac{1}{2} \qquad \csc 210° = \frac{2}{-1} = -2$$

$$\cos 210° = \frac{-\sqrt{3}}{2} = -\frac{\sqrt{3}}{2} \qquad \sec 210° = \frac{2}{-\sqrt{3}} = -\frac{2\sqrt{3}}{3}$$

Rationalize denominators as needed.

$$\tan 210° = \frac{-1}{-\sqrt{3}} = \frac{\sqrt{3}}{3} \qquad \cot 210° = \frac{-\sqrt{3}}{-1} = \sqrt{3}$$

✔ *Now Try Exercise 19.*

Notice in **Example 2** that the trigonometric function values of 210° correspond in absolute value to those of its reference angle 30°. The signs are different for the sine, cosine, secant, and cosecant functions because 210° is a quadrant III angle. These results suggest a shortcut for finding the trigonometric function values of a non-acute angle, using the reference angle. In **Example 2,** the reference angle for 210° is 30°. Using the trigonometric function values of 30°, and choosing the correct signs for a quadrant III angle, we obtain the same results.

We determine the values of the trigonometric functions for any nonquadrantal angle θ as follows.

Finding Trigonometric Function Values for Any Nonquadrantal Angle θ

Step 1 If $\theta > 360°$, or if $\theta < 0°$, then find a coterminal angle by adding or subtracting 360° as many times as needed to get an angle greater than 0° but less than 360°.

Step 2 Find the reference angle θ'.

Step 3 Find the trigonometric function values for reference angle θ'.

Step 4 Determine the correct signs for the values found in Step 3. (Use the table of signs in **Section 1.4,** if necessary.) This gives the values of the trigonometric functions for angle θ.

NOTE To avoid sign errors when finding the trigonometric function values of an angle, sketch it in standard position. Include a reference triangle complete with appropriate values for x, y, and r as done in **Figure 12.**

EXAMPLE 3 Finding Trigonometric Function Values Using Reference Angles

Find the exact value of each expression.

(a) $\cos(-240°)$ **(b)** $\tan 675°$

SOLUTION

(a) Since an angle of $-240°$ is coterminal with an angle of

$$-240° + 360° = 120°, \quad \text{(Section 1.1)}$$

the reference angle is $180° - 120° = 60°$, as shown in **Figure 13(a).** Since the cosine is negative in quadrant II,

$$\cos(-240°) = \underset{\text{Coterminal angle}}{\cos 120°} = \underset{\text{Reference angle}}{-\cos 60°} = -\frac{1}{2}.$$

(b) Begin by subtracting 360° to get a coterminal angle between 0° and 360°.

$$675° - 360° = 315°$$

As shown in **Figure 13(b),** the reference angle is $360° - 315° = 45°$. An angle of 315° is in quadrant IV, so the tangent will be negative.

$$\tan 675° = \tan 315° \quad \text{Coterminal angle}$$
$$= -\tan 45° \quad \text{Reference angle; quadrant-based sign choice}$$
$$= -1 \quad \text{Evaluate.}$$

✔ ***Now Try Exercises 37 and 39.***

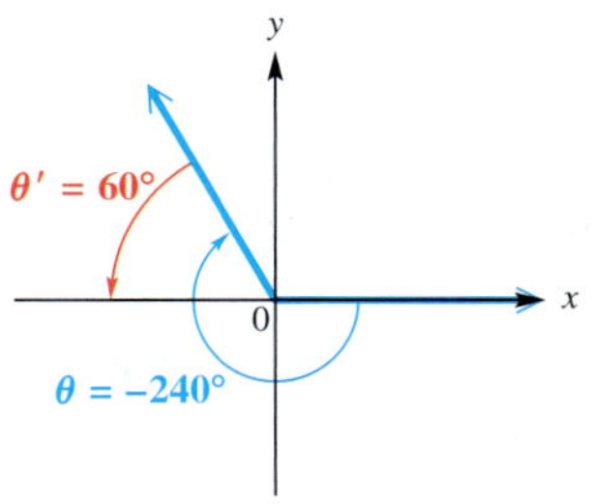

(a)

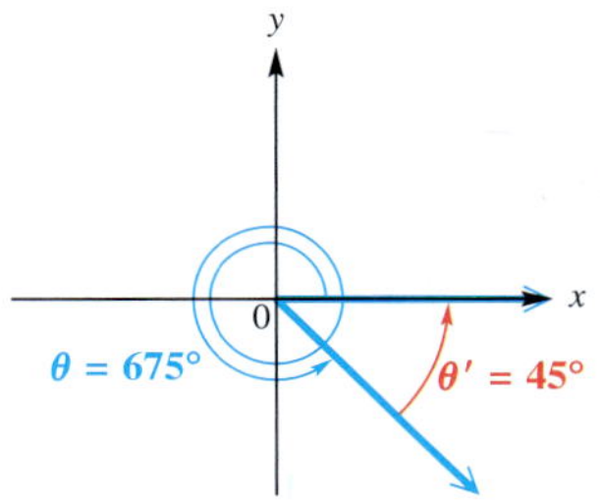

(b)

Figure 13

EXAMPLE 4 Evaluating an Expression with Function Values of Special Angles

Evaluate $\cos 120° + 2\sin^2 60° - \tan^2 30°$.

SOLUTION Use the values $\cos 120° = -\frac{1}{2}$, $\sin 60° = \frac{\sqrt{3}}{2}$, and $\tan 30° = \frac{\sqrt{3}}{3}$.

$$\cos 120° + 2\sin^2 60° - \tan^2 30° = -\frac{1}{2} + 2\left(\frac{\sqrt{3}}{2}\right)^2 - \left(\frac{\sqrt{3}}{3}\right)^2 \quad \text{Substitute values.}$$

$$= -\frac{1}{2} + 2\left(\frac{3}{4}\right) - \frac{3}{9}, \quad \text{or} \quad \frac{2}{3} \quad \text{Simplify.}$$

✔ ***Now Try Exercise 47.***

EXAMPLE 5 Using Coterminal Angles to Find Function Values

Evaluate each function by first expressing in terms of a function of an angle between 0° and 360°.

(a) $\cos 780°$ **(b)** $\cot(-405°)$

SOLUTION

(a) Subtract 360° as many times as necessary to get an angle between 0° and 360°, which gives the following.

$$\cos 780° = \cos(780° - 2 \cdot 360°) \quad \text{Subtract 720°, which is } 2 \cdot 360°.$$

$$= \cos 60°, \quad \text{or} \quad \frac{1}{2} \quad \text{Multiply first, subtract, and evaluate.}$$

(b) Add 360° twice to get $-405° + 2(360°) = 315°$. This angle is located in quadrant IV, and its reference angle is 45°. The cotangent function is negative in quadrant IV.

$$\cot(-405°) = \cot 315° = -\cot 45° = -1$$

Now Try Exercises 27 and 31.

Finding Angle Measures with Special Angles The ideas discussed in this section can also be used to find the measures of certain angles, given a trigonometric function value and an interval in which the angle must lie. We are most often interested in the interval $[0°, 360°)$.

EXAMPLE 6 Finding Angle Measures Given an Interval and a Function Value

Find all values of θ, if θ is in the interval $[0°, 360°)$ and $\cos \theta = -\frac{\sqrt{2}}{2}$.

SOLUTION Since $\cos \theta$ is negative, θ must lie in quadrant II or III. Since the absolute value of $\cos \theta$ is $\frac{\sqrt{2}}{2}$, the reference angle θ' must be 45°. The two possible angles θ are sketched in **Figure 14.**

$$180° - 45° = 135° \quad \text{Quadrant II angle } \theta$$

$$180° + 45° = 225° \quad \text{Quadrant III angle } \theta$$

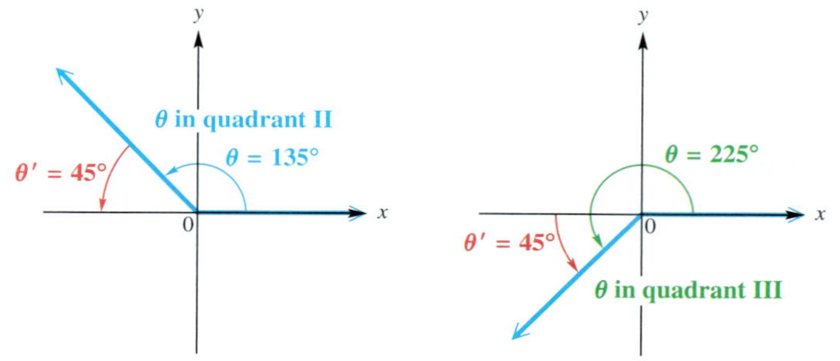

Figure 14

Now Try Exercise 79.

2.2 Exercises

Match each angle in Column I with its reference angle in Column II. Choices may be used once, more than once, or not at all. ***See Example 1.***

I		II	
1. 98°	**2.** 212°	**A.** 45°	**B.** 60°
3. −135°	**4.** −60°	**C.** 82°	**D.** 30°
5. 750°	**6.** 480°	**E.** 38°	**F.** 32°

Give a short explanation in Exercises 7–10.

7. In **Example 2,** why was 2 a good choice for r? Could any other positive number have been used?

8. Explain how the reference angle is used to find values of the trigonometric functions for an angle in quadrant III.

9. Explain why two coterminal angles have the same values for their trigonometric functions.

10. Explain the process for determining the sign of the sine, cosine, and tangent functions of an angle with terminal side in quadrant II.

Complete the table with exact trigonometric function values. Do not use a calculator. ***See Examples 2 and 3.***

	θ	$\sin\theta$	$\cos\theta$	$\tan\theta$	$\cot\theta$	$\sec\theta$	$\csc\theta$
11.	30°	$\frac{1}{2}$	$\frac{\sqrt{3}}{2}$			$\frac{2\sqrt{3}}{3}$	2
12.	45°			1	1		
13.	60°		$\frac{1}{2}$	$\sqrt{3}$		2	
14.	120°	$\frac{\sqrt{3}}{2}$		$-\sqrt{3}$			$\frac{2\sqrt{3}}{3}$
15.	135°	$\frac{\sqrt{2}}{2}$	$-\frac{\sqrt{2}}{2}$			$-\sqrt{2}$	$\sqrt{2}$
16.	150°		$-\frac{\sqrt{3}}{2}$	$-\frac{\sqrt{3}}{3}$			2
17.	210°	$-\frac{1}{2}$		$\frac{\sqrt{3}}{3}$	$\sqrt{3}$		-2
18.	240°	$-\frac{\sqrt{3}}{2}$	$-\frac{1}{2}$			-2	$-\frac{2\sqrt{3}}{3}$

Find exact values of the six trigonometric functions for each angle. Rationalize denominators when applicable. ***See Examples 2, 3, and 5.***

19. 300° **20.** 315° **21.** 405° **22.** 420° **23.** 480° **24.** 495°

25. 570° **26.** 750° **27.** 1305° **28.** 1500° **29.** $-300°$ **30.** $-390°$

31. $-510°$ **32.** $-1020°$ **33.** $-1290°$ **34.** $-855°$ **35.** $-1860°$ **36.** $-2205°$

Find the exact value of each expression. ***See Example 3.***

37. $\sin 1305°$ **38.** $\sin 1500°$ **39.** $\cos(-510°)$ **40.** $\tan(-1020°)$

41. $\csc(-855°)$ **42.** $\sec(-495°)$ **43.** $\tan 3015°$ **44.** $\cot 2280°$

Evaluate each of the following. ***See Example 4.***

45. $\sin^2 120° + \cos^2 120°$

46. $\sin^2 225° + \cos^2 225°$

47. $2\tan^2 120° + 3\sin^2 150° - \cos^2 180°$

48. $\cot^2 135° - \sin 30° + 4\tan 45°$

49. $\sin^2 225° - \cos^2 270° + \tan^2 60°$

50. $\cot^2 90° - \sec^2 180° + \csc^2 135°$

51. $\cos^2 60° + \sec^2 150° - \csc^2 210°$

52. $\cot^2 135° + \tan^4 60° - \sin^4 180°$

Determine whether each statement is true *or* false. *If false, tell why.* ***See Example 4.***

53. $\cos(30° + 60°) = \cos 30° + \cos 60°$

54. $\sin 30° + \sin 60° = \sin(30° + 60°)$

55. $\cos 60° = 2\cos 30°$

56. $\cos 60° = 2\cos^2 30° - 1$

57. $\sin^2 45° + \cos^2 45° = 1$

58. $\tan^2 60° + 1 = \sec^2 60°$

59. $\cos(2 \cdot 45°) = 2\cos 45°$

60. $\sin(2 \cdot 30°) = 2\sin 30° \cdot \cos 30°$

Concept Check *Find the coordinates of the point P on the circumference of each circle. (Hint: Sketch x- and y-axes, and interpret so that the angle is in standard position.)*

61.

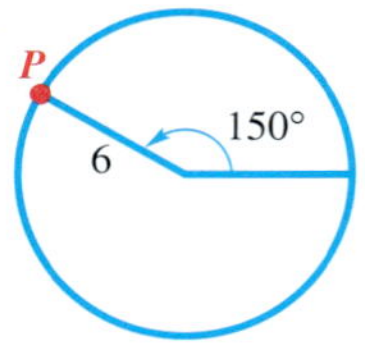

62.

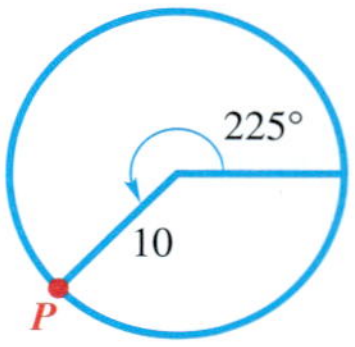

63. *Concept Check* Does there exist an angle θ with the function values $\cos \theta = 0.6$ and $\sin \theta = -0.8$?

64. *Concept Check* Does there exist an angle θ with the function values $\cos \theta = \frac{2}{3}$ and $\sin \theta = \frac{3}{4}$?

Suppose θ is in the interval $(90°, 180°)$. Find the sign of each of the following.

65. $\cos \frac{\theta}{2}$ **66.** $\sin \frac{\theta}{2}$ **67.** $\sec(\theta + 180°)$

68. $\cot(\theta + 180°)$ **69.** $\sin(-\theta)$ **70.** $\cos(-\theta)$

71. Explain why $\sin \theta = \sin(\theta + n \cdot 360°)$ is true for any angle θ and any integer n.

72. Explain why $\cos \theta = \cos(\theta + n \cdot 360°)$ is true for any angle θ and any integer n.

73. Explain why $\tan \theta = \tan(\theta + n \cdot 180°)$ is true for any angle θ and any integer n.

Concept Check *Work Exercises 74–77.*

74. Without using a calculator, determine which of the following numbers is closest to $\cos 115°$: $-0.6, -0.4, 0, 0.4$, or 0.6.

75. Without using a calculator, determine which of the following numbers is closest to $\sin 115°$: $-0.9, -0.1, 0, 0.1$, or 0.9.

76. For what angles θ between 0° and 360° is $\cos \theta = -\sin \theta$ true?

77. For what angles θ between 0° and 360° is $\cos \theta = \sin \theta$ true?

78. *(Modeling) Length of a Sag Curve* When a highway goes downhill and then uphill, it is said to have a **sag curve.** Sag curves are designed so that at night, headlights shine sufficiently far down the road to allow a safe stopping distance. See the figure.

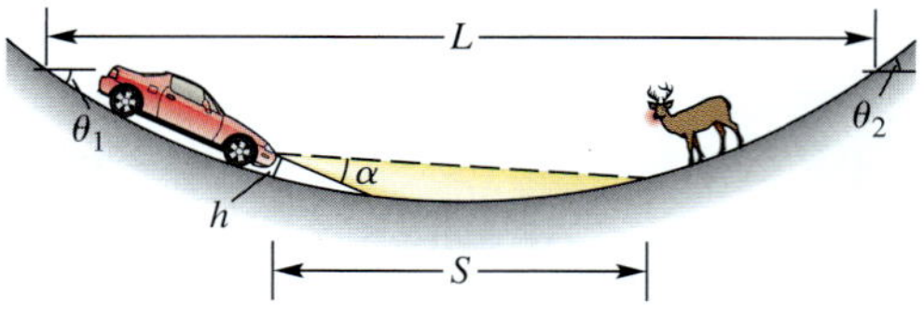

The minimum length L of a sag curve is determined by the height h of the car's headlights above the pavement, the downhill grade $\theta_1 < 0°$, the uphill grade $\theta_2 > 0°$, and the safe stopping distance S for a given speed limit. In addition, L is dependent on the vertical alignment of the headlights. Headlights are usually pointed upward at a slight angle α above the horizontal of the car. Using these quantities, for a 55 mph speed limit, L can be modeled by the formula

$$L = \frac{(\theta_2 - \theta_1)S^2}{200(h + S \tan \alpha)},$$

where $S < L$. (*Source:* Mannering, F. and W. Kilareski, *Principles of Highway Engineering and Traffic Analysis,* Second Edition, John Wiley and Sons.)

(a) Compute L if $h = 1.9$ ft, $\alpha = 0.9°$, $\theta_1 = -3°$, $\theta_2 = 4°$, and $S = 336$ ft.

(b) Repeat part (a) with $\alpha = 1.5°$.

(c) How does the alignment of the headlights affect the value of L?

Find all values of θ, if θ is in the interval $[0°, 360°)$ and has the given function value. ***See Example 6.***

79. $\sin\theta = \frac{1}{2}$ **80.** $\cos\theta = \frac{\sqrt{3}}{2}$ **81.** $\tan\theta = -\sqrt{3}$

82. $\sec\theta = -\sqrt{2}$ **83.** $\cos\theta = \frac{\sqrt{2}}{2}$ **84.** $\cot\theta = -\frac{\sqrt{3}}{3}$

85. $\csc\theta = -2$ **86.** $\sin\theta = -\frac{\sqrt{3}}{2}$ **87.** $\tan\theta = \frac{\sqrt{3}}{3}$

88. $\cos\theta = -\frac{1}{2}$ **89.** $\csc\theta = -\sqrt{2}$ **90.** $\cot\theta = -1$

2.3 Finding Trigonometric Function Values Using a Calculator

- Finding Function Values Using a Calculator
- Finding Angle Measures Using a Calculator

Finding Function Values Using a Calculator Calculators are capable of finding trigonometric function values. For example, the values of $\cos(-240°)$ and $\tan 675°$ in **Example 3** of **Section 2.2** are found with a calculator as shown in **Figure 15.**

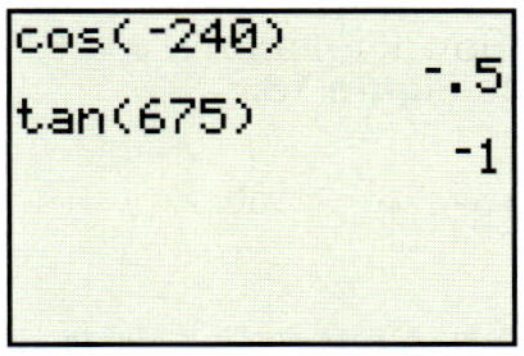

Degree mode

Figure 15

CAUTION ***When evaluating trigonometric functions of angles given in degrees, remember that the calculator must be set in degree mode.*** Get in the habit of always starting work by entering sin 90. If the displayed answer is 1, then the calculator is set for degree measure. Remember that most calculator values of trigonometric functions are *approximations.*

EXAMPLE 1 Finding Function Values with a Calculator

Approximate the value of each expression.

(a) $\sin 49° \, 12'$ **(b)** $\sec 97.977°$ **(c)** $\frac{1}{\cot 51.4283°}$ **(d)** $\sin(-246°)$

SOLUTION

(a) $49° \, 12' = 49\frac{12}{60}° = 49.2°$ Convert 49° 12′ to decimal degrees. (Section 1.1)

$\sin 49° \, 12' = \sin 49.2° \approx 0.75699506$ To eight decimal places

(b) Calculators do not have secant keys. However, $\sec\theta = \frac{1}{\cos\theta}$ for all angles θ where $\cos\theta \neq 0$. Therefore, we use the reciprocal of the cosine function to evaluate the secant function.

$$\sec 97.977° = \frac{1}{\cos 97.977°} \approx -7.20587921$$

(c) Use the reciprocal identity $\tan\theta = \frac{1}{\cot\theta}$ to simplify the expression first.

$$\frac{1}{\cot 51.4283°} = \tan 51.4283° \approx 1.25394815$$

(d) $\sin(-246°) \approx 0.91354546$

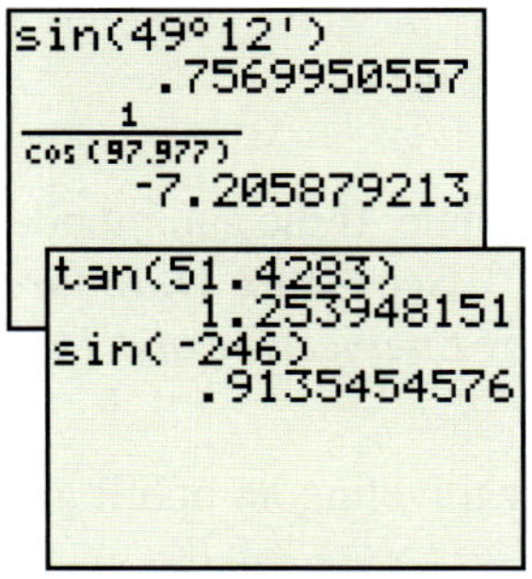

These screens support the results of **Example 1.** We entered the angle measure in degrees and minutes for part (a).

✔ *Now Try Exercises 5, 7, 11, and 15.*

Finding Angle Measures Using a Calculator To find the measure of an angle having a certain trigonometric function value, graphing calculators have three *inverse functions* (denoted $\sin^{-1}$, $\cos^{-1}$, and $\tan^{-1}$). ***If x is an appropriate number, then*** $\sin^{-1} x$, $\cos^{-1} x$, ***or*** $\tan^{-1} x$ ***gives the measure of an angle whose sine, cosine, or tangent, respectively, is x.*** For applications in this chapter, these functions will return angles in quadrant I.

EXAMPLE 2 Using Inverse Trigonometric Functions to Find Angles

Use a calculator to find an angle θ in the interval $[0°, 90°]$ that satisfies each condition.

(a) $\sin \theta \approx 0.96770915$ **(b)** $\sec \theta \approx 1.0545829$

SOLUTION

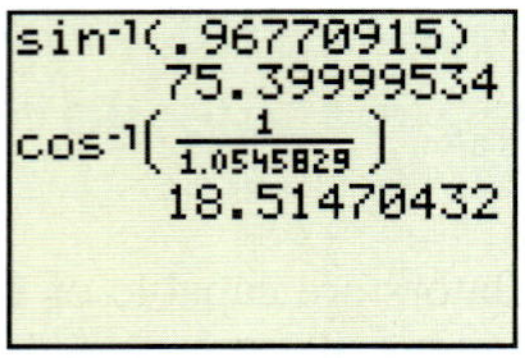

Degree mode

Figure 16

(a) Using degree mode and the inverse sine function, we find that an angle θ having sine value 0.96770915 is about 75.399995°. (There are infinitely many such angles, but the calculator gives only this one.)

$$\theta \approx \sin^{-1} 0.96770915 \approx 75.399995°$$

See **Figure 16.**

(b) Use the identity $\cos \theta = \frac{1}{\sec \theta}$. If $\sec \theta \approx 1.0545829$, then

$$\cos \theta \approx \frac{1}{1.0545829}.$$

Now, find θ using the inverse cosine function. See **Figure 16.**

$$\theta \approx \cos^{-1}\left(\frac{1}{1.0545829}\right) \approx 18.514704°$$

✔ *Now Try Exercises 25 and 29.*

CAUTION Compare **Examples 1(b) and 2(b).** To determine the secant of an angle, as in **Example 1(b),** we find the *reciprocal of the cosine* of the angle. To determine an angle with a given secant value, as in **Example 2(b),** we find the *inverse cosine of the reciprocal* of the value.

EXAMPLE 3 Finding Grade Resistance

When an automobile travels uphill or downhill on a highway, it experiences a force due to gravity. This force F in pounds is the **grade resistance** and is modeled by the equation

$$F = W \sin \theta,$$

where θ is the grade and W is the weight of the automobile. If the automobile is moving uphill, then $\theta > 0°$; if downhill, then $\theta < 0°$. See **Figure 17.** (*Source:* Mannering, F. and W. Kilareski, *Principles of Highway Engineering and Traffic Analysis,* Second Edition, John Wiley and Sons.)

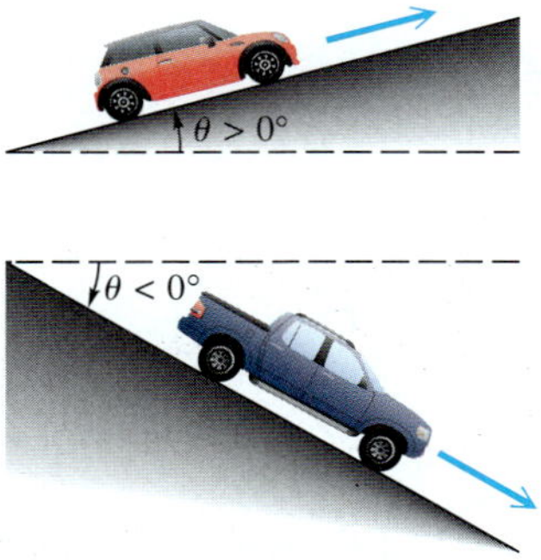

Figure 17

(a) Calculate F to the nearest 10 lb for a 2500-lb car traveling an uphill grade with $\theta = 2.5°$.

(b) Calculate F to the nearest 10 lb for a 5000-lb truck traveling a downhill grade with $\theta = -6.1°$.

(c) Calculate F for $\theta = 0°$ and $\theta = 90°$. Do these answers agree with your intuition?

SOLUTION

(a) $F = W \sin \theta = 2500 \sin 2.5° \approx 110 \text{ lb}$

(b) $F = W \sin \theta = 5000 \sin(-6.1°) \approx -530 \text{ lb}$
F is negative because the truck is moving downhill.

(c) $F = W \sin \theta = W \sin 0° = W(0) = 0 \text{ lb}$

$F = W \sin \theta = W \sin 90° = W(1) = W \text{ lb}$

This agrees with intuition because if $\theta = 0°$, then there is level ground and gravity does not cause the vehicle to roll. If θ were 90°, the road would be vertical and the full weight of the vehicle would be pulled downward by gravity, so $F = W$.

✔ *Now Try Exercises 59 and 61.*

2.3 Exercises

Concept Check Fill in the blanks to complete each statement.

1. The CAUTION at the beginning of this section suggests verifying that a calculator is in degree mode by finding ________ (sin/cos/tan) 90°. If the calculator is in degree mode, then the display should be ________.

2. When a scientific or graphing calculator is used to find a trigonometric function value, in most cases the result is an ________ (exact/approximate) value.

3. To find values of the cotangent, secant, and cosecant functions with a calculator, it is necessary to find the ________ of the ________ function value.

4. To determine the cosecant of an angle, we find the reciprocal of the ________ of the angle, but to determine the angle with a given cosecant value, we find the ________ sine of the reciprocal of the value.

Use a calculator to find a decimal approximation for each value. Give as many digits as your calculator displays. In Exercises 15–22, simplify the expression before using the calculator. ***See Example 1.***

5. $\sin 38° 42'$ **6.** $\cos 41° 24'$ **7.** $\sec 13° 15'$

8. $\csc 145° 45'$ **9.** $\cot 183° 48'$ **10.** $\tan 421° 30'$

11. $\sin(-312° 12')$ **12.** $\tan(-80° 06')$ **13.** $\csc(-317° 36')$

14. $\cot(-512° 20')$ **15.** $\dfrac{1}{\cot 23.4°}$ **16.** $\dfrac{1}{\sec 14.8°}$

17. $\dfrac{\cos 77°}{\sin 77°}$ **18.** $\dfrac{\sin 33°}{\cos 33°}$ **19.** $\cot(90° - 4.72°)$

20. $\cos(90° - 3.69°)$ **21.** $\dfrac{1}{\csc(90° - 51°)}$ **22.** $\dfrac{1}{\tan(90° - 22°)}$

Find a value of θ in the interval $[0°, 90°]$ that satisfies each statement. Write each answer in decimal degrees to six decimal places as needed. ***See Example 2.***

23. $\tan \theta = 1.4739716$ **24.** $\tan \theta = 6.4358841$ **25.** $\sin \theta = 0.27843196$

26. $\sin \theta = 0.84802194$ **27.** $\cot \theta = 1.2575516$ **28.** $\csc \theta = 1.3861147$

29. $\sec \theta = 2.7496222$ **30.** $\sec \theta = 1.1606249$ **31.** $\cos \theta = 0.70058013$

32. $\cos \theta = 0.85536428$ **33.** $\csc \theta = 4.7216543$ **34.** $\cot \theta = 0.21563481$

35. A student, wishing to use a calculator to verify the value of sin 30°, enters the information correctly but gets a display of −0.98803162. He knows that the display should be 0.5, and he also knows that his calculator is in good working order. What do you think is the problem?

36. At one time, a certain make of calculator did not allow the input of angles outside of a particular interval when finding trigonometric function values. For example, trying to find cos 2000° using the methods of this section gave an error message, despite the fact that cos 2000° can be evaluated. Explain how you would use this calculator to find cos 2000°.

37. What value of A between 0° and 90° will produce the output in the graphing calculator screen?

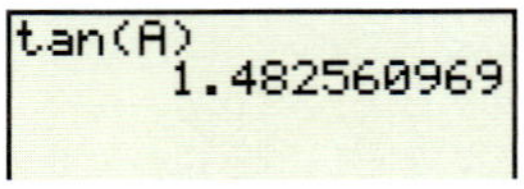

38. What value of A will produce the output (in degrees) in the graphing calculator screen?

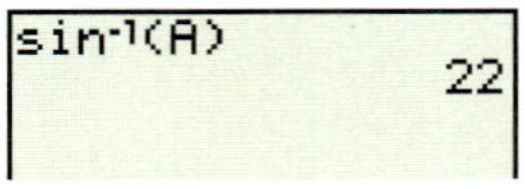

Use a calculator to evaluate each expression.

39. $\sin 35° \cos 55° + \cos 35° \sin 55°$

40. $\cos 100° \cos 80° - \sin 100° \sin 80°$

41. $\sin^2 36° + \cos^2 36°$

42. $2 \sin 25° 13' \cos 25° 13' - \sin 50° 26'$

43. $\cos 75° 29' \cos 14° 31' - \sin 75° 29' \sin 14° 31'$

44. $\sin 28° 14' \cos 61° 46' + \cos 28° 14' \sin 61° 46'$

Work each problem.

45. ***Measuring Speed by Radar*** Any offset between a stationary radar gun and a moving target creates a "cosine effect" that reduces the radar reading by the cosine of the angle between the gun and the vehicle. That is, the radar speed reading is the product of the actual speed and the cosine of the angle. Find the radar readings, to the nearest hundredth, for Auto A and Auto B shown in the figure. (*Source:* Fischetti, M., "Working Knowledge," *Scientific American*.)

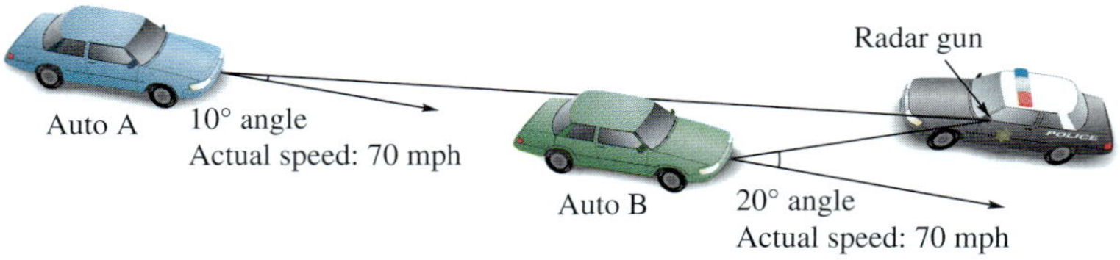

46. ***Measuring Speed by Radar*** In **Exercise 45,** we saw that the speed reported by a radar gun is reduced by the cosine of angle θ, shown in the figure. In the figure, r represents reduced speed and a represents the actual speed. Use the figure to show why this "cosine effect" occurs.

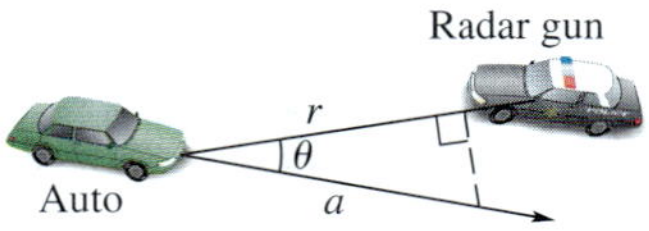

Use a calculator to decide whether each statement is true *or* false. *It may be that a true statement will lead to results that differ in the last decimal place due to rounding error.*

47. $\sin 10° + \sin 10° = \sin 20°$

48. $\cos 40° = 2 \cos 20°$

49. $\sin 50° = 2 \sin 25° \cos 25°$

50. $\cos 70° = 2 \cos^2 35° - 1$

51. $\cos 40° = 1 - 2 \sin^2 80°$

52. $2 \cos 38° 22' = \cos 76° 44'$

53. $\sin 39°48' + \cos 39°48' = 1$

54. $\frac{1}{2}\sin 40° = \sin\left[\frac{1}{2}(40°)\right]$

55. $1 + \cot^2 42.5° = \csc^2 42.5°$

56. $\tan^2 72°25' + 1 = \sec^2 72°25'$

57. $\cos(30° + 20°) = \cos 30° \cos 20° - \sin 30° \sin 20°$

58. $\cos(30° + 20°) = \cos 30° + \cos 20°$

(Modeling) Grade Resistance *See* **Example 3** *to work Exercises 59–65.*

59. Find the grade resistance, to the nearest ten pounds, for a 2100-lb car traveling on a 1.8° uphill grade.

60. Find the grade resistance, to the nearest ten pounds, for a 2400-lb car traveling on a −2.4° downhill grade.

61. A 2600-lb car traveling downhill has a grade resistance of −130 lb. Find the angle of the grade to the nearest tenth of a degree.

62. A 3000-lb car traveling uphill has a grade resistance of 150 lb. Find the angle of the grade to the nearest tenth of a degree.

63. A car traveling on a 2.7° uphill grade has a grade resistance of 120 lb. Determine the weight of the car to the nearest hundred pounds.

64. A car traveling on a −3° downhill grade has a grade resistance of −145 lb. Determine the weight of the car to the nearest hundred pounds.

65. Which has the greater grade resistance: a 2200-lb car on a 2° uphill grade or a 2000-lb car on a 2.2° uphill grade?

66. *Highway Grades* Complete the table for the values of $\sin\theta$, $\tan\theta$, and $\frac{\pi\theta}{180}$ to four decimal places.

θ	0°	0.5°	1°	1.5°	2°	2.5°	3°	3.5°	4°
$\sin\theta$									
$\tan\theta$									
$\frac{\pi\theta}{180}$									

(a) How do $\sin\theta$, $\tan\theta$, and $\frac{\pi\theta}{180}$ compare for small grades θ?

(b) Highway grades are usually small. Give two approximations of the grade resistance $F = W\sin\theta$ that do not use the sine function.

(c) A stretch of highway has a 4-ft vertical rise for every 100 ft of horizontal run. Use an approximation from part (b) to estimate the grade resistance, to the nearest pound, for a 2000-lb car on this stretch of highway.

(d) Without evaluating a trigonometric function, estimate the grade resistance, to the nearest nearest pound, for an 1800-lb car on a stretch of highway that has a 3.75° grade.

(Modeling) Solve each problem.

67. *Design of Highway Curves* When highway curves are designed, the outside of the curve is often slightly elevated or inclined above the inside of the curve. See the figure. This inclination is the **superelevation.** For safety reasons, it is important that both the curve's radius and superelevation be correct for a given speed limit. If an automobile is traveling at velocity V (in feet per second), the safe radius R for a curve with superelevation θ is modeled by the formula

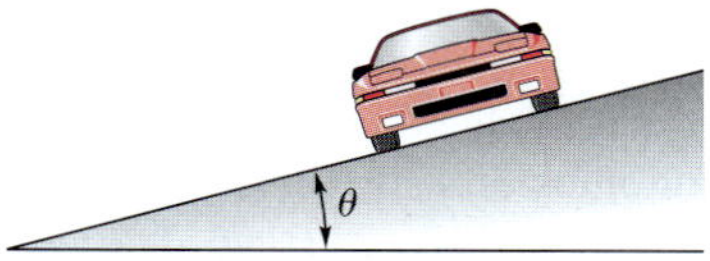

$$R = \frac{V^2}{g(f + \tan\theta)},$$

where f and g are constants. (*Source:* Mannering, F. and W. Kilareski, *Principles of Highway Engineering and Traffic Analysis,* Second Edition, John Wiley and Sons.)

(a) A roadway is being designed for automobiles traveling at 45 mph. If $\theta = 3°$, $g = 32.2$, and $f = 0.14$, calculate R to the nearest foot. (*Hint:* 45 mph = 66 ft per sec)

(b) Determine the radius of the curve, to the nearest foot, if the speed in part (a) is increased to 70 mph.

(c) How would increasing the angle θ affect the results? Verify your answer by repeating parts (a) and (b) with $\theta = 4°$.

68. *Speed Limit on a Curve* Refer to **Exercise 67** and use the same values for f and g. A highway curve has radius $R = 1150$ ft and a superelevation of $\theta = 2.1°$. What should the speed limit (in miles per hour) be for this curve?

(Modeling) Speed of Light When a light ray travels from one medium, such as air, to another medium, such as water or glass, the speed of the light changes, and the light ray is bent, or ***refracted,*** *at the boundary between the two media. (This is why objects under water appear to be in a different position from where they really are.) It can be shown in physics that these changes are related by* ***Snell's law***

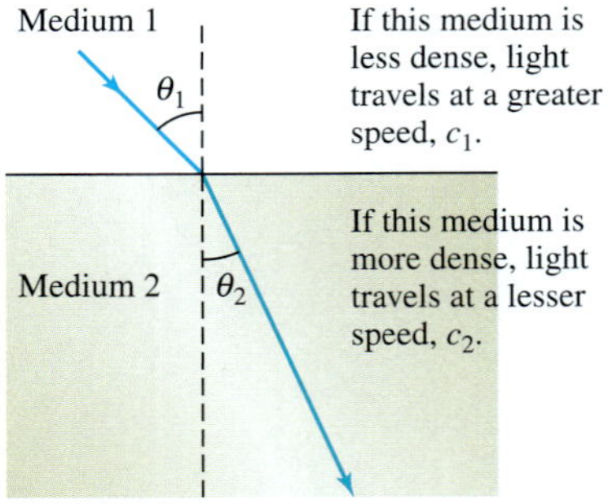

$$\frac{c_1}{c_2} = \frac{\sin \theta_1}{\sin \theta_2},$$

where c_1 is the speed of light in the first medium, c_2 is the speed of light in the second medium, and θ_1 and θ_2 are the angles shown in the figure. In Exercises 69 and 70, assume that $c_1 = 3 \times 10^8$ m per sec.

69. Find the speed of light in the second medium for each of the following.

(a) $\theta_1 = 46°$, $\theta_2 = 31°$ **(b)** $\theta_1 = 39°$, $\theta_2 = 28°$

70. Find θ_2 for each of the following values of θ_1 and c_2. Round to the nearest degree.

(a) $\theta_1 = 40°$, $c_2 = 1.5 \times 10^8$ m per sec **(b)** $\theta_1 = 62°$, $c_2 = 2.6 \times 10^8$ m per sec

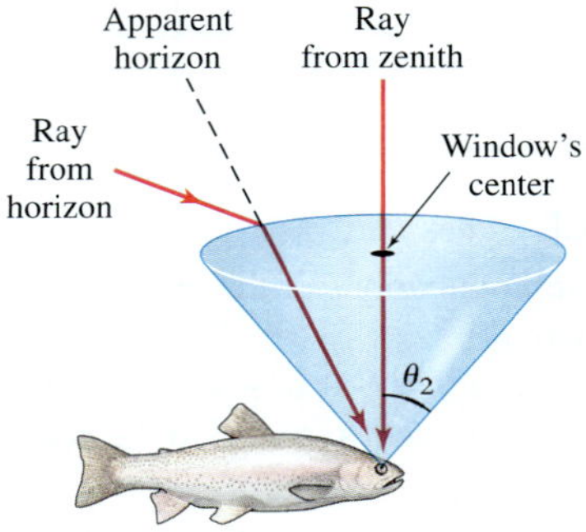

(Modeling) Fish's View of the World The figure in the margin shows a fish's view of the world above the surface of the water. (Source: Walker, J., "The Amateur Scientist," *Scientific American.) Suppose that a light ray comes from the horizon, enters the water, and strikes the fish's eye.*

71. Assume that this ray gives a value of 90° for angle θ_1 in the formula for Snell's law. (In a practical situation, this angle would probably be a little less than 90°.) The speed of light in water is about 2.254×10^8 m per sec. Find angle θ_2 to the nearest tenth.

72. Refer to **Exercise 71.** Suppose an object is located at a true angle of 29.6° above the horizon. Find the apparent angle above the horizon to a fish.

73. *(Modeling) Braking Distance* If aerodynamic resistance is ignored, the braking distance D (in feet) for an automobile to change its velocity from V_1 to V_2 (feet per second) can be modeled using the following equation.

$$D = \frac{1.05(V_1^2 - V_2^2)}{64.4(K_1 + K_2 + \sin \theta)}$$

K_1 is a constant determined by the efficiency of the brakes and tires, K_2 is a constant determined by the rolling resistance of the automobile, and θ is the grade of the highway. (*Source:* Mannering, F. and W. Kilareski, *Principles of Highway Engineering and Traffic Analysis,* Second Edition, John Wiley and Sons.)

(a) Compute the number of feet required to slow a car from 55 mph to 30 mph while traveling uphill with a grade of $\theta = 3.5°$. Let $K_1 = 0.4$ and $K_2 = 0.02$. (*Hint:* Change miles per hour to feet per second.)

(b) Repeat part (a) with $\theta = -2°$.

(c) How is braking distance affected by grade θ? Does this agree with your driving experience?

74. *(Modeling) Car's Speed at Collision* Refer to **Exercise 73.** An automobile is traveling at 90 mph on a highway with a downhill grade of $\theta = -3.5°$. The driver sees a stalled truck in the road 200 ft away and immediately applies the brakes. Assuming that a collision cannot be avoided, how fast (in miles per hour) is the car traveling when it hits the truck? (Use the same values for K_1 and K_2 as in **Exercise 73.**)

Chapter 2 Quiz (Sections 2.1–2.3)

1. Find the exact values of the six trigonometric functions for angle A in the figure.

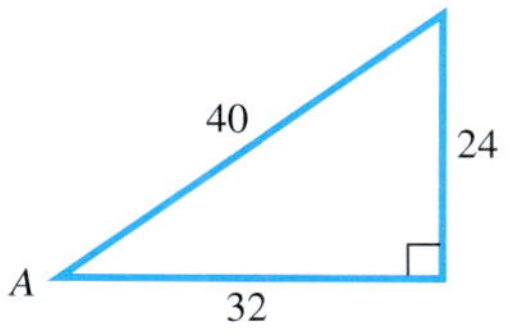

2. Find exact values of the trigonometric functions to complete the table.

θ	$\sin\theta$	$\cos\theta$	$\tan\theta$	$\cot\theta$	$\sec\theta$	$\csc\theta$
30°						
45°						
60°						

3. Find the exact value of each variable in the figure.

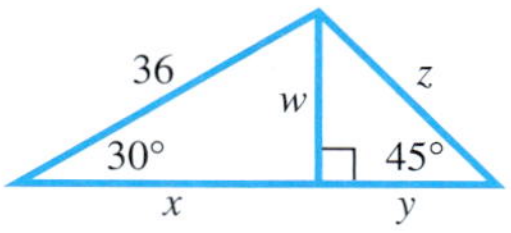

4. *Area of a Solar Cell* A solar cell converts the energy of sunlight directly into electrical energy. The amount of energy a cell produces depends on its area. Suppose a solar cell is hexagonal, as shown in the figure on the left below. Express its area $\mathcal{A}$ in terms of $\sin\theta$ and any side x. (*Hint:* Consider one of the six equilateral triangles from the hexagon. See the figure on the right below.) (*Source:* Kastner, B., *Space Mathematics,* NASA.)

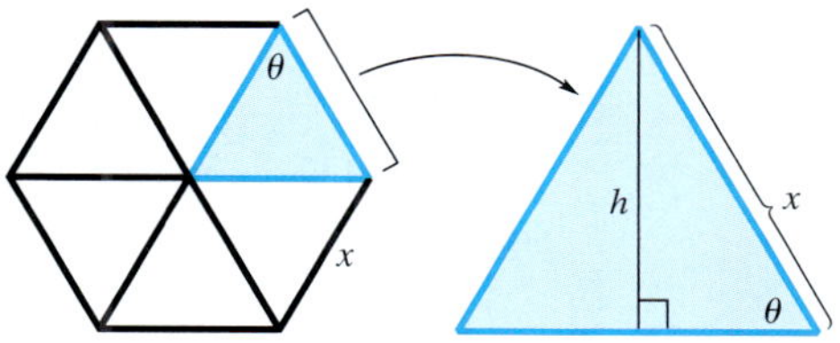

Find exact values of the six trigonometric functions for each angle. Rationalize denominators when applicable.

5. $135°$ **6.** $-150°$ **7.** $1020°$

Find all values of θ in the interval $[0°, 360°)$ that have the given function value.

8. $\sin \theta = \frac{\sqrt{3}}{2}$ **9.** $\sec \theta = -\sqrt{2}$

Use a calculator to approximate each value. Give as many digits as your calculator displays.

10. $\sin 42° \, 18'$ **11.** $\sec(-212° \, 12')$

Use a calculator to find the value of θ in the interval $[0°, 90°]$ that satisfies each statement. Write each answer in decimal degrees to six decimal places as needed.

12. $\tan \theta = 2.6743210$ **13.** $\csc \theta = 2.3861147$

Determine whether each statement is true or false.

14. $\sin(60° + 30°) = \sin 60° + \sin 30°$ **15.** $\tan(90° - 35°) = \cot 35°$

2.4 Solving Right Triangles

- Significant Digits
- Solving Triangles
- Angles of Elevation or Depression

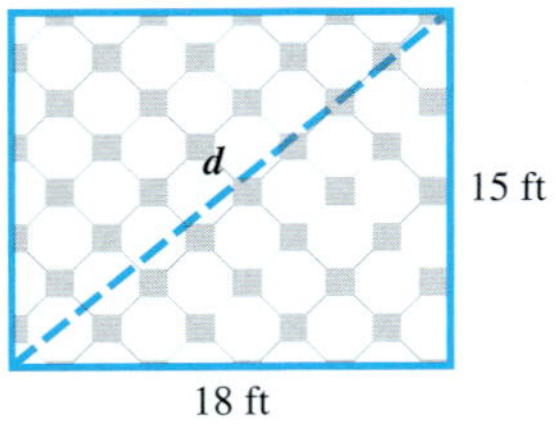

Figure 18

Significant Digits

A number that represents the result of counting, or a number that results from theoretical work and is not the result of measurement, is an **exact number.** There are 50 states in the United States. In this statement, 50 is an exact number.

Most values obtained for trigonometric applications are measured values that are *not* exact. Suppose we quickly measure a room as 15 ft by 18 ft. See **Figure 18.** To calculate the length of a diagonal of the room, we can use the Pythagorean theorem.

$$d^2 = 15^2 + 18^2 \quad \text{(Appendix B)}$$

$$d^2 = 549 \quad \text{Apply the exponents and add.}$$

$$d = \sqrt{549} \quad \text{Square root property (Appendix A); choose the positive root.}$$

$$d \approx 23.430749$$

Should this answer be given as the length of the diagonal of the room? Of course not. The number 23.430749 contains six decimal places, while the original data of 15 ft and 18 ft are accurate only to the nearest foot. In practice, the results of a calculation can be no more accurate than the least accurate number in the calculation. Thus, we should indicate that the diagonal of the 15-by-18-ft room is approximately 23 ft.

If a wall measured to the nearest foot is 18 ft long, this actually means that the wall has length between 17.5 ft and 18.5 ft. If the wall is measured more accurately as 18.3 ft long, then its length is really between 18.25 ft and 18.35 ft. The results of physical measurement are only approximately accurate and depend on the precision of the measuring instrument as well as the aptness of the observer. The digits obtained by actual measurement are called **significant digits.** The measurement 18 ft is said to have two significant digits; 18.3 ft has three significant digits.

In the following numbers, the significant digits are identified in color.

408 21.5 18.00 6.700 0.0025 0.09810 7300

Notice that 18.00 has four significant digits. The zeros in this number represent measured digits accurate to the nearest hundredth. The number 0.0025 has only two significant digits, 2 and 5, because the zeros here are used only to locate the decimal point. The number 7300 causes some confusion because it is impossible to determine whether the zeros are measured values. The number 7300 may have two, three, or four significant digits. When presented with this situation, we assume that the zeros are not significant, unless the context of the problem indicates otherwise.

To determine the number of significant digits for answers in applications of angle measure, use the following table.

Angle Measure to Nearest	Examples	Answer to Number of Significant Digits
Degree	62°, 36°	two
Ten minutes, or nearest tenth of a degree	52° 30′, 60.4°	three
Minute, or nearest hundredth of a degree	81° 48′, 71.25°	four
Ten seconds, or nearest thousandth of a degree	10° 52′ 20″, 21.264°	five

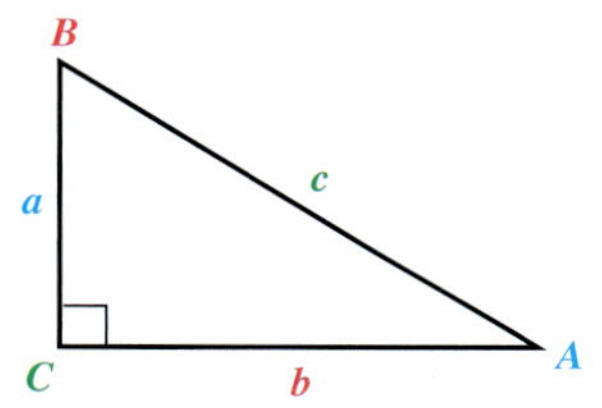

When we are solving triangles, a labeled sketch is an important aid.

Figure 19

To perform calculations with measured numbers, start by identifying the number with the least number of significant digits. Round your final answer to the same number of significant digits as this number. ***Remember that your answer is no more accurate than the least accurate number in your calculation.***

Solving Triangles To *solve a triangle* means to find the measures of all the angles and sides of the triangle. As shown in **Figure 19,** we use a to represent the length of the side opposite angle A, b for the length of the side opposite angle B, and so on. In a right triangle, the letter c is reserved for the hypotenuse.

EXAMPLE 1 Solving a Right Triangle Given an Angle and a Side

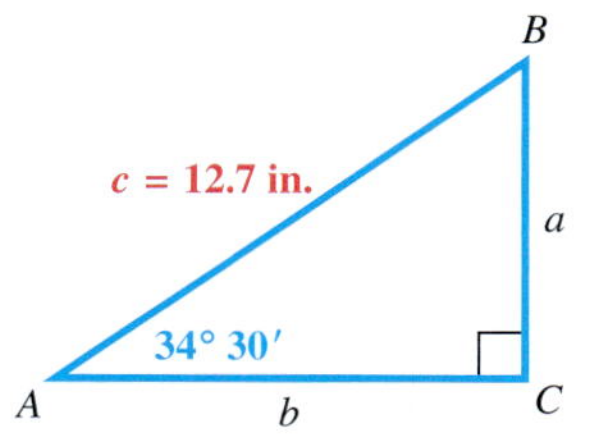

Figure 20

Solve right triangle ABC, if $A = 34° \, 30'$ and $c = 12.7$ in. See **Figure 20.**

SOLUTION To solve the triangle, find the measures of the remaining sides and angles. To find the value of a, use a trigonometric function involving the known values of angle A and side c. Since the sine of angle A is given by the quotient of the side opposite A and the hypotenuse, use sin A.

$$\sin A = \frac{a}{c} \qquad \sin A = \frac{\text{side opposite}}{\text{hypotenuse}} \text{ (Section 2.1)}$$

$$\sin 34° \, 30' = \frac{a}{12.7} \qquad A = 34° \, 30', \; c = 12.7$$

$$a = 12.7 \sin 34° \, 30' \qquad \text{Multiply by 12.7 and rewrite.}$$

$$a = 12.7 \sin 34.5° \qquad \text{Convert to decimal degrees. (Section 1.1)}$$

$$a \approx 12.7(0.56640624) \qquad \text{Use a calculator.}$$

$$a \approx 7.19 \text{ in.} \qquad \text{Three significant digits}$$

Assuming that 34° 30′ is given to the nearest ten minutes, we rounded the answer to three significant digits.

To find the value of b, we could substitute the value of a just calculated and the given value of c in the Pythagorean theorem. It is better, however, to use the information given in the problem rather than a result just calculated.

LOOKING AHEAD TO CALCULUS

The derivatives of the **parametric equations** $x = f(t)$ and $y = g(t)$ often represent the rate of change of physical quantities, such as velocities. When x and y are related by an equation, the derivatives are **related rates** because a change in one causes a related change in the other. Determining these rates in calculus often requires solving a right triangle.

If an error is made in finding a, then b also would be incorrect. And, rounding more than once may cause the result to be less accurate. To find b, use cos A.

$$\cos A = \frac{b}{c} \qquad \cos A = \frac{\text{side adjacent}}{\text{hypotenuse}} \text{ (Section 2.1)}$$

$$\cos 34° 30' = \frac{b}{12.7} \qquad A = 34° 30',\ c = 12.7$$

$$b = 12.7 \cos 34° 30' \qquad \text{Multiply by 12.7 and rewrite.}$$

$$b \approx 10.5 \text{ in.} \qquad \text{Three significant digits}$$

Once b is found, the Pythagorean theorem can be used to verify the results.

All that remains to solve triangle ABC is to find the measure of angle B.

$$B = 90° - A \qquad \text{Use } A + B = 90°, \text{ solved for } B. \text{ (Section 1.1)}$$

$$B = 89° 60' - 34° 30' \qquad \text{Rewrite } 90°. \text{ Substitute } 34° 30' \text{ for } A.$$

$$B = 55° 30' \qquad \text{Subtract degrees and minutes separately.}$$

✔ ***Now Try Exercise 21.***

NOTE In **Example 1,** we could have found the measure of angle B first and then used the trigonometric function values of B to find the unknown sides. A right triangle can usually be solved in several ways, each producing the correct answer. ***To maintain accuracy, always use given information as much as possible, and avoid rounding in intermediate steps.***

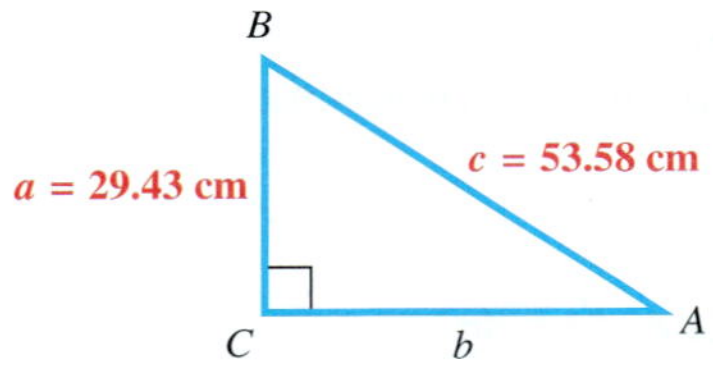

Figure 21

EXAMPLE 2 Solving a Right Triangle Given Two Sides

Solve right triangle ABC, if $a = 29.43$ cm and $c = 53.58$ cm.

SOLUTION We draw a sketch showing the given information, as in **Figure 21.** One way to begin is to find angle A by using the sine function.

$$\sin A = \frac{a}{c} \qquad \sin A = \frac{\text{side opposite}}{\text{hypotenuse}}$$

$$\sin A = \frac{29.43}{53.58} \qquad a = 29.43,\ c = 53.58$$

$$\sin A \approx 0.5492721165 \qquad \text{Use a calculator.}$$

$$A \approx \sin^{-1}(0.5492721165) \qquad \text{Use the inverse sine function. (Section 2.3)}$$

$$A \approx 33.32°, \quad \text{or} \quad 33° 19' \qquad \text{Four significant digits}$$

The measure of B is approximately

$$90° - 33° 19' = 56° 41'. \qquad 90° = 89° 60' \text{ (Section 1.1)}$$

We now find b from the Pythagorean theorem.

$$b^2 = c^2 - a^2 \qquad \text{Pythagorean theorem solved for } b^2 \text{ (Appendix B)}$$

$$b^2 = 53.58^2 - 29.43^2 \qquad c = 53.58,\ a = 29.43$$

$$b = \sqrt{2004.6915} \qquad \text{Simplify on the right; square root property}$$

$$b \approx 44.77 \text{ cm} \qquad \text{Choose the positive square root.}$$

✔ ***Now Try Exercise 31.***

George Polya (1887–1985)

Polya, a native of Budapest, Hungary, wrote more than 250 papers and a number of books. He proposed a general outline for solving applied problems in his classic book *How to Solve It*.

Angles of Elevation or Depression In applications of right triangles, the **angle of elevation** from point X to point Y (above X) is the acute angle formed by ray XY and a horizontal ray with endpoint at X. See **Figure 22(a).** The **angle of depression** from point X to point Y (below X) is the acute angle formed by ray XY and a horizontal ray with endpoint X. See **Figure 22(b).**

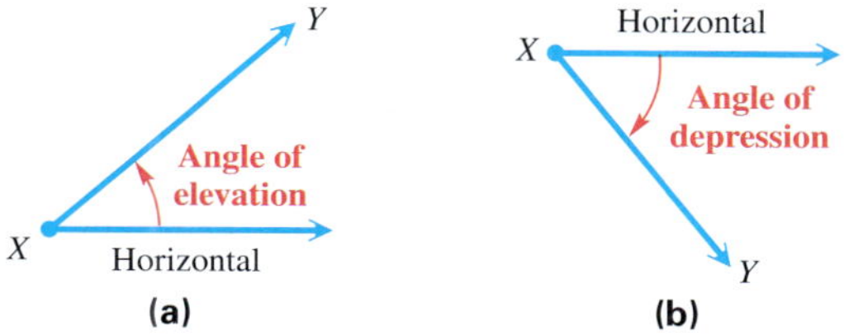

Figure 22

CAUTION Be careful when interpreting the angle of depression. ***Both the angle of elevation and the angle of depression are measured between the line of sight and a horizontal line.***

To solve applied trigonometry problems, follow the same procedure as solving a triangle. ***Drawing a sketch and labeling it correctly in Step 1 is crucial.***

Solving an Applied Trigonometry Problem

Step 1 Draw a sketch, and label it with the given information. Label the quantity to be found with a variable.

Step 2 Use the sketch to write an equation relating the given quantities to the variable.

Step 3 Solve the equation, and check that your answer makes sense.

EXAMPLE 3 Finding a Length Given the Angle of Elevation

Pat Porterfield knows that when she stands 123 ft from the base of a flagpole, the angle of elevation to the top of the flagpole is $26° 40'$. If her eyes are 5.30 ft above the ground, find the height of the flagpole.

SOLUTION

Step 1 The length of the side adjacent to Pat is known, and the length of the side opposite her must be found. See **Figure 23.**

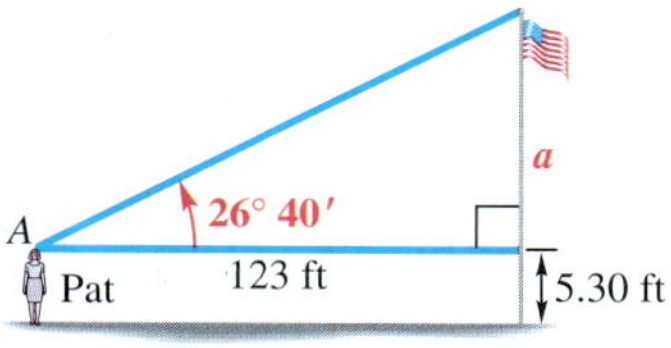

Figure 23

Step 2 The tangent ratio involves the given values. Write an equation.

$$\tan A = \frac{\text{side opposite}}{\text{side adjacent}} \qquad \text{Tangent ratio (Section 2.1)}$$

$$\tan 26° 40' = \frac{a}{123} \qquad A = 26° 40', \text{ side adjacent} = 123$$

Step 3

$$a = 123 \tan 26° 40' \qquad \text{Multiply by 123 and rewrite.}$$

$$a \approx 123(0.50221888) \qquad \text{Use a calculator.}$$

$$a \approx 61.8 \text{ ft} \qquad \text{Three significant digits}$$

The height of the flagpole is

$$61.8 + 5.30 = 67.1 \text{ ft.} \qquad \text{Pat's eyes are 5.30 ft above the ground.}$$

✔ ***Now Try Exercise 51.***

EXAMPLE 4 Finding an Angle of Depression

From the top of a 210-ft cliff, David observes a lighthouse that is 430 ft offshore. Find the angle of depression from the top of the cliff to the base of the lighthouse.

SOLUTION As shown in **Figure 24,** the angle of depression is measured from a horizontal line down to the base of the lighthouse. The angle of depression and angle B, in the right triangle shown, are alternate interior angles whose measures are equal. We use the tangent ratio to solve for angle B.

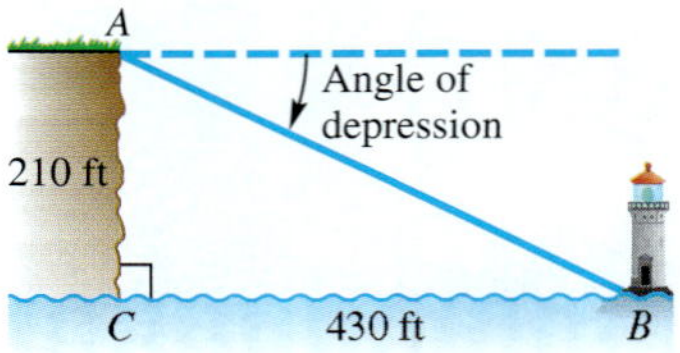

Figure 24

$$\tan B = \frac{210}{430}, \quad \text{so} \quad B = \tan^{-1}\frac{210}{430} \approx 26^\circ \quad \text{Angle of depression}$$

✔ *Now Try Exercise 53.*

2.4 Exercises

Concept Check Refer to the discussion of accuracy and significant digits in this section to work Exercises 1–8.

1. *Leading NFL Receiver* As of the end of the 2009 National Football League season, Jerry Rice was the leading career receiver with 22,895 yd. State the range represented by this number. (*Source:* www.nfl.com)

2. *Height of Mt. Everest* When Mt. Everest was first surveyed, the surveyors obtained a height of 29,000 ft to the nearest foot. State the range represented by this number. (The surveyors thought no one would believe a measurement of 29,000 ft, so they reported it as 29,002.) (*Source:* Dunham, W., *The Mathematical Universe,* John Wiley and Sons.)

3. *Longest Vehicular Tunnel* The E. Johnson Memorial Tunnel in Colorado, which measures 8959 ft, is one of the longest land vehicular tunnels in the United States. What is the range of this number? (*Source: World Almanac and Book of Facts.*)

4. *Top WNBA Scorer* Women's National Basketball Association player Cappie Pondexter of the New York Liberty received the 2010 award for most points scored, 729. Is it appropriate to consider this number as between 728.5 and 729.5? Why or why not? (*Source:* www.wnba.com)

5. *Circumference of a Circle* The formula for the circumference of a circle is $C = 2\pi r$. Suppose you use the $\boxed{\pi}$ key on your calculator to find the circumference of a circle with radius 54.98 cm, and get 345.44953. Since 2 has only one significant digit, should the answer be given as 3×10^2, or 300 cm? If not, explain how the answer should be given.

6. Explain the distinction between a measurement of 23.0 ft and a measurement of 23.00 ft.

7. If h is the actual height of a building and the height is measured as 58.6 ft, then $|h - 58.6| \leq$ _____.

8. If w is the actual weight of a car and the weight is measured as 1542 lb, then $|w - 1542| \leq$ _____.

Solve each right triangle. When two sides are given, give angles in degrees and minutes. ***See Examples 1 and 2.***

9.
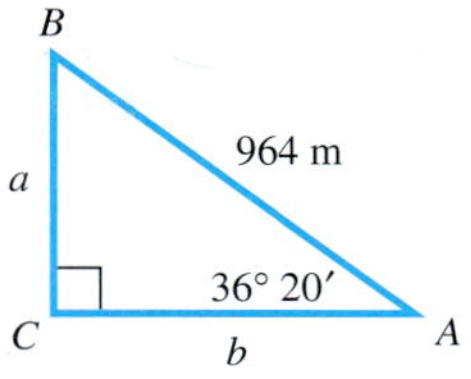

10.
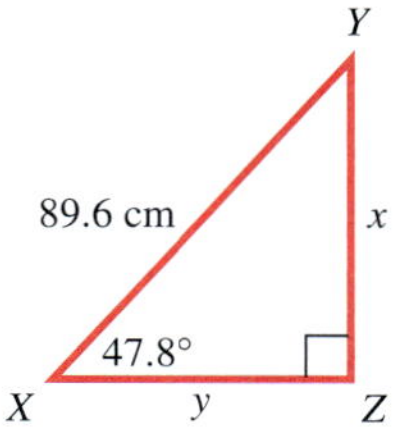

11.
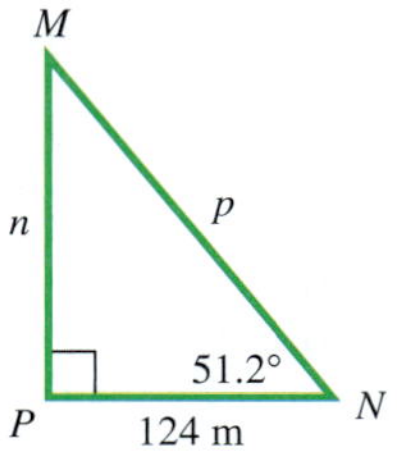

12.
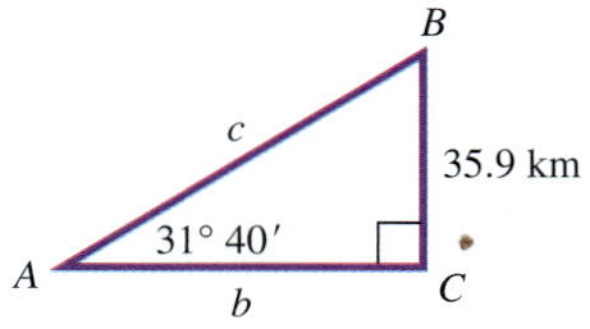

13.
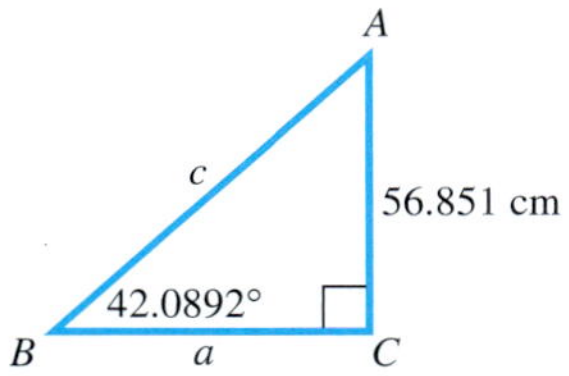

14.
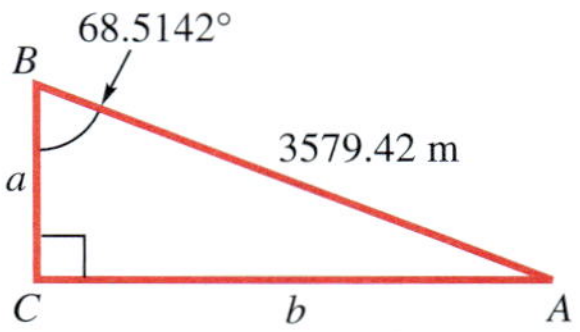

15.
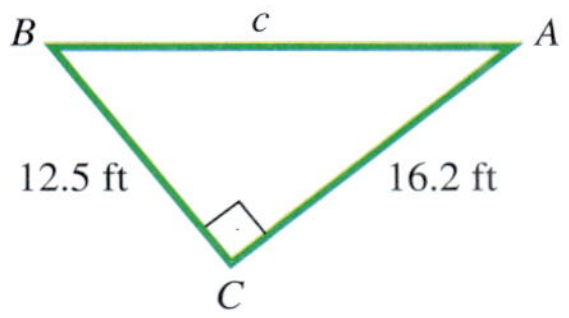

16.
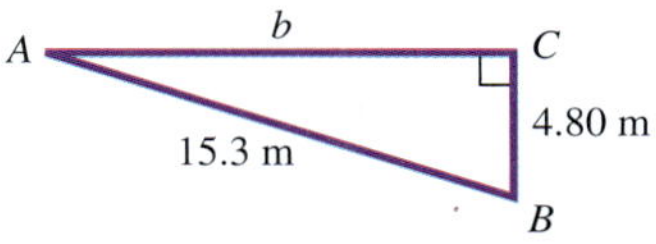

17. Can a right triangle be solved if we are given measures of its two acute angles and no side lengths? Explain.

18. *Concept Check* If we are given an acute angle and a side in a right triangle, what unknown part of the triangle requires the least work to find?

19. Explain why you can always solve a right triangle if you know the measures of one side and one acute angle.

20. Explain why you can always solve a right triangle if you know the lengths of two sides.

Solve each right triangle. In each case, $C = 90°$. *If angle information is given in degrees and minutes, give answers in the same way. If angle information is given in decimal degrees, do likewise in answers. When two sides are given, give angles in degrees and minutes.* ***See Examples 1 and 2.***

21. $A = 28.0°, c = 17.4$ ft

22. $B = 46.0°, c = 29.7$ m

23. $B = 73.0°, b = 128$ in.

24. $A = 62.5°, a = 12.7$ m

25. $A = 61.0°, b = 39.2$ cm

26. $B = 51.7°, a = 28.1$ ft

27. $a = 13$ m, $c = 22$ m

28. $b = 32$ ft, $c = 51$ ft

29. $a = 76.4$ yd, $b = 39.3$ yd

30. $a = 958$ m, $b = 489$ m

31. $a = 18.9$ cm, $c = 46.3$ cm

32. $b = 219$ m, $c = 647$ m

33. $A = 53° 24', c = 387.1$ ft

34. $A = 13° 47', c = 1285$ m

35. $B = 39° 09', c = 0.6231$ m

36. $B = 82° 51', c = 4.825$ cm

37. Explain the meaning of the term *angle of elevation.*

38. *Concept Check* Can an angle of elevation be more than 90°?

39. Explain why the angle of depression *DAB* has the same measure as the angle of elevation *ABC* in the figure.

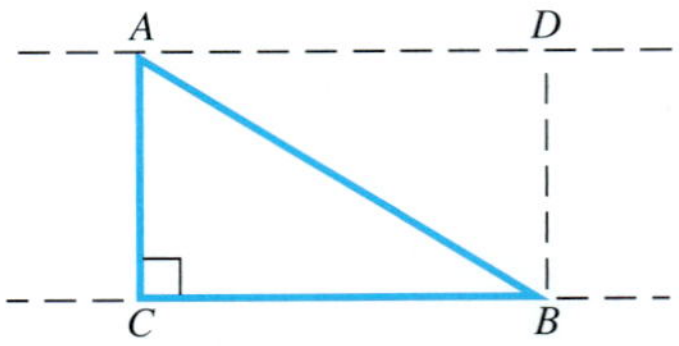

AD is parallel to *BC*.

40. Why is angle *CAB* *not* an angle of depression in the figure for **Exercise 39?**

Solve each problem involving triangles. ***See Examples 1–4.***

41. *Height of a Ladder on a Wall* A 13.5-m fire truck ladder is leaning against a wall. Find the distance *d* the ladder goes up the wall (above the top of the fire truck) if the ladder makes an angle of 43° 50′ with the horizontal.

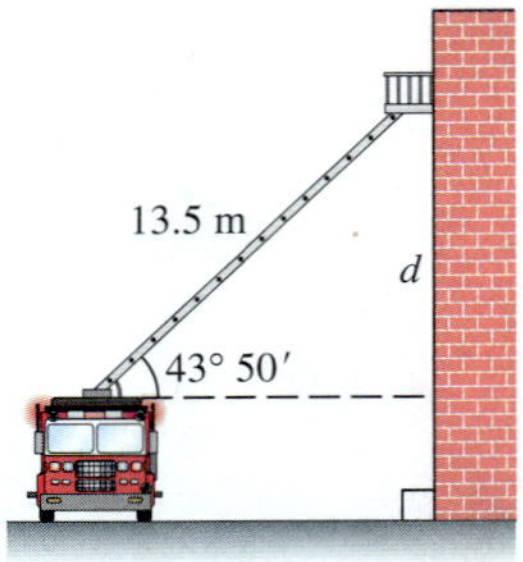

42. *Distance across a Lake* To find the distance *RS* across a lake, a surveyor lays off length $RT = 53.1$ m, so that angle $T = 32° 10'$ and angle $S = 57° 50'$. Find length *RS*.

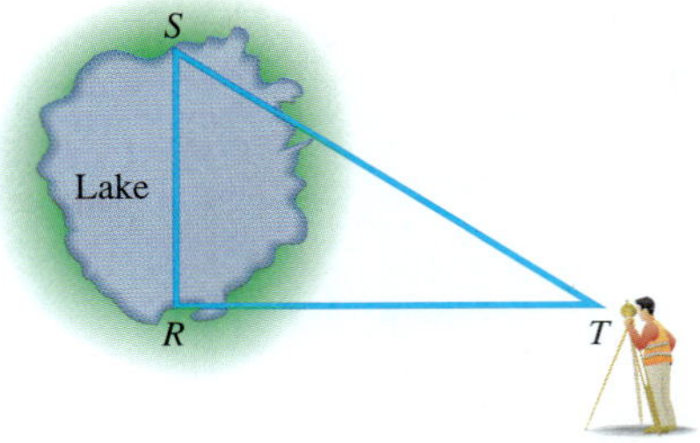

43. *Height of a Building* From a window 30.0 ft above the street, the angle of elevation to the top of the building across the street is 50.0° and the angle of depression to the base of this building is 20.0°. Find the height of the building across the street.

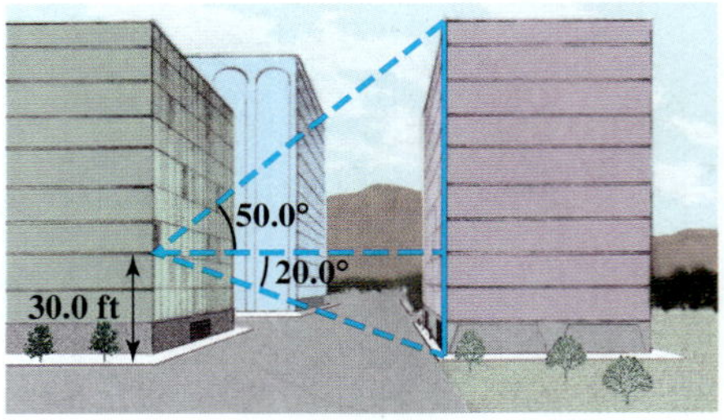

44. *Diameter of the Sun* To determine the diameter of the sun, an astronomer might sight with a **transit** (a device used by surveyors for measuring angles) first to one edge of the sun and then to the other, estimating that the included angle equals 32′. Assuming that the distance *d* from Earth to the sun is 92,919,800 mi, approximate the diameter of the sun.

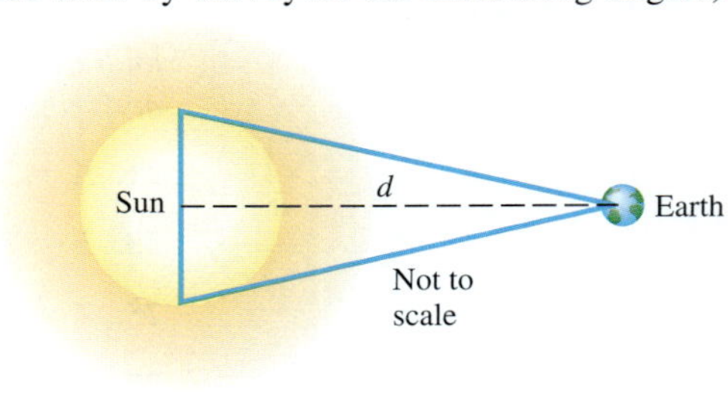

45. *Side Lengths of a Triangle* The length of the base of an isosceles triangle is 42.36 in. Each base angle is 38.12°. Find the length of each of the two equal sides of the triangle. (*Hint:* Divide the triangle into two right triangles.)

46. *Altitude of a Triangle* Find the altitude of an isosceles triangle having base 184.2 cm if the angle opposite the base is 68° 44′.

Solve each problem involving an angle of elevation or depression. ***See Examples 3 and 4.***

47. *Angle of Elevation of the Pyramid of the Sun* The Pyramid of the Sun in the ancient Mexican city of Teotihuacan was the largest and most important structure in the city. The base is a square with sides about 700 ft long, and the height of the pyramid is about 200 ft. Find the angle of elevation of the edge indicated in the figure to two significant digits. (*Hint:* The base of the triangle in the figure is half the diagonal of the square base of the pyramid.) (*Source*: www.britannica.com)

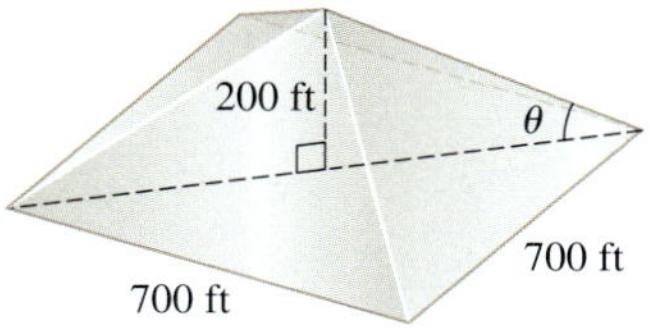

48. *Cloud Ceiling* The U.S. Weather Bureau defines a **cloud ceiling** as the altitude of the lowest clouds that cover more than half the sky. To determine a cloud ceiling, a powerful searchlight projects a circle of light vertically on the bottom of the cloud. An observer sights the circle of light in the crosshairs of a tube called a **clinometer.** A pendant hanging vertically from the tube and resting on a protractor gives the angle of elevation. Find the cloud ceiling if the searchlight is located 1000 ft from the observer and the angle of elevation is 30.0° as measured with a clinometer at eye-height 6 ft. (Assume three significant digits.)

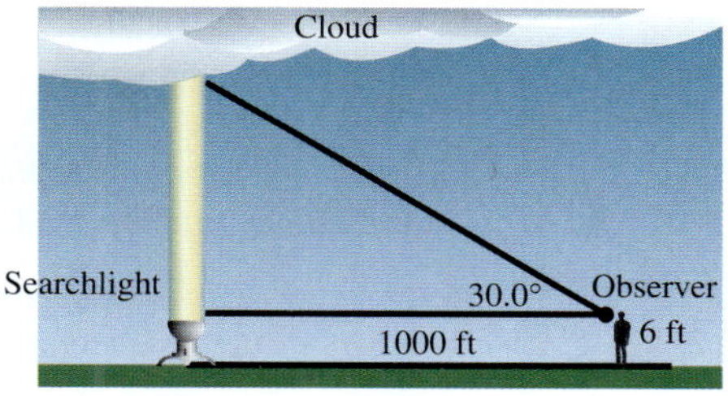

49. *Height of a Tower* The shadow of a vertical tower is 40.6 m long when the angle of elevation of the sun is 34.6°. Find the height of the tower.

50. *Distance from the Ground to the Top of a Building* The angle of depression from the top of a building to a point on the ground is 32° 30′. How far is the point on the ground from the top of the building if the building is 252 m high?

51. *Length of a Shadow* Suppose that the angle of elevation of the sun is 23.4°. Find the length of the shadow cast by Dot Peterson, who is 5.75 ft tall.

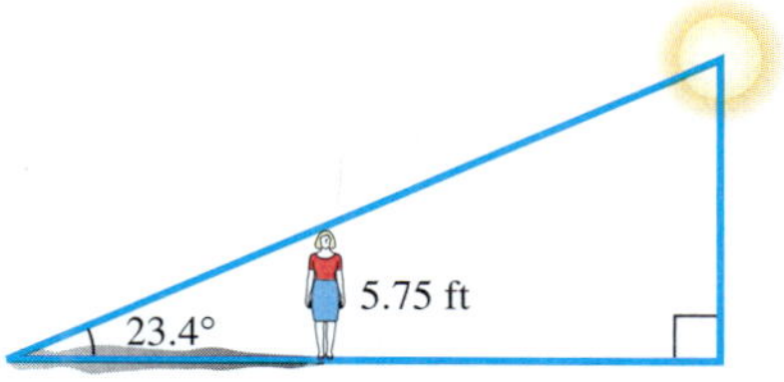

52. *Airplane Distance* An airplane is flying 10,500 ft above level ground. The angle of depression from the plane to the base of a tree is 13° 50′. How far horizontally must the plane fly to be directly over the tree?

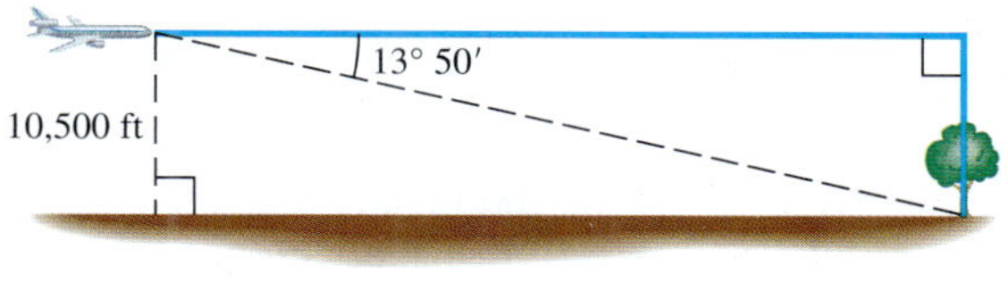

53. *Angle of Depression of a Light* A company safety committee has recommended that a floodlight be mounted in a parking lot so as to illuminate the employee exit. Find the angle of depression of the light to the nearest minute.

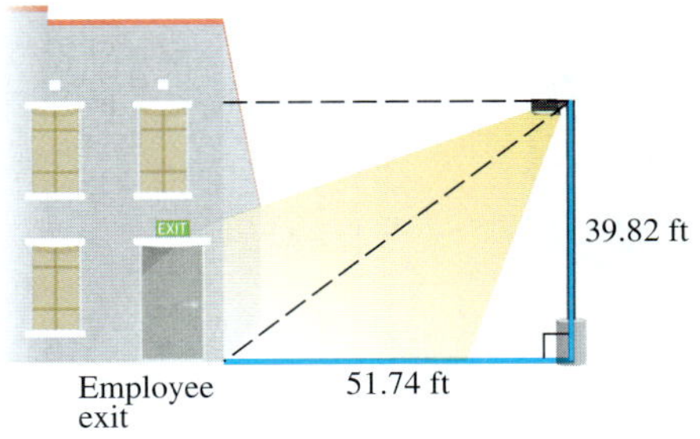

54. *Height of a Building* The angle of elevation from the top of a small building to the top of a nearby taller building is 46° 40′, and the angle of depression to the bottom is 14° 10′. If the shorter building is 28.0 m high, find the height of the taller building.

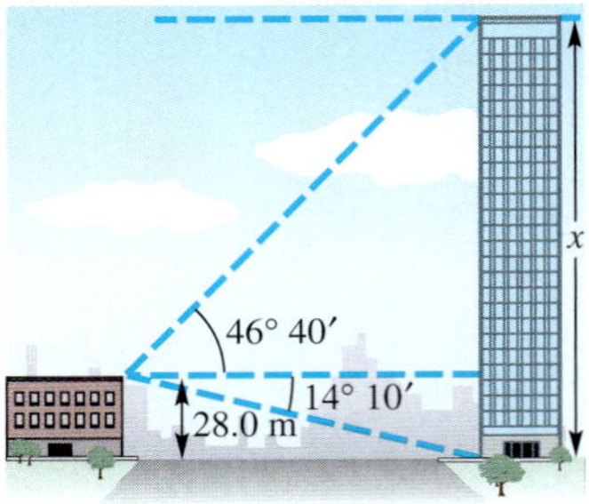

55. *Angle of Elevation of the Sun* The length of the shadow of a building 34.09 m tall is 37.62 m. Find the angle of elevation of the sun to the nearest hundredth of a degree.

56. *Angle of Elevation of the Sun* The length of the shadow of a flagpole 55.20 ft tall is 27.65 ft. Find the angle of elevation of the sun to the nearest hundredth of a degree.

57. *Height of Mt. Everest* The highest mountain peak in the world is Mt. Everest, located in the Himalayas. The height of this enormous mountain was determined in 1856 by surveyors using trigonometry long before it was first climbed in 1953. This difficult measurement had to be done from a great distance. At an altitude of 14,545 ft on a different mountain, the straight-line distance to the peak of Mt. Everest is 27.0134 mi and its angle of elevation is $\theta = 5.82°$. (*Source:* Dunham, W., *The Mathematical Universe,* John Wiley and Sons.)

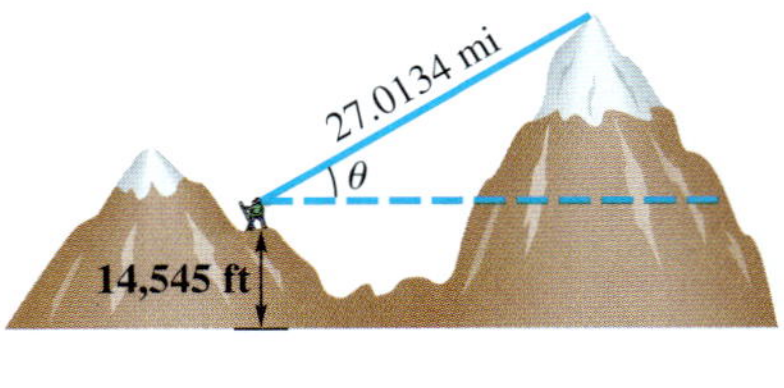

(a) Approximate the height (in feet) of Mt. Everest.

(b) In the actual measurement, Mt. Everest was over 100 mi away and the curvature of Earth had to be taken into account. Would the curvature of Earth make the peak appear taller or shorter than it actually is?

58. *Error in Measurement* A degree may seem like a very small unit, but an error of one degree in measuring an angle may be very significant. For example, suppose a laser beam directed toward the visible center of the moon misses its assigned target by 30 sec. How far is it (in miles) from its assigned target? Take the distance from the surface of Earth to that of the moon to be 234,000 mi. (*Source: A Sourcebook of Applications of School Mathematics* by Donald Bushaw et al.)

2.5 Further Applications of Right Triangles

- Bearing
- Further Applications

Bearing

Other applications of right triangles involve **bearing,** an important concept in navigation. There are two methods for expressing bearing.

Method 1 When a single angle is given, such as 164°, it is understood that the bearing is measured in a clockwise direction from due north.

Several sample bearings using Method 1 are shown in **Figure 25.**

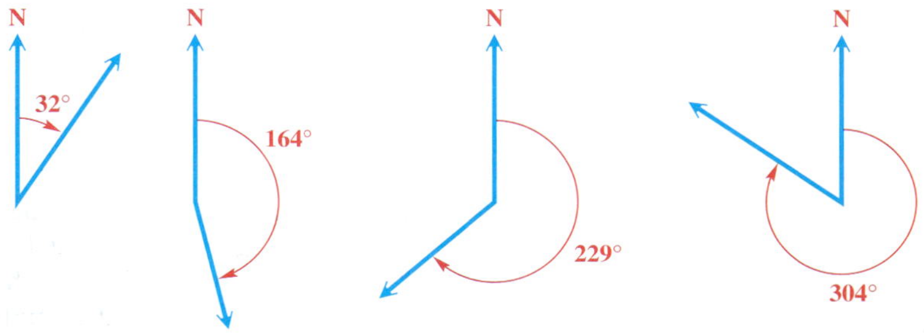

Figure 25

EXAMPLE 1 Solving a Problem Involving Bearing (Method 1)

Radar stations A and B are on an east-west line, 3.7 km apart. Station A detects a plane at C, on a bearing of 61°. Station B simultaneously detects the same plane, on a bearing of 331°. Find the distance from A to C.

SOLUTION Draw a sketch showing the given information, as in **Figure 26.** Since a line drawn due north is perpendicular to an east-west line, right angles are formed at A and B, so angles CAB and CBA can be found as shown in **Figure 26.** Angle C is a right angle because angles CAB and CBA are complementary.

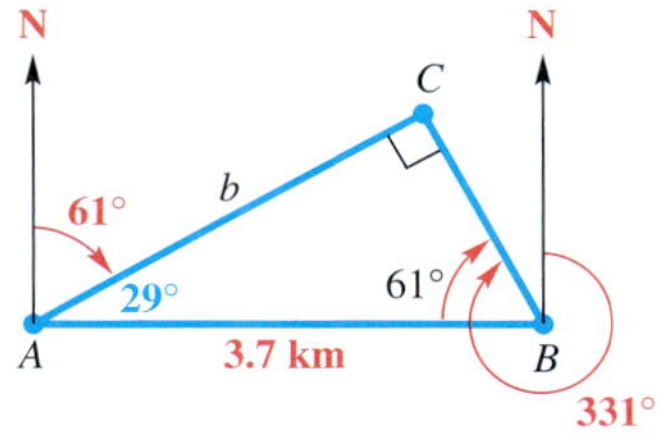

Figure 26

Find distance b by using the cosine function for angle A.

$$\cos 29° = \frac{b}{3.7} \quad \text{Cosine ratio}$$

$$3.7 \cos 29° = b \quad \text{Multiply by 3.7.}$$

$$b \approx 3.2 \text{ km} \quad \text{Use a calculator and round to the nearest tenth.}$$

✔ *Now Try Exercise 19.*

CAUTION ***A correctly labeled sketch is crucial*** when solving applications like that in **Example 1.** Some of the necessary information is often not directly stated in the problem and can be determined only from the sketch.

Method 2 The second method for expressing bearing starts with a north-south line and uses an acute angle to show the direction, either east or west, from this line.

Figure 27 shows several sample bearings using this method. Either N or S always comes first, followed by an acute angle, and then E or W.

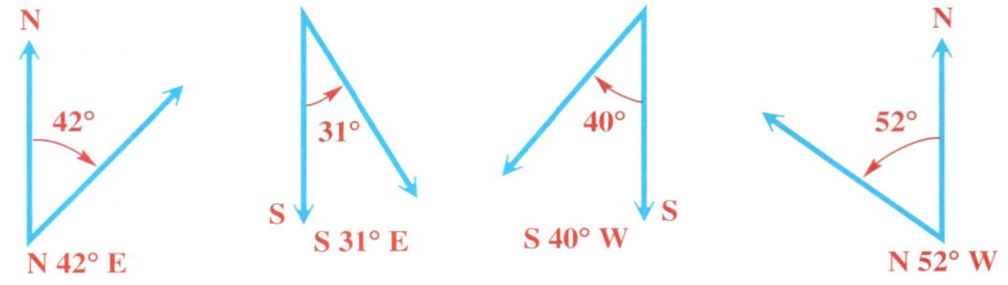

Figure 27

EXAMPLE 2 Solving a Problem Involving Bearing (Method 2)

A ship leaves port and sails on a bearing of N 47° E for 3.5 hr. It then turns and sails on a bearing of S 43° E for 4.0 hr. If the ship's rate of speed is 22 knots (nautical miles per hour), find the distance that the ship is from port.

SOLUTION Draw a sketch as in **Figure 28.** Choose a point C on a bearing of N 47° E from port at point A. Then choose a point B on a bearing of S 43° E from point C. Because north-south lines are parallel, angle ACD is 47° by alternate interior angles. The measure of angle ACB is

$$47^\circ + 43^\circ = 90^\circ,$$

making triangle ABC a right triangle.

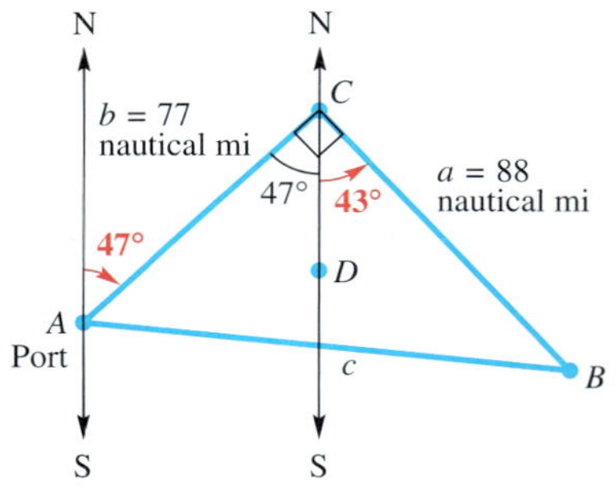

Figure 28

Next, use the formula relating distance, rate, and time to find the distances from A to C and from C to B.

$$b = 22 \times 3.5 = 77 \text{ nautical mi}$$
$$a = 22 \times 4.0 = 88 \text{ nautical mi}$$ distance = rate × time

Now find c, the distance from port at point A to the ship at point B.

$$a^2 + b^2 = c^2$$ Pythagorean theorem (Appendix B)

$$88^2 + 77^2 = c^2$$ $a = 88$, $b = 77$

$$c = \sqrt{88^2 + 77^2}$$ Use the square root property. (Appendix A)

$$c \approx 120 \text{ nautical mi}$$ Two significant digits (Section 2.4)

✔ *Now Try Exercise 25.*

Further Applications

EXAMPLE 3 Using Trigonometry to Measure a Distance

The **subtense bar method** is a method that surveyors use to determine a small distance d between two points P and Q. The subtense bar with length b is centered at Q and situated perpendicular to the line of sight between P and Q. See **Figure 29.** Angle θ is measured, and then the distance d can be determined.

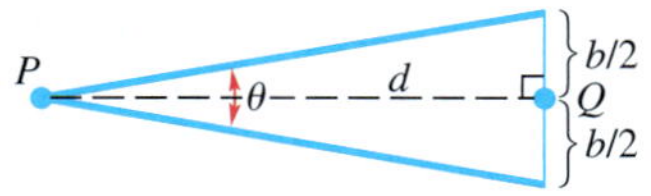

Figure 29

(a) Find d when $\theta = 1° \, 23' \, 12''$ and $b = 2.0000$ cm.

(b) How much change would there be in the value of d if θ measured $1''$ larger?

SOLUTION

(a) From **Figure 29,** we obtain the following.

$$\cot \frac{\theta}{2} = \frac{d}{\frac{b}{2}} \qquad \text{Cotangent ratio}$$

$$d = \frac{b}{2} \cot \frac{\theta}{2} \qquad \text{Multiply and rewrite.}$$

Let $b = 2$. To evaluate $\frac{\theta}{2}$, we change θ to decimal degrees.

$$1° \, 23' \, 12'' \approx 1.386666667°$$

Use $\cot \theta = \frac{1}{\tan \theta}$ to evaluate.

Then $$d = \frac{2}{2} \cot \frac{1.386666667°}{2} \approx 82.634110 \text{ cm.}$$

(b) Since θ is $1''$ larger, use $\theta = 1° \, 23' \, 13'' \approx 1.386944444°$.

$$d = \frac{2}{2} \cot \frac{1.386944444°}{2} \approx 82.617558 \text{ cm}$$

The difference is $82.634110 - 82.617558 = 0.016552$ cm.

✔ *Now Try Exercise 37.*

EXAMPLE 4 Solving a Problem Involving Angles of Elevation

Francisco needs to know the height of a tree. From a given point on the ground, he finds that the angle of elevation to the top of the tree is 36.7°. He then moves back 50 ft. From the second point, the angle of elevation to the top of the tree is 22.2°. See **Figure 30.** Find the height of the tree to the nearest foot.

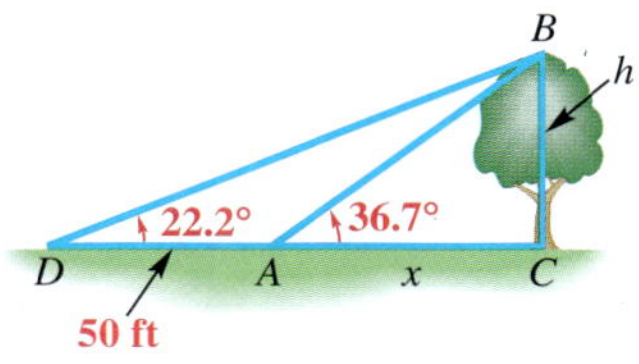

Figure 30

ALGEBRAIC SOLUTION

Figure 30 on the preceding page shows two unknowns: x, the distance from the center of the trunk of the tree to the point where the first observation was made, and h, the height of the tree. See **Figure 31** in the Graphing Calculator Solution. Since nothing is given about the length of the hypotenuse of either triangle ABC or triangle BCD, use a ratio that does not involve the hypotenuse—namely, the tangent.

In triangle ABC, $\tan 36.7^\circ = \frac{h}{x}$ or $h = x \tan 36.7^\circ$.

In triangle BCD, $\tan 22.2^\circ = \frac{h}{50 + x}$ or $h = (50 + x) \tan 22.2^\circ$.

Each expression equals h, so the expressions must be equal.

$$x \tan 36.7^\circ = (50 + x) \tan 22.2^\circ$$
Equate expressions for h.

$$x \tan 36.7^\circ = 50 \tan 22.2^\circ + x \tan 22.2^\circ$$
Distributive property

$$x \tan 36.7^\circ - x \tan 22.2^\circ = 50 \tan 22.2^\circ$$
Write the x-terms on one side.

$$x(\tan 36.7^\circ - \tan 22.2^\circ) = 50 \tan 22.2^\circ$$
Factor out x.

$$x = \frac{50 \tan 22.2^\circ}{\tan 36.7^\circ - \tan 22.2^\circ}$$
Divide by the coefficient of x.

We saw above that $h = x \tan 36.7^\circ$. Substitute for x.

$$h = \left(\frac{50 \tan 22.2^\circ}{\tan 36.7^\circ - \tan 22.2^\circ}\right) \tan 36.7^\circ$$

Use a calculator.

$$\tan 36.7^\circ = 0.74537703 \quad \text{and} \quad \tan 22.2^\circ = 0.40809244$$

Thus,

$$\tan 36.7^\circ - \tan 22.2^\circ = 0.74537703 - 0.40809244 = 0.33728459$$

and $$h = \left(\frac{50(0.40809244)}{0.33728459}\right) 0.74537703 \approx 45.$$

The height of the tree is approximately 45 ft.

GRAPHING CALCULATOR SOLUTION*

In **Figure 31,** we have superimposed **Figure 30** on coordinate axes with the origin at D. By definition, the tangent of the angle between the x-axis and the graph of a line with equation $y = mx + b$ is the slope of the line, m. For line DB, $m = \tan 22.2^\circ$. Since b equals 0, the equation of line DB is

$$y_1 = (\tan 22.2^\circ)x.$$

The equation of line AB is

$$y_2 = (\tan 36.7^\circ)x + b.$$

Since $b \neq 0$ here, we use the point $A(50, 0)$ and the point-slope form to find the equation.

$$y_2 - y_1 = m(x - x_1) \quad \text{Point-slope form}$$
$$y_2 - 0 = m(x - 50) \quad x_1 = 50,\ y_1 = 0$$
$$y_2 = \tan 36.7^\circ(x - 50)$$

Lines y_1 and y_2 are graphed in **Figure 32.** The y-coordinate of the point of intersection of the graphs gives the length of BC, or h. Thus, $h \approx 45$.

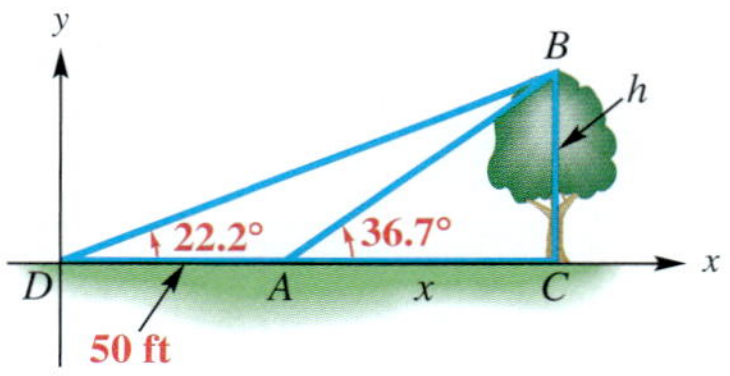

Figure 31

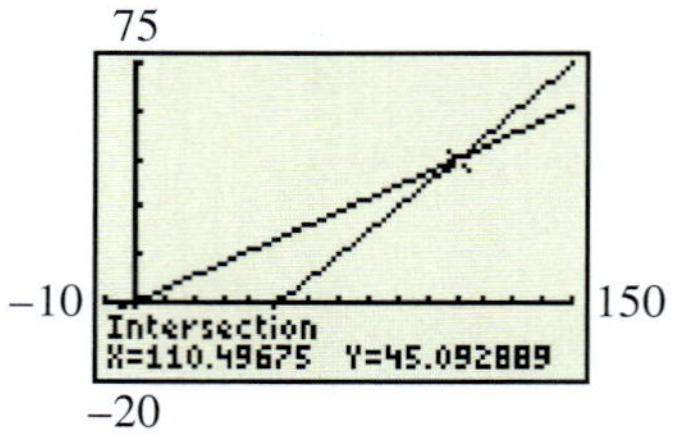

Figure 32

✔ *Now Try Exercise 31.*

NOTE In practice, we usually do not write down intermediate calculator approximation steps. We did in **Example 4** so that you could follow the steps more easily.

**Source:* Reprinted with permission from *Mathematics Teacher,* copyright 1995 by the National Council of Teachers of Mathematics. All rights reserved.

2.5 Exercises

Concept Check Give a short written answer to each question.

1. When bearing is given as a single angle measure, how is the angle represented in a sketch?

2. When bearing is given as N (or S), then an acute angle measure, and then E (or W), how is the angle represented in a sketch?

3. Why is it important to draw a sketch before solving trigonometric problems like those in the last two sections of this chapter?

4. How should the angle of elevation (or depression) from a point X to a point Y be represented?

Concept Check An observer for a radar station is located at the origin of a coordinate system. For each of the points in Exercises 5–12, find the bearing of an airplane located at that point. Express the bearing using both methods.

5. $(-4, 0)$
6. $(5, 0)$
7. $(0, 4)$
8. $(0, -2)$
9. $(-5, 5)$
10. $(-3, -3)$
11. $(2, -2)$
12. $(2, 2)$

13. The ray $y = x$, $x \geq 0$, contains the origin and all points in the coordinate system whose bearing is 45°. Determine the equation of a ray consisting of the origin and all points whose bearing is 240°.

14. Repeat **Exercise 13** for a bearing of 150°.

Work each problem. In these exercises, assume the course of a plane or ship is on the indicated bearing. ***See Examples 1 and 2.***

15. *Distance Flown by a Plane* A plane flies 1.3 hr at 110 mph on a bearing of 38°. It then turns and flies 1.5 hr at the same speed on a bearing of 128°. How far is the plane from its starting point?

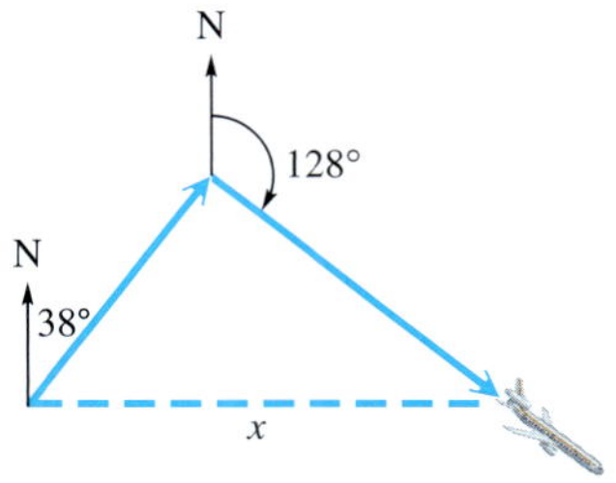

16. *Distance Traveled by a Ship* A ship travels 55 km on a bearing of 27° and then travels on a bearing of 117° for 140 km. Find the distance from the starting point to the ending point.

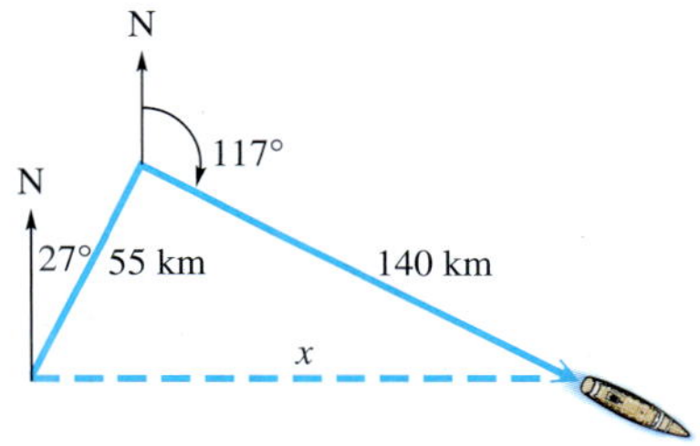

17. *Distance between Two Ships* Two ships leave a port at the same time. The first ship sails on a bearing of 40° at 18 knots (nautical miles per hour) and the second on a bearing of 130° at 26 knots. How far apart are they after 1.5 hr?

18. ***Distance between Two Ships*** Two ships leave a port at the same time. The first ship sails on a bearing of 52° at 17 knots and the second on a bearing of 322° at 22 knots. How far apart are they after 2.5 hr?

19. ***Distance between Two Docks*** Two docks are located on an east-west line 2587 ft apart. From dock A, the bearing of a coral reef is 58° 22′. From dock B, the bearing of the coral reef is 328° 22′. Find the distance from dock A to the coral reef.

20. ***Distance between Two Lighthouses*** Two lighthouses are located on a north-south line. From lighthouse A, the bearing of a ship 3742 m away is 129° 43′. From lighthouse B, the bearing of the ship is 39° 43′. Find the distance between the lighthouses.

21. ***Distance between Two Ships*** A ship leaves its home port and sails on a bearing of S 61° 50′ E. Another ship leaves the same port at the same time and sails on a bearing of N 28° 10′ E. If the first ship sails at 24.0 mph and the second sails at 28.0 mph, find the distance between the two ships after 4 hr.

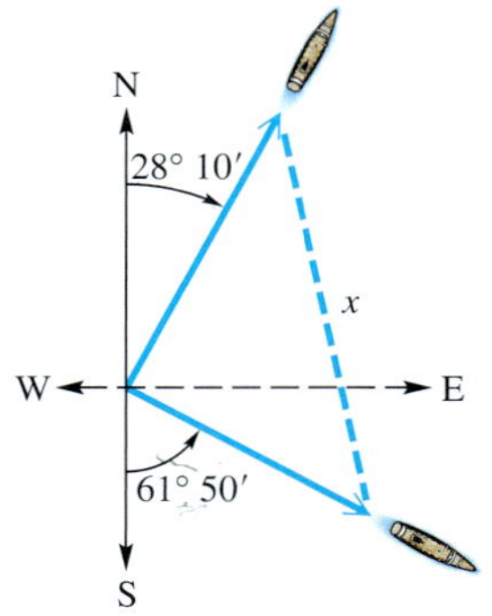

22. ***Distance between Transmitters*** Radio direction finders are set up at two points A and B, which are 2.50 mi apart on an east-west line. From A, it is found that the bearing of a signal from a radio transmitter is N 36° 20′ E, and from B the bearing of the same signal is N 53° 40′ W. Find the distance of the transmitter from B.

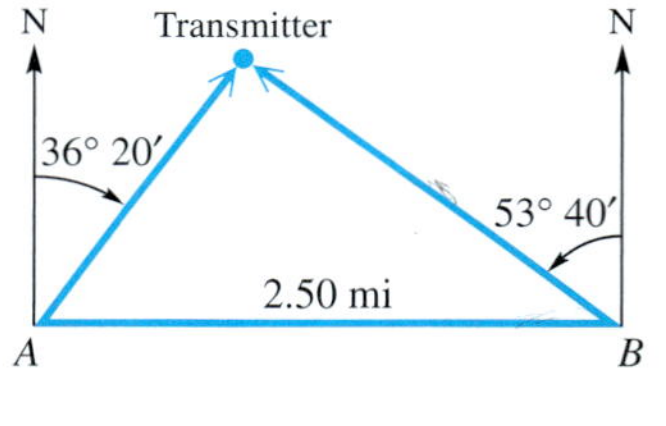

23. ***Flying Distance*** The bearing from A to C is S 52° E. The bearing from A to B is N 84° E. The bearing from B to C is S 38° W. A plane flying at 250 mph takes 2.4 hr to go from A to B. Find the distance from A to C.

24. ***Flying Distance*** The bearing from A to C is N 64° W. The bearing from A to B is S 82° W. The bearing from B to C is N 26° E. A plane flying at 350 mph takes 1.8 hr to go from A to B. Find the distance from B to C.

25. ***Distance between Two Cities*** The bearing from Winston-Salem, North Carolina, to Danville, Virginia, is N 42° E. The bearing from Danville to Goldsboro, North Carolina, is S 48° E. A car driven by Ellen Winchell, traveling at 65 mph, takes 1.1 hr to go from Winston-Salem to Danville and 1.8 hr to go from Danville to Goldsboro. Find the distance from Winston-Salem to Goldsboro.

26. ***Distance between Two Cities*** The bearing from Atlanta to Macon is S 27° E, and the bearing from Macon to Augusta is N 63° E. An automobile traveling at 62 mph needs $1\frac{1}{4}$ hr to go from Atlanta to Macon and $1\frac{3}{4}$ hr to go from Macon to Augusta. Find the distance from Atlanta to Augusta.

27. Solve the equation $ax = b + cx$ for x in terms of a, b, and c. (*Note:* This is essentially the calculation carried out in **Example 4.**)

28. Explain why the line $y = (\tan \theta)(x - a)$ passes through the point $(a, 0)$ and makes an angle θ with the x-axis.

29. Find the equation of the line passing through the point $(25, 0)$ that makes an angle of 35° with the x-axis.

30. Find the equation of the line passing through the point $(5, 0)$ that makes an angle of 15° with the x-axis.

In Exercises 31–36, use the method of ***Example 4.***

31. Find h as indicated in the figure.

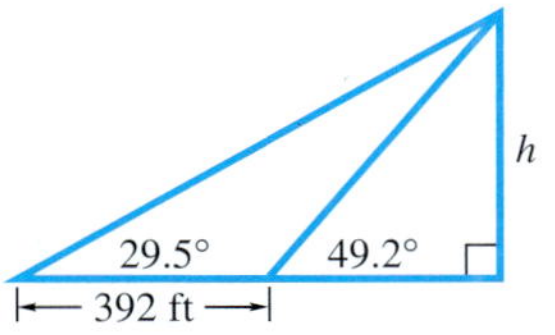

32. Find h as indicated in the figure.

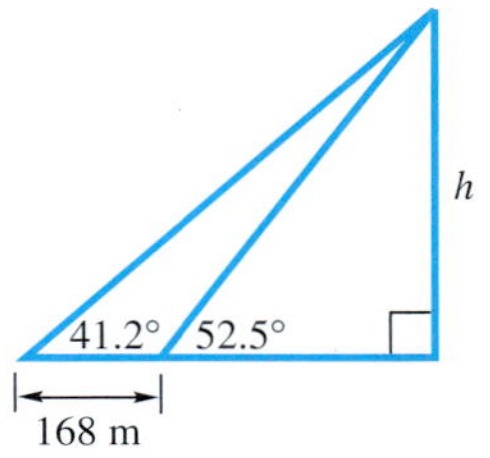

33. ***Height of a Pyramid*** The angle of elevation from a point on the ground to the top of a pyramid is 35° 30′. The angle of elevation from a point 135 ft farther back to the top of the pyramid is 21° 10′. Find the height of the pyramid.

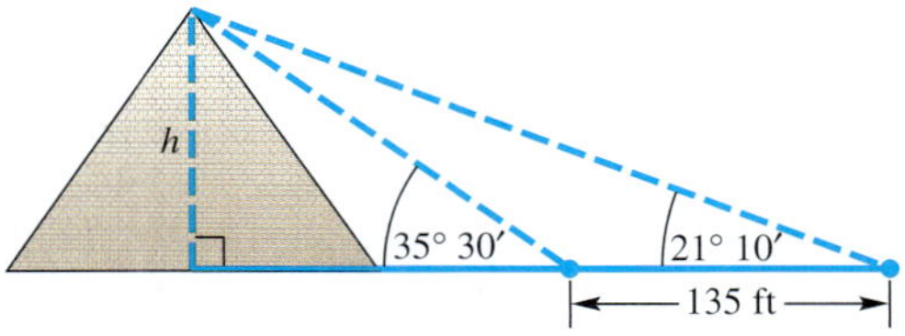

34. ***Distance between a Whale and a Lighthouse*** Debbie Glockner-Ferrari, a whale researcher, is watching a whale approach directly toward a lighthouse as she observes from the top of this lighthouse. When she first begins watching the whale, the angle of depression to the whale is 15° 50′. Just as the whale turns away from the lighthouse, the angle of depression is 35° 40′. If the height of the lighthouse is 68.7 m, find the distance traveled by the whale as it approached the lighthouse.

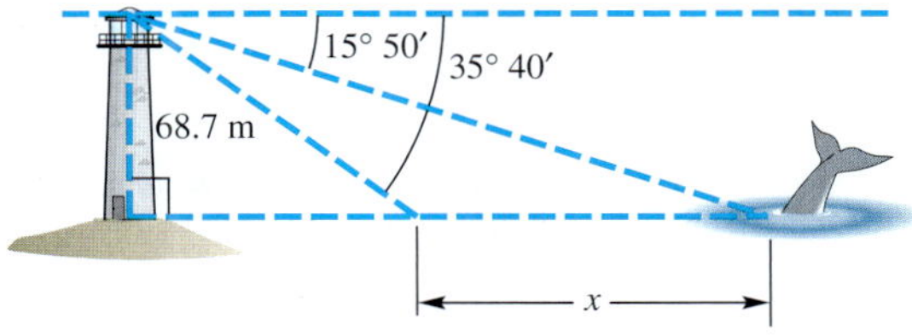

35. ***Height of an Antenna*** A scanner antenna is on top of the center of a house. The angle of elevation from a point 28.0 m from the center of the house to the top of the antenna is 27° 10′, and the angle of elevation to the bottom of the antenna is 18° 10′. Find the height of the antenna.

36. ***Height of Mt. Whitney*** The angle of elevation from Lone Pine to the top of Mt. Whitney is 10° 50′. Van Dong Le, traveling 7.00 km from Lone Pine along a straight, level road toward Mt. Whitney, finds the angle of elevation to be 22° 40′. Find the height of the top of Mt. Whitney above the level of the road.

Solve each problem.

37. *(Modeling) Distance between Two Points* Refer to **Example 3.** A variation of the subtense bar method that surveyors use to determine larger distances d between two points P and Q is shown in the figure. In this case the subtense bar with length b is placed between the points P and Q so that the bar is centered on and perpendicular to the line of sight connecting P and Q. The angles α and β are measured from points P and Q, respectively. (*Source:* Mueller, I. and K. Ramsayer, *Introduction to Surveying,* Frederick Ungar Publishing Co.)

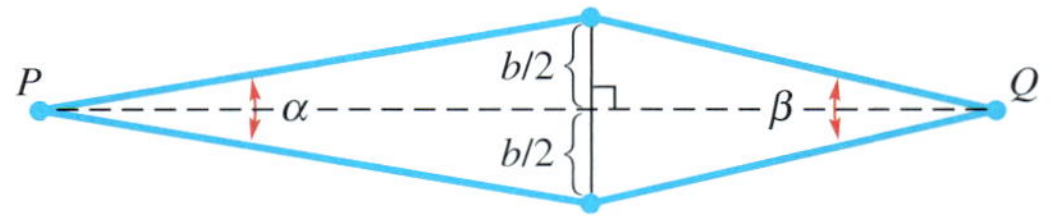

(a) Find a formula for d involving α, β, and b.
(b) Use your formula to determine d if $\alpha = 37' \, 48''$, $\beta = 42' \, 03''$, and $b = 2.000$ cm.

38. *Height of a Plane above Earth* Find the minimum height h above the surface of Earth so that a pilot at point A in the figure can see an object on the horizon at C, 125 mi away. Assume that the radius of Earth is 4.00×10^3 mi.

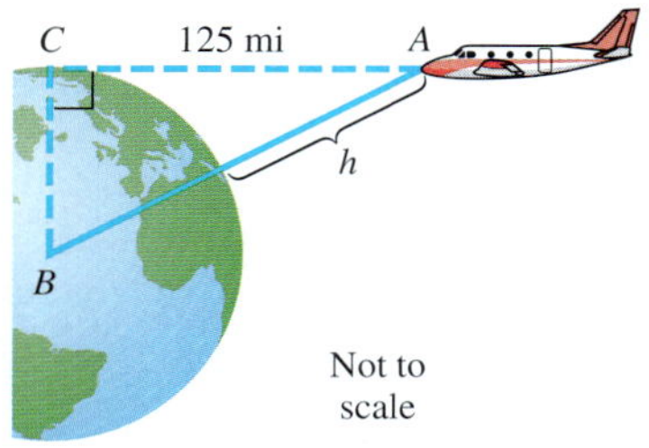

39. *Distance of a Plant from a Fence* In one area, the lowest angle of elevation of the sun in winter is 23° 20′. Find the minimum distance x that a plant needing full sun can be placed from a fence 4.65 ft high.

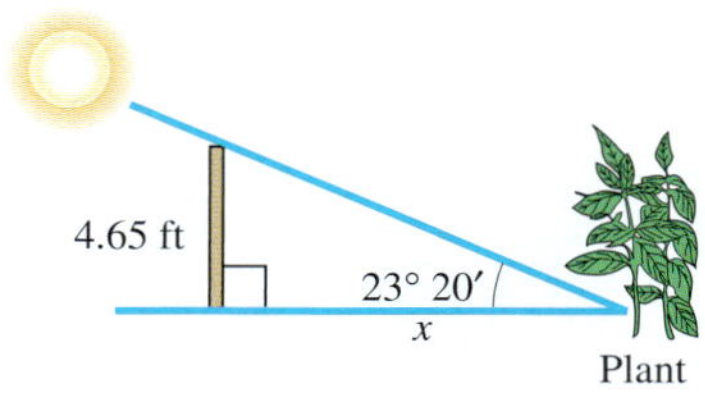

40. *Distance through a Tunnel* A tunnel is to be built from A to B. Both A and B are visible from C. If AC is 1.4923 mi and BC is 1.0837 mi, and if C is 90°, find the measures of angles A and B.

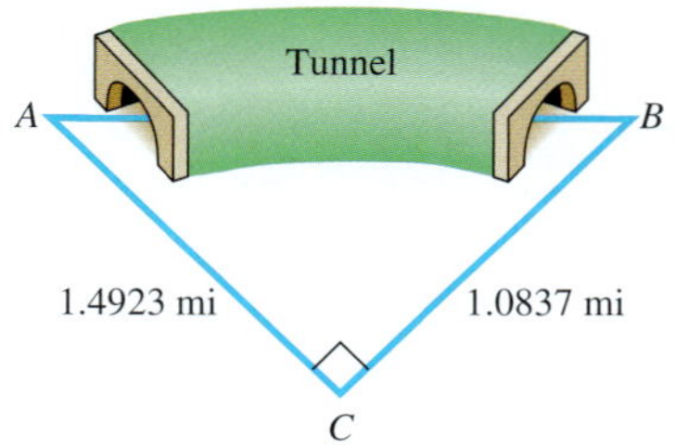

41. *(Modeling) Highway Curves* A basic highway curve connecting two straight sections of road is often circular. In the figure, the points P and S mark the beginning and end of the curve. Let Q be the point of intersection where the two straight sections of highway leading into the curve would meet if extended. The radius of the curve is R, and the central angle θ denotes how many degrees the curve turns. (*Source:* Mannering, F. and W. Kilareski, *Principles of Highway Engineering and Traffic Analysis,* Second Edition, John Wiley and Sons.)

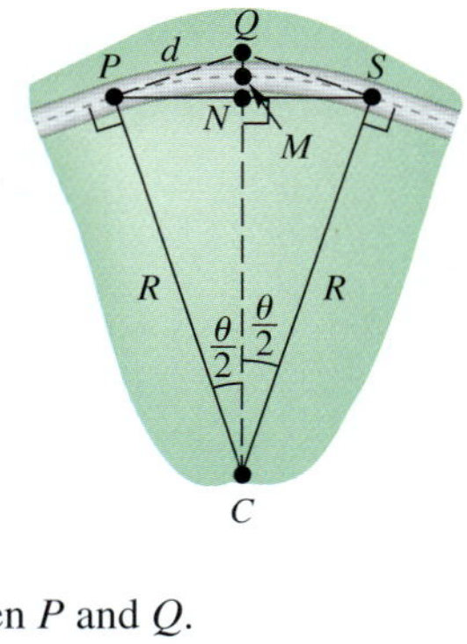

(a) If $R = 965$ ft and $\theta = 37°$, find the distance d between P and Q.
(b) Find an expression in terms of R and θ for the distance between points M and N.

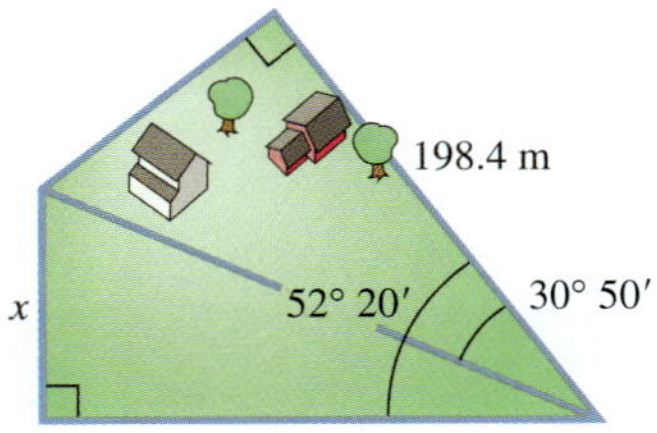

42. *Length of a Side of a Piece of Land* A piece of land has the shape shown in the figure at the left. Find the length x.

43. *(Modeling) Stopping Distance on a Curve* Refer to **Exercise 41.** When an automobile travels along a circular curve, objects like trees and buildings situated on the inside of the curve can obstruct the driver's vision. These obstructions prevent the driver from seeing sufficiently far down the highway to ensure a safe stopping distance. In the figure, the *minimum* distance d that should be cleared on the inside of the highway is modeled by the equation

$$d = R\left(1 - \cos \frac{\theta}{2}\right).$$

(*Source:* Mannering, F. and W. Kilareski, *Principles of Highway Engineering and Traffic Analysis,* Second Edition, John Wiley and Sons.)

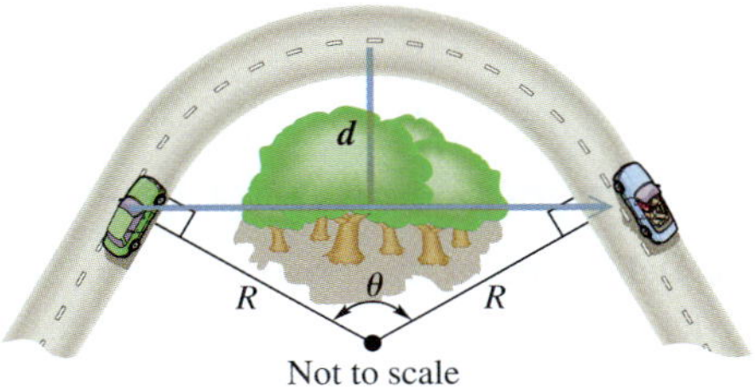

(a) It can be shown that if θ is measured in degrees, then $\theta \approx \frac{57.3S}{R}$, where S is the safe stopping distance for the given speed limit. Compute d to the nearest foot for a 55 mph speed limit if $S = 336$ ft and $R = 600$ ft.

(b) Compute d to the nearest foot for a 65 mph speed limit given $S = 485$ ft and $R = 600$ ft.

(c) How does the speed limit affect the amount of land that should be cleared on the inside of the curve?

44. *(Modeling) Distance of a Shot Put* A shot-putter trying to improve performance may wonder whether there is an optimal angle to aim for, or whether the velocity (speed) at which the ball is thrown is more important. The figure shows the path of a steel ball thrown by a shot-putter. The distance D depends on initial velocity v, height h, and angle θ when the ball is released.

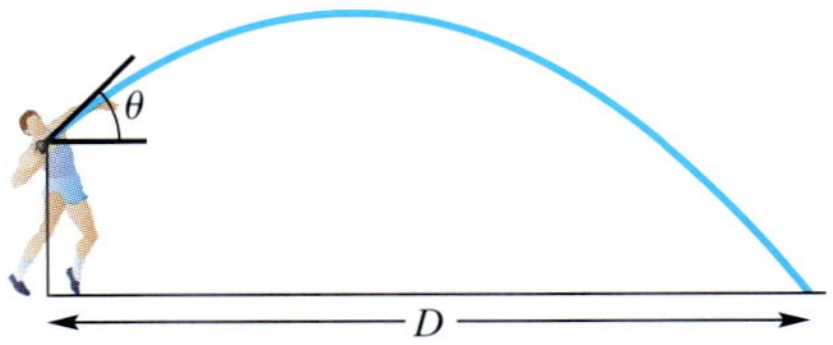

One model developed for this situation gives D as

$$D = \frac{v^2 \sin \theta \cos \theta + v \cos \theta \sqrt{(v \sin \theta)^2 + 64h}}{32}.$$

Typical ranges for the variables are v: 33–46 ft per sec; h: 6–8 ft; and θ: 40°–45°. (*Source:* Kreighbaum, E. and K. Barthels, *Biomechanics*, Allyn & Bacon.)

(a) To see how angle θ affects distance D, let $v = 44$ ft per sec and $h = 7$ ft. Calculate D, to the nearest hundredth, for $\theta = 40°$, $42°$, and $45°$. How does distance D change as θ increases?

(b) To see how velocity v affects distance D, let $h = 7$ and $\theta = 42°$. Calculate D, to the nearest hundredth, for $v = 43$, 44, and 45 ft per sec. How does distance D change as v increases?

(c) Which affects distance D more, v or θ? What should the shot-putter do to improve performance?

Chapter 2 Test Prep

Key Terms

2.1 side opposite
side adjacent
cofunctions

2.2 reference angle

2.4 exact number
significant digits
angle of elevation
angle of depression

2.5 bearing

Quick Review

Concepts | **Examples**

2.1 Trigonometric Functions of Acute Angles

Right-Triangle-Based Definitions of the Trigonometric Functions

Let A represent any acute angle in standard position.

$$\sin A = \frac{y}{r} = \frac{\text{side opposite}}{\text{hypotenuse}} \qquad \csc A = \frac{r}{y} = \frac{\text{hypotenuse}}{\text{side opposite}}$$

$$\cos A = \frac{x}{r} = \frac{\text{side adjacent}}{\text{hypotenuse}} \qquad \sec A = \frac{r}{x} = \frac{\text{hypotenuse}}{\text{side adjacent}}$$

$$\tan A = \frac{y}{x} = \frac{\text{side opposite}}{\text{side adjacent}} \qquad \cot A = \frac{x}{y} = \frac{\text{side adjacent}}{\text{side opposite}}$$

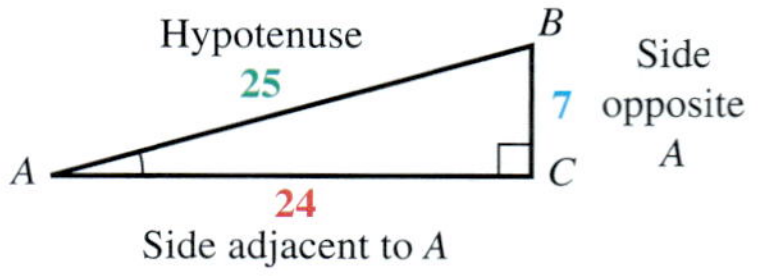

$$\sin A = \frac{7}{25} \qquad \cos A = \frac{24}{25} \qquad \tan A = \frac{7}{24}$$

$$\csc A = \frac{25}{7} \qquad \sec A = \frac{25}{24} \qquad \cot A = \frac{24}{7}$$

Cofunction Identities

For any acute angle A, cofunction values of complementary angles are equal.

$$\sin A = \cos(90° - A) \qquad \cos A = \sin(90° - A)$$

$$\sec A = \csc(90° - A) \qquad \csc A = \sec(90° - A)$$

$$\tan A = \cot(90° - A) \qquad \cot A = \tan(90° - A)$$

$$\sin 55° = \cos(90° - 55°) = \cos 35°$$

$$\sec 48° = \csc(90° - 48°) = \csc 42°$$

$$\tan 72° = \cot(90° - 72°) = \cot 18°$$

Function Values of Special Angles

θ	$\sin\theta$	$\cos\theta$	$\tan\theta$	$\cot\theta$	$\sec\theta$	$\csc\theta$
30°	$\frac{1}{2}$	$\frac{\sqrt{3}}{2}$	$\frac{\sqrt{3}}{3}$	$\sqrt{3}$	$\frac{2\sqrt{3}}{3}$	2
45°	$\frac{\sqrt{2}}{2}$	$\frac{\sqrt{2}}{2}$	1	1	$\sqrt{2}$	$\sqrt{2}$
60°	$\frac{\sqrt{3}}{2}$	$\frac{1}{2}$	$\sqrt{3}$	$\frac{\sqrt{3}}{3}$	2	$\frac{2\sqrt{3}}{3}$

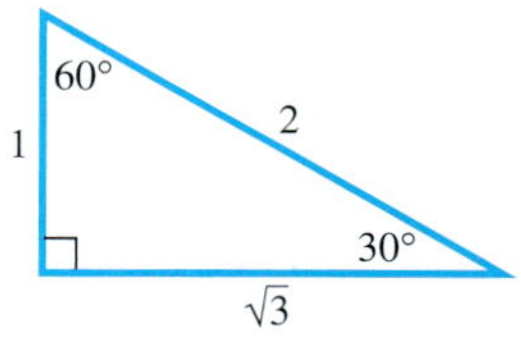

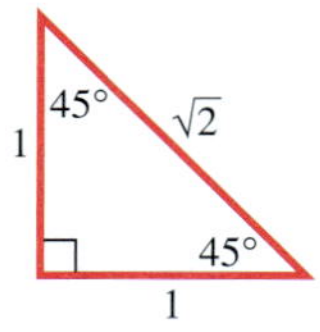

Concepts	Examples

2.2 Trigonometric Functions of Non-Acute Angles

Reference Angle θ' for θ in $(0°, 360°)$

θ in Quadrant	I	II	III	IV
θ' is	θ	$180° - \theta$	$\theta - 180°$	$360° - \theta$

See the figure in **Section 2.2** for illustrations of reference angles.

Quadrant I: For $\theta = 25°$, $\theta' = 25°$
Quadrant II: For $\theta = 152°$, $\theta' = 28°$
Quadrant III: For $\theta = 200°$, $\theta' = 20°$
Quadrant IV: For $\theta = 320°$, $\theta' = 40°$

Finding Trigonometric Function Values for Any Nonquadrantal Angle

Step 1 Add or subtract 360° as many times as needed to get an angle greater than 0° but less than 360°.

Step 2 Find the reference angle θ'.

Step 3 Find the trigonometric function values for θ'.

Step 4 Determine the correct signs for the values found in Step 3.

Find sin 1050°.

$$1050° - 2(360°) = 330° \quad \text{Coterminal angle in quadrant IV}$$

Thus, $\theta' = 30°$.

$$\sin 1050° = -\sin 30° \quad \text{Reference angle}$$
$$= -\frac{1}{2}$$

2.3 Finding Trigonometric Function Values Using a Calculator

To approximate a trigonometric function value of an angle in degrees, make sure your calculator is in degree mode.

Approximate each value.

$$\cos 50° 15' = \cos 50.25° \approx 0.63943900$$

$$\csc 32.5° = \frac{1}{\sin 32.5°} \approx 1.86115900 \quad \csc\theta = \frac{1}{\sin\theta}$$

To find the corresponding angle measure given a trigonometric function value, use an appropriate inverse function.

Find an angle θ in the interval $[0°, 90°]$ that satisfies each condition in color.

$$\cos\theta \approx 0.73677482$$
$$\theta \approx \cos^{-1}(0.73677482)$$
$$\theta \approx 42.542600°$$

$$\csc\theta \approx 1.04766792$$
$$\sin\theta \approx \frac{1}{1.04766792} \quad \sin\theta = \frac{1}{\csc\theta}$$
$$\theta \approx \sin^{-1}\left(\frac{1}{1.04766792}\right)$$
$$\theta \approx 72.65°$$

2.4 Solving Right Triangles

Solving an Applied Trigonometry Problem

Step 1 Draw a sketch, and label it with the given information. Label the quantity to be found with a variable.

Find the angle of elevation of the sun if a 48.6-ft flagpole casts a shadow 63.1 ft long.

Step 1 See the sketch. We must find θ.

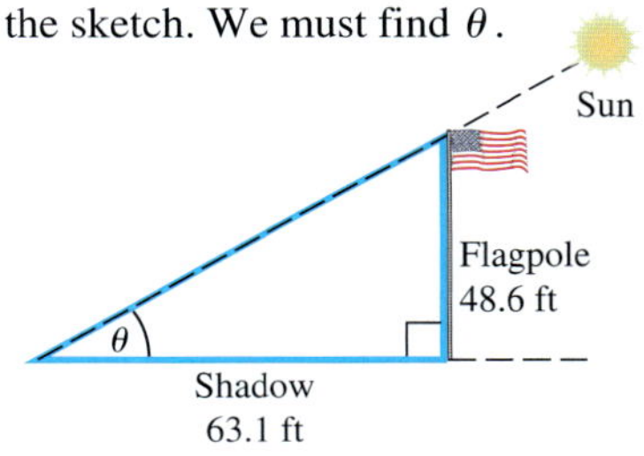

(continued)

Concepts	Examples
Step 2 Use the sketch to write an equation relating the given quantities to the variable.	***Step 2*** $\tan\theta = \dfrac{48.6}{63.1} \approx 0.770206$
Step 3 Solve the equation, and check that your answer makes sense.	***Step 3*** $\theta = \tan^{-1} 0.770206 \approx 37.6°$ The angle of elevation rounded to three significant digits is 37.6°, or 37° 40′.
2.5 Further Applications of Right Triangles	
Expressing Bearing	
Method 1 When a single angle is given, such as 220°, this bearing is measured in a clockwise direction from north.	***Example:*** 220° 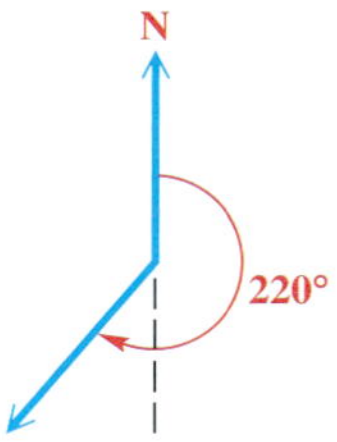
Method 2 Start with a north-south line and use an acute angle to show direction, either east or west, from this line.	***Example:*** S 40° W 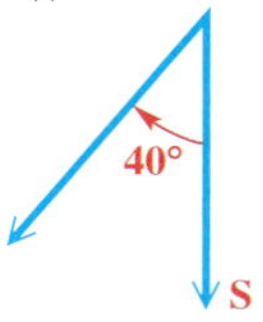

Chapter 2 Review Exercises

Find the values of the six trigonometric functions for each angle A.

1. 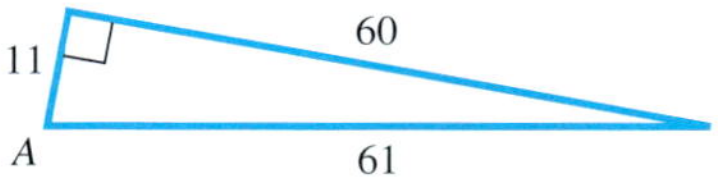

2. 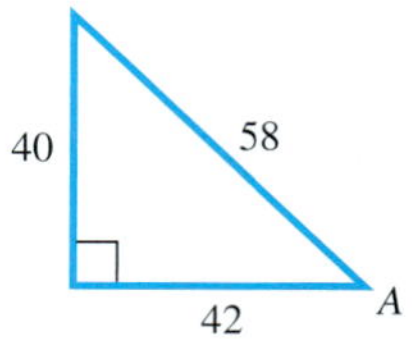

Find one solution for each equation. Assume that all angles are acute angles.

3. $\sin 4\beta = \cos 5\beta$

4. $\sec(2\theta + 10°) = \csc(4\theta + 20°)$

5. $\tan(5x + 11°) = \cot(6x + 2°)$

6. $\cos\left(\dfrac{3\theta}{5} + 11°\right) = \sin\left(\dfrac{7\theta}{10} + 40°\right)$

Tell whether each statement is true *or* false. *If false, tell why.*

7. $\sin 46° < \sin 58°$

8. $\cos 47° < \cos 58°$

9. $\tan 60° \geq \cot 40°$

10. $\csc 22° \leq \csc 68°$

11. Explain why, in the figure, the cosine of angle A is equal to the sine of angle B.

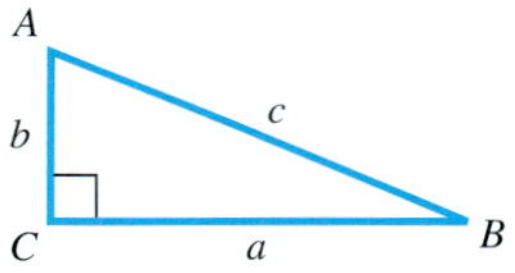

12. *Concept Check* Which one of the following cannot be *exactly* determined using the methods of this chapter?

A. $\cos 135°$ **B.** $\cot(-45°)$ **C.** $\sin 300°$ **D.** $\tan 140°$

Find exact values of the six trigonometric functions for each angle. Do not use a calculator. Rationalize denominators when applicable.

13. $1020°$ **14.** $120°$ **15.** $-1470°$ **16.** $-225°$

Find all values of θ, if θ is in the interval $[0°, 360°)$ and θ has the given function value.

17. $\cos \theta = -\frac{1}{2}$

18. $\sin \theta = -\frac{1}{2}$

19. $\sec \theta = -\frac{2\sqrt{3}}{3}$

20. $\cot \theta = -1$

Evaluate each expression. Give exact values.

21. $\tan^2 120° - 2 \cot 240°$

22. $\cos 60° + 2 \sin^2 30°$

23. $\sec^2 300° - 2 \cos^2 150° + \tan 45°$

24. Find the sine, cosine, and tangent function values for each angle.

(a)

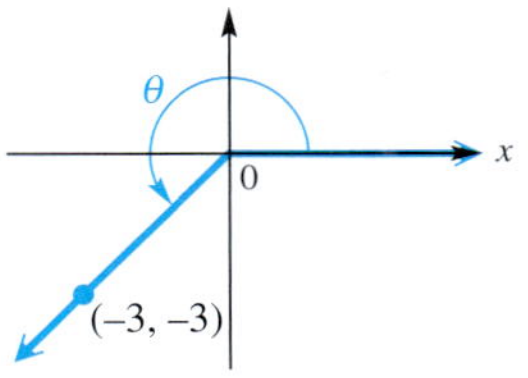

(b)

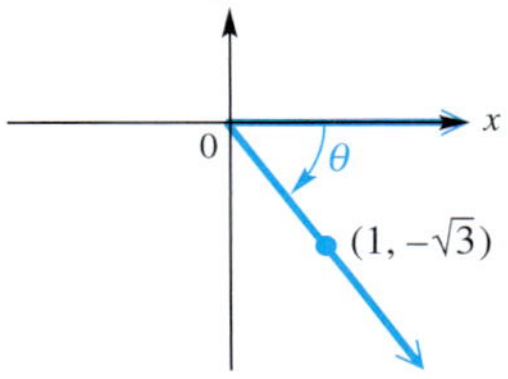

Use a calculator to find each value.

25. $\sec 222° 30'$ **26.** $\sin 72° 30'$ **27.** $\csc 78° 21'$

28. $\cot 305.6°$ **29.** $\tan 11.7689°$ **30.** $\sec 58.9041°$

Use a calculator to find each value of θ, where θ is in the interval $[0°, 90°)$. Give answers in decimal degrees.

31. $\sin \theta = 0.82584121$ **32.** $\cot \theta = 1.1249386$ **33.** $\cos \theta = 0.97540415$

34. $\sec \theta = 1.2637891$ **35.** $\tan \theta = 1.9633124$ **36.** $\csc \theta = 9.5670466$

Find two angles in the interval $[0°, 360°)$ that satisfy each of the following. Leave answers in decimal degrees rounded to the nearest tenth.

37. $\sin \theta = 0.73254290$ **38.** $\tan \theta = 1.3865342$

Determine whether each statement is true *or* false. *If false, tell why. Use a calculator for Exercises 39 and 42.*

39. $\sin 50° + \sin 40° = \sin 90°$

40. $1 + \tan^2 60° = \sec^2 60°$

41. $\sin 240° = 2 \sin 120° \cdot \cos 120°$

42. $\sin 42° + \sin 42° = \sin 84°$

43. A student wants to use a calculator to find the value of $\cot 25°$. However, instead of entering $\frac{1}{\tan 25}$, he enters $\tan^{-1} 25$. Assuming the calculator is in degree mode, will this produce the correct answer? Explain.

44. Explain the process for using a calculator to find $\sec^{-1} 10$.

For each angle θ, use a calculator to find $\cos \theta$ and $\sin \theta$. Use your results to decide in which quadrant the angle lies.

45. $\theta = 1997°$ **46.** $\theta = 2976°$ **47.** $\theta = -3485°$ **48.** $\theta = 4000°$

Solve each right triangle. In Exercise 50, give angles to the nearest minute. In Exercises 51 and 52, label the triangle ABC as in Exercises 49 and 50.

49.

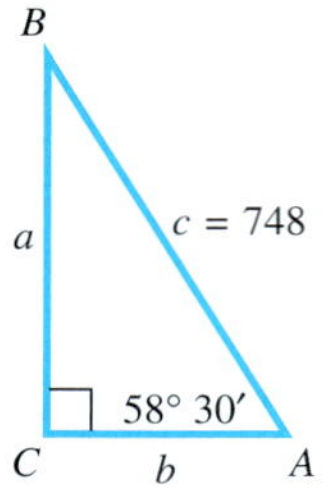

50.

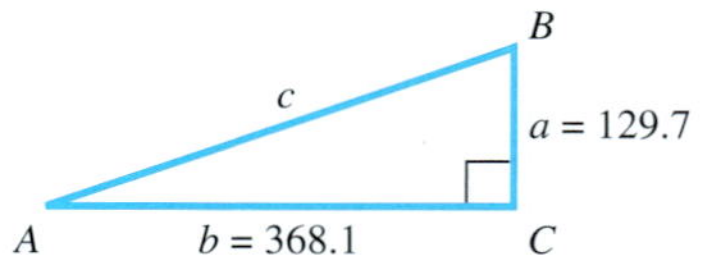

51. $A = 39.72°$, $b = 38.97$ m

52. $B = 47° 53'$, $b = 298.6$ m

Solve each problem. (Source for Exercises 53 and 54: Parker, M., Editor, *She Does Math,* Mathematical Association of America.)

53. ***Height of a Tree*** A civil engineer must determine the vertical height of the tree shown in the figure. The given angle was measured with a **clinometer.** Find the height of the leaning tree to the nearest whole number.

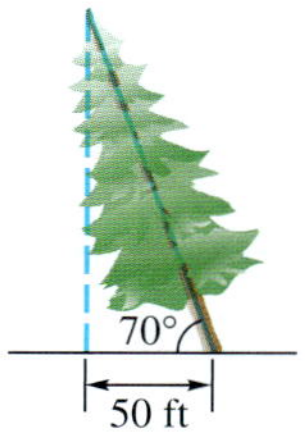

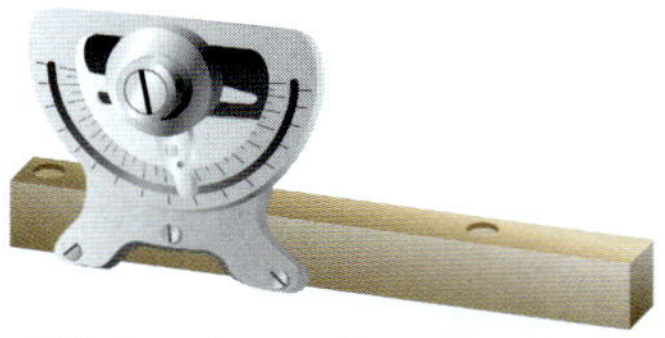

This is a picture of one type of clinometer, called an Abney hand level and clinometer. (Courtesy of Keuffel & Esser Co.)

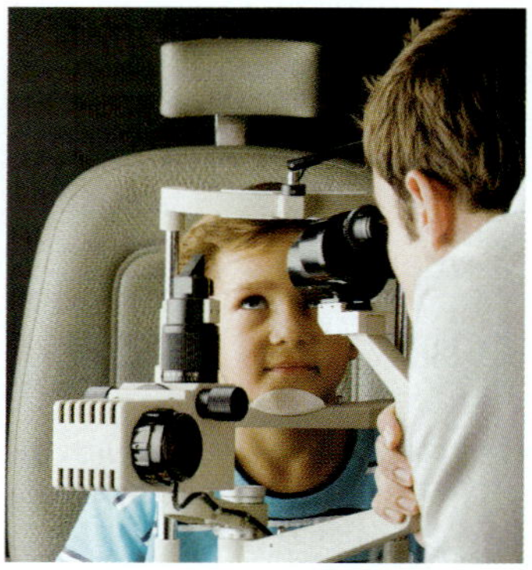

54. ***(Modeling) Double Vision*** To correct mild double vision, a small amount of prism is added to a patient's eyeglasses. The amount of light shift this causes is measured in **prism diopters.** A patient needs 12 prism diopters horizontally and 5 prism diopters vertically. A prism that corrects for both requirements should have length r and be set at angle θ. Find the values of r and θ in the figure.

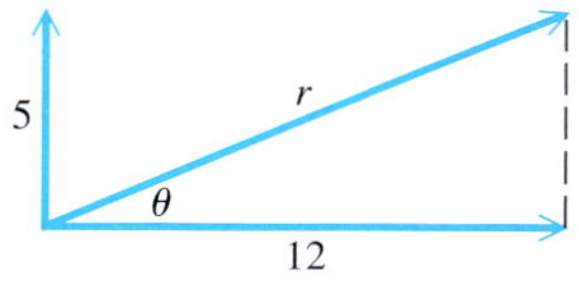

55. ***Height of a Tower*** The angle of elevation from a point 93.2 ft from the base of a tower to the top of the tower is $38° 20'$. Find the height of the tower.

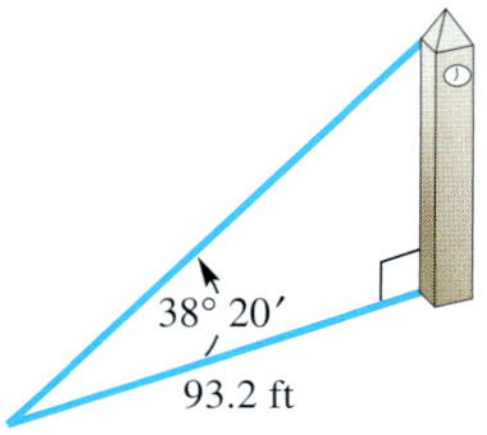

56. ***Height of a Tower*** The angle of depression from a television tower to a point on the ground 36.0 m from the bottom of the tower is 29.5°. Find the height of the tower.

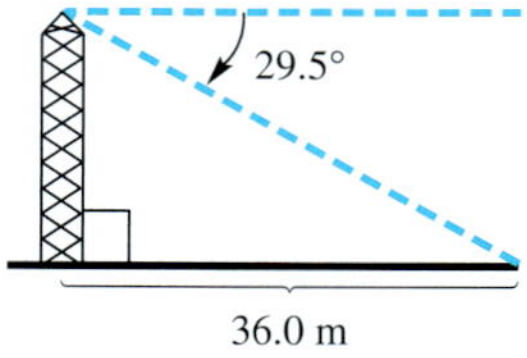

57. ***Length of a Diagonal*** One side of a rectangle measures 15.24 cm. The angle between the diagonal and that side is 35.65°. Find the length of the diagonal.

58. ***Length of Sides of an Isosceles Triangle*** An isosceles triangle has a base of length 49.28 m. The angle opposite the base is 58.746°. Find the length of each of the two equal sides.

59. ***Distance between Two Points*** The bearing of point B from point C is 254°. The bearing of point A from point C is 344°. The bearing of point A from point B is 32°. If the distance from A to C is 780 m, find the distance from A to B.

60. ***Distance a Ship Sails*** The bearing from point A to point B is S 55° E, and the bearing from point B to point C is N 35° E. If a ship sails from A to B, a distance of 81 km, and then from B to C, a distance of 74 km, how far is it from A to C?

61. ***Distance between Two Points*** Two cars leave an intersection at the same time. One heads due south at 55 mph. The other travels due west. After 2 hr, the bearing of the car headed west from the car headed south is 324°. How far apart are they at that time?

62. Find a formula for h in terms of k, A, and B. Assume $A < B$.

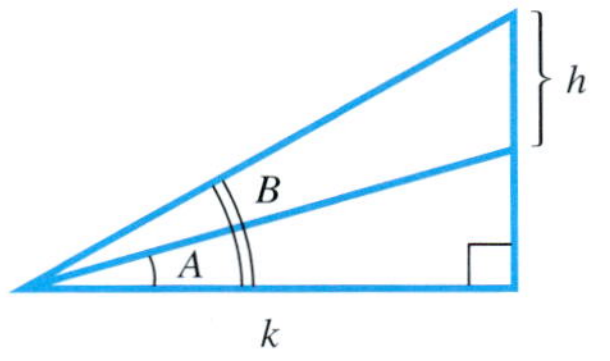

63. Create a right triangle problem whose solution is $3 \tan 25°$.

64. Create a right triangle problem whose solution is found from $\sin \theta = \frac{3}{4}$.

65. ***(Modeling) Height of a Satellite*** Artificial satellites that orbit Earth often use VHF signals to communicate with the ground. VHF signals travel in straight lines. The height h of the satellite above Earth and the time T that the satellite can communicate with a fixed location on the ground are related by the model

$$h = R\left(\frac{1}{\cos \frac{180T}{P}} - 1\right),$$

where $R = 3955$ mi is the radius of Earth and P is the period for the satellite to orbit Earth. (*Source:* Schlosser, W., T. Schmidt-Kaler, and E. Milone, *Challenges of Astronomy,* Springer-Verlag.)

(a) Find h to the nearest mile when $T = 25$ min and $P = 140$ min. (Evaluate the cosine function in degree mode.)

(b) What is the value of h to the nearest mile if T is increased to 30 min?

66. ***(Modeling) Fundamental Surveying Problem*** The first fundamental problem of surveying is to determine the coordinates of a point Q given the coordinates of a point P, the distance between P and Q, and the bearing θ from P to Q. See the figure. (*Source:* Mueller, I. and K. Ramsayer, *Introduction to Surveying,* Frederick Ungar Publishing Co.)

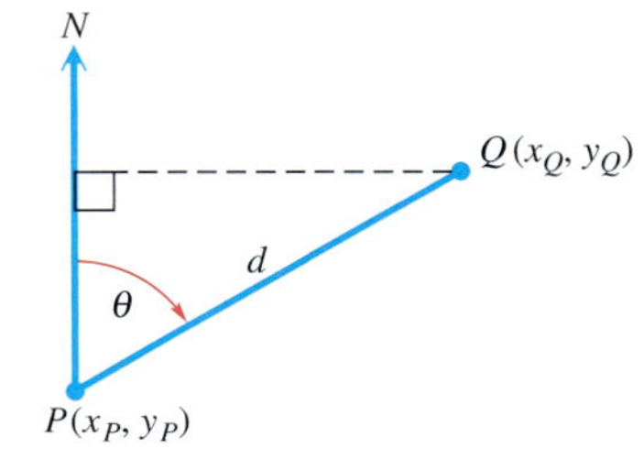

(a) Find a formula for the coordinates (x_Q, y_Q) of the point Q given θ, the coordinates (x_P, y_P) of P, and the distance d between P and Q.

(b) Use your formula to determine (x_Q, y_Q) if $(x_P, y_P) = (123.62, 337.95)$, $\theta = 17° \, 19' \, 22''$, and $d = 193.86$ ft.

Chapter 2 Test

1. Give the six trigonometric function values of angle A.

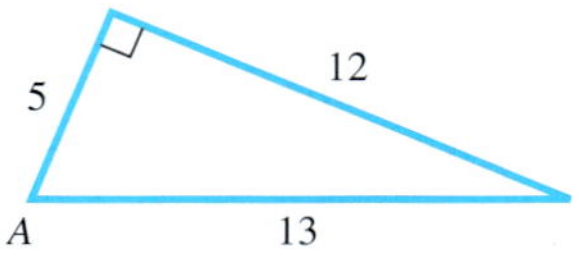

2. Find the exact value of each part labeled with a letter.

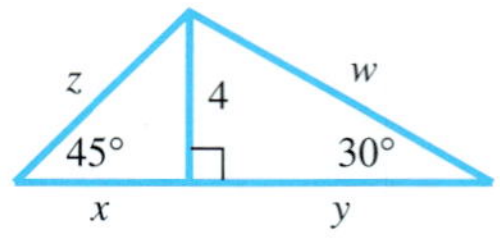

3. Find a solution for $\sin(\theta + 15°) = \cos(2\theta + 30°)$.

4. Determine whether each statement is *true* or *false*. If false, tell why.

(a) $\sin 24° < \sin 48°$ **(b)** $\cos 24° < \cos 48°$

(c) $\cos(60° + 30°) = \cos 60° \cdot \cos 30° - \sin 60° \cdot \sin 30°$

Find the exact values of the six trigonometric functions for each angle. Rationalize denominators when applicable.

5. $240°$ **6.** $-135°$ **7.** $990°$

Find all values of θ in the interval $[0°, 360°)$ that have the given function value.

8. $\cos \theta = -\frac{\sqrt{2}}{2}$ **9.** $\csc \theta = -\frac{2\sqrt{3}}{3}$ **10.** $\tan \theta = 1$

11. How would you find $\cot \theta$ using a calculator, if $\tan \theta = 1.6778490$? Give $\cot \theta$.

12. Use a calculator to approximate each value.

(a) $\sin 78° \, 21'$ **(b)** $\tan 117.689°$ **(c)** $\sec 58.9041°$

13. Find a value of θ in the interval $[0°, 90°)$ in decimal degrees, if

$$\sin \theta = 0.27843196.$$

14. Solve the triangle.

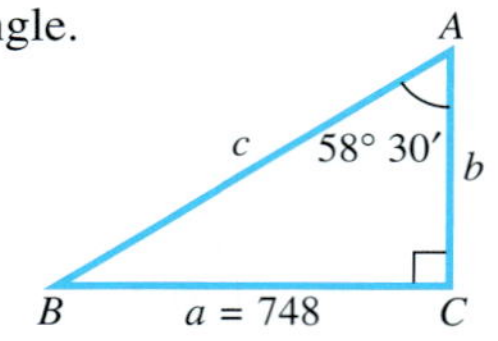

15. *Antenna Mast Guy Wire* A guy wire 77.4 m long is attached to the top of an antenna mast that is 71.3 m high. Find the angle that the wire makes with the ground.

16. *Height of a Flagpole* To measure the height of a flagpole, Amado Carillo found that the angle of elevation from a point 24.7 ft from the base to the top is $32° \, 10'$. What is the height of the flagpole?

17. *Altitude of a Mountain* The highest point in Texas is Guadalupe Peak. The angle of depression from the top of this peak to a small miner's cabin at an approximate elevation of 2000 ft is 26°. The cabin is located 14,000 ft horizontally from a point directly under the top of the mountain. Find the altitude of the top of the mountain to the nearest hundred feet.

18. *Distance between Two Points* Two ships leave a port at the same time. The first ship sails on a bearing of 32° at 16 knots (nautical miles per hour) and the second on a bearing of 122° at 24 knots. How far apart are they after 2.5 hr?

19. *Distance of a Ship from a Pier* A ship leaves a pier on a bearing of S 62° E and travels for 75 km. It then turns and continues on a bearing of N 28° E for 53 km. How far is the ship from the pier?

20. Find h as indicated in the figure.

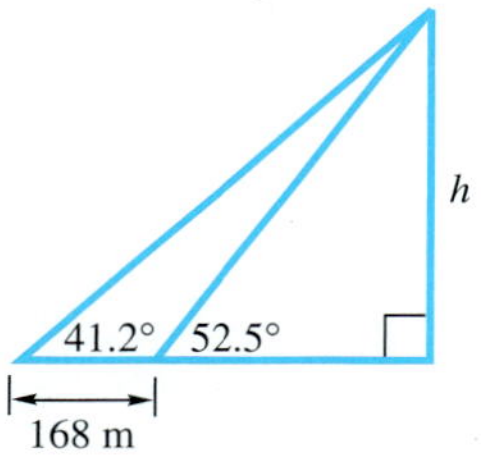

3 Radian Measure and the Unit Circle

The speed of a planet revolving around its sun can be measured in *linear* and *angular speed*, both of which are discussed in this chapter covering *radian measure* of angles.

3.1 Radian Measure

- Radian Measure
- Converting between Degrees and Radians
- Finding Function Values for Angles in Radians

Radian Measure We have seen that angles can be measured in degrees. In more theoretical work in mathematics, *radian measure* of angles is preferred. Radian measure enables us to treat the trigonometric functions as functions with domains of *real numbers,* rather than angles.

Figure 1 shows an angle θ in standard position, along with a circle of radius r. The vertex of θ is at the center of the circle. Because angle θ intercepts an arc on the circle equal in length to the radius of the circle, we say that angle θ has a measure of *1 radian*.

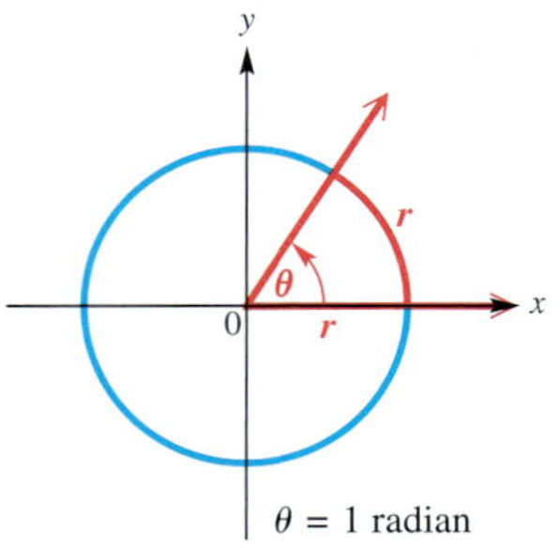

Figure 1

Radian

An angle with its vertex at the center of a circle that intercepts an arc on the circle equal in length to the radius of the circle has a measure of **1 radian.**

It follows that an angle of measure 2 radians intercepts an arc equal in length to twice the radius of the circle, an angle of measure $\frac{1}{2}$ radian intercepts an arc equal in length to half the radius of the circle, and so on. ***In general, if θ is a central angle of a circle of radius r, and θ intercepts an arc of length s, then the radian measure of θ is $\frac{s}{r}$. See* Figure 2.**

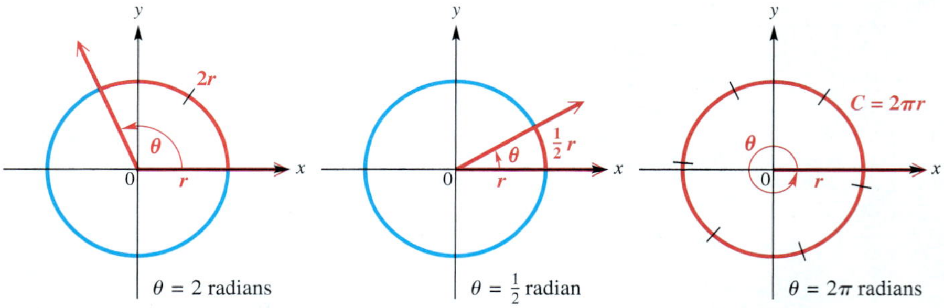

Figure 2

The ratio $\frac{s}{r}$ is a pure number, where s and r are expressed in the same units. ***Thus, "radians" is not a unit of measure like feet or centimeters.***

Converting between Degrees and Radians The **circumference** of a circle—the distance around the circle—is given by $C = 2\pi r$, where r is the radius of the circle. The formula $C = 2\pi r$ shows that the radius can be measured off 2π times around a circle. Therefore, an angle of 360°, which corresponds to a complete circle, intercepts an arc equal in length to 2π times the radius of the circle. Thus, an angle of 360° has a measure of 2π radians.

$$360° = 2\pi \text{ radians}$$

An angle of 180° is half the size of an angle of 360°, so an angle of 180° has half the radian measure of an angle of 360°.

$$180° = \frac{1}{2}(2\pi) \text{ radians} = \pi \text{ radians}$$ Degree/radian relationship

We can use the relationship $180° = \pi$ radians to develop a method for converting between degrees and radians as follows.

$$180° = \pi \text{ radians}$$

$$1° = \frac{\pi}{180} \text{ radian} \quad \text{Divide by 180.} \qquad \text{or} \qquad 1 \text{ radian} = \frac{180°}{\pi} \quad \text{Divide by } \pi.$$

Converting between Degrees and Radians

1. Multiply a degree measure by $\frac{\pi}{180}$ radian and simplify to convert to radians.
2. Multiply a radian measure by $\frac{180°}{\pi}$ and simplify to convert to degrees.

EXAMPLE 1 Converting Degrees to Radians

Convert each degree measure to radians.

(a) $45°$ **(b)** $-270°$ **(c)** $249.8°$

SOLUTION

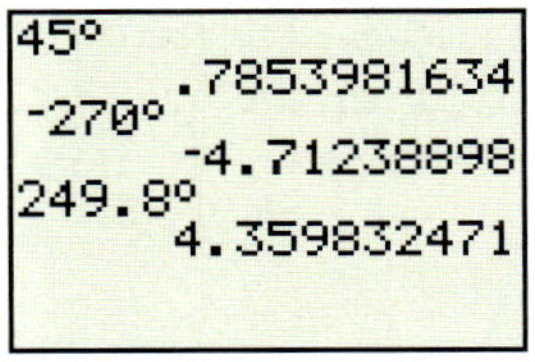

This radian mode screen shows TI-83/84 Plus conversions for **Example 1.** Verify that the first two results are *approximations* for the *exact* values of $\frac{\pi}{4}$ and $-\frac{3\pi}{2}$.

(a) $45° = 45\left(\frac{\pi}{180} \text{ radian}\right) = \frac{\pi}{4} \text{ radian}$ Multiply by $\frac{\pi}{180}$ radian.

(b) $-270° = -270\left(\frac{\pi}{180} \text{ radian}\right)$ Multiply by $\frac{\pi}{180}$ radian.

$$= -\frac{270\pi}{180} \text{ radians}$$

$$= -\frac{3\pi}{2} \text{ radians} \quad \text{Write in lowest terms.}$$

(c) $249.8° = 249.8\left(\frac{\pi}{180} \text{ radian}\right) \approx 4.360 \text{ radians}$ Nearest thousandth

✔ ***Now Try Exercises 7, 13, and 47.***

EXAMPLE 2 Converting Radians to Degrees

Convert each radian measure to degrees.

(a) $\frac{9\pi}{4}$ **(b)** $-\frac{5\pi}{6}$ **(c)** 4.25

SOLUTION

(a) $\frac{9\pi}{4} \text{ radians} = \frac{9\pi}{4}\left(\frac{180°}{\pi}\right) = 405°$ Multiply by $\frac{180°}{\pi}$.

(b) $-\frac{5\pi}{6} \text{ radians} = -\frac{5\pi}{6}\left(\frac{180°}{\pi}\right) = -150°$ Multiply by $\frac{180°}{\pi}$.

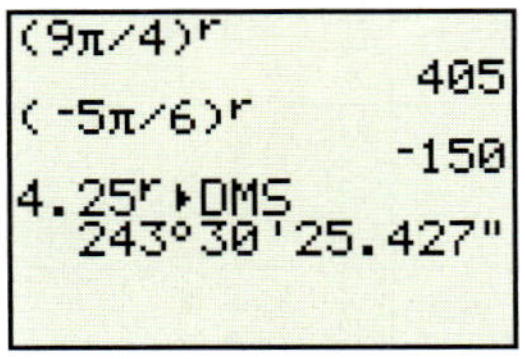

This degree mode screen shows how a TI-83/84 Plus calculator converts the radian measures in **Example 2** to degree measures.

(c) $4.25 \text{ radians} = 4.25\left(\frac{180°}{\pi}\right)$

$$\approx 243.5°, \quad \text{or} \quad 243°\,30' \qquad 0.50706(60') \approx 30'$$

✔ ***Now Try Exercises 31, 35, and 59.***

NOTE Another way to convert a radian measure that is a rational multiple of π, such as $\frac{9\pi}{4}$, to degrees is to just substitute 180° for π. In **Example 2(a),** this would be

$$\frac{9(180°)}{4} = 405°.$$

One of the most important facts to remember when working with angles and their measures is summarized in the following statement.

Agreement on Angle Measurement Units

If no unit of angle measure is specified, then the angle is understood to be measured in radians.

For example, **Figure 3(a)** shows an angle of 30°, and **Figure 3(b)** shows an angle of 30 (which means 30 radians).

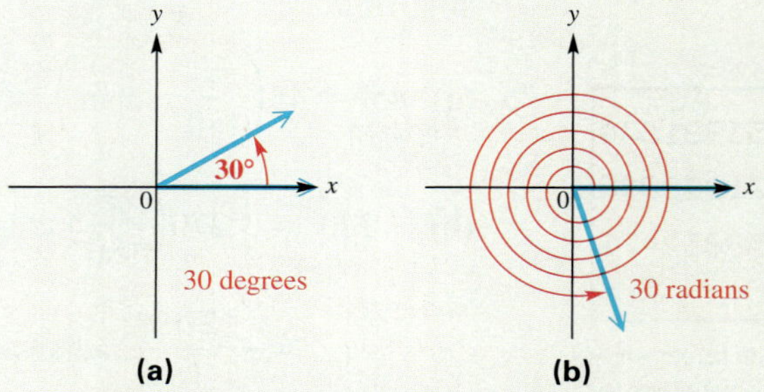

Note the difference between an angle of 30 *degrees* and an angle of 30 *radians*.

Figure 3

The following table and **Figure 4** on the next page give some equivalent angle measures in degrees and radians. Keep in mind that

$$\mathbf{180° = \pi \text{ radians.}}$$

Degrees	Radians		Degrees	Radians	
	Exact	Approximate		Exact	Approximate
0°	0	0	90°	$\frac{\pi}{2}$	1.57
30°	$\frac{\pi}{6}$	0.52	180°	π	3.14
45°	$\frac{\pi}{4}$	0.79	270°	$\frac{3\pi}{2}$	4.71
60°	$\frac{\pi}{3}$	1.05	360°	2π	6.28

These exact values are *rational multiples of* π.

LOOKING AHEAD TO CALCULUS

In calculus, radian measure is much easier to work with than degree measure. If x is measured in radians, then the derivative of $f(x) = \sin x$ is

$$f'(x) = \cos x.$$

However, if x is measured in degrees, then the derivative of $f(x) = \sin x$ is

$$f'(x) = \frac{\pi}{180} \cos x.$$

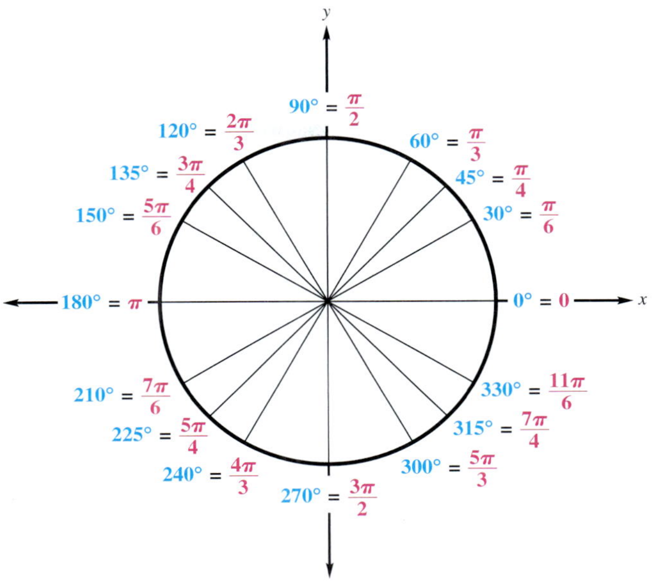

Figure 4

The angles marked in **Figure 4** are extremely important in the study of trigonometry. ***You should learn these equivalences. They will appear often in the chapters to follow.***

Finding Function Values for Angles in Radians Trigonometric function values for angles measured in radians can be found by first converting radian measure to degrees. *(Try to skip this intermediate step as soon as possible, however, and find the function values directly from radian measure.)*

EXAMPLE 3 Finding Function Values of Angles in Radian Measure

Find each function value.

(a) $\tan \frac{2\pi}{3}$ **(b)** $\sin \frac{3\pi}{2}$ **(c)** $\cos\left(-\frac{4\pi}{3}\right)$

SOLUTION

(a) First convert $\frac{2\pi}{3}$ radians to degrees.

$$\tan \frac{2\pi}{3} = \tan\left(\frac{2\pi}{3} \cdot \frac{180^\circ}{\pi}\right) \quad \text{Multiply by } \tfrac{180^\circ}{\pi}.$$

$$= \tan 120^\circ \quad \text{Multiply.}$$

$$= -\sqrt{3} \quad \text{(Section 2.2)}$$

(b) From our earlier observations, $\frac{3\pi}{2}$ radians $= 270^\circ$.

$$\sin \frac{3\pi}{2} = \sin 270^\circ = -1$$

(c) $$\cos\left(-\frac{4\pi}{3}\right) = \cos\left(-\frac{4\pi}{3} \cdot \frac{180^\circ}{\pi}\right) \quad \text{Convert radians to degrees.}$$

$$= -\cos 60^\circ \quad \text{(Section 2.2)}$$

Write in terms of the reference angle.

$$= -\frac{1}{2} \quad \cos 60^\circ = \tfrac{1}{2}$$

✔ ***Now Try Exercises 69, 79, and 83.***

3.1 Exercises

Concept Check In Exercises 1–6, each angle θ is an integer (e.g., 0, ± 1, ± 2, ...) when measured in radians. Give the radian measure of the angle. (It helps to remember that $\pi \approx 3$.)

1.

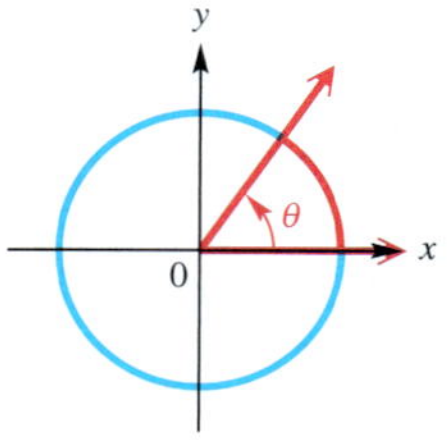

2.

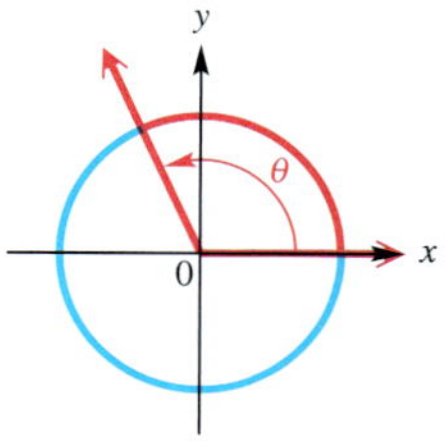

3.

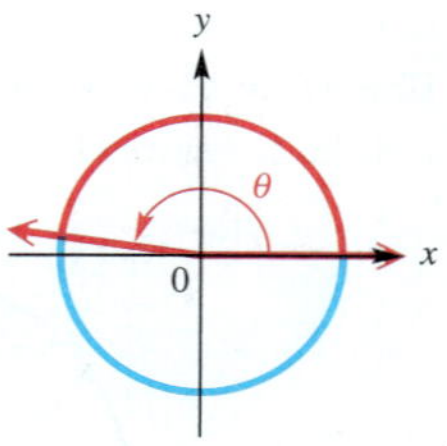

4.

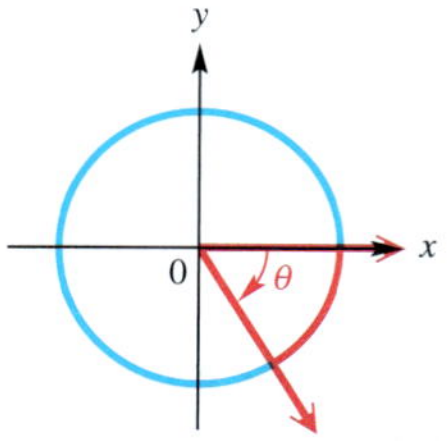

5.

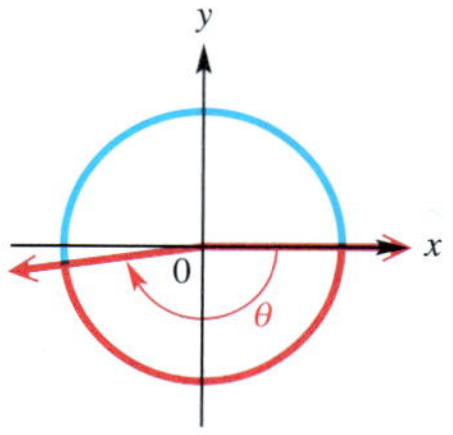

6.

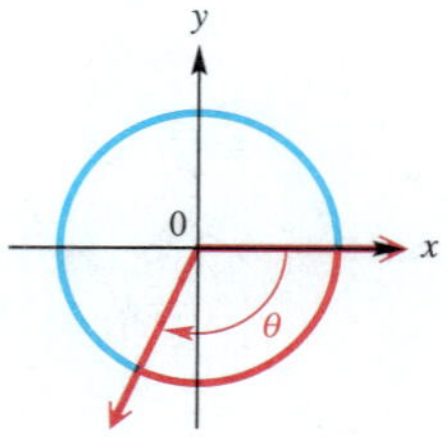

Convert each degree measure to radians. Leave answers as multiples of π. ***See Examples 1(a) and 1(b).***

7. 60° **8.** 30° **9.** 90° **10.** 120°

11. 150° **12.** 270° **13.** −300° **14.** −315°

15. 450° **16.** 480° **17.** 1800° **18.** 3600°

19. 0° **20.** 180° **21.** −900° **22.** −1800°

Give a short explanation in Exercises 23–28.

23. Explain how to convert degree measure to radian measure.

24. Explain how to convert radian measure to degree measure.

25. Explain the meaning of radian measure.

26. Explain the difference between degree measure and radian measure.

27. Use an example to show that you can convert from radian measure to degree measure by multiplying by $\frac{180^\circ}{\pi}$.

28. Explain why an angle of radian measure t in standard position intercepts an arc of length t on a circle of radius 1.

Convert each radian measure to degrees. ***See Examples 2(a) and 2(b).***

29. $\frac{\pi}{3}$ **30.** $\frac{8\pi}{3}$ **31.** $\frac{7\pi}{4}$ **32.** $\frac{2\pi}{3}$

33. $\frac{11\pi}{6}$ **34.** $\frac{15\pi}{4}$ **35.** $-\frac{\pi}{6}$ **36.** $-\frac{8\pi}{5}$

37. $\frac{7\pi}{10}$ **38.** $\frac{11\pi}{15}$ **39.** $-\frac{4\pi}{15}$ **40.** $-\frac{7\pi}{20}$

41. $\frac{17\pi}{20}$ **42.** $\frac{11\pi}{30}$ **43.** -5π **44.** 15π

Convert each degree measure to radians. **See Example 1(c).**

45. 39° **46.** 74° **47.** 42.5° **48.** 264.9°

49. 139° 10′ **50.** 174° 50′ **51.** 64.29° **52.** 85.04°

53. 56° 25′ **54.** 122° 37′ **55.** −47.6925° **56.** −23.0143°

Convert each radian measure to degrees. Write answers to the nearest minute. **See Example 2(c).**

57. 2 **58.** 5 **59.** 1.74 **60.** 3.06

61. 0.3417 **62.** 9.84763 **63.** −5.01095 **64.** −3.47189

65. *Concept Check* The value of sin 30 is not $\frac{1}{2}$. Why is this true?

66. Explain what is meant by an angle of one radian.

Find the exact value of each expression without using a calculator. **See Example 3.**

67. $\sin \frac{\pi}{3}$ **68.** $\cos \frac{\pi}{6}$ **69.** $\tan \frac{\pi}{4}$ **70.** $\cot \frac{\pi}{3}$

71. $\sec \frac{\pi}{6}$ **72.** $\csc \frac{\pi}{4}$ **73.** $\sin \frac{\pi}{2}$ **74.** $\csc \frac{\pi}{2}$

75. $\tan \frac{5\pi}{3}$ **76.** $\cot \frac{2\pi}{3}$ **77.** $\sin \frac{5\pi}{6}$ **78.** $\tan \frac{5\pi}{6}$

79. $\cos 3\pi$ **80.** $\sec \pi$ **81.** $\sin\left(-\frac{8\pi}{3}\right)$ **82.** $\cot\left(-\frac{2\pi}{3}\right)$

83. $\sin\left(-\frac{7\pi}{6}\right)$ **84.** $\cos\left(-\frac{\pi}{6}\right)$ **85.** $\tan\left(-\frac{14\pi}{3}\right)$ **86.** $\csc\left(-\frac{13\pi}{3}\right)$

87. *Concept Check* The figure shows the same angles measured in both degrees and radians. Complete the missing measures.

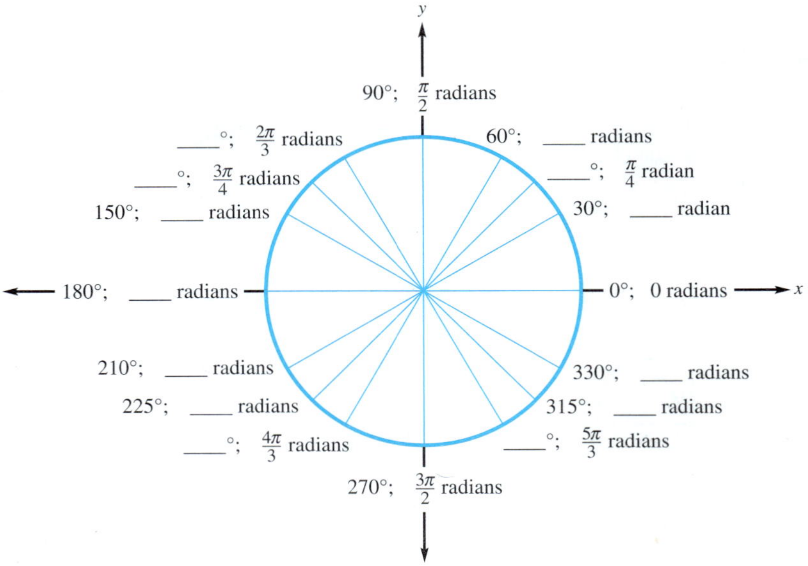

88. *Concept Check* What would be the exact radian measure of an angle that measures π degrees?

Solve each problem.

89. *Rotating Hour Hand on a Clock* Through how many radians does the hour hand on a clock rotate in **(a)** 24 hr and **(b)** 4 hr?

90. *Rotating Minute Hand on a Clock* Through how many radians does the minute hand on a clock rotate in **(a)** 12 hr and **(b)** 3 hr?

91. *Orbits of a Space Vehicle* A space vehicle is orbiting Earth in a circular orbit. What radian measure corresponds to **(a)** 2.5 orbits and **(b)** $\frac{4}{3}$ orbit?

92. *Rotating Pulley* A circular pulley is rotating about its center. Through how many radians does it turn in **(a)** 8 rotations and **(b)** 30 rotations?

93. *Revolutions of a Carousel* A stationary horse on a carousel makes 12 complete revolutions. Through what radian measure angle does the horse revolve?

94. *Railroad Engineering* The term **grade** has several different meanings in construction work. Some engineers use the term **grade** to represent $\frac{1}{100}$ of a right angle and express grade as a percent. For instance, an angle of 0.9° would be referred to as a 1% grade. (*Source:* Hay, W., *Railroad Engineering,* John Wiley and Sons.)

(a) By what number should you multiply a grade (disregarding the % symbol) to convert it to radians?

(b) In a rapid-transit rail system, the maximum grade allowed between two stations is 3.5%. Express this angle in degrees and radians.

3.2 Applications of Radian Measure

- Arc Length on a Circle
- Area of a Sector of a Circle

Arc Length on a Circle

The formula for finding the length of an arc of a circle follows directly from the definition of an angle θ in radians, where $\theta = \frac{s}{r}$.

In **Figure 5,** we see that angle QOP has measure 1 radian and intercepts an arc of length r on the circle. Angle ROT has measure θ radians and intercepts an arc of length s on the circle. From plane geometry, we know that the lengths of the arcs are proportional to the measures of their central angles.

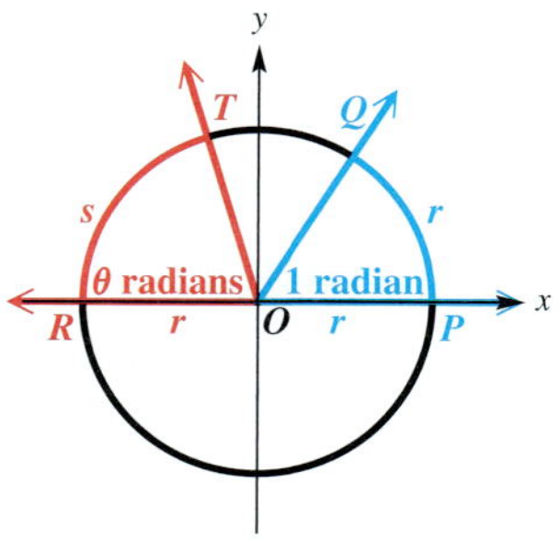

Figure 5

$$\frac{s}{r} = \frac{\theta}{1} \quad \text{Set up a proportion.}$$

Multiplying each side by r gives

$$s = r\theta. \quad \text{Solve for } s.$$

Arc Length

The length s of the arc intercepted on a circle of radius r by a central angle of measure θ radians is given by the product of the radius and the radian measure of the angle.

$$s = r\theta, \quad \textbf{where } \theta \textbf{ is in radians}$$

CAUTION ***When the formula $s = r\theta$ is applied, the value of θ MUST be expressed in radians, not degrees.***

EXAMPLE 1 Finding Arc Length Using $s = r\theta$

A circle has radius 18.20 cm. Find the length of the arc intercepted by a central angle having each of the following measures.

(a) $\frac{3\pi}{8}$ radians **(b)** 144°

SOLUTION

(a) As shown in **Figure 6**, $r = 18.20$ cm and $\theta = \frac{3\pi}{8}$.

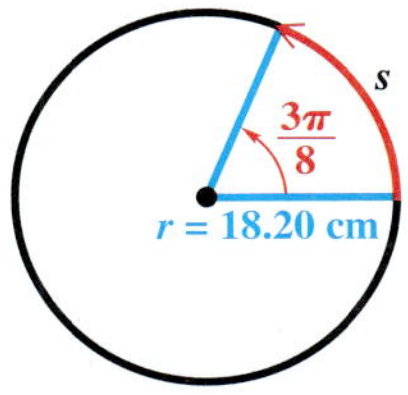

Figure 6

$$s = r\theta \quad \text{Arc length formula}$$

$$s = 18.20\left(\frac{3\pi}{8}\right) \text{ cm} \quad \text{Substitute for } r \text{ and } \theta.$$

$$s \approx 21.44 \text{ cm} \quad \text{Use a calculator.}$$

(b) The formula $s = r\theta$ requires that θ be measured in radians. First, convert θ to radians by multiplying 144° by $\frac{\pi}{180}$ radian.

$$144° = 144\left(\frac{\pi}{180}\right) = \frac{4\pi}{5} \text{ radians} \quad \text{Convert from degrees to radians. (Section 3.1)}$$

The length s is found by using $s = r\theta$.

$$s = r\theta = 18.20\left(\frac{4\pi}{5}\right) \approx 45.74 \text{ cm} \quad \text{Let } r = 18.20 \text{ cm and } \theta = \tfrac{4\pi}{5}.$$

Be sure to use radians for θ in $s = r\theta$.

✔ ***Now Try Exercises 1, 11, and 15.***

EXAMPLE 2 Finding the Distance between Two Cities

Latitude gives the measure of a central angle with vertex at Earth's center whose initial side goes through the equator and whose terminal side goes through the given location. Reno, Nevada, is approximately due north of Los Angeles. The latitude of Reno is 40° N, and that of Los Angeles is 34° N. (The N in 34° N means *north* of the equator.) The radius of Earth is 6400 km. Find the north-south distance between the two cities.

SOLUTION As shown in **Figure 7**, the central angle between Reno and Los Angeles is

$$40° - 34° = 6°.$$

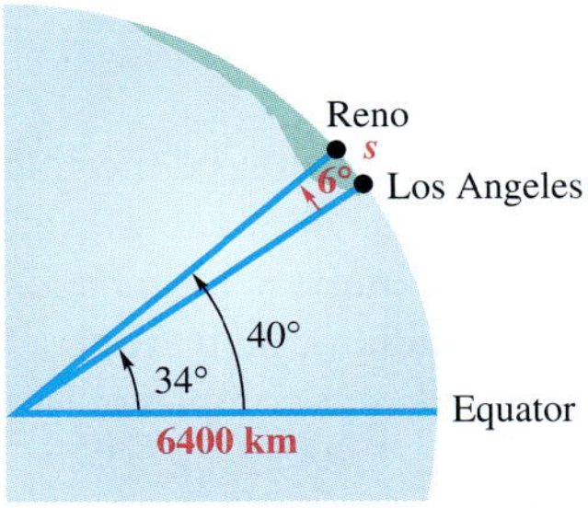

Figure 7

The distance between the two cities can be found by the formula $s = r\theta$, after 6° is converted to radians.

$$6° = 6\left(\frac{\pi}{180}\right) = \frac{\pi}{30} \text{ radian}$$

The distance between the two cities is given by s.

$$s = r\theta = 6400\left(\frac{\pi}{30}\right) \approx 670 \text{ km} \quad \text{Let } r = 6400 \text{ and } \theta = \tfrac{\pi}{30}.$$

✔ ***Now Try Exercise 21.***

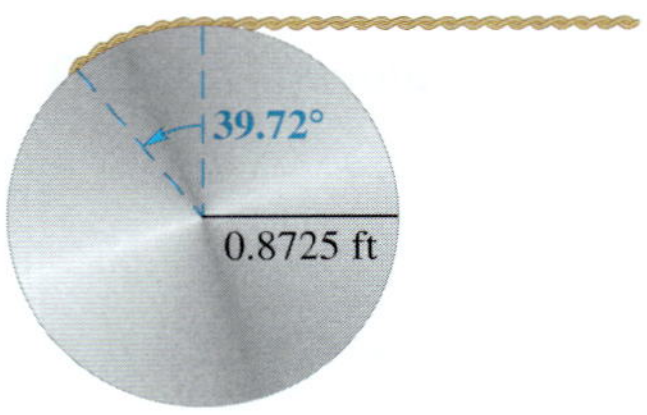

Figure 8

EXAMPLE 3 Finding a Length Using $s = r\theta$

A rope is being wound around a drum with radius 0.8725 ft. (See **Figure 8.**) How much rope will be wound around the drum if the drum is rotated through an angle of 39.72°?

SOLUTION The length of rope wound around the drum is the arc length for a circle of radius 0.8725 ft and a central angle of 39.72°. Use the formula $s = r\theta$, with the angle converted to radian measure. The length of the rope wound around the drum is approximated by s.

$$s = r\theta = 0.8725\left[39.72\left(\frac{\pi}{180}\right)\right] \approx 0.6049 \text{ ft}$$

✔ *Now Try Exercise 33(a).*

Figure 9

EXAMPLE 4 Finding an Angle Measure Using $s = r\theta$

Two gears are adjusted so that the smaller gear drives the larger one, as shown in **Figure 9.** If the smaller gear rotates through an angle of 225°, through how many degrees will the larger gear rotate?

SOLUTION First find the radian measure of the angle of rotation for the smaller gear, and then find the arc length on the smaller gear. This arc length will correspond to the arc length of the motion of the larger gear. Since $225° = \frac{5\pi}{4}$ radians, for the smaller gear,

$$s = r\theta = 2.5\left(\frac{5\pi}{4}\right) = \frac{12.5\pi}{4} = \frac{25\pi}{8} \text{ cm.}$$

The tips of the two mating gear teeth must move at the same linear speed, or the teeth will break. So we must have "equal arc lengths in equal times." An arc with this length s on the larger gear corresponds to an angle measure θ, in radians, where $s = r\theta$.

$$s = r\theta$$

$$\frac{25\pi}{8} = 4.8\theta \qquad \text{Substitute } \tfrac{25\pi}{8} \text{ for } s \text{ and } 4.8 \text{ for } r \text{ (for the larger gear).}$$

$$\frac{125\pi}{192} = \theta \qquad 4.8 = \tfrac{48}{10} = \tfrac{24}{5}. \text{ Multiply by } \tfrac{5}{24} \text{ to solve for } \theta.$$

Converting θ back to degrees shows that the larger gear rotates through

$$\frac{125\pi}{192}\left(\frac{180°}{\pi}\right) \approx 117°. \qquad \text{Convert } \theta = \tfrac{125\pi}{192} \text{ to degrees.}$$

✔ *Now Try Exercise 27.*

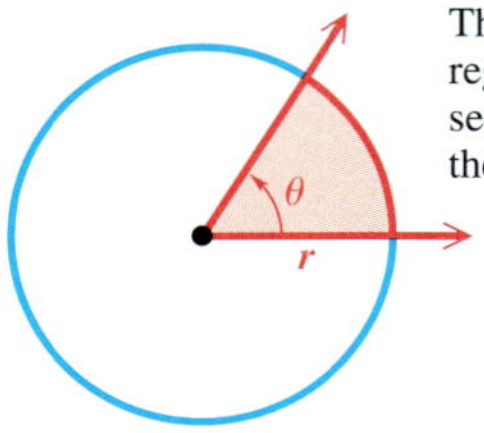

The shaded region is a sector of the circle.

Figure 10

Area of a Sector of a Circle A **sector of a circle** is the portion of the interior of a circle intercepted by a central angle. Think of it as a "piece of pie." See **Figure 10.** A complete circle can be thought of as an angle with measure 2π radians. If a central angle for a sector has measure θ radians, then the sector makes up the fraction $\frac{\theta}{2\pi}$ of a complete circle. The area $\mathscr{A}$ of a complete circle with radius r is $\mathscr{A} = \pi r^2$. Therefore, we have the following.

$$\text{Area } \mathscr{A} \text{ of a sector} = \frac{\theta}{2\pi}(\pi r^2) = \frac{1}{2}r^2\theta, \quad \text{where } \theta \text{ is in radians.}$$

This discussion can be summarized.

Area of a Sector

The area $\mathscr{A}$ of a sector of a circle of radius r and central angle θ is given by the following formula.

$$\mathscr{A} = \frac{1}{2}r^2\theta, \quad \text{where } \theta \text{ is in radians}$$

CAUTION ***As in the formula for arc length, the value of θ must be in radians when this formula is used for the area of a sector.***

EXAMPLE 5 Finding the Area of a Sector-Shaped Field

Center-pivot irrigation system

A center-pivot irrigation system provides water to a sector-shaped field with the measures shown in **Figure 11.** Find the area of the field.

Figure 11

SOLUTION First, convert 15° to radians.

$$15° = 15\left(\frac{\pi}{180}\right) = \frac{\pi}{12} \text{ radian} \qquad \text{Convert to radians.}$$

Now use the formula to find the area of a sector of a circle with radius $r = 321$.

$$\mathscr{A} = \frac{1}{2}r^2\theta$$

$$\mathscr{A} = \frac{1}{2}(321)^2\left(\frac{\pi}{12}\right) \qquad \text{Substitute for } r \text{ and } \theta.$$

$$\mathscr{A} \approx 13{,}500 \text{ m}^2 \qquad \text{Multiply.}$$

✔ ***Now Try Exercise 61.***

3.2 Exercises

Concept Check *Find the exact length of each arc intercepted by the given central angle.*

1.

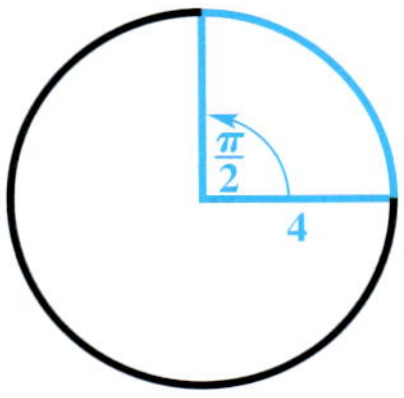

2.

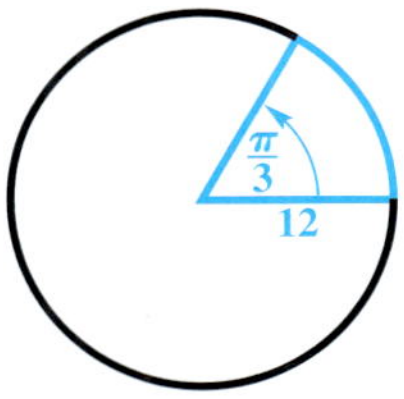

3.

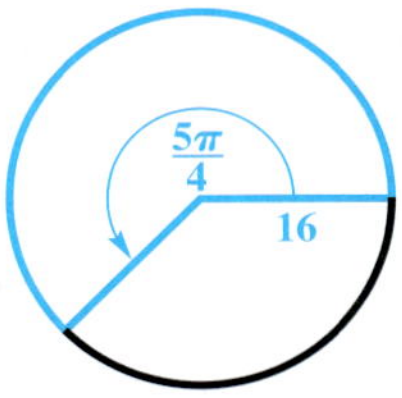

Concept Check *Find the radius of each circle.*

4.

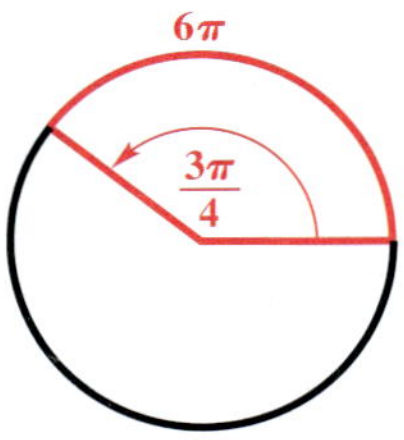

5.

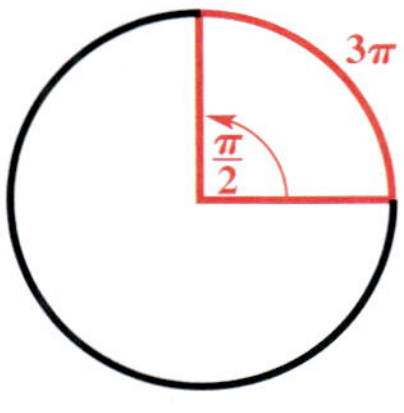

6.

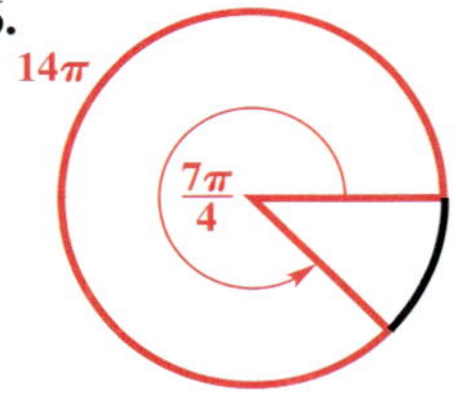

Concept Check Find the measure of each central angle (in radians).

7.

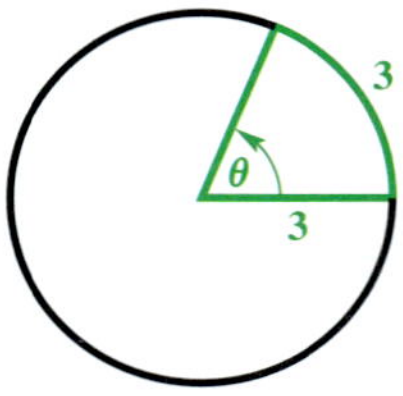

8.

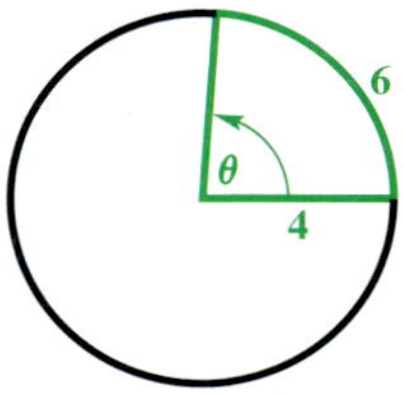

9.

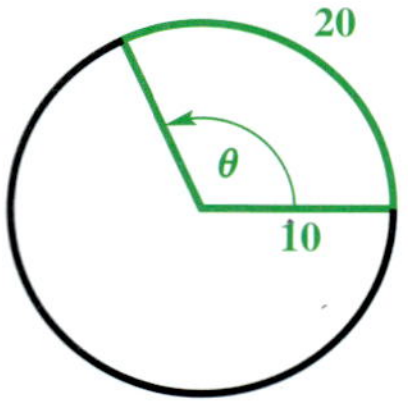

10. Explain how to find the *degree* measure of a central angle in a circle if both the radius and the length of the intercepted arc are known.

Unless otherwise directed, give calculator approximations in your answers in the rest of this exercise set.

Find the length to three significant digits of each arc intercepted by a central angle θ in a circle of radius r. ***See Example 1.***

11. $r = 12.3$ cm, $\theta = \frac{2\pi}{3}$ radians

12. $r = 0.892$ cm, $\theta = \frac{11\pi}{10}$ radians

13. $r = 1.38$ ft, $\theta = \frac{5\pi}{6}$ radians

14. $r = 3.24$ mi, $\theta = \frac{7\pi}{6}$ radians

15. $r = 4.82$ m, $\theta = 60°$

16. $r = 71.9$ cm, $\theta = 135°$

17. $r = 15.1$ in., $\theta = 210°$

18. $r = 12.4$ ft, $\theta = 330°$

19. *Concept Check* If the radius of a circle is doubled, how is the length of the arc intercepted by a fixed central angle changed?

20. *Concept Check* Radian measure simplifies many formulas, such as the formula for arc length, $s = r\theta$. Give the corresponding formula when θ is measured in degrees instead of radians.

Distance between Cities Find the distance in kilometers between each pair of cities, assuming they lie on the same north-south line. Use $r = 6400$ km for the radius of Earth. ***See Example 2.***

21. Panama City, Panama, 9° N, and Pittsburgh, Pennsylvania, 40° N

22. Farmersville, California, 36° N, and Penticton, British Columbia, 49° N

23. New York City, New York, 41° N, and Lima, Peru, 12° S

24. Halifax, Nova Scotia, 45° N, and Buenos Aires, Argentina, 34° S

25. *Latitude of Madison* Madison, South Dakota, and Dallas, Texas, are 1200 km apart and lie on the same north-south line. The latitude of Dallas is 33° N. What is the latitude of Madison?

26. *Latitude of Toronto* Charleston, South Carolina, and Toronto, Canada, are 1100 km apart and lie on the same north-south line. The latitude of Charleston is 33° N. What is the latitude of Toronto?

Work each problem. ***See Examples 3 and 4.***

27. *Gear Movement* Two gears are adjusted so that the smaller gear drives the larger one, as shown in the figure. If the smaller gear rotates through an angle of 300°, through how many degrees does the larger gear rotate?

28. *Gear Movement* Repeat **Exercise 27** for gear radii of 4.8 in. and 7.1 in. and for an angle of 315° for the smaller gear.

29. *Rotating Wheels* The rotation of the smaller wheel in the figure causes the larger wheel to rotate. Through how many degrees does the larger wheel rotate if the smaller one rotates through 60.0°?

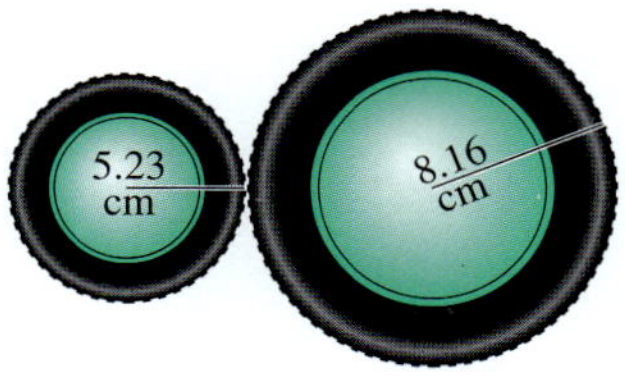

30. *Rotating Wheels* Repeat **Exercise 29** for wheel radii of 6.84 in. and 12.46 in. and an angle of 150° for the smaller wheel.

31. *Rotating Wheels* Find the radius of the larger wheel in the figure if the smaller wheel rotates 80.0° when the larger wheel rotates 50.0°.

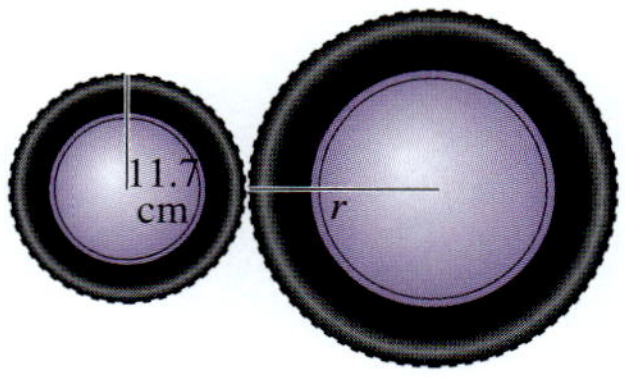

32. *Rotating Wheels* Repeat **Exercise 31** if the smaller wheel of radius 14.6 in. rotates 120° when the larger wheel rotates 60°.

33. *Pulley Raising a Weight* Refer to the figure.

(a) How many inches will the weight in the figure rise if the pulley is rotated through an angle of $71° \, 50'$?

(b) Through what angle, to the nearest minute, must the pulley be rotated to raise the weight 6 in.?

34. *Pulley Raising a Weight* Find the radius of the pulley in the figure if a rotation of 51.6° raises the weight 11.4 cm.

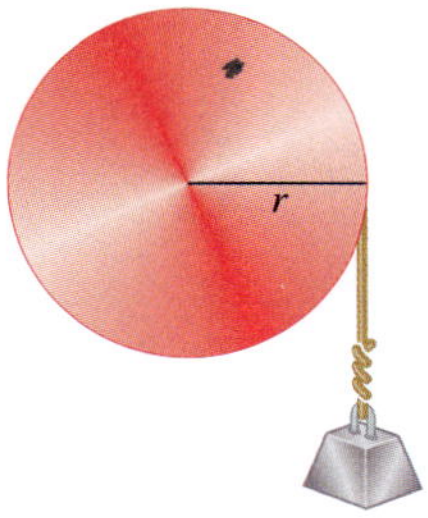

35. *Bicycle Chain Drive* The figure shows the chain drive of a bicycle. How far will the bicycle move if the pedals are rotated through 180°? Assume the radius of the bicycle wheel is 13.6 in.

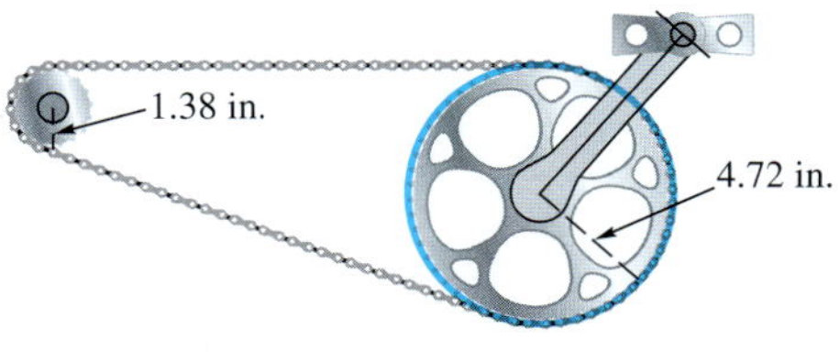

36. *Car Speedometer* The speedometer of Terry's Honda CR-V is designed to be accurate with tires of radius 14 in.

(a) Find the number of rotations of a tire in 1 hr if the car is driven at 55 mph.

(b) Suppose that oversize tires of radius 16 in. are placed on the car. If the car is now driven for 1 hr with the speedometer reading 55 mph, how far has the car gone? If the speed limit is 55 mph, does Terry deserve a speeding ticket?

Suppose the tip of the minute hand of a clock is 3 in. *from the center of the clock. For each duration, determine the distance traveled by the tip of the minute hand.*

37. 30 min

38. 40 min

39. 4.5 hr

40. $6\frac{1}{2}$ hr

If a central angle is very small, there is little difference in length between an arc and the inscribed chord. See the figure. Approximate each of the following lengths by finding the necessary arc length. (Note: When a central angle intercepts an arc, the arc is said to ***subtend*** *the angle.)*

Arc length ≈ length of inscribed chord

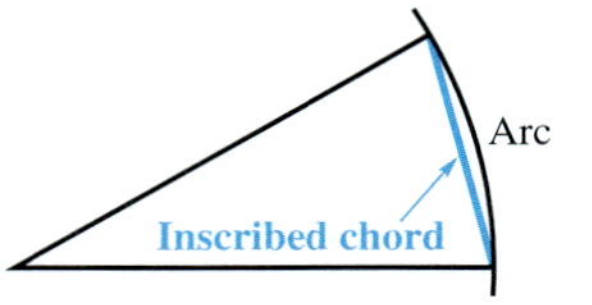

41. *Length of a Train* A railroad track in the desert is 3.5 km away. A train on the track subtends (horizontally) an angle of $3° \, 20'$. Find the length of the train.

42. *Distance to a Boat* The mast of Brent Simon's boat is 32 ft high. If it subtends an angle of $2° \, 10'$, how far away is it?

Concept Check Find the area of each sector.

43.

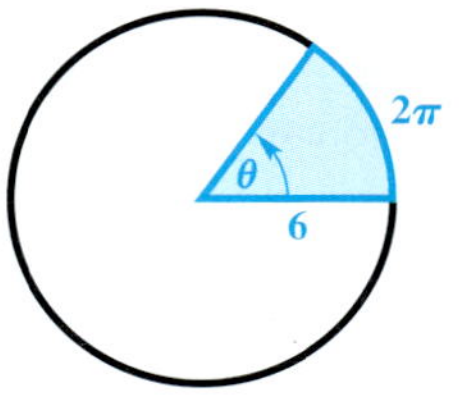

44.

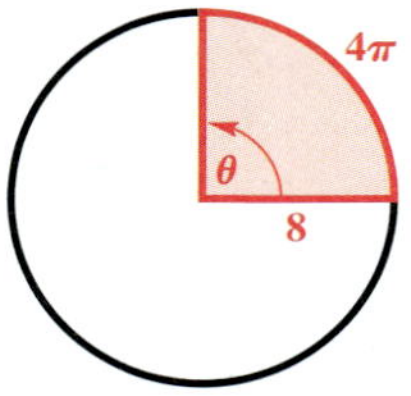

45.

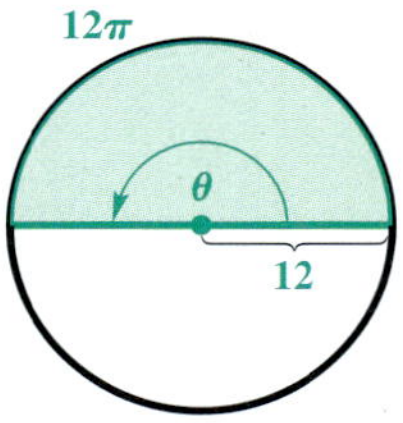

46.

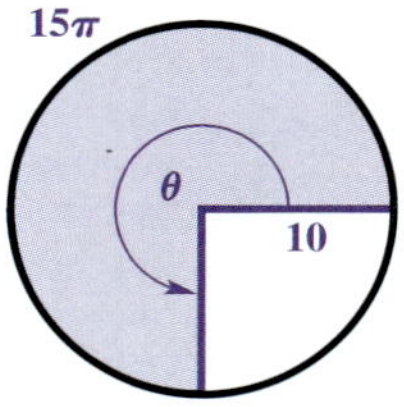

Concept Check Find the measure (in degrees) of each central angle. The number inside the sector is the area.

47.

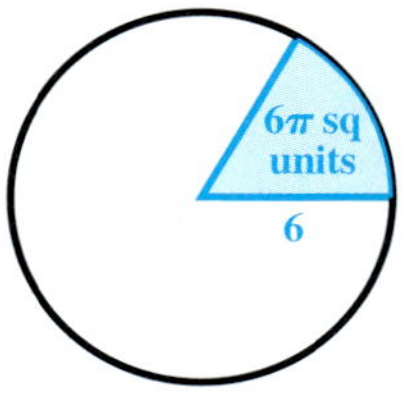

48.

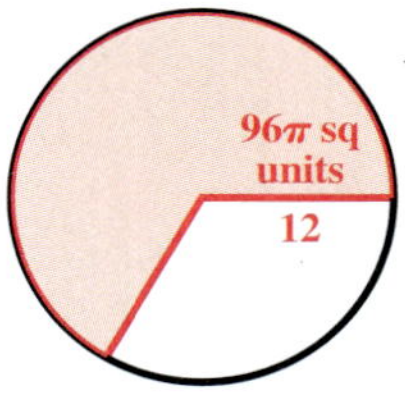

Concept Check Find the measure (in radians) of each central angle. The number inside the sector is the area.

49.

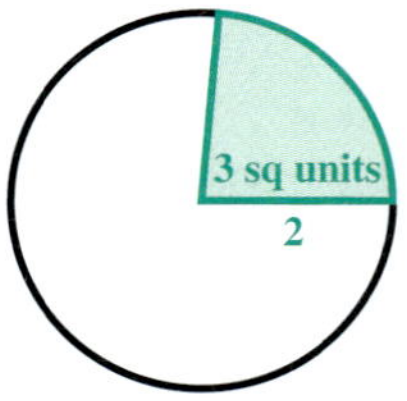

50.

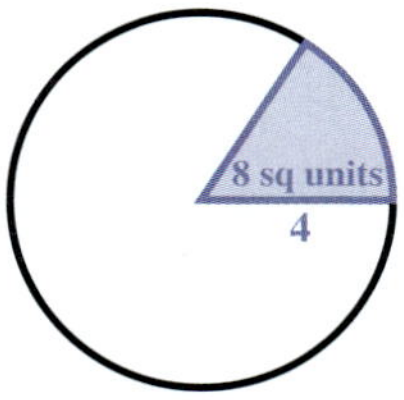

Find the area of a sector of a circle having radius r and central angle θ. Express answers to the nearest tenth. ***See Example 5.***

51. $r = 29.2$ m, $\theta = \frac{5\pi}{6}$ radians

52. $r = 59.8$ km, $\theta = \frac{2\pi}{3}$ radians

53. $r = 30.0$ ft, $\theta = \frac{\pi}{2}$ radians

54. $r = 90.0$ yd, $\theta = \frac{5\pi}{6}$ radians

55. $r = 12.7$ cm, $\theta = 81°$

56. $r = 18.3$ m, $\theta = 125°$

57. $r = 40.0$ mi, $\theta = 135°$

58. $r = 90.0$ km, $\theta = 270°$

Work each problem. ***See Example 5.***

59. *Angle Measure* Find the measure (in radians) of a central angle of a sector of area 16 in.^{2} in a circle of radius 3.0 in.

60. *Radius Length* Find the radius of a circle in which a central angle of $\frac{\pi}{6}$ radian determines a sector of area 64 m^2.

61. *Irrigation Area* A center-pivot irrigation system provides water to a sector-shaped field as shown in the figure. Find the area of the field if $\theta = 40.0°$ and $r = 152$ yd.

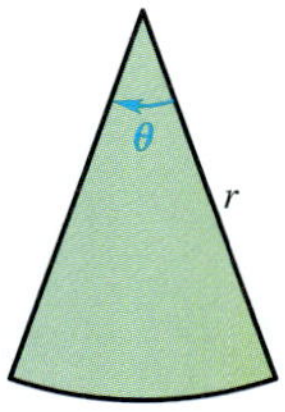

62. *Irrigation Area* Suppose that in **Exercise 61** the angle is halved and the radius length is doubled. How does the new area compare to the original area? Does this result hold in general for any values of θ and r?

63. *Arc Length* A circular sector has an area of 50 in.^{2}. The radius of the circle is 5 in. What is the arc length of the sector?

64. *Angle Measure* In a circle, a sector has an area of 16 cm^2 and an arc length of 6.0 cm. What is the measure of the central angle in degrees?

65. *Measures of a Structure* The figure illustrates Medicine Wheel, a Native American structure in northern Wyoming. There are 27 aboriginal spokes in the wheel, all equally spaced.

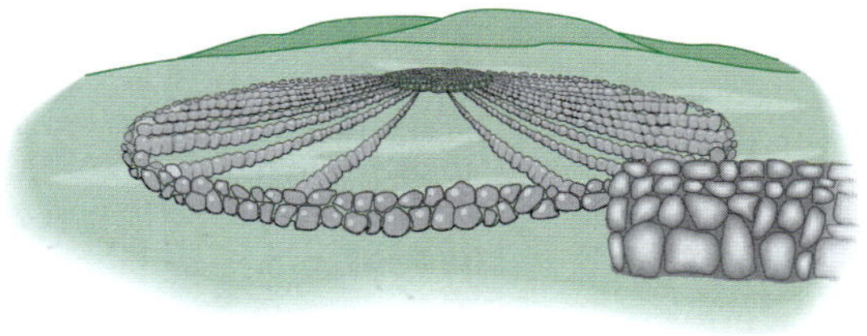

(a) Find the measure of each central angle in degrees and in radians.

(b) If the radius of the wheel is 76.0 ft, find the circumference.

(c) Find the length of each arc intercepted by consecutive pairs of spokes.

(d) Find the area of each sector formed by consecutive spokes.

66. *Area Cleaned by a Windshield Wiper* The Ford Model A, built from 1928 to 1931, had a single windshield wiper on the driver's side. The total arm and blade was 10 in. long and rotated back and forth through an angle of 95°. The shaded region in the figure is the portion of the windshield cleaned by the 7-in. wiper blade. What is the area of the region cleaned?

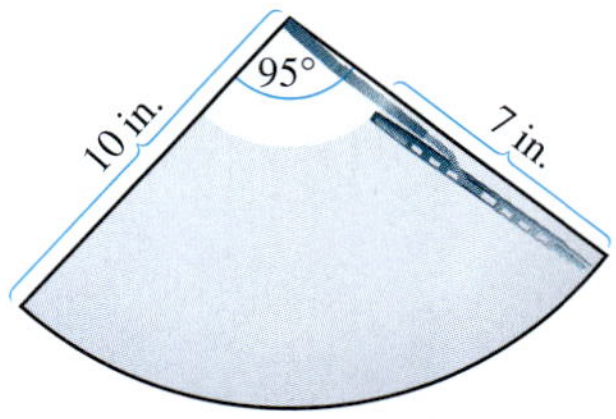

67. *Circular Railroad Curves* In the United States, circular railroad curves are designated by the **degree of curvature**, the central angle subtended by a chord of 100 ft. Suppose a portion of track has curvature 42.0°. (*Source:* Hay, W., *Railroad Engineering,* John Wiley and Sons.)

(a) What is the radius of the curve?

(b) What is the length of the arc determined by the 100-ft chord?

(c) What is the area of the portion of the circle bounded by the arc and the 100-ft chord?

68. *Land Required for a Solar-Power Plant* A 300-megawatt solar-power plant requires approximately 950,000 m^2 of land area to collect the required amount of energy from sunlight. If this land area is circular, what is its radius? If this land area is a 35° sector of a circle, what is its radius?

69. *Area of a Lot* A frequent problem in surveying city lots and rural lands adjacent to curves of highways and railways is that of finding the area when one or more of the boundary lines is the arc of a circle. Find the area (to two significant digits) of the lot shown in the figure. (*Source:* Anderson, J. and E. Michael, *Introduction to Surveying,* McGraw-Hill.)

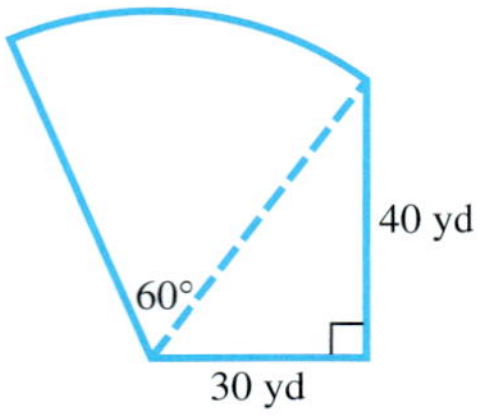

70. *Nautical Miles* **Nautical miles** are used by ships and airplanes. They are different from **statute miles,** which equal 5280 ft. A nautical mile is defined to be the arc length along the equator intercepted by a central angle AOB of 1 min, as illustrated in the figure. If the equatorial radius of Earth is 3963 mi, use the arc length formula to approximate the number of statute miles in 1 nautical mile. Round your answer to two decimal places.

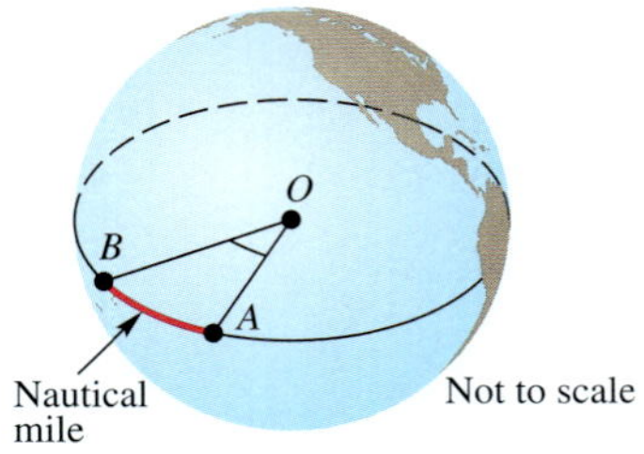

71. *Circumference of Earth* The first accurate estimate of the distance around Earth was done by the Greek astronomer Eratosthenes (276–195 B.C.), who noted that the noontime position of the sun at the summer solstice in the city of Syene differed by 7° 12′ from its noontime position in the city of Alexandria. (See the figure.) The distance between these two cities is 496 mi. Use the arc length formula to estimate the radius of Earth. Then find the circumference of Earth. (*Source:* Zeilik, M., *Introductory Astronomy and Astrophysics,* Third Edition, Saunders College Publishers.)

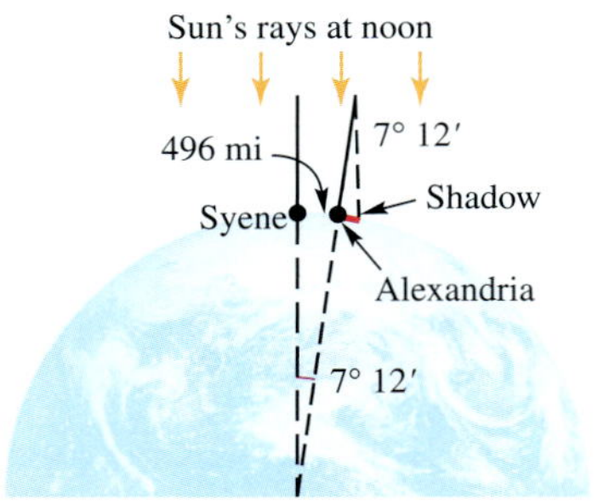

72. *Longitude* **Longitude** is the angular distance (expressed in degrees) East or West of the prime meridian, which goes from the North Pole to the South Pole through Greenwich, England. Arcs of 1° longitude are 110 km apart at the equator, and therefore 15° arcs subtend 15(110) km, or 1650 km, at the equator.

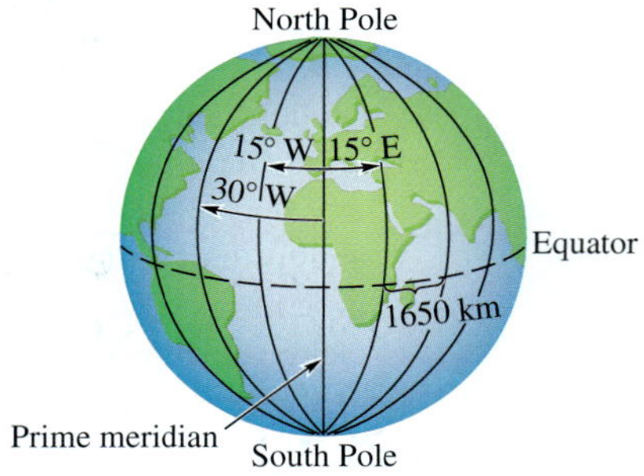

Because Earth rotates 15° per hr, longitude is found by taking the difference between time zones multiplied by 15°. For example, if it is 12 noon where you are (in the United States) and 5 P.M. in Greenwich, you are located at longitude 5(15°), or 75° W.

(a) What is the longitude at Greenwich, England?

(b) Use time zones to determine the longitude where you live.

73. *Concept Check* If the radius of a circle is doubled and the central angle of a sector is unchanged, how is the area of the sector changed?

74. *Concept Check* Give the formula for the area of a sector when the angle is measured in degrees.

Volume of a Solid *Multiply the area of the base by the height to find a formula for the volume V of each solid.*

75.

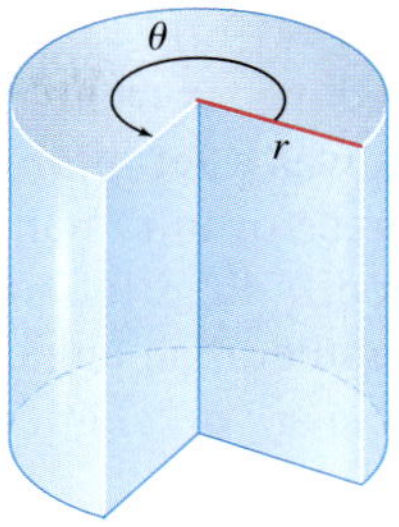

76.

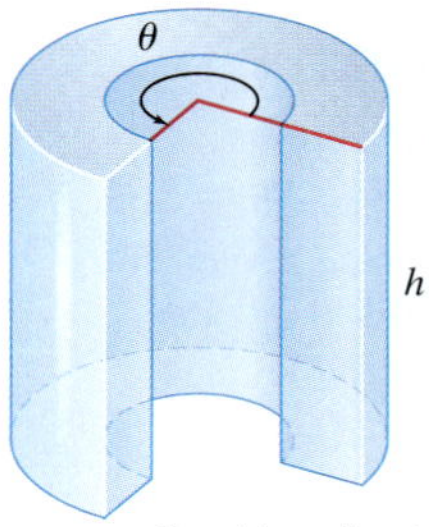

Outside radius is r_1, inside radius is r_2.

Relating Concepts

For individual or collaborative investigation *(Exercises 77–80)*

(Modeling) Measuring Paper Curl *Manufacturers of paper determine its quality by its curl. The curl of a sheet of paper is measured by holding it at the center of one edge and comparing the arc formed by the free end to arcs on a chart lying flat on a table. Each arc in the chart corresponds to a number d that gives the depth of the arc. See the figure. (Source:* Tabakovic, H., J. Paullet, and R. Bertram, "Measuring the Curl of Paper," *The College Mathematics Journal,* Vol. 30, No. 4.*)*

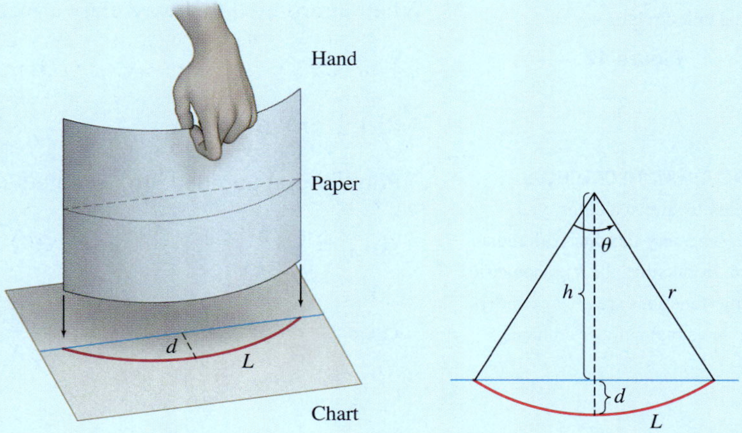

To produce the chart, it is necessary to find a function that relates d to the length of arc L. ***Work Exercises 77–80 in order,*** *to determine that function. Refer to the figure on the right.*

77. Express L in terms of r and θ, and then solve for r.

78. Use a right triangle to relate r, h, and θ. Solve for h.

79. Express d in terms of r and h. Then substitute your answer from **Exercise 78** for h. Factor out r.

80. Use your answer from **Exercise 77** to substitute for r in the result from **Exercise 79.** This result is a formula that gives d for specific values of θ.

3.3 The Unit Circle and Circular Functions

- Circular Functions
- Finding Values of Circular Functions
- Determining a Number with a Given Circular Function Value
- Applying Circular Functions
- Expressing Function Values as Lengths of Line Segments

In **Section 1.3,** we defined the six trigonometric functions in such a way that the domain of each function was a set of *angles* in standard position. These angles can be measured in degrees or in radians. In advanced courses, such as calculus, it is necessary to modify the trigonometric functions so that their domains consist of *real numbers* rather than angles. We do this by using the relationship between an angle θ and an arc of length s on a circle.

Circular Functions

In **Figure 12,** we start at the point $(1, 0)$ and measure an arc of length s along the circle. If $s > 0$, then the arc is measured in a counterclockwise direction, and if $s < 0$, then the direction is clockwise. (If $s = 0$, then no arc is measured.) Let the endpoint of this arc be at the point (x, y). The circle in **Figure 12** is the **unit circle**—it has center at the origin and radius 1 unit (hence the name *unit circle*). Recall from algebra that the equation of this circle is

$$x^2 + y^2 = 1. \qquad \text{(Appendix B)}$$

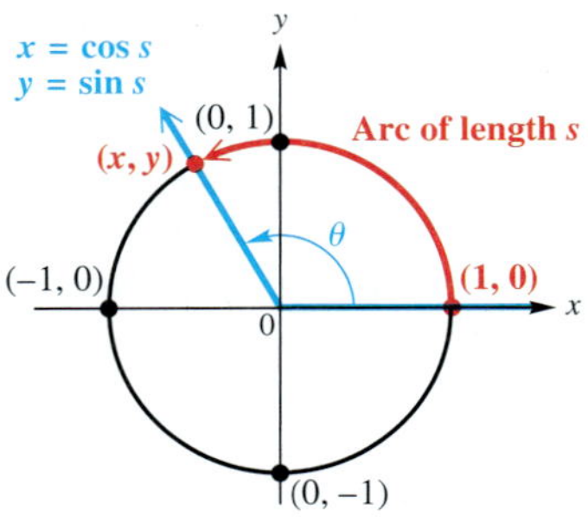

The unit circle $x^2 + y^2 = 1$

Figure 12

The radian measure of θ is related to the arc length s. For θ measured in radians, we know that $s = r\theta$. Here $r = 1$, so s, which is measured in linear units such as inches or centimeters, is equal to θ, measured in radians. Thus, the trigonometric functions of angle θ in radians found by choosing a point (x, y) on the unit circle can be rewritten as functions of the arc length s, a real number. When interpreted this way, they are called **circular functions.**

LOOKING AHEAD TO CALCULUS
If you plan to study calculus, you must become very familiar with radian measure. In calculus, the trigonometric or circular functions are always understood to have real number domains.

Circular Functions

For any real number s represented by a directed arc on the unit circle,

$$\sin s = y \qquad \cos s = x \qquad \tan s = \frac{y}{x} \quad (x \neq 0)$$

$$\csc s = \frac{1}{y} \quad (y \neq 0) \qquad \sec s = \frac{1}{x} \quad (x \neq 0) \qquad \cot s = \frac{x}{y} \quad (y \neq 0).$$

Since x represents the cosine of s and y represents the sine of s, and because of the discussion in **Section 3.1** on converting between degrees and radians, we can summarize a great deal of information in a concise manner, as seen in **Figure 13** on the next page.*

The unit circle is symmetric with respect to the x-axis, the y-axis, and the origin. (See **Appendix D.**) Thus, if a point (a, b) lies on the unit circle, so do $(a, -b)$, $(-a, b)$, and $(-a, -b)$. Furthermore, each of these points has a *reference arc* of equal magnitude. For a point on the unit circle, its **reference arc** is the shortest arc from the point itself to the nearest point on the x-axis. (This concept is analogous to the reference angle concept introduced in **Chapter 2.**) Using the concept of symmetry makes determining sines and cosines of the real numbers identified in **Figure 13** a relatively simple procedure if we know the coordinates of the points labeled in quadrant I.

*The authors thank Professor Marvel Townsend of the University of Florida for her suggestion to include **Figure 13**.

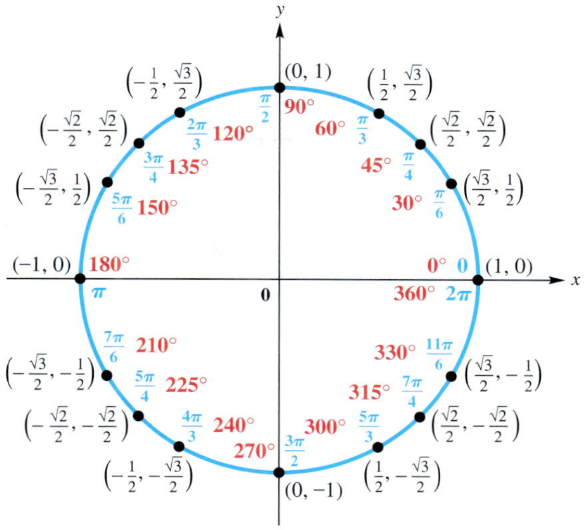

The unit circle $x^2 + y^2 = 1$

Figure 13

For example, the quadrant I real number $\frac{\pi}{3}$ is associated with the point $\left(\frac{1}{2}, \frac{\sqrt{3}}{2}\right)$ on the unit circle. Therefore, we can use symmetry to identify the coordinates of the points associated with

$$\pi - \frac{\pi}{3} = \underset{\text{Quadrant II}}{\frac{2\pi}{3}}, \qquad \pi + \frac{\pi}{3} = \underset{\text{Quadrant III}}{\frac{4\pi}{3}}, \qquad \text{and} \qquad 2\pi - \frac{\pi}{3} = \underset{\text{Quadrant IV}}{\frac{5\pi}{3}}.$$

The following chart summarizes this information.

s	Quadrant of s	Symmetry Type and Corresponding Point	$\cos s$	$\sin s$
$\boldsymbol{\frac{\pi}{3}}$	I	not applicable; $\left(\frac{1}{2}, \frac{\sqrt{3}}{2}\right)$	$\frac{1}{2}$	$\frac{\sqrt{3}}{2}$
$\pi - \frac{\pi}{3} = \boldsymbol{\frac{2\pi}{3}}$	II	y-axis; $\left(-\frac{1}{2}, \frac{\sqrt{3}}{2}\right)$	$-\frac{1}{2}$	$\frac{\sqrt{3}}{2}$
$\pi + \frac{\pi}{3} = \boldsymbol{\frac{4\pi}{3}}$	III	origin; $\left(-\frac{1}{2}, -\frac{\sqrt{3}}{2}\right)$	$-\frac{1}{2}$	$-\frac{\sqrt{3}}{2}$
$2\pi - \frac{\pi}{3} = \boldsymbol{\frac{5\pi}{3}}$	IV	x-axis; $\left(\frac{1}{2}, -\frac{\sqrt{3}}{2}\right)$	$\frac{1}{2}$	$-\frac{\sqrt{3}}{2}$

NOTE Because $\cos s = x$ and $\sin s = y$, we can replace x and y in the equation of the unit circle $x^2 + y^2 = 1$ and obtain the following.

$$\mathbf{\cos^2 s + \sin^2 s = 1} \qquad \text{Pythagorean identity (Section 1.4)}$$

The ordered pair (x, y) represents a point on the unit circle, and therefore

$$-1 \le x \le 1 \quad \text{and} \quad -1 \le y \le 1,$$

$$\mathbf{-1 \le \cos s \le 1} \quad \text{and} \quad \mathbf{-1 \le \sin s \le 1.}$$

For any value of s, both $\sin s$ and $\cos s$ exist, so the domain of these functions is the set of all real numbers.

For tan s, defined as $\frac{y}{x}$, x must not equal 0. The only way x can equal 0 is when the arc length s is $\frac{\pi}{2}$, $-\frac{\pi}{2}$, $\frac{3\pi}{2}$, $-\frac{3\pi}{2}$, and so on. To avoid a 0 denominator, the domain of the tangent function must be restricted to those values of s that satisfy

$$s \neq (2n + 1)\frac{\pi}{2}, \quad \textbf{where } n \textbf{ is any integer.}$$

The definition of secant also has x in the denominator, so the domain of secant is the same as the domain of tangent. Both cotangent and cosecant are defined with a denominator of y. To guarantee that $y \neq 0$, the domain of these functions must be the set of all values of s that satisfy

$$s \neq n\pi, \quad \textbf{where } n \textbf{ is any integer.}$$

Domains of the Circular Functions

The domains of the circular functions are as follows.

Sine and Cosine Functions: $(-\infty, \infty)$

Tangent and Secant Functions:

$$\{s \,|\, s \neq (2n + 1)\frac{\pi}{2}, \quad \textbf{where } n \textbf{ is any integer}\}$$

Cotangent and Cosecant Functions:

$$\{s \,|\, s \neq n\pi, \quad \textbf{where } n \textbf{ is any integer}\}$$

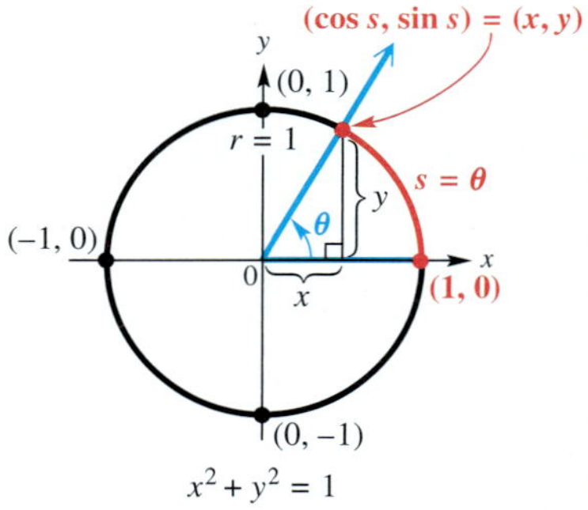

Figure 14

Finding Values of Circular Functions The circular functions of real numbers correspond to the trigonometric functions of angles measured in radians. Let us assume that angle θ is in standard position, superimposed on the unit circle. See **Figure 14**. Suppose that θ is the *radian* measure of this angle. Using the arc length formula

$$s = r\theta \quad \text{with } r = 1, \quad \text{we have} \quad s = \theta.$$

Thus, the length of the intercepted arc is the real number that corresponds to the radian measure of θ. Use the trigonometric function definitions from **Section 1.3** to obtain the following.

$$\sin\theta = \frac{y}{r} = \frac{y}{1} = y = \sin s, \quad \cos\theta = \frac{x}{r} = \frac{x}{1} = x = \cos s, \quad \text{and so on.}$$

As shown here, the trigonometric functions and the circular functions lead to the same function values, provided that we think of the angles as being in radian measure. This leads to the following important result.

Evaluating a Circular Function

Circular function values of real numbers are obtained in the same manner as trigonometric function values of angles measured in radians. This applies both to methods of finding exact values (such as reference angle analysis) and to calculator approximations. ***Calculators must be in radian mode when finding circular function values.***

EXAMPLE 1 Finding Exact Circular Function Values

Find the exact values of $\sin \frac{3\pi}{2}$, $\cos \frac{3\pi}{2}$, and $\tan \frac{3\pi}{2}$.

SOLUTION Evaluating a circular function at the real number $\frac{3\pi}{2}$ is equivalent to evaluating it at $\frac{3\pi}{2}$ radians. An angle of $\frac{3\pi}{2}$ radians intersects the unit circle at the point $(0, -1)$, as shown in **Figure 15.** Since

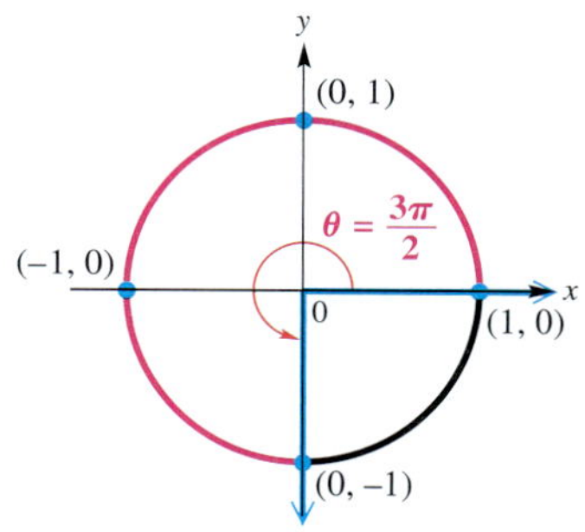

Figure 15

$$\sin s = y, \quad \cos s = x, \quad \text{and} \quad \tan s = \frac{y}{x},$$

it follows that

$$\sin \frac{3\pi}{2} = -1, \quad \cos \frac{3\pi}{2} = 0, \quad \text{and} \quad \tan \frac{3\pi}{2} \text{ is undefined.}$$

✔ *Now Try Exercises 1 and 3.*

EXAMPLE 2 Finding Exact Circular Function Values

Find each exact value using the specified method.

(a) Use **Figure 13** to find the exact values of $\cos \frac{7\pi}{4}$ and $\sin \frac{7\pi}{4}$.

(b) Use **Figure 13** and the definition of the tangent to find the exact value of $\tan\left(-\frac{5\pi}{3}\right)$.

(c) Use reference angles and radian-to-degree conversion to find the exact value of $\cos \frac{2\pi}{3}$.

SOLUTION

(a) In **Figure 13,** we see that the real number $\frac{7\pi}{4}$ corresponds to the unit circle point $\left(\frac{\sqrt{2}}{2}, -\frac{\sqrt{2}}{2}\right)$.

$$\cos \frac{7\pi}{4} = \frac{\sqrt{2}}{2} \quad \text{and} \quad \sin \frac{7\pi}{4} = -\frac{\sqrt{2}}{2}$$

(b) Moving around the unit circle $\frac{5\pi}{3}$ units in the *negative* direction yields the same ending point as moving around $\frac{\pi}{3}$ units in the positive direction. Thus, $-\frac{5\pi}{3}$ corresponds to $\left(\frac{1}{2}, \frac{\sqrt{3}}{2}\right)$.

$\tan s = \frac{y}{x}$

$$\tan\left(-\frac{5\pi}{3}\right) = \tan \frac{\pi}{3} = \frac{\frac{\sqrt{3}}{2}}{\frac{1}{2}} = \frac{\sqrt{3}}{2} \div \frac{1}{2} = \frac{\sqrt{3}}{2} \cdot \frac{2}{1} = \sqrt{3}$$

(c) An angle of $\frac{2\pi}{3}$ radians corresponds to an angle of 120°. In standard position, 120° lies in quadrant II with a reference angle of 60°.

Cosine is negative in quadrant II.

$$\cos \frac{2\pi}{3} = \cos 120^\circ = -\cos 60^\circ = -\frac{1}{2}$$

Reference angle (Section 2.2)

✔ *Now Try Exercises 7, 13, 17, and 21.*

EXAMPLE 3 Approximating Circular Function Values

Find a calculator approximation for each circular function value.

(a) $\cos 1.85$ **(b)** $\cos 0.5149$ **(c)** $\cot 1.3209$ **(d)** $\sec(-2.9234)$

SOLUTION

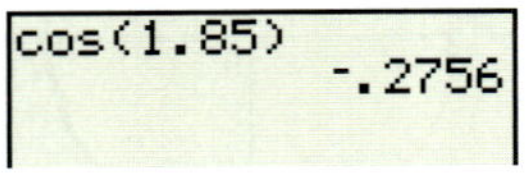

Radian mode

This is how the TI-83/84 Plus calculator displays the result of **Example 3(a)**, fixed to four decimal digits.

(a) $\cos 1.85 \approx -0.2756$ Use a calculator in radian mode.

(b) $\cos 0.5149 \approx 0.8703$ Use a calculator in radian mode.

(c) As before, to find cotangent, secant, and cosecant function values, we must use the appropriate reciprocal functions. To find $\cot 1.3209$, first find $\tan 1.3209$ and then find the reciprocal.

$$\cot 1.3209 = \frac{1}{\tan 1.3209} \approx 0.2552$$ Tangent and cotangent are reciprocals.

(d) $$\sec(-2.9234) = \frac{1}{\cos(-2.9234)} \approx -1.0243$$ Cosine and secant are reciprocals.

✔ ***Now Try Exercises 23, 29, and 33.***

CAUTION A common error is using a calculator in degree mode when radian mode should be used. ***Remember, when finding a circular function value of a real number, the calculator must be in radian mode.***

Determining a Number with a Given Circular Function Value Recall from **Section 2.3** how we used a calculator to determine an angle measure, given a trigonometric function value of the angle. ***Remember that the keys marked $\sin^{-1}$, $\cos^{-1}$, and $\tan^{-1}$ do not represent reciprocal functions. They enable us to find inverse function values.***

For reasons explained in **Chapter 6,** the following statements are true.

- For all x in $[-1, 1]$, a calculator in radian mode returns a single value in $\left[-\frac{\pi}{2}, \frac{\pi}{2}\right]$ for $\sin^{-1} x$.
- For all x in $[-1, 1]$, a calculator in radian mode returns a single value in $[0, \pi]$ for $\cos^{-1} x$.
- For all real numbers x, a calculator in radian mode returns a single value in $\left(-\frac{\pi}{2}, \frac{\pi}{2}\right)$ for $\tan^{-1} x$.

EXAMPLE 4 Finding a Number Given Its Circular Function Value

Find each value as specified.

(a) Approximate the value of s in the interval $\left[0, \frac{\pi}{2}\right]$ if $\cos s = 0.9685$.

(b) Find the exact value of s in the interval $\left[\pi, \frac{3\pi}{2}\right]$ if $\tan s = 1$.

SOLUTION

(a) Since we are given a cosine value and want to determine the real number in $\left[0, \frac{\pi}{2}\right]$ that has this cosine value, we use the *inverse cosine* function of a calculator. With the calculator in radian mode, we find

$$\cos^{-1}(0.9685) \approx 0.2517.$$ (Section 2.3)

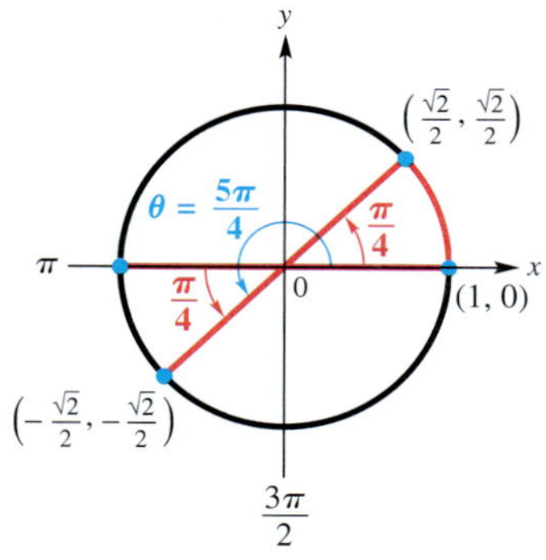

Figure 17

See **Figure 16.** The screen indicates that the real number in $\left[0, \frac{\pi}{2}\right]$ whose cosine is 0.9685 is 0.2517.

Radian mode

Figure 16

(b) Recall that $\tan \frac{\pi}{4} = 1$, and in quadrant III $\tan s$ is positive.

$$\tan\left(\pi + \frac{\pi}{4}\right) = \tan \frac{5\pi}{4} = 1$$

Thus, $s = \frac{5\pi}{4}$. See **Figure 17.**

✔ *Now Try Exercises 55 and 65.*

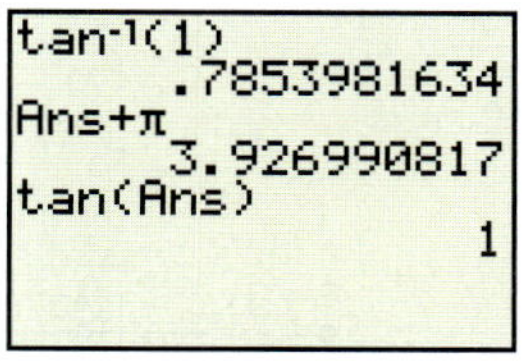

This screen supports the result in **Example 4(b)** with calculator approximations.

Applying Circular Functions

EXAMPLE 5 Modeling the Angle of Elevation of the Sun

The angle of elevation θ of the sun in the sky at any latitude L is calculated with the formula

$$\sin \theta = \cos D \cos L \cos \omega + \sin D \sin L,$$

where $\theta = 0$ corresponds to sunrise and $\theta = \frac{\pi}{2}$ occurs if the sun is directly overhead. ω (the Greek letter *omega*) is the number of radians that Earth has rotated through since noon, when $\omega = 0$. D is the declination of the sun, which varies because Earth is tilted on its axis. (*Source:* Winter, C., R. Sizmann, and L. L. Vant-Hull, Editors, *Solar Power Plants,* Springer-Verlag.)

Sacramento, California, has latitude $L = 38.5°$, or 0.6720 radian. Find the angle of elevation θ of the sun at 3 P.M. on February 29, 2012, where at that time $D \approx -0.1425$ and $\omega \approx 0.7854$.

SOLUTION Use the given formula for $\sin \theta$.

$$\begin{aligned}\sin \theta &= \cos D \cos L \cos \omega + \sin D \sin L \\ &= \cos(-0.1425) \cos(0.6720) \cos(0.7854) + \sin(-0.1425) \sin(0.6720) \\ &\approx 0.4593426188\end{aligned}$$

Thus, $\theta \approx 0.4773$ radian, or 27.3°. Use inverse sine.

✔ *Now Try Exercise 83.*

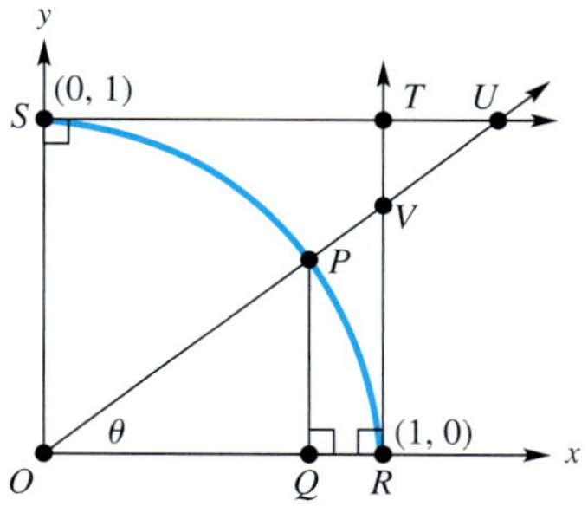

Figure 18

Expressing Function Values as Lengths of Line Segments The diagram shown in **Figure 18** illustrates a correspondence that ties together the right triangle ratio definitions of the trigonometric functions introduced in **Chapter 2** and the unit circle interpretation. The arc SR is the first-quadrant portion of the unit circle, and the standard-position angle POQ is designated θ. By definition, the coordinates of P are $(\cos \theta, \sin \theta)$. The six trigonometric functions of θ can be interpreted as lengths of line segments found in **Figure 18.**

For $\cos \theta$ and $\sin \theta$, use right triangle POQ and right triangle ratios.

$$\boldsymbol{\cos \theta} = \frac{\text{side adjacent to } \theta}{\text{hypotenuse}} = \frac{OQ}{OP} = \frac{OQ}{1} = \boldsymbol{OQ}$$

$$\boldsymbol{\sin \theta} = \frac{\text{side opposite } \theta}{\text{hypotenuse}} = \frac{PQ}{OP} = \frac{PQ}{1} = \boldsymbol{PQ}$$

For $\tan \theta$ and $\sec \theta$, use right triangle VOR in **Figure 18** and right triangle ratios.

$$\mathbf{\tan \theta} = \frac{\text{side opposite } \theta}{\text{side adjacent to } \theta} = \frac{VR}{OR} = \frac{VR}{1} = \mathbf{VR}$$

$$\mathbf{\sec \theta} = \frac{\text{hypotenuse}}{\text{side adjacent to } \theta} = \frac{OV}{OR} = \frac{OV}{1} = \mathbf{OV}$$

For $\csc \theta$ and $\cot \theta$, first note that US and OR are parallel. Thus angle SUO is equal to θ because it is an alternate interior angle to angle POQ, which is equal to θ. Use right triangle USO and right triangle ratios.

$$\csc SUO = \mathbf{\csc \theta} = \frac{\text{hypotenuse}}{\text{side opposite } \theta} = \frac{OU}{OS} = \frac{OU}{1} = \mathbf{OU}$$

$$\cot SUO = \mathbf{\cot \theta} = \frac{\text{side adjacent to } \theta}{\text{side opposite } \theta} = \frac{US}{OS} = \frac{US}{1} = \mathbf{US}$$

Figure 19 uses color to illustrate the results found above.

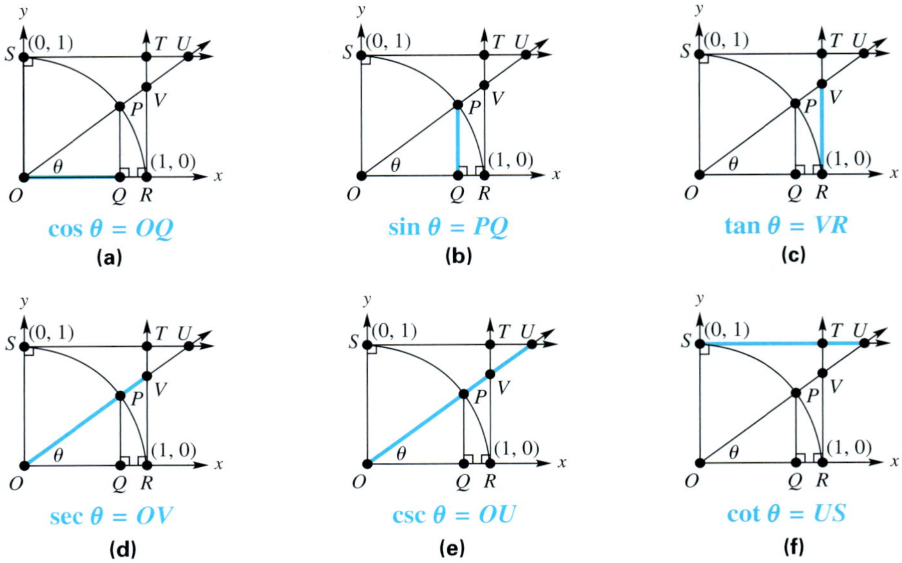

Figure 19

EXAMPLE 6 Finding Lengths of Line Segments

Figure 18 is repeated in the margin. Suppose that angle TVU measures 60°. Find the exact lengths of segments OQ, PQ, VR, OV, OU, and US.

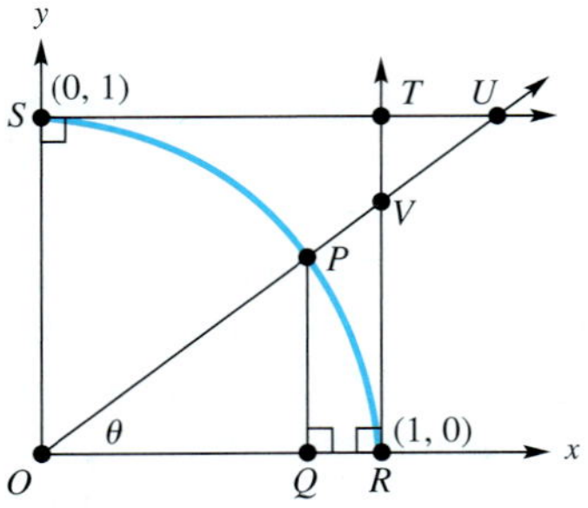

Figure 18 (repeated)

SOLUTION Angle TVU has the same measure as angle OVR because they are vertical angles. Therefore, angle OVR measures 60°. Because it is one of the acute angles in right triangle VOR, θ must be its complement, measuring 30°. Now use the equations found in **Figure 19,** with $\theta = 30°$.

$$OQ = \cos 30° = \frac{\sqrt{3}}{2} \qquad OV = \sec 30° = \frac{2\sqrt{3}}{3}$$

$$PQ = \sin 30° = \frac{1}{2} \qquad OU = \csc 30° = 2$$

$$VR = \tan 30° = \frac{\sqrt{3}}{3} \qquad US = \cot 30° = \sqrt{3}$$

✔ *Now Try Exercise 87.*

3.3 Exercises

For each value of the real number s, find **(a)** $\sin s$, **(b)** $\cos s$, *and* **(c)** $\tan s$. **See Example 1.**

1. $s = \frac{\pi}{2}$ **2.** $s = \pi$ **3.** $s = 2\pi$

4. $s = 3\pi$ **5.** $s = -\pi$ **6.** $s = -\frac{3\pi}{2}$

Find the exact circular function value for each of the following. **See Example 2.**

7. $\sin \frac{7\pi}{6}$ **8.** $\cos \frac{5\pi}{3}$ **9.** $\tan \frac{3\pi}{4}$ **10.** $\sec \frac{2\pi}{3}$

11. $\csc \frac{11\pi}{6}$ **12.** $\cot \frac{5\pi}{6}$ **13.** $\cos\left(-\frac{4\pi}{3}\right)$ **14.** $\tan\left(-\frac{17\pi}{3}\right)$

15. $\cos \frac{7\pi}{4}$ **16.** $\sec \frac{5\pi}{4}$ **17.** $\sin\left(-\frac{4\pi}{3}\right)$ **18.** $\sin\left(-\frac{5\pi}{6}\right)$

19. $\sec \frac{23\pi}{6}$ **20.** $\csc \frac{13\pi}{3}$ **21.** $\tan \frac{5\pi}{6}$ **22.** $\cos \frac{3\pi}{4}$

Find a calculator approximation for each circular function value. **See Example 3.**

23. sin 0.6109 **24.** sin 0.8203 **25.** cos(−1.1519)

26. cos(−5.2825) **27.** tan 4.0203 **28.** tan 6.4752

29. csc(−9.4946) **30.** csc 1.3875 **31.** sec 2.8440

32. sec(−8.3429) **33.** cot 6.0301 **34.** cot 3.8426

Concept Check The figure displays a unit circle and an angle of 1 *radian. The tick marks on the circle are spaced at every two-tenths radian. Use the figure to estimate each value.*

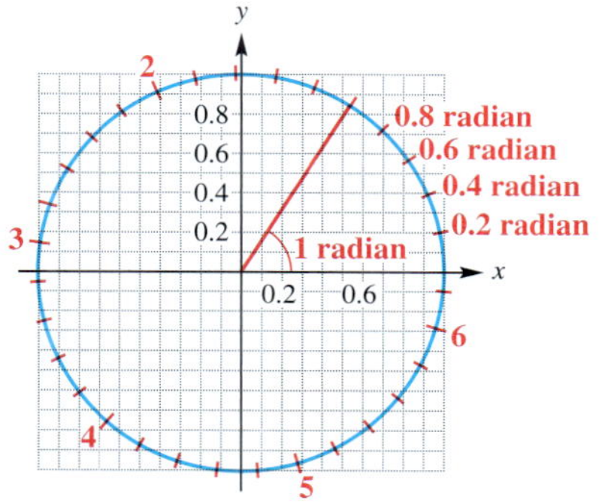

35. cos 0.8 **36.** cos 0.6 **37.** sin 2 **38.** sin 4 **39.** sin 3.8 **40.** cos 3.2

41. a positive angle whose cosine is −0.65

42. a positive angle whose sine is −0.95

43. a positive angle whose sine is 0.7

44. a positive angle whose cosine is 0.3

Concept Check Without using a calculator, decide whether each function value is positive or negative. (Hint: Consider the radian measures of the quadrantal angles, and remember that $\pi \approx 3.14$.*)*

45. cos 2 **46.** sin(−1) **47.** sin 5 **48.** cos 6 **49.** tan 6.29 **50.** tan(−6.29)

Concept Check *Each figure in Exercises 51–54 shows an angle θ in standard position with its terminal side intersecting the unit circle. Evaluate the six circular function values of θ.*

51.

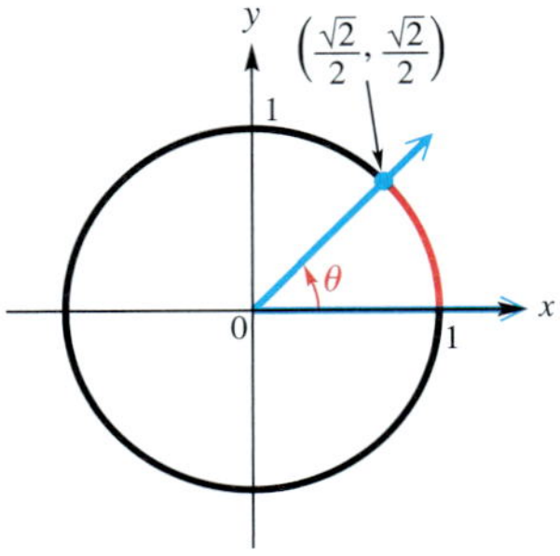

52.

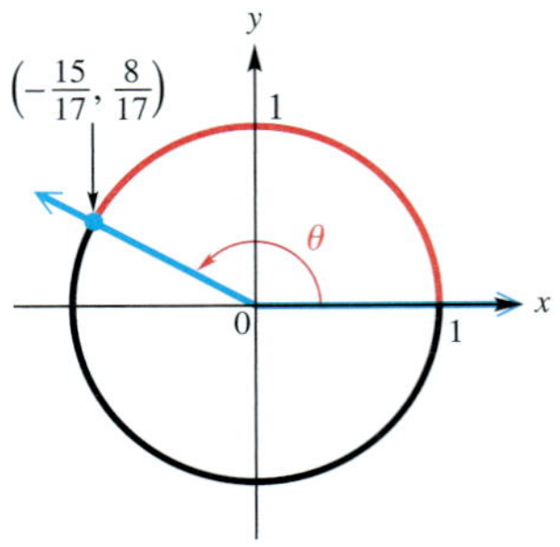

53.

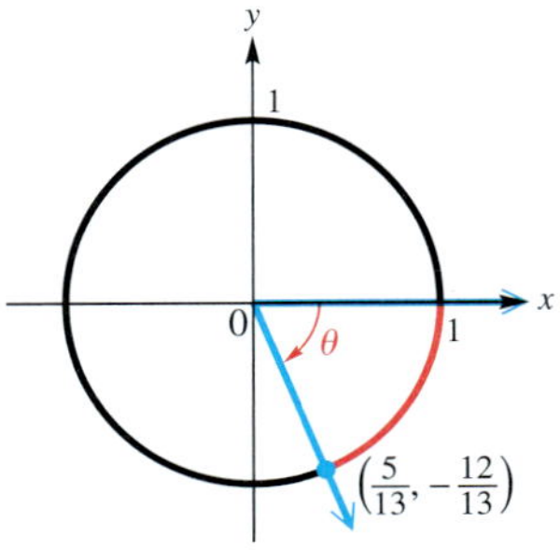

54.

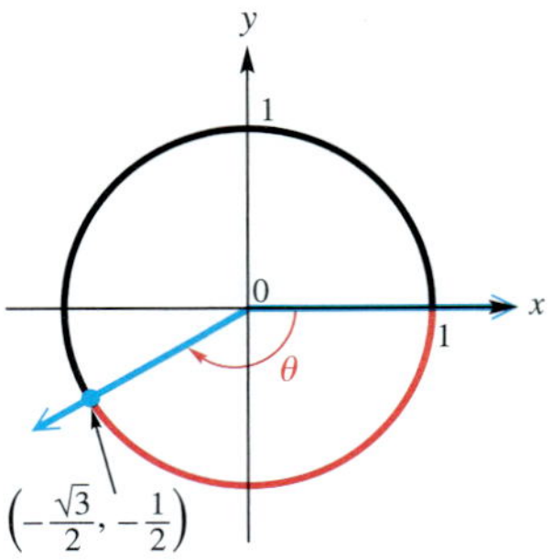

Find the value of s in the interval $\left[0, \frac{\pi}{2}\right]$ *that makes each statement true.* ***See Example 4(a).***

55. $\tan s = 0.2126$ **56.** $\cos s = 0.7826$ **57.** $\sin s = 0.9918$

58. $\cot s = 0.2994$ **59.** $\sec s = 1.0806$ **60.** $\csc s = 1.0219$

Find the exact value of s in the given interval that has the given circular function value. Do not use a calculator. ***See Example 4(b).***

61. $\left[\frac{\pi}{2}, \pi\right]; \quad \sin s = \frac{1}{2}$

62. $\left[\frac{\pi}{2}, \pi\right]; \quad \cos s = -\frac{1}{2}$

63. $\left[\pi, \frac{3\pi}{2}\right]; \quad \tan s = \sqrt{3}$

64. $\left[\pi, \frac{3\pi}{2}\right]; \quad \sin s = -\frac{1}{2}$

65. $\left[\frac{3\pi}{2}, 2\pi\right]; \quad \tan s = -1$

66. $\left[\frac{3\pi}{2}, 2\pi\right]; \quad \cos s = \frac{\sqrt{3}}{2}$

Find the exact values of s in the given interval that satisfy the given condition.

67. $[0, 2\pi); \quad \sin s = -\frac{\sqrt{3}}{2}$

68. $[0, 2\pi); \quad \cos s = -\frac{1}{2}$

69. $[0, 2\pi); \quad \cos^2 s = \frac{1}{2}$

70. $[0, 2\pi); \quad \tan^2 s = 3$

71. $[-2\pi, \pi); \quad 3 \tan^2 s = 1$

72. $[-\pi, \pi); \quad \sin^2 s = \frac{1}{2}$

Suppose an arc of length s lies on the unit circle $x^2 + y^2 = 1$, starting at the point $(1, 0)$ and terminating at the point (x, y). (See **Figure 12.***) Use a calculator to find the approximate coordinates for (x, y). (Hint: $x = \cos s$ and $y = \sin s$.)*

73. $s = 2.5$ **74.** $s = 3.4$ **75.** $s = -7.4$ **76.** $s = -3.9$

Concept Check For each value of s, use a calculator to find sin s and cos s, and then use the results to decide in which quadrant an angle of s radians lies.

77. $s = 51$ **78.** $s = 49$ **79.** $s = 65$ **80.** $s = 79$

Concept Check In Exercises 81 and 82, each graphing calculator screen shows a point on the unit circle. What is the length of the shortest arc of the circle from $(1, 0)$ to the point?

81.

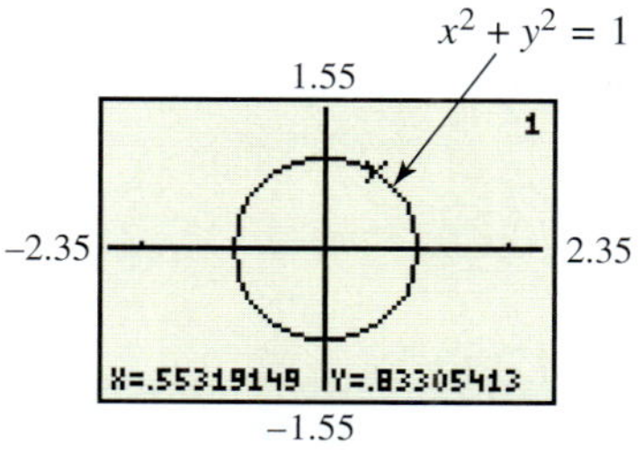

82.

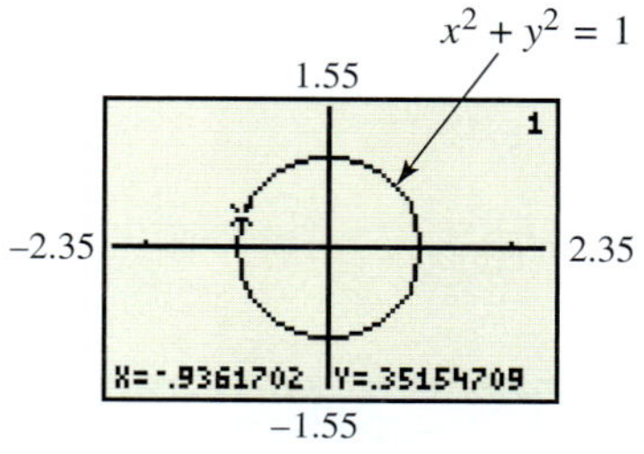

(Modeling) Solve each problem. ***See Example 5.***

83. *Elevation of the Sun* Refer to **Example 5.**

(a) Repeat the example for New Orleans, which has latitude $L = 30°$.

(b) Compare your answers. Do they agree with your intuition?

84. *Length of a Day* The number of daylight hours H at any location can be calculated using the formula

$$\cos(0.1309H) = -\tan D \tan L,$$

where D and L are defined in **Example 5.** Use this trigonometric equation to calculate the shortest and longest days in Minneapolis, Minnesota, if its latitude $L = 44.88°$, the shortest day occurs when $D = -23.44°$, and the longest day occurs when $D = 23.44°$. Remember to convert degrees to radians. (*Source:* Winter, C., R. Sizmann, and L. L. Vant-Hull, Editors, *Solar Power Plants,* Springer-Verlag.)

85. *Maximum Temperatures* Because the values of the circular functions repeat every 2π, they are used to describe things that repeat periodically. For example, the maximum afternoon temperature in a given city might be modeled by

$$t = 60 - 30 \cos\left(\frac{\pi}{6}x\right),$$

where t represents the maximum afternoon temperature in month x, with $x = 0$ representing January, $x = 1$ representing February, and so on. Find the maximum afternoon temperature for each of the following months.

(a) January **(b)** April

(c) May **(d)** June

(e) August **(f)** October

86. *Temperature in Fairbanks* Suppose the temperature in Fairbanks is modeled by

$$T(x) = 37 \sin\left[\frac{2\pi}{365}(x - 101)\right] + 25,$$

where $T(x)$ is the temperature in degrees Fahrenheit on day x, with $x = 1$ corresponding to January 1 and $x = 365$ corresponding to December 31. Use a calculator to estimate the temperature on the following days. (*Source:* Lando, B. and C. Lando, "Is the Graph of Temperature Variation a Sine Curve?," *The Mathematics Teacher,* vol. 70.)

(a) March 1 (day 60) **(b)** April 1 (day 91)

(c) Day 150 **(d)** June 15

(e) September 1 **(f)** October 31

In Exercises 87 and 88, ***see Example 6.***

87. Refer to **Figures 18 and 19.** Suppose that angle θ measures $60°$. Find the exact length of each segment.

(a) OQ **(b)** PQ **(c)** VR

(d) OV **(e)** OU **(f)** US

88. Refer to **Figures 18 and 19.** Repeat **Exercise 87** for $\theta = 38°$, but give lengths as approximations to four significant digits.

Chapter 3 Quiz (Sections 3.1–3.3)

Convert each degree measure to radians.

1. $225°$ **2.** $-330°$

Convert each radian measure to degrees.

3. $\frac{5\pi}{3}$ **4.** $-\frac{7\pi}{6}$

A central angle of a circle with radius 300 in. *intercepts an arc of* 450 in. *(These measures are accurate to the nearest inch.) Find each measure.*

5. the radian measure of the angle **6.** the area of the sector

Find each circular function value. Give exact values.

7. $\cos \frac{7\pi}{4}$ **8.** $\sin\left(-\frac{5\pi}{6}\right)$ **9.** $\tan 3\pi$

10. Find the exact value of s in the interval $\left[\frac{\pi}{2}, \pi\right]$ if $\sin s = \frac{\sqrt{3}}{2}$.

3.4 Linear and Angular Speed

- Linear Speed
- Angular Speed

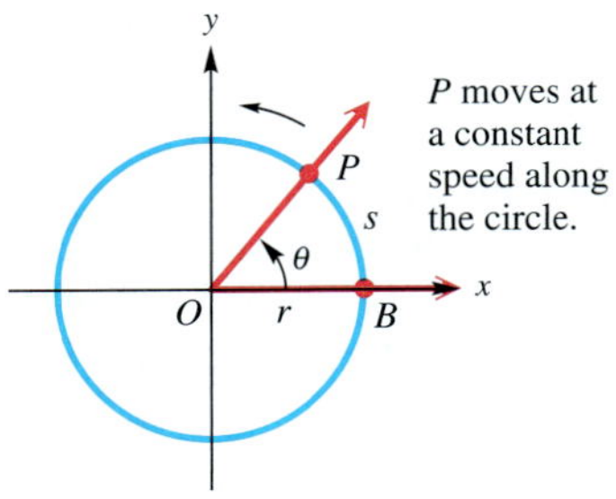

Figure 20

Linear Speed There are situations when we need to know how fast a point on a circular disk is moving or how fast the central angle of such a disk is changing. Some examples occur with machinery involving gears or pulleys or the speed of a car around a curved portion of highway.

Suppose that point P moves at a constant speed along a circle of radius r and center O. See **Figure 20.** The measure of how fast the position of P is changing is the **linear speed.** If v represents linear speed, then

$$\text{speed} = \frac{\text{distance}}{\text{time}}, \quad \text{or} \quad \boldsymbol{v = \frac{s}{t}},$$

where s is the length of the arc traced by point P at time t. (This formula is just a restatement of $r = \frac{d}{t}$ with s as distance, v as rate (speed), and t as time.)

Angular Speed Refer to **Figure 20** at the bottom of the preceding page. As point P in the figure moves along the circle, ray OP rotates around the origin. Since ray OP is the terminal side of angle POB, the measure of the angle changes as P moves along the circle. The measure of how fast angle POB is changing is its **angular speed.** Angular speed, symbolized ω, is given as

$$\boldsymbol{\omega = \frac{\theta}{t},} \quad \textbf{where } \boldsymbol{\theta} \textbf{ is in radians.}$$

Here θ is the measure of angle POB at time t. ***As with earlier formulas in this chapter, θ must be measured in radians, with ω expressed in radians per unit of time.***

In **Section 3.2,** the length s of the arc intercepted on a circle of radius r by a central angle of measure θ radians was found to be $s = r\theta$. Using this formula, the formula for linear speed, $v = \frac{s}{t}$, becomes

$$v = \frac{s}{t} \qquad \text{Formula for linear speed}$$

$$= \frac{r\theta}{t} \qquad s = r\theta$$

$$= r \cdot \frac{\theta}{t}$$

$$v = r\omega. \qquad \omega = \frac{\theta}{t}$$

The formulas for angular and linear speed are summarized in the table.

Angular Speed ω	Linear Speed v
$\omega = \frac{\theta}{t}$ (ω in radians per unit time t, θ in radians)	$v = \frac{s}{t}$ $v = \frac{r\theta}{t}$ $v = r\omega$

As an example of linear and angular speeds, consider the following. The human joint that can be flexed the fastest is the wrist, which can rotate through 90°, or $\frac{\pi}{2}$ radians, in 0.045 sec while holding a tennis racket. The angular speed of a human wrist swinging a tennis racket is

$$\omega = \frac{\theta}{t} \qquad \text{Formula for angular speed}$$

$$= \frac{\frac{\pi}{2}}{0.045} \qquad \text{Substitute.}$$

$$\omega \approx 35 \text{ radians per sec.} \qquad \text{Use a calculator.}$$

If the radius (distance) from the tip of the racket to the wrist joint is 2 ft, then the speed at the tip of the racket is

$$v = r\omega \qquad \text{Formula for linear speed}$$

$$\approx 2(35) \qquad \text{Substitute.}$$

$$v = 70 \text{ ft per sec,} \quad \text{or} \quad \text{about 48 mph.} \qquad \text{Use a calculator.}$$

In a tennis serve the arm rotates at the shoulder, so the final speed of the racket is considerably greater. (*Source:* Cooper, J. and R. Glassow, *Kinesiology,* Second Edition, C.V. Mosby.)

EXAMPLE 1 Using Linear and Angular Speed Formulas

Suppose that point P is on a circle with radius 10 cm, and ray OP is rotating with angular speed $\frac{\pi}{18}$ radian per sec.

(a) Find the angle generated by P in 6 sec.

(b) Find the distance traveled by P along the circle in 6 sec.

(c) Find the linear speed of P in centimeters per second.

SOLUTION

(a) The speed of ray OP is $\omega = \frac{\pi}{18}$ radian per sec. Use $\omega = \frac{\theta}{t}$ and $t = 6$ sec.

$$\frac{\pi}{18} = \frac{\theta}{6}$$ Let $\omega = \frac{\pi}{18}$ and $t = 6$ in the angular speed formula.

$$\theta = \frac{6\pi}{18}, \quad \text{or} \quad \frac{\pi}{3} \text{ radians}$$ Solve for θ.

(b) From part (a), P generates an angle of $\frac{\pi}{3}$ radians in 6 sec. The distance traveled by P along the circle is found as follows.

$$s = r\theta = 10\left(\frac{\pi}{3}\right) = \frac{10\pi}{3} \text{ cm}$$ (Section 3.2)

(c) From part (b), $s = \frac{10\pi}{3}$ cm for 6 sec, so for 1 sec we divide by 6.

$$v = \frac{s}{t} = \frac{\frac{10\pi}{3}}{6} = \frac{10\pi}{3} \div 6 = \frac{10\pi}{3} \cdot \frac{1}{6} = \frac{5\pi}{9} \text{ cm per sec}$$

Be careful simplifying this complex fraction.

✔ *Now Try Exercise 3.*

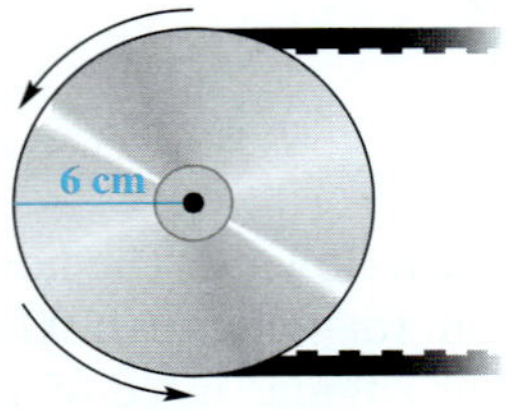

Figure 21

EXAMPLE 2 Finding Angular Speed of a Pulley and Linear Speed of a Belt

A belt runs a pulley of radius 6 cm at 80 revolutions per min. See **Figure 21.**

(a) Find the angular speed of the pulley in radians per second.

(b) Find the linear speed of the belt in centimeters per second.

SOLUTION

(a) In 1 min, the pulley makes 80 revolutions. Each revolution is 2π radians.

$$80(2\pi) = 160\pi \text{ radians per min}$$

Since there are 60 sec in 1 min, we find ω, the angular speed in radians per second, by dividing 160π by 60.

$$\omega = \frac{160\pi}{60} = \frac{8\pi}{3} \text{ radians per sec}$$

(b) The linear speed v of the belt will be the same as that of a point on the circumference of the pulley.

$$v = r\omega = 6\left(\frac{8\pi}{3}\right)$$ Let $r = 6$ and $\omega = \frac{8\pi}{3}$.

$$= 16\pi$$ Multiply.

$$\approx 50 \text{ cm per sec}$$ Approximate.

✔ *Now Try Exercise 41.*

EXAMPLE 3 Finding Linear Speed and Distance Traveled by a Satellite

A satellite traveling in a circular orbit 1600 km above the surface of Earth takes 2 hr to make an orbit. The radius of Earth is approximately 6400 km. See **Figure 22.**

(a) Approximate the linear speed of the satellite in kilometers per hour.

(b) Approximate the distance the satellite travels in 4.5 hr.

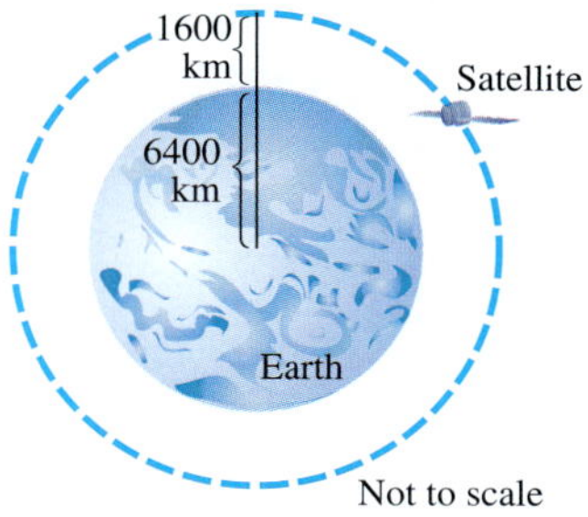

Figure 22

SOLUTION

(a) The distance of the satellite from the center of Earth is approximately

$$r = 1600 + 6400 = 8000 \text{ km.}$$

For one orbit, $\theta = 2\pi$, and

$$s = r\theta = 8000(2\pi) \text{ km.} \qquad \text{Let } r = 8000.$$

Since it takes 2 hr to complete an orbit, the linear speed is approximated as follows.

$$v = \frac{s}{t}$$

$$v = \frac{8000(2\pi)}{2} \qquad \text{Let } s = 8000(2\pi) \text{ and } t = 2.$$

$$v = 8000\pi \qquad \text{Simplify.}$$

$$v \approx 25{,}000 \text{ km per hr} \qquad \text{Approximate.}$$

(b) To approximate the distance traveled by the satellite, we use $s = vt$, which is similar to the distance formula $d = rt$.

$$s = vt$$

$$s = 8000\pi(4.5) \qquad \text{Let } v = 8000\pi \text{ and } t = 4.5.$$

$$s = 36{,}000\pi \qquad \text{Multiply.}$$

$$s \approx 110{,}000 \text{ km} \qquad \text{Approximate.}$$

✔ ***Now Try Exercise 39.***

3.4 Exercises

Concept Check Refer to the figure and answer Exercises 1 and 2.

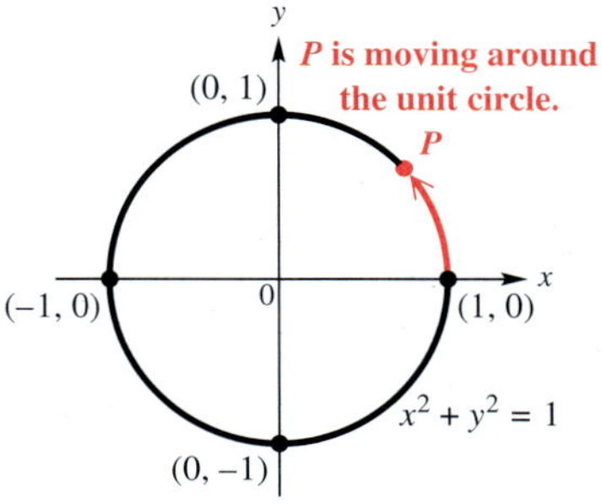

1. If the point P moves around the circumference of the unit circle at an angular velocity of 1 radian per sec, how long will it take for P to move around the entire circle?

2. If the point P moves around the circumference of the unit circle at a speed of 1 unit per sec, how long will it take for P to move around the entire circle?

Suppose that point P is on a circle with radius r, and ray OP is rotating with angular speed ω. For the given values of r, ω, and t, find each of the following. **See Example 1.**

(a) *the angle generated by P in time t*
(b) *the distance traveled by P along the circle in time t*
(c) *the linear speed of P*

3. $r = 20$ cm, $\omega = \frac{\pi}{12}$ radian per sec, $t = 6$ sec

4. $r = 30$ cm, $\omega = \frac{\pi}{10}$ radian per sec, $t = 4$ sec

Use the formula $\omega = \frac{\theta}{t}$ to find the value of the missing variable.

5. $\omega = \frac{2\pi}{3}$ radians per sec, $t = 3$ sec

6. $\omega = \frac{\pi}{4}$ radian per min, $t = 5$ min

7. $\theta = \frac{3\pi}{4}$ radians, $t = 8$ sec

8. $\theta = \frac{2\pi}{5}$ radians, $t = 10$ sec

9. $\theta = \frac{2\pi}{9}$ radian, $\omega = \frac{5\pi}{27}$ radian per min

10. $\theta = \frac{3\pi}{8}$ radians, $\omega = \frac{\pi}{24}$ radian per min

11. $\theta = 3.871$ radians, $t = 21.47$ sec

12. $\theta = 5.225$ radians, $t = 2.515$ sec

13. $\omega = 0.9067$ radian per min, $t = 11.88$ min

14. $\omega = 4.316$ radians per min, $t = 4.752$ min

Use the formula $v = r\omega$ to find the value of the missing variable.

15. $r = 12$ m, $\omega = \frac{2\pi}{3}$ radians per sec

16. $r = 8$ cm, $\omega = \frac{9\pi}{5}$ radians per sec

17. $v = 9$ m per sec, $r = 5$ m

18. $v = 18$ ft per sec, $r = 3$ ft

19. $v = 107.7$ m per sec, $r = 58.74$ m

20. $r = 24.93$ cm, $\omega = 0.3729$ radian per sec

The formula $\omega = \frac{\theta}{t}$ can be rewritten as $\theta = \omega t$. Substituting ωt for θ converts $s = r\theta$ to $s = r\omega t$. Use the formula $s = r\omega t$ to find the value of the missing variable.

21. $r = 6$ cm, $\omega = \frac{\pi}{3}$ radians per sec, $t = 9$ sec

22. $r = 9$ yd, $\omega = \frac{2\pi}{5}$ radians per sec, $t = 12$ sec

23. $s = 6\pi$ cm, $r = 2$ cm, $\omega = \frac{\pi}{4}$ radian per sec

24. $s = \frac{12\pi}{5}$ m, $r = \frac{3}{2}$ m, $\omega = \frac{2\pi}{5}$ radians per sec

25. $s = \frac{3\pi}{4}$ km, $r = 2$ km, $t = 4$ sec

26. $s = \frac{8\pi}{9}$ m, $r = \frac{4}{3}$ m, $t = 12$ sec

Find the angular speed ω for each of the following.

27. the hour hand of a clock

28. the second hand of a clock

29. the minute hand of a clock

30. a line from the center to the edge of a CD revolving 300 times per min

Find the linear speed v for each of the following.

31. the tip of the minute hand of a clock, if the hand is 7 cm long

32. the tip of the second hand of a clock, if the hand is 28 mm long

33. a point on the edge of a flywheel of radius 2 m, rotating 42 times per min

34. a point on the tread of a tire of radius 18 cm, rotating 35 times per min

35. the tip of a propeller 3 m long, rotating 500 times per min (*Hint:* $r = 1.5$ m)

36. a point on the edge of a gyroscope of radius 83 cm, rotating 680 times per min

Solve each problem. ***See Examples 1–3.***

37. ***Speed of a Bicycle*** The tires of a bicycle have radius 13.0 in. and are turning at the rate of 215 revolutions per min. See the figure. How fast is the bicycle traveling in miles per hour? (*Hint:* 5280 ft = 1 mi)

38. ***Hours in a Martian Day*** Mars rotates on its axis at the rate of about 0.2552 radian per hr. Approximately how many hours are in a Martian day (or *sol*)? (*Source: World Almanac and Book of Facts.*)

Opposite sides of Mars

39. ***Angular and Linear Speeds of Earth*** The orbit of Earth about the sun is almost circular. Assume that the orbit is a circle with radius 93,000,000 mi. Its angular and linear speeds are used in designing solar-power facilities.

(a) Assume that a year is 365 days, and find the angle formed by Earth's movement in one day.
(b) Give the angular speed in radians per hour.
(c) Find the linear speed of Earth in miles per hour.

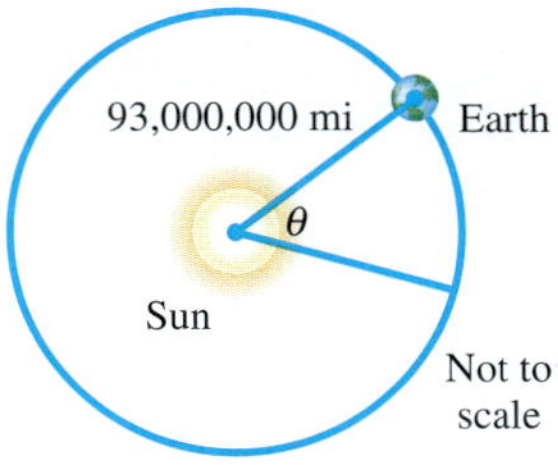

40. ***Angular and Linear Speeds of Earth*** Earth revolves on its axis once every 24 hr. Assuming that Earth's radius is 6400 km, find the following.

(a) angular speed of Earth in radians per day and radians per hour
(b) linear speed at the North Pole or South Pole
(c) linear speed at Quito, Ecuador, a city on the equator
(d) linear speed at Salem, Oregon (halfway from the equator to the North Pole)

41. ***Speeds of a Pulley and a Belt*** The pulley shown has a radius of 12.96 cm. Suppose it takes 18 sec for 56 cm of belt to go around the pulley.

(a) Find the linear speed of the belt in centimeters per second.
(b) Find the angular speed of the pulley in radians per second.

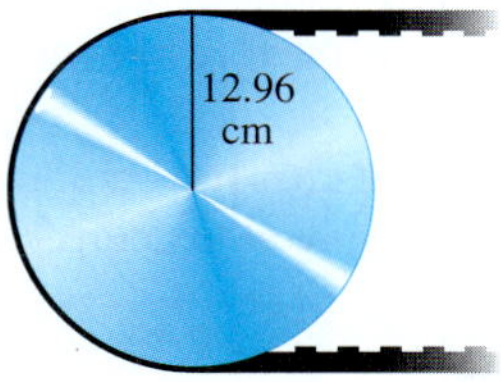

42. *Angular Speeds of Pulleys* The two pulleys in the figure have radii of 15 cm and 8 cm, respectively. The larger pulley rotates 25 times in 36 sec. Find the angular speed of each pulley in radians per second.

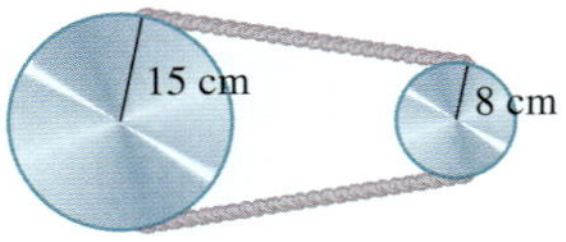

43. *Radius of a Spool of Thread* A thread is being pulled off a spool at the rate of 59.4 cm per sec. Find the radius of the spool if it makes 152 revolutions per min.

44. *Time to Move along a Railroad Track* A railroad track is laid along the arc of a circle of radius 1800 ft. The circular part of the track subtends a central angle of 40°. How long (in seconds) will it take a point on the front of a train traveling 30.0 mph to go around this portion of the track?

45. *Angular Speed of a Motor Propeller* The propeller of a 90-horsepower outboard motor at full throttle rotates at exactly 5000 revolutions per min. Find the angular speed of the propeller in radians per second.

46. *Linear Speed of a Golf Club* The shoulder joint can rotate at 25.0 radians per sec. If a golfer's arm is straight and the distance from the shoulder to the club head is 5.00 ft, find the linear speed of the club head from shoulder rotation. (*Source:* Cooper, J. and R. Glassow, *Kinesiology,* Second Edition, C.V. Mosby.)

Chapter 3 Test Prep

Key Terms

3.1 radian circumference **3.2** latitude sector of a circle longitude	subtend degree of curvature nautical mile statute mile	**3.3** unit circle circular functions reference arc	**3.4** linear speed v angular speed ω

Quick Review

Concepts | **Examples**

3.1 Radian Measure

An angle with its vertex at the center of a circle that intercepts an arc on the circle equal in length to the radius of the circle has a measure of **1 radian.**

Degree/Radian Relationship $\mathbf{180^\circ = \pi}$ **radians**

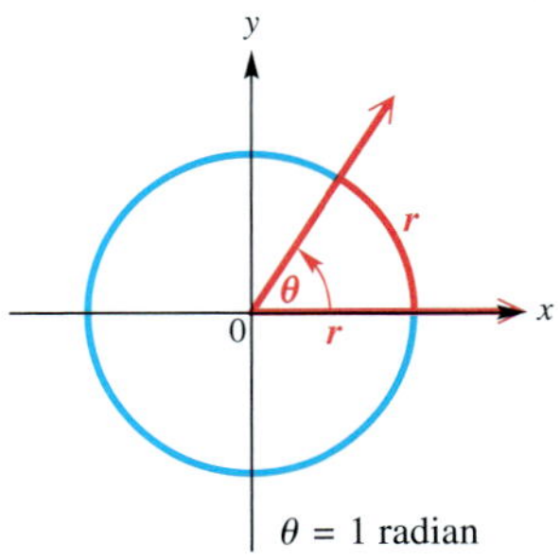

$\theta = 1$ radian

Converting between Degrees and Radians

1. Multiply a degree measure by $\frac{\pi}{180}$ radian and simplify to convert to radians.

2. Multiply a radian measure by $\frac{180^\circ}{\pi}$ and simplify to convert to degrees.

Convert 135° to radians.

$$135^\circ = 135\left(\frac{\pi}{180}\text{ radian}\right) = \frac{3\pi}{4}\text{ radians}$$

Convert $-\frac{5\pi}{3}$ radians to degrees.

$$-\frac{5\pi}{3}\text{ radians} = -\frac{5\pi}{3}\left(\frac{180^\circ}{\pi}\right) = -300^\circ$$

Concepts | **Examples**

3.2 Applications of Radian Measure

Arc Length

The length s of the arc intercepted on a circle of radius r by a central angle of measure θ radians is given by the product of the radius and the radian measure of the angle.

$$s = r\theta, \quad \text{where } \theta \text{ is in radians}$$

Find the central angle θ in the figure.

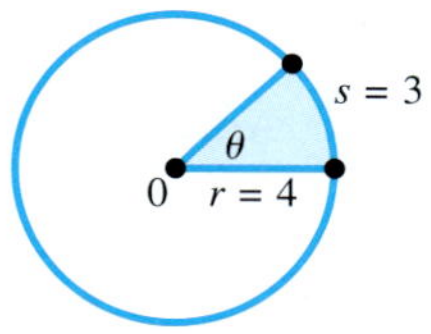

$$\theta = \frac{s}{r} = \frac{3}{4} \text{ radian}$$

Area of a Sector

The area $\mathscr{A}$ of a sector of a circle of radius r and central angle θ is given by the following formula.

$$\mathscr{A} = \frac{1}{2}r^2\theta, \quad \text{where } \theta \text{ is in radians}$$

Find the area $\mathscr{A}$ of the sector in the figure above.

$$\mathscr{A} = \frac{1}{2}(4)^2\left(\frac{3}{4}\right) = 6 \text{ sq units}$$

3.3 The Unit Circle and Circular Functions

Circular Functions

Start at the point $(1, 0)$ on the unit circle $x^2 + y^2 = 1$ and measure off an arc of length $|s|$ along the circle, going counterclockwise if s is positive and clockwise if s is negative. Let the endpoint of the arc be at the point (x, y). The six circular functions of s are defined as follows. (Assume that no denominators are 0.)

$$\sin s = y \qquad \cos s = x \qquad \tan s = \frac{y}{x}$$

$$\csc s = \frac{1}{y} \qquad \sec s = \frac{1}{x} \qquad \cot s = \frac{x}{y}$$

The Unit Circle

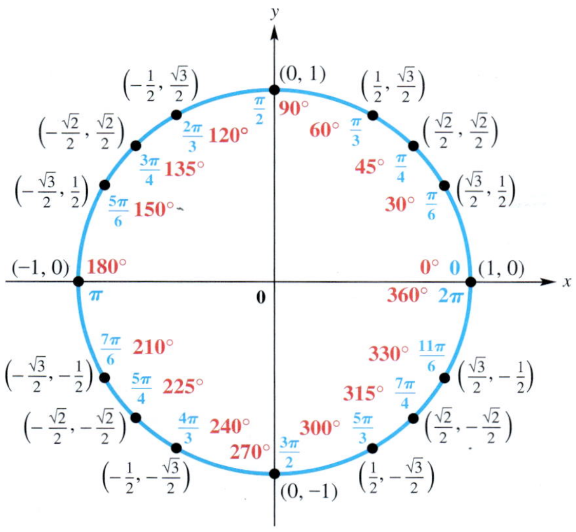

The unit circle $x^2 + y^2 = 1$

Use the unit circle to find each value.

$$\sin\frac{5\pi}{6} = \frac{1}{2}$$

$$\cos\frac{3\pi}{2} = 0$$

$$\tan\frac{\pi}{4} = \frac{\frac{\sqrt{2}}{2}}{\frac{\sqrt{2}}{2}} = 1$$

$$\csc\frac{7\pi}{4} = \frac{1}{-\frac{\sqrt{2}}{2}} = -\sqrt{2}$$

$$\sec\frac{7\pi}{6} = \frac{1}{-\frac{\sqrt{3}}{2}} = -\frac{2\sqrt{3}}{3}$$

$$\cot\frac{\pi}{3} = \frac{\frac{1}{2}}{\frac{\sqrt{3}}{2}} = \frac{\sqrt{3}}{3}$$

$$\sin 0 = 0$$

$$\cos\frac{\pi}{2} = 0$$

Find the value of s in $\left[0, \frac{\pi}{2}\right]$ that makes $\cos s = \frac{\sqrt{3}}{2}$ true.

In $\left[0, \frac{\pi}{2}\right]$, the arc length $s = \frac{\pi}{6}$ is associated with the point $\left(\frac{\sqrt{3}}{2}, \frac{1}{2}\right)$. The first coordinate is

$$\cos s = \cos\frac{\pi}{6} = \frac{\sqrt{3}}{2},$$

so $s = \frac{\pi}{6}$ makes the statement true.

(continued)

Concepts	Examples

3.4 Linear and Angular Speed

Formulas for Angular and Linear Speed

Angular Speed ω	Linear Speed v
$\omega = \frac{\theta}{t}$ (ω in radians per unit time t, θ in radians)	$v = \frac{s}{t}$ $v = \frac{r\theta}{t}$ $v = r\omega$

A belt runs a machine pulley of radius 8 in. at 60 revolutions per min. Find each of the following.

(a) the angular speed ω in radians per minute

$$\omega = 60(2\pi)$$
$$= 120\pi \text{ radians per min}$$

(b) the linear speed v of the belt in inches per minute

$$v = r\omega$$
$$= 8(120\pi)$$
$$= 960\pi \text{ in. per min}$$

Chapter 3 Review Exercises

1. *Concept Check* What is the meaning of "an angle with measure 2 radians"?

2. *Concept Check* Consider each angle in standard position having the given radian measure. In what quadrant does the terminal side lie?

(a) 3 **(b)** 4 **(c)** -2 **(d)** 7

3. Find three angles coterminal with an angle of 1 radian.

4. Give an expression that generates all angles coterminal with an angle of $\frac{\pi}{6}$ radian. Let n represent any integer.

Convert each degree measure to radians. Leave answers as multiples of π.

5. 45° **6.** 120° **7.** 175°

8. 330° **9.** 800° **10.** 1020°

Convert each radian measure to degrees.

11. $\frac{5\pi}{4}$ **12.** $\frac{9\pi}{10}$ **13.** $\frac{8\pi}{3}$

14. $\frac{6\pi}{5}$ **15.** $-\frac{11\pi}{18}$ **16.** $-\frac{21\pi}{5}$

Suppose the tip of the minute hand of a clock is 2 in. *from the center of the clock. For each duration, determine the distance traveled by the tip of the minute hand.*

17. 15 min

18. 20 min

19. 3 hr

Solve each problem. Use a calculator as necessary.

20. *Diameter of the Moon* The distance to the moon is approximately 238,900 mi. Use the arc length formula to estimate the diameter d of the moon if angle θ in the figure is measured to be $0.5170°$.

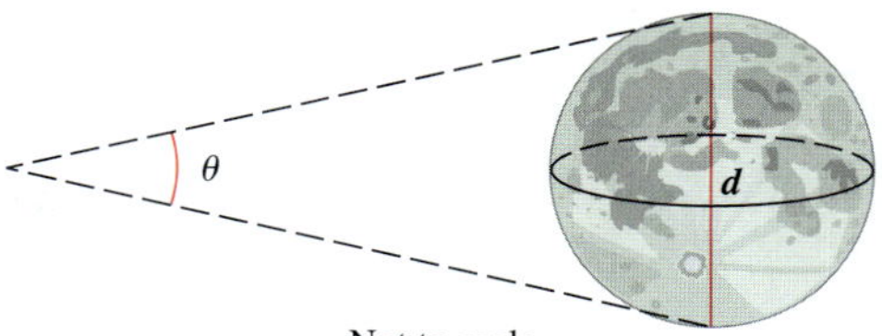

Not to scale

21. *Arc Length* The radius of a circle is 15.2 cm. Find the length of an arc of the circle intercepted by a central angle of $\frac{3\pi}{4}$ radians.

22. *Arc Length* Find the length of an arc intercepted by a central angle of 0.769 radian on a circle with radius 11.4 cm.

23. *Angle Measure* Find the measure (in degrees) of a central angle that intercepts an arc of length 7.683 cm in a circle of radius 8.973 cm.

24. *Area of a Sector* A central angle of $\frac{7\pi}{4}$ radians forms a sector of a circle. Find the area of the sector if the radius of the circle is 28.69 in.

25. *Area of a Sector* Find the area of a sector of a circle having a central angle of $21° 40'$ in a circle of radius 38.0 m.

26. *Concept Check* Use the formulas $s = r\theta$ and $\mathscr{A} = \frac{1}{2}r^2\theta$ to express $\mathscr{A}$ in terms of s and θ.

Distance between Cities *Assume that the radius of Earth is* 6400 km.

27. Find the distance in kilometers between cities on a north-south line that are on latitudes 28° N and 12° S, respectively.

28. Two cities on the equator have longitudes of 72° E and 35° W, respectively. Find the distance between the cities.

Concept Check *Find the measure of the central angle θ (in radians) and the area of the sector.*

29.

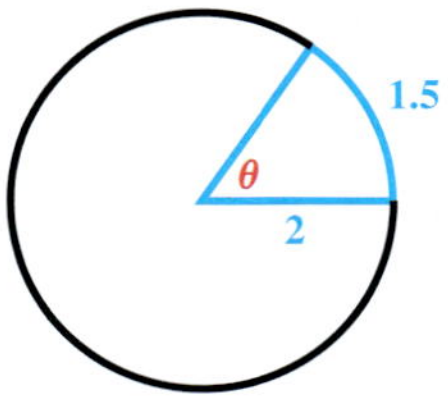

30.

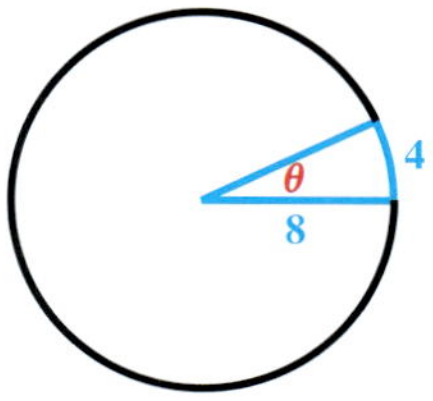

31. *Concept Check* The hour hand of a wall clock measures 6 in. from its tip to the center of the clock.

(a) Through what angle (in radians) does the hour hand pass between 1 o'clock and 3 o'clock?

(b) What distance does the tip of the hour hand travel during the time period from 1 o'clock to 3 o'clock?

32. Describe what would happen to the central angle for a given arc length of a circle if the circle's radius were doubled. (Assume everything else is unchanged.)

Find each exact function value. Do not use a calculator.

33. $\tan \frac{\pi}{3}$ **34.** $\cos \frac{2\pi}{3}$ **35.** $\sin\left(-\frac{5\pi}{6}\right)$

36. $\tan\left(-\frac{7\pi}{3}\right)$ **37.** $\csc\left(-\frac{11\pi}{6}\right)$ **38.** $\cot(-13\pi)$

Without using a calculator, determine which of the following is greater.

39. tan 1 or tan 2 **40.** sin 1 or tan 1 **41.** cos 2 or sin 2

42. ***Concept Check*** Match each domain in Column II with the appropriate circular function pair in Column I.

I	II
(a) sine and cosine	**A.** $(-\infty, \infty)$
(b) tangent and secant	**B.** $\{s \mid s \neq n\pi$, where n is any integer$\}$
(c) cotangent and cosecant	**C.** $\{s \mid s \neq (2n+1)\frac{\pi}{2}$, where n is any integer$\}$

Use a calculator to find an approximation for each circular function value. Be sure your calculator is set in radian mode.

43. sin 1.0472 **44.** tan 1.2275 **45.** $\cos(-0.2443)$

46. cot 3.0543 **47.** sec 7.3159 **48.** csc 4.8386

Find the value of s in the interval $\left[0, \frac{\pi}{2}\right]$ *that makes each statement true.*

49. $\cos s = 0.9250$ **50.** $\tan s = 4.0112$ **51.** $\sin s = 0.4924$

52. $\csc s = 1.2361$ **53.** $\cot s = 0.5022$ **54.** $\sec s = 4.5600$

Find the exact value of s in the given interval that has the given circular function value. Do not use a calculator.

55. $\left[0, \frac{\pi}{2}\right]$; $\cos s = \frac{\sqrt{2}}{2}$ **56.** $\left[\frac{\pi}{2}, \pi\right]$; $\tan s = -\sqrt{3}$

57. $\left[\pi, \frac{3\pi}{2}\right]$; $\sec s = -\frac{2\sqrt{3}}{3}$ **58.** $\left[\frac{3\pi}{2}, 2\pi\right]$; $\sin s = -\frac{1}{2}$

Solve each problem, where t, ω, θ, and s are as defined in ***Section 3.4.***

59. Find t if $\theta = \frac{5\pi}{12}$ radians and $\omega = \frac{8\pi}{9}$ radians per sec.

60. Find θ if $t = 12$ sec and $\omega = 9$ radians per sec.

61. Find ω if $t = 8$ sec and $\theta = \frac{2\pi}{5}$ radians.

62. Find s if $r = 11.46$ cm, $\omega = 4.283$ radians per sec, and $t = 5.813$ sec.

Solve each problem.

63. ***Linear Speed of a Flywheel*** Find the linear speed of a point on the edge of a flywheel of radius 7 cm if the flywheel is rotating 90 times per sec.

64. ***Angular Speed of a Ferris Wheel*** A Ferris wheel has radius 25 ft. If it takes 30 sec for the wheel to turn $\frac{5\pi}{6}$ radians, what is the angular speed of the wheel?

65. *(Modeling) Archaeology* An archaeology professor believes that an unearthed fragment is a piece of the edge of a circular ceremonial plate and uses a formula that will give the radius of the original plate using measurements from the fragment, shown in **Figure A.** Measurements are in inches.

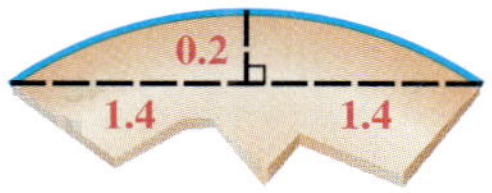

Figure A

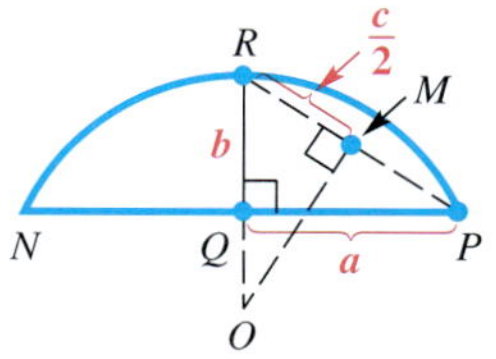

Figure B

In **Figure B,** a is $\frac{1}{2}$ the length of chord NP, and b is the distance from the midpoint of chord NP to the circle. According to the formula, the radius r of the circle, OR, is given by

$$r = \frac{a^2 + b^2}{2b}.$$

What is the radius of the original plate from which the fragment came?

66. *(Modeling) Phase Angle of the Moon* Because the moon orbits Earth, we observe different phases of the moon during the period of a month. In the figure, t is the **phase angle.**

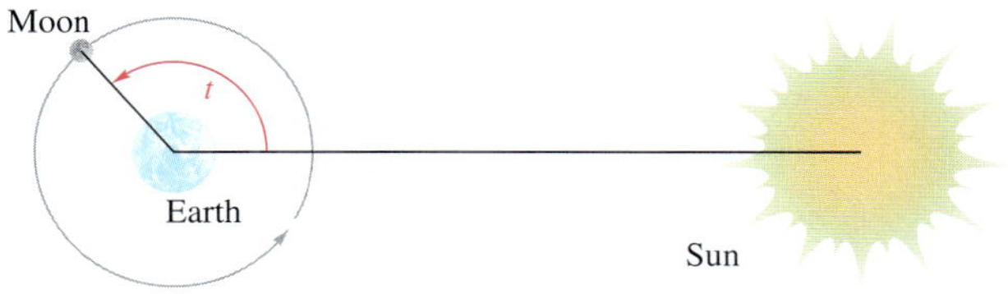

The **phase** F of the moon is modeled by

$$F(t) = \frac{1}{2}(1 - \cos t)$$

and gives the fraction of the moon's face that is illuminated by the sun. (*Source:* Duffet-Smith, P., *Practical Astronomy with Your Calculator,* Cambridge University Press.) Evaluate each expression and interpret the result.

(a) $F(0)$ **(b)** $F\left(\frac{\pi}{2}\right)$ **(c)** $F(\pi)$ **(d)** $F\left(\frac{3\pi}{2}\right)$

Chapter 3 Test

Convert each degree measure to radians.

1. $120°$ **2.** $-45°$ **3.** $5°$ (to the nearest hundredth)

Convert each radian measure to degrees.

4. $\frac{3\pi}{4}$ **5.** $-\frac{7\pi}{6}$ **6.** 4 (to the nearest hundredth)

7. A central angle of a circle with radius 150 cm intercepts an arc of 200 cm. Find each measure.

(a) the radian measure of the angle

(b) the area of a sector with that central angle

8. *Rotation of Gas Gauge Arrow* The arrow on a car's gasoline gauge is $\frac{1}{2}$ in. long. See the figure. Through what angle does the arrow rotate when it moves 1 in. on the gauge?

Find each circular function value.

9. $\sin \frac{3\pi}{4}$ **10.** $\cos\left(-\frac{7\pi}{6}\right)$ **11.** $\tan \frac{3\pi}{2}$

12. $\sec \frac{8\pi}{3}$ **13.** $\tan \pi$ **14.** $\cos \frac{3\pi}{2}$

15. Determine the six exact circular function values of s in the figure.

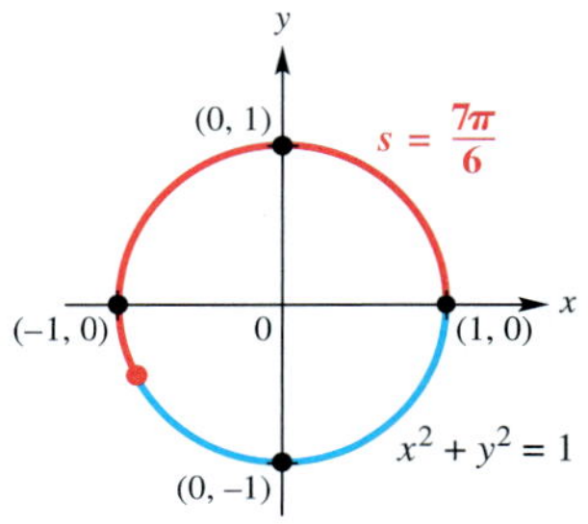

16. Give the domains of the six circular functions.

17. **(a)** Use a calculator to approximate s in the interval $\left[0, \frac{\pi}{2}\right]$ if $\sin s = 0.8258$.

(b) Find the exact value of s in the interval $\left[0, \frac{\pi}{2}\right]$ if $\cos s = \frac{1}{2}$.

18. *Angular and Linear Speed of a Point* Suppose that point P is on a circle with radius 60 cm, and ray OP is rotating with angular speed $\frac{\pi}{12}$ radian per sec.

(a) Find the angle generated by P in 8 sec.

(b) Find the distance traveled by P along the circle in 8 sec.

(c) Find the linear speed of P.

19. *Orbital Speed of Jupiter* It takes Jupiter 11.86 yr to complete one orbit around the sun. See the figure. If Jupiter's average distance from the sun is 483,800,000 mi, find its orbital speed (speed along its orbital path) in miles per second. (*Source: World Almanac and Book of Facts.*)

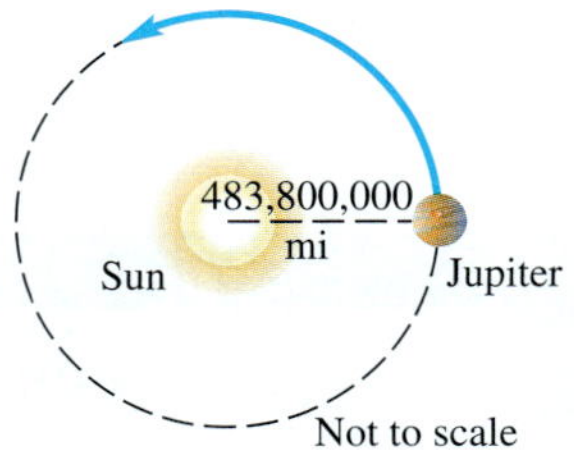

20. *Ferris Wheel* A Ferris wheel has radius 50.0 ft. A person takes a seat and then the wheel turns $\frac{2\pi}{3}$ radians.

(a) How far is the person above the ground?

(b) If it takes 30 sec for the wheel to turn $\frac{2\pi}{3}$ radians, what is the angular speed of the wheel?

4 Graphs of the Circular Functions

Phenomena that repeat in a regular pattern, such as average monthly temperature, rotation of a planet on its axis, and high and low tides, can be modeled by *periodic functions*.

4.1 Graphs of the Sine and Cosine Functions

- Periodic Functions
- Graph of the Sine Function
- Graph of the Cosine Function
- Graphing Techniques, Amplitude, and Period
- Connecting Graphs with Equations
- Using a Trigonometric Model

Periodic Functions Many things in daily life repeat with a predictable pattern, such as weather, tides, and hours of daylight. Because the sine and cosine functions repeat their values in a regular pattern, they are *periodic functions.* **Figure 1** shows a periodic graph that represents a normal heartbeat.

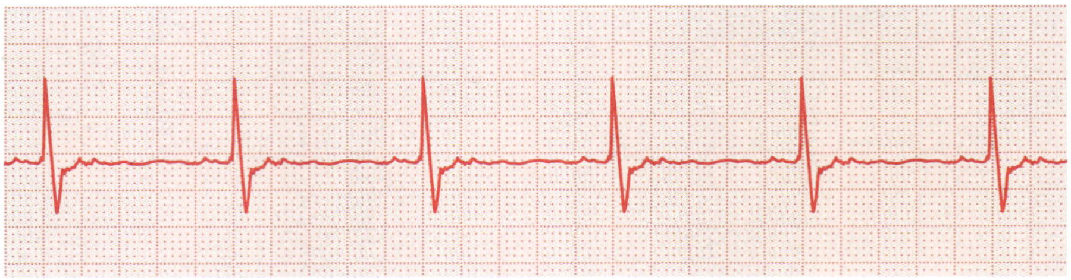

Figure 1

Periodic Function

A **periodic function** is a function f such that

$$f(x) = f(x + np),$$

for every real number x in the domain of f, every integer n, and some positive real number p. The least possible positive value of p is the **period** of the function.

LOOKING AHEAD TO CALCULUS
Periodic functions are used throughout calculus, so you will need to know their characteristics. One use of these functions is to describe the location of a point in the plane using **polar coordinates,** an alternative to rectangular coordinates. (See **Chapter 8.**)

The circumference of the unit circle is 2π, so the least value of p for which the sine and cosine functions repeat is 2π. ***Therefore, the sine and cosine functions are periodic functions with period 2π,*** and the following statements are true for every integer n.

$$\sin x = \sin(x + n \cdot 2\pi) \quad \text{and} \quad \cos x = \cos(x + n \cdot 2\pi)$$

Graph of the Sine Function In **Section 3.3** we saw that for a real number s, the point on the unit circle corresponding to s has coordinates $(\cos s, \sin s)$. See **Figure 2.** Trace along the circle to verify the results shown in the table.

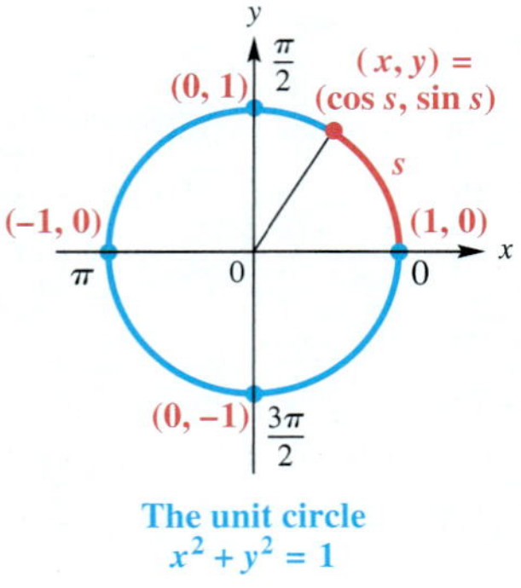

Figure 2

As s Increases from	$\sin s$	$\cos s$
0 to $\frac{\pi}{2}$	Increases from 0 to 1	Decreases from 1 to 0
$\frac{\pi}{2}$ to π	Decreases from 1 to 0	Decreases from 0 to -1
π to $\frac{3\pi}{2}$	Decreases from 0 to -1	Increases from -1 to 0
$\frac{3\pi}{2}$ to 2π	Increases from -1 to 0	Increases from 0 to 1

To avoid confusion when graphing the sine function, we use x rather than s; this corresponds to the letters in the xy-coordinate system. Selecting key values of x and finding the corresponding values of $\sin x$ leads to the table in **Figure 3.**

To obtain the traditional graph in **Figure 3,** we plot the points from the table, use symmetry, and join them with a smooth curve. Since $y = \sin x$ is periodic with period 2π and has domain $(-\infty, \infty)$, the graph continues in the same pattern in both directions. This graph is called a **sine wave,** or **sinusoid.**

Sine Function $f(x) = \sin x$

Domain: $(-\infty, \infty)$ Range: $[-1, 1]$

x	y
0	0
$\frac{\pi}{6}$	$\frac{1}{2}$
$\frac{\pi}{4}$	$\frac{\sqrt{2}}{2}$
$\frac{\pi}{3}$	$\frac{\sqrt{3}}{2}$
$\frac{\pi}{2}$	1
π	0
$\frac{3\pi}{2}$	-1
2π	0

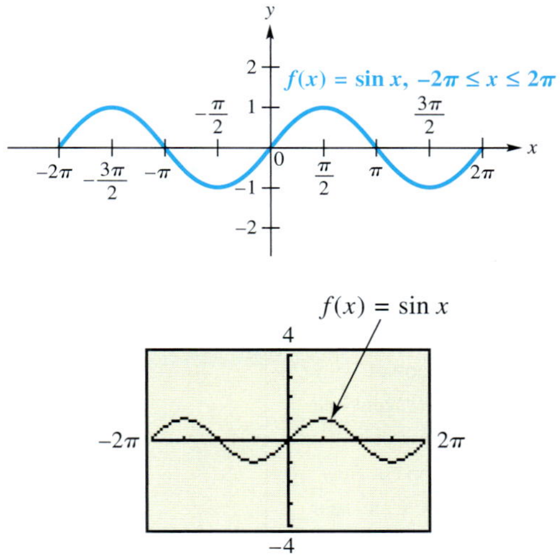

Figure 3

- The graph is continuous over its entire domain, $(-\infty, \infty)$.
- Its x-intercepts are of the form $n\pi$, where n is an integer.
- Its period is 2π.
- The graph is symmetric with respect to the origin, so the function is an odd function. For all x in the domain, $\sin(-x) = -\sin x$.

NOTE A function f is an **odd function** if for all x in the domain of f,

$$f(-x) = -f(x). \quad \text{(Appendix D)}$$

The graph of an odd function is symmetric with respect to the origin. This means that if (x, y) belongs to the function, then $(-x, -y)$ also belongs to the function. For example, $\left(\frac{\pi}{2}, 1\right)$ and $\left(-\frac{\pi}{2}, -1\right)$ are points on the graph of $y = \sin x$, illustrating the property $\sin(-x) = -\sin x$.

The sine function is closely related to the unit circle. **Its domain consists of real numbers corresponding to angle measures (or arc lengths) of the unit circle, and its range corresponds to the *y*-coordinates (or sine values) of the unit circle.**

Consider the unit circle in **Figure 2** and assume that the line from the origin to some point on the circle is part of the pedal of a bicycle, with a foot placed on the circle itself. As the pedal is rotated from 0 radians on the horizontal axis through various angles, the angle (or arc length) giving the pedal's location and its corresponding height from the horizontal axis given by $\sin x$ are used to create points on the sine graph. See **Figure 4** on the next page.

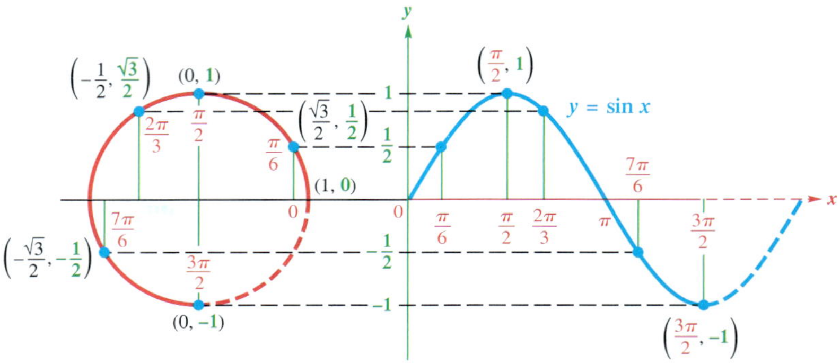

Figure 4

LOOKING AHEAD TO CALCULUS

The discussion of the derivative of a function in calculus shows that for the sine function, the slope of the tangent line at any point x is given by $\cos x$. For example, look at the graph of $y = \sin x$ and notice that a tangent line at $x = \pm\frac{\pi}{2}, \pm\frac{3\pi}{2}, \pm\frac{5\pi}{2}, \ldots$ will be horizontal and thus have slope 0. Now look at the graph of $y = \cos x$ and see that for these values, $\cos x = 0$.

Graph of the Cosine Function The graph of $y = \cos x$ in **Figure 5** has the same shape as the graph of $y = \sin x$. ***The graph of the cosine function is, in fact, the graph of the sine function shifted, or translated, $\frac{\pi}{2}$ units to the left.***

Cosine Function $f(x) = \cos x$

Domain: $(-\infty, \infty)$ Range: $[-1, 1]$

x	y
0	1
$\frac{\pi}{6}$	$\frac{\sqrt{3}}{2}$
$\frac{\pi}{4}$	$\frac{\sqrt{2}}{2}$
$\frac{\pi}{3}$	$\frac{1}{2}$
$\frac{\pi}{2}$	0
π	-1
$\frac{3\pi}{2}$	0
2π	1

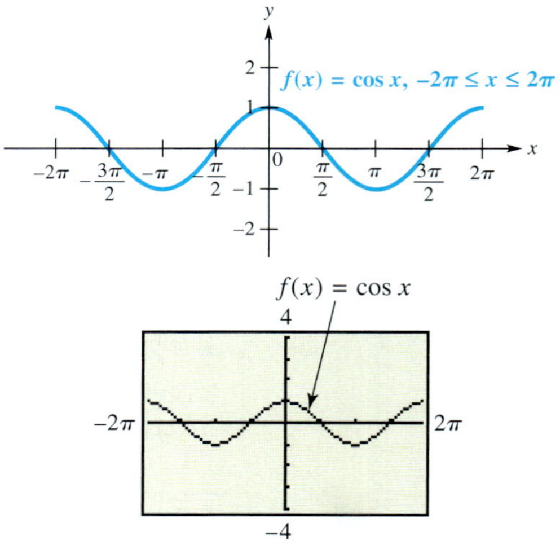

Figure 5

- The graph is continuous over its entire domain, $(-\infty, \infty)$.
- Its x-intercepts are of the form $(2n + 1)\frac{\pi}{2}$, where n is an integer.
- Its period is 2π.
- The graph is symmetric with respect to the y-axis, so the function is an even function. For all x in the domain, $\cos(-x) = \cos x$.

NOTE A function f is an **even function** if for all x in the domain of f,

$$f(-x) = f(x).$$ (Appendix D)

The graph of an even function is symmetric with respect to the y-axis. This means that if (x, y) belongs to the function, then $(-x, y)$ also belongs to the function. For example, $\left(\frac{\pi}{2}, 0\right)$ and $\left(-\frac{\pi}{2}, 0\right)$ are points on the graph of $y = \cos x$, illustrating the property $\cos(-x) = \cos x$.

The calculator graphs of $f(x) = \sin x$ in **Figure 3** and $f(x) = \cos x$ in **Figure 5** are graphed in the window approximately $[-2\pi, 2\pi]$ by $[-4, 4]$, with Xscl $= \frac{\pi}{2}$ and Yscl $= 1$. This is the **trig viewing window.** (Your model may use a different "standard" trig window. Consult your owner's manual.) ■

Graphing Techniques, Amplitude, and Period The examples that follow show graphs that are "stretched" or "compressed" (shrunk) either vertically, horizontally, or both when compared with the graphs of $y = \sin x$ or $y = \cos x$.

EXAMPLE 1 Graphing $y = a \sin x$

Graph $y = 2 \sin x$, and compare to the graph of $y = \sin x$.

SOLUTION For a given value of x, the value of y is twice what it would be for $y = \sin x$, as shown in the table of values. The only change in the graph is the range, which becomes $[-2, 2]$. See **Figure 6,** which includes a graph of $y = \sin x$ for comparison.

x	0	$\frac{\pi}{2}$	π	$\frac{3\pi}{2}$	2π
$\sin x$	0	1	0	-1	0
$2 \sin x$	0	2	0	-2	0

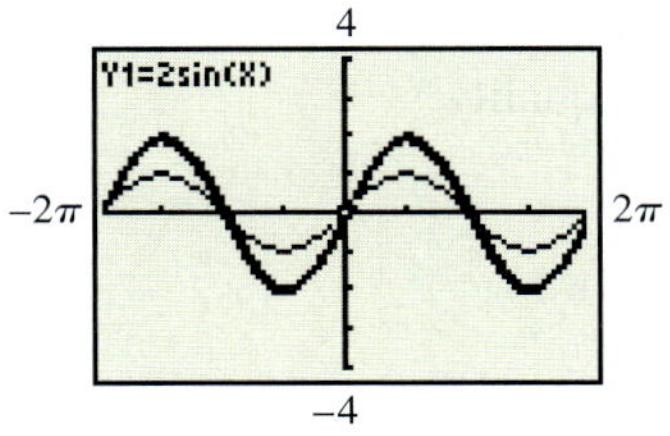

The thick graph style represents the function $y = 2 \sin x$ in **Example 1.**

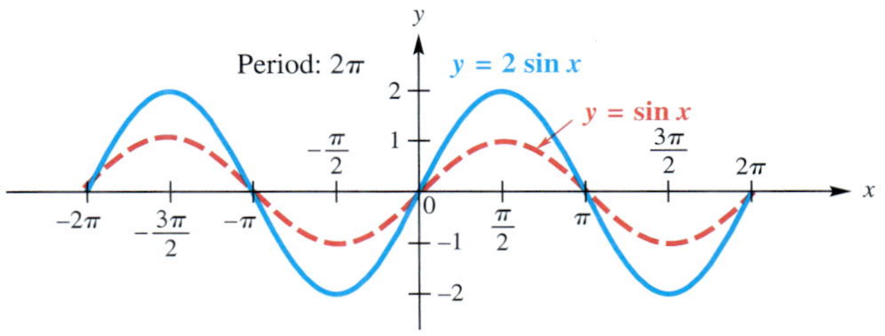

Figure 6

The **amplitude** of a periodic function is half the difference between the maximum and minimum values. It describes the height of the graph both above and below a horizontal line passing through the "middle" of the graph. Thus, for the basic sine function $y = \sin x$ (and also for the basic cosine function $y = \cos x$) the amplitude is computed as follows.

$$\frac{1}{2}[1 - (-1)] = \frac{1}{2}(2) = 1 \quad \text{Amplitude of } y = \sin x$$

For $y = 2 \sin x$, the amplitude is

$$\frac{1}{2}[2 - (-2)] = \frac{1}{2}(4) = 2. \quad \text{Amplitude of } y = 2\sin x$$

We can think of the graph of $y = a \sin x$ as a vertical stretching of the graph of $y = \sin x$ when $a > 1$ and a vertical shrinking when $0 < a < 1$.

✔ *Now Try Exercise 15.*

Generalizing from **Example 1** gives the following.

Amplitude

The graph of $y = a \sin x$ or $y = a \cos x$, with $a \neq 0$, will have the same shape as the graph of $y = \sin x$ or $y = \cos x$, respectively, except with range $[-|a|, |a|]$. The amplitude is $|a|$.

While the coefficient a in $y = a \sin x$ or $y = a \cos x$ affects the amplitude of the graph, the coefficient of x in the argument affects the period. Consider $y = \sin 2x$. We can complete a table of values for the interval $[0, 2\pi]$.

x	0	$\frac{\pi}{4}$	$\frac{\pi}{2}$	$\frac{3\pi}{4}$	π	$\frac{5\pi}{4}$	$\frac{3\pi}{2}$	$\frac{7\pi}{4}$	2π
sin 2*x*	0	1	0	−1	0	1	0	−1	0

Note that one complete cycle occurs in π units, not 2π units. Therefore, the period here is π, which equals $\frac{2\pi}{2}$. Now consider $y = \sin 4x$. Look at the next table.

x	0	$\frac{\pi}{8}$	$\frac{\pi}{4}$	$\frac{3\pi}{8}$	$\frac{\pi}{2}$	$\frac{5\pi}{8}$	$\frac{3\pi}{4}$	$\frac{7\pi}{8}$	π
sin 4*x*	0	1	0	−1	0	1	0	−1	0

These values suggest that one complete cycle is achieved in $\frac{\pi}{2}$ or $\frac{2\pi}{4}$ units, which is reasonable since

$$\sin\left(4 \cdot \frac{\pi}{2}\right) = \sin 2\pi = 0.$$

In general, the graph of a function of the form $y = \sin bx$ or $y = \cos bx$, for $b > 0$, will have a period different from 2π when $b \neq 1$. To see why this is so, remember that the values of $\sin bx$ or $\cos bx$ will take on all possible values as bx ranges from 0 to 2π. Therefore, to find the period of either of these functions, we must solve the following three-part inequality.

$$0 \le bx \le 2\pi \quad \text{(Appendix A)}$$

$$0 \le x \le \frac{2\pi}{b} \quad \text{Divide each part by the positive number } b.$$

Thus, the period is $\frac{2\pi}{b}$. By dividing the interval $\left[0, \frac{2\pi}{b}\right]$ into four equal parts, we obtain the values for which $\sin bx$ or $\cos bx$ is -1, 0, or 1. These values will give minimum points, x-intercepts, and maximum points on the graph. Once these points are determined, we can sketch the graph by joining the points with a smooth sinusoidal curve. (If a function has $b < 0$, then the identities of the next chapter can be used to rewrite the function so that $b > 0$.)

NOTE One method to divide an interval into four equal parts is as follows.

Step 1 Find the midpoint of the interval by adding the x-values of the endpoints and dividing by 2. (See **Appendix B.**)

Step 2 Find the quarter points (the midpoints of the two intervals found in Step 1) using the same procedure.

EXAMPLE 2 Graphing $y = \sin bx$

Graph $y = \sin 2x$, and compare to the graph of $y = \sin x$.

SOLUTION In this function the coefficient of x is 2, so $b = 2$ and the period is $\frac{2\pi}{2} = \pi$. Therefore, the graph will complete one period over the interval $[0, \pi]$.

We can divide the interval $[0, \pi]$ into four equal parts by first finding its midpoint: $\frac{1}{2}(0 + \pi) = \frac{\pi}{2}$. The quarter points are found next by determining the midpoints of the two intervals $\left[0, \frac{\pi}{2}\right]$ and $\left[\frac{\pi}{2}, \pi\right]$.

$\frac{1}{2}\left(\frac{\pi}{2} + \pi\right) = \frac{1}{2}\left(\frac{3\pi}{2}\right) = \frac{3\pi}{4}$

$$\frac{1}{2}\left(0 + \frac{\pi}{2}\right) = \frac{\pi}{4} \quad \text{and} \quad \frac{1}{2}\left(\frac{\pi}{2} + \pi\right) = \frac{3\pi}{4} \quad \text{Quarter points}$$

The interval $[0, \pi]$ is divided into four equal parts using these x-values.

$0,$	$\frac{\pi}{4},$	$\frac{\pi}{2},$	$\frac{3\pi}{4},$	π
↑	↑	↑	↑	↑
Left endpoint	First-quarter point	Midpoint	Third-quarter point	Right endpoint

We plot the points from the table of values given at the top of the previous page, and join them with a smooth sinusoidal curve. More of the graph can be sketched by repeating this cycle, as shown in **Figure 7.** The amplitude is not changed.

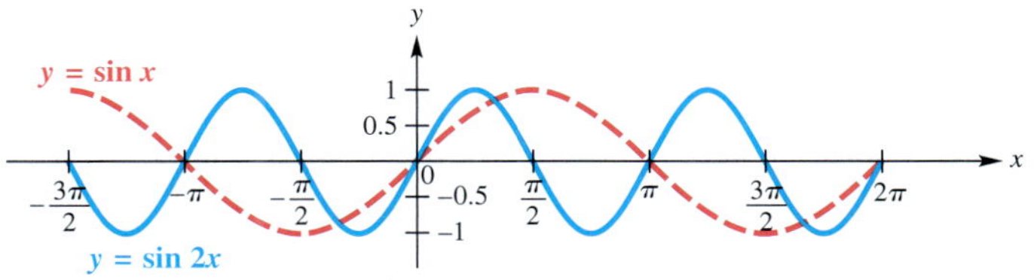

Figure 7

We can think of the graph of $y = \sin bx$ as a horizontal stretching of the graph of $y = \sin x$ when $0 < b < 1$ and as a horizontal shrinking when $b > 1$.

✔ *Now Try Exercise 27.*

Period

For $b > 0$, the graph of $y = \sin bx$ will resemble that of $y = \sin x$, but with period $\frac{2\pi}{b}$. Also, the graph of $y = \cos bx$ will resemble that of $y = \cos x$, but with period $\frac{2\pi}{b}$.

EXAMPLE 3 Graphing $y = \cos bx$

Graph $y = \cos \frac{2}{3}x$ over one period.

SOLUTION The period is

$$\frac{2\pi}{\frac{2}{3}} = 2\pi \div \frac{2}{3} = 2\pi \cdot \frac{3}{2} = 3\pi.$$

To divide by a number, multiply by its reciprocal.

We divide the interval $[0, 3\pi]$ into four equal parts to get the x-values 0, $\frac{3\pi}{4}$, $\frac{3\pi}{2}$, $\frac{9\pi}{4}$, and 3π that yield minimum points, maximum points, and x-intercepts. We use these values to obtain a table of key points for one period.

x	0	$\frac{3\pi}{4}$	$\frac{3\pi}{2}$	$\frac{9\pi}{4}$	3π
$\frac{2}{3}x$	0	$\frac{\pi}{2}$	π	$\frac{3\pi}{2}$	2π
$\cos \frac{2}{3}x$	1	0	-1	0	1

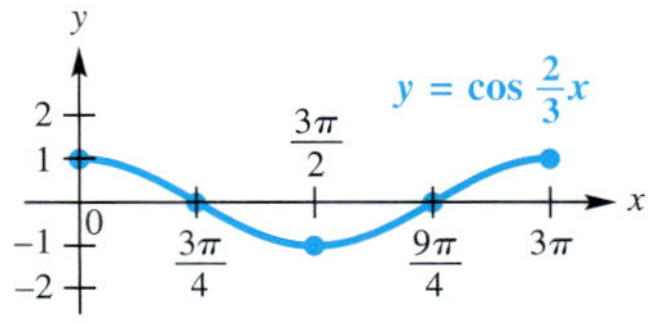

Figure 8

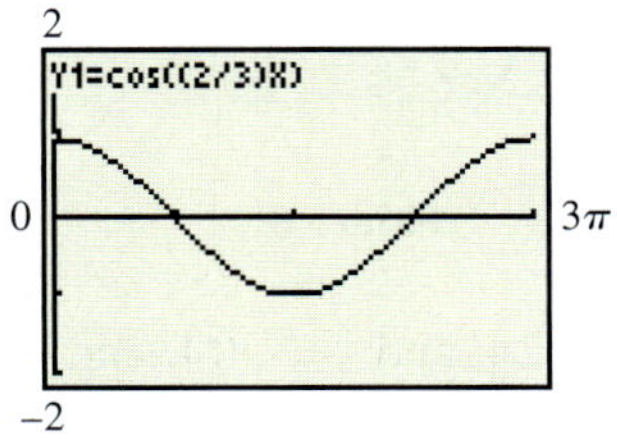

This screen shows a graph of the function in **Example 3.** By our choosing Xscl $= \frac{3\pi}{4}$, the x-intercepts, maxima, and minima coincide with tick marks on the x-axis.

The amplitude is 1 because the maximum value is 1, the minimum value is -1, and $\frac{1}{2}[1 - (-1)] = \frac{1}{2}(2) = 1$. We plot these points and join them with a smooth curve. The graph is shown in **Figure 8.**

✔ *Now Try Exercise 25.*

NOTE Look back at the middle row of the table in **Example 3.** Dividing the interval $\left[0, \frac{2\pi}{b}\right]$ into four equal parts will always give the values 0, $\frac{\pi}{2}$, π, $\frac{3\pi}{2}$, and 2π for this row, in this case resulting in values of -1, 0, or 1. These values lead to key points on the graph, which can then be easily sketched.

Guidelines for Sketching Graphs of Sine and Cosine Functions

To graph $y = a \sin bx$ or $y = a \cos bx$, with $b > 0$, follow these steps.

Step 1 Find the period, $\frac{2\pi}{b}$. Start at 0 on the x-axis, and lay off a distance of $\frac{2\pi}{b}$.

Step 2 Divide the interval into four equal parts. (See the Note preceding **Example 2.**)

Step 3 Evaluate the function for each of the five x-values resulting from Step 2. The points will be maximum points, minimum points, and x-intercepts.

Step 4 Plot the points found in Step 3, and join them with a sinusoidal curve having amplitude $|a|$.

Step 5 Draw the graph over additional periods as needed.

EXAMPLE 4 Graphing $y = a \sin bx$

Graph $y = -2 \sin 3x$ over one period using the preceding guidelines.

SOLUTION

Step 1 For this function, $b = 3$, so the period is $\frac{2\pi}{3}$. The function will be graphed over the interval $\left[0, \frac{2\pi}{3}\right]$.

Step 2 Divide the interval $\left[0, \frac{2\pi}{3}\right]$ into four equal parts to get the x-values 0, $\frac{\pi}{6}$, $\frac{\pi}{3}$, $\frac{\pi}{2}$, and $\frac{2\pi}{3}$.

Step 3 Make a table of values determined by the x-values from Step 2.

x	0	$\frac{\pi}{6}$	$\frac{\pi}{3}$	$\frac{\pi}{2}$	$\frac{2\pi}{3}$
$3x$	0	$\frac{\pi}{2}$	π	$\frac{3\pi}{2}$	2π
$\sin 3x$	0	1	0	-1	0
$-2 \sin 3x$	0	-2	0	2	0

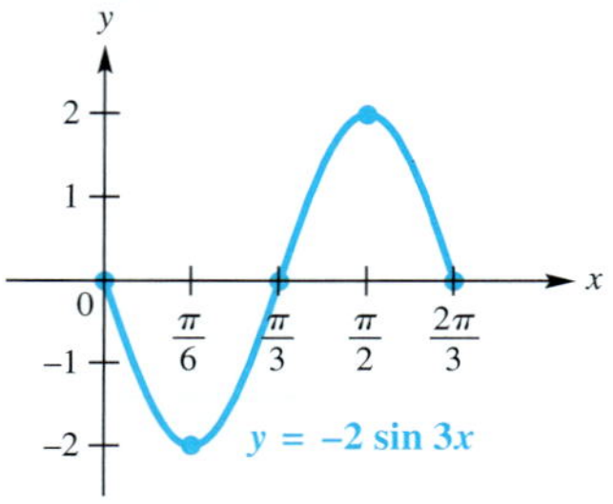

Figure 9

Step 4 Plot the points $(0, 0)$, $\left(\frac{\pi}{6}, -2\right)$, $\left(\frac{\pi}{3}, 0\right)$, $\left(\frac{\pi}{2}, 2\right)$, and $\left(\frac{2\pi}{3}, 0\right)$, and join them with a sinusoidal curve with amplitude 2. See **Figure 9.**

Step 5 The graph can be extended by repeating the cycle.

Notice that when a is negative, the graph of $y = a \sin bx$ is the reflection across the x-axis of the graph of $y = |a| \sin bx$.

✔ *Now Try Exercise 29.*

EXAMPLE 5 Graphing $y = a \cos bx$ for b That Is a Multiple of π

Graph $y = -3 \cos \pi x$ over one period.

SOLUTION

Step 1 Since $b = \pi$, the period is $\frac{2\pi}{\pi} = 2$, so we will graph the function over the interval $[0, 2]$.

Step 2 Dividing $[0, 2]$ into four equal parts yields the x-values $0, \frac{1}{2}, 1, \frac{3}{2}$, and 2.

Step 3 Make a table using these x-values.

x	0	$\frac{1}{2}$	1	$\frac{3}{2}$	2
πx	0	$\frac{\pi}{2}$	π	$\frac{3\pi}{2}$	2π
$\cos \pi x$	1	0	-1	0	1
$-3 \cos \pi x$	-3	0	3	0	-3

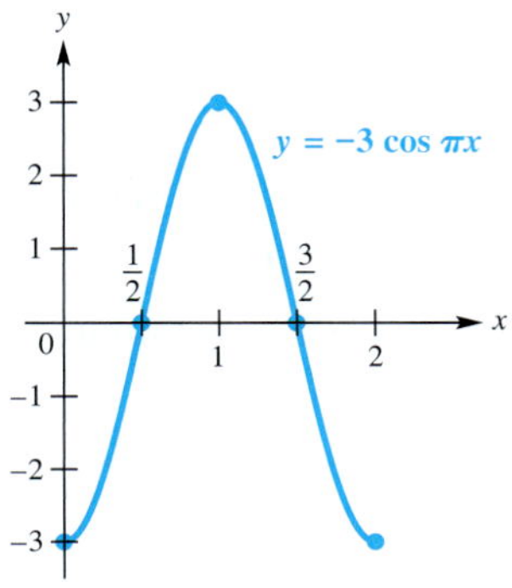

Figure 10

Step 4 Plot the points $(0, -3)$, $\left(\frac{1}{2}, 0\right)$, $(1, 3)$, $\left(\frac{3}{2}, 0\right)$, and $(2, -3)$, and join them with a sinusoidal curve having amplitude $|-3| = 3$. See **Figure 10.**

Step 5 The graph can be extended by repeating the cycle.

Notice that when b is an integer multiple of π, the x-intercepts of the graph are rational numbers.

✔ *Now Try Exercise 37.*

Connecting Graphs with Equations

EXAMPLE 6 Determining an Equation for a Graph

Determine an equation of the form $y = a \cos bx$ or $y = a \sin bx$, where $b > 0$, for the given graph.

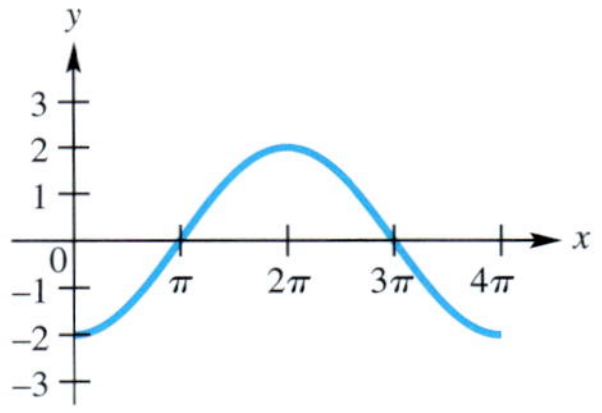

SOLUTION This graph is that of a cosine function that is reflected across its horizontal axis, the x-axis. The amplitude is half the distance between the maximum and minimum values.

$$\frac{1}{2}[2 - (-2)] = \frac{1}{2}(4) = 2 \quad \text{The amplitude } |a| \text{ is 2.}$$

Because the graph completes a cycle on the interval $[0, 4\pi]$, the period is 4π. We use this fact to solve for b.

$$4\pi = \frac{2\pi}{b} \quad \text{Period} = \frac{2\pi}{b}$$

$$4\pi b = 2\pi \quad \text{Multiply each side by } b. \text{ (Appendix A)}$$

$$b = \frac{1}{2} \quad \text{Divide each side by } 4\pi.$$

An equation for the graph is

$$y = -2 \cos \frac{1}{2} x.$$

(-2: x-axis reflection; $\frac{1}{2}$: Horizontal stretch)

✔ *Now Try Exercise 41.*

Using a Trigonometric Model Sine and cosine functions may be used to model many real-life phenomena that repeat their values in a cyclical, or periodic, manner. Average temperature in a certain geographic location is one such example.

EXAMPLE 7 Interpreting a Sine Function Model

The average temperature (in °F) at Mould Bay, Canada, can be approximated by the function

$$f(x) = 34 \sin\left[\frac{\pi}{6}(x - 4.3)\right],$$

where x is the month and $x = 1$ corresponds to January, $x = 2$ to February, and so on.

(a) To observe the graph over a two-year interval and to see the maximum and minimum points, graph f in the window $[0, 25]$ by $[-45, 45]$.

(b) According to this model, what is the average temperature during the month of May?

(c) What would be an approximation for the average *yearly* temperature at Mould Bay?

SOLUTION

(a) The graph of $f(x) = 34 \sin\left[\frac{\pi}{6}(x - 4.3)\right]$ is shown in **Figure 11.** Its amplitude is 34, and the period is

$$\frac{2\pi}{\frac{\pi}{6}} = 2\pi \div \frac{\pi}{6} = 2\pi \cdot \frac{6}{\pi} = 12. \quad \text{Simplify the complex fraction.}$$

The function f has a period of 12 months, or 1 year, which agrees with the changing of the seasons.

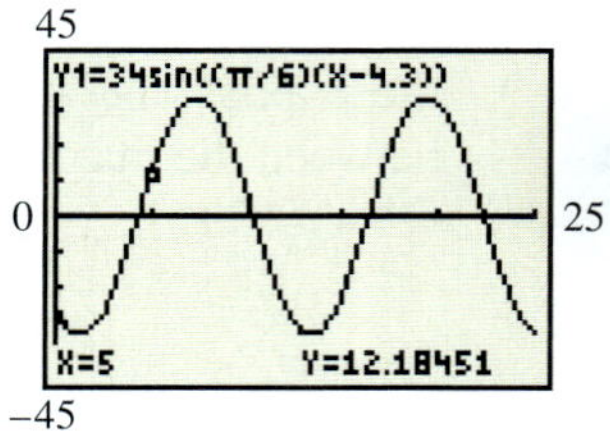

Figure 11

(b) May is the fifth month, so the average temperature during May is

$$f(5) = 34 \sin\left[\frac{\pi}{6}(5 - 4.3)\right] \approx 12°\text{F}. \quad \text{Let } x = 5. \text{ (Appendix C)}$$

See the display at the bottom of the screen in **Figure 11.**

(c) From the graph, it appears that the average yearly temperature is about 0°F since the graph is centered vertically about the line $y = 0$.

✔ *Now Try Exercise 59.*

4.1 Exercises

Concept Check In Exercises 1–8, match each function with its graph in choices A–I. (One choice will not be used.)

1. $y = \sin x$ **2.** $y = \cos x$ **3.** $y = -\sin x$ **4.** $y = -\cos x$

5. $y = \sin 2x$ **6.** $y = \cos 2x$ **7.** $y = 2\sin x$ **8.** $y = 2\cos x$

A.

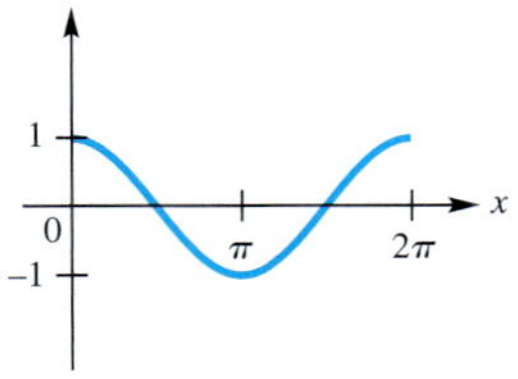

B.

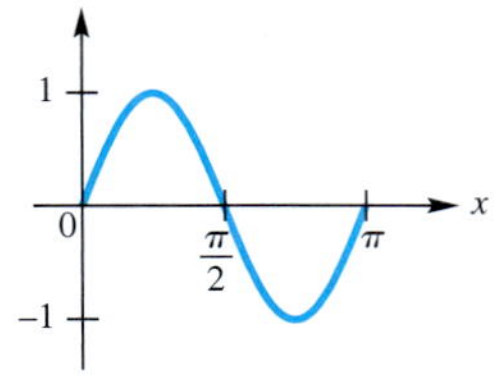

C.

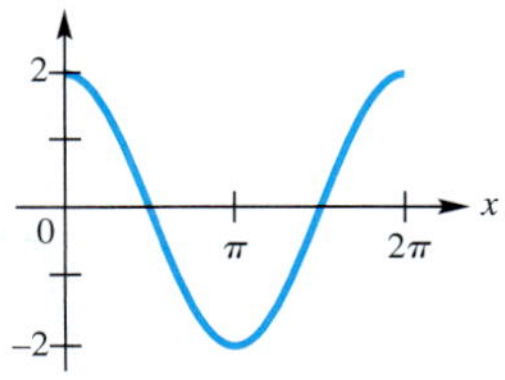

D.

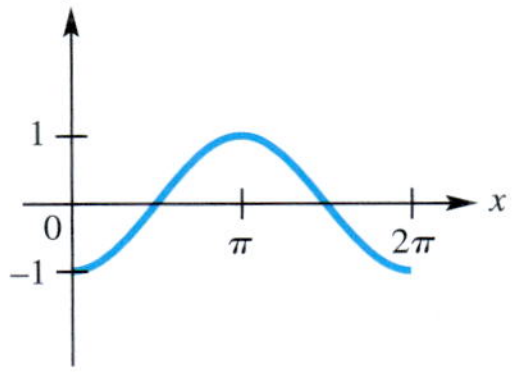

E.

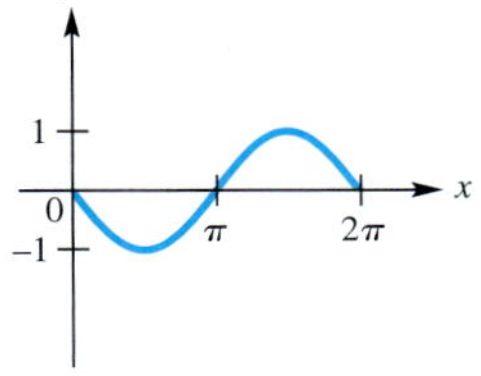

F.

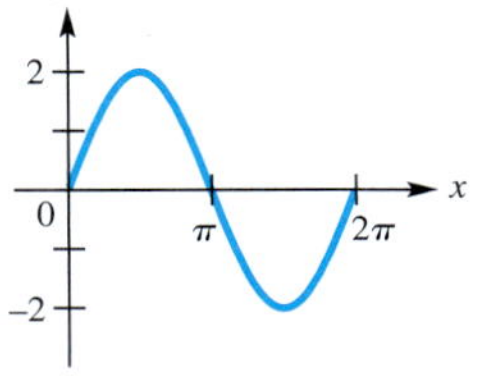

G.

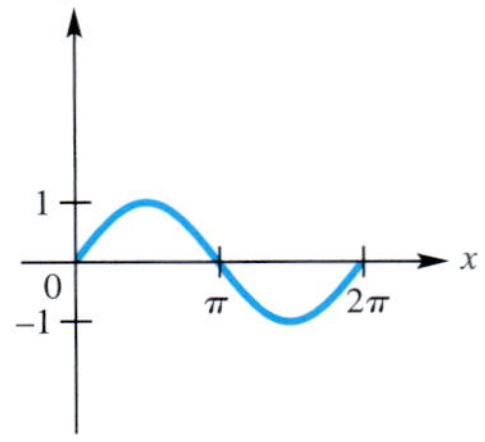

H.

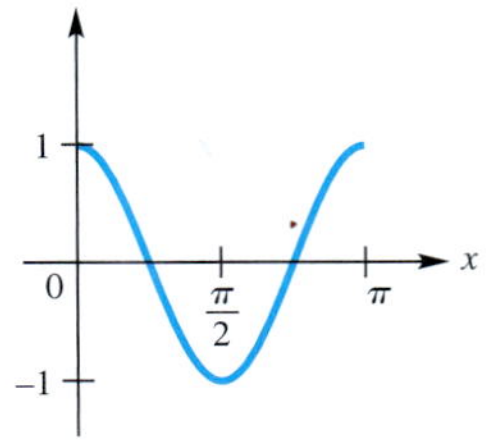

I.

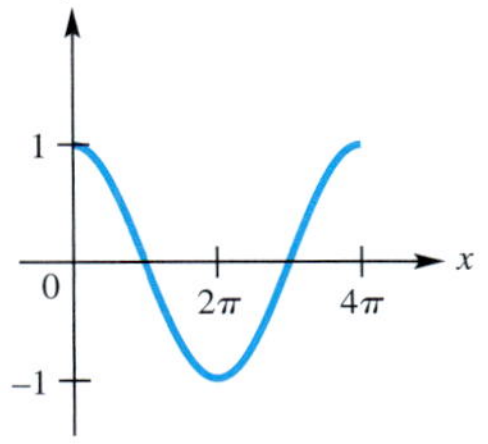

Concept Check In Exercises 9–12, match each function with its calculator graph.

9. $y = \sin 3x$ **10.** $y = \cos 3x$ **11.** $y = 3\cos x$ **12.** $y = 3\sin x$

A.

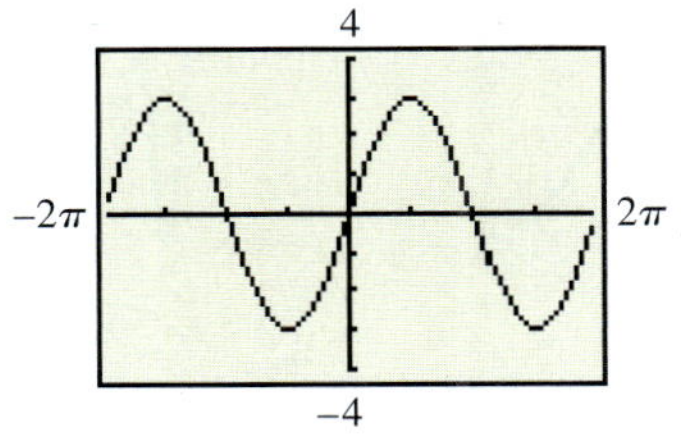

B.

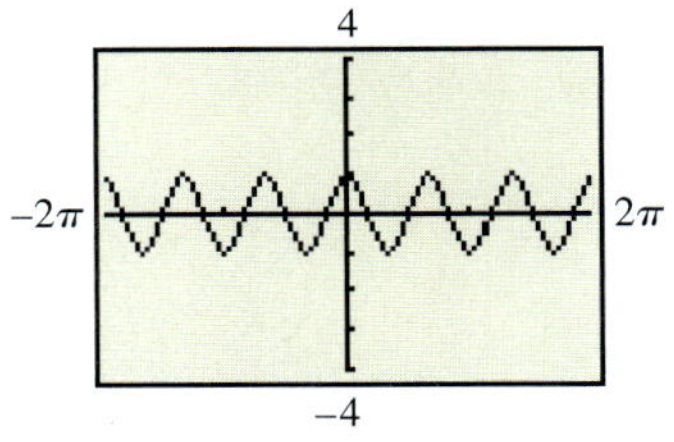

C.

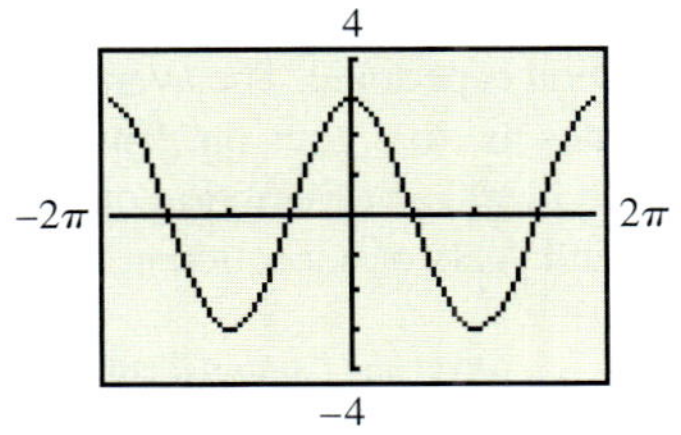

D.

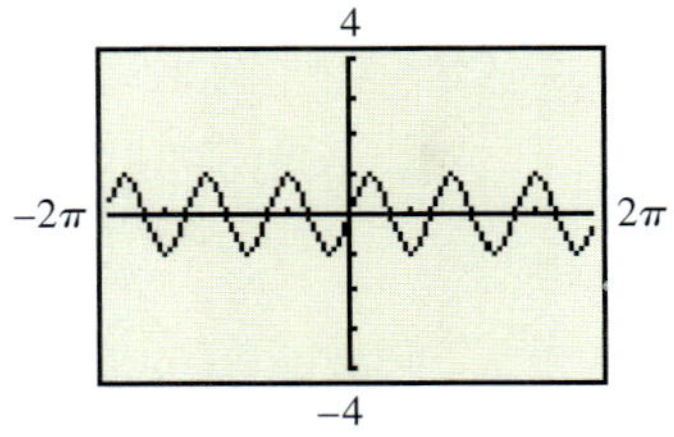

Graph each function over the interval $[-2\pi, 2\pi]$. Give the amplitude. ***See Example 1.***

13. $y = 2\cos x$ **14.** $y = 3\sin x$ **15.** $y = \frac{2}{3}\sin x$

16. $y = \frac{3}{4}\cos x$ **17.** $y = -\cos x$ **18.** $y = -\sin x$

19. $y = -2\sin x$ **20.** $y = -3\cos x$ **21.** $y = \sin(-x)$

22. *Concept Check* In **Exercise 21,** why is the graph the same as that of $y = -\sin x$?

Graph each function over a two-period interval. Give the period and amplitude. ***See Examples 2–5.***

23. $y = \sin \frac{1}{2}x$ **24.** $y = \sin \frac{2}{3}x$ **25.** $y = \cos \frac{3}{4}x$

26. $y = \cos \frac{1}{3}x$ **27.** $y = \sin 3x$ **28.** $y = \cos 2x$

29. $y = 2 \sin \frac{1}{4}x$ **30.** $y = 3 \sin 2x$ **31.** $y = -2 \cos 3x$

32. $y = -5 \cos 2x$ **33.** $y = \cos \pi x$ **34.** $y = -\sin \pi x$

35. $y = -2 \sin 2\pi x$ **36.** $y = 3 \cos 2\pi x$ **37.** $y = \frac{1}{2} \cos \frac{\pi}{2}x$

38. $y = -\frac{2}{3} \sin \frac{\pi}{4}x$ **39.** $y = \pi \sin \pi x$ **40.** $y = -\pi \cos \pi x$

Connecting Graphs with Equations *Each function graphed is of the form $y = a \sin bx$ or $y = a \cos bx$, where $b > 0$. Determine the equation of the graph.* ***See Example 6.***

41.

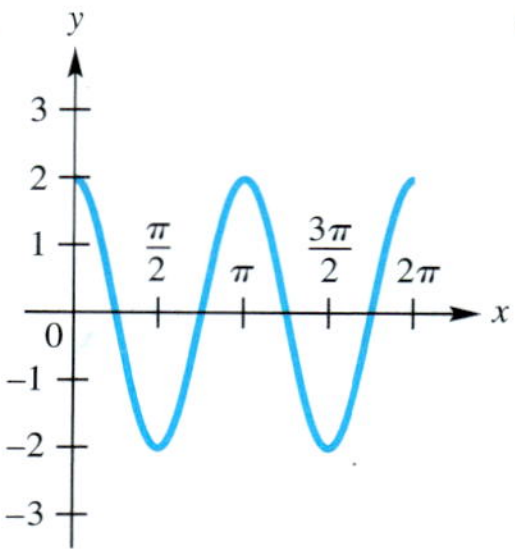

42.

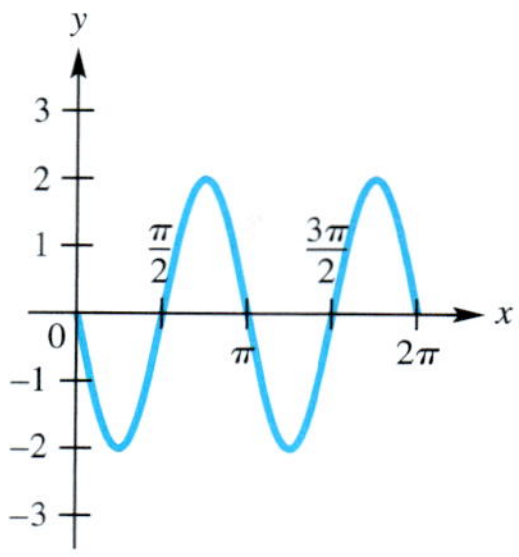

43.

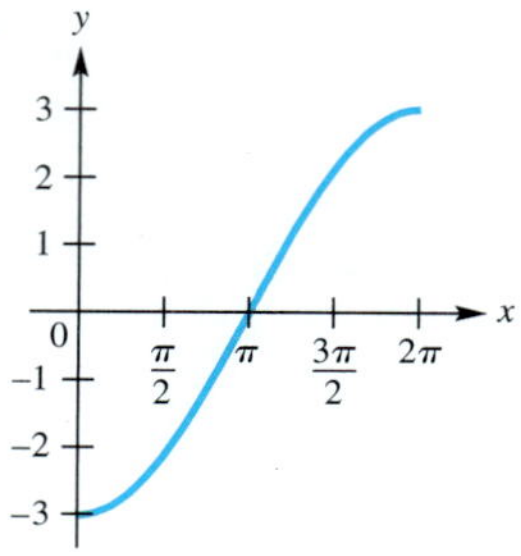

44.

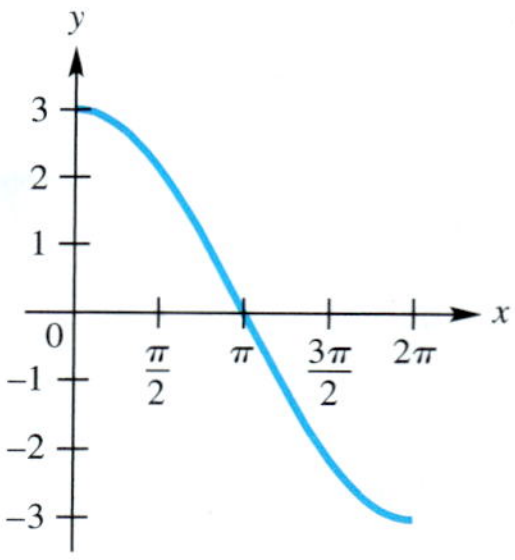

45.

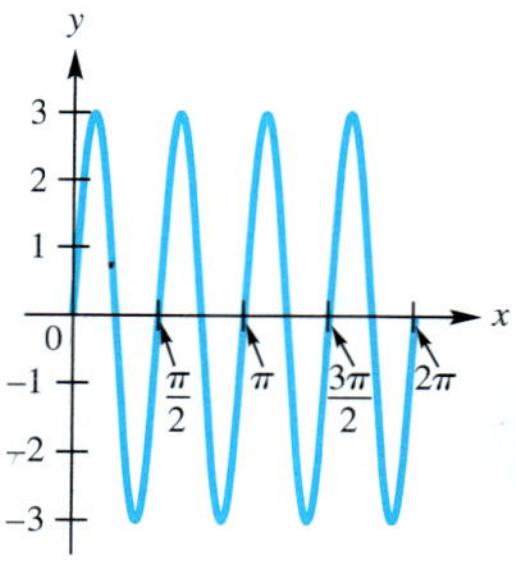

46.

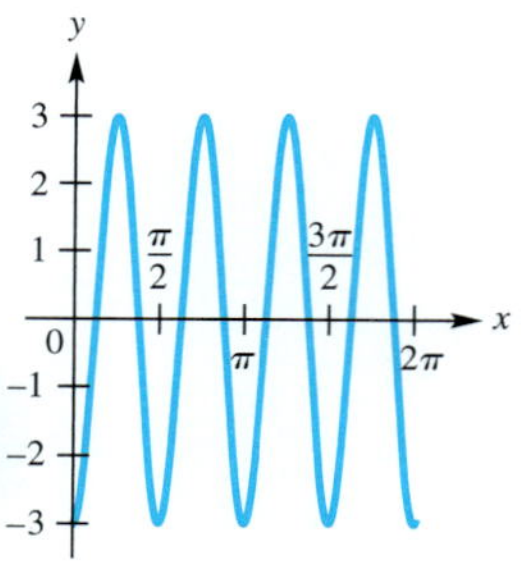

(Modeling) Solve each problem.

47. *Average Annual Temperature* Scientists believe that the average annual temperature in a given location is periodic. The average temperature at a given place during a given season fluctuates as time goes on, from colder to warmer, and back to colder. The graph shows an idealized description of the temperature (in °F) for approximately the last 150 thousand years of a location at the same latitude as Anchorage, Alaska.

Average Annual Temperature (Idealized)

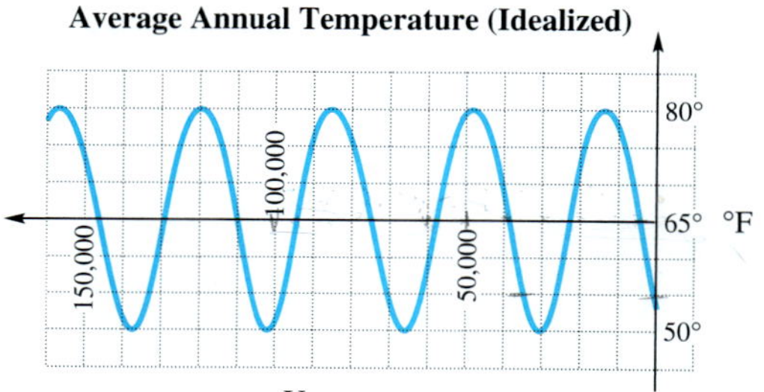

Years ago

(a) Find the highest and lowest temperatures recorded.
(b) Use these two numbers to find the amplitude.
(c) Find the period of the function.
(d) What is the trend of the temperature now?

48. *Blood Pressure Variation* The graph gives the variation in blood pressure for a typical person. **Systolic** and **diastolic pressures** are the upper and lower limits of the periodic changes in pressure that produce the pulse. The length of time between peaks is called the period of the pulse.

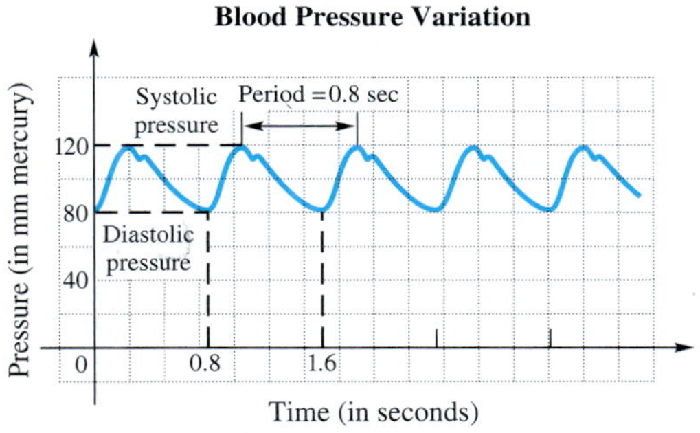

(a) Find the systolic and diastolic pressures.
(b) Find the amplitude of the graph.
(c) Find the pulse rate (the number of pulse beats in 1 min) for this person.

Tides for Kahului Harbor The chart shows the tides for Kahului Harbor (on the island of Maui, Hawaii). To identify high and low tides and times for other Maui areas, the following adjustments must be made.

Hana:	High, +40 min, +0.1 ft; Low, +18 min, −0.2 ft	Makena:	High, +1:21, −0.5 ft; Low, +1:09, −0.2 ft
Maalaea:	High, +1:52, −0.1 ft; Low, +1:19, −0.2 ft	Lahaina:	High, +1:18, −0.2 ft; Low, +1:01, −0.1 ft

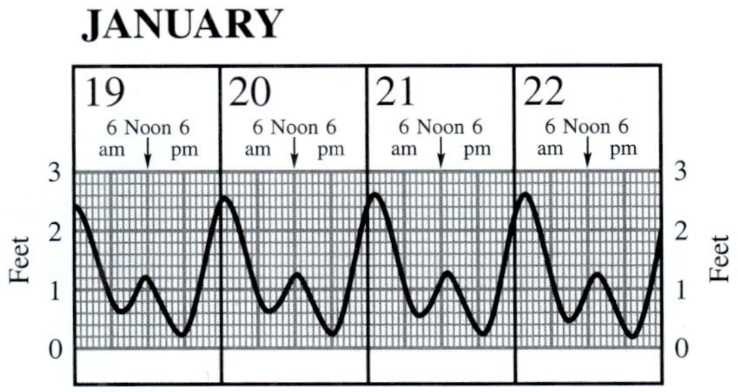

Source: Maui News. Original chart prepared by Edward K. Noda and Associates.

Use the graph to work Exercises 49–53.

49. The graph is an example of a periodic function. What is the period (in hours)?

50. What is the amplitude?

51. At what time on January 20 was low tide at Kahului? What was the height then?

52. Repeat **Exercise 51** for Maalaea.

53. At what time on January 22 was high tide at Lahaina? What was the height then?

54. *Activity of a Nocturnal Animal* Many of the activities of living organisms are periodic. For example, the graph at the right shows the time that a certain nocturnal animal begins its evening activity.

(a) Find the amplitude of this graph.
(b) Find the period.

Activity of a Nocturnal Animal

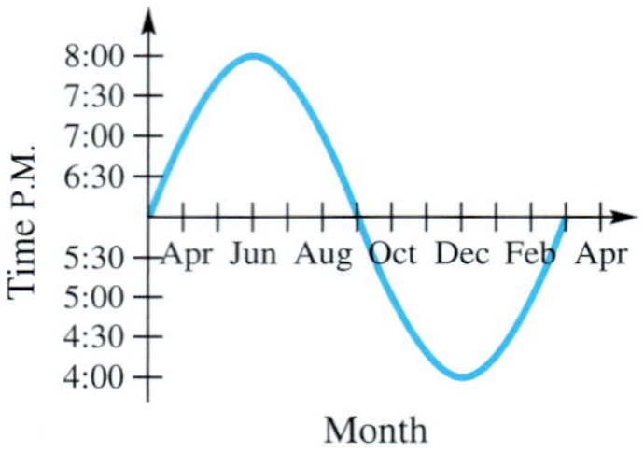

55. *Voltage of an Electrical Circuit* The voltage E in an electrical circuit is modeled by

$$E = 5 \cos 120\pi t,$$

where t is time measured in seconds.

(a) Find the amplitude and the period.
(b) How many cycles are completed in 1 sec? (The number of cycles, or periods, completed in 1 sec is the **frequency** of the function.)
(c) Find E when $t = 0, 0.03, 0.06, 0.09, 0.12$.
(d) Graph E for $0 \le t \le \frac{1}{30}$.

56. *Voltage of an Electrical Circuit* For another electrical circuit, the voltage E is modeled by

$$E = 3.8 \cos 40\pi t,$$

where t is time measured in seconds.

(a) Find the amplitude and the period.
(b) Find the frequency. See **Exercise 55(b).**
(c) Find E when $t = 0.02, 0.04, 0.08, 0.12, 0.14$.
(d) Graph one period of E.

57. *Atmospheric Carbon Dioxide* At Mauna Loa, Hawaii, atmospheric carbon dioxide levels in parts per million (ppm) were measured regularly from 1958 to 2004. The function

$$L(x) = 0.022x^2 + 0.55x + 316 + 3.5 \sin 2\pi x$$

can be used to model these levels, where x is in years and $x = 0$ corresponds to 1960. (*Source:* Nilsson, A., *Greenhouse Earth,* John Wiley and Sons.)

(a) Graph L in the window $[15, 45]$ by $[325, 385]$.
(b) When do the seasonal maximum and minimum carbon dioxide levels occur?
(c) L is the sum of a quadratic function and a sine function. What is the significance of each of these functions? Discuss what physical phenomena may be responsible for each function.

58. *Atmospheric Carbon Dioxide* Refer to **Exercise 57.** The carbon dioxide content in the atmosphere at Barrow, Alaska, in parts per million (ppm) can be modeled using the function

$$C(x) = 0.04x^2 + 0.6x + 330 + 7.5 \sin 2\pi x,$$

where $x = 0$ corresponds to 1970. (*Source:* Zeilik, M. and S. Gregory, *Introductory Astronomy and Astrophysics,* Brooks/Cole.)

(a) Graph C in the window $[5, 40]$ by $[320, 420]$.

(b) Discuss possible reasons why the amplitude of the oscillations in the graph of C is larger than the amplitude of the oscillations in the graph of L in **Exercise 57,** which models Hawaii.

(c) Define a new function C that is valid if x represents the actual year, where $1970 \le x \le 2010$. (See horizontal translations in **Appendix D.**)

59. *Average Daily Temperature* The temperature in Anchorage, Alaska, is modeled by

$$T(x) = 37 + 21 \sin\left[\frac{2\pi}{365}(x - 91)\right],$$

where $T(x)$ is the temperature in degrees Fahrenheit on day x, with $x = 1$ corresponding to January 1 and $x = 365$ corresponding to December 31. Use a calculator to estimate the temperature on the following days. (*Source: World Almanac and Book of Facts.*)

(a) March 15 (day 74) **(b)** April 5 (day 95) **(c)** Day 200
(d) June 25 **(e)** October 1 **(f)** December 31

60. *Fluctuation in the Solar Constant* The **solar constant** S is the amount of energy per unit area that reaches Earth's atmosphere from the sun. It is equal to 1367 watts per m^2 but varies slightly throughout the seasons. This fluctuation ΔS in S can be calculated using the formula

$$\Delta S = 0.034S \sin\left[\frac{2\pi(82.5 - N)}{365.25}\right].$$

In this formula, N is the day number covering a four-year period, where $N = 1$ corresponds to January 1 of a leap year and $N = 1461$ corresponds to December 31 of the fourth year. (*Source:* Winter, C., R. Sizmann, and L. L.Vant-Hull, Editors, *Solar Power Plants,* Springer-Verlag.)

(a) Calculate ΔS for $N = 80$, which is the spring equinox in the first year.

(b) Calculate ΔS for $N = 1268$, which is the summer solstice in the fourth year.

(c) What is the maximum value of ΔS?

(d) Find a value for N where ΔS is equal to 0.

Musical Sound Waves *Pure sounds produce single sine waves on an oscilloscope. Find the amplitude and period of each sine wave graph in Exercises 61 and 62. On the vertical scale, each square represents* 0.5; *on the horizontal scale, each square represents* 30° *or* $\frac{\pi}{6}$.

61.

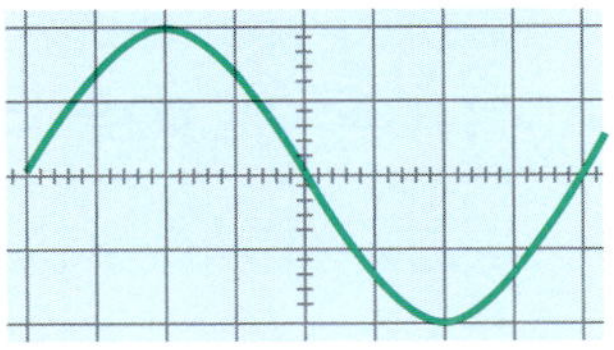

62.

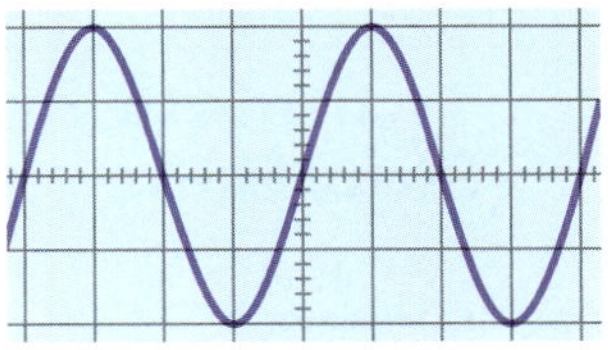

63. Compare the graphs of $y = \sin 2x$ and $y = 2 \sin x$ over the interval $[0, 2\pi]$. Can we say that, in general, $\sin bx = b \sin x$? Explain.

64. Compare the graphs of $y = \cos 3x$ and $y = 3 \cos x$ over the interval $[0, 2\pi]$. Can we say that, in general, $\cos bx = b \cos x$? Explain.

Relating Concepts

For individual or collaborative investigation *(Exercises 65–68)*

Connecting the Unit Circle and Sine Graph *Using a TI-83/84 Plus calculator, adjust the settings to correspond to the following screens.*

MODE FORMAT Y = editor

Tmax is 2π,
Tstep is $\frac{\pi}{40}$,
Xmax is 2π,
Xscl is $\frac{\pi}{2}$.

Graph the two equations (which are in **parametric form**), *and watch as the unit circle and the sine function are graphed simultaneously. Press the* TRACE *key once to get the screen shown on the left below, and then press the up-arrow key to get the screen shown on the right below. The screen on the left gives a unit circle interpretation of* $\cos 0 = 1$ *and* $\sin 0 = 0$. *The screen on the right gives a rectangular coordinate graph interpretation of* $\sin 0 = 0$.

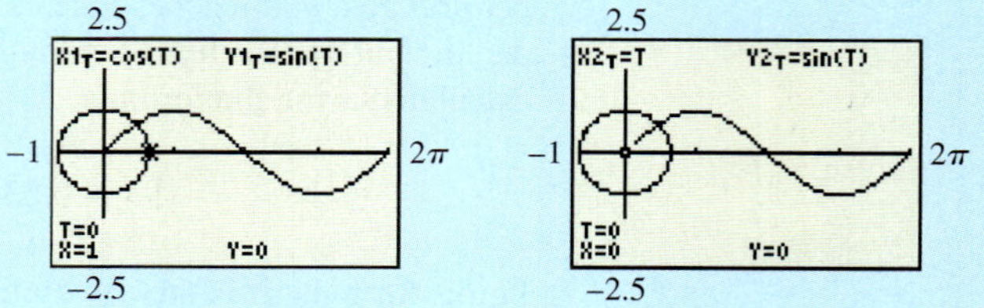

65. On the unit circle graph, let $T = 2$. Find X and Y, and interpret their values.

66. On the sine graph, let $T = 2$. What values of X and Y are displayed? Interpret these values with an equation in X and Y.

67. Now go back and redefine Y_{2T} as cos(T). Graph both equations. On the cosine graph, let $T = 2$. What values of X and Y are displayed? Interpret these values with an equation in X and Y.

68. Explain the relationship between the coordinates of the unit circle and the coordinates of the sine and cosine graphs.

4.2 Translations of the Graphs of the Sine and Cosine Functions

- Horizontal Translations
- Vertical Translations
- Combinations of Translations
- Determining a Trigonometric Model

Horizontal Translations

The graph of the function

$$y = f(x - d)$$

is translated *horizontally* compared to the graph of $y = f(x)$. The translation is d units to the right if $d > 0$ and is $|d|$ units to the left if $d < 0$. See **Figure 12** on the next page.

With circular functions, a horizontal translation is called a **phase shift.** In the function $y = f(x - d)$, the expression $x - d$ is the **argument.**

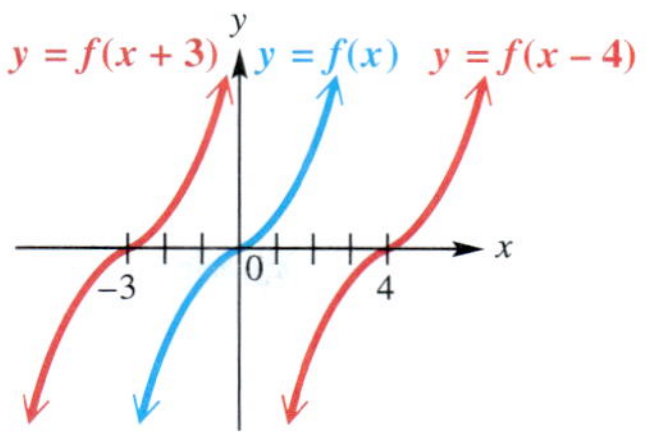

Horizontal translations of $y = f(x)$

(Appendix D)

Figure 12

In **Examples 1–3,** we give two methods that can be used to sketch the graph of a circular function involving a phase shift.

EXAMPLE 1 Graphing $y = \sin(x - d)$

Graph $y = \sin\left(x - \frac{\pi}{3}\right)$ over one period.

SOLUTION ***Method 1*** For the argument $x - \frac{\pi}{3}$ to result in all possible values throughout one period, it must take on all values between 0 and 2π, inclusive. To find an interval of one period, we solve the following three-part inequality.

$$0 \leq x - \frac{\pi}{3} \leq 2\pi$$ (Appendix A)

$$\frac{\pi}{3} \leq x \leq \frac{7\pi}{3} \quad \text{Add } \frac{\pi}{3} \text{ to each part.}$$

Use the method described in the Note preceding **Example 2** in **Section 4.1** to divide the interval $\left[\frac{\pi}{3}, \frac{7\pi}{3}\right]$ into four equal parts, obtaining the following x-values.

$$\frac{\pi}{3}, \quad \frac{5\pi}{6}, \quad \frac{4\pi}{3}, \quad \frac{11\pi}{6}, \quad \frac{7\pi}{3}$$

These are *key* x-values.

A table of values using these x-values follows.

x	$\frac{\pi}{3}$	$\frac{5\pi}{6}$	$\frac{4\pi}{3}$	$\frac{11\pi}{6}$	$\frac{7\pi}{3}$
$x - \frac{\pi}{3}$	0	$\frac{\pi}{2}$	π	$\frac{3\pi}{2}$	2π
$\sin\left(x - \frac{\pi}{3}\right)$	0	1	0	−1	0

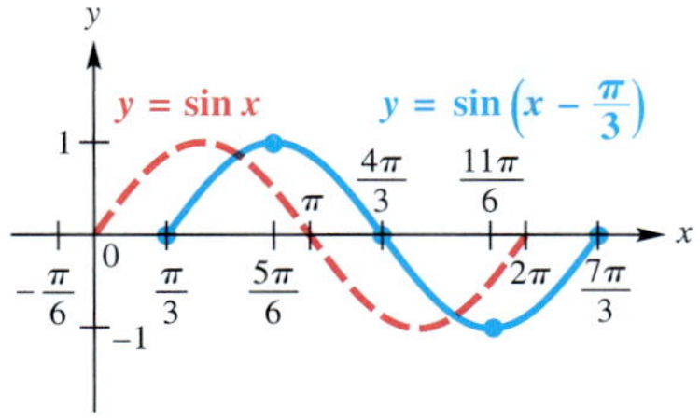

Figure 13

We join the corresponding points with a smooth curve to get the solid blue graph shown in **Figure 13.** The period is 2π, and the amplitude is 1.

Method 2 We can also graph $y = \sin\left(x - \frac{\pi}{3}\right)$ by using a horizontal translation of the graph of $y = \sin x$. The argument $x - \frac{\pi}{3}$ indicates that the graph will be translated $\frac{\pi}{3}$ units to the *right* (the phase shift) compared to the graph of $y = \sin x$. See **Figure 13.**

Therefore, to graph a function using this method, first graph the basic circular function, and then graph the desired function by using the appropriate translation.

✔ ***Now Try Exercise 35.***

NOTE The graph in **Figure 13** of **Example 1** can be extended through additional periods by repeating the given portion of the graph, as necessary.

EXAMPLE 2 Graphing $y = a\cos(x - d)$

Graph $y = 3\cos\left(x + \frac{\pi}{4}\right)$ over one period.

SOLUTION ***Method 1*** First solve the following three-part inequality.

$$0 \le x + \frac{\pi}{4} \le 2\pi$$

$$-\frac{\pi}{4} \le x \le \frac{7\pi}{4} \quad \text{Subtract } \frac{\pi}{4} \text{ from each part.}$$

Dividing this interval into four equal parts gives these x-values.

$$-\frac{\pi}{4}, \quad \frac{\pi}{4}, \quad \frac{3\pi}{4}, \quad \frac{5\pi}{4}, \quad \frac{7\pi}{4} \quad \text{Key } x\text{-values}$$

Use these x-values to make a table of points.

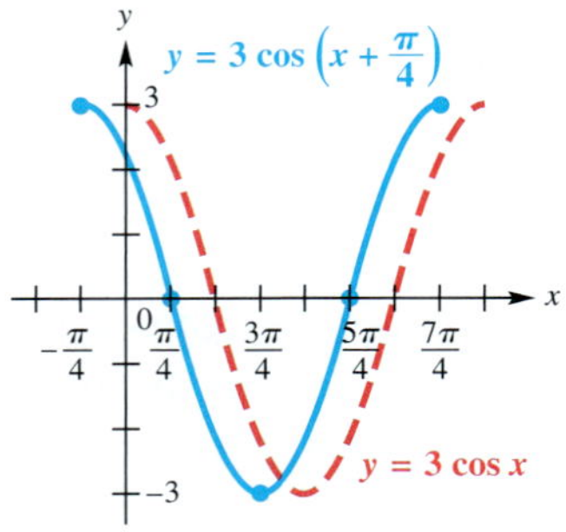

Figure 14

x	$-\frac{\pi}{4}$	$\frac{\pi}{4}$	$\frac{3\pi}{4}$	$\frac{5\pi}{4}$	$\frac{7\pi}{4}$
$x + \frac{\pi}{4}$	0	$\frac{\pi}{2}$	π	$\frac{3\pi}{2}$	2π
$\cos\left(x + \frac{\pi}{4}\right)$	1	0	-1	0	1
$3\cos\left(x + \frac{\pi}{4}\right)$	3	0	-3	0	3

These x-values lead to maximum points, minimum points, and x-intercepts.

We join the corresponding points with a smooth curve to get the solid blue graph shown in **Figure 14.** The period is 2π, and the amplitude is 3.

Method 2 Write $y = 3\cos\left(x + \frac{\pi}{4}\right)$ in the form $y = a\cos(x - d)$.

$$y = 3\cos\left(x + \frac{\pi}{4}\right), \quad \text{or} \quad y = 3\cos\left[x - \left(-\frac{\pi}{4}\right)\right] \quad \text{Rewrite to subtract } -\frac{\pi}{4}.$$

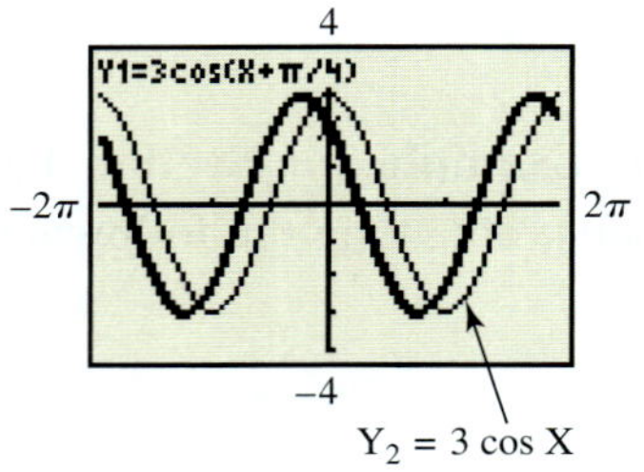

This result shows that $d = -\frac{\pi}{4}$. Since $-\frac{\pi}{4}$ is negative, the phase shift is $\left|-\frac{\pi}{4}\right| = \frac{\pi}{4}$ unit to the left. The graph is the same as that of $y = 3\cos x$ (the thin-lined graph in the margin calculator screen), except that it is translated $\frac{\pi}{4}$ unit to the left (the thick-lined graph).

✔ *Now Try Exercise 37.*

EXAMPLE 3 Graphing $y = a\cos[b(x - d)]$

Graph $y = -2\cos(3x + \pi)$ over two periods.

SOLUTION ***Method 1*** The function can be sketched over one period by solving the three-part inequality

$$0 \le 3x + \pi \le 2\pi$$

to find the interval $\left[-\frac{\pi}{3}, \frac{\pi}{3}\right]$. Divide this interval into four equal parts to find the points $\left(-\frac{\pi}{3}, -2\right)$, $\left(-\frac{\pi}{6}, 0\right)$, $(0, 2)$, $\left(\frac{\pi}{6}, 0\right)$, and $\left(\frac{\pi}{3}, -2\right)$. Plot these points and join them with a smooth curve. By graphing an additional half period to the left and to the right, we obtain the graph shown in **Figure 15.**

Method 2 First write the equation in the form $y = a\cos[b(x - d)]$.

$$y = -2\cos(3x + \pi), \quad \text{or} \quad y = -2\cos\left[3\left(x + \frac{\pi}{3}\right)\right] \quad \text{Rewrite by factoring out 3.}$$

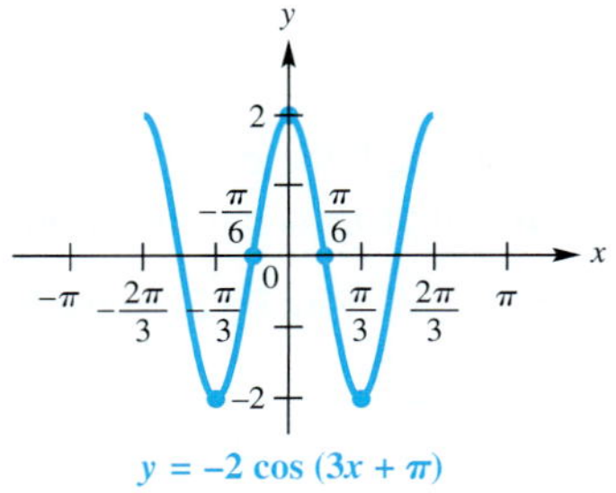

Figure 15

Then $a = -2$, $b = 3$, and $d = -\frac{\pi}{3}$. The amplitude is $|-2| = 2$, and the period is $\frac{2\pi}{3}$ (since the value of b is 3). The phase shift is $\left|-\frac{\pi}{3}\right| = \frac{\pi}{3}$ units to the left compared to the graph of $y = -2\cos 3x$. Again, see **Figure 15.**

✔ *Now Try Exercise 43.*

Vertical Translations The graph of a function of the form

$$y = c + f(x)$$

is translated *vertically* compared to the graph of $y = f(x)$. See **Figure 16.** The translation is c units up if $c > 0$ and is $|c|$ units down if $c < 0$.

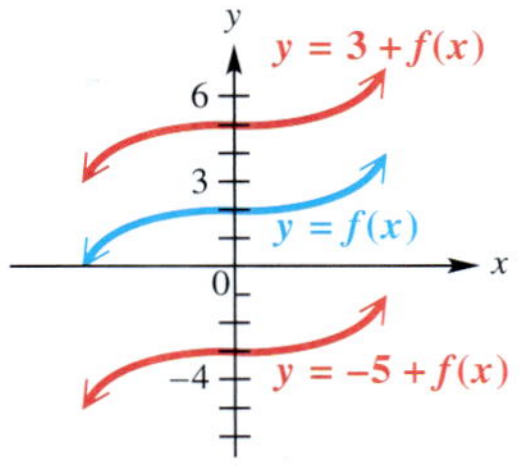

Vertical translations of $y = f(x)$

(Appendix D)

Figure 16

EXAMPLE 4 Graphing $y = c + a \cos bx$

Graph $y = 3 - 2 \cos 3x$ over two periods.

SOLUTION The values of y will be 3 greater than the corresponding values of y in $y = -2 \cos 3x$. This means that the graph of $y = 3 - 2 \cos 3x$ is the same as the graph of $y = -2 \cos 3x$, vertically translated 3 units up. Since the period of $y = -2 \cos 3x$ is $\frac{2\pi}{3}$, the key points have these x-values.

$$0, \frac{\pi}{6}, \frac{\pi}{3}, \frac{\pi}{2}, \frac{2\pi}{3} \quad \text{Key } x\text{-values}$$

Use these x-values to make a table of points.

x	0	$\frac{\pi}{6}$	$\frac{\pi}{3}$	$\frac{\pi}{2}$	$\frac{2\pi}{3}$
cos 3*x*	1	0	−1	0	1
2 cos 3*x*	2	0	−2	0	2
3 − 2 cos 3*x*	1	3	5	3	1

The key points are shown on the graph in **Figure 17,** along with more of the graph, which is sketched using the fact that the function is periodic.

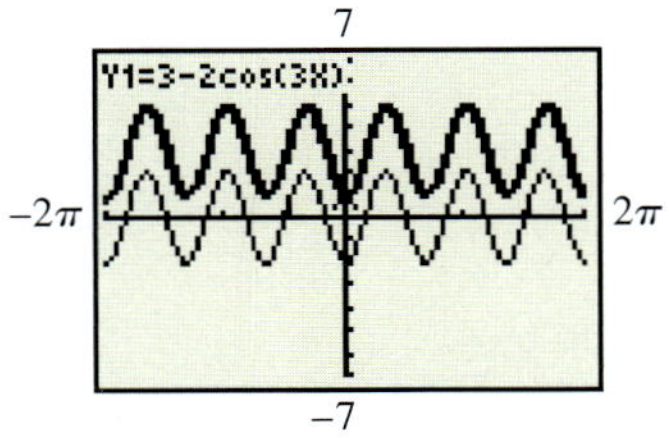

The function in **Example 4** is shown using the thick graph style. Notice also the thin graph style for $y = -2 \cos 3x$.

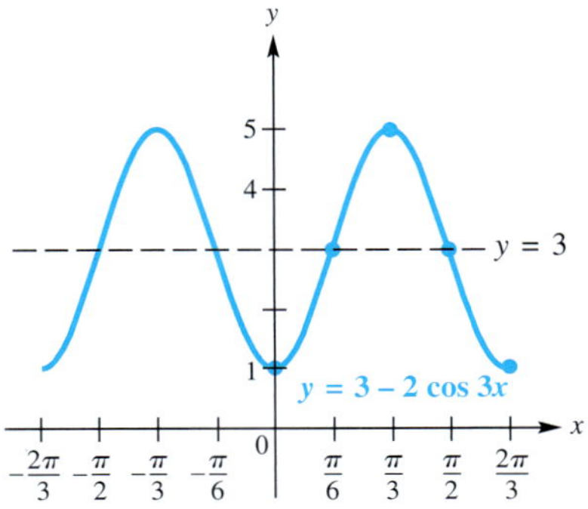

Figure 17

✔ *Now Try Exercise 47.*

Combinations of Translations A function of the form

$$y = c + a\sin[b(x - d)] \quad \text{or} \quad y = c + a\cos[b(x - d)], \quad \text{where } b > 0,$$

which involves stretching, shrinking, and translating, can be graphed according to the following guidelines.

Further Guidelines for Sketching Graphs of Sine and Cosine Functions

Method 1 Follow these steps.

Step 1 Find an interval whose length is one period $\frac{2\pi}{b}$ by solving the three-part inequality $0 \le b(x - d) \le 2\pi$. (See **Appendix A.**)

Step 2 Divide the interval into four equal parts. (See the Note preceding **Example 2** in **Section 4.1.**)

Step 3 Evaluate the function for each of the five x-values resulting from Step 2. The points will be maximum points, minimum points, and points that intersect the line $y = c$ ("middle" points of the wave).

Step 4 Plot the points found in Step 3, and join them with a sinusoidal curve having amplitude $|a|$.

Step 5 Draw the graph over additional periods, as needed.

Method 2 Follow these steps.

Step 1 Graph $y = a \sin bx$ or $y = a \cos bx$. The amplitude of the function is $|a|$, and the period is $\frac{2\pi}{b}$.

Step 2 Use translations to graph the desired function. The vertical translation is c units up if $c > 0$ and is $|c|$ units down if $c < 0$. The horizontal translation (phase shift) is d units to the right if $d > 0$ and is $|d|$ units to the left if $d < 0$.

EXAMPLE 5 **Graphing $y = c + a\sin[b(x - d)]$**

Graph $y = -1 + 2\sin(4x + \pi)$ over two periods.

SOLUTION We use Method 1. First write the expression on the right side of the equation in the form $c + a\sin[b(x - d)]$.

$$y = -1 + 2\sin(4x + \pi), \quad \text{or} \quad y = -1 + 2\sin\left[4\left(x + \frac{\pi}{4}\right)\right]$$ Rewrite by factoring out 4.

Step 1 Find an interval whose length is one period.

$$0 \le 4\left(x + \frac{\pi}{4}\right) \le 2\pi$$

$$0 \le x + \frac{\pi}{4} \le \frac{\pi}{2}$$ Divide each part by 4.

$$-\frac{\pi}{4} \le x \le \frac{\pi}{4}$$ Subtract $\frac{\pi}{4}$ from each part.

Step 2 Divide the interval $\left[-\frac{\pi}{4}, \frac{\pi}{4}\right]$ into four equal parts to get these x-values.

$$-\frac{\pi}{4}, \quad -\frac{\pi}{8}, \quad 0, \quad \frac{\pi}{8}, \quad \frac{\pi}{4}$$ Key x-values

Step 3 Make a table of values.

x	$-\frac{\pi}{4}$	$-\frac{\pi}{8}$	0	$\frac{\pi}{8}$	$\frac{\pi}{4}$
$x + \frac{\pi}{4}$	0	$\frac{\pi}{8}$	$\frac{\pi}{4}$	$\frac{3\pi}{8}$	$\frac{\pi}{2}$
$4(x + \frac{\pi}{4})$	0	$\frac{\pi}{2}$	π	$\frac{3\pi}{2}$	2π
$\sin[4(x + \frac{\pi}{4})]$	0	1	0	-1	0
$2\sin[4(x + \frac{\pi}{4})]$	0	2	0	-2	0
$-1 + 2\sin(4x + \pi)$	-1	1	-1	-3	-1

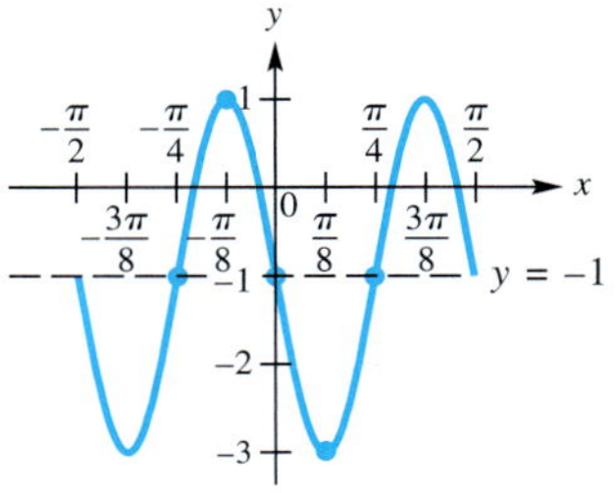

$y = -1 + 2\sin(4x + \pi)$

Figure 18

Steps 4 and 5 Plot the points found in the table and join them with a sinusoidal curve. **Figure 18** shows the graph, extended to the right and left to include two full periods.

✔ ***Now Try Exercise 53.***

Determining a Trigonometric Model A sinusoidal function is often a good approximation of a set of real data points.

EXAMPLE 6 Modeling Temperature with a Sine Function

The maximum average monthly temperature in New Orleans is 83°F, and the minimum is 53°F. The table shows the average monthly temperatures. The scatter diagram for a two-year interval in **Figure 19** strongly suggests that the temperatures can be modeled with a sine curve.

Month	°F	Month	°F
Jan	53	July	83
Feb	56	Aug	83
Mar	62	Sept	79
Apr	68	Oct	70
May	76	Nov	61
June	81	Dec	55

Source: World Almanac and Book of Facts.

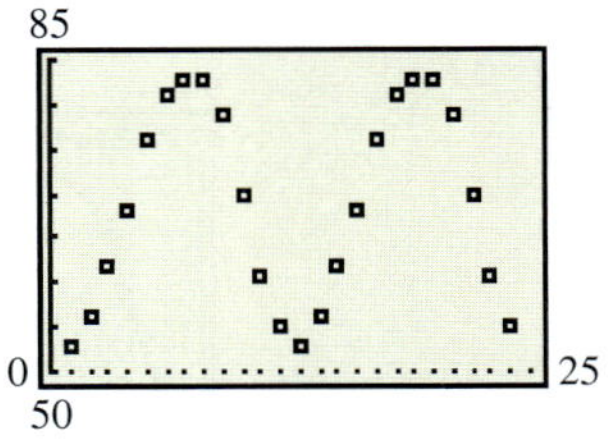

Figure 19

(a) Using only the maximum and minimum temperatures, determine a function of the form

$$f(x) = a\sin[b(x - d)] + c, \quad \text{where } a, b, c, \text{ and } d \text{ are constants,}$$

that models the average monthly temperature in New Orleans. Let x represent the month, with January corresponding to $x = 1$.

(b) On the same coordinate axes, graph f for a two-year period together with the actual data values found in the table.

(c) Use the **sine regression** feature of a graphing calculator to determine a second model for these data.

SOLUTION

(a) We use the maximum and minimum average monthly temperatures to find the amplitude a.

$$a = \frac{83 - 53}{2} = 15 \quad \text{Amplitude}$$

The average of the maximum and minimum temperatures is a good choice for c. The average is

$$\frac{83 + 53}{2} = 68. \quad \text{Vertical translation}$$

Since temperatures repeat every 12 months, b can be found as follows.

$$12 = \frac{2\pi}{b} \quad \text{Period} = \frac{2\pi}{b}$$

$$b = \frac{\pi}{6} \quad \text{Solve for } b. \text{ (Appendix A)}$$

The coldest month is January, when $x = 1$, and the hottest month is July, when $x = 7$. A good choice for d is 4 because April, when $x = 4$, is located at the midpoint between January and July. Also, notice that the average monthly temperature in April is 68°F, which is the value of the vertical translation, c. The average monthly temperature in New Orleans is modeled closely by the following equation.

$$f(x) = a \sin[b(x - d)] + c$$

$$f(x) = 15 \sin\left[\frac{\pi}{6}(x - 4)\right] + 68 \quad \text{Substitute.}$$

(b) **Figure 20** shows the data points from the table, along with the graph of $y = 15 \sin\left[\frac{\pi}{6}(x - 4)\right] + 68$ and the graph of $y = 15 \sin \frac{\pi}{6}x + 68$ for comparison.

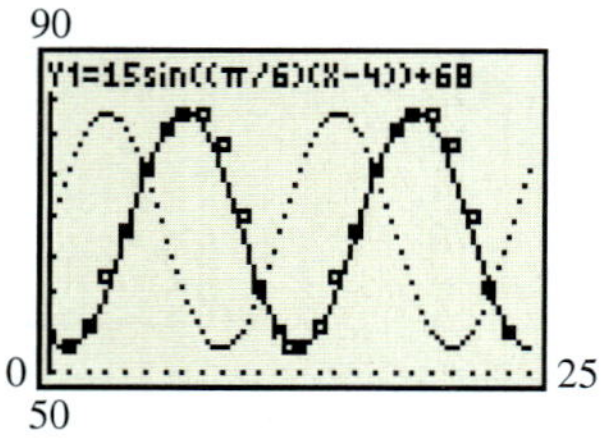

Figure 20

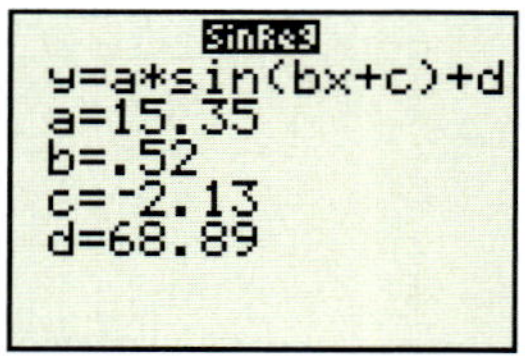

Values are rounded to the nearest hundredth.

(a)

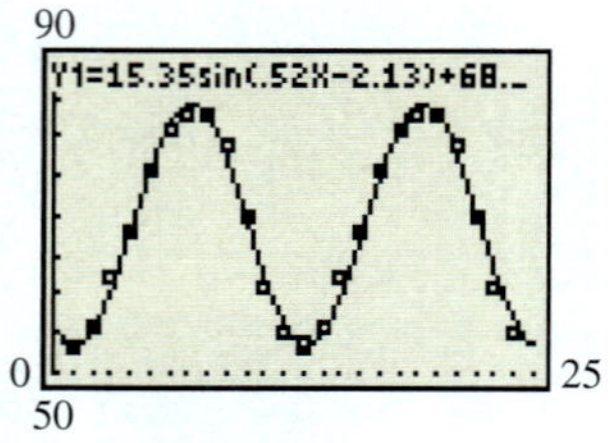

(b)

Figure 21

(c) We used the given data for a two-year period and the sine regression capability of a graphing calculator to produce the model

$$f(x) = 15.35 \sin(0.52x - 2.13) + 68.89$$

described in **Figure 21(a).** Its graph along with the data points is shown in **Figure 21(b).**

✔ ***Now Try Exercise 57.***

4.2 Exercises

Concept Check *In Exercises 1–8, match each function with its graph in choices A–I. (One choice will not be used.)*

1. $y = \sin\left(x - \frac{\pi}{4}\right)$ **2.** $y = \sin\left(x + \frac{\pi}{4}\right)$ **3.** $y = \cos\left(x - \frac{\pi}{4}\right)$

4. $y = \cos\left(x + \frac{\pi}{4}\right)$ **5.** $y = 1 + \sin x$ **6.** $y = -1 + \sin x$

7. $y = 1 + \cos x$ **8.** $y = -1 + \cos x$

A.
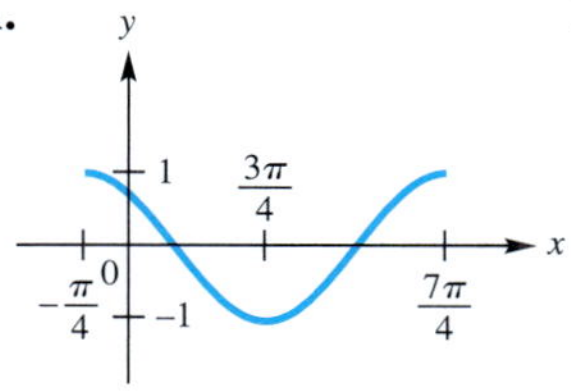

B.
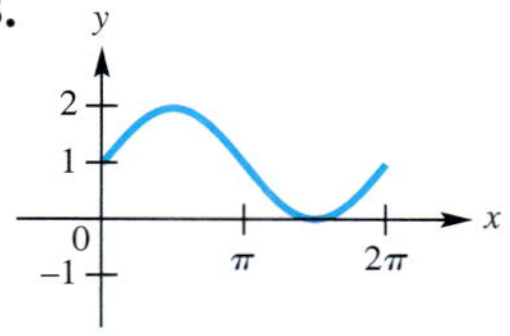

C.
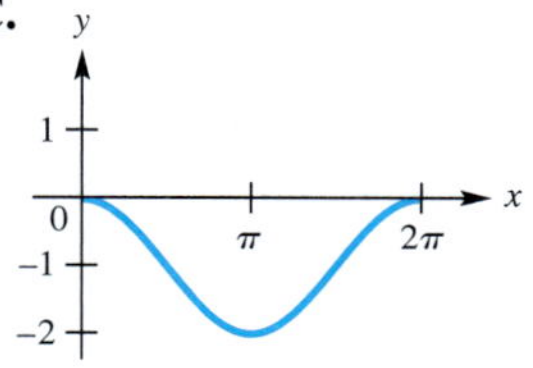

D.
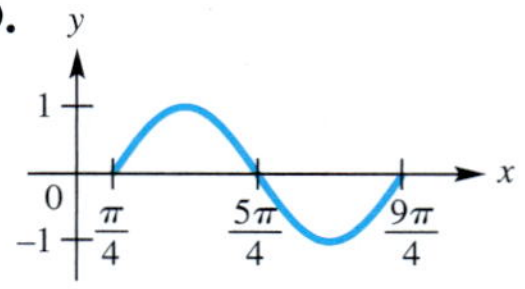

E.
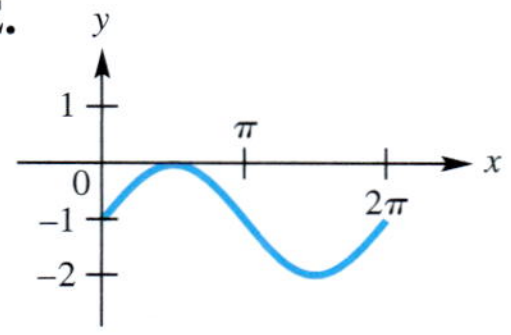

F.
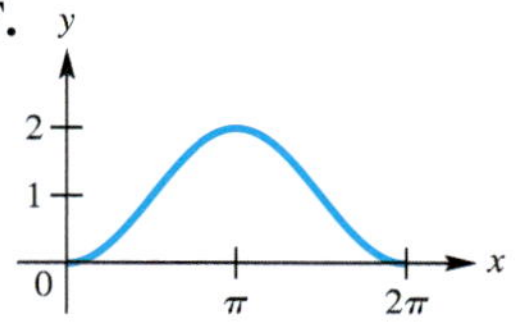

G.
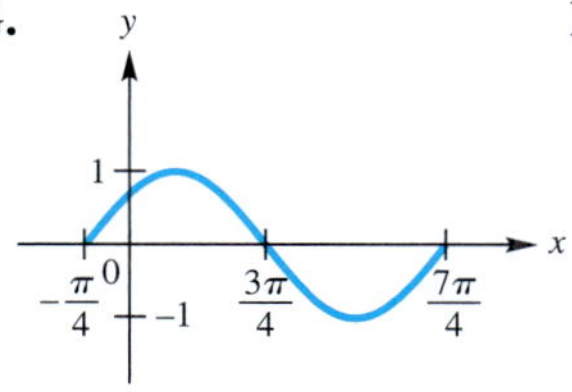

H.
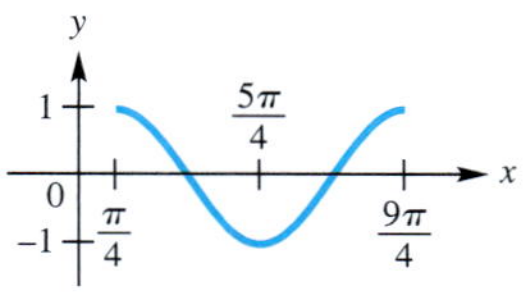

I.
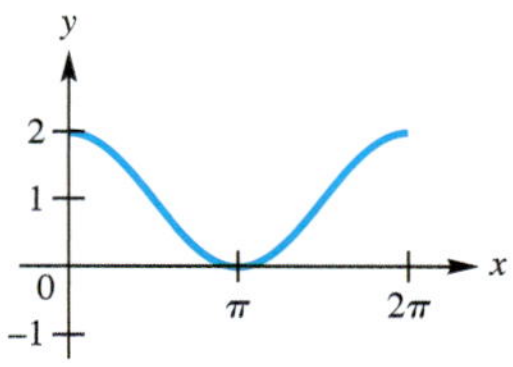

Concept Check *In Exercises 9–12, match each function with its calculator graph in the standard trig window in choices A–D.*

9. $y = \cos\left(x - \frac{\pi}{4}\right)$ **10.** $y = \sin\left(x - \frac{\pi}{4}\right)$

11. $y = 1 + \sin x$ **12.** $y = -1 + \cos x$

A.
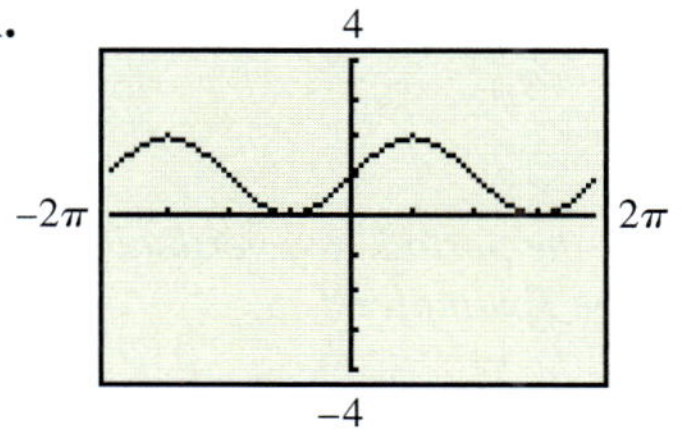

B.
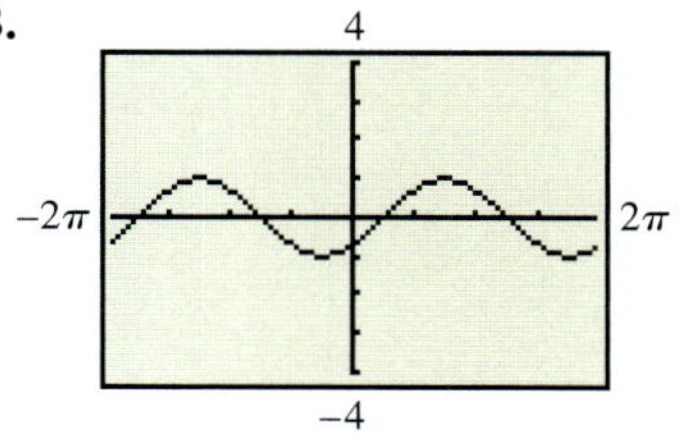

C.
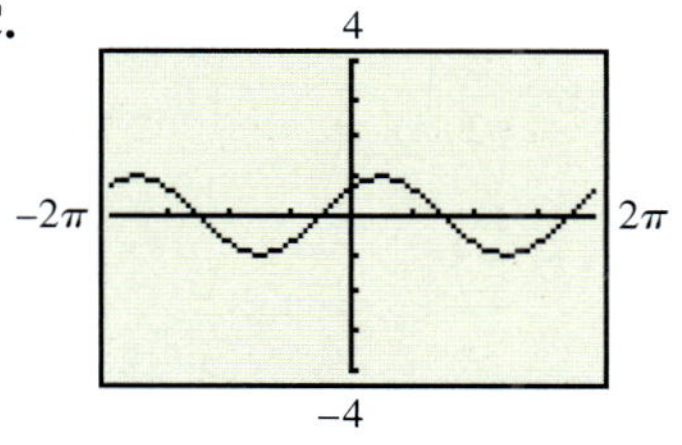

D.
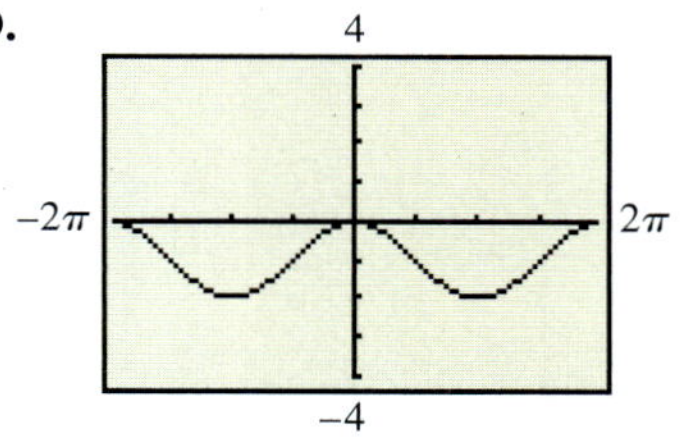

13. The graphs of $y = \sin x + 1$ and $y = \sin(x + 1)$ are **NOT** the same. Explain why this is so.

14. *Concept Check* Refer to **Exercise 13.** Which one of the two graphs is the same as that of $y = 1 + \sin x$?

Concept Check Match each function in Column I with the appropriate description in Column II.

I	II
15. $y = 3\sin(2x - 4)$	**A.** amplitude $= 2$, period $= \frac{\pi}{2}$, phase shift $= \frac{3}{4}$
16. $y = 2\sin(3x - 4)$	**B.** amplitude $= 3$, period $= \pi$, phase shift $= 2$
17. $y = -4\sin(3x - 2)$	**C.** amplitude $= 4$, period $= \frac{2\pi}{3}$, phase shift $= \frac{2}{3}$
18. $y = -2\sin(4x - 3)$	**D.** amplitude $= 2$, period $= \frac{2\pi}{3}$, phase shift $= \frac{4}{3}$

Concept Check In Exercises 19 and 20, fill in the blanks with the word right *or the word* left.

19. If the graph of $y = \cos x$ is translated $\frac{\pi}{2}$ units horizontally to the ____________, it will coincide with the graph of $y = \sin x$.

20. If the graph of $y = \sin x$ is translated $\frac{\pi}{2}$ units horizontally to the ____________, it will coincide with the graph of $y = \cos x$.

Connecting Graphs with Equations Each function graphed in Exercises 21–24 is of the form $y = c + \cos x$, $y = c + \sin x$, $y = \cos(x - d)$, or $y = \sin(x - d)$, where d is the least possible positive value. Determine the equation of the graph.

21.

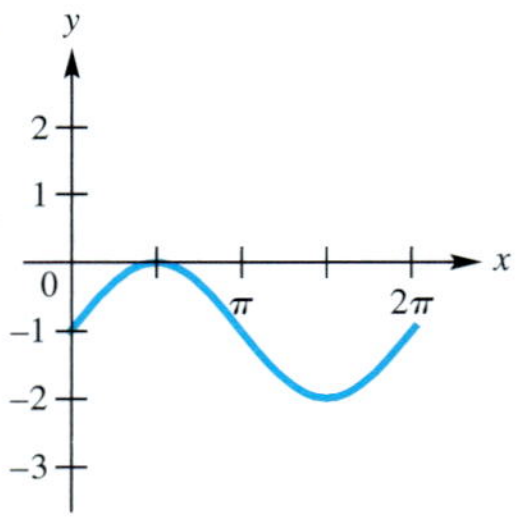

22.

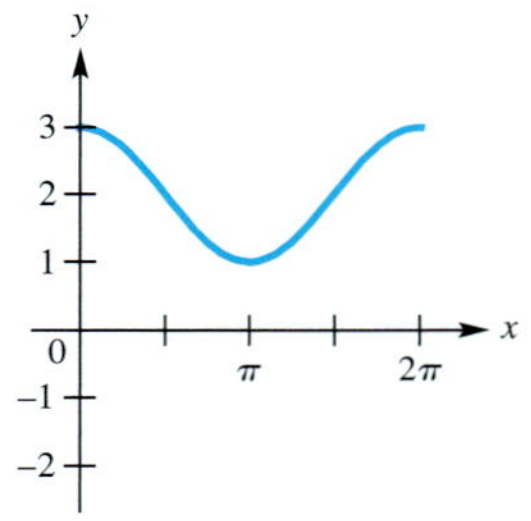

23.

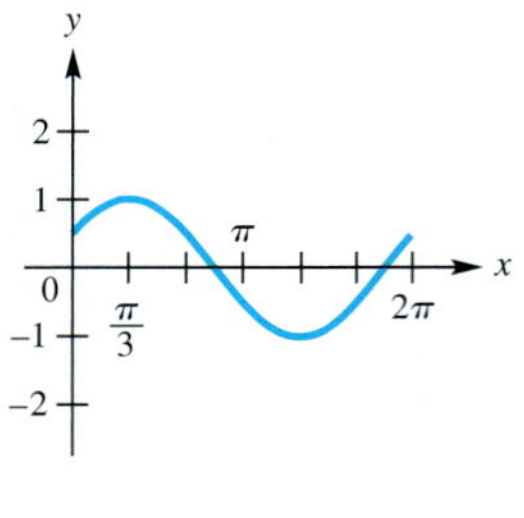

24.

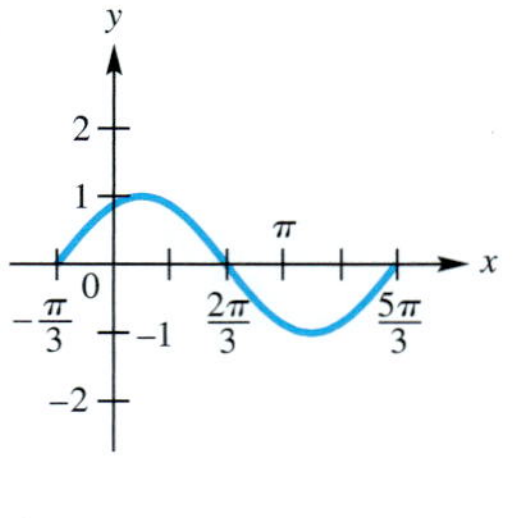

Find the amplitude, the period, any vertical translation, and any phase shift of the graph of each function. ***See Examples 1–5.***

25. $y = 2\sin(x + \pi)$

26. $y = 3\sin\left(x + \dfrac{\pi}{2}\right)$

27. $y = -\dfrac{1}{4}\cos\left(\dfrac{1}{2}x + \dfrac{\pi}{2}\right)$

28. $y = -\dfrac{1}{2}\sin\left(\dfrac{1}{2}x + \pi\right)$

29. $y = 3\cos\left[\dfrac{\pi}{2}\left(x - \dfrac{1}{2}\right)\right]$

30. $y = -\cos\left[\pi\left(x - \dfrac{1}{3}\right)\right]$

31. $y = 2 - \sin\left(3x - \dfrac{\pi}{5}\right)$

32. $y = -1 + \dfrac{1}{2}\cos(2x - 3\pi)$

Graph each function over a two-period interval. **See Examples 1 and 2.**

33. $y = \cos\left(x - \frac{\pi}{2}\right)$

34. $y = \sin\left(x - \frac{\pi}{4}\right)$

35. $y = \sin\left(x + \frac{\pi}{4}\right)$

36. $y = \cos\left(x + \frac{\pi}{3}\right)$

37. $y = 2\cos\left(x - \frac{\pi}{3}\right)$

38. $y = 3\sin\left(x - \frac{3\pi}{2}\right)$

Graph each function over a one-period interval. **See Example 3.**

39. $y = \frac{3}{2}\sin\left[2\left(x + \frac{\pi}{4}\right)\right]$

40. $y = -\frac{1}{2}\cos\left[4\left(x + \frac{\pi}{2}\right)\right]$

41. $y = -4\sin(2x - \pi)$

42. $y = 3\cos(4x + \pi)$

43. $y = \frac{1}{2}\cos\left(\frac{1}{2}x - \frac{\pi}{4}\right)$

44. $y = -\frac{1}{4}\sin\left(\frac{3}{4}x + \frac{\pi}{8}\right)$

Graph each function over a two-period interval. **See Example 4.**

45. $y = -3 + 2\sin x$

46. $y = 2 - 3\cos x$

47. $y = -1 - 2\cos 5x$

48. $y = 1 - \frac{2}{3}\sin\frac{3}{4}x$

49. $y = 1 - 2\cos\frac{1}{2}x$

50. $y = -3 + 3\sin\frac{1}{2}x$

51. $y = -2 + \frac{1}{2}\sin 3x$

52. $y = 1 + \frac{2}{3}\cos\frac{1}{2}x$

Graph each function over a one-period interval. **See Example 5.**

53. $y = -3 + 2\sin\left(x + \frac{\pi}{2}\right)$

54. $y = 4 - 3\cos(x - \pi)$

55. $y = \frac{1}{2} + \sin\left[2\left(x + \frac{\pi}{4}\right)\right]$

56. $y = -\frac{5}{2} + \cos\left[3\left(x - \frac{\pi}{6}\right)\right]$

(Modeling) Solve each problem. **See Example 6.**

57. ***Average Monthly Temperature*** The average monthly temperature (in °F) in Seattle, Washington, is shown in the table.

Month	°F	Month	°F
Jan	41	July	65
Feb	43	Aug	66
Mar	46	Sept	61
Apr	50	Oct	53
May	56	Nov	45
June	61	Dec	41

Source: World Almanac and Book of Facts.

(a) Plot the average monthly temperature over a two-year period, letting $x = 1$ correspond to January during the first year. Do the data seem to indicate a translated sine graph?

(b) The highest average monthly temperature is 66°F in August, and the lowest average monthly temperature is 41°F in January. Their average is 53.5°F. Graph the data together with the line $y = 53.5$. What does this line represent with regard to temperature in Seattle?

(c) Approximate the amplitude, period, and phase shift of the translated sine wave.

(d) Determine a function of the form $f(x) = a\sin[b(x - d)] + c$, where a, b, c, and d are constants, that models the data.

(e) Graph f together with the data on the same coordinate axes. How well does f model the given data?

(f) Use the sine regression capability of a graphing calculator to find the equation of a sine curve that fits these data.

58. ***Average Monthly Temperature*** The average monthly temperature (in °F) in Phoenix, Arizona, is shown in the table.

(a) Predict the average yearly temperature.

(b) Plot the average monthly temperature over a two-year period, letting $x = 1$ correspond to January of the first year.

(c) Determine a function of the form $f(x) = a \cos[b(x - d)] + c$, where a, b, c, and d are constants, that models the data.

(d) Graph f together with the data on the same coordinate axes. How well does f model the data?

(e) Use the sine regression capability of a graphing calculator to find the equation of a sine curve that fits these data (two years).

Month	°F	Month	°F
Jan	54	July	93
Feb	58	Aug	91
Mar	63	Sept	86
Apr	70	Oct	75
May	79	Nov	62
June	89	Dec	54

Source: World Almanac and Book of Facts.

(Modeling) Utility Bills *In an article entitled "I Found Sinusoids in My Gas Bill" (*Mathematics Teacher, *January 2000), Cathy G. Schloemer presents the following graph* that accompanied her gas bill.*

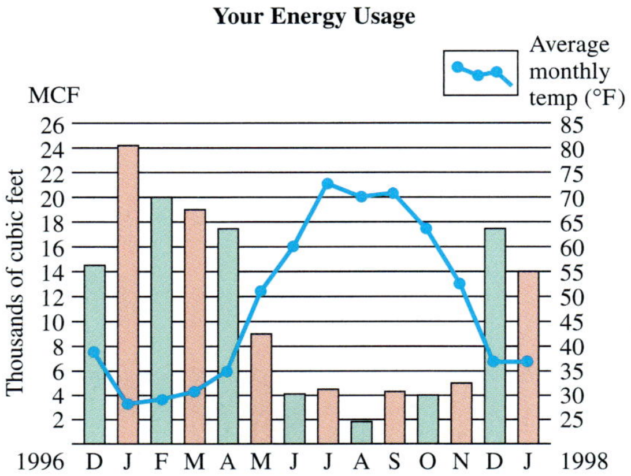

Notice that two sinusoids are suggested here: one for the behavior of the average monthly temperature and another for gas use in MCF (thousands of cubic feet). Use this information in Exercises 59 and 60.

59. If January 1997 is represented by $x = 1$, the data of estimated ordered pairs (month, temperature) are given in the list shown on the two graphing calculator screens below.

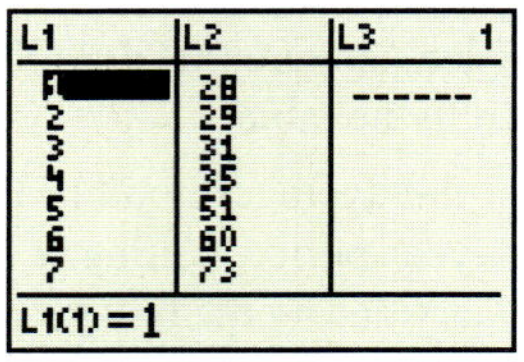

Use the sine regression feature of a graphing calculator to find a sine function that fits these data points. Then make a scatter diagram, and graph the function.

60. If January 1997 is again represented by $x = 1$, the data of estimated ordered pairs (month, gas use in thousands of cubic feet (MCF)) are given in the list shown on the two graphing calculator screens below.

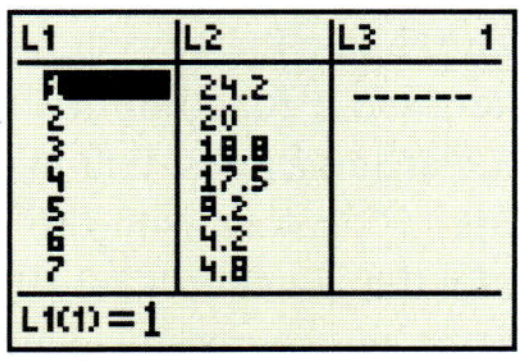

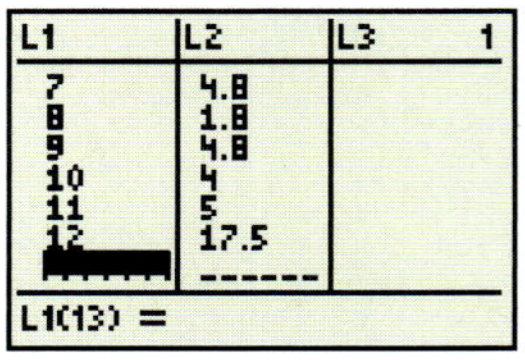

Use the sine regression feature of a graphing calculator to find a sine function that fits these data points. Then make a scatter diagram, and graph the function.

Chapter 4 Quiz (Sections 4.1–4.2)

1. Give the amplitude, period, vertical translation, and phase shift of the function $y = 3 - 4\sin\left(2x + \frac{\pi}{2}\right)$.

Graph each function over a two-period interval. Give the period and amplitude.

2. $y = -4\sin x$

3. $y = -\dfrac{1}{2}\cos 2x$

4. $y = 3\sin \pi x$

5. $y = -2\cos\left(x + \dfrac{\pi}{4}\right)$

6. $y = 2 + \sin(2x - \pi)$

7. $y = -1 + \dfrac{1}{2}\sin x$

Connecting Graphs with Equations *Each function graphed is of the form $y = a\cos bx$ or $y = a\sin bx$, where $b > 0$. Determine the equation of the graph.*

8.

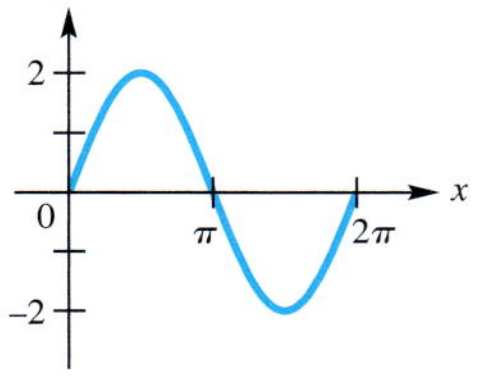

9.

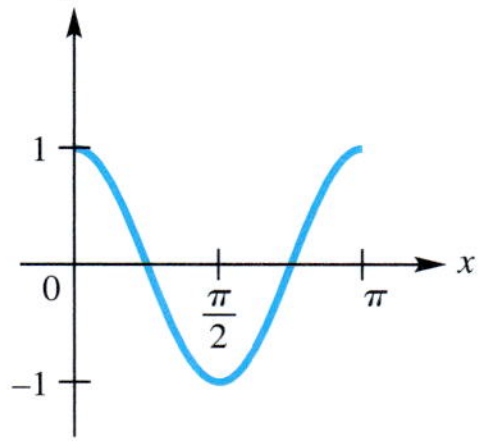

10.

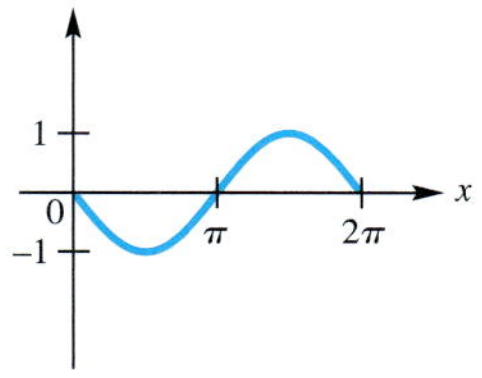

Average Monthly Temperature *The average temperature (in °F) at a certain location can be approximated by the function*

$$f(x) = 12\sin\left[\frac{\pi}{6}(x - 3.9)\right] + 72,$$

where $x = 1$ represents January, $x = 2$ represents February, and so on.

11. What is the average temperature in April?

12. What is the lowest average monthly temperature? What is the highest?

4.3 Graphs of the Tangent and Cotangent Functions

- Graph of the Tangent Function
- Graph of the Cotangent Function
- Graphing Techniques
- Connecting Graphs with Equations

Graph of the Tangent Function Consider the table of selected points accompanying the graph of the tangent function in **Figure 22** on the next page. These points include special values between $-\frac{\pi}{2}$ and $\frac{\pi}{2}$. The tangent function is undefined for odd multiples of $\frac{\pi}{2}$ and, thus, has *vertical asymptotes* for such values. A **vertical asymptote** is a vertical line that the graph approaches but does not intersect. As the x-values get closer and closer to the line, the function values increase or decrease without bound. Furthermore, since

$$\tan(-x) = -\tan x, \quad \text{(See Exercise 45.)}$$

the graph of the tangent function is symmetric with respect to the origin.

x	$y = \tan x$
$-\frac{\pi}{3}$	$-\sqrt{3} \approx -1.7$
$-\frac{\pi}{4}$	-1
$-\frac{\pi}{6}$	$-\frac{\sqrt{3}}{3} \approx -0.6$
0	0
$\frac{\pi}{6}$	$\frac{\sqrt{3}}{3} \approx 0.6$
$\frac{\pi}{4}$	1
$\frac{\pi}{3}$	$\sqrt{3} \approx 1.7$

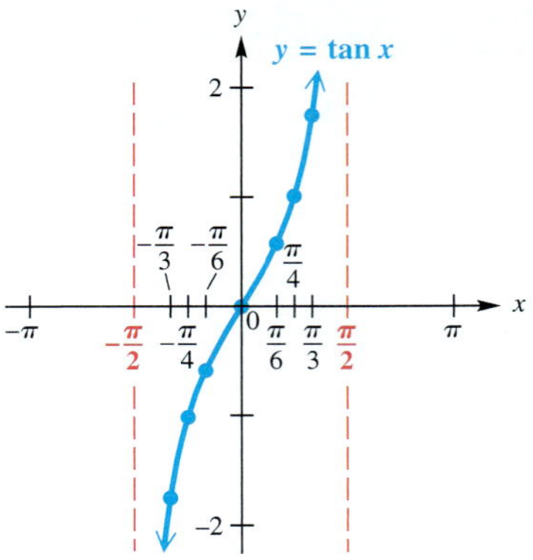

Figure 22

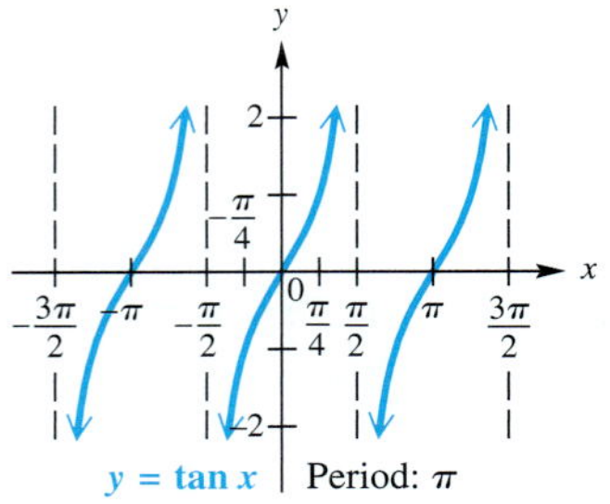

Figure 23

The tangent function has period π. Because $\tan x = \frac{\sin x}{\cos x}$, tangent values are 0 when sine values are 0, and are undefined when cosine values are 0. As x-values increase from $-\frac{\pi}{2}$ to $\frac{\pi}{2}$, tangent values range from $-\infty$ to ∞ and increase throughout the interval. Those same values are repeated as x increases from $\frac{\pi}{2}$ to $\frac{3\pi}{2}$, from $\frac{3\pi}{2}$ to $\frac{5\pi}{2}$, and so on. The graph of $y = \tan x$ from $-\frac{3\pi}{2}$ to $\frac{3\pi}{2}$ is shown in **Figure 23.** The graph continues in this pattern.

Tangent Function $f(x) = \tan x$

Domain: $\{x \mid x \neq (2n + 1)\frac{\pi}{2}$, where n is any integer$\}$ Range: $(-\infty, \infty)$

x	y
$-\frac{\pi}{2}$	undefined
$-\frac{\pi}{4}$	-1
0	0
$\frac{\pi}{4}$	1
$\frac{\pi}{2}$	undefined

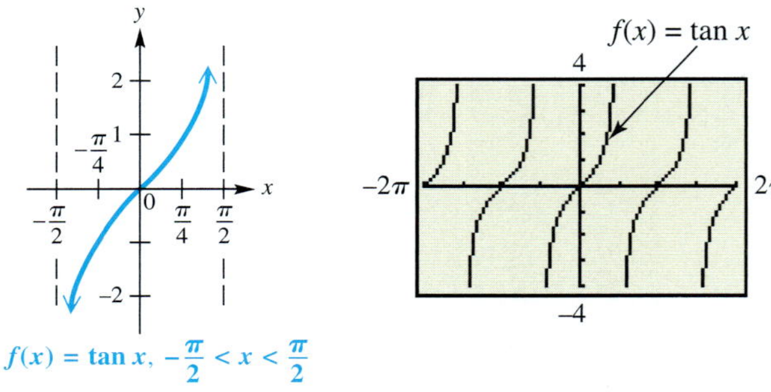

Figure 24

- The graph is discontinuous at values of x of the form $x = (2n + 1)\frac{\pi}{2}$ and has vertical asymptotes at these values.
- Its x-intercepts are of the form $x = n\pi$.
- Its period is π.
- Its graph has no amplitude, since there are no minimum or maximum values.
- The graph is symmetric with respect to the origin, so the function is an odd function. For all x in the domain, $\tan(-x) = -\tan x$.

Graph of the Cotangent Function A similar analysis for selected points between 0 and π for the graph of the cotangent function yields the graph in **Figure 25** on the next page. Here the vertical asymptotes are at x-values that are integer multiples of π. Because

$$\cot(-x) = -\cot x, \quad \text{(See Exercise 46.)}$$

this graph is also symmetric with respect to the origin. (This can be seen when more of the graph is plotted.)

x	$y = \cot x$
$\frac{\pi}{6}$	$\sqrt{3} \approx 1.7$
$\frac{\pi}{4}$	1
$\frac{\pi}{3}$	$\frac{\sqrt{3}}{3} \approx 0.6$
$\frac{\pi}{2}$	0
$\frac{2\pi}{3}$	$-\frac{\sqrt{3}}{3} \approx -0.6$
$\frac{3\pi}{4}$	-1
$\frac{5\pi}{6}$	$-\sqrt{3} \approx -1.7$

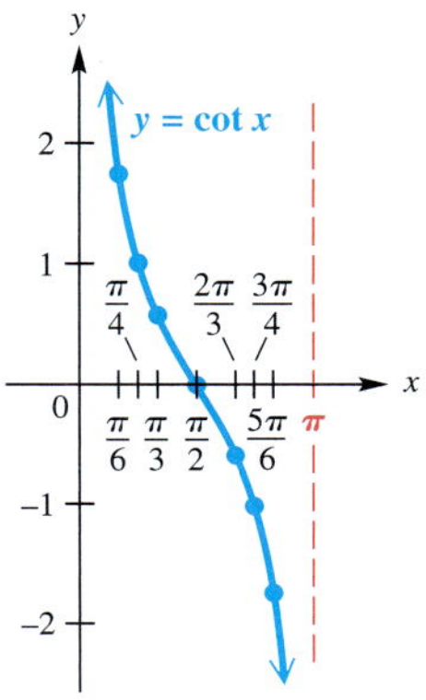

Figure 25

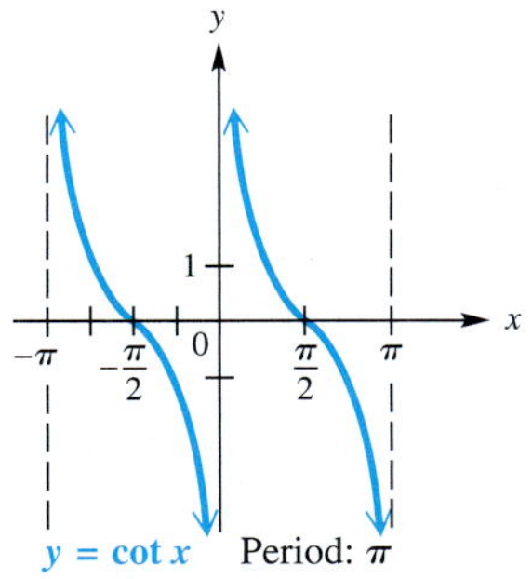

Figure 26

The cotangent function also has period π. Cotangent values are 0 when cosine values are 0, and are undefined when sine values are 0. As x-values increase from 0 to π, cotangent values range from ∞ to $-\infty$ and decrease throughout the interval. Those same values are repeated as x increases from π to 2π, from 2π to 3π, and so on. The graph of $y = \cot x$ from $-\pi$ to π is shown in **Figure 26.** The graph continues in this pattern.

Cotangent Function $f(x) = \cot x$

Domain: $\{x \mid x \neq n\pi$, where n is any integer$\}$ Range: $(-\infty, \infty)$

x	y
0	undefined
$\frac{\pi}{4}$	1
$\frac{\pi}{2}$	0
$\frac{3\pi}{4}$	-1
π	undefined

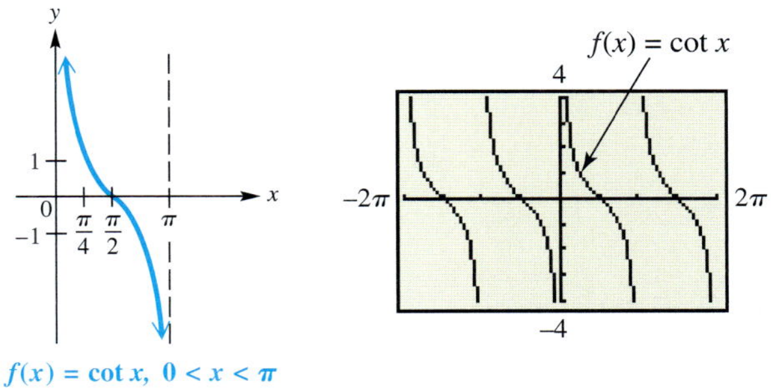

Figure 27

- The graph is discontinuous at values of x of the form $x = n\pi$ and has vertical asymptotes at these values.
- Its x-intercepts are of the form $x = (2n + 1)\frac{\pi}{2}$.
- Its period is π.
- Its graph has no amplitude, since there are no minimum or maximum values.
- The graph is symmetric with respect to the origin, so the function is an odd function. For all x in the domain, $\cot(-x) = -\cot x$.

The tangent function can be graphed directly with a graphing calculator, using the tangent key. To graph the cotangent function, however, we must use one of the identities

$$\cot x = \frac{1}{\tan x} \quad \text{or} \quad \cot x = \frac{\cos x}{\sin x},$$

because graphing calculators generally do not have cotangent keys. ■

Graphing Techniques

Guidelines for Sketching Graphs of Tangent and Cotangent Functions

To graph $y = a \tan bx$ or $y = a \cot bx$, with $b > 0$, follow these steps.

Step 1 Determine the period, $\frac{\pi}{b}$. To locate two adjacent vertical asymptotes, solve the following equations for x:

For $y = a \tan bx$: $\quad bx = -\frac{\pi}{2}$ and $bx = \frac{\pi}{2}$.

For $y = a \cot bx$: $\quad bx = 0$ and $bx = \pi$.

Step 2 Sketch the two vertical asymptotes found in Step 1.

Step 3 Divide the interval formed by the vertical asymptotes into four equal parts.

Step 4 Evaluate the function for the first-quarter point, midpoint, and third-quarter point, using the x-values found in Step 3.

Step 5 Join the points with a smooth curve, approaching the vertical asymptotes. Indicate additional asymptotes and periods of the graph as necessary.

EXAMPLE 1 Graphing $y = \tan bx$

Graph $y = \tan 2x$.

SOLUTION

Step 1 The period of this function is $\frac{\pi}{2}$. To locate two adjacent vertical asymptotes, solve $2x = -\frac{\pi}{2}$ and $2x = \frac{\pi}{2}$ (because this is a tangent function). The two asymptotes have equations $x = -\frac{\pi}{4}$ and $x = \frac{\pi}{4}$.

Step 2 Sketch the two vertical asymptotes $x = \pm\frac{\pi}{4}$, as shown in **Figure 28.**

Step 3 Divide the interval $\left(-\frac{\pi}{4}, \frac{\pi}{4}\right)$ into four equal parts. This gives the following key x-values.

first-quarter value: $-\frac{\pi}{8}$, middle value: 0, third-quarter value: $\frac{\pi}{8}$ Key x-values

Step 4 Evaluate the function for the x-values found in Step 3.

x	$-\frac{\pi}{8}$	0	$\frac{\pi}{8}$
$2x$	$-\frac{\pi}{4}$	0	$\frac{\pi}{4}$
$\tan 2x$	-1	0	1

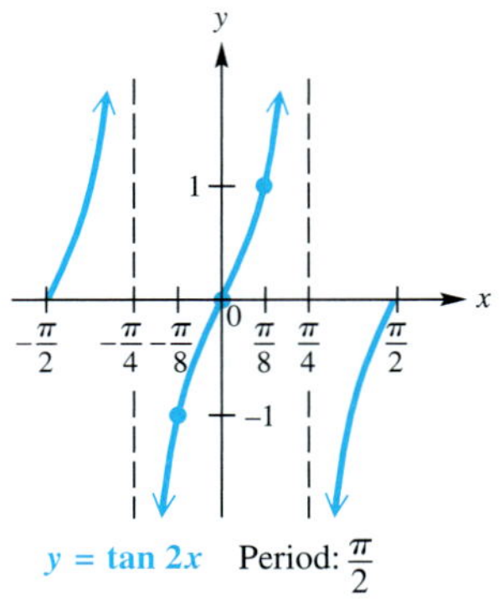

Figure 28

Step 5 Join these points with a smooth curve, approaching the vertical asymptotes. See **Figure 28.** Another period has been graphed, one half period to the left and one half period to the right.

✔ *Now Try Exercise 7.*

EXAMPLE 2 **Graphing $y = a \tan bx$**

Graph $y = -3 \tan \frac{1}{2}x$.

SOLUTION The period is $\frac{\pi}{\frac{1}{2}} = \pi \div \frac{1}{2} = \pi \cdot \frac{2}{1} = 2\pi$. Adjacent asymptotes are at $x = -\pi$ and $x = \pi$. Dividing the interval $(-\pi, \pi)$ into four equal parts gives key x-values of $-\frac{\pi}{2}$, 0, and $\frac{\pi}{2}$. Evaluating the function at these x-values gives the following key points.

$$\left(-\frac{\pi}{2}, 3\right), \quad (0, 0), \quad \left(\frac{\pi}{2}, -3\right) \qquad \text{Key points}$$

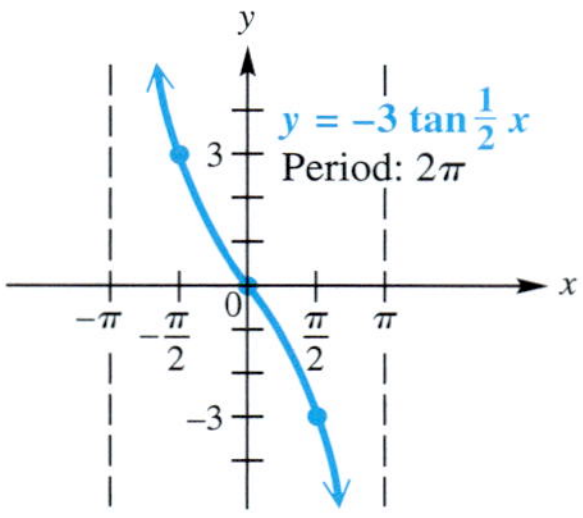

Figure 29

By plotting these points and joining them with a smooth curve, we obtain the graph shown in **Figure 29.** Because the coefficient -3 is negative, the graph is reflected across the x-axis compared to the graph of $y = 3 \tan \frac{1}{2}x$.

✔ *Now Try Exercise 15.*

NOTE The function $y = -3 \tan \frac{1}{2}x$ in **Example 2,** graphed in **Figure 29,** has a graph that compares to the graph of $y = \tan x$ as follows.

1. The period is larger because $b = \frac{1}{2}$, and $\frac{1}{2} < 1$.
2. The graph is "stretched" vertically because $a = -3$, and $|-3| > 1$.
3. Each branch of the graph falls from left to right (that is, the function decreases) between each pair of adjacent asymptotes because $a = -3$, and $-3 < 0$. When $a < 0$, the graph is reflected across the x-axis compared to the graph of $y = |a| \tan bx$.

EXAMPLE 3 **Graphing $y = a \cot bx$**

Graph $y = \frac{1}{2} \cot 2x$.

SOLUTION Because this function involves the cotangent, we can locate two adjacent asymptotes by solving the equations $2x = 0$ and $2x = \pi$. The lines $x = 0$ (the y-axis) and $x = \frac{\pi}{2}$ are two such asymptotes. We divide the interval $\left(0, \frac{\pi}{2}\right)$ into four equal parts, getting key x-values of $\frac{\pi}{8}$, $\frac{\pi}{4}$, and $\frac{3\pi}{8}$. Evaluating the function at these x-values gives the key points $\left(\frac{\pi}{8}, \frac{1}{2}\right)$, $\left(\frac{\pi}{4}, 0\right)$, $\left(\frac{3\pi}{8}, -\frac{1}{2}\right)$. We plot these points and join them with a smooth curve approaching the asymptotes to obtain the graph shown in **Figure 30.**

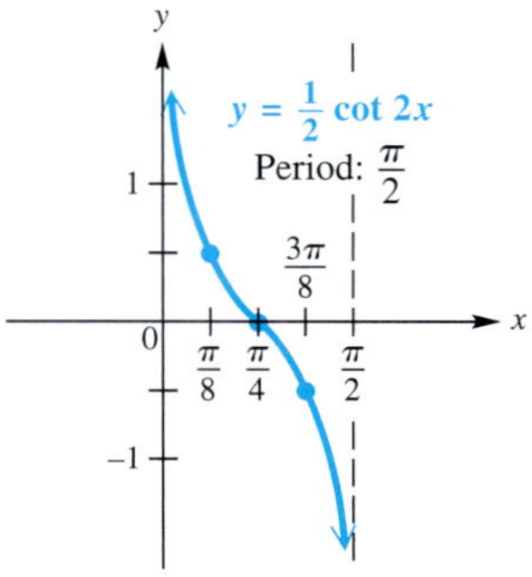

Figure 30

✔ *Now Try Exercise 17.*

Like the other circular functions, the graphs of the tangent and cotangent functions may be translated horizontally and vertically.

EXAMPLE 4 Graphing $y = c + \tan x$

Graph $y = 2 + \tan x$.

ANALYTIC SOLUTION

Every value of y for this function will be 2 units more than the corresponding value of y in $y = \tan x$, causing the graph of $y = 2 + \tan x$ to be translated 2 units up compared to the graph of $y = \tan x$. See **Figure 31.**

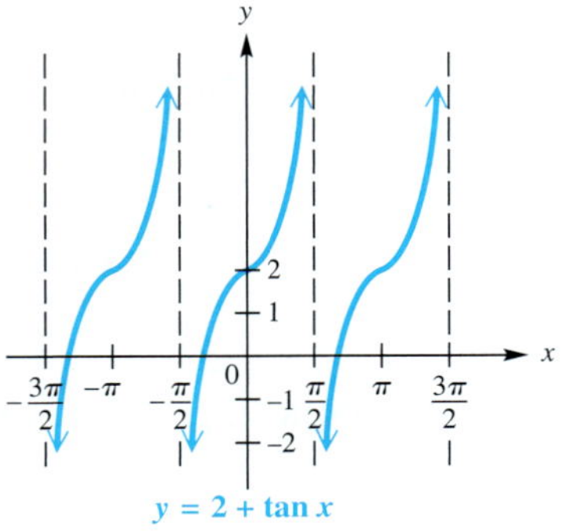

Figure 31

Three periods of the function are shown in **Figure 31.** Because the period of $y = 2 + \tan x$ is π, additional asymptotes and periods of the function can be drawn by repeating the basic graph every π units on the x-axis to the left or to the right of the graph shown.

GRAPHING CALCULATOR SOLUTION

To see the vertical translation, observe the coordinates displayed at the bottoms of the screens in **Figures 32 and 33.** For $X = \frac{\pi}{4} \approx 0.78539816$,

$$Y_1 = \tan X = 1,$$

while for the same X-value,

$$Y_2 = 2 + \tan X = 2 + 1 = 3.$$

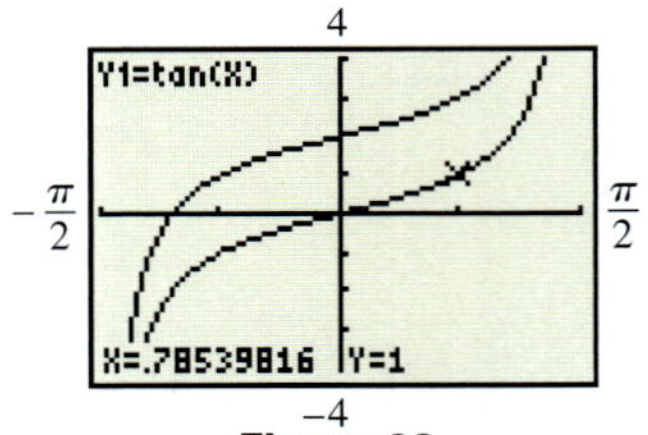

Figure 32

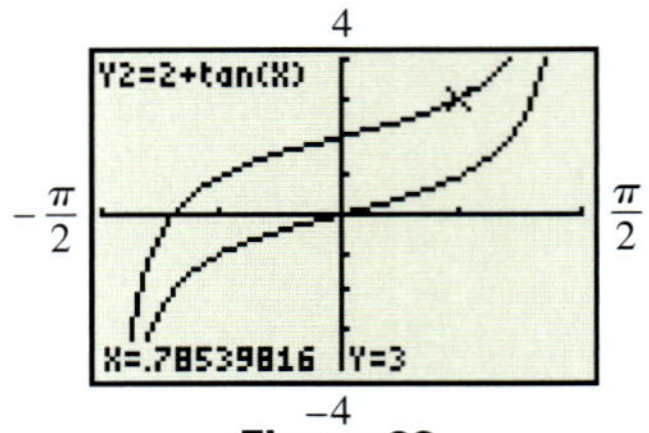

Figure 33

✔ *Now Try Exercise 23.*

EXAMPLE 5 Graphing $y = c + a\cot(x - d)$

Graph $y = -2 - \cot\left(x - \frac{\pi}{4}\right)$.

SOLUTION Here $b = 1$, so the period is π. The negative sign in front of the cotangent will cause the graph to be reflected across the x-axis, and the argument $\left(x - \frac{\pi}{4}\right)$ indicates a phase shift (horizontal shift) $\frac{\pi}{4}$ unit to the right. Because $c = -2$, the graph will then be translated down 2 units. To locate adjacent asymptotes, since this function involves the cotangent, we solve the following equations.

$$x - \frac{\pi}{4} = 0 \quad \text{and} \quad x - \frac{\pi}{4} = \pi$$

$$x = \frac{\pi}{4} \quad \text{and} \quad x = \frac{5\pi}{4} \qquad \text{Add } \tfrac{\pi}{4}.$$

Dividing the interval $\left(\frac{\pi}{4}, \frac{5\pi}{4}\right)$ into four equal parts and evaluating the function at the three key x-values within the interval give these points.

$$\left(\frac{\pi}{2}, -3\right), \quad \left(\frac{3\pi}{4}, -2\right), \quad (\pi, -1) \qquad \text{Key points}$$

We join these points with a smooth curve. This period of the graph, along with the one in the domain interval $\left(-\frac{3\pi}{4}, \frac{\pi}{4}\right)$, is shown in **Figure 34** on the next page.

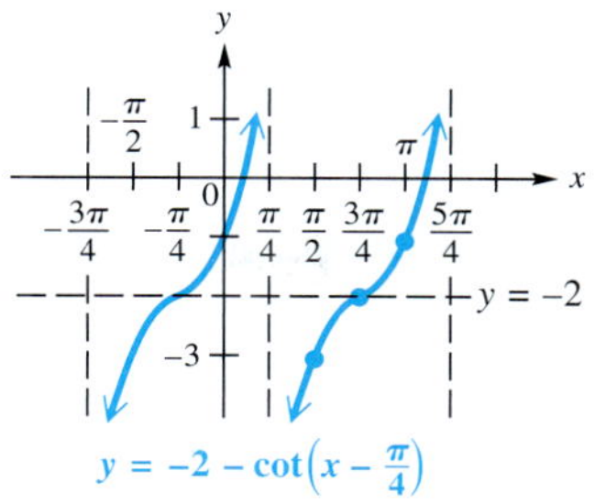

Figure 34

✔ ***Now Try Exercise 31.***

Connecting Graphs with Equations

EXAMPLE 6 Determining an Equation for a Graph

Determine an equation for each graph.

(a)

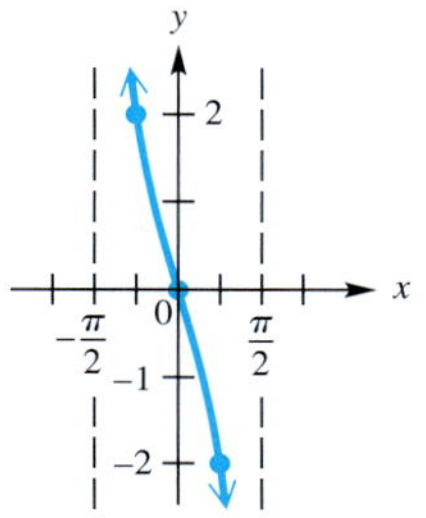

(b)

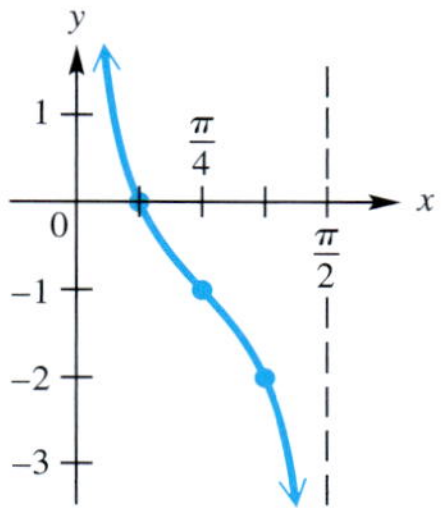

SOLUTION

(a) This graph is that of $y = \tan x$ but reflected across the x-axis and stretched vertically by a factor of 2. Therefore, an equation for this graph is

$$y = -2 \tan x.$$

Vertical stretch

x-axis reflection

(b) This is the graph of a cotangent function, but the period is $\frac{\pi}{2}$ rather than π. Therefore, the coefficient of x is 2. This graph is vertically translated 1 unit down compared to the graph of $y = \cot 2x$. An equation for this graph is

$$y = -1 + \cot 2x.$$

Period is $\frac{\pi}{2}$.

Vertical translation
1 unit down

✔ ***Now Try Exercises 33 and 37.***

NOTE Because the circular functions are periodic, there are infinitely many equations that correspond to each graph in **Example 6.** Confirm that both

$$y = -1 - \cot(-2x) \quad \text{and} \quad y = -1 - \tan\left(2x - \frac{\pi}{2}\right)$$

are equations for the graph in **Example 6(b).** When writing the equation from a graph, it is practical to write the simplest form. Therefore, we choose values of b where $b > 0$ and write the function without a phase shift when possible.

4.3 Exercises

Concept Check *In Exercises 1–6, match each function with its graph from choices A–F.*

1. $y = -\tan x$

2. $y = -\cot x$

3. $y = \tan\left(x - \frac{\pi}{4}\right)$

4. $y = \cot\left(x - \frac{\pi}{4}\right)$

5. $y = \cot\left(x + \frac{\pi}{4}\right)$

6. $y = \tan\left(x + \frac{\pi}{4}\right)$

A.

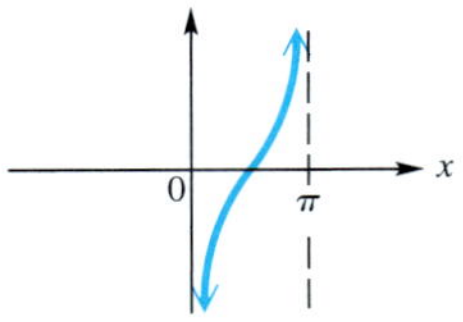

B.

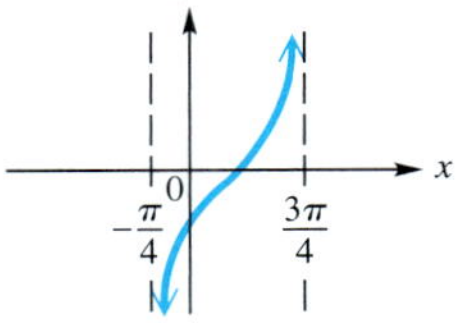

C.

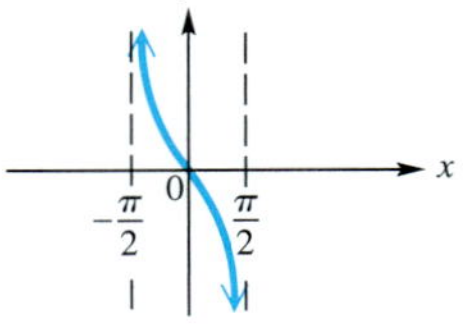

D.

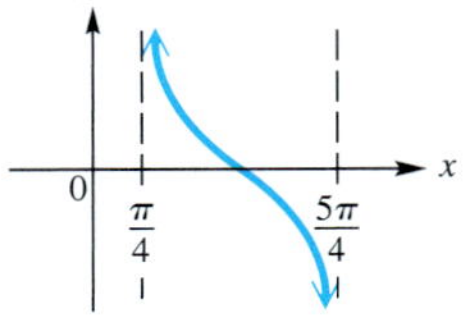

E.

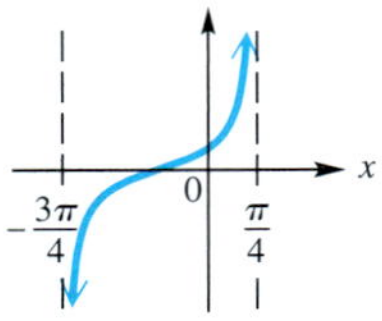

F.

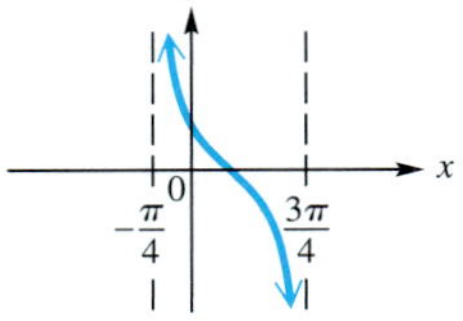

Graph each function over a one-period interval. **See Examples 1–3.**

7. $y = \tan 4x$

8. $y = \tan \frac{1}{2}x$

9. $y = 2 \tan x$

10. $y = 2 \cot x$

11. $y = 2 \tan \frac{1}{4}x$

12. $y = \frac{1}{2} \cot x$

13. $y = \cot 3x$

14. $y = -\cot \frac{1}{2}x$

15. $y = -2 \tan \frac{1}{4}x$

16. $y = 3 \tan \frac{1}{2}x$

17. $y = \frac{1}{2} \cot 4x$

18. $y = -\frac{1}{2} \cot 2x$

Graph each function over a two-period interval. **See Examples 4 and 5.**

19. $y = \tan(2x - \pi)$

20. $y = \tan\left(\frac{x}{2} + \pi\right)$

21. $y = \cot\left(3x + \frac{\pi}{4}\right)$

22. $y = \cot\left(2x - \frac{3\pi}{2}\right)$

23. $y = 1 + \tan x$

24. $y = 1 - \tan x$

25. $y = 1 - \cot x$

26. $y = -2 - \cot x$

27. $y = -1 + 2 \tan x$

28. $y = 3 + \frac{1}{2} \tan x$

29. $y = -1 + \frac{1}{2} \cot(2x - 3\pi)$

30. $y = -2 + 3 \tan(4x + \pi)$

31. $y = 1 - 2 \cot\left[2\left(x + \frac{\pi}{2}\right)\right]$

32. $y = -2 + \frac{2}{3} \tan\left(\frac{3}{4}x - \pi\right)$

Connecting Graphs with Equations *Determine the simplest form of an equation for each graph. Choose $b > 0$, and include no phase shifts. (Midpoints and quarter-points are identified by dots.)* ***See Example 6.***

33.

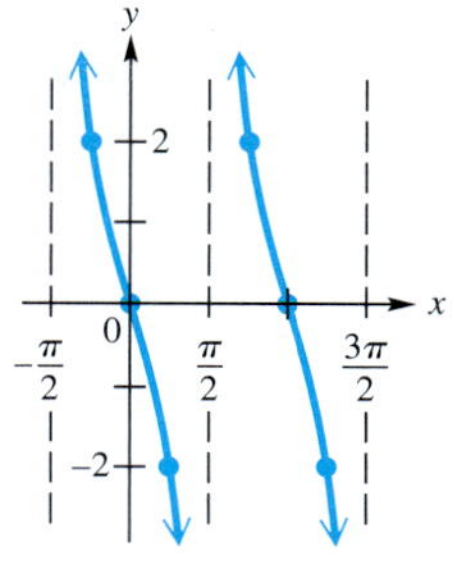

34.

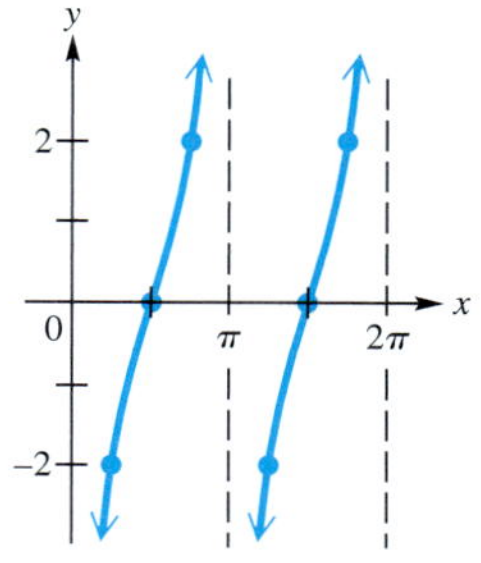

35.

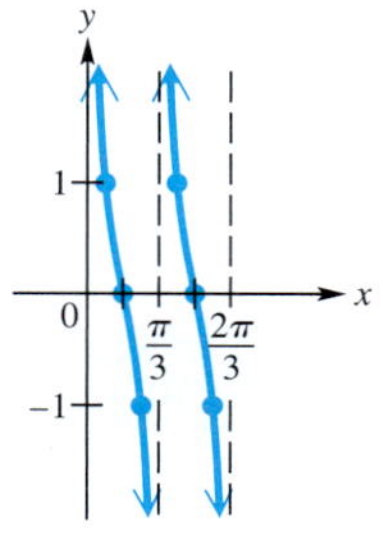

36.

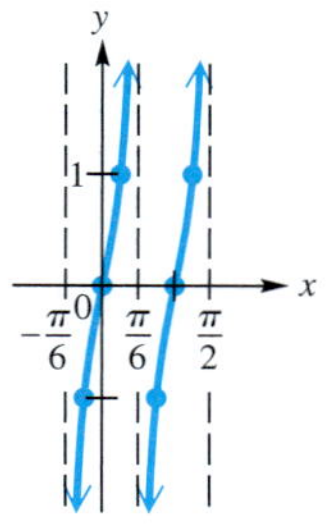

37.

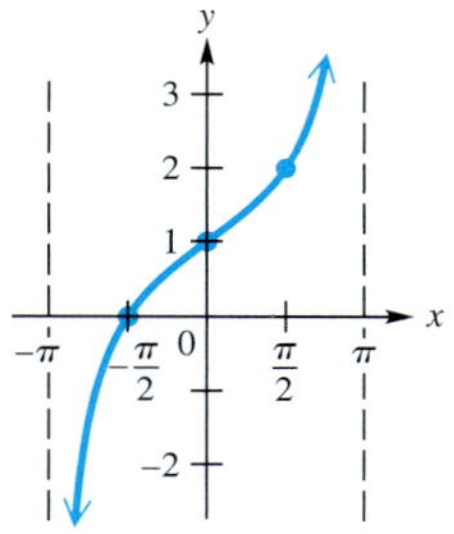

38.

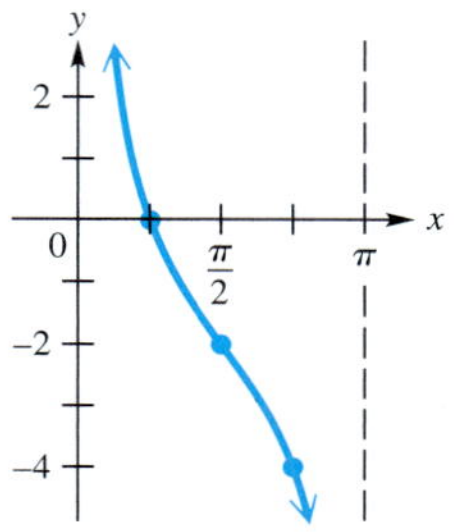

Concept Check *In Exercises 39–42, tell whether each statement is* true *or* false. *If* false, *tell why.*

39. The least positive number k for which $x = k$ is an asymptote for the tangent function is $\frac{\pi}{2}$.

40. The least positive number k for which $x = k$ is an asymptote for the cotangent function is $\frac{\pi}{2}$.

41. The graph of $y = \tan x$ in **Figure 23** suggests that $\tan(-x) = \tan x$ for all x in the domain of $\tan x$.

42. The graph of $y = \cot x$ in **Figure 26** suggests that $\cot(-x) = -\cot x$ for all x in the domain of $\cot x$.

Work each exercise.

43. *Concept Check* If c is any number, then how many solutions does the equation $c = \tan x$ have in the interval $(-2\pi, 2\pi]$?

44. *Concept Check* Consider the function defined by $f(x) = -4\tan(2x + \pi)$. What is the domain of f? What is its range?

45. Show that $\tan(-x) = -\tan x$ by writing $\tan(-x)$ as $\frac{\sin(-x)}{\cos(-x)}$ and then using the relationships for $\sin(-x)$ and $\cos(-x)$.

46. Show that $\cot(-x) = -\cot x$ by writing $\cot(-x)$ as $\frac{\cos(-x)}{\sin(-x)}$ and then using the relationships for $\cos(-x)$ and $\sin(-x)$.

47. *(Modeling) Distance of a Rotating Beacon* A rotating beacon is located at point A next to a long wall. The beacon is 4 m from the wall. The distance d is given by

$$d = 4 \tan 2\pi t,$$

where t is time measured in seconds since the beacon started rotating. (When $t = 0$, the beacon is aimed at point R. When the beacon is aimed to the right of R, the value of d is positive; d is negative when the beacon is aimed to the left of R.) Find d for each time.

(a) $t = 0$
(b) $t = 0.4$
(c) $t = 0.8$
(d) $t = 1.2$
(e) Why is 0.25 a meaningless value for t?

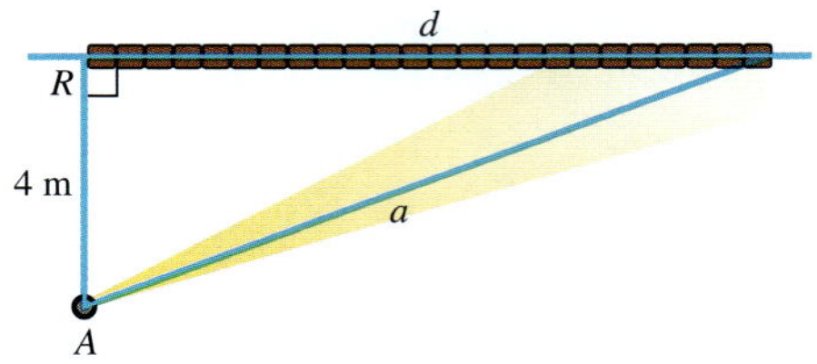

48. Simultaneously graph $y = \tan x$ and $y = x$ in the window $[-1, 1]$ by $[-1, 1]$ with a graphing calculator. Write a short description of the relationship between $\tan x$ and x for small x-values.

Relating Concepts

For individual or collaborative investigation *(Exercises 49–54)*

Consider the following function from ***Example 5. Work these exercises in order.***

$$y = -2 - \cot\left(x - \frac{\pi}{4}\right)$$

49. What is the least positive number for which $y = \cot x$ is undefined?

50. Let k represent the number you found in **Exercise 49.** Set $x - \frac{\pi}{4}$ equal to k, and solve to find a positive number for which $\cot\left(x - \frac{\pi}{4}\right)$ is undefined.

51. Based on your answer in **Exercise 50** and the fact that the cotangent function has period π, give the general form of the equations of the asymptotes of the graph of $y = -2 - \cot\left(x - \frac{\pi}{4}\right)$. Let n represent any integer.

52. Use the capabilities of your calculator to find the least positive x-intercept of the graph of this function.

53. Use the fact that the period of this function is π to find the next positive x-intercept.

54. Give the solution set of the equation $-2 - \cot\left(x - \frac{\pi}{4}\right) = 0$ over all real numbers. Let n represent any integer.

4.4 Graphs of the Secant and Cosecant Functions

- Graph of the Secant Function
- Graph of the Cosecant Function
- Graphing Techniques
- Connecting Graphs with Equations
- Addition of Ordinates

Graph of the Secant Function Consider the table of selected points accompanying the graph of the secant function in **Figure 35** on the next page. These points include special values from $-\pi$ to π. The secant function is undefined for odd multiples of $\frac{\pi}{2}$ and thus, like the tangent function, has vertical asymptotes for such values. Furthermore, since

$$\sec(-x) = \sec x, \quad \text{(See Exercise 31.)}$$

the graph of the secant function is symmetric with respect to the y-axis.

x	$y = \sec x$
0	1
$\pm\frac{\pi}{6}$	$\frac{2\sqrt{3}}{3} \approx 1.2$
$\pm\frac{\pi}{4}$	$\sqrt{2} \approx 1.4$
$\pm\frac{\pi}{3}$	2
$\pm\frac{2\pi}{3}$	-2
$\pm\frac{3\pi}{4}$	$-\sqrt{2} \approx -1.4$
$\pm\frac{5\pi}{6}$	$-\frac{2\sqrt{3}}{3} \approx -1.2$
$\pm\pi$	-1

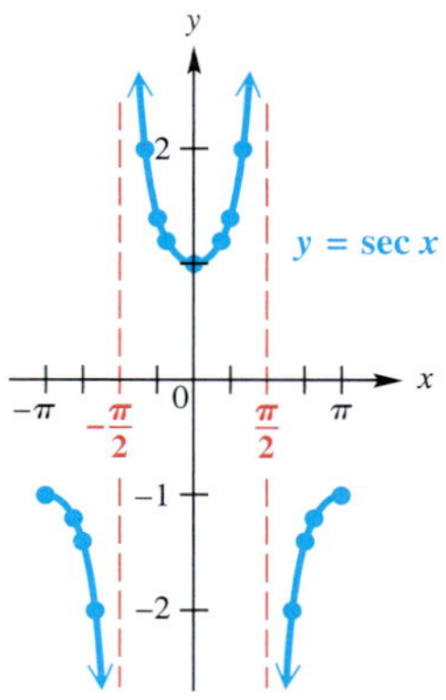

Figure 35

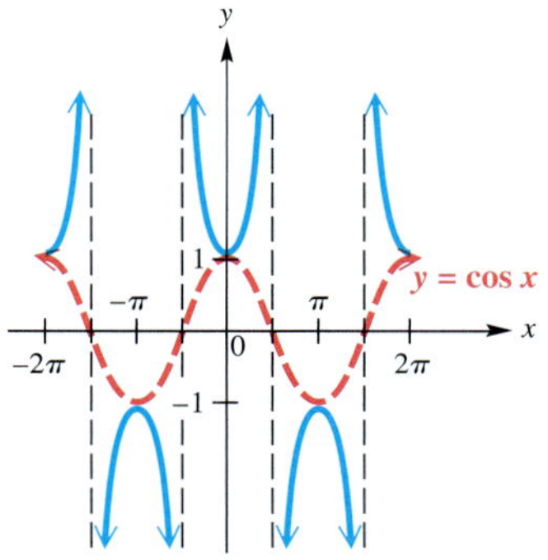

$y = \sec x$ Period: 2π

Figure 36

Because secant values are reciprocals of corresponding cosine values, the period of the secant function is 2π, the same as for $y = \cos x$. When $\cos x = 1$, the value of $\sec x$ is also 1. Likewise, when $\cos x = -1$, $\sec x = -1$. For all x, $-1 \leq \cos x \leq 1$, and thus, $|\sec x| \geq 1$ for all x in its domain. **Figure 36** shows how the graphs of $y = \cos x$ and $y = \sec x$ are related.

Secant Function $f(x) = \sec x$

Domain: $\{x \mid x \neq (2n+1)\frac{\pi}{2}$, where n is any integer$\}$ Range: $(-\infty, -1] \cup [1, \infty)$

x	y
$-\frac{\pi}{2}$	undefined
$-\frac{\pi}{4}$	$\sqrt{2}$
0	1
$\frac{\pi}{4}$	$\sqrt{2}$
$\frac{\pi}{2}$	undefined
$\frac{3\pi}{4}$	$-\sqrt{2}$
π	-1
$\frac{3\pi}{2}$	undefined

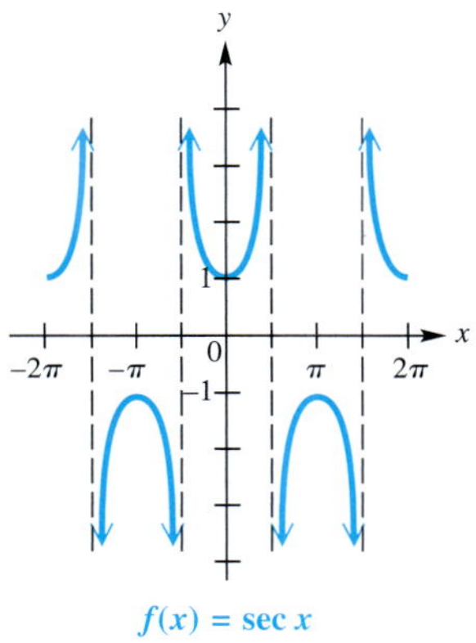

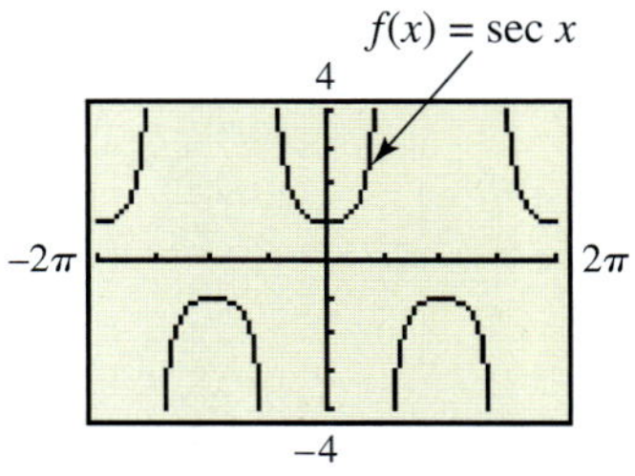

Figure 37

- The graph is discontinuous at values of x of the form $x = (2n+1)\frac{\pi}{2}$ and has vertical asymptotes at these values.
- There are no x-intercepts.
- Its period is 2π.
- Its graph has no amplitude, since there are no minimum or maximum values.
- The graph is symmetric with respect to the y-axis, so the function is an even function. For all x in the domain, $\sec(-x) = \sec x$.

Graph of the Cosecant Function A similar analysis for selected points between $-\pi$ and π for the graph of the cosecant function yields the graph in **Figure 38.** The vertical asymptotes are at x-values that are integer multiples of π. Because

$$\csc(-x) = -\csc x, \quad \text{(See Exercise 32.)}$$

this graph is symmetric with respect to the origin.

x	$y = \csc x$	x	$y = \csc x$
$\frac{\pi}{6}$	2	$-\frac{\pi}{6}$	-2
$\frac{\pi}{4}$	$\sqrt{2} \approx 1.4$	$-\frac{\pi}{4}$	$-\sqrt{2} \approx -1.4$
$\frac{\pi}{3}$	$\frac{2\sqrt{3}}{3} \approx 1.2$	$-\frac{\pi}{3}$	$-\frac{2\sqrt{3}}{3} \approx -1.2$
$\frac{\pi}{2}$	1	$-\frac{\pi}{2}$	-1
$\frac{2\pi}{3}$	$\frac{2\sqrt{3}}{3} \approx 1.2$	$-\frac{2\pi}{3}$	$-\frac{2\sqrt{3}}{3} \approx -1.2$
$\frac{3\pi}{4}$	$\sqrt{2} \approx 1.4$	$-\frac{3\pi}{4}$	$-\sqrt{2} \approx -1.4$
$\frac{5\pi}{6}$	2	$-\frac{5\pi}{6}$	-2

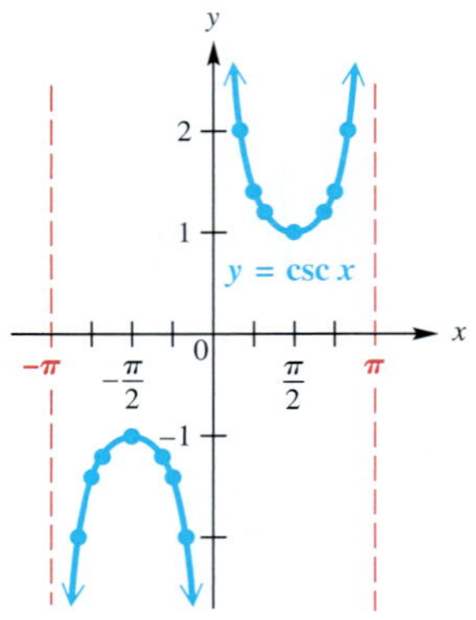

Figure 38

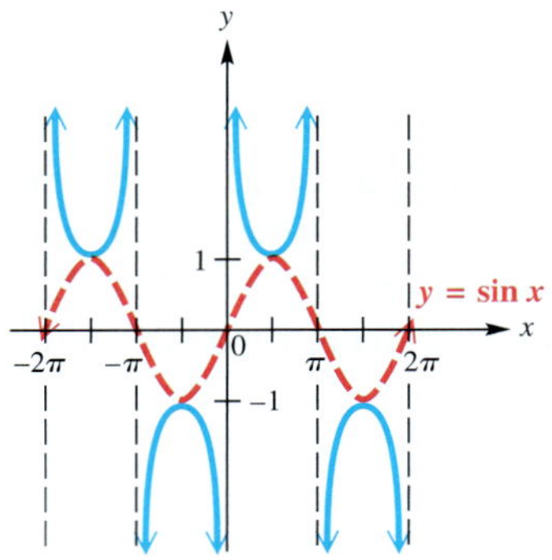

$y = \csc x$ Period: 2π

Figure 39

Because cosecant values are reciprocals of corresponding sine values, the period of the cosecant function is 2π, the same as for $y = \sin x$. When $\sin x = 1$, the value of $\csc x$ is also 1. Likewise, when $\sin x = -1$, $\csc x = -1$. For all x, $-1 \le \sin x \le 1$, and thus $|\csc x| \ge 1$ for all x in its domain. **Figure 39** shows how the graphs of $y = \sin x$ and $y = \csc x$ are related.

Cosecant Function $f(x) = \csc x$

Domain: $\{x \mid x \neq n\pi$, where n is any integer$\}$ Range: $(-\infty, -1] \cup [1, \infty)$

x	y
0	undefined
$\frac{\pi}{6}$	2
$\frac{\pi}{3}$	$\frac{2\sqrt{3}}{3}$
$\frac{\pi}{2}$	1
$\frac{2\pi}{3}$	$\frac{2\sqrt{3}}{3}$
π	undefined
$\frac{3\pi}{2}$	-1
2π	undefined

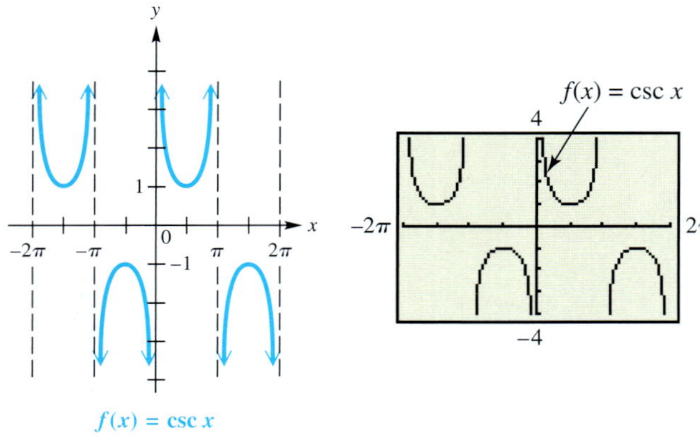

Figure 40

- The graph is discontinuous at values of x of the form $x = n\pi$ and has vertical asymptotes at these values.
- There are no x-intercepts.
- Its period is 2π.
- Its graph has no amplitude, since there are no minimum or maximum values.
- The graph is symmetric with respect to the origin, so the function is an odd function. For all x in the domain, $\csc(-x) = -\csc x$.

Typically, calculators do not have keys for the cosecant and secant functions. To graph $y = \csc x$ with a graphing calculator, use

$$\csc x = \frac{1}{\sin x}. \quad \text{Reciprocal identity}$$

Figure 41 shows the graph of $Y_1 = \sin X$ as a thin graph and that of $Y_2 = \csc X$ as a thick graph. Although this calculator screen does not show the vertical asymptotes, they occur at each x-intercept of the guide function $Y_1 = \sin X$. **Figure 42** shows the graph of the secant function, graphed in a similar manner, using the identity

$$\sec x = \frac{1}{\cos x}. \quad \text{Reciprocal identity}$$

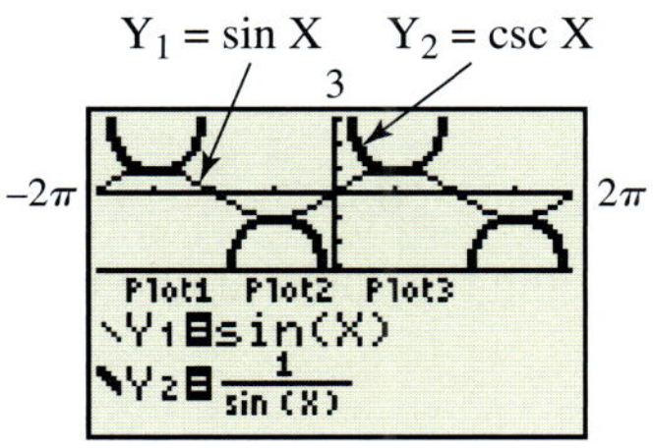

Connected mode

Figure 41

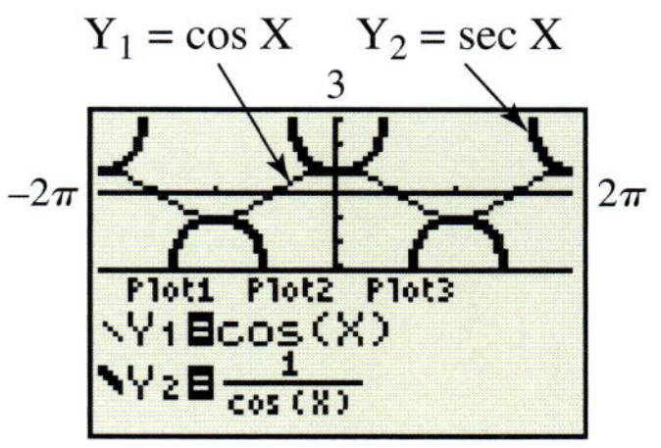

Connected mode

Figure 42

Graphing Techniques In the previous section, we gave guidelines for sketching graphs of tangent and cotangent functions. We now present similar guidelines for graphing cosecant and secant functions.

Guidelines for Sketching Graphs of Cosecant and Secant Functions

To graph $\boldsymbol{y = a \csc bx}$ or $\boldsymbol{y = a \sec bx}$, with $b > 0$, follow these steps.

Step 1 Graph the corresponding reciprocal function as a guide, using a dashed curve.

To Graph	Use as a Guide
$y = a \csc bx$	$y = a \sin bx$
$y = a \sec bx$	$y = a \cos bx$

Step 2 Sketch the vertical asymptotes. They will have equations of the form $x = k$, where k is an x-intercept of the graph of the guide function.

Step 3 Sketch the graph of the desired function by drawing the typical U-shaped branches between the adjacent asymptotes. The branches will be above the graph of the guide function when the guide function values are positive and below the graph of the guide function when the guide function values are negative. The graph will resemble those in **Figures 37 and 40** in the function boxes given earlier in this section.

Like graphs of the sine and cosine functions, graphs of the secant and cosecant functions may be translated vertically and horizontally. The period of both basic functions is 2π.

EXAMPLE 1 Graphing $y = a \sec bx$

Graph $y = 2 \sec \frac{1}{2}x$.

SOLUTION

Step 1 This function involves the secant, so the corresponding reciprocal function will involve the cosine. The guide function to graph is

$$y = 2 \cos \frac{1}{2}x.$$

Using the guidelines of **Section 4.1,** we find that this guide function has amplitude 2 and that one period of the graph lies along the interval that satisfies the following inequality.

$$0 \le \frac{1}{2}x \le 2\pi$$

$$0 \le x \le 4\pi, \quad \text{or} \quad [0, 4\pi] \quad \text{(Appendix A)}$$

Dividing this interval into four equal parts gives these key points.

$$(0, 2), \quad (\pi, 0), \quad (2\pi, -2), \quad (3\pi, 0), \quad (4\pi, 2) \quad \text{Key points}$$

These points are plotted and joined with a dashed red curve to indicate that this graph is only a guide. An additional period is graphed as shown in **Figure 43(a).**

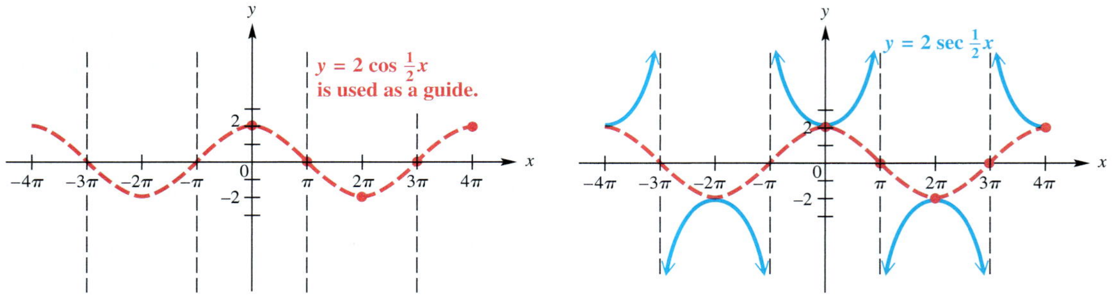

(a) (b)

Figure 43

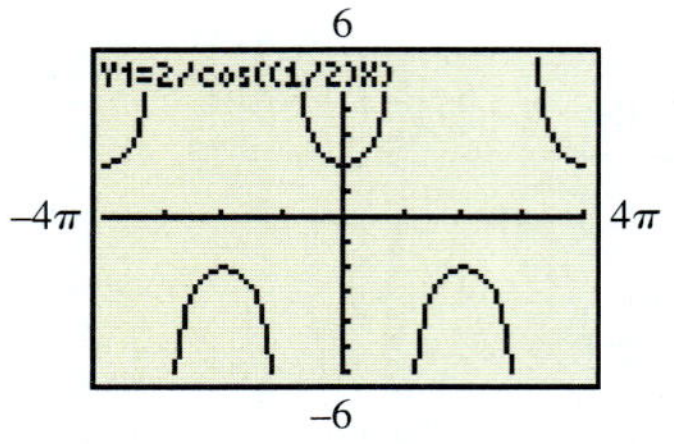

This is a calculator graph of the function in **Example 1.**

Step 2 Sketch the vertical asymptotes as shown in **Figure 43(a).** These occur at x-values for which the guide function equals 0, such as

$$x = -3\pi, \quad x = -\pi, \quad x = \pi, \quad x = 3\pi.$$

Step 3 Sketch the graph of $y = 2 \sec \frac{1}{2}x$ by drawing the typical U-shaped branches, approaching the asymptotes. See the solid blue graph in **Figure 43(b).**

✔ *Now Try Exercise 5.*

EXAMPLE 2 Graphing $y = a \csc(x - d)$

Graph $y = \frac{3}{2} \csc\left(x - \frac{\pi}{2}\right)$.

SOLUTION

Step 1 Use the guidelines of **Section 4.2** to graph the corresponding reciprocal function defined by

$$y = \frac{3}{2} \sin\left(x - \frac{\pi}{2}\right),$$

shown as a red dashed curve in **Figure 44** on the next page.

Step 2 Sketch the vertical asymptotes through the x-intercepts of the graph of $y = \frac{3}{2}\sin\left(x - \frac{\pi}{2}\right)$. These have the form $x = (2n + 1)\frac{\pi}{2}$, where n is any integer. See the black dashed lines in **Figure 44.**

Step 3 Sketch the graph of $y = \frac{3}{2}\csc\left(x - \frac{\pi}{2}\right)$ by drawing the typical U-shaped branches between adjacent asymptotes. See the solid blue graph in **Figure 44.**

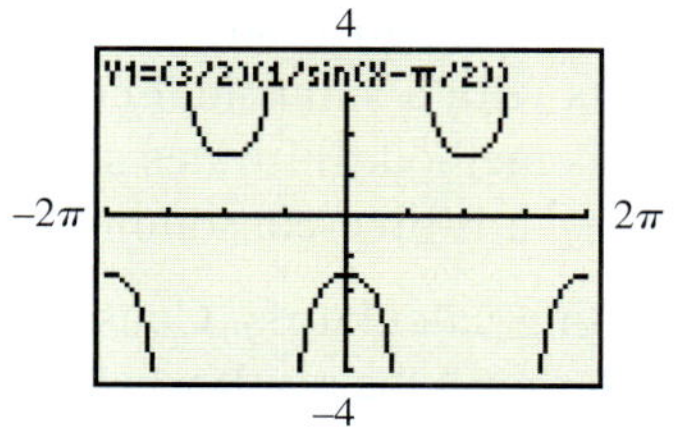

This is a calculator graph of the function in **Example 2.**

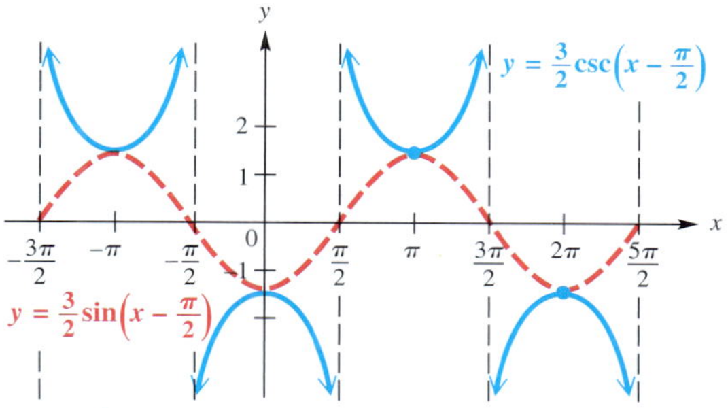

Figure 44

✔ ***Now Try Exercise 7.***

Connecting Graphs with Equations

EXAMPLE 3 Determining an Equation for a Graph

Determine an equation for each graph.

(a)

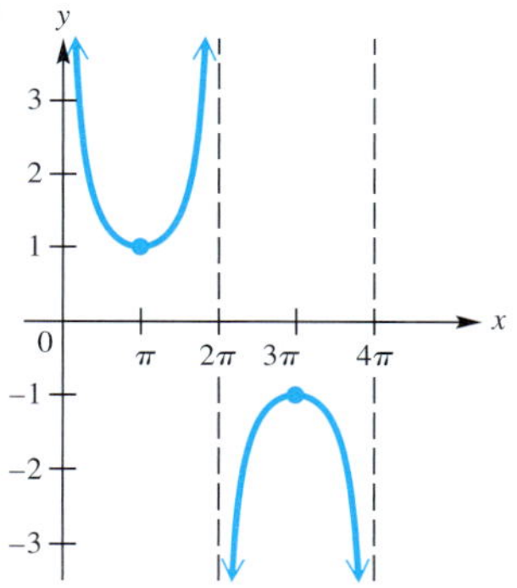

(b)

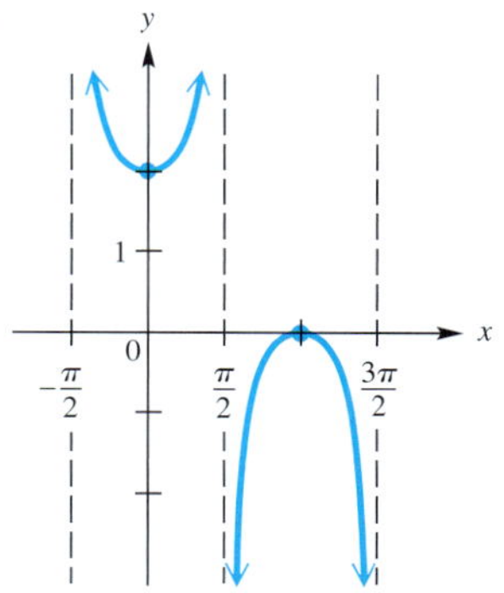

SOLUTION

(a) This graph is that of a cosecant function that is stretched horizontally having period 4π. Therefore, if $y = \csc bx$, where $b > 0$, we must have $b = \frac{1}{2}$. An equation for this graph is

$$y = \csc \frac{1}{2}x.$$

↑ Horizontal stretch

(b) This is the graph of $y = \sec x$, translated 1 unit upward. An equation is

$$y = 1 + \sec x.$$

↑ Vertical translation

✔ ***Now Try Exercises 19 and 21.***

Addition of Ordinates New functions can be formed by adding or subtracting other functions. A function formed by combining two other functions, such as

$$y = \cos x + \sin x,$$

has historically been graphed using a method known as **addition of ordinates.** (The x-value of a point is sometimes called its **abscissa,** while its y-value is called its **ordinate.**)

To apply this method to this function, we graph the functions $y = \cos x$ and $y = \sin x$. Then, for selected values of x, we add $\cos x$ and $\sin x$, and plot the points $(x, \cos x + \sin x)$. Joining the resulting points with a sinusoidal curve gives the graph of the desired function. Although this method illustrates some valuable concepts involving the arithmetic of functions, it is time-consuming.

This technique is easily illustrated with graphing calculators. Consider $Y_1 = \cos X$, $Y_2 = \sin X$, and $Y_3 = Y_1 + Y_2$. **Figure 45** shows the result when Y_1 and Y_2 are graphed in thin graph style, and $Y_3 = \cos X + \sin X$ is graphed in thick graph style. Notice that for $X = \frac{\pi}{6} \approx 0.52359878$, $Y_1 + Y_2 = Y_3$.

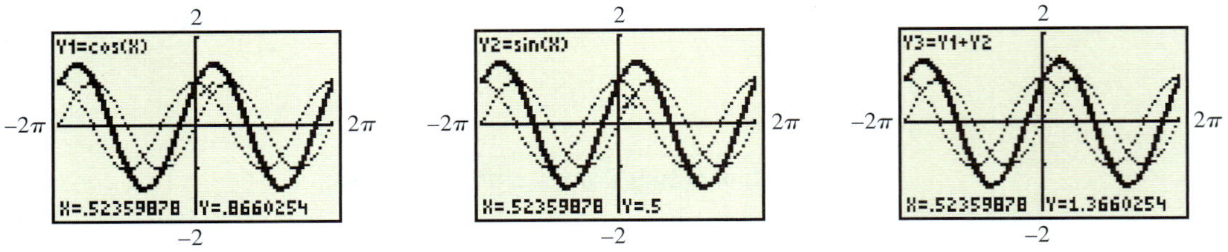

Figure 45

4.4 Exercises

Concept Check In Exercises 1–4, match each function with its graph from choices A–D.

1. $y = -\csc x$ **2.** $y = -\sec x$ **3.** $y = \sec\left(x - \frac{\pi}{2}\right)$ **4.** $y = \csc\left(x + \frac{\pi}{2}\right)$

A.

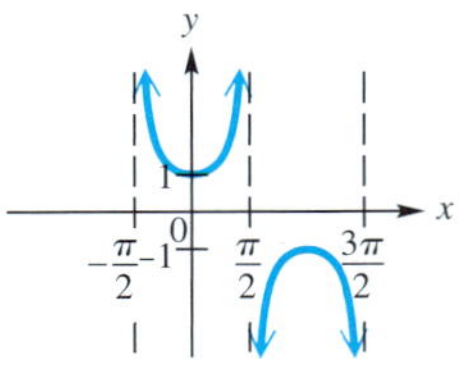

B.

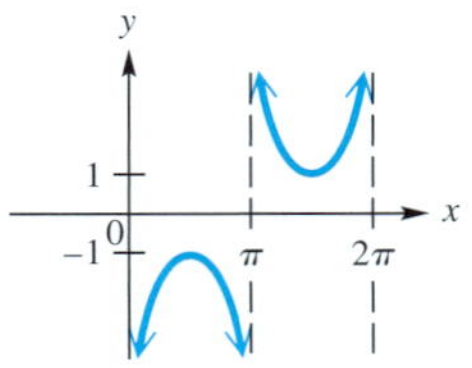

C.

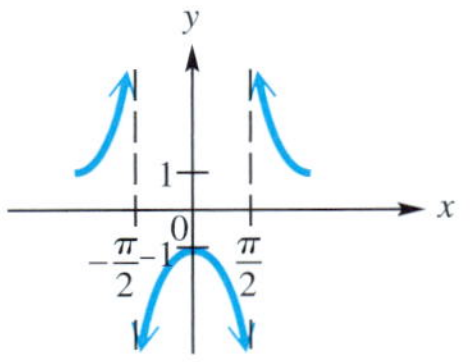

D.

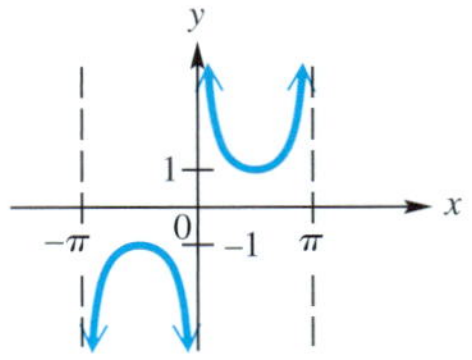

Graph each function over a one-period interval. ***See Examples 1 and 2.***

5. $y = 3 \sec \frac{1}{4}x$ **6.** $y = -2 \sec \frac{1}{2}x$ **7.** $y = -\frac{1}{2} \csc\left(x + \frac{\pi}{2}\right)$

8. $y = \frac{1}{2} \csc\left(x - \frac{\pi}{2}\right)$ **9.** $y = \csc\left(x - \frac{\pi}{4}\right)$ **10.** $y = \sec\left(x + \frac{3\pi}{4}\right)$

11. $y = \sec\left(x + \frac{\pi}{4}\right)$

12. $y = \csc\left(x + \frac{\pi}{3}\right)$

13. $y = \csc\left(\frac{1}{2}x - \frac{\pi}{4}\right)$

14. $y = \sec\left(\frac{1}{2}x + \frac{\pi}{3}\right)$

15. $y = 2 + 3\sec(2x - \pi)$

16. $y = 1 - 2\csc\left(x + \frac{\pi}{2}\right)$

17. $y = 1 - \frac{1}{2}\csc\left(x - \frac{3\pi}{4}\right)$

18. $y = 2 + \frac{1}{4}\sec\left(\frac{1}{2}x - \pi\right)$

Connecting Graphs with Equations *Determine an equation for each graph.* ***See Example 3.***

19.

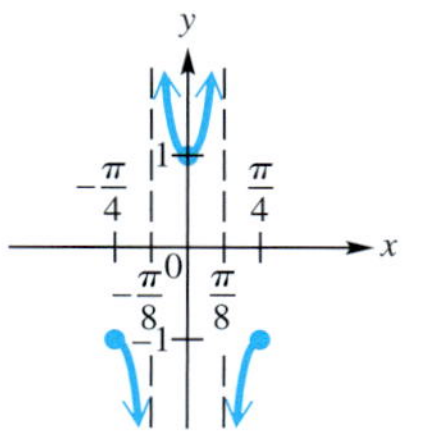

20.

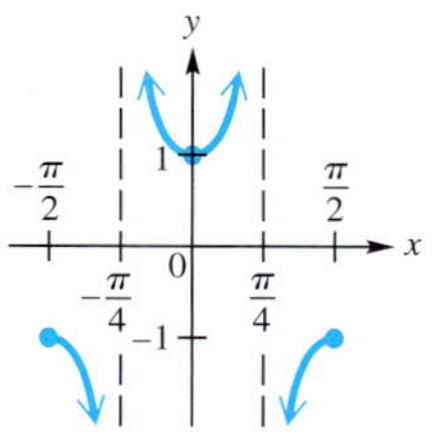

21.

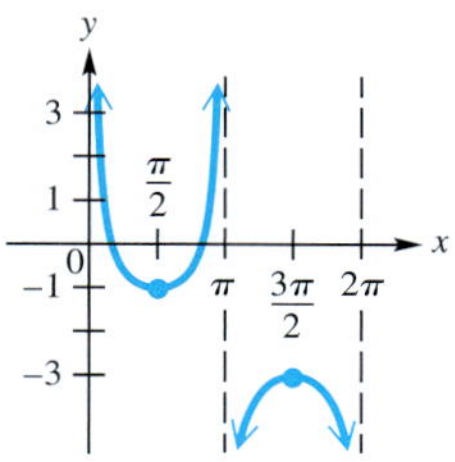

22.

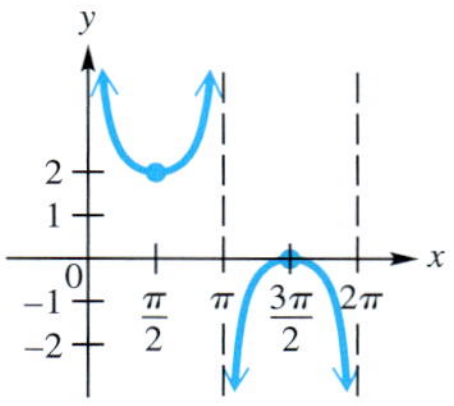

23.

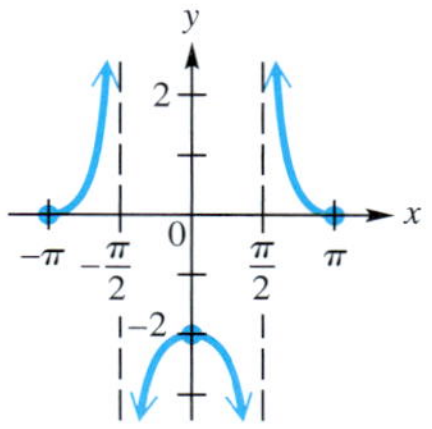

24.

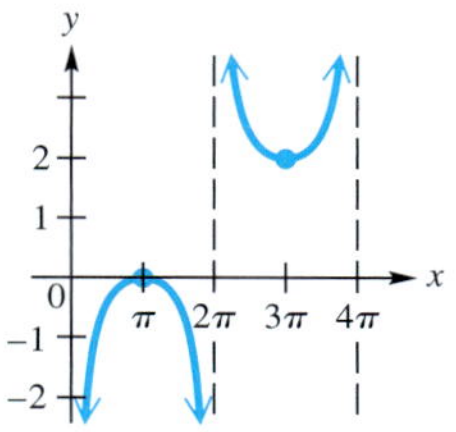

Concept Check *In Exercises 25–28, tell whether each statement is* true *or* false. *If* false, *tell why.*

25. The tangent and secant functions are undefined for the same values.

26. The secant and cosecant functions are undefined for the same values.

27. The graph of $y = \sec x$ in **Figure 37** suggests that $\sec(-x) = \sec x$ for all x in the domain of $\sec x$.

28. The graph of $y = \csc x$ in **Figure 40** suggests that $\csc(-x) = -\csc x$ for all x in the domain of $\csc x$.

Work each exercise.

29. *Concept Check* If c is any number such that $-1 < c < 1$, then how many solutions does the equation $c = \sec x$ have over the entire domain of the secant function?

30. *Concept Check* Consider the function $g(x) = -2\csc(4x + \pi)$. What is the domain of g? What is its range?

31. Show that $\sec(-x) = \sec x$ by writing $\sec(-x)$ as $\frac{1}{\cos(-x)}$ and then using the relationship between $\cos(-x)$ and $\cos x$.

32. Show that $\csc(-x) = -\csc x$ by writing $\csc(-x)$ as $\frac{1}{\sin(-x)}$ and then using the relationship between $\sin(-x)$ and $\sin x$.

33. *(Modeling) Distance of a Rotating Beacon* In the figure for **Exercise 47** in **Section 4.3,** the distance a is given by

$$a = 4|\sec 2\pi t|.$$

Find a for each time.

(a) $t = 0$ **(b)** $t = 0.86$ **(c)** $t = 1.24$

34. Between each pair of successive asymptotes, a portion of the graph of $y = \sec x$ or $y = \csc x$ resembles a parabola. Can each of these portions actually be a parabola? Explain.

Use a graphing calculator to graph Y_1, Y_2, *and* $Y_1 + Y_2$ *on the same screen. Evaluate each of the three functions at* $X = \frac{\pi}{6}$, *and verify that* $Y_1\left(\frac{\pi}{6}\right) + Y_2\left(\frac{\pi}{6}\right) = (Y_1 + Y_2)\left(\frac{\pi}{6}\right)$. *See the discussion on addition of ordinates.*

35. $Y_1 = \sin X, \quad Y_2 = \sin 2X$

36. $Y_1 = \cos X, \quad Y_2 = \sec X$

Summary Exercises on Graphing Circular Functions

These summary exercises provide practice with the various graphing techniques presented in this chapter. Graph each function over a one-period interval.

1. $y = 2 \sin \pi x$

2. $y = 4 \cos \frac{3}{2}x$

3. $y = -2 + \frac{1}{2} \cos \frac{\pi}{4}x$

4. $y = 3 \sec \frac{\pi}{2}x$

5. $y = -4 \csc \frac{1}{2}x$

6. $y = 3 \tan\left(\frac{\pi}{2}x + \pi\right)$

Graph each function over a two-period interval.

7. $y = -5 \sin \frac{x}{3}$

8. $y = 10 \cos\left(\frac{x}{4} + \frac{\pi}{2}\right)$

9. $y = 3 - 4 \sin\left(\frac{5}{2}x + \pi\right)$

10. $y = 2 - \sec[\pi(x - 3)]$

4.5 Harmonic Motion

- Simple Harmonic Motion
- Damped Oscillatory Motion

Simple Harmonic Motion In part A of **Figure 46,** a spring with a weight attached to its free end is in equilibrium (or rest) position. If the weight is pulled down a units and released (part B of the figure), the spring's elasticity causes the weight to rise a units $(a > 0)$ above the equilibrium position, as seen in part C, and then to oscillate about the equilibrium position.

If friction is neglected, this oscillatory motion is described mathematically by a sinusoid. Other applications of this type of motion include sound, electric current, and electromagnetic waves.

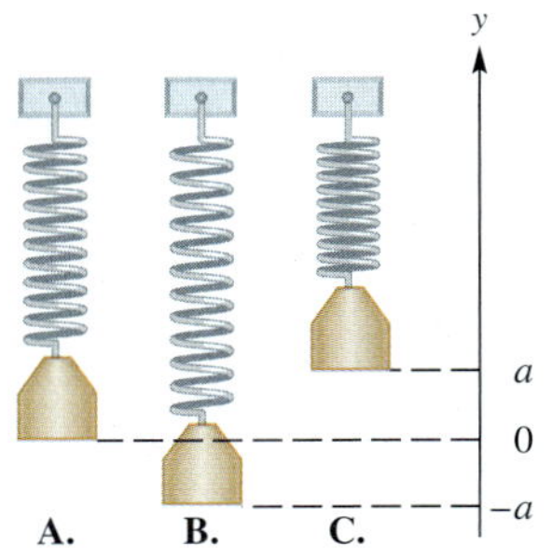

Figure 46

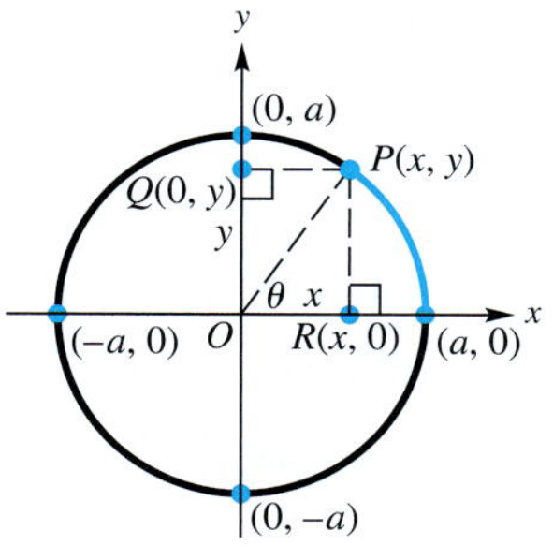

Figure 47

To develop a general equation for such motion, consider **Figure 47.** Suppose the point $P(x, y)$ moves around the circle counterclockwise at a uniform angular speed ω. Assume that at time $t = 0$, P is at $(a, 0)$. The angle swept out by ray OP at time t is given by $\theta = \omega t$. The coordinates of point P at time t are

$$x = a \cos \theta = a \cos \omega t \quad \text{and} \quad y = a \sin \theta = a \sin \omega t.$$

As P moves around the circle from the point $(a, 0)$, the point $Q(0, y)$ oscillates back and forth along the y-axis between the points $(0, a)$ and $(0, -a)$. Similarly, the point $R(x, 0)$ oscillates back and forth between $(a, 0)$ and $(-a, 0)$. This oscillatory motion is called **simple harmonic motion.**

The amplitude of the motion is $|a|$, and the period is $\frac{2\pi}{\omega}$. The moving points P and Q or P and R complete one oscillation or cycle per period. The number of cycles per unit of time, called the **frequency,** is the reciprocal of the period, $\frac{\omega}{2\pi}$, where $\omega > 0$.

Simple Harmonic Motion

The position of a point oscillating about an equilibrium position at time t is modeled by either

$$s(t) = a \cos \omega t \quad \text{or} \quad s(t) = a \sin \omega t,$$

where a and ω are constants, with $\omega > 0$. The amplitude of the motion is $|a|$, the period is $\frac{2\pi}{\omega}$, and the frequency is $\frac{\omega}{2\pi}$ oscillations per time unit.

EXAMPLE 1 Modeling the Motion of a Spring

Suppose that an object is attached to a coiled spring such as the one in **Figure 46** on the preceding page. It is pulled down a distance of 5 in. from its equilibrium position and then released. The time for one complete oscillation is 4 sec.

(a) Give an equation that models the position of the object at time t.

(b) Determine the position at $t = 1.5$ sec.

(c) Find the frequency.

SOLUTION

(a) When the object is released at $t = 0$, the distance of the object from the equilibrium position is 5 in. below equilibrium. If $s(t)$ is to model the motion, then $s(0)$ must equal -5. We use

$$s(t) = a \cos \omega t, \quad \text{with } a = -5.$$

We choose the cosine function because $\cos \omega(0) = \cos 0 = 1$, and $-5 \cdot 1 = -5$. (Had we chosen the sine function, a phase shift would have been required.) Use the fact that the period is 4 to solve for ω.

$$\frac{2\pi}{\omega} = 4 \qquad \text{The period is } \tfrac{2\pi}{\omega}.$$

$$\omega = \frac{\pi}{2} \qquad \text{Solve for } \omega. \text{ (Appendix A)}$$

Thus, the motion is modeled by

$$s(t) = -5 \cos \frac{\pi}{2} t.$$

(b) $s(1.5) = -5 \cos\left[\frac{\pi}{2}(1.5)\right]$ Let $t = 1.5$ in the equation from part (a). (Appendix C)

≈ 3.54 in.

Because $3.54 > 0$, the object is above the equilibrium position.

(c) The frequency is the reciprocal of the period, or $\frac{1}{4}$ oscillation per sec.

✔ *Now Try Exercise 9.*

EXAMPLE 2 Analyzing Harmonic Motion

Suppose that an object oscillates according to the model

$$s(t) = 8 \sin 3t,$$

where t is in seconds and $s(t)$ is in feet. Analyze the motion.

SOLUTION The motion is harmonic because the model is $s(t) = a \sin \omega t$. Because $a = 8$, the object oscillates 8 ft in either direction from its starting point. The period $\frac{2\pi}{3} \approx 2.1$ is the time, in seconds, it takes for one complete oscillation. The frequency is the reciprocal of the period, so the object completes $\frac{3}{2\pi} \approx 0.48$ oscillation per sec.

✔ *Now Try Exercise 17.*

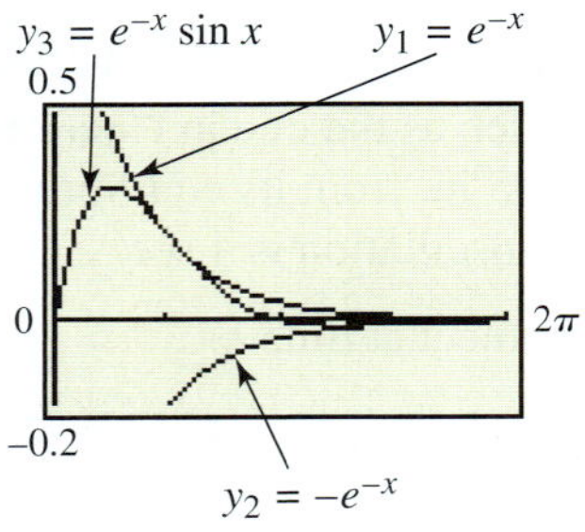

Figure 48

Damped Oscillatory Motion In the example of the stretched spring, we disregard the effect of friction. Friction causes the amplitude of the motion to diminish gradually until the weight comes to rest. In this situation, we say that the motion has been *damped* by the force of friction. Most oscillatory motions are damped, and the decrease in amplitude follows the pattern of exponential decay. An example of **damped oscillatory motion** is provided by the function

$$\mathbf{s(t) = e^{-t} \sin t.}$$

(The number $e \approx 2.718$ is the base of the natural logarithmic function, first studied in college algebra courses.) **Figure 48** shows how the graph of $y_3 = e^{-x} \sin x$ is bounded above by the graph of $y_1 = e^{-x}$ and below by the graph of $y_2 = -e^{-x}$. The damped motion curve dips below the x-axis at $x = \pi$ but stays above the graph of y_2. **Figure 49** shows a traditional graph of $s(t) = e^{-t} \sin t$, along with the graph of $y = \sin t$.

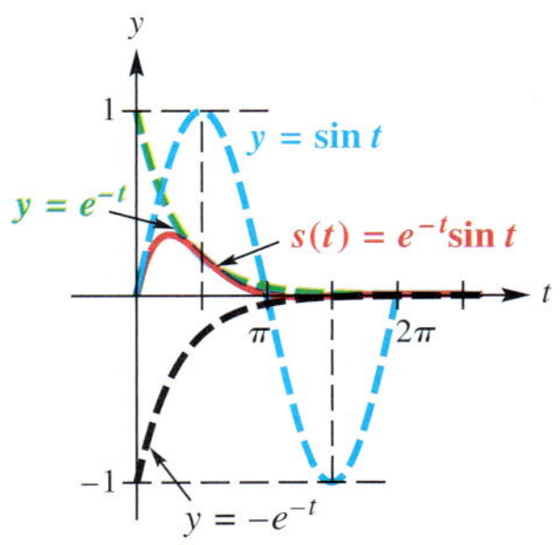

Figure 49

Shock absorbers are put on an automobile in order to damp oscillatory motion. Instead of the car oscillating up and down for a long while after hitting a bump or pothole, the oscillations of the car are quickly damped out for a smoother ride.

4.5 Exercises

(Modeling) Springs A weight on a spring has initial position $s(0)$ and period P.

(a) *Find a function s given by $s(t) = a \cos \omega t$ that models the displacement of the weight.*

(b) *Evaluate $s(1)$. Is the weight moving upward, downward, or neither when $t = 1$? Support your results graphically or numerically.*

1. $s(0) = 2$ in.; $P = 0.5$ sec

2. $s(0) = 5$ in.; $P = 1.5$ sec

3. $s(0) = -3$ in.; $P = 0.8$ sec

4. $s(0) = -4$ in.; $P = 1.2$ sec

(Modeling) Music A note on the piano has given frequency F. Suppose the maximum displacement at the center of the piano wire is given by $s(0)$. Find constants a and ω so that the equation

$$s(t) = a \cos \omega t$$

models this displacement. Graph s in the viewing window $[0, 0.05]$ by $[-0.3, 0.3]$.

5. $F = 27.5$; $s(0) = 0.21$

6. $F = 110$; $s(0) = 0.11$

7. $F = 55$; $s(0) = 0.14$

8. $F = 220$; $s(0) = 0.06$

(Modeling) Solve each problem. ***See Examples 1 and 2.***

9. *Spring Motion* An object is attached to a coiled spring, as in **Figure 46.** It is pulled down a distance of 4 units from its equilibrium position and then released. The time for one complete oscillation is 3 sec.

(a) Give an equation that models the position of the object at time t.
(b) Determine the position at $t = 1.25$ sec.
(c) Find the frequency.

10. *Spring Motion* Repeat **Exercise 9,** but assume that the object is pulled down 6 units and that the time for one complete oscillation is 4 sec.

11. *Particle Movement* Write the equation and then determine the amplitude, period, and frequency of the simple harmonic motion of a particle moving uniformly around a circle of radius 2 units, with the given angular speed.

(a) 2 radians per sec **(b)** 4 radians per sec

12. *Spring Motion* The height attained by a weight attached to a spring set in motion is

$$s(t) = -4 \cos 8\pi t \text{ inches after } t \text{ seconds.}$$

(a) Find the maximum height that the weight rises above the equilibrium position of $s(t) = 0$.
(b) When does the weight first reach its maximum height if $t \geq 0$?
(c) What are the frequency and the period?

13. *Pendulum Motion* What are the period P and frequency T of oscillation of a pendulum of length $\frac{1}{2}$ ft? $\left(\textit{Hint: } P = 2\pi\sqrt{\frac{L}{32}},\right.$ where L is the length of the pendulum in feet and the period P is in seconds.$\left.\right)$

14. *Pendulum Motion* In **Exercise 13,** how long should the pendulum be to have a period of 1 sec?

15. *Spring Motion* The formula for the up and down motion of a weight on a spring is given by

$$s(t) = a \sin \sqrt{\frac{k}{m}}\, t.$$

If the spring constant k is 4, what mass m must be used to produce a period of 1 sec?

16. *Spring Motion* (See **Exercise 15.**) A spring with spring constant $k = 2$ and a 1-unit mass m attached to it is stretched and then allowed to come to rest.

(a) If the spring is stretched $\frac{1}{2}$ ft and released, what are the amplitude, period, and frequency of the resulting oscillatory motion?
(b) What is the equation of the motion?

17. *Spring Motion* The position of a weight attached to a spring is

$$s(t) = -5 \cos 4\pi t \text{ inches after } t \text{ seconds.}$$

(a) What is the maximum height that the weight rises above the equilibrium position?
(b) What are the frequency and period?
(c) When does the weight first reach its maximum height?
(d) Calculate and interpret $s(1.3)$.

18. *Spring Motion* The position of a weight attached to a spring is

$$s(t) = -4 \cos 10t \text{ inches after } t \text{ seconds.}$$

(a) What is the maximum height that the weight rises above the equilibrium position?
(b) What are the frequency and period?
(c) When does the weight first reach its maximum height?
(d) Calculate and interpret $s(1.466)$.

19. *Spring Motion* A weight attached to a spring is pulled down 3 in. below the equilibrium position.

(a) Assuming that the frequency is $\frac{6}{\pi}$ cycles per sec, determine a model that gives the position of the weight at time t seconds.
(b) What is the period?

20. *Spring Motion* A weight attached to a spring is pulled down 2 in. below the equilibrium position.

(a) Assuming that the period is $\frac{1}{3}$ sec, determine a model that gives the position of the weight at time t seconds.
(b) What is the frequency?

Damped Oscillatory Motion *Use a graphing calculator to graph*

$$y_1 = e^{-t} \sin t, \quad y_2 = e^{-t}, \quad \text{and} \quad y_3 = -e^{-t}$$

in the viewing window $[0, \pi]$ *by* $[-0.5, 0.5]$.

21. Find the t-intercepts of the graph of y_1. Explain the relationship of these intercepts to the x-intercepts of the graph of $y = \sin x$.

22. Find any points of intersection of y_1 and y_2 or y_1 and y_3. How are these points related to the graph of $y = \sin x$?

Chapter 4 Test Prep

Key Terms

4.1 periodic function
period
sine wave (sinusoid)
amplitude

4.2 phase shift
argument

4.3 vertical asymptote

4.4 addition of ordinates

4.5 simple harmonic motion
frequency
damped oscillatory motion

Quick Review

Concepts	Examples

4.1 Graphs of the Sine and Cosine Functions

4.2 Translations of the Graphs of the Sine and Cosine Functions

Sine and Cosine Functions

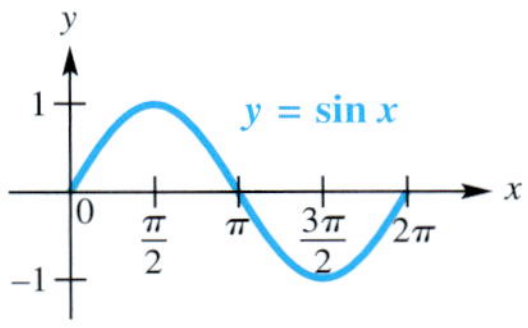

Domain: $(-\infty, \infty)$
Range: $[-1, 1]$
Amplitude: 1
Period: 2π

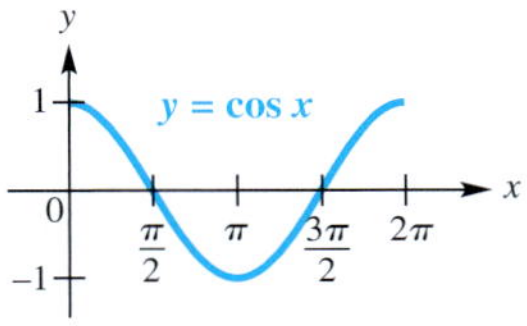

Domain: $(-\infty, \infty)$
Range: $[-1, 1]$
Amplitude: 1
Period: 2π

The graph of

$$y = c + a \sin[b(x - d)] \quad \text{or} \quad y = c + a \cos[b(x - d)],$$

with $b > 0$, has the following characteristics.

1. amplitude $|a|$
2. period $\frac{2\pi}{b}$
3. vertical translation c units up if $c > 0$ or $|c|$ units down if $c < 0$
4. phase shift d units to the right if $d > 0$ or $|d|$ units to the left if $d < 0$

See **Sections 4.1 and 4.2** for a summary of graphing techniques.

Graph $y = 1 + \sin 3x$.

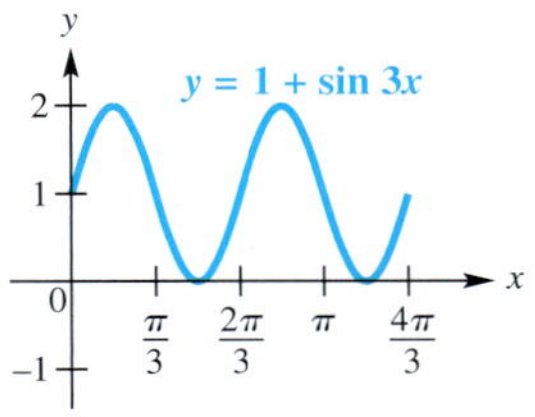

amplitude: 1 domain: $(-\infty, \infty)$
period: $\frac{2\pi}{3}$ range: $[0, 2]$
vertical translation: 1 unit up

Graph $y = -2 \cos\left(x + \frac{\pi}{2}\right)$.

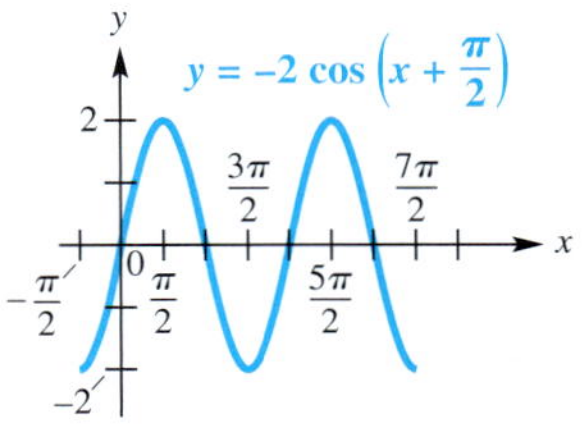

amplitude: 2 domain: $(-\infty, \infty)$
period: 2π range: $[-2, 2]$
phase shift: $\frac{\pi}{2}$ left

(continued)

Concepts	Examples

4.3 Graphs of the Tangent and Cotangent Functions

Tangent and Cotangent Functions

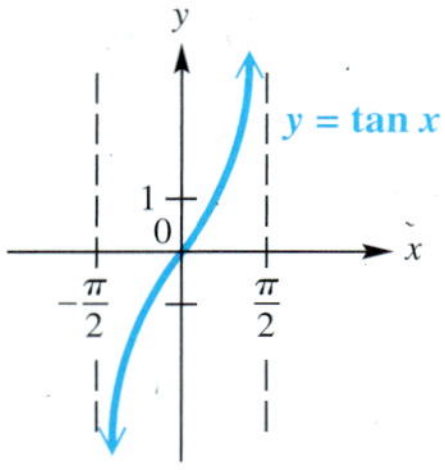

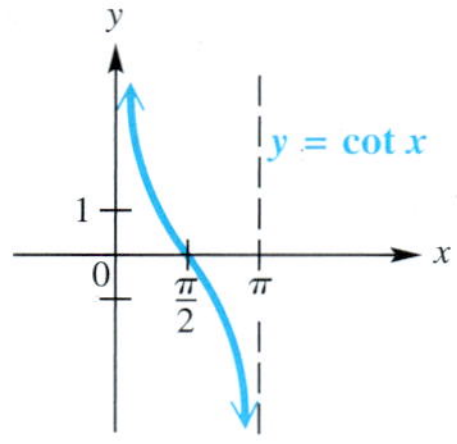

Domain: $\left\{x \mid x \neq (2n+1)\frac{\pi}{2},\right.$ where n is any integer$\left.\right\}$
Range: $(-\infty, \infty)$
Period: π

Domain: $\{x \mid x \neq n\pi,$ where n is any integer$\}$
Range: $(-\infty, \infty)$
Period: π

See **Section 4.3** for a summary of graphing techniques.

Graph one period of $y = 2 \tan x$.

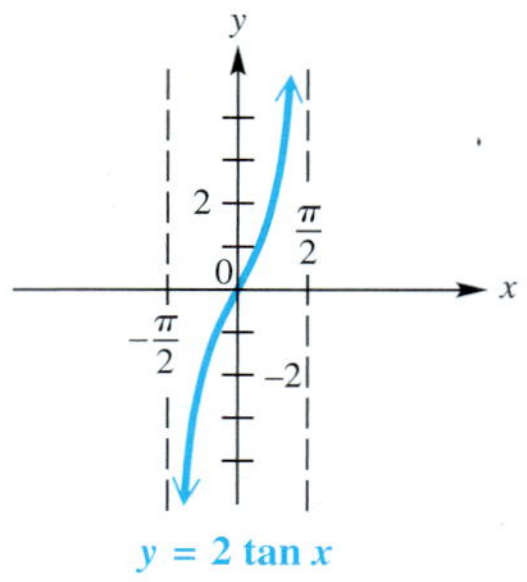

period: π
domain: $\left\{x \mid x \neq (2n+1)\frac{\pi}{2},\right.$ where n is any integer$\left.\right\}$
range: $(-\infty, \infty)$

4.4 Graphs of the Secant and Cosecant Functions

Secant and Cosecant Functions

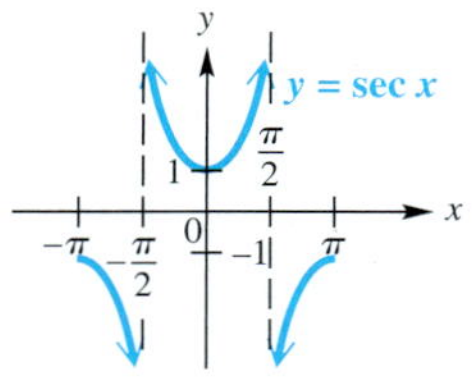

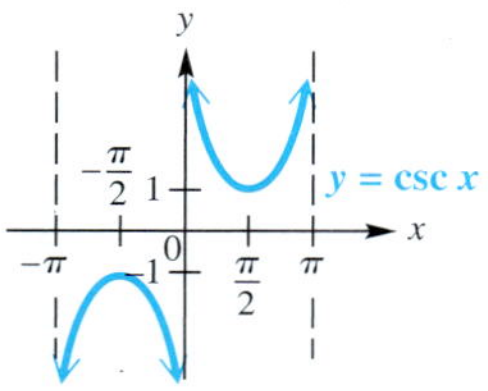

Domain: $\left\{x \mid x \neq (2n+1)\frac{\pi}{2},\right.$ where n is any integer$\left.\right\}$
Range: $(-\infty, -1] \cup [1, \infty)$
Period: 2π

Domain: $\{x \mid x \neq n\pi,$ where n is any integer$\}$
Range: $(-\infty, -1] \cup [1, \infty)$
Period: 2π

See **Section 4.4** for a summary of graphing techniques.

Graph one period of $y = \sec\left(x + \frac{\pi}{4}\right)$.

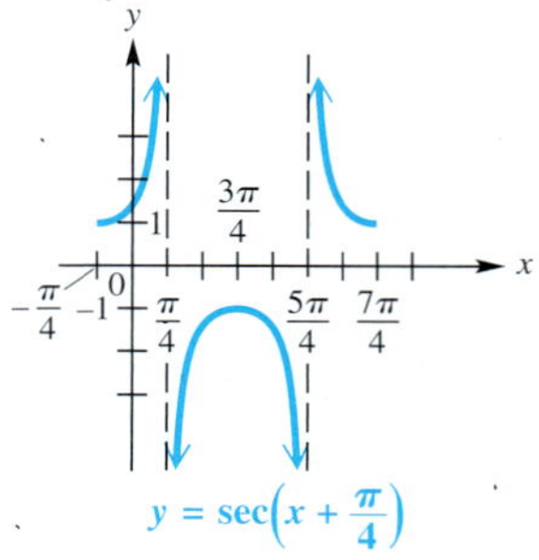

period: 2π
phase shift: $\frac{\pi}{4}$ left
domain: $\left\{x \mid x \neq \frac{\pi}{4} + n\pi,\right.$ where n is any integer$\left.\right\}$
range: $(-\infty, -1] \cup [1, \infty)$

4.5 Harmonic Motion

Simple Harmonic Motion

The position of a point oscillating about an equilibrium position at time t is modeled by either

$$s(t) = a \cos \omega t \quad \text{or} \quad s(t) = a \sin \omega t,$$

where a and ω are constants, with $\omega > 0$. The amplitude of the motion is $|a|$, the period is $\frac{2\pi}{\omega}$, and the frequency is $\frac{\omega}{2\pi}$ oscillations per time unit.

A spring oscillates according to

$$s(t) = -5 \cos 6t,$$

where t is in seconds and $s(t)$ is in inches. Find the amplitude, period, and frequency.

$$\text{amplitude} = |-5| = 5 \text{ in.} \qquad \text{period} = \frac{2\pi}{6} = \frac{\pi}{3} \text{ sec}$$

$$\text{frequency} = \frac{3}{\pi} \text{ oscillation per sec}$$

Chapter 4 Review Exercises

1. *Concept Check* Which one of the following is true about the graph of $y = 4 \sin 2x$?

A. It has amplitude 2 and period $\frac{\pi}{2}$.

B. It has amplitude 4 and period π.

C. Its range is $[0, 4]$.

D. Its range is $[-4, 0]$.

2. *Concept Check* Which one of the following is false about the graph of $y = -3 \cos \frac{1}{2}x$?

A. Its range is $[-3, 3]$.

B. Its domain is $(-\infty, \infty)$.

C. Its amplitude is 3, and its period is 4π.

D. Its amplitude is -3, and its period is π.

3. *Concept Check* Which of the basic circular functions can have y-value $\frac{1}{2}$?

4. *Concept Check* Which of the basic circular functions can have y-value 2?

For each function, give the amplitude, period, vertical translation, and phase shift, as applicable.

5. $y = 2 \sin x$

6. $y = \tan 3x$

7. $y = -\frac{1}{2} \cos 3x$

8. $y = 2 \sin 5x$

9. $y = 1 + 2 \sin \frac{1}{4}x$

10. $y = 3 - \frac{1}{4} \cos \frac{2}{3}x$

11. $y = 3 \cos\left(x + \frac{\pi}{2}\right)$

12. $y = -\sin\left(x - \frac{3\pi}{4}\right)$

13. $y = \frac{1}{2} \csc\left(2x - \frac{\pi}{4}\right)$

14. $y = 2 \sec(\pi x - 2\pi)$

15. $y = \frac{1}{3} \tan\left(3x - \frac{\pi}{3}\right)$

16. $y = \cot\left(\frac{x}{2} + \frac{3\pi}{4}\right)$

Concept Check *Identify the circular function that satisfies each description.*

17. period is π, x-intercepts are of the form $n\pi$, where n is any integer

18. period is 2π, graph passes through the origin

19. period is 2π, graph passes through the point $\left(\frac{\pi}{2}, 0\right)$

20. period is 2π, domain is $\{x \mid x \neq n\pi$, where n is any integer$\}$

21. period is π, function is decreasing on the interval $(0, \pi)$

22. period is 2π, has vertical asymptotes of the form $x = (2n + 1)\frac{\pi}{2}$, where n is any integer

23. Suppose that f is a sine function with period 10 and $f(5) = 2$. Explain why $f(25) = 2$.

24. Suppose that f is a sine function with period π and $f\left(\frac{6\pi}{5}\right) = 1$. Explain why $f\left(-\frac{4\pi}{5}\right) = 1$.

Graph each function over a one-period interval.

25. $y = 3 \sin x$

26. $y = \frac{1}{2} \sec x$

27. $y = -\tan x$

28. $y = -2 \cos x$

29. $y = 2 + \cot x$

30. $y = -1 + \csc x$

31. $y = \sin 2x$

32. $y = \tan 3x$

33. $y = 3 \cos 2x$

34. $y = \frac{1}{2} \cot 3x$

35. $y = \cos\left(x - \frac{\pi}{4}\right)$

36. $y = \tan\left(x - \frac{\pi}{2}\right)$

37. $y = \sec\left(2x + \frac{\pi}{3}\right)$

38. $y = \sin\left(3x + \frac{\pi}{2}\right)$

39. $y = 1 + 2 \cos 3x$

40. $y = -1 - 3 \sin 2x$

41. $y = 2 \sin \pi x$

42. $y = -\frac{1}{2} \cos(\pi x - \pi)$

43. *Concept Check* Determine the range of a function of the form $f(x) = 2 \sin(bx + c)$.

44. *Concept Check* Determine the range of a function of the form $f(x) = 2 \csc(bx + c)$.

Connecting Graphs with Equations *Determine the simplest form of an equation for each graph. Choose $b > 0$, and include no phase shifts.*

45.

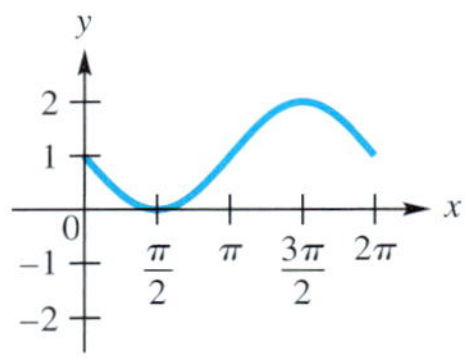

46.

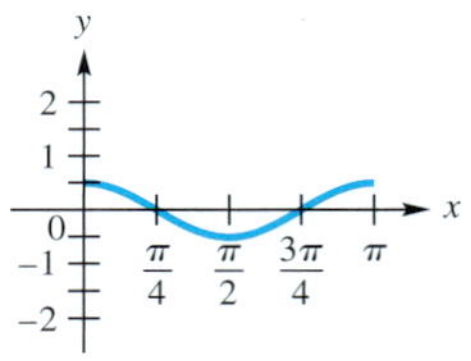

47.

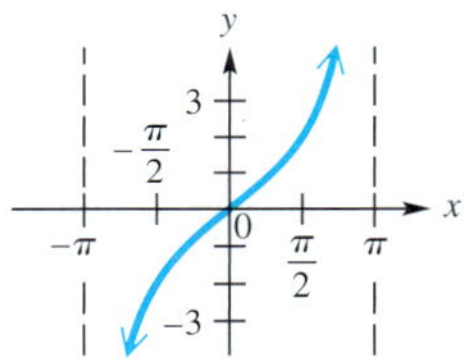

48.

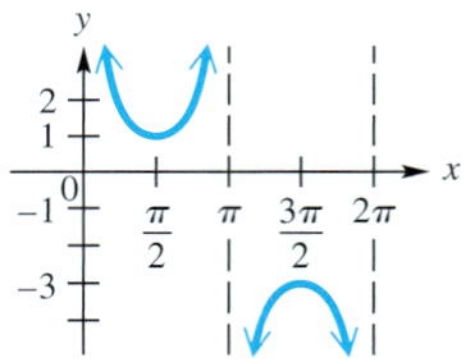

Solve each problem.

49. *Viewing Angle to an Object* Let a person whose eyes are h_1 feet from the ground stand d feet from an object h_2 feet tall, where $h_2 > h_1$. Let θ be the angle of elevation to the top of the object. See the figure.

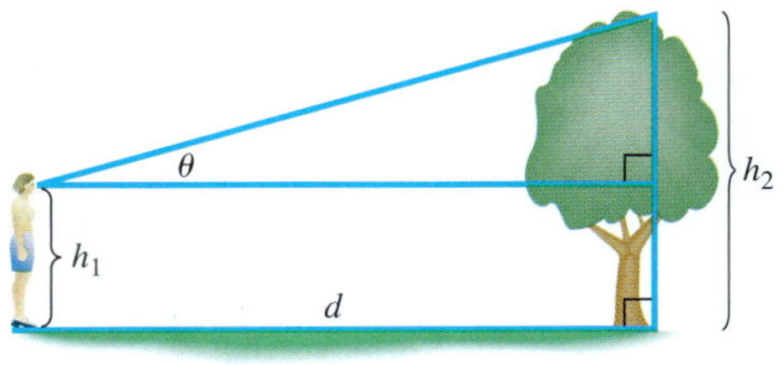

(a) Show that $d = (h_2 - h_1) \cot \theta$.

(b) Let $h_2 = 55$ and $h_1 = 5$. Graph d for the interval $0 < \theta \leq \frac{\pi}{2}$.

50. *(Modeling) Tides* The figure shows a function f that models the tides in feet at Clearwater Beach, Florida, x hours after midnight. (*Source:* Pentcheff, D., *WWW Tide and Current Predictor.*)

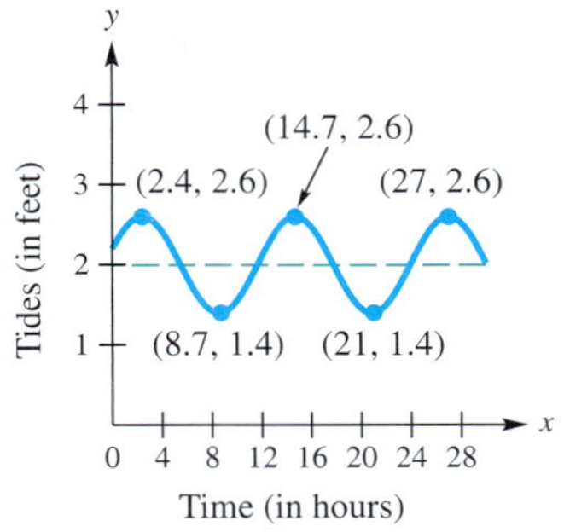

(a) Find the time between high tides.
(b) What is the difference in water levels between high tide and low tide?
(c) The tides can be modeled by

$$f(x) = 0.6 \cos[0.511(x - 2.4)] + 2.$$

Estimate the tides when $x = 10$.

51. *(Modeling) Maximum Temperatures* The maximum afternoon temperature (in °F) in a given city might be modeled by

$$t = 60 - 30 \cos \frac{x\pi}{6},$$

where t represents the maximum afternoon temperature in month x, with $x = 0$ representing January, $x = 1$ representing February, and so on. Find the maximum afternoon temperature to the nearest degree for each month.

(a) January **(b)** April **(c)** May
(d) June **(e)** August **(f)** October

52. *(Modeling) Average Monthly Temperature* The average monthly temperature (in °F) in Chicago, Illinois, is shown in the table.

Month	°F	Month	°F
Jan	22	July	73
Feb	27	Aug	72
Mar	37	Sept	64
Apr	48	Oct	52
May	59	Nov	39
June	68	Dec	27

Source: World Almanac and Book of Facts.

(a) Plot the average monthly temperature over a two-year period. Let $x = 1$ correspond to January of the first year.
(b) Determine a model function of the form $f(x) = a \sin[b(x - d)] + c$, where a, b, c, and d are constants.
(c) Explain the significance of each constant.
(d) Graph f together with the data on the same coordinate axes. How well does f model the data?
(e) Use the sine regression capability of a graphing calculator to find the equation of a sine curve that fits these data.

53. *(Modeling) Pollution Trends* The amount of pollution in the air is lower after heavy spring rains and higher after periods of little rain. In addition to this seasonal fluctuation, the long-term trend is upward. An idealized graph of this situation is shown in the figure.

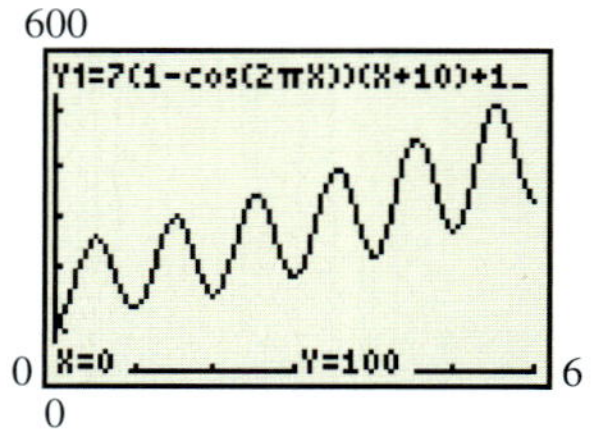

Circular functions can be used to model the fluctuating part of the pollution levels. Powers of the number e (e is the base of the natural logarithm; $e \approx 2.718282$) can be used to model long-term growth. The pollution level in a certain area might be given by

$$y = 7(1 - \cos 2\pi x)(x + 10) + 100e^{0.2x},$$

where x is the time in years, with $x = 0$ representing January 1 of the base year. July 1 of the same year would be represented by $x = 0.5$, October 1 of the following year would be represented by $x = 1.75$, and so on. Find the pollution levels on each date.

(a) January 1, base year
(b) July 1, base year
(c) January 1, following year
(d) July 1, following year

54. *(Modeling) Lynx and Hare Populations* The figure shows the populations of lynx and hares in Canada for the years 1847–1903. The hares are food for the lynx. An increase in hare population causes an increase in lynx population some time later. The increasing lynx population then causes a decline in hare population. The two graphs have the same period.

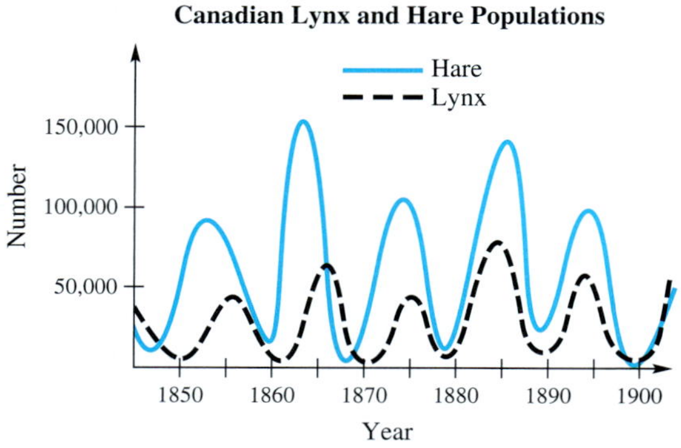

(a) Estimate the length of one period.
(b) Estimate the maximum and minimum hare populations.

An object in simple harmonic motion has position function $s(t)$ inches from an equilibrium point, where t is the time in seconds. Find the amplitude, period, and frequency.

55. $s(t) = 4 \sin \pi t$

56. $s(t) = 3 \cos 2t$

57. In **Exercise 55,** what does the frequency represent? Find the position of the object relative to the equilibrium point at 1.5 sec, 2 sec, and 3.25 sec.

58. In **Exercise 56,** what does the period represent? What does the amplitude represent?

Chapter 4 Test

1. Identify each of the following basic circular function graphs.

(a)

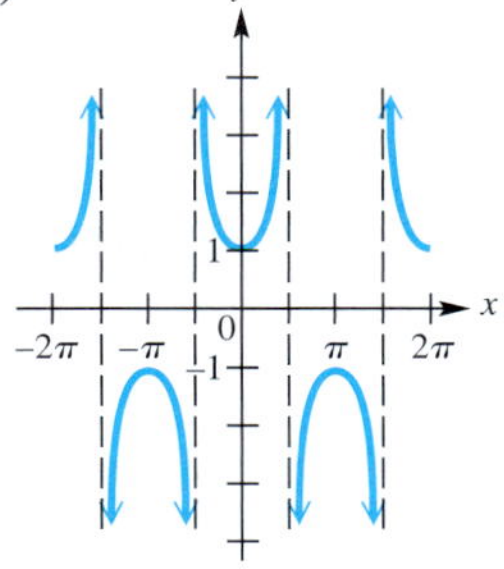

(b)

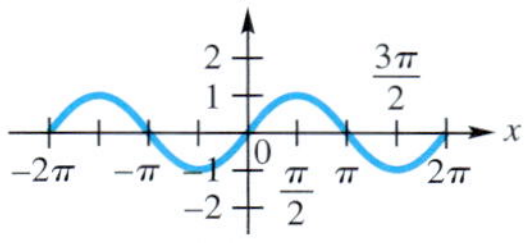

(c)

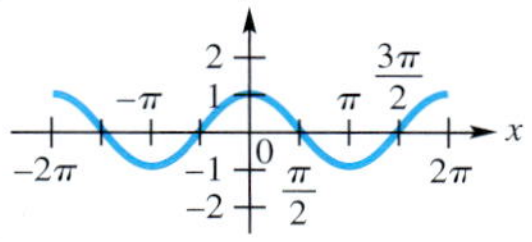

(d)

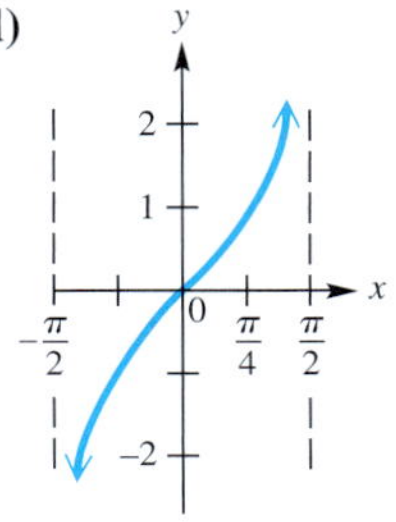

(e)

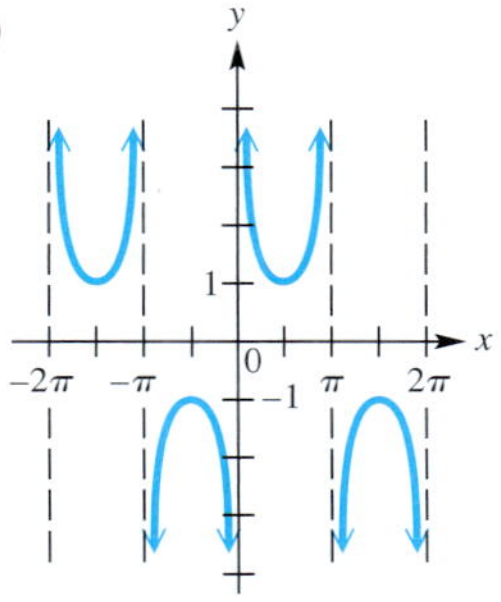

(f)

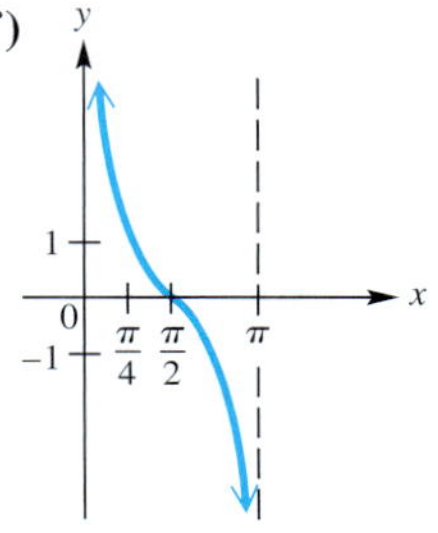

2. ***Connecting Graphs with Equations*** Determine the simplest form of an equation for each graph. Choose $b > 0$, and include no phase shifts.

(a)

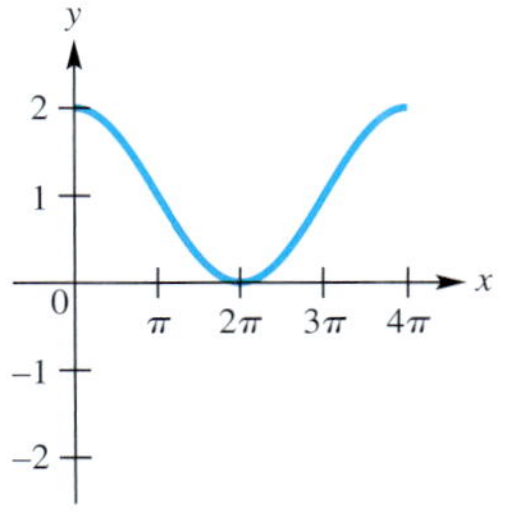

(b)

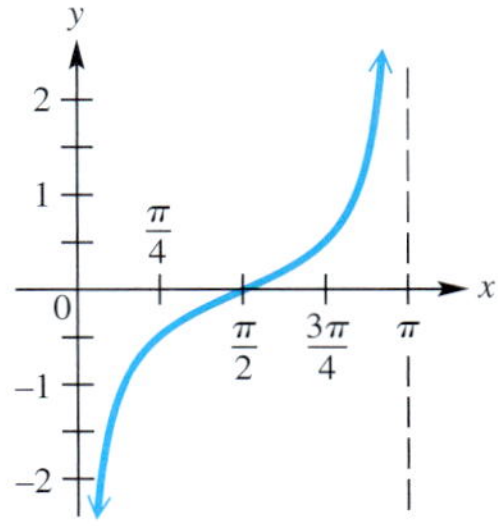

3. Give a short answer to each of the following.

(a) What is the domain of the cosine function?
(b) What is the range of the sine function?
(c) What is the least positive value for which the tangent function is undefined?
(d) What is the range of the secant function?

4. Consider the function $y = 3 - 6\sin\left(2x + \frac{\pi}{2}\right)$.

(a) What is its period?
(b) What is the amplitude of its graph?
(c) What is its range?
(d) What is the y-intercept of its graph?
(e) What is its phase shift?

Graph each function over a two-period interval. Identify asymptotes when applicable.

5. $y = \sin(2x + \pi)$

6. $y = -\cos 2x$

7. $y = 2 + \cos x$

8. $y = -1 + 2\sin(x + \pi)$

9. $y = \tan\left(x - \frac{\pi}{2}\right)$

10. $y = -2 - \cot\left(x - \frac{\pi}{2}\right)$

11. $y = -\csc 2x$

12. $y = 3 \csc \pi x$

(Modeling) Solve each problem.

13. ***Average Monthly Temperature*** The average monthly temperature (in °F) in San Antonio, Texas, can be modeled using the circular function

$$f(x) = 16.5 \sin\left[\frac{\pi}{6}(x - 4)\right] + 67.5,$$

where x is the month and $x = 1$ corresponds to January. (*Source: World Almanac and Book of Facts.*)

(a) Graph f in the window $[0, 25]$ by $[40, 90]$.
(b) Determine the amplitude, period, phase shift, and vertical translation of f.
(c) What is the average monthly temperature for the month of December?
(d) Determine the minimum and maximum average monthly temperatures and the months when they occur.
(e) What would be an approximation for the average *yearly* temperature in San Antonio? How is this related to the vertical translation of the sine function in the formula for f?

14. ***Spring Motion*** The height of a weight attached to a spring is

$$s(t) = -4 \cos 8\pi t \text{ inches after } t \text{ seconds.}$$

(a) Find the maximum height that the weight rises above the equilibrium position of $s(t) = 0$.
(b) When does the weight first reach its maximum height if $t \geq 0$?
(c) What are the frequency and period?

15. Explain why the domains of the tangent and secant functions are the same, and then give a similar explanation for the cotangent and cosecant functions.

5 Trigonometric Identities

Electricity that passes through wires to homes and businesses alternates its direction on those wires and is modeled by *sine* and *cosine functions*.

5.1 Fundamental Identities

Fundamental Identities

As suggested by the circle in **Figure 1,** an angle θ having the point (x, y) on its terminal side has a corresponding angle $-\theta$ with the point $(x, -y)$ on its terminal side.

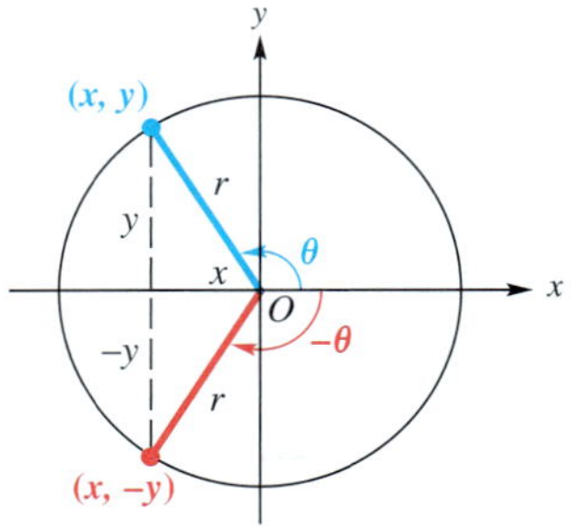

$\sin(-\theta) = -\frac{y}{r} = -\sin\theta$

Figure 1

From the definition of sine,

$$\sin(-\theta) = \frac{-y}{r} \quad \text{and} \quad \sin\theta = \frac{y}{r}, \qquad \text{(Section 1.3)}$$

so $\sin(-\theta)$ and $\sin\theta$ are negatives of each other.

$$\boldsymbol{\sin(-\theta) = -\sin\theta}$$

This is an example of an **identity,** an equation that is satisfied by *every* value in the domain of its variable. (See **Appendix A.**) Examples from algebra are

$$x^2 - y^2 = (x + y)(x - y),$$

$$x(x + y) = x^2 + xy, \qquad \text{Identities}$$

and $$x^2 + 2xy + y^2 = (x + y)^2.$$

Figure 1 shows an angle θ in quadrant II, but the same result holds for θ in any quadrant. The figure also suggests the following identity.

$$\cos(-\theta) = \frac{x}{r} \quad \text{and} \quad \cos\theta = \frac{x}{r} \qquad \text{(Section 1.3)}$$

$$\boldsymbol{\cos(-\theta) = \cos\theta}$$

We use the identities for $\sin(-\theta)$ and $\cos(-\theta)$ to find $\tan(-\theta)$ in terms of $\tan\theta$.

$$\tan(-\theta) = \frac{\sin(-\theta)}{\cos(-\theta)} = \frac{-\sin\theta}{\cos\theta} = -\frac{\sin\theta}{\cos\theta}$$

$$\boldsymbol{\tan(-\theta) = -\tan\theta}$$

Similar reasoning gives the remaining three **negative-angle** or **negative-number identities,** which, together with the reciprocal, quotient, and Pythagorean identities from **Chapter 1,** make up the **fundamental identities.** For reference, we summarize these identities in the box at the top of the next page.

NOTE In trigonometric identities, θ can be an angle in degrees, a real number, or a variable.

Fundamental Identities

Reciprocal Identities

$$\cot\theta = \frac{1}{\tan\theta} \qquad \sec\theta = \frac{1}{\cos\theta} \qquad \csc\theta = \frac{1}{\sin\theta}$$

Quotient Identities

$$\tan\theta = \frac{\sin\theta}{\cos\theta} \qquad \cot\theta = \frac{\cos\theta}{\sin\theta}$$

Pythagorean Identities

$$\sin^2\theta + \cos^2\theta = 1 \qquad \tan^2\theta + 1 = \sec^2\theta \qquad 1 + \cot^2\theta = \csc^2\theta$$

Negative-Angle Identities

$$\sin(-\theta) = -\sin\theta \qquad \cos(-\theta) = \cos\theta \qquad \tan(-\theta) = -\tan\theta$$

$$\csc(-\theta) = -\csc\theta \qquad \sec(-\theta) = \sec\theta \qquad \cot(-\theta) = -\cot\theta$$

NOTE We will also use alternative forms of the fundamental identities. *For example, two other forms of* $\sin^2\theta + \cos^2\theta = 1$ *are*

$$\sin^2\theta = 1 - \cos^2\theta \quad \textit{and} \quad \cos^2\theta = 1 - \sin^2\theta.$$

Using the Fundamental Identities We can use these identities to find the values of other trigonometric functions from the value of a given trigonometric function.

EXAMPLE 1 Finding Trigonometric Function Values Given One Value and the Quadrant

If $\tan\theta = -\frac{5}{3}$ and θ is in quadrant II, find each function value.

(a) $\sec\theta$ **(b)** $\sin\theta$ **(c)** $\cot(-\theta)$

SOLUTION

(a) We use an identity that relates the tangent and secant functions. Remember that $\sec\theta$ will be negative because θ is in quadrant II.

$$\tan^2\theta + 1 = \sec^2\theta \quad \text{Pythagorean identity}$$

$$\left(-\frac{5}{3}\right)^2 + 1 = \sec^2\theta \quad \tan\theta = -\tfrac{5}{3}$$

$$\frac{25}{9} + 1 = \sec^2\theta \quad \text{Square } -\tfrac{5}{3}.$$

$$\frac{34}{9} = \sec^2\theta \quad \text{Add; } 1 = \tfrac{9}{9}$$

Choose the correct sign.

$$-\sqrt{\frac{34}{9}} = \sec\theta \quad \text{Take the negative square root. (Appendix A)}$$

$$-\frac{\sqrt{34}}{3} = \sec\theta \quad \text{Simplify the radical: } -\sqrt{\tfrac{34}{9}} = -\tfrac{\sqrt{34}}{\sqrt{9}} = -\tfrac{\sqrt{34}}{3}.$$

(b)

$$\tan\theta = \frac{\sin\theta}{\cos\theta}$$ Quotient identity

$$\cos\theta\tan\theta = \sin\theta$$ Multiply each side by $\cos\theta$.

$$\left(\frac{1}{\sec\theta}\right)\tan\theta = \sin\theta$$ Reciprocal identity

$$\left(-\frac{3\sqrt{34}}{34}\right)\left(-\frac{5}{3}\right) = \sin\theta$$ $\frac{1}{\sec\theta} = \frac{1}{-\frac{\sqrt{34}}{3}} = -\frac{3}{\sqrt{34}} = -\frac{3}{\sqrt{34}} \cdot \frac{\sqrt{34}}{\sqrt{34}} = -\frac{3\sqrt{34}}{34}$; From part (a)

$\tan\theta = -\frac{5}{3}$, as given

$$\sin\theta = \frac{5\sqrt{34}}{34}$$ Multiply and rewrite.

(c)

$$\cot(-\theta) = \frac{1}{\tan(-\theta)}$$ Reciprocal identity

$$\cot(-\theta) = \frac{1}{-\tan\theta}$$ Negative-angle identity

$$\cot(-\theta) = \frac{1}{-\left(-\frac{5}{3}\right)} = \frac{3}{5}$$ Use $\tan\theta = -\frac{5}{3}$, and simplify the complex fraction.

✔ ***Now Try Exercises 7, 15, and 31.***

CAUTION ***To avoid a common error, when taking the square root, be sure to choose the sign based on the quadrant of θ and the function being evaluated.***

EXAMPLE 2 Writing One Trigonometric Function in Terms of Another

Write $\cos x$ in terms of $\tan x$.

SOLUTION Since $\sec x$ is related to both $\cos x$ and $\tan x$ by identities, we start with $1 + \tan^2 x = \sec^2 x$.

$$1 + \tan^2 x = \sec^2 x$$ Pythagorean identity

$$\frac{1}{1+\tan^2 x} = \frac{1}{\sec^2 x}$$ Take reciprocals.

$$\frac{1}{1+\tan^2 x} = \cos^2 x$$ The reciprocal of $\sec^2 x$ is $\cos^2 x$.

Remember both the positive and negative roots.

$$\pm\sqrt{\frac{1}{1+\tan^2 x}} = \cos x$$ Take the square root of each side.

$$\cos x = \frac{\pm 1}{\sqrt{1+\tan^2 x}}$$ Quotient rule for radicals: $\sqrt[n]{\frac{a}{b}} = \frac{\sqrt[n]{a}}{\sqrt[n]{b}}$; rewrite.

$$\cos x = \frac{\pm\sqrt{1+\tan^2 x}}{1+\tan^2 x}$$ Rationalize the denominator.

The choice of the + sign or the − sign is made depending on the quadrant of x.

✔ ***Now Try Exercise 53.***

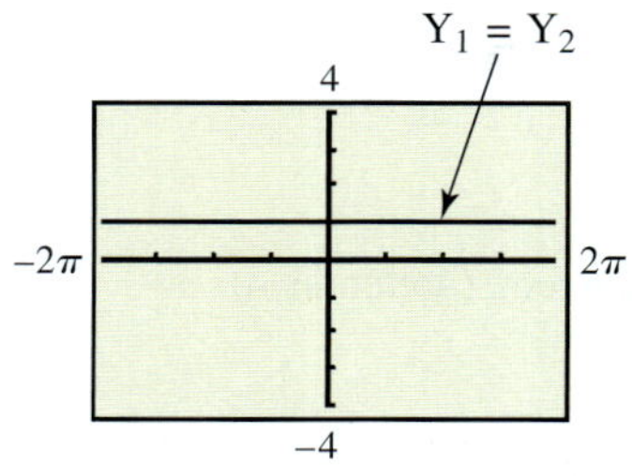

Figure 2

We can use a graphing calculator to decide whether two functions are identical. See **Figure 2,** which supports the identity $\sin^2 x + \cos^2 x = 1$. Y_1 is defined as $\sin^2 x + \cos^2 x$, and Y_2 is defined as 1. With an identity, you should see no difference between the two graphs. ■

Each of the functions $\tan\theta$, $\cot\theta$, $\sec\theta$, and $\csc\theta$ can easily be expressed in terms of $\sin\theta$, $\cos\theta$, or both. We often make such substitutions in an expression to simplify it.

EXAMPLE 3 **Rewriting an Expression in Terms of Sine and Cosine**

Write $\frac{1+\cot^2\theta}{1-\csc^2\theta}$ in terms of $\sin\theta$ and $\cos\theta$, and then simplify the expression so that no quotients appear.

SOLUTION

$$\frac{1+\cot^2\theta}{1-\csc^2\theta}$$

$$=\frac{1+\frac{\cos^2\theta}{\sin^2\theta}}{1-\frac{1}{\sin^2\theta}} \qquad \text{Quotient identities}$$

$$=\frac{\left(1+\frac{\cos^2\theta}{\sin^2\theta}\right)\sin^2\theta}{\left(1-\frac{1}{\sin^2\theta}\right)\sin^2\theta} \qquad \text{Simplify the complex fraction by multiplying both numerator and denominator by the LCD.}$$

$$=\frac{\sin^2\theta+\cos^2\theta}{\sin^2\theta-1} \qquad \text{Distributive property: } a(b+c)=ab+ac$$

$$=\frac{1}{-\cos^2\theta} \qquad \text{Pythagorean identities}$$

$$=-\sec^2\theta \qquad \text{Reciprocal identity}$$

✔ *Now Try Exercise 65.*

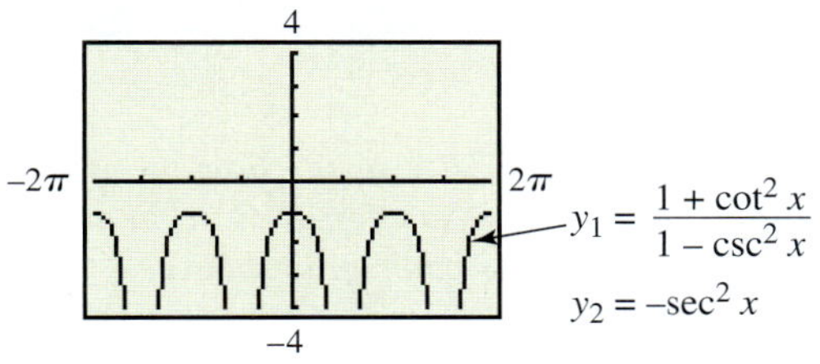

The graph supports the result in **Example 3.** The graphs of y_1 and y_2 coincide.

CAUTION ***When working with trigonometric expressions and identities, be sure to write the argument of the function.*** For example, we would *not* write $\sin^2 + \cos^2 = 1$. An argument such as θ is necessary in this identity.

5.1 Exercises

Concept Check In Exercises 1–6, use identities to fill in the blanks.

1. If $\tan\theta = 2.6$, then $\tan(-\theta) =$ ______.

2. If $\cos\theta = -0.65$, then $\cos(-\theta) =$ ______.

3. If $\tan\theta = 1.6$, then $\cot\theta =$ ______.

4. If $\cos\theta = 0.8$ and $\sin\theta = 0.6$, then $\tan(-\theta) =$ ______.

5. If $\sin\theta = \frac{2}{3}$, then $-\sin(-\theta) =$ ______.

6. If $\cos\theta = -\frac{1}{5}$, then $-\cos(-\theta) =$ ______.

Find $\sin\theta$. ***See Example 1.***

7. $\cos\theta = \frac{3}{4}$, θ in quadrant I

8. $\cos\theta = \frac{5}{6}$, θ in quadrant I

9. $\cot\theta = -\frac{1}{5}$, θ in quadrant IV

10. $\cot\theta = -\frac{1}{3}$, θ in quadrant IV

11. $\cos(-\theta) = \frac{\sqrt{5}}{5}$, $\tan\theta < 0$

12. $\cos(-\theta) = \frac{\sqrt{3}}{6}$, $\cot\theta < 0$

13. $\tan\theta = -\frac{\sqrt{6}}{2}$, $\cos\theta > 0$

14. $\tan\theta = -\frac{\sqrt{7}}{2}$, $\sec\theta > 0$

15. $\sec\theta = \frac{11}{4}$, $\cot\theta < 0$

16. $\sec\theta = \frac{7}{2}$, $\tan\theta < 0$

17. $\csc\theta = -\frac{9}{4}$

18. $\csc\theta = -\frac{8}{5}$

19. Why is it unnecessary to give the quadrant of θ in **Exercises 17 and 18?**

20. *Concept Check* What is **WRONG** with the statement of this problem?

Find $\cos(-\theta)$ *if* $\cos\theta = 3$.

Relating Concepts

For individual or collaborative investigation *(Exercises 21–26)*

A function is an ***even function*** *if* $f(-x) = f(x)$ *for all x in the domain of f. Similarly, a function is an* ***odd function*** *if* $f(-x) = -f(x)$ *for all x in the domain of f.* ***Work Exercises 21–26 in order,*** *to see the connection between the negative-angle identities and even and odd functions.*

21. Complete the statement: $\sin(-x) =$ ______.

22. Is the function $f(x) = \sin x$ *even* or *odd?*

23. Complete the statement: $\cos(-x) =$ ______.

24. Is the function $f(x) = \cos x$ *even* or *odd?*

25. Complete the statement: $\tan(-x) =$ ______.

26. Is the function $f(x) = \tan x$ *even* or *odd?*

Concept Check *For each graph of a circular function* $y = f(x)$, *determine whether* $f(-x) = f(x)$ *or* $f(-x) = -f(x)$ *is true.*

27.

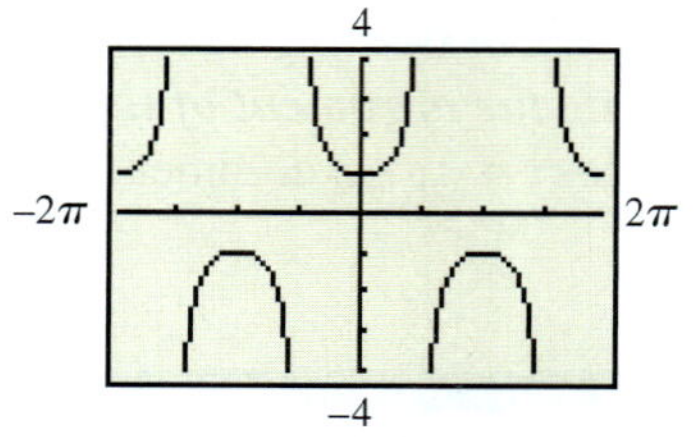

28.

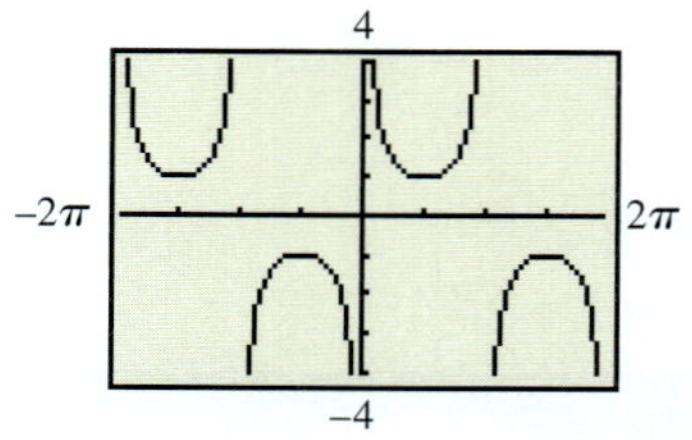

29.

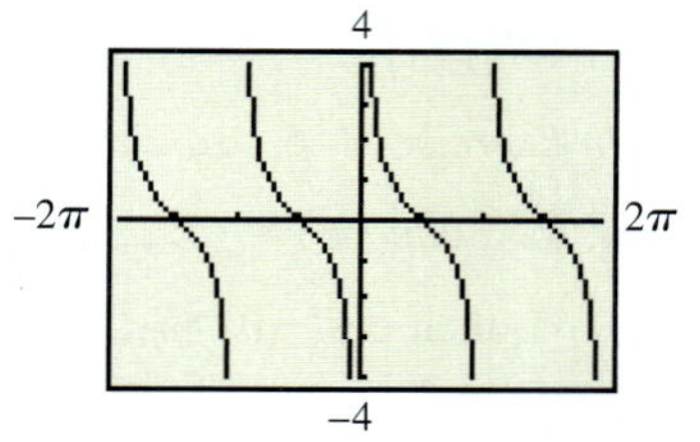

30.

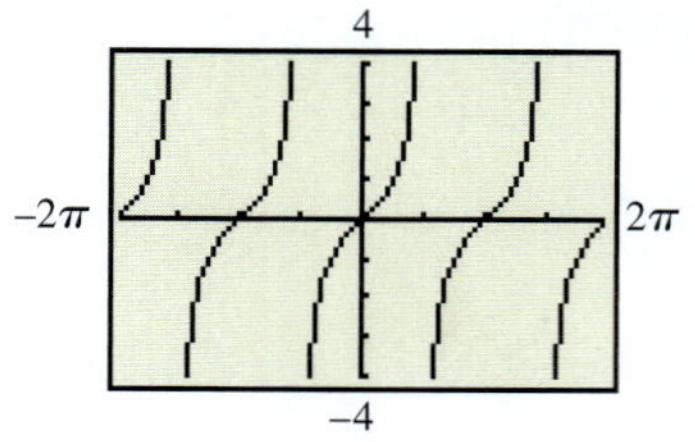

Find the remaining five trigonometric functions of θ. ***See Example 1.***

31. $\sin\theta = \frac{2}{3}$, θ in quadrant II

32. $\cos\theta = \frac{1}{5}$, θ in quadrant I

33. $\tan\theta = -\frac{1}{4}$, θ in quadrant IV

34. $\csc\theta = -\frac{5}{2}$, θ in quadrant III

35. $\cot\theta = \frac{4}{3}$, $\sin\theta > 0$

36. $\sin\theta = -\frac{4}{5}$, $\cos\theta < 0$

37. $\sec\theta = \frac{4}{3}$, $\sin\theta < 0$

38. $\cos\theta = -\frac{1}{4}$, $\sin\theta > 0$

Concept Check *For each expression in Column I, choose the expression from Column II that completes an identity.*

I	II
39. $\frac{\cos x}{\sin x} =$ ______	**A.** $\sin^2 x + \cos^2 x$
40. $\tan x =$ ______	**B.** $\cot x$
41. $\cos(-x) =$ ______	**C.** $\sec^2 x$
42. $\tan^2 x + 1 =$ ______	**D.** $\frac{\sin x}{\cos x}$
43. $1 =$ ______	**E.** $\cos x$

Concept Check *For each expression in Column I, choose the expression from Column II that completes an identity. You may have to rewrite one or both expressions.*

I	II
44. $-\tan x \cos x =$ ______	**A.** $\frac{\sin^2 x}{\cos^2 x}$
45. $\sec^2 x - 1 =$ ______	**B.** $\frac{1}{\sec^2 x}$
46. $\frac{\sec x}{\csc x} =$ ______	**C.** $\sin(-x)$
47. $1 + \sin^2 x =$ ______	**D.** $\csc^2 x - \cot^2 x + \sin^2 x$
48. $\cos^2 x =$ ______	**E.** $\tan x$

49. A student writes "$1 + \cot^2 = \csc^2$." Comment on this student's work.

50. A student makes the following claim: "Since $\sin^2\theta + \cos^2\theta = 1$, I should be able to also say that $\sin\theta + \cos\theta = 1$ if I take the square root of each side." Comment on this student's statement.

51. *Concept Check* Suppose that $\cos\theta = \frac{x}{x+1}$. Find an expression in x for $\sin\theta$.

52. *Concept Check* Suppose that $\sec\theta = \frac{x+4}{x}$. Find an expression in x for $\tan\theta$.

Perform each transformation. ***See Example 2.***

53. Write $\sin x$ in terms of $\cos x$.

54. Write $\cot x$ in terms of $\sin x$.

55. Write $\tan x$ in terms of $\sec x$.

56. Write $\cot x$ in terms of $\csc x$.

57. Write $\csc x$ in terms of $\cos x$.

58. Write $\sec x$ in terms of $\sin x$.

Write each expression in terms of sine and cosine, and simplify so that no quotients appear in the final expression and all functions are of θ only. ***See Example 3.***

59. $\cot\theta \sin\theta$

60. $\tan\theta \cos\theta$

61. $\sec\theta \cot\theta \sin\theta$

62. $\csc\theta \cos\theta \tan\theta$

63. $\cos\theta \csc\theta$

64. $\sin\theta \sec\theta$

65. $\sin^2\theta(\csc^2\theta - 1)$

66. $\cot^2\theta(1 + \tan^2\theta)$

67. $(1 - \cos\theta)(1 + \sec\theta)$

68. $(\sec\theta - 1)(\sec\theta + 1)$

69. $\frac{1 + \tan(-\theta)}{\tan(-\theta)}$

70. $\frac{1 + \cot\theta}{\cot\theta}$

71. $\frac{1 - \cos^2(-\theta)}{1 + \tan^2(-\theta)}$

72. $\frac{1 - \sin^2(-\theta)}{1 + \cot^2(-\theta)}$

73. $\sec\theta - \cos\theta$

74. $\csc\theta - \sin\theta$

75. $(\sec\theta + \csc\theta)(\cos\theta - \sin\theta)$

76. $(\sin\theta - \cos\theta)(\csc\theta + \sec\theta)$

77. $\sin\theta(\csc\theta - \sin\theta)$

78. $\cos\theta(\cos\theta - \sec\theta)$

79. $\dfrac{1 + \tan^2\theta}{1 + \cot^2\theta}$

80. $\dfrac{\sec^2\theta - 1}{\csc^2\theta - 1}$

81. $\sin^2(-\theta) + \tan^2(-\theta) + \cos^2(-\theta)$

82. $-\sec^2(-\theta) + \sin^2(-\theta) + \cos^2(-\theta)$

83. $\dfrac{\csc\theta}{\cot(-\theta)}$

84. $\dfrac{\tan(-\theta)}{\sec\theta}$

Work each problem.

85. Let $\cos x = \frac{1}{5}$. Find all possible values of $\frac{\sec x - \tan x}{\sin x}$.

86. Let $\csc x = -3$. Find all possible values of $\frac{\sin x + \cos x}{\sec x}$.

Relating Concepts

For individual or collaborative investigation *(Exercises 87–92)*

*In **Chapter 4** we graphed functions defined by*

$$y = c + a \cdot f[b(x - d)]$$

*with the assumption that $b > 0$. To see what happens when $b < 0$, **work Exercises 87–92 in order.***

87. Use a negative-angle identity to write $y = \sin(-2x)$ as a function of $2x$.

88. How is your answer to **Exercise 87** related to $y = \sin(2x)$?

89. Use a negative-angle identity to write $y = \cos(-4x)$ as a function of $4x$.

90. How is your answer to **Exercise 89** related to $y = \cos(4x)$?

91. Use your results from **Exercises 87–90** to rewrite the following with a positive value of b.

(a) $y = \sin(-4x)$ **(b)** $y = \cos(-2x)$ **(c)** $y = -5\sin(-3x)$

92. Write a short response to this statement, which is often used by one of the authors of this text in trigonometry classes: *Students who tend to ignore negative signs should enjoy graphing functions involving the cosine and the secant.*

Use a graphing calculator to make a conjecture about whether each equation is an identity.

93. $\cos 2x = 1 - 2\sin^2 x$

94. $2\sin x = \sin 2x$

95. $\sin x = \sqrt{1 - \cos^2 x}$

96. $\cos 2x = \cos^2 x - \sin^2 x$

5.2 Verifying Trigonometric Identities

- Strategies
- Verifying Identities by Working with One Side
- Verifying Identities by Working with Both Sides

Strategies

One of the skills required for more advanced work in mathematics, especially in calculus, is the ability to use identities to write expressions in alternative forms. We develop this skill by using the fundamental identities to verify that a trigonometric equation is an identity (for those values of the variable for which it is defined). Here are some helpful hints.

LOOKING AHEAD TO CALCULUS

Trigonometric identities are used in calculus to simplify trigonometric expressions, determine derivatives of trigonometric functions, and change the form of some integrals.

Hints for Verifying Identities

1. ***Learn the fundamental identities given in Section 5.1.*** Whenever you see either side of a fundamental identity, the other side should come to mind. ***Also, be aware of equivalent forms of the fundamental identities.*** For example,

$$\sin^2\theta = 1 - \cos^2\theta \quad \text{is an alternative form of} \quad \sin^2\theta + \cos^2\theta = 1.$$

2. ***Try to rewrite the more complicated side*** of the equation so that it is identical to the simpler side.
3. ***It is sometimes helpful to express all trigonometric functions in the equation in terms of sine and cosine*** and then simplify the result.
4. ***Usually, any factoring or indicated algebraic operations should be performed.*** These *algebraic* identities are often used in verifying trigonometric identities.

$$(a+b)^2 = a^2 + 2ab + b^2 \qquad (a-b)^2 = a^2 - 2ab + b^2$$

$$a^3 - b^3 = (a-b)(a^2 + ab + b^2) \qquad a^3 + b^3 = (a+b)(a^2 - ab + b^2)$$

$$a^2 - b^2 = (a+b)(a-b)$$

For example, the expression

$$\sin^2 x + 2\sin x + 1 \quad \text{can be factored as} \quad (\sin x + 1)^2.$$

The sum or difference of two trigonometric expressions can be found in the same way as any other rational expression. For example,

$$\frac{1}{\sin\theta} + \frac{1}{\cos\theta} = \frac{1\cdot\cos\theta}{\sin\theta\cos\theta} + \frac{1\cdot\sin\theta}{\cos\theta\sin\theta} \qquad \text{Write with the LCD.}$$

$$= \frac{\cos\theta + \sin\theta}{\sin\theta\cos\theta}. \qquad \frac{a}{c} + \frac{b}{c} = \frac{a+b}{c}$$

5. ***As you select substitutions, keep in mind the side you are not changing, because it represents your goal.*** For example, to verify the identity

$$\tan^2 x + 1 = \frac{1}{\cos^2 x},$$

try to think of an identity that relates $\tan x$ to $\cos x$. In this case, since $\sec x = \frac{1}{\cos x}$ and $\sec^2 x = \tan^2 x + 1$, the secant function is the best link between the two sides.

6. If an expression contains $1 + \sin x$, ***multiplying both numerator and denominator*** by $1 - \sin x$ would give $1 - \sin^2 x$, which could be replaced with $\cos^2 x$. Similar procedures apply for $1 - \sin x$, $1 + \cos x$, and $1 - \cos x$.

CAUTION ***The procedure for verifying identities is not the same as that of solving equations.*** Techniques used in solving equations, such as adding the same term to each side, and multiplying each side by the same term, should not be used when working with identities.

Verifying Identities by Working with One Side To avoid the temptation to use algebraic properties of equations to verify identities, ***one strategy is to work with only one side and rewrite it to match the other side.***

EXAMPLE 1 Verifying an Identity (Working with One Side)

Verify that the following equation is an identity.

$$\cot \theta + 1 = \csc \theta(\cos \theta + \sin \theta)$$

SOLUTION We use the fundamental identities from **Section 5.1** to rewrite one side of the equation so that it is identical to the other side. Since the right side is more complicated, we work with it, as suggested in Hint 2, and use Hint 3 to change all functions to expressions involving sine or cosine.

For $\theta = x$,
$y_1 = \cot x + 1$
$y_2 = \csc x (\cos x + \sin x)$

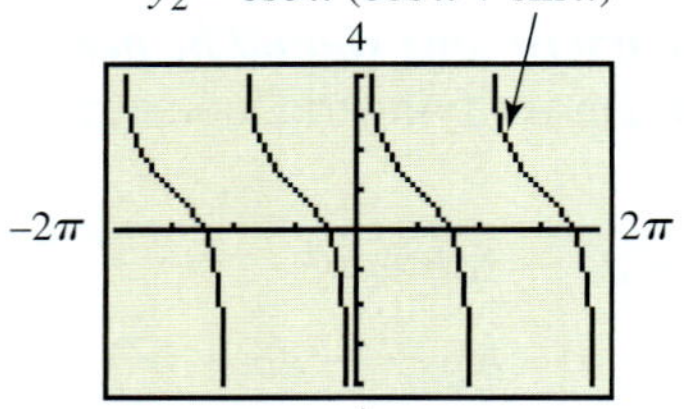

The graphs coincide, which supports the conclusion in **Example 1.**

Steps	Reasons
$\underbrace{\csc \theta(\cos \theta + \sin \theta)}_{\text{Right side of given equation}} = \frac{1}{\sin \theta}(\cos \theta + \sin \theta)$	$\csc \theta = \frac{1}{\sin \theta}$
$= \frac{\cos \theta}{\sin \theta} + \frac{\sin \theta}{\sin \theta}$	Distributive property: $a(b + c) = ab + ac$
$= \underbrace{\cot \theta + 1}_{\text{Left side of given equation}}$	$\frac{\cos \theta}{\sin \theta} = \cot \theta$; $\frac{\sin \theta}{\sin \theta} = 1$

The given equation is an identity. The right side of the equation is identical to the left side.

✔ *Now Try Exercise 35.*

EXAMPLE 2 Verifying an Identity (Working with One Side)

Verify that the following equation is an identity.

$$\tan^2 x(1 + \cot^2 x) = \frac{1}{1 - \sin^2 x}$$

SOLUTION We work with the more complicated left side, as suggested in Hint 2. Again, we use the fundamental identities from **Section 5.1.**

$\underbrace{\tan^2 x(1 + \cot^2 x)}_{\text{Left side of given equation}} = \tan^2 x + \tan^2 x \cot^2 x$	Distributive property
$= \tan^2 x + \tan^2 x \cdot \frac{1}{\tan^2 x}$	$\cot^2 x = \frac{1}{\tan^2 x}$
$= \tan^2 x + 1$	$\tan^2 x \cdot \frac{1}{\tan^2 x} = 1$
$= \sec^2 x$	Pythagorean identity
$= \frac{1}{\cos^2 x}$	$\sec^2 x = \frac{1}{\cos^2 x}$
$= \frac{1}{\underbrace{1 - \sin^2 x}_{\text{Right side of given equation}}}$	Pythagorean identity

$y_1 = \tan^2 x(1 + \cot^2 x)$
$y_2 = \frac{1}{1 - \sin^2 x}$

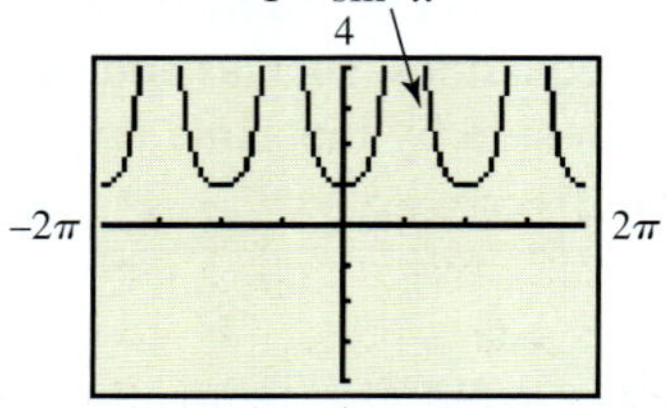

The screen supports the conclusion in **Example 2.**

Since the left side of the equation is identical to the right side, the given equation is an identity.

✔ *Now Try Exercise 39.*

EXAMPLE 3 Verifying an Identity (Working with One Side)

Verify that the following equation is an identity.

$$\frac{\tan t - \cot t}{\sin t \cos t} = \sec^2 t - \csc^2 t$$

SOLUTION We transform the more complicated left side to match the right side.

$$\begin{aligned}
\frac{\tan t - \cot t}{\sin t \cos t} &= \frac{\tan t}{\sin t \cos t} - \frac{\cot t}{\sin t \cos t} && \frac{a-b}{c} = \frac{a}{c} - \frac{b}{c} \\
&= \tan t \cdot \frac{1}{\sin t \cos t} - \cot t \cdot \frac{1}{\sin t \cos t} && \frac{a}{b} = a \cdot \frac{1}{b} \\
&= \frac{\sin t}{\cos t} \cdot \frac{1}{\sin t \cos t} - \frac{\cos t}{\sin t} \cdot \frac{1}{\sin t \cos t} && \tan t = \frac{\sin t}{\cos t};\ \cot t = \frac{\cos t}{\sin t} \\
&= \frac{1}{\cos^2 t} - \frac{1}{\sin^2 t} && \text{Multiply.} \\
&= \sec^2 t - \csc^2 t && \frac{1}{\cos^2 t} = \sec^2 t;\ \frac{1}{\sin^2 t} = \csc^2 t
\end{aligned}$$

Hint 3 about writing all trigonometric functions in terms of sine and cosine was used in the third line of the solution.

✔ ***Now Try Exercise 43.***

EXAMPLE 4 Verifying an Identity (Working with One Side)

Verify that the following equation is an identity.

$$\frac{\cos x}{1 - \sin x} = \frac{1 + \sin x}{\cos x}$$

SOLUTION We work on the right side, using Hint 6 in the list given earlier to multiply the numerator and denominator on the right by $1 - \sin x$.

$$\begin{aligned}
\frac{1 + \sin x}{\cos x} &= \frac{(1 + \sin x)(1 - \sin x)}{\cos x(1 - \sin x)} && \text{Multiply by 1 in the form } \frac{1 - \sin x}{1 - \sin x}. \\
&= \frac{1 - \sin^2 x}{\cos x(1 - \sin x)} && (x+y)(x-y) = x^2 - y^2 \\
&= \frac{\cos^2 x}{\cos x(1 - \sin x)} && 1 - \sin^2 x = \cos^2 x \\
&= \frac{\cos x \cdot \cos x}{\cos x(1 - \sin x)} && a^2 = a \cdot a \\
&= \frac{\cos x}{1 - \sin x} && \text{Write in lowest terms.}
\end{aligned}$$

✔ ***Now Try Exercise 49.***

Verifying Identities by Working with Both Sides If both sides of an identity appear to be equally complex, the identity can be verified by working independently on the left side and on the right side, until each side is changed into some common third result. ***Each step, on each side, must be reversible.*** With all steps reversible, the procedure is as shown in the margin. The left side leads to a common third expression, which leads back to the right side.

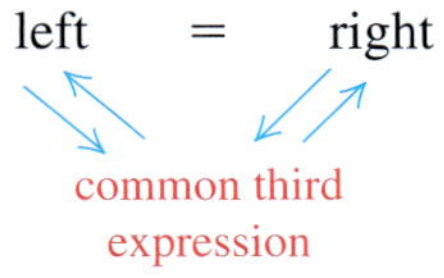

NOTE Working with both sides is often a good alternative for identities that are difficult. In practice, if working with one side does not seem to be effective, switch to the other side. Somewhere along the way it may happen that the same expression occurs on both sides.

EXAMPLE 5 Verifying an Identity (Working with Both Sides)

Verify that the following equation is an identity.

$$\frac{\sec\alpha + \tan\alpha}{\sec\alpha - \tan\alpha} = \frac{1 + 2\sin\alpha + \sin^2\alpha}{\cos^2\alpha}$$

SOLUTION Both sides appear equally complex, so we verify the identity by changing each side into a common third expression. We work first on the left, multiplying the numerator and denominator by $\cos\alpha$.

$$\underbrace{\frac{\sec\alpha + \tan\alpha}{\sec\alpha - \tan\alpha}}_{\text{Left side of given equation}} = \frac{(\sec\alpha + \tan\alpha)\cos\alpha}{(\sec\alpha - \tan\alpha)\cos\alpha} \qquad \text{Multiply by 1 in the form } \tfrac{\cos\alpha}{\cos\alpha}.$$

$$= \frac{\sec\alpha\cos\alpha + \tan\alpha\cos\alpha}{\sec\alpha\cos\alpha - \tan\alpha\cos\alpha} \qquad \text{Distributive property}$$

$$= \frac{1 + \tan\alpha\cos\alpha}{1 - \tan\alpha\cos\alpha} \qquad \sec\alpha\cos\alpha = 1$$

$$= \frac{1 + \frac{\sin\alpha}{\cos\alpha}\cdot\cos\alpha}{1 - \frac{\sin\alpha}{\cos\alpha}\cdot\cos\alpha} \qquad \tan\alpha = \tfrac{\sin\alpha}{\cos\alpha}$$

$$= \frac{1 + \sin\alpha}{1 - \sin\alpha} \qquad \text{Simplify.}$$

On the right side of the original equation, begin by factoring.

$$\underbrace{\frac{1 + 2\sin\alpha + \sin^2\alpha}{\cos^2\alpha}}_{\text{Right side of given equation}} = \frac{(1 + \sin\alpha)^2}{\cos^2\alpha} \qquad \text{Factor the numerator; } x^2 + 2xy + y^2 = (x + y)^2.$$

$$= \frac{(1 + \sin\alpha)^2}{1 - \sin^2\alpha} \qquad \cos^2\alpha = 1 - \sin^2\alpha$$

$$= \frac{(1 + \sin\alpha)^2}{(1 + \sin\alpha)(1 - \sin\alpha)} \qquad \text{Factor the denominator; } x^2 - y^2 = (x + y)(x - y).$$

$$= \frac{1 + \sin\alpha}{1 - \sin\alpha} \qquad \text{Write in lowest terms.}$$

We have shown that

$$\overbrace{\frac{\sec\alpha + \tan\alpha}{\sec\alpha - \tan\alpha}}^{\text{Left side of given equation}} = \overbrace{\frac{1 + \sin\alpha}{1 - \sin\alpha}}^{\text{Common third expression}} = \overbrace{\frac{1 + 2\sin\alpha + \sin^2\alpha}{\cos^2\alpha}}^{\text{Right side of given equation}},$$

and thus have verified that the given equation is an identity.

✔ *Now Try Exercise 65.*

CAUTION Use the method of **Example 5** *only* if the steps are reversible.

There are usually several ways to verify a given identity. For instance, another way to begin verifying the identity in **Example 5** is to work on the left as follows.

$$\underbrace{\frac{\sec\alpha + \tan\alpha}{\sec\alpha - \tan\alpha}}_{\substack{\text{Left side of} \\ \text{given equation} \\ \text{in Example 5}}} = \frac{\frac{1}{\cos\alpha} + \frac{\sin\alpha}{\cos\alpha}}{\frac{1}{\cos\alpha} - \frac{\sin\alpha}{\cos\alpha}} \qquad \text{Fundamental identities (Section 5.1)}$$

$$= \frac{\frac{1+\sin\alpha}{\cos\alpha}}{\frac{1-\sin\alpha}{\cos\alpha}} \qquad \text{Add and subtract fractions.}$$

$$= \frac{1+\sin\alpha}{\cos\alpha} \div \frac{1-\sin\alpha}{\cos\alpha} \qquad \text{Simplify the complex fraction. Use the definition of division.}$$

$$= \frac{1+\sin\alpha}{\cos\alpha} \cdot \frac{\cos\alpha}{1-\sin\alpha} \qquad \text{Multiply by the reciprocal.}$$

$$= \frac{1+\sin\alpha}{1-\sin\alpha} \qquad \text{Multiply and write in lowest terms.}$$

Compare this with the result shown in **Example 5** for the right side to see that the two sides indeed agree.

EXAMPLE 6 Applying a Pythagorean Identity to Electronics

Tuners in radios select a radio station by adjusting the frequency. A tuner may contain an inductor L and a capacitor C, as illustrated in **Figure 3.** The energy stored in the inductor at time t is given by

$$L(t) = k\sin^2(2\pi Ft)$$

and the energy stored in the capacitor is given by

$$C(t) = k\cos^2(2\pi Ft),$$

where F is the frequency of the radio station and k is a constant. The total energy E in the circuit is given by

$$E(t) = L(t) + C(t).$$

Show that E is a constant function. (*Source:* Weidner, R. and R. Sells, *Elementary Classical Physics,* Vol. 2, Allyn & Bacon.)

An Inductor and a Capacitor

Figure 3

SOLUTION

$$\begin{aligned} E(t) &= L(t) + C(t) && \text{Given equation} \\ &= k\sin^2(2\pi Ft) + k\cos^2(2\pi Ft) && \text{Substitute.} \\ &= k[\sin^2(2\pi Ft) + \cos^2(2\pi Ft)] && \text{Factor out } k. \\ &= k(1) && \sin^2\theta + \cos^2\theta = 1 \text{ (Here } \theta = 2\pi Ft.) \\ &= k && \text{Identity property} \end{aligned}$$

Since k is a constant, $E(t)$ is a constant function.

✔ ***Now Try Exercise 95.***

5.2 Exercises

To the student: **Exercises 1–34** are designed for practice in applying algebraic techniques to trigonometric expressions. These techniques are essential in verifying the identities that follow.

Perform each indicated operation and simplify the result so that there are no quotients.

1. $\cot\theta + \dfrac{1}{\cot\theta}$

2. $\dfrac{\sec x}{\csc x} + \dfrac{\csc x}{\sec x}$

3. $\tan x(\cot x + \csc x)$

4. $\cos\beta(\sec\beta + \csc\beta)$

5. $\dfrac{1}{\csc^2\theta} + \dfrac{1}{\sec^2\theta}$

6. $\dfrac{\cos x}{\sec x} + \dfrac{\sin x}{\csc x}$

7. $(\sin\alpha - \cos\alpha)^2$

8. $(\tan x + \cot x)^2$

9. $(1 + \sin t)^2 + \cos^2 t$

10. $(1 + \tan\theta)^2 - 2\tan\theta$

11. $\dfrac{1}{1 + \cos x} - \dfrac{1}{1 - \cos x}$

12. $\dfrac{1}{\sin\alpha - 1} - \dfrac{1}{\sin\alpha + 1}$

Factor each trigonometric expression.

13. $\sin^2\theta - 1$

14. $\sec^2\theta - 1$

15. $(\sin x + 1)^2 - (\sin x - 1)^2$

16. $(\tan x + \cot x)^2 - (\tan x - \cot x)^2$

17. $2\sin^2 x + 3\sin x + 1$

18. $4\tan^2\beta + \tan\beta - 3$

19. $\cos^4 x + 2\cos^2 x + 1$

20. $\cot^4 x + 3\cot^2 x + 2$

21. $\sin^3 x - \cos^3 x$

22. $\sin^3\alpha + \cos^3\alpha$

Each expression simplifies to a constant, a single function, or a power of a function. Use fundamental identities to simplify each expression.

23. $\tan\theta\cos\theta$

24. $\cot\alpha\sin\alpha$

25. $\sec r\cos r$

26. $\cot t\tan t$

27. $\dfrac{\sin\beta\tan\beta}{\cos\beta}$

28. $\dfrac{\csc\theta\sec\theta}{\cot\theta}$

29. $\sec^2 x - 1$

30. $\csc^2 t - 1$

31. $\dfrac{\sin^2 x}{\cos^2 x} + \sin x\csc x$

32. $\dfrac{1}{\tan^2\alpha} + \cot\alpha\tan\alpha$

33. $1 - \dfrac{1}{\csc^2 x}$

34. $1 - \dfrac{1}{\sec^2 x}$

In Exercises 35–78, verify that each trigonometric equation is an identity. ***See Examples 1–5.***

35. $\dfrac{\cot\theta}{\csc\theta} = \cos\theta$

36. $\dfrac{\tan\alpha}{\sec\alpha} = \sin\alpha$

37. $\dfrac{1 - \sin^2\beta}{\cos\beta} = \cos\beta$

38. $\dfrac{\tan^2\alpha + 1}{\sec\alpha} = \sec\alpha$

39. $\cos^2\theta(\tan^2\theta + 1) = 1$

40. $\sin^2\beta(1 + \cot^2\beta) = 1$

41. $\cot\theta + \tan\theta = \sec\theta\csc\theta$

42. $\sin^2\alpha + \tan^2\alpha + \cos^2\alpha = \sec^2\alpha$

43. $\dfrac{\cos\alpha}{\sec\alpha} + \dfrac{\sin\alpha}{\csc\alpha} = \sec^2\alpha - \tan^2\alpha$

44. $\dfrac{\sin^2\theta}{\cos\theta} = \sec\theta - \cos\theta$

45. $\sin^4\theta - \cos^4\theta = 2\sin^2\theta - 1$

46. $\sec^4 x - \sec^2 x = \tan^4 x + \tan^2 x$

47. $\dfrac{1 - \cos x}{1 + \cos x} = (\cot x - \csc x)^2$

48. $(\sec\alpha - \tan\alpha)^2 = \dfrac{1 - \sin\alpha}{1 + \sin\alpha}$

49. $\dfrac{\cos\theta + 1}{\tan^2\theta} = \dfrac{\cos\theta}{\sec\theta - 1}$

50. $\dfrac{(\sec\theta - \tan\theta)^2 + 1}{\sec\theta\csc\theta - \tan\theta\csc\theta} = 2\tan\theta$

51. $\dfrac{1}{1 - \sin\theta} + \dfrac{1}{1 + \sin\theta} = 2\sec^2\theta$

52. $\dfrac{1}{\sec\alpha - \tan\alpha} = \sec\alpha + \tan\alpha$

53. $\dfrac{\cot\alpha + 1}{\cot\alpha - 1} = \dfrac{1 + \tan\alpha}{1 - \tan\alpha}$

54. $\dfrac{\csc\theta + \cot\theta}{\tan\theta + \sin\theta} = \cot\theta\csc\theta$

55. $\dfrac{\cos\theta}{\sin\theta\cot\theta} = 1$

56. $\sin^2\theta(1 + \cot^2\theta) - 1 = 0$

57. $\dfrac{\sec^4\theta - \tan^4\theta}{\sec^2\theta + \tan^2\theta} = \sec^2\theta - \tan^2\theta$

58. $\dfrac{\sin^4\alpha - \cos^4\alpha}{\sin^2\alpha - \cos^2\alpha} = 1$

59. $\dfrac{\tan^2 t - 1}{\sec^2 t} = \dfrac{\tan t - \cot t}{\tan t + \cot t}$

60. $\dfrac{\cot^2 t - 1}{1 + \cot^2 t} = 1 - 2\sin^2 t$

61. $\sin^2\alpha\sec^2\alpha + \sin^2\alpha\csc^2\alpha = \sec^2\alpha$

62. $\tan^2\alpha\sin^2\alpha = \tan^2\alpha + \cos^2\alpha - 1$

63. $\dfrac{\tan x}{1 + \cos x} + \dfrac{\sin x}{1 - \cos x} = \cot x + \sec x\csc x$

64. $\dfrac{\sin\theta}{1 - \cos\theta} - \dfrac{\sin\theta\cos\theta}{1 + \cos\theta} = \csc\theta(1 + \cos^2\theta)$

65. $\dfrac{1 + \cos x}{1 - \cos x} - \dfrac{1 - \cos x}{1 + \cos x} = 4\cot x\csc x$

66. $\dfrac{1 + \sin\theta}{1 - \sin\theta} - \dfrac{1 - \sin\theta}{1 + \sin\theta} = 4\tan\theta\sec\theta$

67. $\dfrac{1 - \sin\theta}{1 + \sin\theta} = \sec^2\theta - 2\sec\theta\tan\theta + \tan^2\theta$

68. $\sin\theta + \cos\theta = \dfrac{\sin\theta}{1 - \cot\theta} + \dfrac{\cos\theta}{1 - \tan\theta}$

69. $\dfrac{-1}{\tan\alpha - \sec\alpha} + \dfrac{-1}{\tan\alpha + \sec\alpha} = 2\tan\alpha$

70. $(1 + \sin x + \cos x)^2 = 2(1 + \sin x)(1 + \cos x)$

71. $(1 - \cos^2\alpha)(1 + \cos^2\alpha) = 2\sin^2\alpha - \sin^4\alpha$

72. $(\sec\alpha + \csc\alpha)(\cos\alpha - \sin\alpha) = \cot\alpha - \tan\alpha$

73. $\dfrac{1 - \cos x}{1 + \cos x} = \csc^2 x - 2\csc x\cot x + \cot^2 x$

74. $\dfrac{1 - \cos\theta}{1 + \cos\theta} = 2\csc^2\theta - 2\csc\theta\cot\theta - 1$

75. $(2\sin x + \cos x)^2 + (2\cos x - \sin x)^2 = 5$

76. $\sin^2 x(1 + \cot x) + \cos^2 x(1 - \tan x) + \cot^2 x = \csc^2 x$

77. $\sec x - \cos x + \csc x - \sin x - \sin x\tan x = \cos x\cot x$

78. $\sin^3\theta + \cos^3\theta = (\cos\theta + \sin\theta)(1 - \cos\theta\sin\theta)$

Graph each expression and use the graph to make a conjecture, predicting what might be an identity. Then verify your conjecture algebraically.

79. $(\sec\theta + \tan\theta)(1 - \sin\theta)$

80. $(\csc\theta + \cot\theta)(\sec\theta - 1)$

81. $\dfrac{\cos\theta + 1}{\sin\theta + \tan\theta}$

82. $\tan\theta\sin\theta + \cos\theta$

Graph the expressions on each side of the equals symbol to determine whether the equation might be an identity. (Note: Use a domain whose length is at least 2π.) If the equation looks like an identity, verify it algebraically. ***See Example 1.***

83. $\dfrac{2 + 5\cos x}{\sin x} = 2\csc x + 5\cot x$

84. $1 + \cot^2 x = \dfrac{\sec^2 x}{\sec^2 x - 1}$

85. $\dfrac{\tan x - \cot x}{\tan x + \cot x} = 2\sin^2 x$

86. $\dfrac{1}{1 + \sin x} + \dfrac{1}{1 - \sin x} = \sec^2 x$

By substituting a number for t, show that the equation is not an identity.

87. $\sin(\csc t) = 1$

88. $\sqrt{\cos^2 t} = \cos t$

89. $\csc t = \sqrt{1 + \cot^2 t}$

90. $\cos t = \sqrt{1 - \sin^2 t}$

91. *Concept Check* When is $\sin x = -\sqrt{1 - \cos^2 x}$ a true statement?

92. *Concept Check* When is $\cos x = -\sqrt{1 - \sin^2 x}$ a true statement?

(Modeling) Work each problem.

93. *Intensity of a Lamp* According to **Lambert's law,** the intensity of light from a single source on a flat surface at point P is given by

$$I = k\cos^2\theta,$$

where k is a constant. (*Source:* Winter, C., *Solar Power Plants,* Springer-Verlag.)

(a) Write I in terms of the sine function.
(b) Why does the maximum value of I occur when $\theta = 0$?

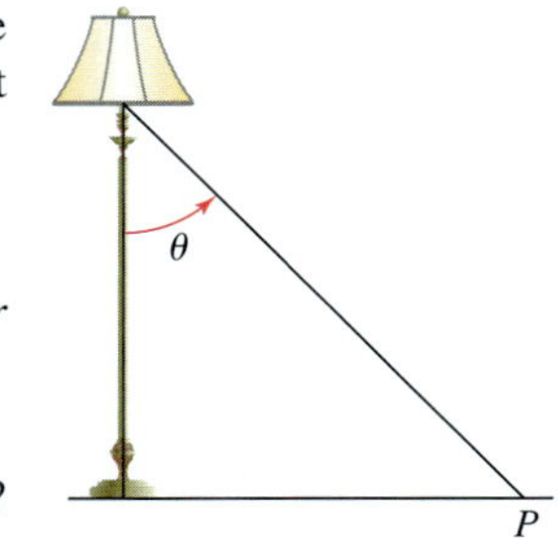

94. *Oscillating Spring* The distance or displacement y of a weight attached to an oscillating spring from its natural position is modeled by

$$y = 4\cos(2\pi t),$$

where t is time in seconds. Potential energy is the energy of position and is given by

$$P = ky^2,$$

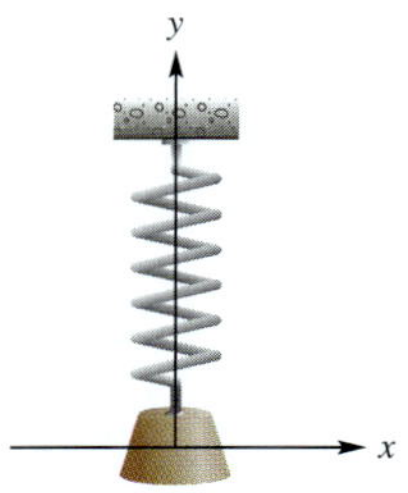

where k is a constant. The weight has the greatest potential energy when the spring is stretched the most. (*Source:* Weidner, R. and R. Sells, *Elementary Classical Physics,* Vol. 2, Allyn & Bacon.)

(a) Write an expression for P that involves the cosine function.
(b) Use a fundamental identity to write P in terms of $\sin(2\pi t)$.

95. *Radio Tuners* Refer to **Example 6.** Let the energy stored in the inductor be given by

$$L(t) = 3\cos^2(6{,}000{,}000t)$$

and let the energy in the capacitor be given by

$$C(t) = 3\sin^2(6{,}000{,}000t),$$

where t is time in seconds. The total energy E in the circuit is given by $E(t) = L(t) + C(t)$.

(a) Graph L, C, and E in the window $[0, 10^{-6}]$ by $[-1, 4]$, with Xscl $= 10^{-7}$ and Yscl $= 1$. Interpret the graph.
(b) Make a table of values for L, C, and E starting at $t = 0$, incrementing by 10^{-7}. Interpret your results.
(c) Use a fundamental identity to derive a simplified expression for $E(t)$.

5.3 Sum and Difference Identities for Cosine

- Difference Identity for Cosine
- Sum Identity for Cosine
- Cofunction Identities
- Applying the Sum and Difference Identities
- Verifying an Identity

Difference Identity for Cosine

Several examples presented earlier should have convinced you by now that

$$\cos(A - B) \textit{ does not equal } \cos A - \cos B.$$

For example, if $A = \frac{\pi}{2}$ and $B = 0$, then

$$\cos(A - B) = \cos\left(\frac{\pi}{2} - 0\right) = \cos\frac{\pi}{2} = 0,$$

while $$\cos A - \cos B = \cos\frac{\pi}{2} - \cos 0 = 0 - 1 = -1.$$

To derive a formula for $\cos(A - B)$, we start by locating angles A and B in standard position on a unit circle, with $B < A$. Let S and Q be the points where the terminal sides of angles A and B, respectively, intersect the circle. Let P be the point $(1, 0)$, and locate point R on the unit circle so that angle POR equals the difference $A - B$. See **Figure 4.**

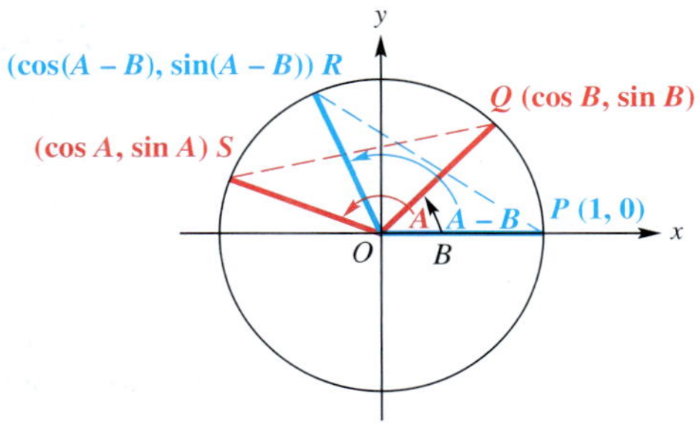

Figure 4

Because point Q is on the unit circle, the x-coordinate of Q is the cosine of angle B, while the y-coordinate of Q is the sine of angle B.

$$Q \text{ has coordinates } (\cos B, \sin B).$$

In the same way,

$$S \text{ has coordinates } (\cos A, \sin A),$$

and $$R \text{ has coordinates } (\cos(A - B), \sin(A - B)).$$

Angle SOQ also equals $A - B$. Since the central angles SOQ and POR are equal, chords PR and SQ are equal. By the distance formula, since $PR = SQ$,

$$\sqrt{[\cos(A - B) - 1]^2 + [\sin(A - B) - 0]^2}$$
$$= \sqrt{(\cos A - \cos B)^2 + (\sin A - \sin B)^2}. \quad \text{(Appendix B)}$$

Square each side and clear parentheses.

$$\cos^2(A - B) - 2\cos(A - B) + 1 + \sin^2(A - B)$$
$$= \cos^2 A - 2\cos A \cos B + \cos^2 B + \sin^2 A - 2\sin A \sin B + \sin^2 B$$

Since $\sin^2 x + \cos^2 x = 1$ for any value of x, we can rewrite the equation, as shown on the next page.

$2 - 2\cos(A - B) = 2 - 2\cos A \cos B - 2\sin A \sin B$ — Use $\sin^2 x + \cos^2 x = 1$ three times and add like terms.

$\mathbf{\cos(A - B) = \cos A \cos B + \sin A \sin B}$ — Subtract 2 and divide by -2.

This is the identity for $\cos(A - B)$. Although **Figure 4** shows angles A and B in the second and first quadrants, respectively, this result is the same for any values of these angles.

Sum Identity for Cosine To find a similar expression for $\cos(A + B)$, rewrite $A + B$ as $A - (-B)$ and use the identity for $\cos(A - B)$.

$\cos(A + B) = \cos[A - (-B)]$	Definition of subtraction
$= \cos A \cos(-B) + \sin A \sin(-B)$	Cosine difference identity
$= \cos A \cos B + \sin A(-\sin B)$	Negative-angle identities (Section 5.1)
$\mathbf{\cos(A + B) = \cos A \cos B - \sin A \sin B}$	Multiply.

Cosine of a Sum or Difference

$$\cos(A + B) = \cos A \cos B - \sin A \sin B$$
$$\cos(A - B) = \cos A \cos B + \sin A \sin B$$

These identities are important in calculus and useful in certain applications. For example, the method shown in **Example 1** can be applied to find an exact value for $\cos 15°$.

EXAMPLE 1 Finding Exact Cosine Function Values

Find the *exact* value of each expression.

(a) $\cos 15°$ **(b)** $\cos \frac{5\pi}{12}$ **(c)** $\cos 87° \cos 93° - \sin 87° \sin 93°$

SOLUTION

(a) To find $\cos 15°$, we write $15°$ as the sum or difference of two angles with known function values, such as $45°$ and $30°$, since

$$15° = 45° - 30°.$$

(We could also use $60° - 45°$.) Then we use the cosine difference identity.

$\cos 15°$	
$= \cos(45° - 30°)$	$15° = 45° - 30°$
$= \cos 45° \cos 30° + \sin 45° \sin 30°$	Cosine difference identity
$= \frac{\sqrt{2}}{2} \cdot \frac{\sqrt{3}}{2} + \frac{\sqrt{2}}{2} \cdot \frac{1}{2}$	Substitute known values. (Section 2.1)
$= \frac{\sqrt{6} + \sqrt{2}}{4}$	Multiply and then add fractions.

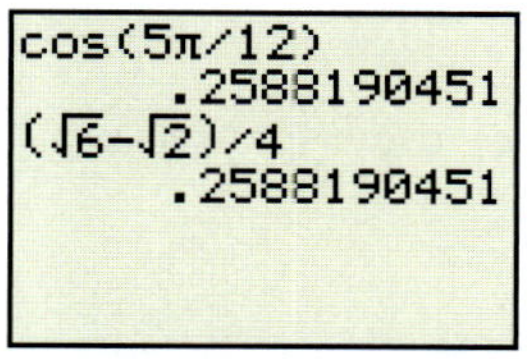

This screen supports the solution in **Example 1(b)** by showing that the decimal approximations for $\cos \frac{5\pi}{12}$ and $\frac{\sqrt{6}-\sqrt{2}}{4}$ agree.

(b) $\cos \frac{5\pi}{12}$

$$= \cos\left(\frac{\pi}{6} + \frac{\pi}{4}\right) \qquad \frac{\pi}{6} = \frac{2\pi}{12} \text{ and } \frac{\pi}{4} = \frac{3\pi}{12}$$

$$= \cos \frac{\pi}{6} \cos \frac{\pi}{4} - \sin \frac{\pi}{6} \sin \frac{\pi}{4} \qquad \text{Cosine sum identity}$$

$$= \frac{\sqrt{3}}{2} \cdot \frac{\sqrt{2}}{2} - \frac{1}{2} \cdot \frac{\sqrt{2}}{2} \qquad \text{Substitute known values. (Section 3.1)}$$

$$= \frac{\sqrt{6} - \sqrt{2}}{4} \qquad \text{Multiply and then subtract fractions.}$$

(c) $\cos 87° \cos 93° - \sin 87° \sin 93°$

$$= \cos(87° + 93°) \qquad \text{Cosine sum identity}$$

$$= \cos 180° \qquad \text{Add.}$$

$$= -1 \qquad \text{(Section 1.3)}$$

✔ ***Now Try Exercises 7, 11, and 15.***

Cofunction Identities We can use the identity for the cosine of the difference of two angles and the fundamental identities to derive *cofunction identities,* presented originally in **Section 2.1** for values of θ in the interval $[0°, 90°]$.

Cofunction Identities

The following identities hold for any angle θ for which the functions are defined.

$$\cos(90° - \theta) = \sin \theta \qquad \cot(90° - \theta) = \tan \theta$$

$$\sin(90° - \theta) = \cos \theta \qquad \sec(90° - \theta) = \csc \theta$$

$$\tan(90° - \theta) = \cot \theta \qquad \csc(90° - \theta) = \sec \theta$$

The same identities can be obtained for a real number domain by replacing 90° with $\frac{\pi}{2}$.

Substituting 90° for A and θ for B in the identity for $\cos(A - B)$ gives the following.

$$\cos(90° - \theta) = \cos 90° \cos \theta + \sin 90° \sin \theta \qquad \text{Cosine difference identity}$$

$$= 0 \cdot \cos \theta + 1 \cdot \sin \theta \qquad \text{Substitute.}$$

$$= \sin \theta \qquad \text{Simplify.}$$

This result is true for *any* value of θ since the identity for $\cos(A - B)$ is true for any values of A and B.

NOTE Because trigonometric (circular) functions are periodic, the solutions that follow in **Example 2** on the next page are not unique. We give only one of infinitely many possibilities.

EXAMPLE 2 **Using Cofunction Identities to Find θ**

Find one value of θ or x that satisfies each of the following.

(a) $\cot \theta = \tan 25°$ **(b)** $\sin \theta = \cos(-30°)$ **(c)** $\csc \frac{3\pi}{4} = \sec x$

SOLUTION

(a) Since tangent and cotangent are cofunctions, $\tan(90° - \theta) = \cot \theta$.

$$\cot \theta = \tan 25°$$

$$\tan(90° - \theta) = \tan 25° \quad \text{Cofunction identity}$$

$$90° - \theta = 25° \quad \text{Set angle measures equal.}$$

$$\theta = 65° \quad \text{Solve for } \theta.$$

(b)

$$\sin \theta = \cos(-30°)$$

$$\cos(90° - \theta) = \cos(-30°) \quad \text{Cofunction identity}$$

$$90° - \theta = -30° \quad \text{Set angle measures equal.}$$

$$\theta = 120° \quad \text{Solve for } \theta.$$

(c) $\csc \frac{3\pi}{4} = \sec x$

$$\csc \frac{3\pi}{4} = \csc\left(\frac{\pi}{2} - x\right) \quad \text{Cofunction identity}$$

$$\frac{3\pi}{4} = \frac{\pi}{2} - x \quad \text{Set angle measures equal.}$$

$$x = -\frac{\pi}{4} \quad \text{Solve for } x;\ \tfrac{\pi}{2} - \tfrac{3\pi}{4} = \tfrac{2\pi}{4} - \tfrac{3\pi}{4} = -\tfrac{\pi}{4}$$

✔ *Now Try Exercises 37 and 41.*

Applying the Sum and Difference Identities If either angle A or angle B in the identities for $\cos(A + B)$ and $\cos(A - B)$ is a quadrantal angle, then the identity allows us to write the expression in terms of a single function of A or B.

EXAMPLE 3 **Reducing $\cos(A - B)$ to a Function of a Single Variable**

Write $\cos(180° - \theta)$ as a trigonometric function of θ alone.

SOLUTION

$$\cos(180° - \theta)$$

$$= \cos 180° \cos \theta + \sin 180° \sin \theta \quad \text{Cosine difference identity}$$

$$= (-1) \cos \theta + (0) \sin \theta \quad \text{(Section 1.3)}$$

$$= -\cos \theta \quad \text{Simplify.}$$

✔ *Now Try Exercise 49.*

A standard problem in trigonometry involves being given information about two angles (or numbers)—say s and t—and being asked to find a function value of their sum $s + t$ or their difference $s - t$. These problems can be solved either by using angles in standard position or by using the Pythagorean identities. In **Example 4,** we show both methods.

EXAMPLE 4 Finding $\cos(s + t)$ Given Information about s and t

Suppose that $\sin s = \frac{3}{5}$, $\cos t = -\frac{12}{13}$, and both s and t are in quadrant II. Find $\cos(s + t)$.

SOLUTION By the cosine sum identity,

$$\cos(s + t) = \cos s \cos t - \sin s \sin t.$$

The values of $\sin s$ and $\cos t$ are given, so we can find $\cos(s + t)$ if we know the values of $\cos s$ and $\sin t$.

Method 1 We use angles in standard position. To find $\cos s$ and $\sin t$, we sketch two reference triangles in the second quadrant, one with $\sin s = \frac{3}{5}$ and the other with $\cos t = -\frac{12}{13}$. Be sure to include any negative signs when labeling the sides of the reference triangles. Notice that for angle t, we use -12 to denote the length of the side that lies along the x-axis. See **Figure 5.**

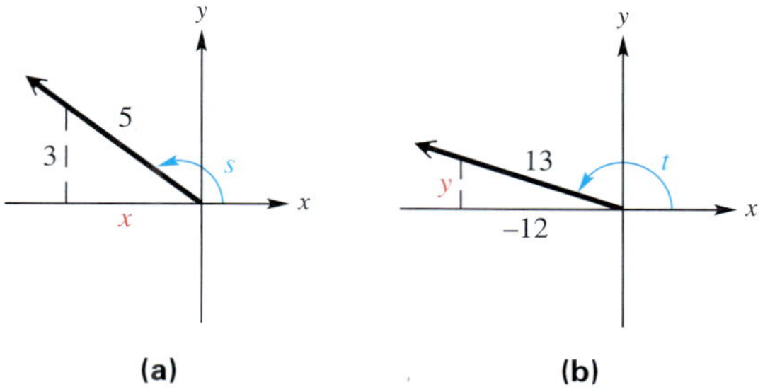

Figure 5

In **Figure 5(a),** $y = 3$ and $r = 5$. We find x using the Pythagorean theorem.

$$x^2 + y^2 = r^2 \quad \text{(Appendix B)}$$
$$x^2 + 3^2 = 5^2 \quad \text{Substitute.}$$
$$x^2 = 16$$
$$x = -4$$

Choose the negative square root here.

Thus, $\cos s = \frac{x}{r} = -\frac{4}{5}$.

In **Figure 5(b),** $x = -12$ and $r = 13$. We must find y.

$$x^2 + y^2 = r^2 \quad \text{Pythagorean theorem}$$
$$(-12)^2 + y^2 = 13^2 \quad \text{Substitute.}$$
$$y^2 = 25$$
$$y = 5$$

Choose the positive square root here.

Thus, $\sin t = \frac{y}{r} = \frac{5}{13}$.

Now we can find $\cos(s + t)$.

$$\cos(s + t) = \cos s \cos t - \sin s \sin t \quad \text{Cosine sum identity (1)}$$
$$= -\frac{4}{5}\left(-\frac{12}{13}\right) - \frac{3}{5} \cdot \frac{5}{13} \quad \text{Substitute.}$$
$$= \frac{48}{65} - \frac{15}{65} \quad \text{Multiply.}$$
$$= \frac{33}{65} \quad \text{Subtract.}$$

Method 2 We use Pythagorean identities here. To find $\cos s$, recall that $\sin^2 s + \cos^2 s = 1$, where s is in quadrant II.

$$\left(\frac{3}{5}\right)^2 + \cos^2 s = 1 \qquad \sin s = \frac{3}{5}$$

$$\frac{9}{25} + \cos^2 s = 1 \qquad \text{Square.}$$

$$\cos^2 s = \frac{16}{25} \qquad \text{Subtract } \frac{9}{25}.$$

$$\cos s = -\frac{4}{5}$$

$\cos s < 0$ because s is in quadrant II.

To find $\sin t$, we use $\sin^2 t + \cos^2 t = 1$, where t is in quadrant II.

$$\sin^2 t + \left(-\frac{12}{13}\right)^2 = 1 \qquad \cos t = -\frac{12}{13}$$

$$\sin^2 t + \frac{144}{169} = 1 \qquad \text{Square.}$$

$$\sin^2 t = \frac{25}{169} \qquad \text{Subtract } \frac{144}{169}.$$

$$\sin t = \frac{5}{13}$$

$\sin t > 0$ because t is in quadrant II.

From this point, the problem is solved by using the same steps beginning with the equation marked (1) in Method 1 on the previous page. The result is

$$\cos(s + t) = \frac{33}{65}. \qquad \text{Same result as in Method 1}$$

✔ ***Now Try Exercise 51.***

EXAMPLE 5 Applying the Cosine Difference Identity to Voltage

Common household electric current is called **alternating current** because the current alternates direction within the wires. The voltage V in a typical 115-volt outlet can be expressed by the function

$$V(t) = 163 \sin \omega t,$$

where ω is the angular speed (in radians per second) of the rotating generator at the electrical plant and t is time measured in seconds. (*Source:* Bell, D., *Fundamentals of Electric Circuits,* Fourth Edition, Prentice-Hall.)

(a) It is essential for electric generators to rotate at precisely 60 cycles per sec so household appliances and computers will function properly. Determine ω for these electric generators.

(b) Graph V in the window $[0, 0.05]$ by $[-200, 200]$.

(c) Determine a value of ϕ so that the graph of

$$V(t) = 163 \cos(\omega t - \phi)$$

is the same as the graph of $V(t) = 163 \sin \omega t$.

SOLUTION

(a) Each cycle is 2π radians at 60 cycles per sec, so the angular speed is

$$\omega = 60(2\pi) = 120\pi \text{ radians per sec.}$$

(b) $V(t) = 163 \sin \omega t$

$V(t) = 163 \sin 120\pi t$ From part (a), $\omega = 120\pi$ radians per sec.

Because the amplitude of the function $V(t)$ is 163 (from **Section 4.1**), $[-200, 200]$ is an appropriate interval for the range, as shown in the graph in **Figure 6.**

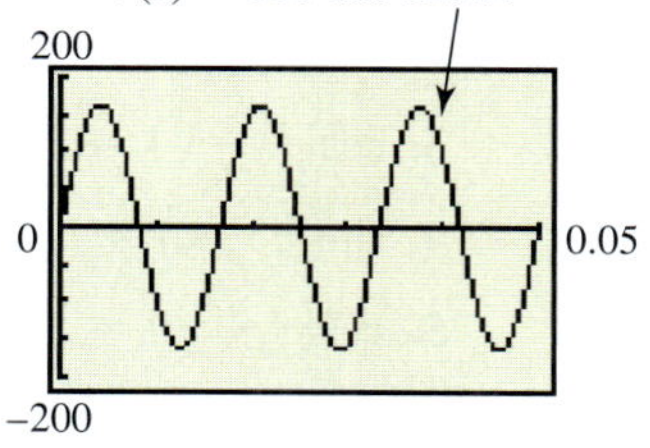

Figure 6

(c) Using the negative-angle identity for cosine and a cofunction identity gives

$$\cos\left(x - \frac{\pi}{2}\right) = \cos\left[-\left(\frac{\pi}{2} - x\right)\right] = \cos\left(\frac{\pi}{2} - x\right) = \sin x.$$

Therefore, if $\phi = \frac{\pi}{2}$, then

$$V(t) = 163 \cos(\omega t - \phi) = 163 \cos\left(\omega t - \frac{\pi}{2}\right) = 163 \sin \omega t.$$

✔ *Now Try Exercise 75.*

Verifying an Identity

EXAMPLE 6 **Verifying an Identity**

Verify that the following equation is an identity.

$$\sec\left(\frac{3\pi}{2} - x\right) = -\csc x$$

SOLUTION We work with the more complicated left side.

$$\sec\left(\frac{3\pi}{2} - x\right) = \frac{1}{\cos\left(\frac{3\pi}{2} - x\right)} \quad \text{Reciprocal identity}$$

$$= \frac{1}{\cos \frac{3\pi}{2} \cos x + \sin \frac{3\pi}{2} \sin x} \quad \text{Cosine difference identity}$$

$$= \frac{1}{0 \cdot \cos x + (-1)\sin x} \quad \cos \tfrac{3\pi}{2} = 0 \text{ and } \sin \tfrac{3\pi}{2} = -1$$

$$= \frac{1}{-\sin x} \quad \text{Simplify.}$$

$$= -\csc x \quad \text{Reciprocal identity}$$

The left side is identical to the right side, so the given equation is an identity.

✔ *Now Try Exercise 67.*

5.3 Exercises

Concept Check *Match each expression in Column I with the correct expression in Column II to form an identity. Choices may be used once, more than once, or not at all.*

I	II
1. $\cos(x + y) =$ ______	**A.** $\cos x \cos y + \sin x \sin y$
2. $\cos(x - y) =$ ______	**B.** $\cos x$
3. $\cos\left(\frac{\pi}{2} - x\right) =$ ______	**C.** $-\cos x$
4. $\sin\left(\frac{\pi}{2} - x\right) =$ ______	**D.** $-\sin x$
5. $\cos\left(x - \frac{\pi}{2}\right) =$ ______	**E.** $\sin x$
6. $\sin\left(x - \frac{\pi}{2}\right) =$ ______	**F.** $\cos x \cos y - \sin x \sin y$

Use identities to find each exact value. (Do not use a calculator.) **See Example 1.**

7. $\cos 75°$

8. $\cos(-15°)$

9. $\cos(-105°)$
(*Hint:* $-105° = -60° + (-45°)$)

10. $\cos 105°$
(*Hint:* $105° = 60° + 45°$)

11. $\cos \frac{7\pi}{12}$

12. $\cos\left(\frac{\pi}{12}\right)$

13. $\cos\left(-\frac{\pi}{12}\right)$

14. $\cos\left(-\frac{7\pi}{12}\right)$

15. $\cos 40° \cos 50° - \sin 40° \sin 50°$

16. $\cos \frac{7\pi}{9} \cos \frac{2\pi}{9} - \sin \frac{7\pi}{9} \sin \frac{2\pi}{9}$

Use a graphing or scientific calculator to support your answer for each of the following. **See Example 1.**

17. Exercise 15

18. Exercise 16

Write each function value in terms of the cofunction of a complementary angle. **See Example 2.**

19. $\tan 87°$

20. $\sin 15°$

21. $\cos \frac{\pi}{12}$

22. $\sin \frac{2\pi}{5}$

23. $\csc(14° 24')$

24. $\sin 142° 14'$

25. $\sin \frac{5\pi}{8}$

26. $\cot \frac{9\pi}{10}$

27. $\sec 146° 42'$

28. $\tan 174° 03'$

29. $\cot 176.9814°$

30. $\sin 98.0142°$

Use identities to fill in each blank with the appropriate trigonometric function name. **See Example 2.**

31. $\cot \frac{\pi}{3} =$ ______ $\frac{\pi}{6}$

32. $\sin \frac{2\pi}{3} =$ ______ $\left(-\frac{\pi}{6}\right)$

33. ______ $33° = \sin 57°$

34. ______ $72° = \cot 18°$

35. $\cos 70° = \frac{1}{\text{\_\_\_\_\_\_ } 20°}$

36. $\tan 24° = \frac{1}{\text{\_\_\_\_\_\_ } 66°}$

Find one angle θ that satisfies each of the following. **See Example 2.**

37. $\tan\theta = \cot(45° + 2\theta)$

38. $\sin\theta = \cos(2\theta + 30°)$

39. $\sec\theta = \csc\left(\frac{\theta}{2} + 20°\right)$

40. $\cos\theta = \sin\left(\frac{\theta}{4} + 3°\right)$

41. $\sin(3\theta - 15°) = \cos(\theta + 25°)$

42. $\cot(\theta - 10°) = \tan(2\theta - 20°)$

Use the identities for the cosine of a sum or difference to write each expression as a function of θ. **See Example 3.**

43. $\cos(0° - \theta)$

44. $\cos(90° - \theta)$

45. $\cos(\theta - 180°)$

46. $\cos(\theta - 270°)$

47. $\cos(0° + \theta)$

48. $\cos(90° + \theta)$

49. $\cos(180° + \theta)$

50. $\cos(270° + \theta)$

Find $\cos(s + t)$ *and* $\cos(s - t)$. **See Example 4.**

51. $\sin s = \frac{3}{5}$ and $\sin t = -\frac{12}{13}$, s in quadrant I and t in quadrant III

52. $\cos s = -\frac{8}{17}$ and $\cos t = -\frac{3}{5}$, s and t in quadrant III

53. $\cos s = -\frac{1}{5}$ and $\sin t = \frac{3}{5}$, s and t in quadrant II

54. $\sin s = \frac{2}{3}$ and $\sin t = -\frac{1}{3}$, s in quadrant II and t in quadrant IV

55. $\sin s = \frac{\sqrt{5}}{7}$ and $\sin t = \frac{\sqrt{6}}{8}$, s and t in quadrant I

56. $\cos s = \frac{\sqrt{2}}{4}$ and $\sin t = -\frac{\sqrt{5}}{6}$, s and t in quadrant IV

Concept Check *Determine whether each statement is* true *or* false.

57. $\cos 42° = \cos(30° + 12°)$

58. $\cos(-24°) = \cos 16° - \cos 40°$

59. $\cos 74° = \cos 60° \cos 14° + \sin 60° \sin 14°$

60. $\cos 140° = \cos 60° \cos 80° - \sin 60° \sin 80°$

61. $\cos\frac{\pi}{3} = \cos\frac{\pi}{12}\cos\frac{\pi}{4} - \sin\frac{\pi}{12}\sin\frac{\pi}{4}$

62. $\cos\frac{2\pi}{3} = \cos\frac{11\pi}{12}\cos\frac{\pi}{4} + \sin\frac{11\pi}{12}\sin\frac{\pi}{4}$

63. $\cos 70° \cos 20° - \sin 70° \sin 20° = 0$

64. $\cos 85° \cos 40° + \sin 85° \sin 40° = \frac{\sqrt{2}}{2}$

65. $\tan\left(x - \frac{\pi}{2}\right) = \cot x$

66. $\sin\left(x - \frac{\pi}{2}\right) = \cos x$

Verify that each equation is an identity. **See Example 6.**

67. $\cos\left(\frac{\pi}{2} + x\right) = -\sin x$

68. $\sec(\pi - x) = -\sec x$

69. $\cos 2x = \cos^2 x - \sin^2 x$ (*Hint:* $\cos 2x = \cos(x + x)$.)

70. $1 + \cos 2x - \cos^2 x = \cos^2 x$ (*Hint:* Use the result from **Exercise 69.**)

Relating Concepts

For individual or collaborative investigation *(Exercises 71–74)*

The identities for $\cos(A + B)$ *and* $\cos(A - B)$ *can be used to find exact values of expressions like* $\cos 195°$ *and* $\cos 255°$*, where the angle is not in the first quadrant.* ***Work Exercises 71–74 in order,*** *to see how this is done.*

71. By writing $195°$ as $180° + 15°$, use the identity for $\cos(A + B)$ to express $\cos 195°$ as $-\cos 15°$.

72. Use the identity for $\cos(A - B)$ to find $-\cos 15°$.

73. By the results of **Exercises 71 and 72,** $\cos 195° =$ ______.

74. Find each exact value using the method shown in **Exercises 71–73.**

(a) $\cos 255°$ **(b)** $\cos \frac{11\pi}{12}$

(Modeling) Solve each problem. ***See Example 5.***

75. *Electric Current* Refer to **Example 5.**

(a) How many times does the current oscillate in 0.05 sec?

(b) What are the maximum and minimum voltages in this outlet? Is the voltage always equal to 115 volts?

76. *Sound Waves* Sound is a result of waves applying pressure to a person's eardrum. For a pure sound wave radiating outward in a spherical shape, the trigonometric function

$$P = \frac{a}{r}\cos\left(\frac{2\pi r}{\lambda} - ct\right)$$

can be used to model the sound pressure at a radius of r feet from the source, where t is time in seconds, λ is length of the sound wave in feet, c is speed of sound in feet per second, and a is maximum sound pressure at the source measured in pounds per square foot. (*Source:* Beranek, L., *Noise and Vibration Control,* Institute of Noise Control Engineering, Washington, D.C.) Let $\lambda = 4.9$ ft and $c = 1026$ ft per sec.

(a) Let $a = 0.4$ lb per ft^2. Graph the sound pressure at distance $r = 10$ ft from its source in the window $[0, 0.05]$ by $[-0.05, 0.05]$. Describe P at this distance.

(b) Now let $a = 3$ and $t = 10$. Graph the sound pressure in the window $[0, 20]$ by $[-2, 2]$. What happens to pressure P as radius r increases?

(c) Suppose a person stands at a radius r so that $r = n\lambda$, where n is a positive integer. Use the difference identity for cosine to simplify P in this situation.

Relating Concepts

For individual or collaborative investigation *(Exercises 77–82)*

(This discussion applies to functions of both angles and real numbers.) The result of **Example 3** in this section can be written as an identity.

$$\mathbf{\cos(180° - \theta) = -\cos\theta}$$

This is an example of a **reduction formula,** which is an identity that *reduces* a function of a quadrantal angle plus or minus θ to a function of θ alone. Another example of a reduction formula is

$$\cos(270° + \theta) = \sin\theta.$$

Here is an interesting method for quickly determining a reduction formula for a trigonometric function f of the form $f(Q \pm \theta)$, where Q is a quadrantal angle. ***There are two cases to consider, and in each case, think of θ as a small positive angle*** in order to determine the quadrant in which $Q \pm \theta$ will lie.

Case 1 **Suppose that Q is a quadrantal angle whose terminal side lies along the x-axis.** Determine the quadrant in which $Q \pm \theta$ will lie for a small positive angle θ. If the given function f is positive in that quadrant, use a + sign on the reduced form. If f is negative in that quadrant, use a − sign. The reduced form will have that sign, f as the function, and θ as the argument. For example:

Terminates on the x-axis

Cosine is negative in quadrant II.

$$\cos(180° - \theta) = -\cos\theta.$$

This is in quadrant II for small θ.

Same function

Case 2 **Suppose that Q is a quadrantal angle whose terminal side lies along the y-axis.** Determine the quadrant in which $Q \pm \theta$ will lie for a small positive angle θ. If the given function f is positive in that quadrant, use a + sign on the reduced form. If f is negative in that quadrant, use a − sign. The reduced form will have that sign, the ***cofunction of f*** as the function, and θ as the argument. For example:

Terminates on the y-axis

Cosine is positive in quadrant IV.

$$\cos(270° + \theta) = +\sin\theta \text{ (or } \sin\theta\text{, as it is usually written).}$$

This is in quadrant IV for small θ.

Cofunctions

Use these ideas to write reduction formulas for the following. (Those involving sine and tangent can be verified after the introduction of the identities in ***Section 5.4.****)*

77. $\cos(90° + \theta)$ **78.** $\cos(270° - \theta)$ **79.** $\cos(180° + \theta)$

80. $\cos(270° + \theta)$ **81.** $\sin(180° + \theta)$ **82.** $\tan(270° - \theta)$

5.4 Sum and Difference Identities for Sine and Tangent

- **Sum and Difference Identities for Sine**
- **Sum and Difference Identities for Tangent**
- **Applying the Sum and Difference Identities**
- **Verifying an Identity**

Sum and Difference Identities for Sine We can use the cosine sum and difference identities from the previous section to derive similar identities for sine and tangent. In $\sin\theta = \cos(90° - \theta)$, replace θ with $A + B$.

$\sin(A + B) = \cos[90° - (A + B)]$ Cofunction identity (Section 5.3)

$= \cos[(90° - A) - B]$ Distribute negative sign and regroup.

$= \cos(90° - A)\cos B + \sin(90° - A)\sin B$ Cosine difference identity (Section 5.3)

$\mathbf{\sin(A + B) = \sin A \cos B + \cos A \sin B}$ Cofunction identities

Now we write $\sin(A - B)$ as $\sin[A + (-B)]$ and use the identity just found for $\sin(A + B)$.

$$\sin(A - B) = \sin[A + (-B)] \qquad \text{Definition of subtraction}$$

$$= \sin A \cos(-B) + \cos A \sin(-B) \qquad \text{Sine sum identity}$$

$$\mathbf{\sin(A - B) = \sin A \cos B - \cos A \sin B} \qquad \text{Negative-angle identities (Section 5.1)}$$

Sine of a Sum or Difference

$$\sin(A + B) = \sin A \cos B + \cos A \sin B$$

$$\sin(A - B) = \sin A \cos B - \cos A \sin B$$

Sum and Difference Identities for Tangent We can now derive the identity for $\tan(A + B)$ as follows.

$$\tan(A + B) = \frac{\sin(A + B)}{\cos(A + B)} \qquad \text{Fundamental identity (Section 5.1)}$$

We express this result in terms of the tangent function.

$$= \frac{\sin A \cos B + \cos A \sin B}{\cos A \cos B - \sin A \sin B} \qquad \text{Sum identities}$$

$$= \frac{\dfrac{\sin A \cos B + \cos A \sin B}{1}}{\dfrac{\cos A \cos B - \sin A \sin B}{1}} \cdot \frac{\dfrac{1}{\cos A \cos B}}{\dfrac{1}{\cos A \cos B}} \qquad \text{Multiply by 1, where } 1 = \frac{\frac{1}{\cos A \cos B}}{\frac{1}{\cos A \cos B}}.$$

$$= \frac{\dfrac{\sin A \cos B}{\cos A \cos B} + \dfrac{\cos A \sin B}{\cos A \cos B}}{\dfrac{\cos A \cos B}{\cos A \cos B} - \dfrac{\sin A \sin B}{\cos A \cos B}} \qquad \text{Multiply numerators and multiply denominators.}$$

$$= \frac{\dfrac{\sin A}{\cos A} + \dfrac{\sin B}{\cos B}}{1 - \dfrac{\sin A}{\cos A} \cdot \dfrac{\sin B}{\cos B}} \qquad \text{Simplify.}$$

$$\mathbf{\tan(A + B) = \frac{\tan A + \tan B}{1 - \tan A \tan B}} \qquad \frac{\sin\theta}{\cos\theta} = \tan\theta$$

We can replace B with $-B$ and use the fact that $\tan(-B) = -\tan B$ to obtain the identity for the tangent of the difference of two angles, as seen below.

Tangent of a Sum or Difference

$$\tan(A + B) = \frac{\tan A + \tan B}{1 - \tan A \tan B} \qquad \tan(A - B) = \frac{\tan A - \tan B}{1 + \tan A \tan B}$$

Applying the Sum and Difference Identities

EXAMPLE 1 Finding Exact Sine and Tangent Function Values

Find the *exact* value of each expression.

(a) $\sin 75^\circ$ **(b)** $\tan \frac{7\pi}{12}$ **(c)** $\sin 40^\circ \cos 160^\circ - \cos 40^\circ \sin 160^\circ$

SOLUTION

(a) $\sin 75^\circ$

$= \sin(45^\circ + 30^\circ)$ $75^\circ = 45^\circ + 30^\circ$

$= \sin 45^\circ \cos 30^\circ + \cos 45^\circ \sin 30^\circ$ Sine sum identity

$= \frac{\sqrt{2}}{2} \cdot \frac{\sqrt{3}}{2} + \frac{\sqrt{2}}{2} \cdot \frac{1}{2}$ Substitute known values. (Section 2.1)

$= \frac{\sqrt{6} + \sqrt{2}}{4}$ Multiply and then add fractions.

(b) $\tan \frac{7\pi}{12}$

$= \tan\left(\frac{\pi}{3} + \frac{\pi}{4}\right)$ $\frac{\pi}{3} = \frac{4\pi}{12}$ and $\frac{\pi}{4} = \frac{3\pi}{12}$

$= \frac{\tan \frac{\pi}{3} + \tan \frac{\pi}{4}}{1 - \tan \frac{\pi}{3} \tan \frac{\pi}{4}}$ Tangent sum identity

$= \frac{\sqrt{3} + 1}{1 - \sqrt{3} \cdot 1}$ Substitute known values. (Section 3.1)

$= \frac{\sqrt{3} + 1}{1 - \sqrt{3}} \cdot \frac{1 + \sqrt{3}}{1 + \sqrt{3}}$ Rationalize the denominator.

$= \frac{\sqrt{3} + 3 + 1 + \sqrt{3}}{1 - 3}$ $(a + b)(c + d) = ac + ad + bc + bd$; $(a - b)(a + b) = a^2 - b^2$

$= \frac{4 + 2\sqrt{3}}{-2}$ Combine like terms.

$= \frac{2(2 + \sqrt{3})}{2(-1)}$ Factor out 2.

Factor first. Then divide out the common factor.

$= -2 - \sqrt{3}$ Write in lowest terms.

(c) $\sin 40^\circ \cos 160^\circ - \cos 40^\circ \sin 160^\circ$

$= \sin(40^\circ - 160^\circ)$ Sine difference identity

$= \sin(-120^\circ)$ Subtract.

$= -\sin 120^\circ$ Negative-angle identity

$= -\frac{\sqrt{3}}{2}$ (Section 2.2)

✔ *Now Try Exercises 9, 11, and 19.*

EXAMPLE 2 Writing Functions as Expressions Involving Functions of θ

Write each function as an expression involving functions of θ.

(a) $\sin(30^\circ + \theta)$ **(b)** $\tan(45^\circ - \theta)$ **(c)** $\sin(180^\circ - \theta)$

SOLUTION

(a) $\sin(30^\circ + \theta)$

$= \sin 30^\circ \cos\theta + \cos 30^\circ \sin\theta$ Sine sum identity

$= \frac{1}{2}\cos\theta + \frac{\sqrt{3}}{2}\sin\theta$ $\sin 30^\circ = \frac{1}{2}$ and $\cos 30^\circ = \frac{\sqrt{3}}{2}$

$= \frac{\cos\theta + \sqrt{3}\sin\theta}{2}$ Add.

(b) $\tan(45^\circ - \theta)$

$= \frac{\tan 45^\circ - \tan\theta}{1 + \tan 45^\circ \tan\theta}$ Tangent difference identity

$= \frac{1 - \tan\theta}{1 + 1 \cdot \tan\theta}$ $\tan 45^\circ = 1$

$= \frac{1 - \tan\theta}{1 + \tan\theta}$ Multiply.

(c) $\sin(180^\circ - \theta)$

$= \sin 180^\circ \cos\theta - \cos 180^\circ \sin\theta$ Sine difference identity

$= 0 \cdot \cos\theta - (-1)\sin\theta$ $\sin 180^\circ = 0$ and $\cos 180^\circ = -1$

$= \sin\theta$ Simplify.

✔ *Now Try Exercises 27, 33, and 37.*

EXAMPLE 3 Finding Function Values and the Quadrant of $A + B$

Suppose that A and B are angles in standard position, with $\sin A = \frac{4}{5}$, $\frac{\pi}{2} < A < \pi$, and $\cos B = -\frac{5}{13}$, $\pi < B < \frac{3\pi}{2}$. Find each of the following.

(a) $\sin(A + B)$ **(b)** $\tan(A + B)$ **(c)** the quadrant of $A + B$

SOLUTION

(a) The identity for $\sin(A + B)$ involves $\sin A$, $\cos A$, $\sin B$, and $\cos B$. We are given values of $\sin A$ and $\cos B$. We must find values of $\cos A$ and $\sin B$.

$\sin^2 A + \cos^2 A = 1$ Fundamental identity (Section 5.1)

$\left(\frac{4}{5}\right)^2 + \cos^2 A = 1$ $\sin A = \frac{4}{5}$

$\frac{16}{25} + \cos^2 A = 1$ Square.

$\cos^2 A = \frac{9}{25}$ Subtract $\frac{16}{25}$.

Pay attention to signs. $\cos A = -\frac{3}{5}$ Take square roots (Appendix A). Since A is in quadrant II, $\cos A < 0$.

In the same way, $\sin B = -\frac{12}{13}$. Now find $\sin(A + B)$.

$$\sin(A+B) = \sin A \cos B + \cos A \sin B \qquad \text{Sine sum identity}$$

$$= \frac{4}{5}\left(-\frac{5}{13}\right) + \left(-\frac{3}{5}\right)\left(-\frac{12}{13}\right) \qquad \text{Substitute the given values for } \sin A \text{ and } \cos B \text{ and the values found for } \cos A \text{ and } \sin B.$$

$$= -\frac{20}{65} + \frac{36}{65} \qquad \text{Multiply.}$$

$$\sin(A+B) = \frac{16}{65} \qquad \text{Add.}$$

(b) To find $\tan(A + B)$, use the values of sine and cosine from part (a), $\sin A = \frac{4}{5}$, $\cos A = -\frac{3}{5}$, $\sin B = -\frac{12}{13}$, and $\cos B = -\frac{5}{13}$, to get $\tan A$ and $\tan B$.

$$\tan A = \frac{\sin A}{\cos A} = \frac{\frac{4}{5}}{-\frac{3}{5}} = \frac{4}{5} \div \left(-\frac{3}{5}\right)$$

$$\tan A = \frac{4}{5} \cdot \left(-\frac{5}{3}\right), \quad \text{or} \quad -\frac{4}{3}$$

$$\tan B = \frac{\sin B}{\cos B} = \frac{-\frac{12}{13}}{-\frac{5}{13}} = -\frac{12}{13} \div \left(-\frac{5}{13}\right)$$

$$\tan B = -\frac{12}{13} \cdot \left(-\frac{13}{5}\right), \quad \text{or} \quad \frac{12}{5}$$

Now use the identity for $\tan(A + B)$.

$$\tan(A+B) = \frac{\tan A + \tan B}{1 - \tan A \tan B} \qquad \text{Tangent sum identity}$$

$$= \frac{\left(-\frac{4}{3}\right) + \frac{12}{5}}{1 - \left(-\frac{4}{3}\right)\left(\frac{12}{5}\right)} \qquad \text{Substitute.}$$

$$= \frac{\frac{16}{15}}{1 + \frac{48}{15}} \qquad \text{Perform the indicated operations.}$$

$$= \frac{\frac{16}{15}}{\frac{63}{15}} \qquad \text{Add terms in the denominator.}$$

$$= \frac{16}{15} \div \frac{63}{15} \qquad \text{Simplify the complex fraction.}$$

$$= \frac{16}{15} \cdot \frac{15}{63} \qquad \text{Definition of division}$$

$$\tan(A+B) = \frac{16}{63} \qquad \text{Multiply.}$$

(c) From parts (a) and (b),

$$\sin(A+B) = \frac{16}{65} \quad \text{and} \quad \tan(A+B) = \frac{16}{63},$$

and thus both are positive. Therefore, $A + B$ must be in quadrant I, since it is the only quadrant in which both sine and tangent are positive.

✔ ***Now Try Exercise 45.***

Verifying an Identity

EXAMPLE 4 Verifying an Identity Using Sum and Difference Identities

Verify that the equation is an identity.

$$\sin\left(\frac{\pi}{6} + \theta\right) + \cos\left(\frac{\pi}{3} + \theta\right) = \cos\theta$$

SOLUTION Work on the left side, using the sum identities for $\sin(A + B)$ and $\cos(A + B)$.

$$\sin\left(\frac{\pi}{6} + \theta\right) + \cos\left(\frac{\pi}{3} + \theta\right)$$

$$= \left(\sin\frac{\pi}{6}\cos\theta + \cos\frac{\pi}{6}\sin\theta\right) + \left(\cos\frac{\pi}{3}\cos\theta - \sin\frac{\pi}{3}\sin\theta\right)$$

Sine sum identity; cosine sum identity (Section 5.3)

$$= \left(\frac{1}{2}\cos\theta + \frac{\sqrt{3}}{2}\sin\theta\right) + \left(\frac{1}{2}\cos\theta - \frac{\sqrt{3}}{2}\sin\theta\right)$$

$\sin\frac{\pi}{6} = \frac{1}{2}$; $\cos\frac{\pi}{6} = \frac{\sqrt{3}}{2}$; $\cos\frac{\pi}{3} = \frac{1}{2}$; $\sin\frac{\pi}{3} = \frac{\sqrt{3}}{2}$

$$= \frac{1}{2}\cos\theta + \frac{1}{2}\cos\theta \qquad \text{Simplify.}$$

$$= \cos\theta \qquad \text{Add.}$$

Now Try Exercise 63.

5.4 Exercises

Concept Check Match each expression in Column I with its value in Column II. ***See Example 1.***

I		II	
1. $\sin 15°$	**2.** $\sin 105°$	**A.** $\frac{\sqrt{6}+\sqrt{2}}{4}$	**B.** $\frac{-\sqrt{6}-\sqrt{2}}{4}$
3. $\tan 15°$	**4.** $\tan 105°$	**C.** $\frac{\sqrt{6}-\sqrt{2}}{4}$	**D.** $2+\sqrt{3}$
5. $\sin(-105°)$	**6.** $\tan(-105°)$	**E.** $2-\sqrt{3}$	**F.** $-2-\sqrt{3}$

7. Compare the formulas for $\sin(A - B)$ and $\sin(A + B)$. How do they differ? How are they alike?

8. Compare the formulas for $\tan(A - B)$ and $\tan(A + B)$. How do they differ? How are they alike?

Use identities to find each exact value. ***See Example 1.***

9. $\sin\frac{5\pi}{12}$ **10.** $\sin\frac{13\pi}{12}$ **11.** $\tan\frac{\pi}{12}$ **12.** $\tan\frac{5\pi}{12}$

13. $\sin\frac{7\pi}{12}$ **14.** $\sin\frac{\pi}{12}$ **15.** $\sin\left(-\frac{7\pi}{12}\right)$

16. $\sin\left(-\frac{5\pi}{12}\right)$ **17.** $\tan\left(-\frac{5\pi}{12}\right)$ **18.** $\tan\left(-\frac{7\pi}{12}\right)$

19. $\sin 76^\circ \cos 31^\circ - \cos 76^\circ \sin 31^\circ$

20. $\sin 40^\circ \cos 50^\circ + \cos 40^\circ \sin 50^\circ$

21. $\dfrac{\tan 80^\circ + \tan 55^\circ}{1 - \tan 80^\circ \tan 55^\circ}$

22. $\dfrac{\tan 80^\circ - \tan(-55^\circ)}{1 + \tan 80^\circ \tan(-55^\circ)}$

23. $\dfrac{\tan 100^\circ + \tan 80^\circ}{1 - \tan 100^\circ \tan 80^\circ}$

24. $\dfrac{\tan \frac{5\pi}{12} + \tan \frac{\pi}{4}}{1 - \tan \frac{5\pi}{12} \tan \frac{\pi}{4}}$

25. $\sin \dfrac{\pi}{5} \cos \dfrac{3\pi}{10} + \cos \dfrac{\pi}{5} \sin \dfrac{3\pi}{10}$

26. $\sin 100^\circ \cos 10^\circ - \cos 100^\circ \sin 10^\circ$

Use identities to write each expression as a single function of x or θ. ***See Example 2.***

27. $\cos(30^\circ + \theta)$

28. $\cos(\theta - 30^\circ)$

29. $\cos(60^\circ + \theta)$

30. $\cos(45^\circ - \theta)$

31. $\cos\left(\dfrac{3\pi}{4} - x\right)$

32. $\sin(45^\circ + \theta)$

33. $\tan(\theta + 30^\circ)$

34. $\tan\left(\dfrac{\pi}{4} + x\right)$

35. $\sin\left(\dfrac{\pi}{4} + x\right)$

36. $\sin\left(\dfrac{3\pi}{4} - x\right)$

37. $\sin(270^\circ - \theta)$

38. $\tan(180^\circ + \theta)$

39. $\tan(2\pi - x)$

40. $\sin(\pi + x)$

41. $\tan(\pi - x)$

42. Why is it not possible to use the method of **Example 2** to find a formula for $\tan(270^\circ - \theta)$?

43. Why is it that standard trigonometry texts usually do not develop formulas for the cotangent, secant, and cosecant of the sum and difference of two numbers or angles?

44. Show that if A, B, and C are the angles of a triangle, then

$$\sin(A + B + C) = 0.$$

Use the given information to find ***(a)*** $\sin(s + t)$, ***(b)*** $\tan(s + t)$, *and* ***(c)*** *the quadrant of* $s + t$. ***See Example 3.***

45. $\cos s = \frac{3}{5}$ and $\sin t = \frac{5}{13}$, s and t in quadrant I

46. $\sin s = \frac{3}{5}$ and $\sin t = -\frac{12}{13}$, s in quadrant I and t in quadrant III

47. $\cos s = -\frac{8}{17}$ and $\cos t = -\frac{3}{5}$, s and t in quadrant III

48. $\cos s = -\frac{15}{17}$ and $\sin t = \frac{4}{5}$, s in quadrant II and t in quadrant I

49. $\sin s = \frac{2}{3}$ and $\sin t = -\frac{1}{3}$, s in quadrant II and t in quadrant IV

50. $\cos s = -\frac{1}{5}$ and $\sin t = \frac{3}{5}$, s and t in quadrant II

Find each exact value. Use an appropriate sum or difference identity.

51. $\sin 165^\circ$

52. $\sin 255^\circ$

53. $\tan 165^\circ$

54. $\tan 285^\circ$

55. $\tan \dfrac{11\pi}{12}$

56. $\sin\left(-\dfrac{13\pi}{12}\right)$

Graph each expression and use the graph to make a conjecture, predicting what might be an identity. Then verify your conjecture algebraically.

57. $\sin\left(\dfrac{\pi}{2} + \theta\right)$

58. $\sin\left(\dfrac{3\pi}{2} + \theta\right)$

59. $\tan\left(\dfrac{\pi}{2} + \theta\right)$

60. $\tan\left(\dfrac{\pi}{2} - \theta\right)$

Verify that each equation is an identity. ***See Example 4.***

61. $\sin 2x = 2 \sin x \cos x \quad$ (*Hint:* $\sin 2x = \sin(x + x)$)

62. $\sin(x + y) + \sin(x - y) = 2 \sin x \cos y$

63. $\sin\left(\dfrac{7\pi}{6} + x\right) - \cos\left(\dfrac{2\pi}{3} + x\right) = 0$

64. $\tan(x - y) - \tan(y - x) = \dfrac{2(\tan x - \tan y)}{1 + \tan x \tan y}$

65. $\dfrac{\cos(\alpha - \beta)}{\cos \alpha \sin \beta} = \tan \alpha + \cot \beta$

66. $\dfrac{\sin(s + t)}{\cos s \cos t} = \tan s + \tan t$

67. $\dfrac{\sin(x - y)}{\sin(x + y)} = \dfrac{\tan x - \tan y}{\tan x + \tan y}$

68. $\dfrac{\sin(x + y)}{\cos(x - y)} = \dfrac{\cot x + \cot y}{1 + \cot x \cot y}$

69. $\dfrac{\sin(s - t)}{\sin t} + \dfrac{\cos(s - t)}{\cos t} = \dfrac{\sin s}{\sin t \cos t}$

70. $\dfrac{\tan(\alpha + \beta) - \tan \beta}{1 + \tan(\alpha + \beta) \tan \beta} = \tan \alpha$

Relating Concepts

For individual or collaborative investigation *(Exercises 71–76)*

Refer to the figure on the left below. By the definition of $\tan \theta$,

$m = \tan \theta$, *where m is the slope and* θ *is the angle of inclination of the line.*

The following exercises, which depend on properties of triangles, refer to triangle ABC in the figure on the right below. ***Work Exercises 71–76 in order.*** *Assume that all angles are measured in degrees.*

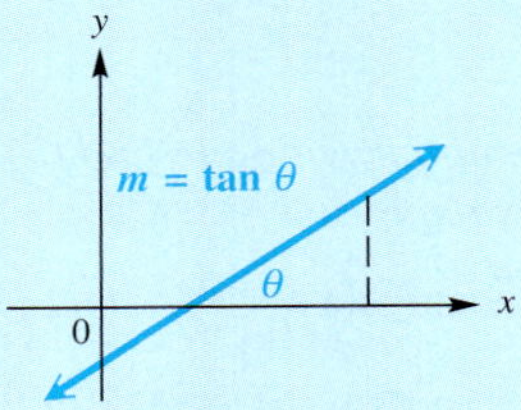

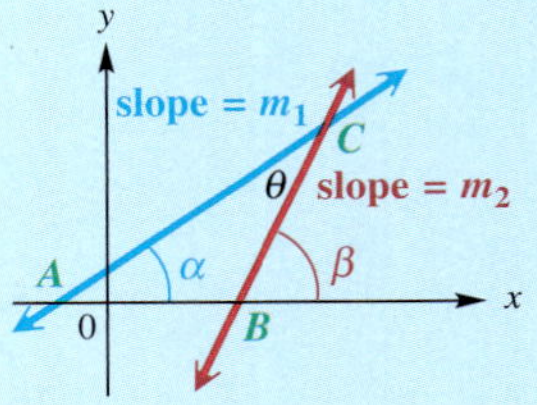

71. In terms of β, what is the measure of angle ABC?

72. Use the fact that the sum of the angles in a triangle is 180° to express θ in terms of α and β.

73. Apply the formula for $\tan(A - B)$ to obtain an expression for $\tan \theta$ in terms of $\tan \alpha$ and $\tan \beta$.

74. Replace $\tan \alpha$ with m_1 and $\tan \beta$ with m_2 to obtain

$$\tan \theta = \frac{m_2 - m_1}{1 + m_1 m_2}.$$

In Exercises 75 and 76, use the result from ***Exercise 74*** *to find the acute angle between each pair of lines. (Note that the tangent of the angle will be positive.) Use a calculator and round to the nearest tenth of a degree.*

75. $x + y = 9, \quad 2x + y = -1$

76. $5x - 2y + 4 = 0, \quad 3x + 5y = 6$

(Modeling) Solve each problem.

77. *Back Stress* If a person bends at the waist with a straight back making an angle of θ degrees with the horizontal, then the force F exerted on the back muscles can be modeled by the equation

$$F = \frac{0.6W \sin(\theta + 90^\circ)}{\sin 12^\circ},$$

where W is the weight of the person. (*Source:* Metcalf, H., *Topics in Classical Biophysics,* Prentice-Hall.)

(a) Calculate force F for $W = 170$ lb and $\theta = 30^\circ$.
(b) Use an identity to show that F is approximately equal to $2.9W \cos \theta$.
(c) For what value of θ is F maximum?

78. *Back Stress* Refer to **Exercise 77.**

(a) Suppose a 200-lb person bends at the waist so that $\theta = 45^\circ$. Estimate the force exerted on the person's back muscles.
(b) Approximate graphically the value of θ that results in the back muscles exerting a force of 400 lb.

79. *Voltage* A coil of wire rotating in a magnetic field induces a voltage

$$E = 20 \sin\left(\frac{\pi t}{4} - \frac{\pi}{2}\right).$$

Use an identity from this section to express this in terms of $\cos \frac{\pi t}{4}$.

80. *Voltage of a Circuit* When the two voltages

$$V_1 = 30 \sin 120\pi t \quad \text{and} \quad V_2 = 40 \cos 120\pi t$$

are applied to the same circuit, the resulting voltage V will be equal to their sum. (*Source:* Bell, D., *Fundamentals of Electric Circuits,* Second Edition, Reston Publishing Company.)

(a) Graph the sum in the window $[0, 0.05]$ by $[-60, 60]$.
(b) Use the graph to estimate values for a and ϕ so that $V = a \sin(120\pi t + \phi)$.
(c) Use identities to verify that your expression for V is valid.

(Modeling) Roll of a Spacecraft The figure on the left below shows the three quantities that determine the motion of a spacecraft. A conventional three-dimensional spacecraft coordinate system is shown on the right.

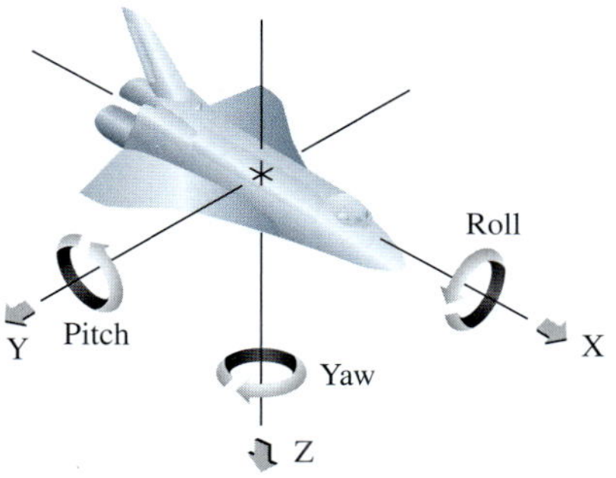

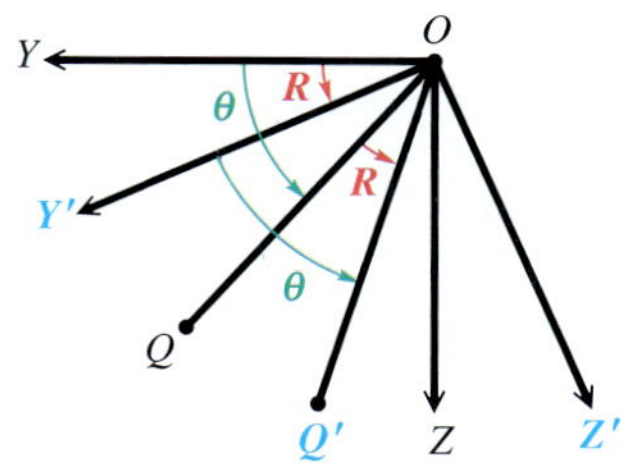

Angle $YOQ = \theta$ and $OQ = r$. The coordinates of Q are (x, y, z), where

$$y = r \cos \theta \quad \text{and} \quad z = r \sin \theta.$$

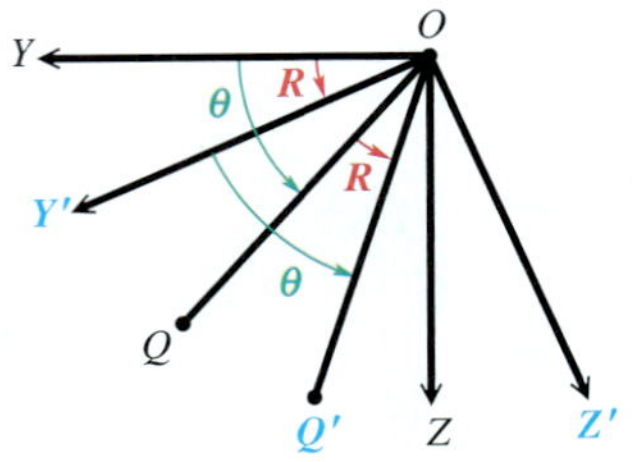

When the spacecraft performs a rotation, it is necessary to find the coordinates in the spacecraft system after the rotation takes place. For example, suppose the spacecraft undergoes roll through angle R. The coordinates (x, y, z) of point Q become (x', y', z'), the coordinates of the corresponding point Q′. In the new reference system, $OQ' = r$ and, since the roll is around the x-axis and angle $Y'OQ' = YOQ = \theta$,

$$x' = x, \quad y' = r\cos(\theta + R), \quad \text{and} \quad z' = r\sin(\theta + R).$$

(Source: Kastner, B., Space Mathematics, *NASA.)*

81. Write y' in terms of y, R, and z.

82. Write z' in terms of y, R, and z.

Chapter 5 Quiz (Sections 5.1–5.4)

1. If $\sin\theta = -\frac{7}{25}$ and θ is in quadrant IV, find the remaining five trigonometric function values of θ.

2. Express $\cot^2 x + \csc^2 x$ in terms of $\sin x$ and $\cos x$, and simplify.

3. Find the exact value of $\sin\left(-\frac{7\pi}{12}\right)$.

4. Express $\cos(180° - \theta)$ as a function of θ alone.

5. If $\cos A = \frac{3}{5}$, $\sin B = -\frac{5}{13}$, $0 < A < \frac{\pi}{2}$, and $\pi < B < \frac{3\pi}{2}$, find each of the following.

(a) $\cos(A + B)$ **(b)** $\sin(A + B)$ **(c)** the quadrant of $A + B$

6. Express $\tan\left(\frac{3\pi}{4} + x\right)$ as a function of x alone.

Verify each identity.

7. $\dfrac{1 + \sin\theta}{\cot^2\theta} = \dfrac{\sin\theta}{\csc\theta - 1}$

8. $\sin\left(\dfrac{\pi}{3} + \theta\right) - \sin\left(\dfrac{\pi}{3} - \theta\right) = \sin\theta$

9. $\dfrac{\sin^2\theta - \cos^2\theta}{\sin^4\theta - \cos^4\theta} = 1$

10. $\dfrac{\cos(x + y) + \cos(x - y)}{\sin(x - y) + \sin(x + y)} = \cot x$

5.5 Double-Angle Identities

- Double-Angle Identities
- An Application
- Product-to-Sum and Sum-to-Product Identities

Double-Angle Identities

When $A = B$ in the identities for the sum of two angles, the **double-angle identities** result. To derive an expression for $\cos 2A$, we let $B = A$ in the identity $\cos(A + B) = \cos A\cos B - \sin A\sin B$.

$$\cos 2A = \cos(A + A)$$

$$= \cos A\cos A - \sin A\sin A \qquad \text{Cosine sum identity (Section 5.3)}$$

$$\mathbf{\cos 2A = \cos^2 A - \sin^2 A} \qquad a \cdot a = a^2$$

Two other useful forms of this identity can be obtained by substituting either $\cos^2 A = 1 - \sin^2 A$ or $\sin^2 A = 1 - \cos^2 A$. Replacing $\cos^2 A$ with the expression $1 - \sin^2 A$ gives the following.

$$\cos 2A = \cos^2 A - \sin^2 A \qquad \text{From above}$$

$$= (1 - \sin^2 A) - \sin^2 A \qquad \text{Fundamental identity (Section 5.1)}$$

$$\mathbf{\cos 2A = 1 - 2\sin^2 A} \qquad \text{Subtract.}$$

LOOKING AHEAD TO CALCULUS

The identities

$$\cos 2A = 1 - 2\sin^2 A$$

and $$\cos 2A = 2\cos^2 A - 1$$

can be rewritten as

$$\sin^2 A = \frac{1}{2}(1 - \cos 2A)$$

and $$\cos^2 A = \frac{1}{2}(1 + \cos 2A).$$

These identities are used to integrate the functions $f(A) = \sin^2 A$ and $g(A) = \cos^2 A$.

Replacing $\sin^2 A$ with $1 - \cos^2 A$ gives a third form.

$$\cos 2A = \cos^2 A - \sin^2 A$$

$$= \cos^2 A - (1 - \cos^2 A) \quad \text{Fundamental identity}$$

$$= \cos^2 A - 1 + \cos^2 A \quad \text{Distributive property}$$

$$\mathbf{\cos 2A = 2\cos^2 A - 1} \quad \text{Add.}$$

We find $\sin 2A$ using $\sin(A + B) = \sin A \cos B + \cos A \sin B$, with $B = A$.

$$\sin 2A = \sin(A + A)$$

$$= \sin A \cos A + \cos A \sin A \quad \text{Sine sum identity (Section 5.4)}$$

$$\mathbf{\sin 2A = 2\sin A \cos A} \quad \text{Add.}$$

Using the identity for $\tan(A + B)$, we find $\tan 2A$.

$$\tan 2A = \tan(A + A)$$

$$= \frac{\tan A + \tan A}{1 - \tan A \tan A} \quad \text{Tangent sum identity (Section 5.4)}$$

$$\mathbf{\tan 2A = \frac{2\tan A}{1 - \tan^2 A}} \quad \text{Simplify.}$$

NOTE In general, for a trigonometric function f,

$$f(2A) \neq 2f(A).$$

Double-Angle Identities

$$\cos 2A = \cos^2 A - \sin^2 A \qquad \cos 2A = 1 - 2\sin^2 A$$

$$\cos 2A = 2\cos^2 A - 1 \qquad \sin 2A = 2\sin A \cos A$$

$$\tan 2A = \frac{2\tan A}{1 - \tan^2 A}$$

EXAMPLE 1 Finding Function Values of 2θ Given Information about θ

Given $\cos\theta = \frac{3}{5}$ and $\sin\theta < 0$, find $\sin 2\theta$, $\cos 2\theta$, and $\tan 2\theta$.

SOLUTION To find $\sin 2\theta$, we must first find the value of $\sin\theta$.

$$\sin^2\theta + \left(\frac{3}{5}\right)^2 = 1 \quad \sin^2\theta + \cos^2\theta = 1 \text{ and } \cos\theta = \frac{3}{5}$$

$$\sin^2\theta = \frac{16}{25} \quad \left(\frac{3}{5}\right)^2 = \frac{9}{25}\text{; subtract } \frac{9}{25}.$$

$$\sin\theta = -\frac{4}{5} \quad \text{Take square roots (Appendix A). Choose the negative square root since } \sin\theta < 0.$$

Pay attention to signs here.

Now use the double-angle identity for sine.

$$\sin 2\theta = 2\sin\theta\cos\theta = 2\left(-\frac{4}{5}\right)\left(\frac{3}{5}\right) = -\frac{24}{25} \quad \sin\theta = -\frac{4}{5} \text{ and } \cos\theta = \frac{3}{5}$$

Now we find cos 2θ, using the first of the double-angle identities for cosine.

Any of the three forms may be used.

$$\cos 2\theta = \cos^2\theta - \sin^2\theta = \frac{9}{25} - \frac{16}{25} = -\frac{7}{25} \qquad \cos\theta = \tfrac{3}{5} \text{ and } \sin\theta = -\tfrac{4}{5}$$

The value of tan 2θ can be found in either of two ways. We can use the double-angle identity and the fact that $\tan\theta = \frac{\sin\theta}{\cos\theta} = \frac{-\frac{4}{5}}{\frac{3}{5}} = -\frac{4}{5} \div \frac{3}{5} = -\frac{4}{5} \cdot \frac{5}{3} = -\frac{4}{3}$.

$$\tan 2\theta = \frac{2\tan\theta}{1-\tan^2\theta} = \frac{2\left(-\frac{4}{3}\right)}{1-\left(-\frac{4}{3}\right)^2} = \frac{-\frac{8}{3}}{-\frac{7}{9}} = \frac{24}{7}$$

Alternatively, we can find tan 2θ by finding the quotient of sin 2θ and cos 2θ.

$$\tan 2\theta = \frac{\sin 2\theta}{\cos 2\theta} = \frac{-\frac{24}{25}}{-\frac{7}{25}} = \frac{24}{7} \qquad \text{Same result as above}$$

✔ **Now Try Exercise 11.**

EXAMPLE 2 Finding Function Values of θ Given Information about 2θ

Find the values of the six trigonometric functions of θ if $\cos 2\theta = \frac{4}{5}$ and $90° < \theta < 180°$.

SOLUTION We must obtain a trigonometric function value of θ alone.

$$\cos 2\theta = 1 - 2\sin^2\theta \qquad \text{Double-angle identity}$$

$$\frac{4}{5} = 1 - 2\sin^2\theta \qquad \cos 2\theta = \tfrac{4}{5}$$

$$-\frac{1}{5} = -2\sin^2\theta \qquad \text{Subtract 1 from each side.}$$

$$\frac{1}{10} = \sin^2\theta \qquad \text{Multiply by } -\tfrac{1}{2}.$$

$$\sin\theta = \sqrt{\frac{1}{10}} \qquad \text{Take square roots and choose the positive square root since } \theta \text{ terminates in quadrant II.}$$

$$\sin\theta = \frac{1}{\sqrt{10}} \cdot \frac{\sqrt{10}}{\sqrt{10}} \qquad \text{Use the quotient rule and rationalize the denominator.}$$

$$\sin\theta = \frac{\sqrt{10}}{10} \qquad \sqrt{a}\cdot\sqrt{a} = a$$

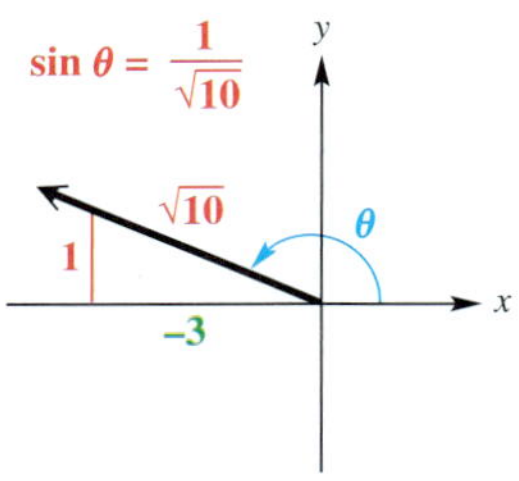

Figure 7

Now find values of cos θ and tan θ by sketching and labeling a right triangle in quadrant II. Since $\sin\theta = \frac{1}{\sqrt{10}}$, the triangle in **Figure 7** is labeled accordingly. The Pythagorean theorem is used to find the remaining leg. Now,

$$\cos\theta = \frac{-3}{\sqrt{10}} = -\frac{3\sqrt{10}}{10}, \quad \text{and} \quad \tan\theta = \frac{1}{-3} = -\frac{1}{3}. \qquad \text{(Section 1.3)}$$

Find the other three functions using reciprocals.

$$\csc\theta = \frac{1}{\sin\theta} = \sqrt{10}, \quad \sec\theta = \frac{1}{\cos\theta} = -\frac{\sqrt{10}}{3}, \quad \cot\theta = \frac{1}{\tan\theta} = -3$$

✔ **Now Try Exercise 15.**

EXAMPLE 3 Verifying a Double-Angle Identity

Verify that the following equation is an identity.

$$\cot x \sin 2x = 1 + \cos 2x$$

SOLUTION We start by working on the left side, using Hint 3 from **Section 5.2** about writing all functions in terms of sine and cosine.

$$\cot x \sin 2x = \frac{\cos x}{\sin x} \cdot \sin 2x \quad \text{Quotient identity}$$

$$= \frac{\cos x}{\sin x}(2 \sin x \cos x) \quad \text{Double-angle identity}$$

$$= 2\cos^2 x \quad \text{Multiply.}$$

$$= 1 + \cos 2x \quad \cos 2x = 2\cos^2 x - 1, \text{ so } 2\cos^2 x = 1 + \cos 2x$$

Be able to recognize alternative forms of identities.

✔ Now Try Exercise 17.

EXAMPLE 4 Simplifying Expressions Using Double-Angle Identities

Simplify each expression.

(a) $\cos^2 7x - \sin^2 7x$ **(b)** $\sin 15° \cos 15°$

SOLUTION

(a) This expression suggests one of the double-angle identities for cosine: $\cos 2A = \cos^2 A - \sin^2 A$. Substitute $7x$ for A.

$$\cos^2 7x - \sin^2 7x = \cos 2(7x) = \cos 14x$$

(b) If the expression $\sin 15° \cos 15°$ were

$$2 \sin 15° \cos 15°,$$

we could apply the identity for $\sin 2A$ directly because $\sin 2A = 2 \sin A \cos A$.

$$\sin 15° \cos 15°$$

$$= \frac{1}{2}(2) \sin 15° \cos 15° \quad \text{Multiply by 1 in the form } \tfrac{1}{2}(2).$$

This is not an obvious way to begin, but it is indeed valid.

$$= \frac{1}{2}(2 \sin 15° \cos 15°) \quad \text{Associative property}$$

$$= \frac{1}{2}\sin(2 \cdot 15°) \quad 2 \sin A \cos A = \sin 2A, \text{ with } A = 15°$$

$$= \frac{1}{2}\sin 30° \quad \text{Multiply.}$$

$$= \frac{1}{2} \cdot \frac{1}{2} \quad \sin 30° = \tfrac{1}{2} \text{ (Section 2.1)}$$

$$= \frac{1}{4} \quad \text{Multiply.}$$

✔ Now Try Exercises 37 and 39.

Identities involving larger multiples of the variable can be derived by repeated use of the double-angle identities and other identities.

EXAMPLE 5 Deriving a Multiple-Angle Identity

Write $\sin 3x$ in terms of $\sin x$.

SOLUTION

$\sin 3x$

$= \sin(2x + x)$ — Use the simple fact that 3 = 2 + 1 here.

$= \sin 2x \cos x + \cos 2x \sin x$ — Sine sum identity (Section 5.4)

$= (2 \sin x \cos x)\cos x + (\cos^2 x - \sin^2 x)\sin x$ — Double-angle identities

$= 2 \sin x \cos^2 x + \cos^2 x \sin x - \sin^3 x$ — Multiply.

$= 2 \sin x(1 - \sin^2 x) + (1 - \sin^2 x)\sin x - \sin^3 x$ — $\cos^2 x = 1 - \sin^2 x$

$= 2 \sin x - 2 \sin^3 x + \sin x - \sin^3 x - \sin^3 x$ — Distributive property

$= 3 \sin x - 4 \sin^3 x$ — Combine like terms.

✔ *Now Try Exercise 49.*

An Application

EXAMPLE 6 Determining Wattage Consumption

If a toaster is plugged into a common household outlet, the wattage consumed is not constant. Instead, it varies at a high frequency according to the model

$$W = \frac{V^2}{R},$$

where V is the voltage and R is a constant that measures the resistance of the toaster in ohms. (*Source:* Bell, D., *Fundamentals of Electric Circuits,* Fourth Edition, Prentice-Hall.) Graph the wattage W consumed by a typical toaster with $R = 15$ and $V = 163 \sin 120\pi t$ in the window $[0, 0.05]$ by $[-500, 2000]$. How many oscillations are there?

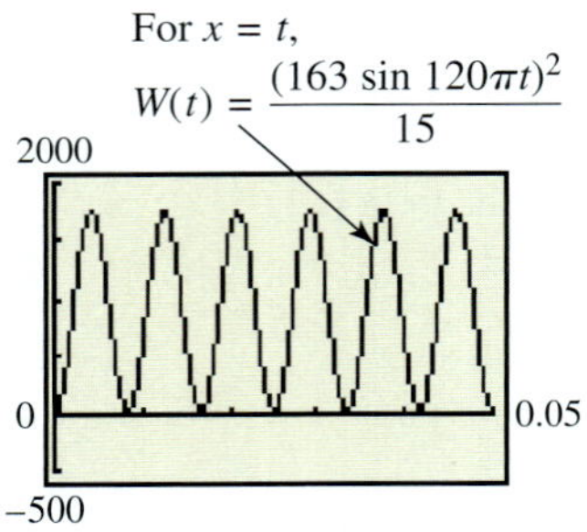

Figure 8

SOLUTION Substituting the given values into the wattage equation gives

$$W = \frac{V^2}{R} = \frac{(163 \sin 120\pi t)^2}{15}.$$

To determine the range of W, we note that $\sin 120\pi t$ has maximum value 1, so the expression for W has maximum value $\frac{163^2}{15} \approx 1771$. The minimum value is 0. The graph in **Figure 8** shows that there are six oscillations.

✔ *Now Try Exercise 69.*

Product-to-Sum and Sum-to-Product Identities

We can add the identities for $\cos(A + B)$ and $\cos(A - B)$ to derive an identity useful in calculus.

$$\begin{aligned} \cos(A + B) &= \cos A \cos B - \sin A \sin B \\ \cos(A - B) &= \cos A \cos B + \sin A \sin B \\ \hline \cos(A + B) + \cos(A - B) &= 2 \cos A \cos B \end{aligned}$$ Add.

or

$$\mathbf{\cos A \cos B = \frac{1}{2}[\cos(A + B) + \cos(A - B)]}$$

LOOKING AHEAD TO CALCULUS

The product-to-sum identities are used in calculus to find **integrals** of functions that are products of trigonometric functions. The classic calculus text by Earl Swokowski includes the following example:

Evaluate $\int \cos 5x \cos 3x \, dx$.

The first solution line reads:
"We may write

$\cos 5x \cos 3x = \frac{1}{2}[\cos 8x + \cos 2x]$."

Similarly, subtracting $\cos(A + B)$ from $\cos(A - B)$ gives

$$\sin A \sin B = \frac{1}{2}[\cos(A - B) - \cos(A + B)].$$

Using the identities for $\sin(A + B)$ and $\sin(A - B)$ in the same way, we obtain two more identities. Those and the previous ones are now summarized.

Product-to-Sum Identities

$$\cos A \cos B = \frac{1}{2}[\cos(A + B) + \cos(A - B)]$$

$$\sin A \sin B = \frac{1}{2}[\cos(A - B) - \cos(A + B)]$$

$$\sin A \cos B = \frac{1}{2}[\sin(A + B) + \sin(A - B)]$$

$$\cos A \sin B = \frac{1}{2}[\sin(A + B) - \sin(A - B)]$$

EXAMPLE 7 Using a Product-to-Sum Identity

Write $4 \cos 75° \sin 25°$ as the sum or difference of two functions.

SOLUTION

$4 \cos 75° \sin 25°$

$= 4\left[\frac{1}{2}\left(\sin(75° + 25°) - \sin(75° - 25°)\right)\right]$ Use the identity for $\cos A \sin B$, with $A = 75°$ and $B = 25°$.

$= 2 \sin 100° - 2 \sin 50°$ Simplify.

✔ *Now Try Exercise 57.*

We can convert the product-to-sum identities into equivalent useful forms that enable us to write sums as products.

Sum-to-Product Identities

$$\sin A + \sin B = 2 \sin\left(\frac{A + B}{2}\right) \cos\left(\frac{A - B}{2}\right)$$

$$\sin A - \sin B = 2 \cos\left(\frac{A + B}{2}\right) \sin\left(\frac{A - B}{2}\right)$$

$$\cos A + \cos B = 2 \cos\left(\frac{A + B}{2}\right) \cos\left(\frac{A - B}{2}\right)$$

$$\cos A - \cos B = -2 \sin\left(\frac{A + B}{2}\right) \sin\left(\frac{A - B}{2}\right)$$

EXAMPLE 8 Using a Sum-to-Product Identity

Write $\sin 2\theta - \sin 4\theta$ as a product of two functions.

SOLUTION

$$\sin 2\theta - \sin 4\theta$$

$$= 2\cos\left(\frac{2\theta + 4\theta}{2}\right)\sin\left(\frac{2\theta - 4\theta}{2}\right)$$ Use the identity for $\sin A - \sin B$, with $A = 2\theta$ and $B = 4\theta$.

$$= 2\cos\frac{6\theta}{2}\sin\left(\frac{-2\theta}{2}\right)$$ Simplify the numerators.

$$= 2\cos 3\theta \sin(-\theta)$$ Divide.

$$= -2\cos 3\theta \sin\theta$$ $\sin(-\theta) = -\sin\theta$ (Section 5.1)

✔ *Now Try Exercise 63.*

5.5 Exercises

Concept Check Match each expression in Column I with its value in Column II.

I		II	
1. $2\cos^2 15° - 1$	**2.** $\dfrac{2\tan 15°}{1 - \tan^2 15°}$	**A.** $\dfrac{1}{2}$	**B.** $\dfrac{\sqrt{2}}{2}$
3. $2\sin 22.5° \cos 22.5°$	**4.** $\cos^2\dfrac{\pi}{6} - \sin^2\dfrac{\pi}{6}$	**C.** $\dfrac{\sqrt{3}}{2}$	**D.** $-\sqrt{3}$
5. $4\sin\dfrac{\pi}{3}\cos\dfrac{\pi}{3}$	**6.** $\dfrac{2\tan\frac{\pi}{3}}{1 - \tan^2\frac{\pi}{3}}$	**E.** $\dfrac{\sqrt{3}}{3}$	**F.** $\sqrt{3}$

Use identities to find values of the sine and cosine functions for each angle measure. ***See Examples 1 and 2.***

7. 2θ, given $\sin\theta = \frac{2}{5}$ and $\cos\theta < 0$

8. 2θ, given $\cos\theta = -\frac{12}{13}$ and $\sin\theta > 0$

9. $2x$, given $\tan x = 2$ and $\cos x > 0$

10. $2x$, given $\tan x = \frac{5}{3}$ and $\sin x < 0$

11. 2θ, given $\sin\theta = -\frac{\sqrt{5}}{7}$ and $\cos\theta > 0$

12. 2θ, given $\cos\theta = \frac{\sqrt{3}}{5}$ and $\sin\theta > 0$

13. θ, given $\cos 2\theta = \frac{3}{5}$ and θ terminates in quadrant I

14. θ, given $\cos 2\theta = \frac{3}{4}$ and θ terminates in quadrant III

15. θ, given $\cos 2\theta = -\frac{5}{12}$ and $90° < \theta < 180°$

16. θ, given $\cos 2\theta = \frac{2}{3}$ and $90° < \theta < 180°$

Verify that each equation is an identity. ***See Example 3.***

17. $(\sin x + \cos x)^2 = \sin 2x + 1$

18. $\sec 2x = \dfrac{\sec^2 x + \sec^4 x}{2 + \sec^2 x - \sec^4 x}$

19. $(\cos 2x + \sin 2x)^2 = 1 + \sin 4x$

20. $(\cos 2x - \sin 2x)^2 = 1 - \sin 4x$

21. $\tan 8\theta - \tan 8\theta \tan^2 4\theta = 2\tan 4\theta$

22. $\sin 2x = \dfrac{2\tan x}{1 + \tan^2 x}$

23. $\cos 2\theta = \dfrac{2 - \sec^2\theta}{\sec^2\theta}$

24. $\tan 2\theta = \dfrac{-2\tan\theta}{\sec^2\theta - 2}$

25. $\sin 4x = 4 \sin x \cos x \cos 2x$

26. $\dfrac{1 + \cos 2x}{\sin 2x} = \cot x$

27. $\dfrac{2 \cos 2\theta}{\sin 2\theta} = \cot \theta - \tan \theta$

28. $\cot 4\theta = \dfrac{1 - \tan^2 2\theta}{2 \tan 2\theta}$

29. $\tan x + \cot x = 2 \csc 2x$

30. $\cos 2x = \dfrac{1 - \tan^2 x}{1 + \tan^2 x}$

31. $1 + \tan x \tan 2x = \sec 2x$

32. $\dfrac{\cot A - \tan A}{\cot A + \tan A} = \cos 2A$

33. $\sin 2A \cos 2A = \sin 2A - 4 \sin^3 A \cos A$

34. $\sin 4x = 4 \sin x \cos x - 8 \sin^3 x \cos x$

35. $\tan(\theta - 45°) + \tan(\theta + 45°) = 2 \tan 2\theta$

36. $\cot \theta \tan(\theta + \pi) - \sin(\pi - \theta) \cos\left(\dfrac{\pi}{2} - \theta\right) = \cos^2 \theta$

Use an identity to write each expression as a single trigonometric function value or as a single number. ***See Example 4.***

37. $\cos^2 15° - \sin^2 15°$

38. $\dfrac{2 \tan 15°}{1 - \tan^2 15°}$

39. $1 - 2 \sin^2 15°$

40. $1 - 2 \sin^2 22\dfrac{1}{2}°$

41. $2 \cos^2 67\dfrac{1}{2}° - 1$

42. $\cos^2 \dfrac{\pi}{8} - \dfrac{1}{2}$

43. $\dfrac{\tan 51°}{1 - \tan^2 51°}$

44. $\dfrac{\tan 34°}{2(1 - \tan^2 34°)}$

45. $\dfrac{1}{4} - \dfrac{1}{2} \sin^2 47.1°$

46. $\dfrac{1}{8} \sin 29.5° \cos 29.5°$

47. $\sin^2 \dfrac{2\pi}{5} - \cos^2 \dfrac{2\pi}{5}$

48. $\cos^2 2x - \sin^2 2x$

Express each function as a trigonometric function of x. ***See Example 5.***

49. $\sin 4x$

50. $\cos 3x$

51. $\tan 3x$

52. $\cos 4x$

Graph each expression and use the graph to make a conjecture, predicting what might be an identity. Then verify your conjecture algebraically.

53. $\cos^4 x - \sin^4 x$

54. $\dfrac{4 \tan x \cos^2 x - 2 \tan x}{1 - \tan^2 x}$

55. $\dfrac{2 \tan x}{2 - \sec^2 x}$

56. $\dfrac{\cot^2 x - 1}{2 \cot x}$

Write each expression as a sum or difference of trigonometric functions. ***See Example 7.***

57. $2 \sin 58° \cos 102°$

58. $2 \cos 85° \sin 140°$

59. $2 \sin \dfrac{\pi}{6} \cos \dfrac{\pi}{3}$

60. $5 \cos 3x \cos 2x$

61. $6 \sin 4x \sin 5x$

62. $8 \sin 7x \sin 9x$

Write each expression as a product of trigonometric functions. ***See Example 8.***

63. $\cos 4x - \cos 2x$

64. $\cos 5x + \cos 8x$

65. $\sin 25° + \sin(-48°)$

66. $\sin 102° - \sin 95°$

67. $\cos 4x + \cos 8x$

68. $\sin 9x - \sin 3x$

(Modeling) Solve each problem. ***See Example 6.***

69. *Wattage Consumption* Use an identity to determine values of a, c, and ω in **Example 6** so that

$$W = a \cos(\omega t) + c.$$

Check your answer by graphing both expressions for W on the same coordinate axes.

70. *Amperage, Wattage, and Voltage* Amperage is a measure of the amount of electricity that is moving through a circuit, whereas voltage is a measure of the force pushing the electricity. The wattage W consumed by an electrical device can be determined by calculating the product of the amperage I and voltage V. (*Source:* Wilcox, G. and C. Hesselberth, *Electricity for Engineering Technology,* Allyn & Bacon.)

(a) A household circuit has voltage

$$V = 163 \sin 120\pi t$$

when an incandescent light bulb is turned on with amperage

$$I = 1.23 \sin 120\pi t.$$

Graph the wattage $W = VI$ consumed by the light bulb in the window $[0, 0.05]$ by $[-50, 300]$.

(b) Determine the maximum and minimum wattages used by the light bulb.

(c) Use identities to determine values for a, c, and ω so that $W = a \cos(\omega t) + c$.

(d) Check your answer by graphing both expressions for W on the same coordinate axes.

(e) Use the graph to estimate the average wattage used by the light. For how many watts (to the nearest integer) do you think this incandescent light bulb is rated?

5.6 Half-Angle Identities

- Half-Angle Identities
- Applying the Half-Angle Identities
- Verifying an Identity

Half-Angle Identities From the alternative forms of the identity for cos $2A$, we derive identities for $\sin \frac{A}{2}$, $\cos \frac{A}{2}$, and $\tan \frac{A}{2}$. These are known as **half-angle identities.**

We derive the identity for $\sin \frac{A}{2}$ as follows.

$$\cos 2x = 1 - 2 \sin^2 x$$ Cosine double-angle identity (Section 5.5)

$$2 \sin^2 x = 1 - \cos 2x$$ Add $2 \sin^2 x$ and subtract $\cos 2x$.

$$\sin x = \pm\sqrt{\frac{1 - \cos 2x}{2}}$$ Divide by 2 and take square roots. (Appendix A)

Remember both the positive and negative square roots.

$$\sin \frac{A}{2} = \pm\sqrt{\frac{1 - \cos A}{2}}$$ Let $2x = A$, so $x = \frac{A}{2}$. Substitute.

The $\pm$ sign in this identity indicates that the appropriate sign is chosen depending on the quadrant of $\frac{A}{2}$. For example, if $\frac{A}{2}$ is a quadrant III angle, we choose the negative sign because the sine function is negative in quadrant III.

We derive the identity for $\cos \frac{A}{2}$ using another double-angle identity.

$$\cos 2x = 2\cos^2 x - 1$$ Cosine double-angle identity (Section 5.5)

$$1 + \cos 2x = 2\cos^2 x$$ Add 1.

$$\cos^2 x = \frac{1 + \cos 2x}{2}$$ Rewrite and divide by 2.

$$\cos x = \pm\sqrt{\frac{1 + \cos 2x}{2}}$$ Take square roots.

$$\mathbf{\cos \frac{A}{2} = \pm\sqrt{\frac{1 + \cos A}{2}}}$$ Replace x with $\frac{A}{2}$.

An identity for $\tan \frac{A}{2}$ comes from the identities for $\sin \frac{A}{2}$ and $\cos \frac{A}{2}$.

$$\tan \frac{A}{2} = \frac{\sin \frac{A}{2}}{\cos \frac{A}{2}} = \frac{\pm\sqrt{\frac{1 - \cos A}{2}}}{\pm\sqrt{\frac{1 + \cos A}{2}}} = \pm\sqrt{\frac{1 - \cos A}{1 + \cos A}}$$

We derive an alternative identity for $\tan \frac{A}{2}$ using double-angle identities.

$$\tan \frac{A}{2} = \frac{\sin \frac{A}{2}}{\cos \frac{A}{2}} = \frac{2\sin \frac{A}{2}\cos \frac{A}{2}}{2\cos^2 \frac{A}{2}}$$ Multiply by $2\cos \frac{A}{2}$ in numerator and denominator.

$$= \frac{\sin 2\left(\frac{A}{2}\right)}{1 + \cos 2\left(\frac{A}{2}\right)}$$ Double-angle identities (Section 5.5)

$$\mathbf{\tan \frac{A}{2} = \frac{\sin A}{1 + \cos A}}$$ Simplify.

From the identity $\tan \frac{A}{2} = \frac{\sin A}{1 + \cos A}$, we can also derive an equivalent identity.

$$\mathbf{\tan \frac{A}{2} = \frac{1 - \cos A}{\sin A}}$$

Half-Angle Identities

In the following identities, the symbol $\pm$ indicates that the sign is chosen based on the function under consideration and the quadrant of $\frac{A}{2}$.

$$\cos \frac{A}{2} = \pm\sqrt{\frac{1 + \cos A}{2}} \qquad \sin \frac{A}{2} = \pm\sqrt{\frac{1 - \cos A}{2}}$$

$$\tan \frac{A}{2} = \pm\sqrt{\frac{1 - \cos A}{1 + \cos A}} \qquad \tan \frac{A}{2} = \frac{\sin A}{1 + \cos A} \qquad \tan \frac{A}{2} = \frac{1 - \cos A}{\sin A}$$

The final two identities for $\tan \frac{A}{2}$ do not require a sign choice. When using the other half-angle identities, select the plus or minus sign according to the quadrant in which $\frac{A}{2}$ terminates. For example, if an angle $A = 324°$, then $\frac{A}{2} = 162°$, which lies in quadrant II. So when $A = 324°$, $\cos \frac{A}{2}$ and $\tan \frac{A}{2}$ are negative, and $\sin \frac{A}{2}$ is positive.

Applying the Half-Angle Identities

EXAMPLE 1 **Using a Half-Angle Identity to Find an Exact Value**

Find the exact value of cos 15° using the half-angle identity for cosine.

SOLUTION

$$\cos 15° = \cos \frac{1}{2}(30°) = \sqrt{\frac{1 + \cos 30°}{2}}$$

Choose the positive square root.

$$= \sqrt{\frac{1 + \frac{\sqrt{3}}{2}}{2}} = \sqrt{\frac{\left(1 + \frac{\sqrt{3}}{2}\right) \cdot 2}{2 \cdot 2}} = \frac{\sqrt{2 + \sqrt{3}}}{2}$$

Simplify the radicals.

✔ *Now Try Exercise 11.*

EXAMPLE 2 **Using a Half-Angle Identity to Find an Exact Value**

Find the exact value of tan 22.5° using the identity $\tan \frac{A}{2} = \frac{\sin A}{1 + \cos A}$.

SOLUTION Since $22.5° = \frac{1}{2}(45°)$, replace A with 45°.

$$\tan 22.5° = \tan \frac{45°}{2} = \frac{\sin 45°}{1 + \cos 45°} = \frac{\frac{\sqrt{2}}{2}}{1 + \frac{\sqrt{2}}{2}} = \frac{\frac{\sqrt{2}}{2}}{1 + \frac{\sqrt{2}}{2}} \cdot \frac{2}{2}$$

$$= \frac{\sqrt{2}}{2 + \sqrt{2}} = \frac{\sqrt{2}}{2 + \sqrt{2}} \cdot \frac{2 - \sqrt{2}}{2 - \sqrt{2}} = \frac{2\sqrt{2} - 2}{2}$$

Rationalize the denominator.

$$= \frac{2(\sqrt{2} - 1)}{2} = \sqrt{2} - 1$$

Factor out 2.

Factor first, and then divide out the common factor.

✔ *Now Try Exercise 13.*

EXAMPLE 3 **Finding Function Values of $\frac{s}{2}$ Given Information about s**

Given $\cos s = \frac{2}{3}$, with $\frac{3\pi}{2} < s < 2\pi$, find $\sin \frac{s}{2}$, $\cos \frac{s}{2}$, and $\tan \frac{s}{2}$.

SOLUTION The angle associated with $\frac{s}{2}$ terminates in quadrant II, since

$$\frac{3\pi}{2} < s < 2\pi \quad \text{and} \quad \frac{3\pi}{4} < \frac{s}{2} < \pi. \quad \text{Divide by 2. (Appendix A)}$$

See **Figure 9.** In quadrant II, the values of $\cos \frac{s}{2}$ and $\tan \frac{s}{2}$ are negative and the value of $\sin \frac{s}{2}$ is positive. Use the appropriate half-angle identities and simplify.

$$\sin \frac{s}{2} = \sqrt{\frac{1 - \frac{2}{3}}{2}} = \sqrt{\frac{1}{6}} = \frac{\sqrt{1}}{\sqrt{6}} \cdot \frac{\sqrt{6}}{\sqrt{6}} = \frac{\sqrt{6}}{6}$$

Rationalize all denominators.

$$\cos \frac{s}{2} = -\sqrt{\frac{1 + \frac{2}{3}}{2}} = -\sqrt{\frac{5}{6}} = -\frac{\sqrt{5}}{\sqrt{6}} \cdot \frac{\sqrt{6}}{\sqrt{6}} = -\frac{\sqrt{30}}{6}$$

$$\tan \frac{s}{2} = \frac{\sin \frac{s}{2}}{\cos \frac{s}{2}} = \frac{\frac{\sqrt{6}}{6}}{-\frac{\sqrt{30}}{6}} = \frac{\sqrt{6}}{-\sqrt{30}} = -\frac{\sqrt{6}}{\sqrt{30}} \cdot \frac{\sqrt{30}}{\sqrt{30}} = -\frac{\sqrt{180}}{30} = -\frac{6\sqrt{5}}{6 \cdot 5} = -\frac{\sqrt{5}}{5}$$

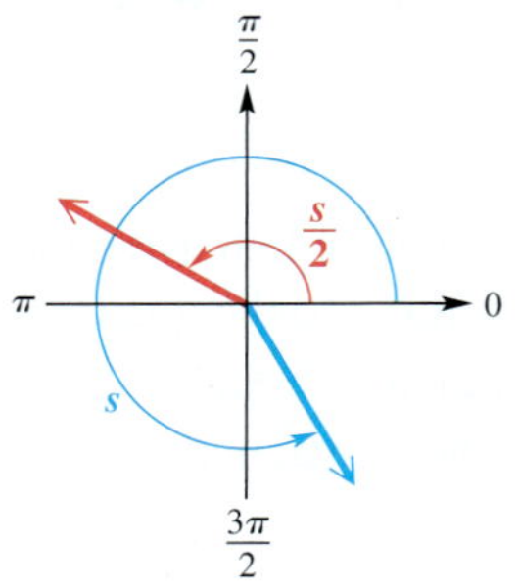

Figure 9

Notice that it is not necessary to use a half-angle identity for $\tan \frac{s}{2}$ once we find $\sin \frac{s}{2}$ and $\cos \frac{s}{2}$. However, using this identity provides an excellent check.

✔ *Now Try Exercise 19.*

EXAMPLE 4 Simplifying Expressions Using the Half-Angle Identities

Simplify each expression.

(a) $\pm\sqrt{\dfrac{1+\cos 12x}{2}}$ **(b)** $\dfrac{1-\cos 5\alpha}{\sin 5\alpha}$

SOLUTION

(a) This matches part of the identity for $\cos \frac{A}{2}$. Replace A with $12x$ to get

$$\cos\frac{A}{2} = \pm\sqrt{\frac{1+\cos A}{2}} = \pm\sqrt{\frac{1+\cos 12x}{2}} = \cos\frac{12x}{2} = \cos 6x.$$

(b) Use the third identity for $\tan \frac{A}{2}$ given earlier with $A = 5\alpha$ to get

$$\frac{1-\cos 5\alpha}{\sin 5\alpha} = \tan\frac{5\alpha}{2}.$$

✔ *Now Try Exercises 37 and 39.*

Verifying an Identity

EXAMPLE 5 Verifying an Identity

Verify that the following equation is an identity.

$$\left(\sin\frac{x}{2}+\cos\frac{x}{2}\right)^2 = 1+\sin x$$

SOLUTION We work on the more complicated left side.

$$\left(\sin\frac{x}{2}+\cos\frac{x}{2}\right)^2$$

Remember the term $2ab$ when squaring a binomial.

$= \sin^2\frac{x}{2} + 2\sin\frac{x}{2}\cos\frac{x}{2} + \cos^2\frac{x}{2}$ $\quad (a+b)^2 = a^2 + 2ab + b^2$

$= 1 + 2\sin\frac{x}{2}\cos\frac{x}{2}$ $\quad \sin^2\frac{x}{2} + \cos^2\frac{x}{2} = 1$

$= 1 + \sin 2\left(\frac{x}{2}\right)$ $\quad 2\sin\frac{x}{2}\cos\frac{x}{2} = \sin 2\left(\frac{x}{2}\right)$

$= 1 + \sin x$ $\quad$ Multiply.

✔ *Now Try Exercise 47.*

5.6 Exercises

Concept Check Determine whether the positive or the negative square root should be selected.

1. $\sin 195^\circ = \pm\sqrt{\dfrac{1-\cos 390^\circ}{2}}$

2. $\cos 58^\circ = \pm\sqrt{\dfrac{1+\cos 116^\circ}{2}}$

3. $\tan 225^\circ = \pm\sqrt{\dfrac{1-\cos 450^\circ}{1+\cos 450^\circ}}$

4. $\sin(-10^\circ) = \pm\sqrt{\dfrac{1-\cos(-20^\circ)}{2}}$

Match each expression in Column I with its value in Column II. **See Examples 1 and 2.**

I		II	
5. $\sin 15°$	**6.** $\tan 15°$	**A.** $2-\sqrt{3}$	**B.** $\dfrac{\sqrt{2-\sqrt{2}}}{2}$
7. $\cos \dfrac{\pi}{8}$	**8.** $\tan\left(-\dfrac{\pi}{8}\right)$	**C.** $\dfrac{\sqrt{2-\sqrt{3}}}{2}$	**D.** $\dfrac{\sqrt{2+\sqrt{2}}}{2}$
9. $\tan 67.5°$	**10.** $\cos 67.5°$	**E.** $1-\sqrt{2}$	**F.** $1+\sqrt{2}$

Use a half-angle identity to find each exact value. **See Examples 1 and 2.**

11. $\sin 67.5°$ **12.** $\sin 195°$ **13.** $\tan 195°$

14. $\cos 195°$ **15.** $\cos 165°$ **16.** $\sin 165°$

17. Explain how you can use an identity of this section to find the exact value of $\sin 7.5°$. (*Hint:* $7.5 = \frac{1}{2}\left(\frac{1}{2}\right)(30)$.)

18. The half-angle identity

$$\tan \frac{A}{2} = \pm\sqrt{\frac{1-\cos A}{1+\cos A}}$$

can be used to find $\tan 22.5° = \sqrt{3-2\sqrt{2}}$, and the half-angle identity

$$\tan \frac{A}{2} = \frac{\sin A}{1+\cos A}$$

can be used to find $\tan 22.5° = \sqrt{2}-1$. Show that these answers are the same, without using a calculator. (*Hint:* If $a > 0$ and $b > 0$ and $a^2 = b^2$, then $a = b$.)

Find each of the following. **See Example 3.**

19. $\cos \frac{x}{2}$, given $\cos x = \frac{1}{4}$, with $0 < x < \frac{\pi}{2}$

20. $\sin \frac{x}{2}$, given $\cos x = -\frac{5}{8}$, with $\frac{\pi}{2} < x < \pi$

21. $\tan \frac{\theta}{2}$, given $\sin \theta = \frac{3}{5}$, with $90° < \theta < 180°$

22. $\cos \frac{\theta}{2}$, given $\sin \theta = -\frac{4}{5}$, with $180° < \theta < 270°$

23. $\sin \frac{x}{2}$, given $\tan x = 2$, with $0 < x < \frac{\pi}{2}$

24. $\cos \frac{x}{2}$, given $\cot x = -3$, with $\frac{\pi}{2} < x < \pi$

25. $\tan \frac{\theta}{2}$, given $\tan \theta = \frac{\sqrt{7}}{3}$, with $180° < \theta < 270°$

26. $\cot \frac{\theta}{2}$, given $\tan \theta = -\frac{\sqrt{5}}{2}$, with $90° < \theta < 180°$

27. $\sin \theta$, given $\cos 2\theta = \frac{3}{5}$ and θ terminates in quadrant I

28. $\cos \theta$, given $\cos 2\theta = \frac{1}{2}$ and θ terminates in quadrant II

29. $\cos x$, given $\cos 2x = -\frac{5}{12}$, with $\frac{\pi}{2} < x < \pi$

30. $\sin x$, given $\cos 2x = \frac{2}{3}$, with $\pi < x < \frac{3\pi}{2}$

31. *Concept Check* If $\cos x \approx 0.9682$ and $\sin x = 0.25$, then $\tan \frac{x}{2} \approx$ ______.

32. *Concept Check* If $\cos x = -0.75$ and $\sin x \approx 0.6614$, then $\tan \frac{x}{2} \approx$ ______.

Use an identity to write each expression as a single trigonometric function. **See Example 4.**

33. $\sqrt{\dfrac{1-\cos 40°}{2}}$ **34.** $\sqrt{\dfrac{1+\cos 76°}{2}}$ **35.** $\sqrt{\dfrac{1-\cos 147°}{1+\cos 147°}}$

36. $\sqrt{\dfrac{1+\cos 165°}{1-\cos 165°}}$ **37.** $\dfrac{1-\cos 59.74°}{\sin 59.74°}$ **38.** $\dfrac{\sin 158.2°}{1+\cos 158.2°}$

39. $\pm\sqrt{\dfrac{1+\cos 18x}{2}}$

40. $\pm\sqrt{\dfrac{1+\cos 20\alpha}{2}}$

41. $\pm\sqrt{\dfrac{1-\cos 8\theta}{1+\cos 8\theta}}$

42. $\pm\sqrt{\dfrac{1-\cos 5A}{1+\cos 5A}}$

43. $\pm\sqrt{\dfrac{1+\cos \frac{x}{4}}{2}}$

44. $\pm\sqrt{\dfrac{1-\cos \frac{3\theta}{5}}{2}}$

Verify that each equation is an identity. ***See Example 5.***

45. $\sec^2 \dfrac{x}{2} = \dfrac{2}{1+\cos x}$

46. $\cot^2 \dfrac{x}{2} = \dfrac{(1+\cos x)^2}{\sin^2 x}$

47. $\sin^2 \dfrac{x}{2} = \dfrac{\tan x - \sin x}{2\tan x}$

48. $\dfrac{\sin 2x}{2\sin x} = \cos^2 \dfrac{x}{2} - \sin^2 \dfrac{x}{2}$

49. $\dfrac{2}{1+\cos x} - \tan^2 \dfrac{x}{2} = 1$

50. $\tan \dfrac{\theta}{2} = \csc\theta - \cot\theta$

51. $1 - \tan^2 \dfrac{\theta}{2} = \dfrac{2\cos\theta}{1+\cos\theta}$

52. $\cos x = \dfrac{1-\tan^2 \frac{x}{2}}{1+\tan^2 \frac{x}{2}}$

53. Use the half-angle identity

$$\tan \frac{A}{2} = \frac{\sin A}{1+\cos A}$$

to derive the equivalent identity

$$\tan \frac{A}{2} = \frac{1-\cos A}{\sin A}$$

by multiplying both the numerator and the denominator by $1 - \cos A$.

54. Use the identity $\tan \frac{A}{2} = \frac{\sin A}{1+\cos A}$ to determine an identity for $\cot \frac{A}{2}$.

Graph each expression and use the graph to make a conjecture, predicting what might be an identity. Then verify your conjecture algebraically.

55. $\dfrac{\sin x}{1+\cos x}$

56. $\dfrac{1-\cos x}{\sin x}$

57. $\dfrac{\tan \frac{x}{2} + \cot \frac{x}{2}}{\cot \frac{x}{2} - \tan \frac{x}{2}}$

58. $1 - 8\sin^2 \dfrac{x}{2}\cos^2 \dfrac{x}{2}$

(Modeling) Mach Number *An airplane flying faster than sound sends out sound waves that form a cone, as shown in the figure. The cone intersects the ground to form a* ***hyperbola.*** *As this hyperbola passes over a particular point on the ground, a sonic boom is heard at that point. If θ is the angle at the vertex of the cone, then*

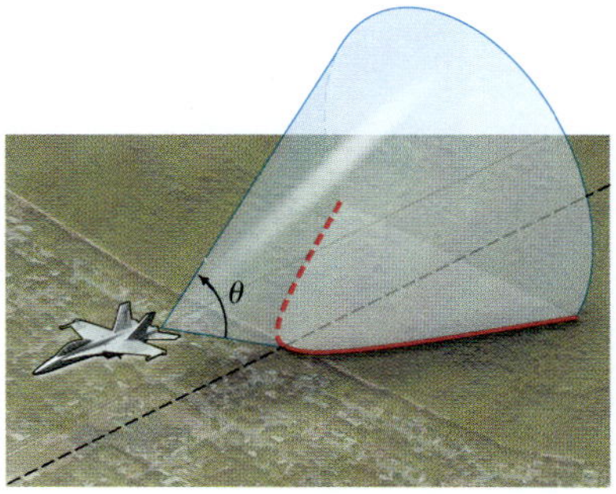

$$\sin \frac{\theta}{2} = \frac{1}{m},$$

where m is the Mach number for the speed of the plane. (We assume $m > 1$.) The Mach number is the ratio of the speed of the plane to the speed of sound. Thus, a speed of Mach 1.4 means that the plane is flying at 1.4 times the speed of sound. In Exercises 59–62, one of the values θ or m is given. Find the other value.

59. $m = \dfrac{5}{4}$

60. $m = \dfrac{3}{2}$

61. $\theta = 60^\circ$

62. $\theta = 30^\circ$

63. *(Modeling) Railroad Curves* In the United States, circular railroad curves are designated by the **degree of curvature,** the central angle subtended by a chord of 100 ft. See the figure. (*Source:* Hay, W. W., *Railroad Engineering,* John Wiley and Sons.)

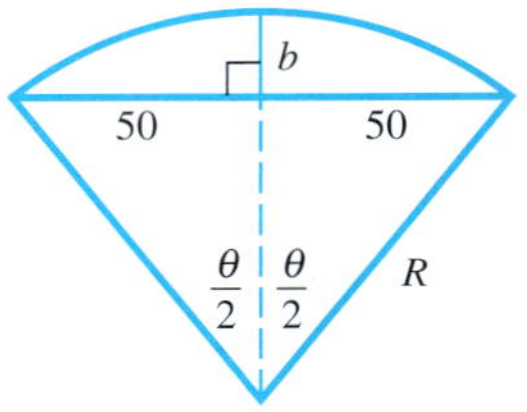

(a) Use the figure to write an expression for $\cos \frac{\theta}{2}$.

(b) Use the result of part (a) and the third half-angle identity for tangent to write an expression for $\tan \frac{\theta}{4}$.

64. In **Exercise 63,** if $b = 12$, what is the measure of angle θ to the nearest degree?

Relating Concepts

For individual or collaborative investigation *(Exercises 65–72)*

These exercises use results from plane geometry to obtain exact values of the trigonometric functions of 15°. *Start with a right triangle ACB having a* 60° *angle at A and a* 30° *angle at B. Let the hypotenuse of this triangle have length* 2. *Extend side BC and draw a semicircle with diameter along BC extended, center at B, and radius AB. Draw segment AE. (See the figure.) Since any angle inscribed in a semicircle is a right angle, triangle EAD is a right triangle.* ***Work Exercises 65–72 in order.***

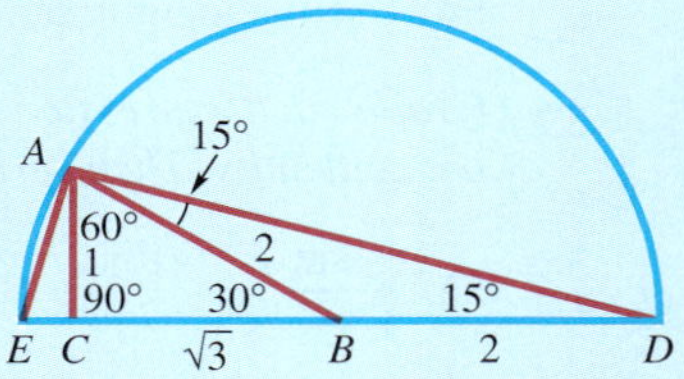

65. Why is $AB = BD$ true? Conclude that triangle ABD is isosceles.

66. Why does angle ABD have measure 150°?

67. Why do angles DAB and ADB both have measures of 15°?

68. What is the length DC?

69. Use the Pythagorean theorem to show that the length AD is $\sqrt{6} + \sqrt{2}$.

70. Use angle ADB of triangle EAD to find cos 15°.

71. Show that AE has length $\sqrt{6} - \sqrt{2}$ and find sin 15°.

72. Use triangle ACD to find tan 15°.

Advanced methods of trigonometry can be used to find the following exact *value. (See, for example, Hobson's* A Treatise on Plane Trigonometry.*)*

$$\sin 18^\circ = \frac{\sqrt{5} - 1}{4}$$

Use this exact value and identities to find each exact value. Support your answers with calculator approximations if you wish.

73. cos 18° **74.** tan 18° **75.** cot 18°

76. $\sec 18^\circ$

77. $\csc 18^\circ$

78. $\cos 72^\circ$

79. $\sin 72^\circ$

80. $\tan 72^\circ$

81. $\cot 72^\circ$

82. $\csc 72^\circ$

83. $\sec 72^\circ$

84. $\sin 162^\circ$

Summary Exercises on Verifying Trigonometric Identities

These summary exercises provide practice with the various types of trigonometric identities presented in this chapter. Verify that each equation is an identity.

1. $\tan\theta + \cot\theta = \sec\theta\csc\theta$

2. $\csc\theta\cos^2\theta + \sin\theta = \csc\theta$

3. $\tan\frac{x}{2} = \csc x - \cot x$

4. $\sec(\pi - x) = -\sec x$

5. $\frac{\sin t}{1+\cos t} = \frac{1-\cos t}{\sin t}$

6. $\frac{1-\sin t}{\cos t} = \frac{1}{\sec t + \tan t}$

7. $\sin 2\theta = \frac{2\tan\theta}{1+\tan^2\theta}$

8. $\frac{2}{1+\cos x} - \tan^2\frac{x}{2} = 1$

9. $\cot\theta - \tan\theta = \frac{2\cos^2\theta - 1}{\sin\theta\cos\theta}$

10. $\frac{1}{\sec t - 1} + \frac{1}{\sec t + 1} = 2\cot t\csc t$

11. $\frac{\sin(x+y)}{\cos(x-y)} = \frac{\cot x + \cot y}{1 + \cot x\cot y}$

12. $1 - \tan^2\frac{\theta}{2} = \frac{2\cos\theta}{1+\cos\theta}$

13. $\frac{\sin\theta + \tan\theta}{1+\cos\theta} = \tan\theta$

14. $\csc^4 x - \cot^4 x = \frac{1+\cos^2 x}{1-\cos^2 x}$

15. $\cos x = \frac{1-\tan^2\frac{x}{2}}{1+\tan^2\frac{x}{2}}$

16. $\cos 2x = \frac{2-\sec^2 x}{\sec^2 x}$

17. $\frac{\tan^2 t + 1}{\tan t\csc^2 t} = \tan t$

18. $\frac{\sin s}{1+\cos s} + \frac{1+\cos s}{\sin s} = 2\csc s$

19. $\tan 4\theta = \frac{2\tan 2\theta}{2-\sec^2 2\theta}$

20. $\tan\left(\frac{x}{2} + \frac{\pi}{4}\right) = \sec x + \tan x$

21. $\frac{\cot s - \tan s}{\cos s + \sin s} = \frac{\cos s - \sin s}{\sin s\cos s}$

22. $\frac{\tan\theta - \cot\theta}{\tan\theta + \cot\theta} = 1 - 2\cos^2\theta$

23. $\frac{\tan(x+y) - \tan y}{1 + \tan(x+y)\tan y} = \tan x$

24. $2\cos^2\frac{x}{2}\tan x = \tan x + \sin x$

25. $\frac{\cos^4 x - \sin^4 x}{\cos^2 x} = 1 - \tan^2 x$

26. $\frac{\csc t + 1}{\csc t - 1} = (\sec t + \tan t)^2$

27. $\frac{2(\sin x - \sin^3 x)}{\cos x} = \sin 2x$

28. $\frac{1}{2}\cot\frac{x}{2} - \frac{1}{2}\tan\frac{x}{2} = \cot x$

29. $\sin(60^\circ + x) + \sin(60^\circ - x) = \sqrt{3}\cos x$

30. $\sin(60^\circ - x) - \sin(60^\circ + x) = -\sin x$

31. $\frac{\cos(x+y) + \cos(y-x)}{\sin(x+y) - \sin(y-x)} = \cot x$

32. $\sin x + \sin 3x + \sin 5x + \sin 7x = 4\cos x\cos 2x\sin 4x$

33. $\sin^3\theta + \cos^3\theta + \sin\theta\cos^2\theta + \sin^2\theta\cos\theta = \sin\theta + \cos\theta$

34. $\frac{\cos x + \sin x}{\cos x - \sin x} - \frac{\cos x - \sin x}{\cos x + \sin x} = 2\tan 2x$

Chapter 5 Test Prep

Quick Review

Concepts	Examples

5.1 Fundamental Identities

Reciprocal Identities

$$\cot\theta = \frac{1}{\tan\theta} \qquad \sec\theta = \frac{1}{\cos\theta} \qquad \csc\theta = \frac{1}{\sin\theta}$$

Quotient Identities

$$\tan\theta = \frac{\sin\theta}{\cos\theta} \qquad \cot\theta = \frac{\cos\theta}{\sin\theta}$$

Pythagorean Identities

$$\sin^2\theta + \cos^2\theta = 1 \qquad \tan^2\theta + 1 = \sec^2\theta$$

$$1 + \cot^2\theta = \csc^2\theta$$

Negative-Angle Identities

$$\sin(-\theta) = -\sin\theta \quad \cos(-\theta) = \cos\theta \quad \tan(-\theta) = -\tan\theta$$

$$\csc(-\theta) = -\csc\theta \quad \sec(-\theta) = \sec\theta \quad \cot(-\theta) = -\cot\theta$$

If θ is in quadrant IV and $\sin\theta = -\frac{3}{5}$, find $\csc\theta$, $\cos\theta$, and $\sin(-\theta)$.

$$\csc\theta = \frac{1}{\sin\theta} = \frac{1}{-\frac{3}{5}} = -\frac{5}{3} \qquad \text{Reciprocal identity}$$

$$\sin^2\theta + \cos^2\theta = 1 \qquad \text{Pythagorean identity}$$

$$\left(-\frac{3}{5}\right)^2 + \cos^2\theta = 1 \qquad \text{Substitute.}$$

$$\cos^2\theta = 1 - \frac{9}{25} = \frac{16}{25} \qquad \text{Subtract } \tfrac{9}{25}.$$

$$\cos\theta = +\sqrt{\frac{16}{25}} = \frac{4}{5} \qquad \cos\theta \text{ is positive in quadrant IV.}$$

$$\sin(-\theta) = -\sin\theta = -\left(-\frac{3}{5}\right) = \frac{3}{5}$$

Negative angle identity

5.2 Verifying Trigonometric Identities

See the box titled Hints for Verifying Identities in **Section 5.2.**

5.3 Sum and Difference Identities for Cosine

5.4 Sum and Difference Identities for Sine and Tangent

Cofunction Identities

$$\cos(90^\circ - \theta) = \sin\theta \qquad \cot(90^\circ - \theta) = \tan\theta$$

$$\sin(90^\circ - \theta) = \cos\theta \qquad \sec(90^\circ - \theta) = \csc\theta$$

$$\tan(90^\circ - \theta) = \cot\theta \qquad \csc(90^\circ - \theta) = \sec\theta$$

Find one value of θ such that $\tan\theta = \cot 78^\circ$.

$$\tan\theta = \cot 78^\circ$$

$$\cot(90^\circ - \theta) = \cot 78^\circ \qquad \text{Cofunction identity}$$

$$90^\circ - \theta = 78^\circ \qquad \text{Set angles equal.}$$

$$\theta = 12^\circ \qquad \text{Solve for } \theta.$$

Sum and Difference Identities

$$\cos(A - B) = \cos A \cos B + \sin A \sin B$$

$$\cos(A + B) = \cos A \cos B - \sin A \sin B$$

$$\sin(A + B) = \sin A \cos B + \cos A \sin B$$

$$\sin(A - B) = \sin A \cos B - \cos A \sin B$$

$$\tan(A + B) = \frac{\tan A + \tan B}{1 - \tan A \tan B}$$

$$\tan(A - B) = \frac{\tan A - \tan B}{1 + \tan A \tan B}$$

Find the exact value of $\cos(-15^\circ)$.

$$\cos(-15^\circ)$$

$$= \cos(30^\circ - 45^\circ)$$

$$= \cos 30^\circ \cos 45^\circ + \sin 30^\circ \sin 45^\circ$$

Cosine difference identity

$$= \frac{\sqrt{3}}{2} \cdot \frac{\sqrt{2}}{2} + \frac{1}{2} \cdot \frac{\sqrt{2}}{2} \qquad \text{Substitute values.}$$

$$= \frac{\sqrt{6} + \sqrt{2}}{4} \qquad \text{Simplify.}$$

Concepts	Examples
	Write $\tan\left(\frac{\pi}{4}+\theta\right)$ in terms of $\tan\theta$. $\tan\left(\frac{\pi}{4}+\theta\right)=\frac{\tan\frac{\pi}{4}+\tan\theta}{1-\tan\frac{\pi}{4}\tan\theta}=\frac{1+\tan\theta}{1-\tan\theta}$ $\quad \tan\frac{\pi}{4}=1$

5.5 Double-Angle Identities

Double-Angle Identities

$$\cos 2A=\cos^2 A-\sin^2 A \qquad \cos 2A = 1-2\sin^2 A$$

$$\cos 2A = 2\cos^2 A - 1 \qquad \sin 2A = 2\sin A\cos A$$

$$\tan 2A=\frac{2\tan A}{1-\tan^2 A}$$

Given $\cos\theta=-\frac{5}{13}$ and $\sin\theta>0$, find $\sin 2\theta$.

Sketch a triangle in quadrant II since $\cos\theta<0$ and $\sin\theta>0$. Use it to find that $\sin\theta=\frac{12}{13}$.

$$\sin 2\theta = 2\sin\theta\cos\theta$$

$$=2\left(\frac{12}{13}\right)\left(-\frac{5}{13}\right)$$

$$=-\frac{120}{169}$$

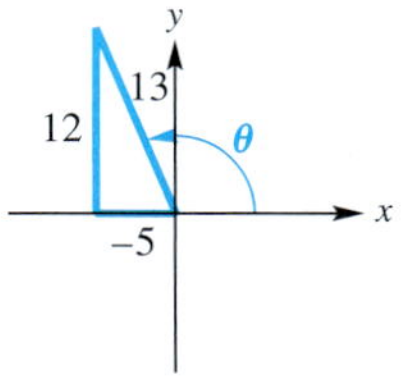

Product-to-Sum Identities

$$\cos A\cos B=\frac{1}{2}[\cos(A+B)+\cos(A-B)]$$

$$\sin A\sin B=\frac{1}{2}[\cos(A-B)-\cos(A+B)]$$

$$\sin A\cos B=\frac{1}{2}[\sin(A+B)+\sin(A-B)]$$

$$\cos A\sin B=\frac{1}{2}[\sin(A+B)-\sin(A-B)]$$

Write $\sin(-\theta)\sin 2\theta$ as the difference of two functions.

$$\sin(-\theta)\sin 2\theta$$

$$=\frac{1}{2}[\cos(-\theta-2\theta)-\cos(-\theta+2\theta)]$$

$$=\frac{1}{2}[\cos(-3\theta)-\cos\theta]$$

$$=\frac{1}{2}\cos(-3\theta)-\frac{1}{2}\cos\theta$$

$$=\frac{1}{2}\cos 3\theta-\frac{1}{2}\cos\theta$$

Sum-to-Product Identities

$$\sin A+\sin B=2\sin\left(\frac{A+B}{2}\right)\cos\left(\frac{A-B}{2}\right)$$

$$\sin A-\sin B=2\cos\left(\frac{A+B}{2}\right)\sin\left(\frac{A-B}{2}\right)$$

$$\cos A+\cos B=2\cos\left(\frac{A+B}{2}\right)\cos\left(\frac{A-B}{2}\right)$$

$$\cos A-\cos B=-2\sin\left(\frac{A+B}{2}\right)\sin\left(\frac{A-B}{2}\right)$$

Write $\cos\theta+\cos 3\theta$ as a product of two functions.

$$\cos\theta+\cos 3\theta$$

$$=2\cos\left(\frac{\theta+3\theta}{2}\right)\cos\left(\frac{\theta-3\theta}{2}\right)$$

$$=2\cos\left(\frac{4\theta}{2}\right)\cos\left(\frac{-2\theta}{2}\right)$$

$$=2\cos 2\theta\cos(-\theta)$$

$$=2\cos 2\theta\cos\theta$$

5.6 Half-Angle Identities

Half-Angle Identities

$$\cos\frac{A}{2}=\pm\sqrt{\frac{1+\cos A}{2}} \qquad \sin\frac{A}{2}=\pm\sqrt{\frac{1-\cos A}{2}}$$

$$\tan\frac{A}{2}=\pm\sqrt{\frac{1-\cos A}{1+\cos A}} \qquad \tan\frac{A}{2}=\frac{\sin A}{1+\cos A}$$

$$\tan\frac{A}{2}=\frac{1-\cos A}{\sin A}$$

(In the identities involving radicals, the sign is chosen on the basis of the function under consideration and the quadrant of $\frac{A}{2}$.)

Find the exact value of $\tan 67.5^\circ$.

We choose the last form with $A=135^\circ$.

$$\tan 67.5^\circ=\tan\frac{135^\circ}{2}=\frac{1-\cos 135^\circ}{\sin 135^\circ}=\frac{1-\left(-\frac{\sqrt{2}}{2}\right)}{\frac{\sqrt{2}}{2}}$$

$$=\frac{1+\frac{\sqrt{2}}{2}}{\frac{\sqrt{2}}{2}}\cdot\frac{2}{2}=\frac{2+\sqrt{2}}{\sqrt{2}},\quad \text{or}\quad \sqrt{2}+1$$

Rationalize the denominator and simplify.

Chapter 5 Review Exercises

Concept Check For each expression in Column I, choose the expression from Column II that completes an identity.

I		II	
1. $\sec x =$ ______	**2.** $\csc x =$ ______	**A.** $\dfrac{1}{\sin x}$	**B.** $\dfrac{1}{\cos x}$
3. $\tan x =$ ______	**4.** $\cot x =$ ______	**C.** $\dfrac{\sin x}{\cos x}$	**D.** $\dfrac{1}{\cot^2 x}$
5. $\tan^2 x =$ ______	**6.** $\sec^2 x =$ ______	**E.** $\dfrac{1}{\cos^2 x}$	**F.** $\dfrac{\cos x}{\sin x}$

Use identities to write each expression in terms of $\sin \theta$ *and* $\cos \theta$, *and simplify.*

7. $\sec^2 \theta - \tan^2 \theta$ **8.** $\dfrac{\cot \theta}{\sec \theta}$ **9.** $\tan^2 \theta(1 + \cot^2 \theta)$

10. $\csc \theta + \cot \theta$ **11.** $\tan \theta - \sec \theta \csc \theta$ **12.** $\csc^2 \theta + \sec^2 \theta$

13. Use the trigonometric identities to find $\sin x$, $\tan x$, and $\cot(-x)$, given $\cos x = \frac{3}{5}$ and x in quadrant IV.

14. Given $\tan x = -\frac{5}{4}$, where $\frac{\pi}{2} < x < \pi$, use the trigonometric identities to find $\cot x$, $\csc x$, and $\sec x$.

15. Find the exact values of the six trigonometric functions of 165°.

16. Find the exact values of $\sin x$, $\cos x$, and $\tan x$, for $x = \frac{\pi}{12}$, using
(a) difference identities **(b)** half-angle identities.

Concept Check For each expression in Column I, use an identity to choose an expression from Column II with the same value. Choices may be used once, more than once, or not at all.

I		II	
17. $\cos 210°$	**18.** $\sin 35°$	**A.** $\sin(-35°)$	**B.** $\cos 55°$
19. $\tan(-35°)$	**20.** $-\sin 35°$	**C.** $\sqrt{\dfrac{1 + \cos 150°}{2}}$	**D.** $2 \sin 150° \cos 150°$
21. $\cos 35°$	**22.** $\cos 75°$	**E.** $\cot(-35°)$	**F.** $\cos^2 150° - \sin^2 150°$
23. $\sin 75°$	**24.** $\sin 300°$	**G.** $\cos(-35°)$	**H.** $\cot 125°$
25. $\cos 300°$	**26.** $\cos(-55°)$	**I.** $\cos 150° \cos 60° - \sin 150° \sin 60°$	
		J. $\sin 15° \cos 60° + \cos 15° \sin 60°$	

For each of the following, find $\sin(x + y)$, $\cos(x - y)$, $\tan(x + y)$, *and the quadrant of* $x + y$.

27. $\sin x = -\frac{3}{5}$, $\cos y = -\frac{7}{25}$, x and y in quadrant III

28. $\sin x = \frac{3}{5}$, $\cos y = \frac{24}{25}$, x in quadrant I, y in quadrant IV

29. $\sin x = -\frac{1}{2}$, $\cos y = -\frac{2}{5}$, x and y in quadrant III

30. $\sin y = -\frac{2}{3}$, $\cos x = -\frac{1}{5}$, x in quadrant II, y in quadrant III

31. $\sin x = \frac{1}{10}$, $\cos y = \frac{4}{5}$, x in quadrant I, y in quadrant IV

32. $\cos x = \frac{2}{9}$, $\sin y = -\frac{1}{2}$, x in quadrant IV, y in quadrant III

Find sine and cosine of each of the following.

33. θ, given $\cos 2\theta = -\frac{3}{4}$, $90° < 2\theta < 180°$

34. B, given $\cos 2B = \frac{1}{8}$, $540° < 2B < 720°$

35. $2x$, given $\tan x = 3$, $\sin x < 0$

36. $2y$, given $\sec y = -\frac{5}{3}$, $\sin y > 0$

Find each of the following.

37. $\cos \frac{\theta}{2}$, given $\cos \theta = -\frac{1}{2}$, $90° < \theta < 180°$

38. $\sin \frac{A}{2}$, given $\cos A = -\frac{3}{4}$, $90° < A < 180°$

39. $\tan x$, given $\tan 2x = 2$, $\pi < x < \frac{3\pi}{2}$

40. $\sin y$, given $\cos 2y = -\frac{1}{3}$, $\frac{\pi}{2} < y < \pi$

41. $\tan \frac{x}{2}$, given $\sin x = 0.8$, $0 < x < \frac{\pi}{2}$

42. $\sin 2x$, given $\sin x = 0.6$, $\frac{\pi}{2} < x < \pi$

Graph each expression and use the graph to make a conjecture, predicting what might be an identity. Then verify your conjecture algebraically.

43. $-\dfrac{\sin 2x + \sin x}{\cos 2x - \cos x}$

44. $\dfrac{1 - \cos 2x}{\sin 2x}$

45. $\dfrac{\sin x}{1 - \cos x}$

46. $\dfrac{\cos x \sin 2x}{1 + \cos 2x}$

47. $\dfrac{2(\sin x - \sin^3 x)}{\cos x}$

48. $\csc x - \cot x$

Verify that each equation is an identity.

49. $\sin^2 x - \sin^2 y = \cos^2 y - \cos^2 x$

50. $2 \cos^3 x - \cos x = \dfrac{\cos^2 x - \sin^2 x}{\sec x}$

51. $\dfrac{\sin^2 x}{2 - 2 \cos x} = \cos^2 \dfrac{x}{2}$

52. $\dfrac{\sin 2x}{\sin x} = \dfrac{2}{\sec x}$

53. $2 \cos A - \sec A = \cos A - \dfrac{\tan A}{\csc A}$

54. $\dfrac{2 \tan B}{\sin 2B} = \sec^2 B$

55. $1 + \tan^2 \alpha = 2 \tan \alpha \csc 2\alpha$

56. $\dfrac{2 \cot x}{\tan 2x} = \csc^2 x - 2$

57. $\tan \theta \sin 2\theta = 2 - 2 \cos^2 \theta$

58. $\csc A \sin 2A - \sec A = \cos 2A \sec A$

59. $2 \tan x \csc 2x - \tan^2 x = 1$

60. $2 \cos^2 \theta - 1 = \dfrac{1 - \tan^2 \theta}{1 + \tan^2 \theta}$

61. $\tan \theta \cos^2 \theta = \dfrac{2 \tan \theta \cos^2 \theta - \tan \theta}{1 - \tan^2 \theta}$

62. $\sec^2 \alpha - 1 = \dfrac{\sec 2\alpha - 1}{\sec 2\alpha + 1}$

63. $\dfrac{\sin^2 x - \cos^2 x}{\csc x} = 2 \sin^3 x - \sin x$

64. $\sin^3 \theta = \sin \theta - \cos^2 \theta \sin \theta$

65. $\tan 4\theta = \dfrac{2 \tan 2\theta}{2 - \sec^2 2\theta}$

66. $2 \cos^2 \dfrac{x}{2} \tan x = \tan x + \sin x$

67. $\tan\left(\dfrac{x}{2} + \dfrac{\pi}{4}\right) = \sec x + \tan x$

68. $\dfrac{1}{2} \cot \dfrac{x}{2} - \dfrac{1}{2} \tan \dfrac{x}{2} = \cot x$

69. $-\cot \dfrac{x}{2} = \dfrac{\sin 2x + \sin x}{\cos 2x - \cos x}$

70. $\dfrac{\sin 3t + \sin 2t}{\sin 3t - \sin 2t} = \dfrac{\tan \frac{5t}{2}}{\tan \frac{t}{2}}$

(Modeling) Solve each problem.

71. ***Distance Traveled by a Stone*** The distance D of an object thrown (or projected) from height h (feet) at angle θ with initial velocity v is modeled by the formula

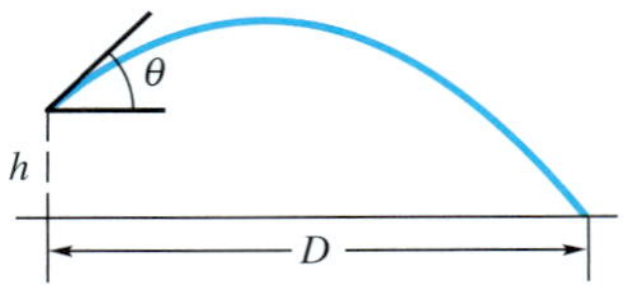

$$D = \frac{v^2 \sin\theta\cos\theta + v\cos\theta\sqrt{(v\sin\theta)^2 + 64h}}{32}.$$

See the figure. (*Source:* Kreighbaum, E. and K. Barthels, *Biomechanics,* Allyn & Bacon.)

(a) Find D when $h = 0$—that is, when the object is projected from the ground.

(b) Suppose a car driving over loose gravel kicks up a small stone at a velocity of 36 ft per sec (about 25 mph) and an angle $\theta = 30°$. How far will the stone travel?

72. ***Amperage, Wattage, and Voltage*** Suppose that for an electric heater, voltage is given by $V = a \sin 2\pi\omega t$ and amperage by $I = b \sin 2\pi\omega t$, where t is time in seconds.

(a) Find the period of the graph for the voltage.

(b) Show that the graph of the wattage $W = VI$ will have half the period of the voltage. Interpret this result.

Chapter 5 Test

1. If $\cos\theta = \frac{24}{25}$ and θ is in quadrant IV, find the five remaining trigonometric function values of θ.

2. Express $\sec\theta - \sin\theta\tan\theta$ as a single function of θ.

3. Express $\tan^2 x - \sec^2 x$ in terms of $\sin x$ and $\cos x$, and simplify.

4. Find the exact value of $\cos\frac{5\pi}{12}$.

5. Express as a function of x alone.

(a) $\cos(270° - x)$ **(b)** $\tan(\pi + x)$

6. Use a half-angle identity to find the exact value of $\sin(-22.5°)$.

7. Graph $y = \cot\frac{1}{2}x - \cot x$, and use the graph to make a conjecture, predicting what might be an identity. Then verify your conjecture algebraically.

8. Given that $\sin A = \frac{5}{13}$, $\cos B = -\frac{3}{5}$, A is a quadrant I angle, and B is a quadrant II angle, find each of the following.

(a) $\sin(A + B)$ **(b)** $\cos(A + B)$ **(c)** $\tan(A - B)$ **(d)** the quadrant of $A + B$

9. Given that $\cos\theta = -\frac{3}{5}$ and $90° < \theta < 180°$, find each of the following.

(a) $\cos 2\theta$ **(b)** $\sin 2\theta$ **(c)** $\tan 2\theta$ **(d)** $\cos\frac{\theta}{2}$ **(e)** $\tan\frac{\theta}{2}$

Verify each identity.

10. $\sec^2 B = \dfrac{1}{1 - \sin^2 B}$

11. $\cos 2A = \dfrac{\cot A - \tan A}{\csc A \sec A}$

12. $\dfrac{\sin 2x}{\cos 2x + 1} = \tan x$

13. $\tan^2 x - \sin^2 x = (\tan x \sin x)^2$

14. $\dfrac{\tan x - \cot x}{\tan x + \cot x} = 2\sin^2 x - 1$

15. ***(Modeling) Voltage*** The voltage in common household current is expressed as $V = 163 \sin \omega t$, where ω is the angular speed (in radians per second) of the generator at the electrical plant and t is time (in seconds).

(a) Use an identity to express V in terms of cosine.

(b) If $\omega = 120\pi$, what is the maximum voltage? Give the least positive value of t when the maximum voltage occurs.

6 Inverse Circular Functions and Trigonometric Equations

Sound waves, such as those initiated by musical instruments, travel in sinusoidal patterns that can be graphed as sine or cosine functions and described by *trigonometric equations*.

6.1 Inverse Circular Functions

Inverse Functions ***Recall that for a function f, every element x in the domain corresponds to one and only one element y, or $f(x)$, in the range.*** **(See Appendix C.)** This means the following:

1. If point (a, b) lies on the graph of f, then there is no other point on the graph that has a as first coordinate.
2. Other points may have b as second coordinate, however, since the definition of function allows range elements to be used more than once.

If a function is defined so that ***each range element is used only once,*** then it is called a **one-to-one function.** For example, the function

$$f(x) = x^3 \text{ is a one-to-one function}$$

because every real number has exactly one real cube root. However,

$$g(x) = x^2 \text{ is not a one-to-one function}$$

because $g(2) = 4$ and $g(-2) = 4$. There are two domain elements, 2 and -2, that correspond to the range element 4.

The **horizontal line test** helps determine graphically whether a function is one-to-one.

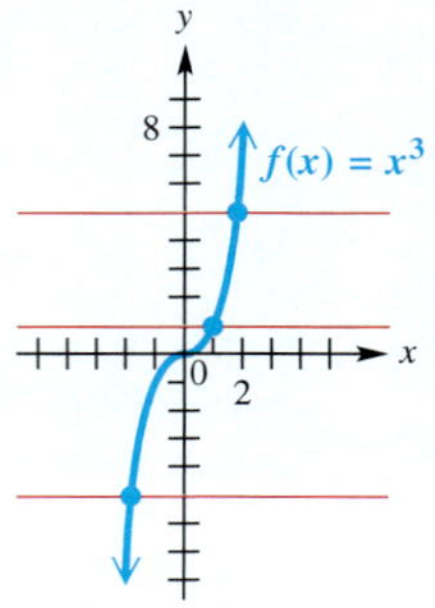

$f(x) = x^3$ is a one-to-one function. It satisfies the conditions of the horizontal line test.

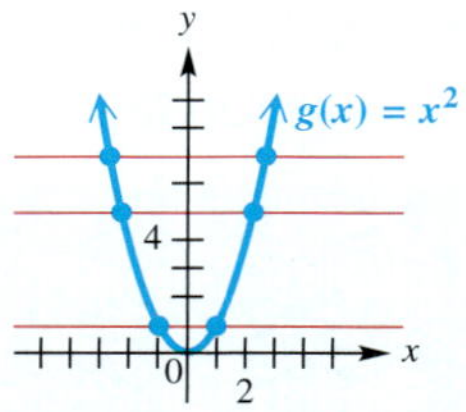

$g(x) = x^2$ is not one-to-one. It does not satisfy the conditions of the horizontal line test.

Figure 1

Horizontal Line Test

A function is one-to-one if every horizontal line intersects the graph of the function at most once.

This test is applied to the graphs of $f(x) = x^3$ and $g(x) = x^2$ in **Figure 1.**

By interchanging the components of the ordered pairs of a one-to-one function f, we obtain a new set of ordered pairs that satisfies the definition of a function. This new function is called the *inverse function,* or *inverse,* of f.

Inverse Function

The **inverse function** of the one-to-one function f is defined as follows.

$$f^{-1} = \{(y, x) \mid (x, y) \text{ belongs to } f\}$$

The special notation used for inverse functions is f^{-1} (read **"f-inverse"**). In simple terms, it represents the function created by interchanging the input (domain) and the output (range) of a one-to-one function.

CAUTION ***Do not confuse the -1 in f^{-1} with a negative exponent.*** The symbol $f^{-1}(x)$ does *not* represent $\frac{1}{f(x)}$. It represents the inverse function of f.

The following statements summarize the concepts of inverse functions.

Summary of Inverse Functions

1. In a one-to-one function, each x-value corresponds to only one y-value and each y-value corresponds to only one x-value.
2. If a function f is one-to-one, then f has an inverse function f^{-1}.
3. The domain of f is the range of f^{-1}, and the range of f is the domain of f^{-1}. That is, if the point (a, b) is on the graph of f, then (b, a) is on the graph of f^{-1}.
4. The graphs of f and f^{-1} are reflections of each other across the line $y = x$.
5. To find $f^{-1}(x)$ from $f(x)$, follow these steps.

   Step 1 Replace $f(x)$ with y and interchange x and y.

   Step 2 Solve for y.

   Step 3 Replace y with $f^{-1}(x)$.

Figure 2 illustrates some of these concepts.

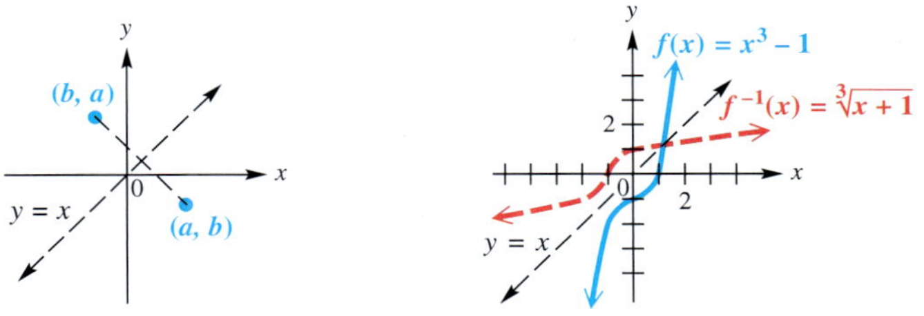

(b, a) is the reflection of (a, b) across the line $y = x$.

The graph of f^{-1} is the reflection of the graph of f across the line $y = x$.

Figure 2

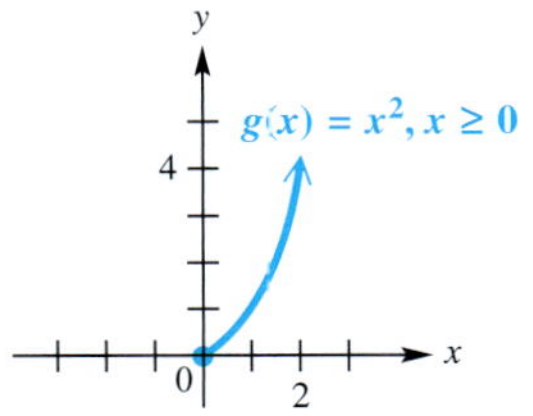

This is a one-to-one function.

Figure 3

We often restrict the domain of a function that is not one-to-one to make it one-to-one, without changing the range. For example, we saw in **Figure 1** that $g(x) = x^2$, with its natural domain $(-\infty, \infty)$, is not one-to-one. However, if we restrict its domain to the set of nonnegative numbers $[0, \infty)$, we obtain a new function f that is one-to-one and has the same range as g, $[0, \infty)$. See **Figure 3.**

NOTE We could have chosen to restrict the domain of $g(x) = x^2$ to $(-\infty, 0]$ to obtain a different one-to-one function. For the trigonometric functions, such choices are made based on general agreement by mathematicians.

LOOKING AHEAD TO CALCULUS

The **inverse circular functions** are used in calculus to solve certain types of related-rates problems and to integrate certain rational functions.

Inverse Sine Function Refer to the graph of the sine function in **Figure 4** on the next page. Applying the horizontal line test, we see that $y = \sin x$ does not define a one-to-one function. If we restrict the domain to the interval $\left[-\frac{\pi}{2}, \frac{\pi}{2}\right]$, which is the part of the graph in **Figure 4** shown in color, this restricted function is one-to-one and has an inverse function. The range of $y = \sin x$ is $[-1, 1]$, so the domain of the inverse function will be $[-1, 1]$, and its range will be $\left[-\frac{\pi}{2}, \frac{\pi}{2}\right]$.

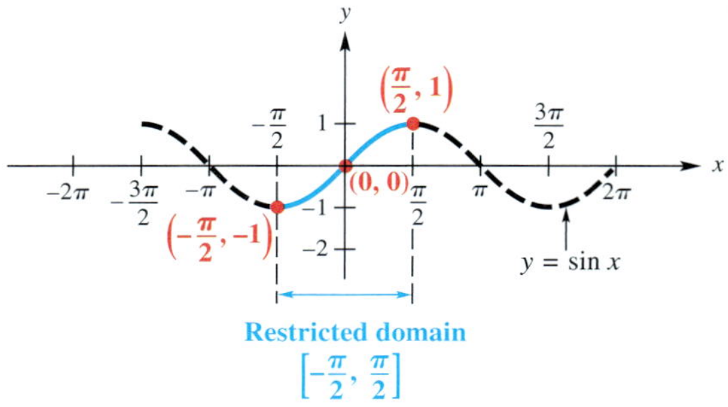

Figure 4

Reflecting the graph of $y = \sin x$ on the restricted domain, shown in **Figure 5(a),** across the line $y = x$ gives the graph of the inverse function, shown in **Figure 5(b).** Some key points are labeled on the graph. The equation of the inverse of $y = \sin x$ is found by interchanging x and y to get

$$x = \sin y.$$

This equation is solved for y by writing

$$\boldsymbol{y = \sin^{-1} x} \quad \text{(read "}\textbf{inverse sine of } \boldsymbol{x}\text{")}.$$

As **Figure 5(b)** shows, the domain of $y = \sin^{-1} x$ is $[-1, 1]$, while the restricted domain of $y = \sin x$, $\left[-\frac{\pi}{2}, \frac{\pi}{2}\right]$, is the range of $y = \sin^{-1} x$. An alternative notation for $\sin^{-1} x$ is $\arcsin x$.

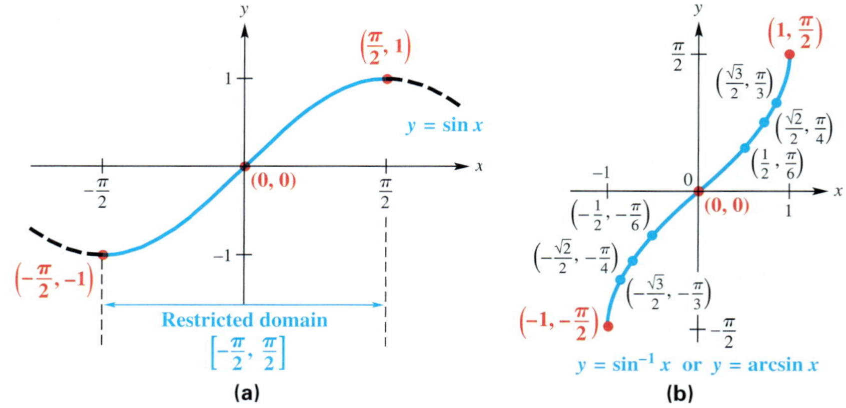

Figure 5

Inverse Sine Function

$\boldsymbol{y = \sin^{-1} x}$ or $\boldsymbol{y = \arcsin x}$ means that $x = \sin y$, for $-\frac{\pi}{2} \le y \le \frac{\pi}{2}$.

We can think of $\boldsymbol{y = \sin^{-1} x}$ *or* $\boldsymbol{y = \arcsin x}$ *as*

"y is the number (angle) in the interval $\left[-\frac{\pi}{2}, \frac{\pi}{2}\right]$ *whose sine is x."*

Thus, we can write $y = \sin^{-1} x$ as $\sin y = x$ to evaluate it. We must pay close attention to the domain and range intervals.

EXAMPLE 1 Finding Inverse Sine Values

Find y in each equation.

(a) $y = \arcsin \frac{1}{2}$ **(b)** $y = \sin^{-1}(-1)$ **(c)** $y = \sin^{-1}(-2)$

ALGEBRAIC SOLUTION

(a) The graph of the function defined by $y = \arcsin x$ (**Figure 5(b)**) includes the point $\left(\frac{1}{2}, \frac{\pi}{6}\right)$. Therefore, $\arcsin \frac{1}{2} = \frac{\pi}{6}$.

Alternatively, we can think of $y = \arcsin \frac{1}{2}$ as "y is the number in $\left[-\frac{\pi}{2}, \frac{\pi}{2}\right]$ whose sine is $\frac{1}{2}$." Then we can write the given equation as $\sin y = \frac{1}{2}$. Since $\sin \frac{\pi}{6} = \frac{1}{2}$ and $\frac{\pi}{6}$ is in the range of the arcsine function, $y = \frac{\pi}{6}$.

(b) Writing the equation $y = \sin^{-1}(-1)$ in the form $\sin y = -1$ shows that $y = -\frac{\pi}{2}$. Notice that the point $\left(-1, -\frac{\pi}{2}\right)$ is on the graph of $y = \sin^{-1} x$.

(c) Because -2 is not in the domain of the inverse sine function, $\sin^{-1}(-2)$ does not exist.

GRAPHING CALCULATOR SOLUTION

We graph the equation $Y_1 = \sin^{-1} X$ and find the points with X-values $\frac{1}{2} = 0.5$ and -1. For these two X-values, **Figure 6** indicates that $Y = \frac{\pi}{6} \approx 0.52359878$ and $Y = -\frac{\pi}{2} \approx -1.570796$.

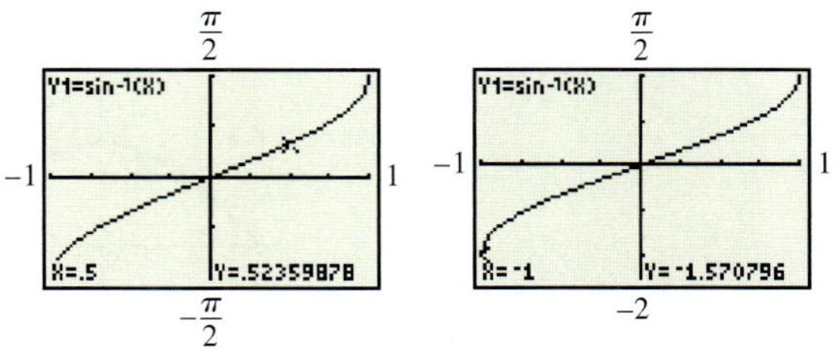

Figure 6

Since $\sin^{-1}(-2)$ does not exist, a calculator will give an error message for this input.

✔ *Now Try Exercises 13, 21, and 25.*

CAUTION In **Example 1(b),** it is tempting to give the value of $\sin^{-1}(-1)$ as $\frac{3\pi}{2}$, since $\sin \frac{3\pi}{2} = -1$. Notice, however, that $\frac{3\pi}{2}$ is not in the range of the inverse sine function. ***Be certain that the number given for an inverse function value is in the range of the particular inverse function being considered.***

We summarize this discussion about the inverse sine function as follows.

Inverse Sine Function $y = \sin^{-1} x$ or $y = \arcsin x$

Domain: $[-1, 1]$ Range: $\left[-\frac{\pi}{2}, \frac{\pi}{2}\right]$

x	y
-1	$-\frac{\pi}{2}$
$-\frac{\sqrt{2}}{2}$	$-\frac{\pi}{4}$
0	0
$\frac{\sqrt{2}}{2}$	$\frac{\pi}{4}$
1	$\frac{\pi}{2}$

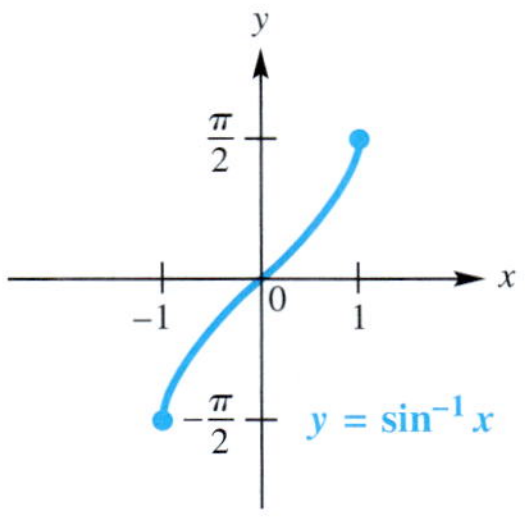

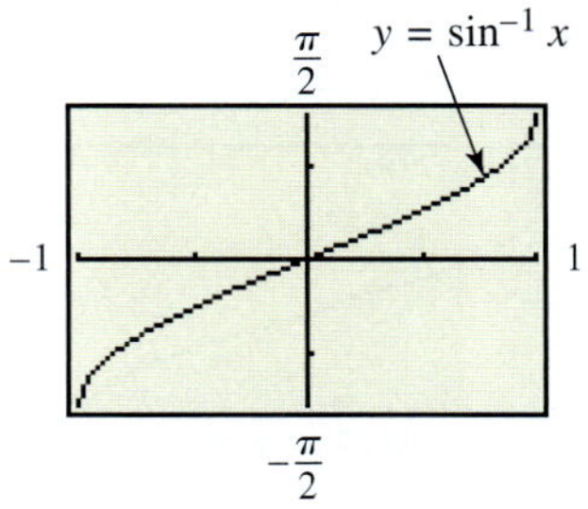

Figure 7

- The inverse sine function is increasing and continuous on its domain $[-1, 1]$.
- Its x-intercept is 0, and its y-intercept is 0.
- Its graph is symmetric with respect to the origin, so the function is an odd function. For all x in the domain, $\sin^{-1}(-x) = -\sin^{-1} x$.

Inverse Cosine Function The function

$$y = \cos^{-1} x \quad (\text{or } y = \arccos x)$$

is defined by restricting the domain of the function $y = \cos x$ to the interval $[0, \pi]$ as in **Figure 8.** This restricted function, which is the part of the graph in **Figure 8** shown in color, is one-to-one and has an inverse function. The inverse function, $y = \cos^{-1} x$, is found by interchanging the roles of x and y. Reflecting the graph of $y = \cos x$ across the line $y = x$ gives the graph of the inverse function shown in **Figure 9.** Some key points are shown on the graph.

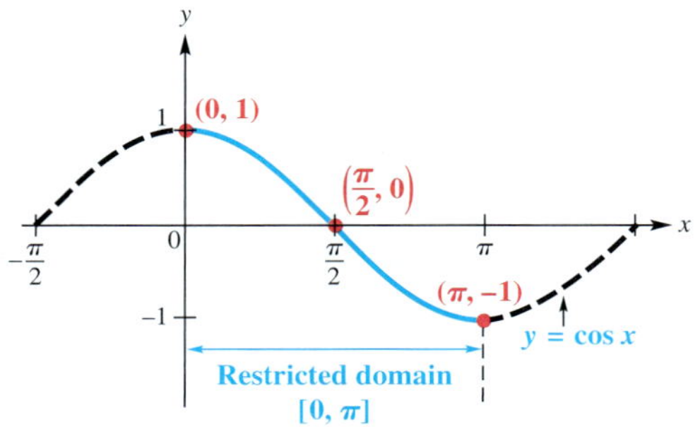

Figure 8

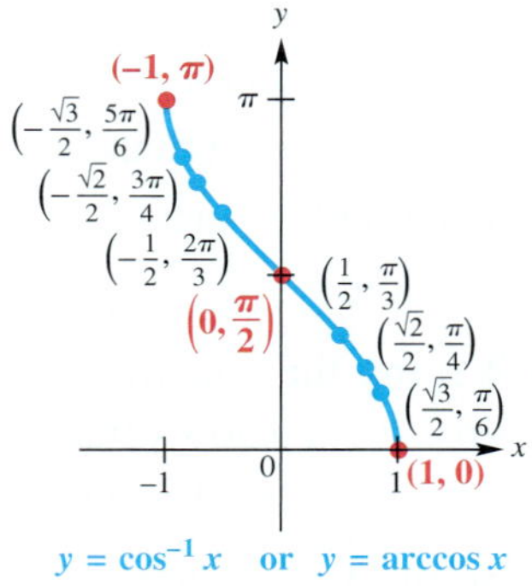

Figure 9

Inverse Cosine Function

$y = \cos^{-1} x$ or $y = \arccos x$ means that $x = \cos y$, for $0 \le y \le \pi$.

We can think of $y = \cos^{-1} x$ or $y = \arccos x$ as

"y is the number (angle) in the interval $[0, \pi]$ whose cosine is x."

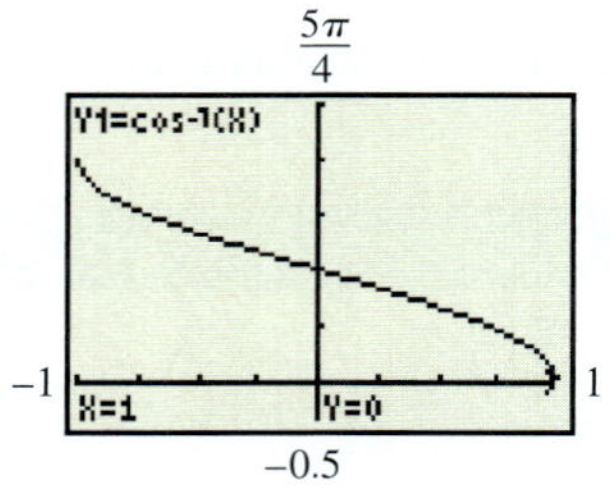

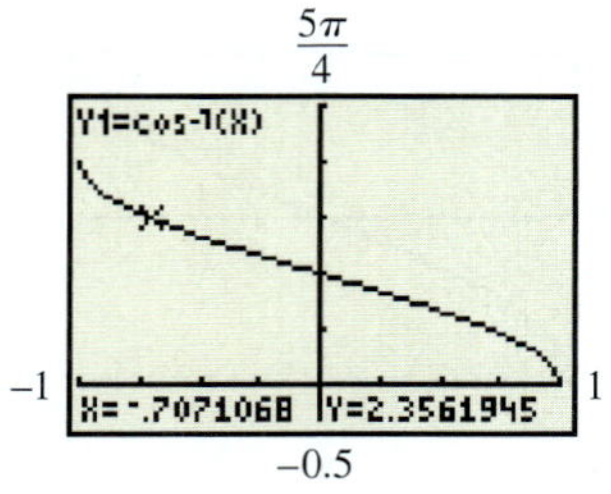

These screens support the results of **Example 2** because $-\frac{\sqrt{2}}{2} \approx -0.7071068$ and $\frac{3\pi}{4} \approx 2.3561945$.

EXAMPLE 2 **Finding Inverse Cosine Values**

Find y in each equation.

(a) $y = \arccos 1$ **(b)** $y = \cos^{-1}\left(-\frac{\sqrt{2}}{2}\right)$

SOLUTION

(a) Since the point $(1, 0)$ lies on the graph of $y = \arccos x$ in **Figure 9,** the value of y, or arccos 1, is 0. Alternatively, we can think of $y = \arccos 1$ as

"y is the number in $[0, \pi]$ whose cosine is 1," or $\cos y = 1$.

Thus, $y = 0$, since $\cos 0 = 1$ and 0 is in the range of the arccosine function.

(b) We must find the value of y that satisfies

$$\cos y = -\frac{\sqrt{2}}{2}, \quad \text{where } y \text{ is in the interval } [0, \pi],$$

which is the range of the function $y = \cos^{-1} x$. The only value for y that satisfies these conditions is $\frac{3\pi}{4}$. Again, this can be verified from the graph in **Figure 9.**

✔ *Now Try Exercises 15 and 23.*

Our observations about the inverse cosine function lead to the following generalizations.

Inverse Cosine Function $y = \cos^{-1} x$ or $y = \arccos x$

Domain: $[-1, 1]$ Range: $[0, \pi]$

x	y
-1	π
$-\frac{\sqrt{2}}{2}$	$\frac{3\pi}{4}$
0	$\frac{\pi}{2}$
$\frac{\sqrt{2}}{2}$	$\frac{\pi}{4}$
1	0

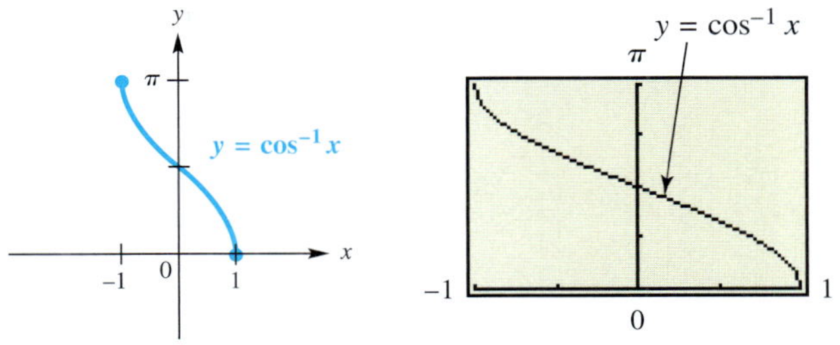

Figure 10

- The inverse cosine function is decreasing and continuous on its domain $[-1, 1]$.
- Its x-intercept is 1, and its y-intercept is $\frac{\pi}{2}$.
- Its graph is not symmetric with respect to either the y-axis or the origin.

Inverse Tangent Function Restricting the domain of the function $y = \tan x$ to the open interval $\left(-\frac{\pi}{2}, \frac{\pi}{2}\right)$ yields a one-to-one function. By interchanging the roles of x and y, we obtain the inverse tangent function given by $y = \tan^{-1} x$ or $y = \arctan x$. **Figure 11** shows the graph of the restricted tangent function. **Figure 12** gives the graph of $y = \tan^{-1} x$.

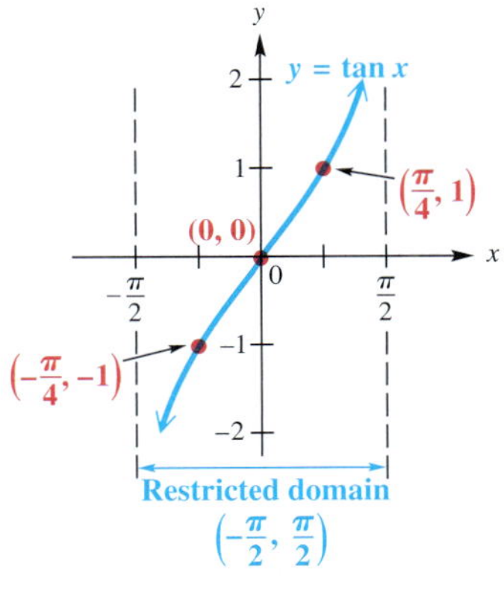

Figure 11

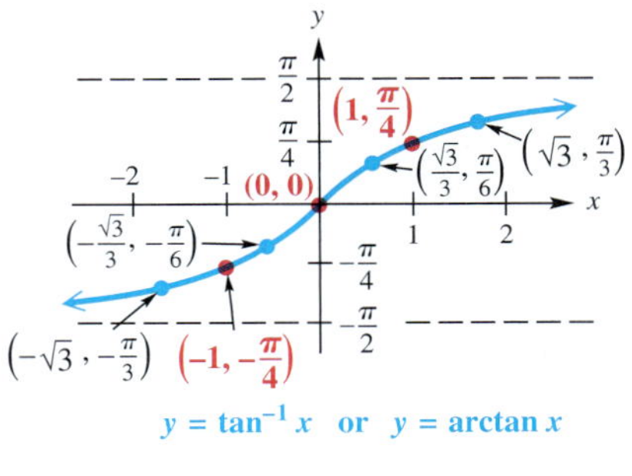

Figure 12

Inverse Tangent Function

$y = \tan^{-1} x$ or $y = \arctan x$ means that $x = \tan y$, for $-\frac{\pi}{2} < y < \frac{\pi}{2}$.

We can think of $y = \tan^{-1} x$ *or* $y = \arctan x$ *as*

"y is the number (angle) in the interval $\left(-\frac{\pi}{2}, \frac{\pi}{2}\right)$ *whose tangent is x."*

We summarize this discussion about the inverse tangent function as follows.

Inverse Tangent Function $y = \tan^{-1} x$ or $y = \arctan x$

Domain: $(-\infty, \infty)$ Range: $\left(-\frac{\pi}{2}, \frac{\pi}{2}\right)$

x	y
-1	$-\frac{\pi}{4}$
$-\frac{\sqrt{3}}{3}$	$-\frac{\pi}{6}$
0	0
$\frac{\sqrt{3}}{3}$	$\frac{\pi}{6}$
1	$\frac{\pi}{4}$

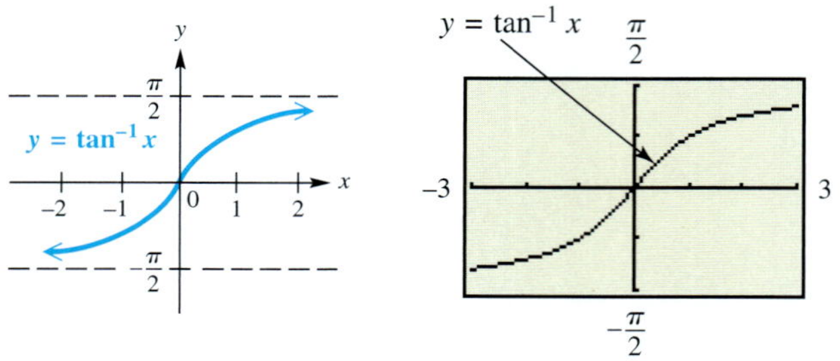

Figure 13

- The inverse tangent function is increasing and continuous on its domain $(-\infty, \infty)$.
- Its x-intercept is 0, and its y-intercept is 0.
- Its graph is symmetric with respect to the origin so the function is an odd function. For all x in the domain, $\tan^{-1}(-x) = -\tan^{-1} x$.
- The lines $y = \frac{\pi}{2}$ and $y = -\frac{\pi}{2}$ are horizontal asymptotes.

Remaining Inverse Circular Functions The remaining three inverse trigonometric functions are defined similarly. Their graphs are shown in **Figure 14.**

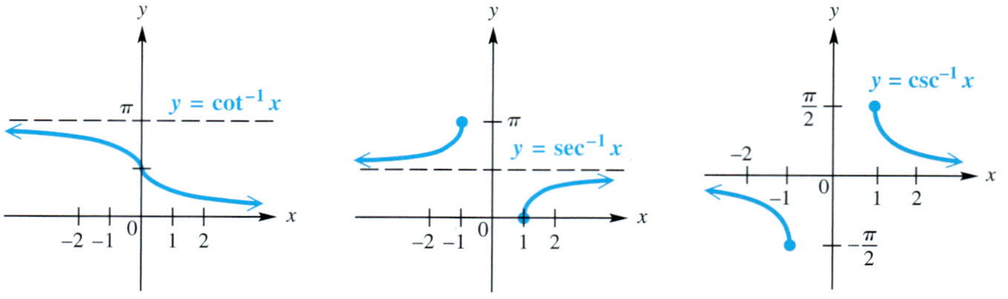

(a) **(b)** **(c)**

Figure 14

Inverse Cotangent, Secant, and Cosecant Functions*

$y = \cot^{-1} x$ or $y = \text{arccot}\, x$ means that $x = \cot y$, for $0 < y < \pi$.

$y = \sec^{-1} x$ or $y = \text{arcsec}\, x$ means that $x = \sec y$, for $0 \le y \le \pi$, $y \ne \frac{\pi}{2}$.

$y = \csc^{-1} x$ or $y = \text{arccsc}\, x$ means that $x = \csc y$, for $-\frac{\pi}{2} \le y \le \frac{\pi}{2}$, $y \ne 0$.

*The inverse secant and inverse cosecant functions are sometimes defined with different ranges. We use intervals that match those of the inverse cosine and inverse sine functions, respectively (except for one missing point).

The table gives all six inverse trigonometric functions with their domains and ranges.

Inverse Function	Domain	Range: Interval	Range: Quadrants of the Unit Circle
$y = \sin^{-1} x$	$[-1, 1]$	$\left[-\frac{\pi}{2}, \frac{\pi}{2}\right]$	I and IV
$y = \cos^{-1} x$	$[-1, 1]$	$[0, \pi]$	I and II
$y = \tan^{-1} x$	$(-\infty, \infty)$	$\left(-\frac{\pi}{2}, \frac{\pi}{2}\right)$	I and IV
$y = \cot^{-1} x$	$(-\infty, \infty)$	$(0, \pi)$	I and II
$y = \sec^{-1} x$	$(-\infty, -1] \cup [1, \infty)$	$\left[0, \frac{\pi}{2}\right) \cup \left(\frac{\pi}{2}, \pi\right]$	I and II
$y = \csc^{-1} x$	$(-\infty, -1] \cup [1, \infty)$	$\left[-\frac{\pi}{2}, 0\right) \cup \left(0, \frac{\pi}{2}\right]$	I and IV

Inverse Function Values The inverse circular functions are formally defined with real number ranges. However, there are times when it may be convenient to find degree-measured angles equivalent to these real number values. It is also often convenient to think in terms of the unit circle and choose the inverse function values on the basis of the quadrants given in the preceding table.

EXAMPLE 3 Finding Inverse Function Values (Degree-Measured Angles)

Find the *degree measure* of θ in the following.

(a) $\theta = \arctan 1$ **(b)** $\theta = \sec^{-1} 2$

SOLUTION

(a) Here θ must be in $(-90°, 90°)$, but since 1 is positive, θ must be in quadrant I. The alternative statement, $\tan \theta = 1$, leads to $\theta = 45°$.

(b) Write the equation as $\sec \theta = 2$. For $\sec^{-1} x$, θ is in quadrant I or II. Because 2 is positive, θ is in quadrant I and $\theta = 60°$, since $\sec 60° = 2$. Note that $60°$ $\left(\text{the degree equivalent of } \frac{\pi}{3}\right)$ is in the range of the inverse secant function.

✔ *Now Try Exercises 37 and 45.*

The inverse trigonometric function keys on a calculator give correct results for the inverse sine, inverse cosine, and inverse tangent functions.

$$\sin^{-1} 0.5 = 30°, \qquad \sin^{-1}(-0.5) = -30°,$$
$$\tan^{-1}(-1) = -45°, \quad \text{and} \quad \cos^{-1}(-0.5) = 120°$$

Degree mode

However, finding $\cot^{-1} x$, $\sec^{-1} x$, and $\csc^{-1} x$ with a calculator is not as straightforward, because these functions must first be expressed in terms of $\tan^{-1} x$, $\cos^{-1} x$, and $\sin^{-1} x$, respectively. If $y = \sec^{-1} x$, for example, then $\sec y = x$, which must be written in terms of cosine as follows.

$$\text{If } \sec y = x, \quad \text{then} \quad \frac{1}{\cos y} = x, \quad \text{or} \quad \cos y = \frac{1}{x}, \quad \text{and} \quad y = \cos^{-1} \frac{1}{x}.$$

Use the following to evaluate these inverse trigonometric functions on a calculator.

$\sec^{-1} x$ *can be evaluated as* $\cos^{-1} \frac{1}{x}$; $\csc^{-1} x$ *can be evaluated as* $\sin^{-1} \frac{1}{x}$;

$$\cot^{-1} x \textit{ can be evaluated as } \begin{cases} \tan^{-1} \frac{1}{x} & \text{if } x > 0 \\ 180° + \tan^{-1} \frac{1}{x} & \text{if } x < 0. \end{cases}$$

Degree mode

EXAMPLE 4 Finding Inverse Function Values with a Calculator

Use a calculator to give each value.

(a) Find y in radians if $y = \csc^{-1}(-3)$.

(b) Find θ in degrees if $\theta = \text{arccot}(-0.3541)$.

SOLUTION

(a) With the calculator in radian mode, enter $\csc^{-1}(-3)$ as $\sin^{-1}\left(\frac{1}{-3}\right)$ to get $y \approx -0.3398369095$. See **Figure 15.**

(b) Now set the calculator to degree mode. A calculator gives the inverse tangent value of a negative number as a quadrant IV angle. The restriction on the range of arccotangent implies that θ must be in quadrant II, so enter

$$\text{arccot}(-0.3541) \quad \text{as} \quad \tan^{-1}\left(\frac{1}{-0.3541}\right) + 180°.$$

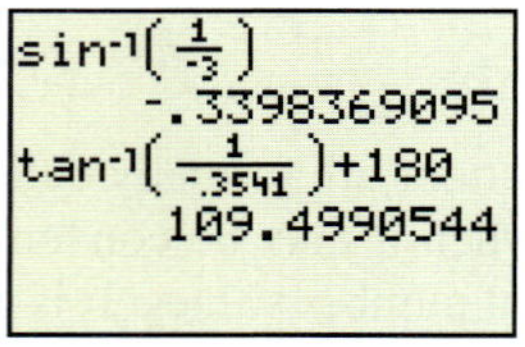

Figure 15

As shown in **Figure 15,**

$$\theta \approx 109.4990544°.$$

✔ *Now Try Exercises 53 and 65.*

CAUTION ***Be careful when using your calculator to evaluate the inverse cotangent of a negative quantity.*** To do this, we must enter the inverse tangent of the *reciprocal* of the negative quantity, which returns an angle in quadrant IV. Since inverse cotangent is negative in quadrant II, adjust your calculator result by adding 180° or π accordingly. Note that $\cot^{-1} 0 = \frac{\pi}{2}$.

EXAMPLE 5 Finding Function Values Using Definitions of the Trigonometric Functions

Evaluate each expression without using a calculator.

(a) $\sin\left(\tan^{-1} \frac{3}{2}\right)$ **(b)** $\tan\left(\cos^{-1}\left(-\frac{5}{13}\right)\right)$

SOLUTION

(a) Let $\theta = \tan^{-1} \frac{3}{2}$, so $\tan \theta = \frac{3}{2}$. The inverse tangent function yields values only in quadrants I and IV, and since $\frac{3}{2}$ is positive, θ is in quadrant I. Sketch θ in quadrant I, and label a triangle, as shown in **Figure 16** on the next page. By the Pythagorean theorem, the hypotenuse is $\sqrt{13}$. The value of sine is the quotient of the side opposite and the hypotenuse.

$$\sin\left(\tan^{-1} \frac{3}{2}\right) = \sin \theta = \frac{3}{\sqrt{13}} = \frac{3}{\sqrt{13}} \cdot \frac{\sqrt{13}}{\sqrt{13}} = \frac{3\sqrt{13}}{13} \quad \text{(Section 2.1)}$$

Rationalize the denominator.

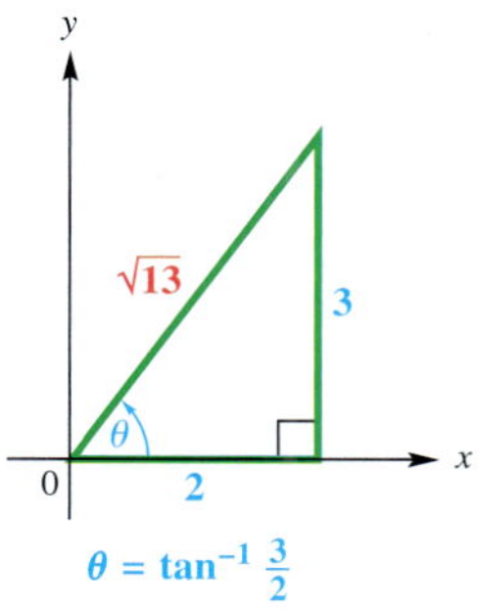

Figure 16

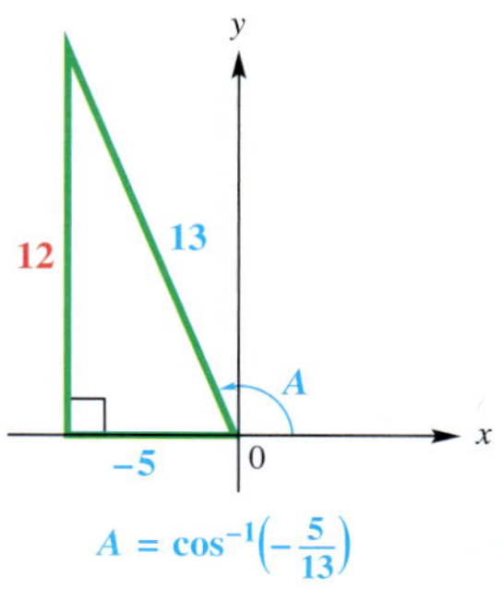

Figure 17

(b) Let $A = \cos^{-1}\left(-\frac{5}{13}\right)$. Then, $\cos A = -\frac{5}{13}$. Since $\cos^{-1} x$ for a negative value of x is in quadrant II, sketch A in quadrant II, as shown in **Figure 17.**

$$\tan\left(\cos^{-1}\left(-\frac{5}{13}\right)\right) = \tan A = -\frac{12}{5}$$

✔ ***Now Try Exercises 79 and 81.***

EXAMPLE 6 Finding Function Values Using Identities

Evaluate each expression without using a calculator.

(a) $\cos\left(\arctan\sqrt{3} + \arcsin\frac{1}{3}\right)$ **(b)** $\tan\left(2\arcsin\frac{2}{5}\right)$

SOLUTION

(a) Let $A = \arctan\sqrt{3}$ and $B = \arcsin\frac{1}{3}$, so $\tan A = \sqrt{3}$ and $\sin B = \frac{1}{3}$. Sketch both A and B in quadrant I, as shown in **Figure 18,** and use the Pythagorean theorem to find the unknown side in each triangle. Then, use the cosine sum identity.

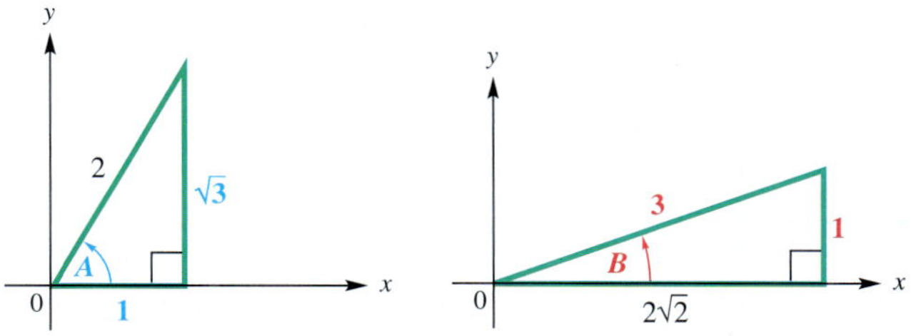

Figure 18

$$\cos\left(\arctan\sqrt{3} + \arcsin\frac{1}{3}\right)$$

$= \cos(A + B)$ Let $A = \arctan\sqrt{3}$ and $B = \arcsin\frac{1}{3}$.

$= \cos A \cos B - \sin A \sin B$ Cosine sum identity **(Section 5.3)**

$= \frac{1}{2} \cdot \frac{2\sqrt{2}}{3} - \frac{\sqrt{3}}{2} \cdot \frac{1}{3}$ Substitute values using **Figure 18.**

$= \frac{2\sqrt{2} - \sqrt{3}}{6}$ Multiply and write as a single fraction.

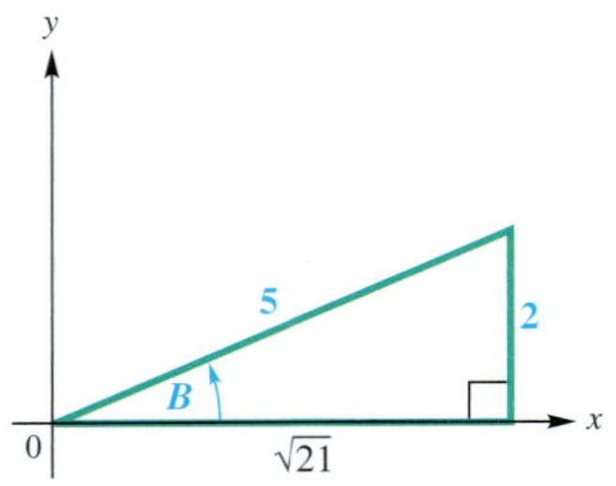

Figure 19

(b) Let $B = \arcsin \frac{2}{5}$, so that $\sin B = \frac{2}{5}$. Sketch angle B in quadrant I, and find the length of the third side of the triangle. Then, use the double-angle tangent identity $\tan 2B = \frac{2 \tan B}{1 - \tan^2 B}$.

$$\tan\left(2 \arcsin \tfrac{2}{5}\right)$$

$$= \frac{2\left(\frac{2}{\sqrt{21}}\right)}{1 - \left(\frac{2}{\sqrt{21}}\right)^2} \quad \tan B = \tfrac{2}{\sqrt{21}} \text{ from Figure 19. (Section 5.5)}$$

$$= \frac{\frac{4}{\sqrt{21}}}{1 - \frac{4}{21}} \quad \text{Multiply and apply the exponent.}$$

$$= \frac{\frac{4}{\sqrt{21}} \cdot \frac{\sqrt{21}}{\sqrt{21}}}{\frac{17}{21}} \quad \text{Rationalize in the numerator. Subtract in the denominator.}$$

$$= \frac{\frac{4\sqrt{21}}{21}}{\frac{17}{21}} \quad \text{Multiply in the numerator.}$$

$$= \frac{4\sqrt{21}}{17} \quad \text{Divide; } \frac{\frac{a}{b}}{\frac{c}{d}} = \frac{a}{b} \div \frac{c}{d} = \frac{a}{b} \cdot \frac{d}{c}.$$

✔ *Now Try Exercises 83 and 91.*

While the work shown in **Examples 5 and 6** does not rely on a calculator, we can support our algebraic work with one. By entering $\cos\left(\arctan \sqrt{3} + \arcsin \frac{1}{3}\right)$ from **Example 6(a)** into a calculator, we get the approximation 0.1827293862, the same approximation as when we enter $\frac{2\sqrt{2} - \sqrt{3}}{6}$ (the exact value we obtained algebraically). Similarly, we obtain the same approximation when we evaluate $\tan\left(2 \arcsin \frac{2}{5}\right)$ and $\frac{4\sqrt{21}}{17}$, supporting our answer in **Example 6(b).**

EXAMPLE 7 Writing Function Values in Terms of *u*

Write each trigonometric expression as an algebraic expression in u.

(a) $\sin(\tan^{-1} u)$ **(b)** $\cos(2 \sin^{-1} u)$

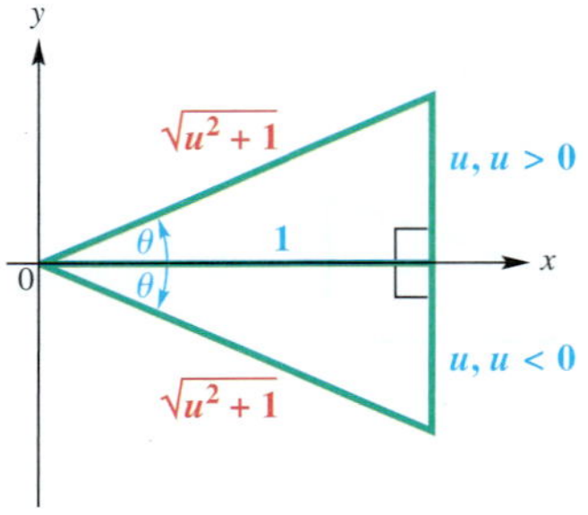

Figure 20

SOLUTION

(a) Let $\theta = \tan^{-1} u$, so $\tan \theta = u$. Here, u may be positive or negative. Since $-\frac{\pi}{2} < \tan^{-1} u < \frac{\pi}{2}$, sketch θ in quadrants I and IV and label two triangles, as shown in **Figure 20.** Because sine is given by the quotient of the side opposite and the hypotenuse, we have the following.

$$\sin(\tan^{-1} u) = \sin \theta = \frac{u}{\sqrt{u^2+1}} = \frac{u}{\sqrt{u^2+1}} \cdot \frac{\sqrt{u^2+1}}{\sqrt{u^2+1}} = \frac{u\sqrt{u^2+1}}{u^2+1}$$

Rationalize the denominator.

The result is positive when u is positive and negative when u is negative.

(b) Let $\theta = \sin^{-1} u$, so $\sin \theta = u$. To find $\cos 2\theta$, use the double-angle identity $\cos 2\theta = 1 - 2 \sin^2 \theta$.

$$\cos(2 \sin^{-1} u) = \cos 2\theta = 1 - 2 \sin^2 \theta = 1 - 2u^2 \quad \text{(Section 5.5)}$$

✔ *Now Try Exercises 99 and 103.*

EXAMPLE 8 **Finding the Optimal Angle of Elevation of a Shot Put**

The optimal angle of elevation θ that a shot-putter should aim for in order to throw the greatest distance depends on the velocity v of the throw and the initial height h of the shot. See **Figure 21.** One model for θ that achieves this greatest distance is

$$\theta = \arcsin\left(\sqrt{\frac{v^2}{2v^2 + 64h}}\right).$$

(*Source:* Townend, M. S., *Mathematics in Sport,* Chichester, Ellis Horwood Limited.)

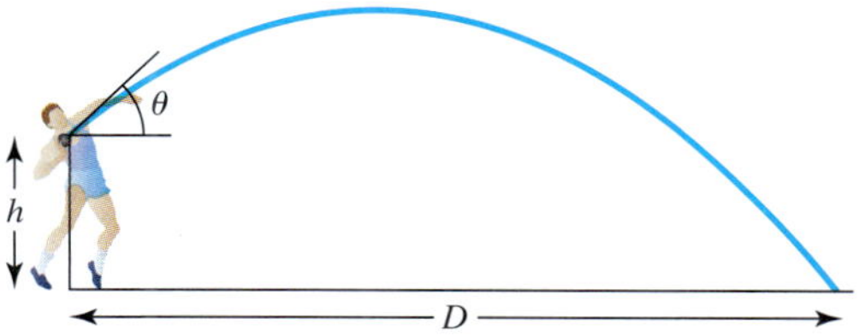

Figure 21

Suppose a shot-putter can consistently throw the steel ball with $h = 6.6$ ft and $v = 42$ ft per sec. At what angle should he release the ball to maximize distance?

SOLUTION To find this angle, substitute and use a calculator in degree mode.

$$\theta = \arcsin\left(\sqrt{\frac{42^2}{2(42^2) + 64(6.6)}}\right) \approx 42^\circ \quad h = 6.6,\ v = 42$$

✔ ***Now Try Exercise 109.***

6.1 Exercises

Concept Check Complete each statement, or answer the question.

1. For a function to have an inverse, it must be ______.

2. The domain of $y = \arcsin x$ equals the ______ of $y = \sin x$.

3. $y = \cos^{-1} x$ means that $x =$ ______, for $0 \leq y \leq \pi$.

4. The point $\left(\frac{\pi}{4}, 1\right)$ lies on the graph of $y = \tan x$. Therefore, the point ______ lies on the graph of ______.

5. If a function f has an inverse and $f(\pi) = -1$, then $f^{-1}(-1) =$ ______.

6. How can the graph of f^{-1} be sketched if the graph of f is known?

Concept Check In Exercises 7–10, write short answers.

7. Consider the inverse sine function, defined by $y = \sin^{-1} x$ or $y = \arcsin x$.

(a) What is its domain?
(b) What is its range?
(c) Is this function increasing or decreasing?
(d) Why is $\arcsin(-2)$ not defined?

8. Consider the inverse cosine function, defined by $y = \cos^{-1} x$, or $y = \arccos x$.

(a) What is its domain?
(b) What is its range?
(c) Is this function increasing or decreasing?
(d) $\arccos\left(-\frac{1}{2}\right) = \frac{2\pi}{3}$. Why is $\arccos\left(-\frac{1}{2}\right)$ not equal to $-\frac{4\pi}{3}$?

9. Consider the inverse tangent function, defined by $y = \tan^{-1} x$, or $y = \arctan x$.

(a) What is its domain?
(b) What is its range?
(c) Is this function increasing or decreasing?
(d) Is there any real number x for which $\arctan x$ is not defined? If so, what is it (or what are they)?

10. Give the domain and range of each inverse trigonometric function, as defined in this section.

(a) inverse cosecant function
(b) inverse secant function
(c) inverse cotangent function

11. *Concept Check* Is $\sec^{-1} a$ calculated as $\cos^{-1} \frac{1}{a}$ or as $\frac{1}{\cos^{-1} a}$?

12. *Concept Check* For positive values of a, $\cot^{-1} a$ is calculated as $\tan^{-1} \frac{1}{a}$. How is $\cot^{-1} a$ calculated for negative values of a?

Find the exact value of each real number y if it exists. Do not use a calculator. ***See Examples 1 and 2.***

13. $y = \sin^{-1} 0$
14. $y = \sin^{-1}(-1)$
15. $y = \cos^{-1}(-1)$
16. $y = \arccos 0$
17. $y = \tan^{-1} 1$
18. $y = \arctan(-1)$
19. $y = \arctan 0$
20. $y = \tan^{-1}(-1)$
21. $y = \arcsin\left(-\frac{\sqrt{3}}{2}\right)$
22. $y = \sin^{-1} \frac{\sqrt{2}}{2}$
23. $y = \arccos\left(-\frac{\sqrt{3}}{2}\right)$
24. $y = \cos^{-1}\left(-\frac{1}{2}\right)$
25. $y = \sin^{-1} \sqrt{3}$
26. $y = \arcsin\left(-\sqrt{2}\right)$
27. $y = \cot^{-1}(-1)$
28. $y = \operatorname{arccot}\left(-\sqrt{3}\right)$
29. $y = \csc^{-1}(-2)$
30. $y = \csc^{-1} \sqrt{2}$
31. $y = \operatorname{arcsec} \frac{2\sqrt{3}}{3}$
32. $y = \sec^{-1}\left(-\sqrt{2}\right)$
33. $y = \sec^{-1} 1$
34. $y = \sec^{-1} 0$
35. $y = \csc^{-1} \frac{\sqrt{2}}{2}$
36. $y = \operatorname{arccsc}\left(-\frac{1}{2}\right)$

Give the degree measure of θ if it exists. Do not use a calculator. ***See Example 3.***

37. $\theta = \arctan(-1)$
38. $\theta = \tan^{-1} \sqrt{3}$
39. $\theta = \arcsin\left(-\frac{\sqrt{3}}{2}\right)$
40. $\theta = \arcsin\left(-\frac{\sqrt{2}}{2}\right)$
41. $\theta = \arccos\left(-\frac{1}{2}\right)$
42. $\theta = \sec^{-1}(-2)$
43. $\theta = \cot^{-1}\left(-\frac{\sqrt{3}}{3}\right)$
44. $\theta = \cot^{-1} \frac{\sqrt{3}}{3}$
45. $\theta = \csc^{-1}(-2)$
46. $\theta = \csc^{-1}(-1)$
47. $\theta = \sin^{-1} 2$
48. $\theta = \cos^{-1}(-2)$

Use a calculator to give each value in decimal degrees. ***See Example 4.***

49. $\theta = \sin^{-1}(-0.13349122)$
50. $\theta = \arcsin 0.77900016$
51. $\theta = \arccos(-0.39876459)$
52. $\theta = \cos^{-1}(-0.13348816)$

53. $\theta = \csc^{-1} 1.9422833$

54. $\theta = \cot^{-1} 1.7670492$

55. $\theta = \cot^{-1}(-0.60724226)$

56. $\theta = \cot^{-1}(-2.7733744)$

57. $\theta = \tan^{-1}(-7.7828641)$

58. $\theta = \sec^{-1}(-5.1180378)$

Use a calculator to give each real number value. (Be sure the calculator is in radian mode.) ***See Example 4.***

59. $y = \arcsin 0.92837781$

60. $y = \arcsin 0.81926439$

61. $y = \cos^{-1}(-0.32647891)$

62. $y = \arccos 0.44624593$

63. $y = \arctan 1.1111111$

64. $y = \cot^{-1} 1.0036571$

65. $y = \cot^{-1}(-0.92170128)$

66. $y = \cot^{-1}(-36.874610)$

67. $y = \sec^{-1}(-1.2871684)$

68. $y = \sec^{-1} 4.7963825$

The screen here shows how to define the inverse secant, cosecant, and cotangent functions in order to graph them using a TI-83/84 Plus graphing calculator.

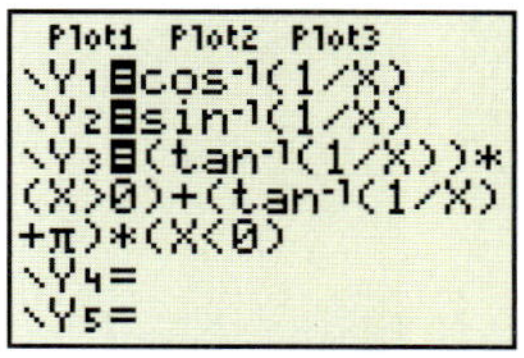

Use this information to graph each inverse circular function and compare your graphs to those in **Figure 14.**

69. $y = \sec^{-1} x$

70. $y = \csc^{-1} x$

71. $y = \cot^{-1} x$

Graph each inverse circular function by hand.

72. $y = \operatorname{arccsc} 2x$

73. $y = \operatorname{arcsec} \frac{1}{2}x$

74. $y = 2 \cot^{-1} x$

75. *Concept Check* Explain why attempting to find $\sin^{-1} 1.003$ on your calculator will result in an error message.

Relating Concepts

For individual or collaborative investigation *(Exercises 76–78)**

76. Consider the function

$$f(x) = 3x - 2 \quad \text{and its inverse} \quad f^{-1}(x) = \frac{1}{3}x + \frac{2}{3}.$$

Simplify $f(f^{-1}(x))$ and $f^{-1}(f(x))$. What do you notice in each case? What would the graph look like in each case?

77. Use a graphing calculator to graph $y = \tan(\tan^{-1} x)$ in the standard viewing window, using radian mode. How does this compare to the graph you described in **Exercise 76?**

78. Use a graphing calculator to graph $y = \tan^{-1}(\tan x)$ in the standard viewing window, using radian and dot modes. Why does this graph not agree with the graph you found in **Exercise 77?**

*The authors wish to thank Carol Walker of Hinds Community College for making a suggestion on which these exercises are based.

Give the exact value of each expression without using a calculator. **See Examples 5 and 6.**

79. $\tan\left(\arccos \frac{3}{4}\right)$

80. $\sin\left(\arccos \frac{1}{4}\right)$

81. $\cos(\tan^{-1}(-2))$

82. $\sec\left(\sin^{-1}\left(-\frac{1}{5}\right)\right)$

83. $\sin\left(2 \tan^{-1} \frac{12}{5}\right)$

84. $\cos\left(2 \sin^{-1} \frac{1}{4}\right)$

85. $\cos\left(2 \arctan \frac{4}{3}\right)$

86. $\tan\left(2 \cos^{-1} \frac{1}{4}\right)$

87. $\sin\left(2 \cos^{-1} \frac{1}{5}\right)$

88. $\cos(2 \tan^{-1}(-2))$

89. $\sec(\sec^{-1} 2)$

90. $\csc(\csc^{-1}\sqrt{2})$

91. $\cos\left(\tan^{-1} \frac{5}{12} - \tan^{-1} \frac{3}{4}\right)$

92. $\cos\left(\sin^{-1} \frac{3}{5} + \cos^{-1} \frac{5}{13}\right)$

93. $\sin\left(\sin^{-1} \frac{1}{2} + \tan^{-1}(-3)\right)$

94. $\tan\left(\cos^{-1} \frac{\sqrt{3}}{2} - \sin^{-1}\left(-\frac{3}{5}\right)\right)$

Use a calculator to find each value. Give answers as real numbers.

95. $\cos(\tan^{-1} 0.5)$

96. $\sin(\cos^{-1} 0.25)$

97. $\tan(\arcsin 0.12251014)$

98. $\cot(\arccos 0.58236841)$

Write each expression as an algebraic (nontrigonometric) expression in u, for $u > 0$. **See Example 7.**

99. $\sin(\arccos u)$

100. $\tan(\arccos u)$

101. $\cos(\arcsin u)$

102. $\cot(\arcsin u)$

103. $\sin\left(2 \sec^{-1} \frac{u}{2}\right)$

104. $\cos\left(2 \tan^{-1} \frac{3}{u}\right)$

105. $\tan\left(\sin^{-1} \frac{u}{\sqrt{u^2 + 2}}\right)$

106. $\sec\left(\cos^{-1} \frac{u}{\sqrt{u^2 + 5}}\right)$

107. $\sec\left(\operatorname{arccot} \frac{\sqrt{4 - u^2}}{u}\right)$

108. $\csc\left(\arctan \frac{\sqrt{9 - u^2}}{u}\right)$

(Modeling) Solve each problem.

109. *Angle of Elevation of a Shot Put* Refer to **Example 8.** Suppose a shot-putter can consistently release the steel ball with velocity v of 32 ft per sec from an initial height h of 5.0 ft. What angle, to the nearest degree, will maximize the distance?

110. *Angle of Elevation of a Shot Put* Refer to **Example 8.**

(a) What is the optimal angle, to the nearest degree, when $h = 0$?

(b) Fix h at 6 ft and regard θ as a function of v. As v increases without bound, the graph approaches an asymptote. Find the equation of that asymptote.

111. *Observation of a Painting* A painting 1 m high and 3 m from the floor will cut off an angle θ to an observer, where

$$\theta = \tan^{-1}\left(\frac{x}{x^2 + 2}\right),$$

assuming that the observer is x meters from the wall where the painting is displayed and that the eyes of the observer are 2 m above the ground. (See the figure.) Find the value of θ for the following values of x. Round to the nearest degree.

(a) 1 **(b)** 2 **(c)** 3

(d) Derive the formula given above. (*Hint:* Use the identity for $\tan(\theta + \alpha)$. Use right triangles.)

(e) Graph the function for θ with a graphing calculator, and determine the distance that maximizes the angle.

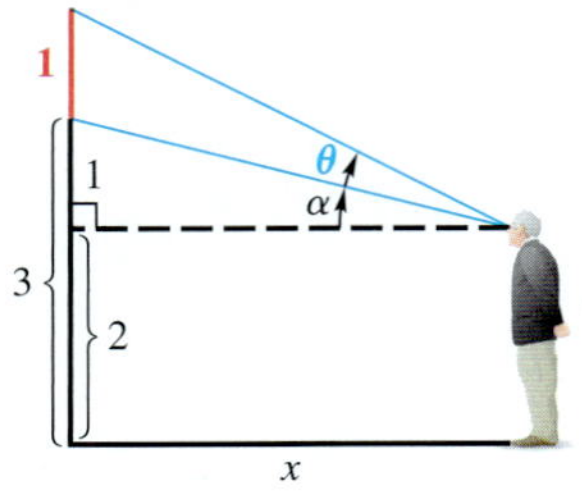

(f) The concept in part (e) was first investigated in 1471 by the astronomer Regiomontanus. (*Source:* Maor, E., *Trigonometric Delights,* Princeton University Press.) If the bottom of the picture is a meters above eye level and the top of the picture is b meters above eye level, then the optimum value of x is $\sqrt{ab}$ meters. Use this result to find the exact answer to part (e).

112. *Landscaping Formula* A shrub is planted in a 100-ft-wide space between buildings measuring 75 ft and 150 ft tall. The location of the shrub determines how much sun it receives each day. Show that if θ is the angle in the figure and x is the distance of the shrub from the taller building, then the value of θ (in radians) is given by

$$\theta = \pi - \arctan\left(\frac{75}{100 - x}\right) - \arctan\left(\frac{150}{x}\right).$$

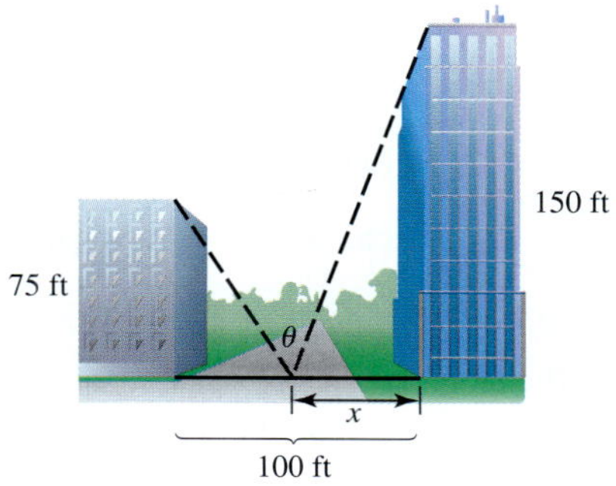

113. *Communications Satellite Coverage* The figure shows a stationary communications satellite positioned 20,000 mi above the equator. What percent, to the nearest tenth, of the equator can be seen from the satellite? The diameter of Earth is 7927 mi at the equator.

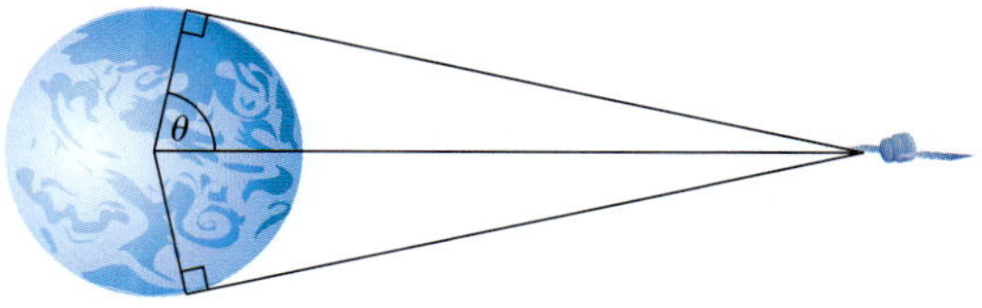

114. *Oil in a Storage Tank* The level of oil in a storage tank buried in the ground can be found in much the same way as a dipstick is used to determine the oil level in an automobile crankcase. Suppose the ends of the cylindrical storage tank in the figure are circles of radius 3 ft and the cylinder is 20 ft long. Determine the volume of oil in the tank to the nearest cubic foot if the rod shows a depth of 2 ft. (*Hint:* The volume will be 20 times the area of the shaded segment of the circle shown in the figure on the right.)

6.2 Trigonometric Equations I

- Solving by Linear Methods
- Solving by Factoring
- Solving by Quadratic Methods
- Solving by Using Trigonometric Identities

In **Chapter 5,** we studied trigonometric equations that were identities. We now consider trigonometric equations that are *conditional.* These equations are satisfied by some values but not others. (See **Appendix A.**)

Solving by Linear Methods

The most basic trigonometric equations are solved by first using properties of equality to isolate a trigonometric expression on one side of the equation.

EXAMPLE 1 Solving a Trigonometric Equation by Linear Methods

Solve the equation $2 \sin \theta + 1 = 0$

(a) over the interval $[0°, 360°)$, and

(b) for all solutions.

ALGEBRAIC SOLUTION

(a) Because $\sin \theta$ is to the first power, we use the same method as we would to solve the linear equation $2x + 1 = 0$.

$$2 \sin \theta + 1 = 0 \quad \text{Original equation}$$

$$2 \sin \theta = -1 \quad \text{Subtract 1. (Appendix A)}$$

$$\sin \theta = -\frac{1}{2} \quad \text{Divide by 2.}$$

To find values of θ that satisfy $\sin \theta = -\frac{1}{2}$, we observe that θ must be in either quadrant III or quadrant IV because the sine function is negative only in these two quadrants. Furthermore, the reference angle must be 30°. The graph of the unit circle in **Figure 22** shows the two possible values of θ. The solution set is $\{210°, 330°\}$.

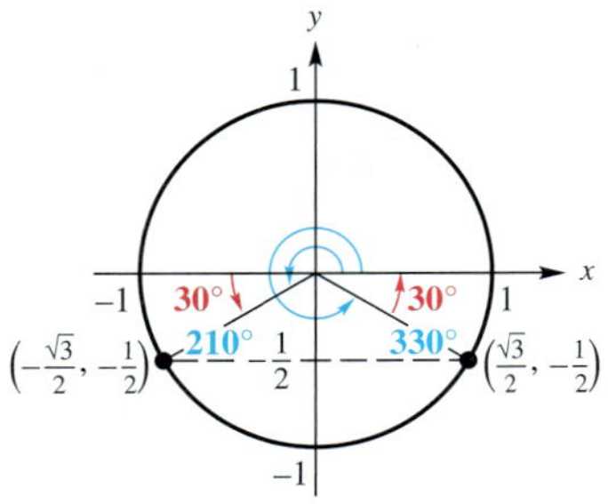

Figure 22

(b) To find all solutions, we add integer multiples of the period of the sine function, 360°, to each solution found in part (a). The solution set is written as follows.

$$\{210° + 360°n, 330° + 360°n, \text{ where } n \text{ is any integer}\}$$

GRAPHING CALCULATOR SOLUTION

(a) Consider the original equation.

$$2 \sin \theta + 1 = 0$$

We can find the solution set of this equation by graphing the function

$$Y_1 = 2 \sin X + 1$$

and then determining its x-intercepts, or zeros. Since we are finding solutions over the interval $[0°, 360°)$, we use degree mode and choose this interval of values for the input X on the graph.

The screen in **Figure 23(a)** indicates that one solution is 210°, and the screen in **Figure 23(b)** indicates that the other solution is 330°. The solution set is $\{210°, 330°\}$, which agrees with the algebraic solution.

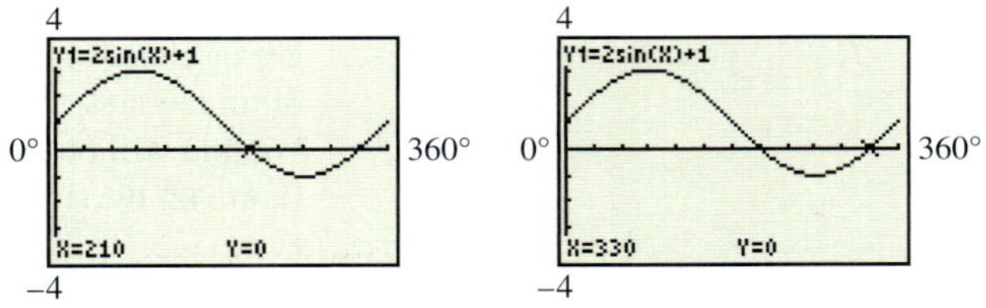

Degree mode
(a)

Degree mode
(b)

Figure 23

(b) Because the graph of

$$Y_1 = 2 \sin X + 1$$

repeats the same y-values every 360°, all solutions are found by adding integer multiples of 360° to the solutions found in part (a). See the algebraic solution.

✔ *Now Try Exercises 11 and 43.*

Solving by Factoring

EXAMPLE 2 Solving a Trigonometric Equation by Factoring

Solve $\sin\theta\tan\theta = \sin\theta$ over the interval $[0°, 360°)$.

SOLUTION

$$\sin\theta\tan\theta = \sin\theta \quad \text{Original equation}$$

$$\sin\theta\tan\theta - \sin\theta = 0 \quad \text{Subtract } \sin\theta.$$

$$\sin\theta(\tan\theta - 1) = 0 \quad \text{Factor out } \sin\theta.$$

$$\sin\theta = 0 \quad \text{or} \quad \tan\theta - 1 = 0 \quad \text{Zero-factor property (Appendix A)}$$

$$\tan\theta = 1$$

$$\theta = 0° \quad \text{or} \quad \theta = 180° \qquad \theta = 45° \quad \text{or} \quad \theta = 225° \quad \text{Apply the inverse function. (Section 6.1)}$$

See **Figure 24.** The solution set is $\{0°, 45°, 180°, 225°\}$.

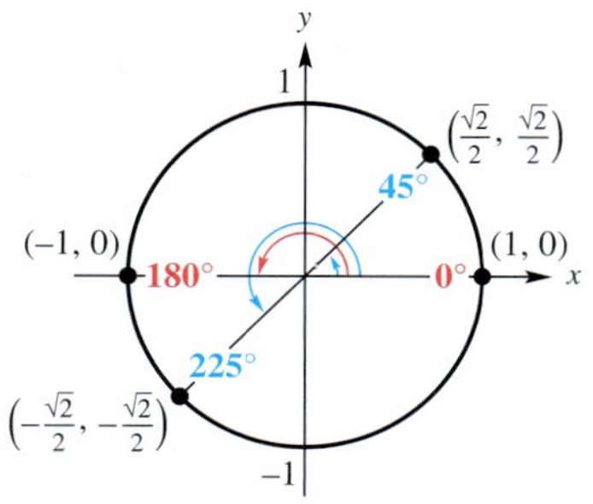

Figure 24

Now Try Exercise 31.

CAUTION Trying to solve the equation in **Example 2** by dividing each side by $\sin\theta$ would lead to $\tan\theta = 1$, which would give $\theta = 45°$ or $\theta = 225°$. The missing two solutions are the ones that make the divisor, $\sin\theta$, equal 0. ***For this reason, we avoid dividing by a variable expression.***

Solving by Quadratic Methods

The equation $au^2 + bu + c = 0$, where u is an algebraic expression, is solved by quadratic methods. The expression u may be a trigonometric function, as in the next example.

EXAMPLE 3 Solving a Trigonometric Equation by Factoring

Solve $\tan^2 x + \tan x - 2 = 0$ over the interval $[0, 2\pi)$.

SOLUTION

$$\tan^2 x + \tan x - 2 = 0 \quad \text{This equation is quadratic in form.}$$

$$(\tan x - 1)(\tan x + 2) = 0 \quad \text{Factor.}$$

$$\tan x - 1 = 0 \quad \text{or} \quad \tan x + 2 = 0 \quad \text{Zero-factor property}$$

$$\tan x = 1 \quad \text{or} \quad \tan x = -2 \quad \text{Solve each equation.}$$

The solutions for $\tan x = 1$ over the interval $[0, 2\pi)$ are $x = \frac{\pi}{4}$ and $x = \frac{5\pi}{4}$.

To solve $\tan x = -2$ over that interval, we use a scientific calculator set in *radian* mode. We find that $\tan^{-1}(-2) \approx -1.1071487$. This is a quadrant IV number, based on the range of the inverse tangent function. However, since we want solutions over the interval $[0, 2\pi)$, we must first add π to -1.1071487, and then add 2π. See **Figure 25.**

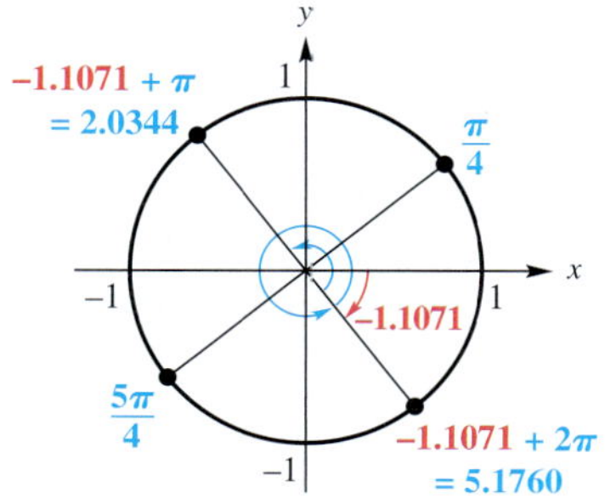

The solutions shown in blue represent angle measures, in radians, ***and*** their intercepted arc lengths on the unit circle.

Figure 25

$$x \approx -1.1071487 + \pi \approx 2.0344439$$

$$x \approx -1.1071487 + 2\pi \approx 5.1760366$$

The solutions over the required interval form the following solution set.

$$\left\{\underbrace{\frac{\pi}{4}, \frac{5\pi}{4}}_{\text{Exact values}}, \underbrace{2.0344, 5.1760}_{\text{Approximate values to four decimal places}}\right\}$$

Now Try Exercise 21.

EXAMPLE 4 Solving a Trigonometric Equation Using the Quadratic Formula

Find all solutions of $\cot x(\cot x + 3) = 1$. Write the solution set.

SOLUTION We multiply the factors on the left and subtract 1 to write the equation in standard quadratic form.

$$\cot x(\cot x + 3) = 1 \quad \text{Original equation}$$

$$\cot^2 x + 3\cot x - 1 = 0 \quad \text{(Appendix A)}$$

This equation is quadratic in form, but cannot be solved by factoring. Therefore, we use the quadratic formula, with $a = 1$, $b = 3$, $c = -1$, and $\cot x$ as the variable.

$$\cot x = \frac{-b \pm \sqrt{b^2 - 4ac}}{2a} \quad \text{Quadratic formula (Appendix A)}$$

$$= \frac{-3 \pm \sqrt{3^2 - 4(1)(-1)}}{2(1)} \quad a = 1, b = 3, c = -1$$

Be careful with signs.

$$= \frac{-3 \pm \sqrt{9 + 4}}{2} \quad \text{Simplify.}$$

$$= \frac{-3 \pm \sqrt{13}}{2} \quad \text{Add.}$$

$\cot x \approx -3.302775638$ or $\cot x \approx 0.3027756377$ Use a calculator. (Section 2.3)

$x \approx \cot^{-1}(-3.302775638)$ or $x \approx \cot^{-1}(0.3027756377)$ Definition of inverse cotangent

$x \approx \tan^{-1}\left(\frac{1}{-3.302775638}\right) + \pi$ or $x \approx \tan^{-1}\left(\frac{1}{0.3027756377}\right)$ Reciprocal identity: $\tan x = \frac{1}{\cot x}$ (Section 1.4)

$x \approx -0.2940013018 + \pi$ or $x \approx 1.276795025$ Use a calculator in radian mode.

$x \approx 2.847591352$

To find *all* solutions, we add integer multiples of the period of the tangent function, which is π, to each solution found previously. Although not unique, a common form of the solution set of the equation, written using the least possible nonnegative angle measures, is given as follows.

$$\{2.8476 + n\pi, 1.2768 + n\pi, \text{ where } n \text{ is any integer}\}$$

Round to four decimal places.

✔ ***Now Try Exercise 53.***

Solving by Using Trigonometric Identities Recall that squaring each side of an equation, such as

$$\sqrt{x + 4} = x + 2,$$

will yield all solutions but may also give extraneous solutions—solutions that satisfy the final equation but *not* the original equation. As a result, all proposed solutions *must* be checked in the original equation as shown in **Example 5.**

EXAMPLE 5 Solving a Trigonometric Equation by Squaring

Solve $\tan x + \sqrt{3} = \sec x$ over the interval $[0, 2\pi)$.

SOLUTION Our first goal is to rewrite the equation in terms of a single trigonometric function. Since the tangent and secant functions are related by the identity $1 + \tan^2 x = \sec^2 x$, square each side and express $\sec^2 x$ in terms of $\tan^2 x$.

$$\left(\tan x + \sqrt{3}\right)^2 = (\sec x)^2 \quad \text{Square each side.}$$

Don't forget the middle term.

$$\tan^2 x + 2\sqrt{3}\tan x + 3 = \sec^2 x \quad (x+y)^2 = x^2 + 2xy + y^2$$

$$\tan^2 x + 2\sqrt{3}\tan x + 3 = 1 + \tan^2 x \quad \text{Pythagorean identity (Section 1.4)}$$

$$2\sqrt{3}\tan x = -2 \quad \text{Subtract } 3 + \tan^2 x.$$

$$\tan x = -\frac{1}{\sqrt{3}}, \text{ or } -\frac{\sqrt{3}}{3} \quad \text{Divide by } 2\sqrt{3}. \text{ Rationalize the denominator.}$$

Solutions of $\tan x = -\frac{\sqrt{3}}{3}$ over $[0, 2\pi)$ are $\frac{5\pi}{6}$ and $\frac{11\pi}{6}$. These possible, or proposed, solutions must be checked to determine whether they are also solutions of the original equation.

CHECK $\tan x + \sqrt{3} = \sec x$ Original equation

Let $x = \frac{5\pi}{6}$.

$$\tan\left(\frac{5\pi}{6}\right) + \sqrt{3} \stackrel{?}{=} \sec\left(\frac{5\pi}{6}\right)$$

$$-\frac{\sqrt{3}}{3} + \frac{3\sqrt{3}}{3} \stackrel{?}{=} -\frac{2\sqrt{3}}{3}$$

$$\frac{2\sqrt{3}}{3} = -\frac{2\sqrt{3}}{3} \quad \text{False}$$

Let $x = \frac{11\pi}{6}$.

$$\tan\left(\frac{11\pi}{6}\right) + \sqrt{3} \stackrel{?}{=} \sec\left(\frac{11\pi}{6}\right)$$

$$-\frac{\sqrt{3}}{3} + \frac{3\sqrt{3}}{3} \stackrel{?}{=} \frac{2\sqrt{3}}{3}$$

$$\frac{2\sqrt{3}}{3} = \frac{2\sqrt{3}}{3} \quad ✓ \text{ True}$$

As the check shows, only $\frac{11\pi}{6}$ is a solution, so the solution set is $\left\{\frac{11\pi}{6}\right\}$.

✔ *Now Try Exercise 41.*

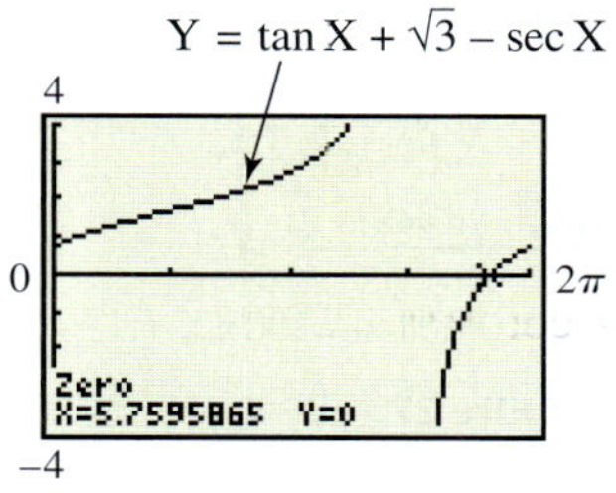

Radian mode

The graph shows that on the interval $[0, 2\pi)$, the only x-intercept of the graph of $Y = \tan X + \sqrt{3} - \sec X$ is 5.7595865, which is an approximation for $\frac{11\pi}{6}$, the solution found in **Example 5.**

Methods for solving trigonometric equations can be summarized as follows.

Solving a Trigonometric Equation

1. Decide whether the equation is linear or quadratic in form, so that you can determine the solution method.
2. If only one trigonometric function is present, solve the equation for that function.
3. If more than one trigonometric function is present, rearrange the equation so that one side equals 0. Then try to factor and set each factor equal to 0 to solve.
4. If the equation is quadratic in form, but not factorable, use the quadratic formula. Check that solutions are in the desired interval.
5. Try using identities to change the form of the equation. It may be helpful to square each side of the equation first. In this case, check for extraneous solutions.

LOOKING AHEAD TO CALCULUS

There are many instances in calculus where it is necessary to solve trigonometric equations. Examples include solving related-rates problems and optimization problems.

EXAMPLE 6 Describing a Musical Tone from a Graph

A basic component of music is a pure tone. The graph in **Figure 26** models the sinusoidal pressure $y = P$ in pounds per square foot from a pure tone at time $x = t$ in seconds.

(a) The frequency of a pure tone is often measured in hertz. One hertz is equal to one cycle per second and is abbreviated Hz. What is the frequency f, in hertz, of the pure tone shown in the graph?

(b) The time for the tone to produce one complete cycle is called the **period.** Approximate the period T, in seconds, of the pure tone.

(c) An equation for the graph is $y = 0.004 \sin 300\pi x$. Use a calculator to estimate all solutions to the equation that make $y = 0.004$ over the interval $[0, 0.02]$.

SOLUTION

(a) From the graph in **Figure 26,** we see that there are 6 cycles in 0.04 sec. This is equivalent to $\frac{6}{0.04} = 150$ cycles per sec. The pure tone has a frequency of $f = 150$ Hz.

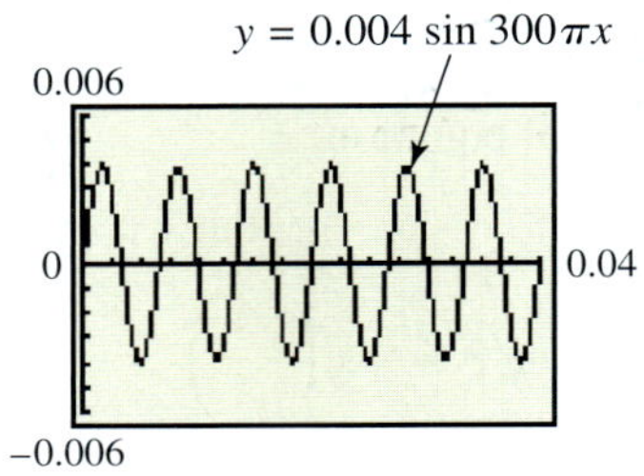

Figure 26

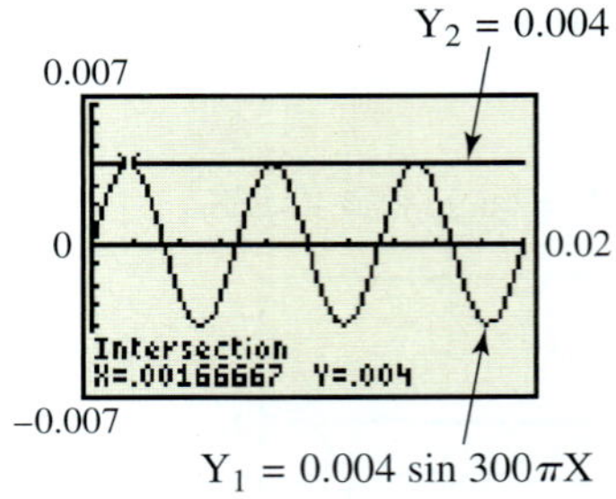

Figure 27

(b) Six periods cover a time interval of 0.04 sec. One period would be equal to $T = \frac{0.04}{6} = \frac{1}{150}$, or $0.00\overline{6}$ sec.

(c) If we reproduce the graph in **Figure 26** on a calculator as Y_1 and also graph a second function as $Y_2 = 0.004$, we can determine that the approximate values of x at the points of intersection of the graphs over the interval $[0, 0.02]$ are

$$0.0017, \quad 0.0083, \quad \text{and} \quad 0.015.$$

The first value is shown in **Figure 27.** These values represent time in seconds.

✔ *Now Try Exercise 61.*

6.2 Exercises

Concept Check Refer to the summary box on solving a trigonometric equation following **Example 5.** *Decide on the appropriate technique to begin the solution of each equation. Do not solve the equation.*

1. $2 \cot x + 1 = -1$

2. $\sin x + 2 = 3$

3. $5 \sec^2 x = 6 \sec x$

4. $2 \cos^2 x - \cos x = 1$

5. $9 \sin^2 x - 5 \sin x = 1$

6. $\tan^2 x - 4 \tan x + 2 = 0$

7. $\tan x - \cot x = 0$

8. $\cos^2 x = \sin^2 x + 1$

9. Suppose that in solving an equation over the interval $[0°, 360°)$, you reach the step $\sin \theta = -\frac{1}{2}$. Why is $-30°$ not a correct answer?

10. Lindsay solved the equation $\sin x = 1 - \cos x$ by squaring each side to get

$$\sin^2 x = 1 - 2 \cos x + \cos^2 x.$$

Several steps later, using correct algebra, she concluded that the solution set for solutions over the interval $[0, 2\pi)$ is $\{0, \frac{\pi}{2}, \frac{3\pi}{2}\}$. Explain why this is not the correct solution set.

Solve each equation for exact solutions over the interval $[0, 2\pi)$. ***See Examples 1–3.***

11. $2 \cot x + 1 = -1$

12. $\sin x + 2 = 3$

13. $2 \sin x + 3 = 4$

14. $2 \sec x + 1 = \sec x + 3$

15. $\tan^2 x + 3 = 0$

16. $\sec^2 x + 2 = -1$

17. $(\cot x - 1)(\sqrt{3} \cot x + 1) = 0$

18. $(\csc x + 2)(\csc x - \sqrt{2}) = 0$

19. $\cos^2 x + 2 \cos x + 1 = 0$

20. $2 \cos^2 x - \sqrt{3} \cos x = 0$

21. $-2 \sin^2 x = 3 \sin x + 1$

22. $2 \cos^2 x - \cos x = 1$

Solve each equation for solutions over the interval $[0°, 360°)$. *Give solutions to the nearest tenth as appropriate.* ***See Examples 2–5.***

23. $(\cot \theta - \sqrt{3})(2 \sin \theta + \sqrt{3}) = 0$

24. $(\tan \theta - 1)(\cos \theta - 1) = 0$

25. $2 \sin \theta - 1 = \csc \theta$

26. $\tan \theta + 1 = \sqrt{3} + \sqrt{3} \cot \theta$

27. $\tan \theta - \cot \theta = 0$

28. $\cos^2 \theta = \sin^2 \theta + 1$

29. $\csc^2 \theta - 2 \cot \theta = 0$

30. $\sin^2 \theta \cos \theta = \cos \theta$

31. $2 \tan^2 \theta \sin \theta - \tan^2 \theta = 0$

32. $\sin^2 \theta \cos^2 \theta = 0$

33. $\sec^2 \theta \tan \theta = 2 \tan \theta$

34. $\cos^2 \theta - \sin^2 \theta = 0$

35. $9 \sin^2 \theta - 6 \sin \theta = 1$

36. $4 \cos^2 \theta + 4 \cos \theta = 1$

37. $\tan^2 \theta + 4 \tan \theta + 2 = 0$

38. $3 \cot^2 \theta - 3 \cot \theta - 1 = 0$

39. $\sin^2 \theta - 2 \sin \theta + 3 = 0$

40. $2 \cos^2 \theta + 2 \cos \theta + 1 = 0$

41. $\cot \theta + 2 \csc \theta = 3$

42. $2 \sin \theta = 1 - 2 \cos \theta$

Solve each equation (x in radians and θ in degrees) for all exact solutions where appropriate. Round approximate answers in radians to four decimal places and approximate answers in degrees to the nearest tenth. Write answers using the least possible nonnegative angle measures. ***See Examples 1–5.***

43. $\cos \theta + 1 = 0$

44. $\tan \theta + 1 = 0$

45. $3 \csc x - 2\sqrt{3} = 0$

46. $\cot x + \sqrt{3} = 0$

47. $6 \sin^2 \theta + \sin \theta = 1$

48. $3 \sin^2 \theta - \sin \theta = 2$

49. $2 \cos^2 x + \cos x - 1 = 0$

50. $4 \cos^2 x - 1 = 0$

51. $\sin \theta \cos \theta - \sin \theta = 0$

52. $\tan \theta \csc \theta - \sqrt{3} \csc \theta = 0$

53. $\sin x (3 \sin x - 1) = 1$

54. $\tan x (\tan x - 2) = 5$

55. $5 + 5 \tan^2 \theta = 6 \sec \theta$

56. $\sec^2 \theta = 2 \tan \theta + 4$

57. $\dfrac{2 \tan \theta}{3 - \tan^2 \theta} = 1$

58. $\dfrac{2 \cot^2 \theta}{\cot \theta + 3} = 1$

The following equations cannot be solved by algebraic methods. Use a graphing calculator to find all solutions over the interval $[0, 2\pi)$. Express solutions to four decimal places.

59. $x^2 + \sin x - x^3 - \cos x = 0$

60. $x^3 - \cos^2 x = \frac{1}{2}x - 1$

(Modeling) Solve each problem.

61. *Pressure on the Eardrum* **See Example 6.** No musical instrument can generate a true pure tone. A pure tone has a unique, constant frequency and amplitude that sounds rather dull and uninteresting. The pressures caused by pure tones on the eardrum are sinusoidal. The change in pressure P in pounds per square foot on a person's eardrum from a pure tone at time t in seconds can be modeled using the equation

$$P = A \sin(2\pi f t + \phi),$$

where f is the frequency in cycles per second, and ϕ is the phase angle. When P is positive, there is an increase in pressure and the eardrum is pushed inward. When P is negative, there is a decrease in pressure and the eardrum is pushed outward. (*Source:* Roederer, J., *Introduction to the Physics and Psychophysics of Music,* Second Edition, Springer-Verlag.) A graph of the tone middle C is shown in the figure.

(a) Determine algebraically the values of t for which $P = 0$ over $[0, 0.005]$.

(b) From the graph and your answer in part (a), determine the interval for which $P \le 0$ over $[0, 0.005]$.

(c) Would an eardrum hearing this tone be vibrating outward or inward when $P < 0$?

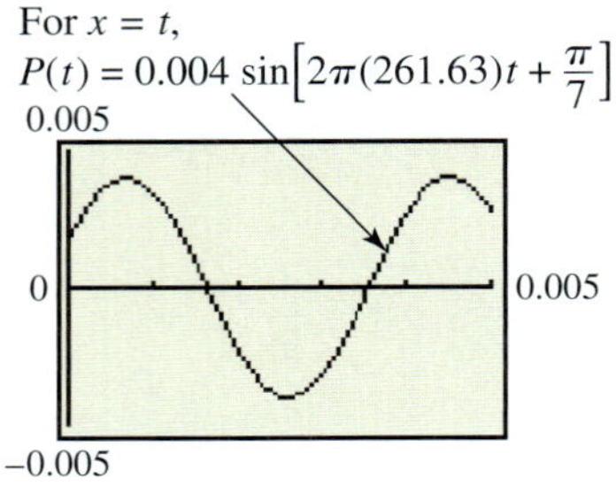

62. *Accident Reconstruction* The model

$$0.342D \cos \theta + h \cos^2 \theta = \frac{16D^2}{{V_0}^2}$$

is used to reconstruct accidents in which a vehicle vaults into the air after hitting an obstruction. V_0 is velocity in feet per second of the vehicle when it hits the obstruction, D is distance (in feet) from the obstruction to the landing point, and h is the difference in height (in feet) between landing point and takeoff point. Angle θ is the takeoff angle, the angle between the horizontal and the path of the vehicle. Find θ to the nearest degree if $V_0 = 60$, $D = 80$, and $h = 2$.

63. *Electromotive Force* In an electric circuit, suppose that

$$V = \cos 2\pi t$$

models the electromotive force in volts at t seconds. Find the least value of t where $0 \le t \le \frac{1}{2}$ for each value of V.

(a) $V = 0$ **(b)** $V = 0.5$ **(c)** $V = 0.25$

64. *Voltage Induced by a Coil of Wire* A coil of wire rotating in a magnetic field induces a voltage modeled by

$$E = 20 \sin\left(\frac{\pi t}{4} - \frac{\pi}{2}\right),$$

where t is time in seconds. Find the least positive time to produce each voltage.

(a) 0 **(b)** $10\sqrt{3}$

65. *Movement of a Particle* A particle moves along a straight line. The distance of the particle from the origin at time t is modeled by

$$s(t) = \sin t + 2 \cos t.$$

Find a value of t that satisfies each equation.

(a) $s(t) = \frac{2 + \sqrt{3}}{2}$ **(b)** $s(t) = \frac{3\sqrt{2}}{2}$

66. Explain what is **WRONG** with the following solution of the trigonometric equation $\sin^2 x - \sin x = 0$ for all x over the interval $[0, 2\pi)$.

$$\sin^2 x - \sin x = 0 \quad \text{Original equation}$$

$$\sin x - 1 = 0 \quad \text{Divide by } \sin x.$$

$$\sin x = 1 \quad \text{Add 1.}$$

$$x = \frac{\pi}{2} \quad \text{Definition of inverse sine}$$

The solution set is $\left\{\frac{\pi}{2}\right\}$.

6.3 Trigonometric Equations II

- Equations with Half-Angles
- Equations with Multiple Angles

In this section, we discuss trigonometric equations that involve functions of half-angles and multiple angles. Solving these equations often requires adjusting solution intervals to fit given domains.

Equations with Half-Angles

EXAMPLE 1 Solving an Equation with a Half-Angle

Solve the equation $2 \sin \frac{x}{2} = 1$

(a) over the interval $[0, 2\pi)$, and **(b)** for all solutions.

SOLUTION

(a) Write the interval $[0, 2\pi)$ as the inequality

$$0 \le x < 2\pi.$$

The corresponding interval for $\frac{x}{2}$ is

$$0 \le \frac{x}{2} < \pi. \quad \text{Divide by 2. (Appendix A)}$$

To find all values of $\frac{x}{2}$ over the interval $[0, \pi)$ that satisfy the given equation, first solve for $\sin \frac{x}{2}$.

$$2 \sin \frac{x}{2} = 1 \quad \text{Original equation}$$

$$\sin \frac{x}{2} = \frac{1}{2} \quad \text{Divide by 2.}$$

The two numbers over the interval $[0, \pi)$ with sine value $\frac{1}{2}$ are $\frac{\pi}{6}$ and $\frac{5\pi}{6}$, so

$$\frac{x}{2} = \frac{\pi}{6} \quad \text{or} \quad \frac{x}{2} = \frac{5\pi}{6} \quad \text{Definition of inverse sine (Section 6.1)}$$

$$x = \frac{\pi}{3} \quad \text{or} \quad x = \frac{5\pi}{3}. \quad \text{Multiply by 2.}$$

The solution set over the given interval is $\left\{\frac{\pi}{3}, \frac{5\pi}{3}\right\}$.

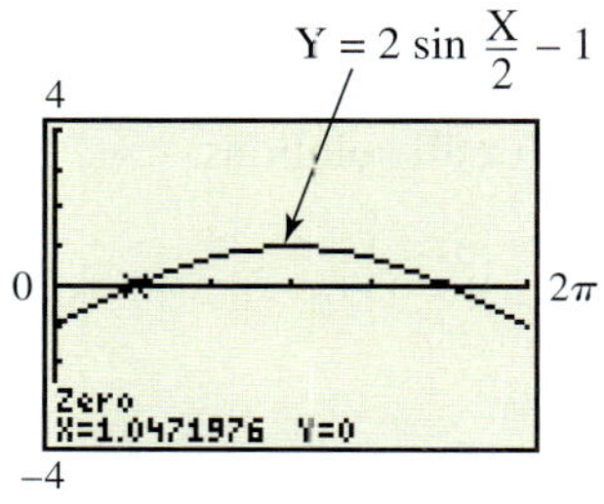

The x-intercepts are the solutions found in **Example 1(a).** Using Xscl $= \frac{\pi}{3}$ makes it possible to support the exact solutions by counting the tick marks from 0 on the graph.

(b) Because this is a sine function with period 4π, all solutions are found by adding integer multiples of 4π.

$$\left\{\frac{\pi}{3} + 4n\pi, \frac{5\pi}{3} + 4n\pi, \text{ where } n \text{ is any integer}\right\}$$

✔ *Now Try Exercises 15 and 29.*

Equations with Multiple Angles

EXAMPLE 2 Solving an Equation Using a Double Angle Identity

Solve $\cos 2x = \cos x$ over the interval $[0, 2\pi)$.

SOLUTION First change $\cos 2x$ to a trigonometric function of x. Use the identity $\cos 2x = 2\cos^2 x - 1$ so that the equation involves only $\cos x$. Then factor.

$$\cos 2x = \cos x \quad \text{Original equation}$$

$$2\cos^2 x - 1 = \cos x \quad \text{Cosine double-angle identity (Section 5.5)}$$

$$2\cos^2 x - \cos x - 1 = 0 \quad \text{Subtract } \cos x.$$

$$(2\cos x + 1)(\cos x - 1) = 0 \quad \text{Factor.}$$

$$2\cos x + 1 = 0 \quad \text{or} \quad \cos x - 1 = 0 \quad \text{Zero-factor property (Appendix A)}$$

$$\cos x = -\frac{1}{2} \quad \text{or} \quad \cos x = 1 \quad \text{Solve each equation for } \cos x.$$

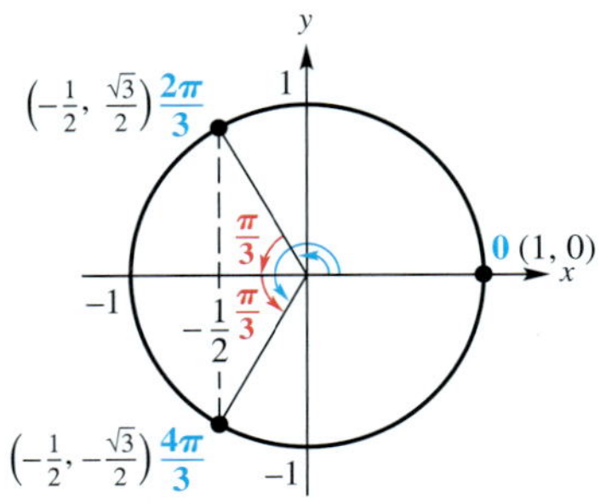

Figure 28

Cosine is $-\frac{1}{2}$ in quadrants II and III with reference angle $\frac{\pi}{3}$, and it has a value of 1 at 0 radians. We can use **Figure 28** to determine that solutions over the required interval are as follows.

$$x = \frac{2\pi}{3} \quad \text{or} \quad x = \frac{4\pi}{3} \quad \text{or} \quad x = 0.$$

The solution set is $\left\{0, \frac{2\pi}{3}, \frac{4\pi}{3}\right\}$.

✔ *Now Try Exercise 17.*

CAUTION In **Example 2,** because 2 is not a factor of $\cos 2x$, $\frac{\cos 2x}{2} \neq \cos x$. The only way to change $\cos 2x$ to a trigonometric function of x is by using one of the identities for $\cos 2x$.

EXAMPLE 3 Solving an Equation Using a Multiple-Angle Identity

Solve the equation $4 \sin\theta \cos\theta = \sqrt{3}$

(a) over the interval $[0°, 360°)$, and **(b)** for all solutions.

SOLUTION

(a) The identity $2\sin\theta\cos\theta = \sin 2\theta$ is useful here.

$$4\sin\theta\cos\theta = \sqrt{3} \quad \text{Original equation}$$

$$2(2\sin\theta\cos\theta) = \sqrt{3} \quad 4 = 2 \cdot 2$$

$$2\sin 2\theta = \sqrt{3} \quad 2\sin\theta\cos\theta = \sin 2\theta \text{ (Section 5.5)}$$

$$\sin 2\theta = \frac{\sqrt{3}}{2} \quad \text{Divide by 2.}$$

From the given interval $0° \le \theta < 360°$, the corresponding interval for 2θ is $0° \le 2\theta < 720°$. Because the sine is positive in quadrants I and II, solutions over this interval are as follows.

$$2\theta = 60°, 120°, 420°, 480°, \quad \text{Reference angle is } 60°.$$

or

$$\theta = 30°, 60°, 210°, 240° \quad \text{Divide by 2.}$$

The final two solutions for 2θ were found by adding 360° to 60° and 120°, respectively, which gives the solution set $\{30°, 60°, 210°, 240°\}$.

(b) All angles 2θ that are solutions of the equation $\sin 2\theta = \frac{\sqrt{3}}{2}$ are found by adding integer multiples of 360° to the basic solution angles, 60° and 120°.

$$2\theta = 60° + 360°n \quad \text{and} \quad 2\theta = 120° + 360°n \quad \text{Add integer multiples of } 360°.$$

$$\theta = 30° + 180°n \quad \text{and} \quad \theta = 60° + 180°n \quad \text{Divide by 2.}$$

All solutions are given by the following set, where 180° represents the period of $\sin 2\theta$.

$$\{30° + 180°n, 60° + 180°n, \text{ where } n \text{ is any integer}\}$$

✔ ***Now Try Exercises 13 and 37.***

EXAMPLE 4 Solving an Equation with a Multiple Angle

Solve $\tan 3x + \sec 3x = 2$ over the interval $[0, 2\pi)$.

SOLUTION Since the tangent and secant functions are related by the identity $1 + \tan^2\theta = \sec^2\theta$, one way to begin is to express everything in terms of secant.

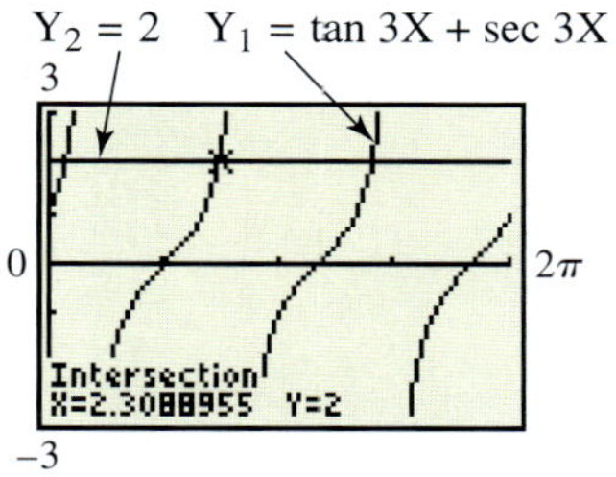

Connected mode; radian mode

The screen shows that one solution is approximately 2.3089. An advantage of using a graphing calculator is that extraneous values do not appear.

$$\tan 3x + \sec 3x = 2$$

Don't forget the middle term.

$$\tan 3x = 2 - \sec 3x \quad \text{Subtract } \sec 3x.$$

$$\tan^2 3x = 4 - 4\sec 3x + \sec^2 3x \quad \text{Square each side. } (x - y)^2 = x^2 - 2xy + y^2$$

$$\sec^2 3x - 1 = 4 - 4\sec 3x + \sec^2 3x \quad \text{Replace } \tan^2 3x \text{ with } \sec^2 3x - 1. \text{ (Section 5.1)}$$

$$4\sec 3x = 5 \quad \text{Simplify.}$$

$$\sec 3x = \frac{5}{4} \quad \text{Divide by 4.}$$

$$\frac{1}{\cos 3x} = \frac{5}{4} \quad \sec\theta = \frac{1}{\cos\theta} \text{ (Section 1.4)}$$

$$\cos 3x = \frac{4}{5} \quad \text{Use reciprocals.}$$

Multiply each term of the inequality $0 \le x < 2\pi$ by 3 to find the interval for $3x$: $[0, 6\pi)$. Use a calculator and the fact that cosine is positive in quadrants I and IV.

$$3x \approx 0.6435, 5.6397, 6.9267, 11.9229, 13.2099, 18.2061 \quad \text{Definition of inverse cosine}$$

$$x \approx 0.2145, 1.8799, 2.3089, 3.9743, 4.4033, 6.0687 \quad \text{Divide by 3.}$$

Since both sides of the equation were squared, each proposed solution must be checked. Verify by substitution in the given equation that the solution set is $\{0.2145, 2.3089, 4.4033\}$.

✔ ***Now Try Exercise 43.***

A piano string can vibrate at more than one frequency when it is struck. It produces a complex wave that can mathematically be modeled by a sum of several pure tones. When a piano key with a frequency of f_1 is played, the corresponding string vibrates not only at f_1 but also at the higher frequencies of $2f_1, 3f_1, 4f_1, \ldots, nf_1$. f_1 is the **fundamental frequency** of the string, and higher frequencies are the **upper harmonics.** The human ear will hear the sum of these frequencies as one complex tone. (*Source:* Roederer, J., *Introduction to the Physics and Psychophysics of Music,* Second Edition, Springer-Verlag.)

EXAMPLE 5 Analyzing Pressures of Upper Harmonics

Suppose that the A key above middle C is played on a piano. Its fundamental frequency is $f_1 = 440$ Hz, and its associated pressure is expressed as

$$P_1 = 0.002 \sin 880\pi t.$$

The string will also vibrate at

$$f_2 = 880, \quad f_3 = 1320, \quad f_4 = 1760, \quad f_5 = 2200, \ldots \text{ Hz}.$$

The corresponding pressures of these upper harmonics are as follows.

$$P_2 = \frac{0.002}{2} \sin 1760\pi t, \qquad P_3 = \frac{0.002}{3} \sin 2640\pi t,$$

$$P_4 = \frac{0.002}{4} \sin 3520\pi t, \qquad \text{and} \qquad P_5 = \frac{0.002}{5} \sin 4400\pi t$$

The graph of

$$P = P_1 + P_2 + P_3 + P_4 + P_5,$$

shown in **Figure 29,** is "saw-toothed."

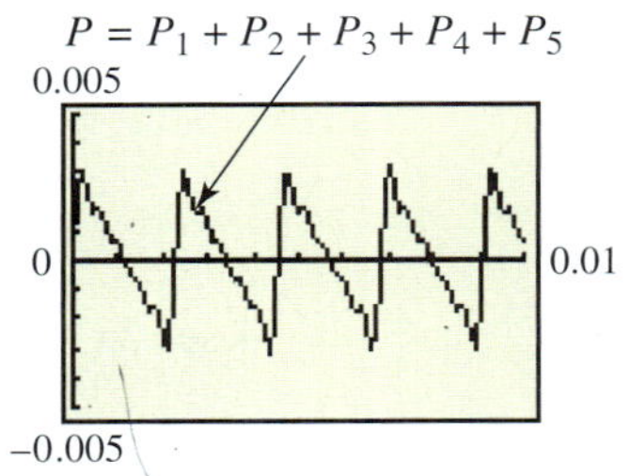

Figure 29

(a) What is the maximum value of P?

(b) At what values of $t = x$ does this maximum occur over the interval $[0, 0.01]$?

SOLUTION

(a) A graphing calculator shows that the maximum value of P is approximately 0.00317. See **Figure 30.**

(b) The maximum occurs at

$$t = x \approx 0.000191, 0.00246, 0.00474, 0.00701, \text{ and } 0.00928.$$

Figure 30 shows how the second value is found. The other values are found similarly.

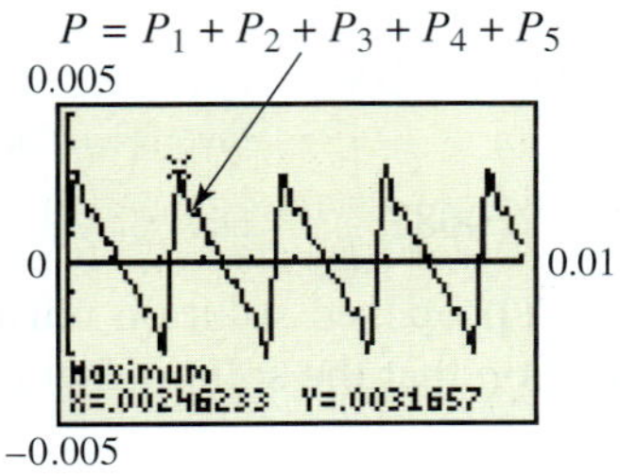

Figure 30

✔ *Now Try Exercise 47.*

6.3 Exercises

Concept Check Answer each question.

1. Suppose you are solving a trigonometric equation for solutions over the interval $[0, 2\pi)$, and your work leads to $2x = \frac{2\pi}{3}, 2\pi, \frac{8\pi}{3}$. What are the corresponding values of x?

2. Suppose you are solving a trigonometric equation for solutions over the interval $[0, 2\pi)$, and your work leads to $\frac{1}{2}x = \frac{\pi}{16}, \frac{5\pi}{12}, \frac{5\pi}{8}$. What are the corresponding values of x?

3. Suppose you are solving a trigonometric equation for solutions over the interval $[0°, 360°)$, and your work leads to $3\theta = 180°, 630°, 720°, 930°$. What are the corresponding values of θ?

4. Suppose you are solving a trigonometric equation for solutions over the interval $[0°, 360°)$, and your work leads to $\frac{1}{3}\theta = 45°, 60°, 75°, 90°$. What are the corresponding values of θ?

5. Explain what is **WRONG** with the following solution.
Solve $\tan 2\theta = 2$ over the interval $[0, 2\pi)$.

$\tan 2\theta = 2$	Original equation
$\frac{\tan 2\theta}{2} = \frac{2}{2}$	Divide by 2.
$\tan \theta = 1$	Perform the division.
$\theta = \frac{\pi}{4}$ or $\theta = \frac{5\pi}{4}$	Definition of inverse tangent

The solution set is $\left\{\frac{\pi}{4}, \frac{5\pi}{4}\right\}$.

6. The equation $\cot \frac{x}{2} - \csc \frac{x}{2} - 1 = 0$ has no solution over the interval $[0, 2\pi)$. Using this information, what can we say about the graph of

$$y = \cot \frac{x}{2} - \csc \frac{x}{2} - 1$$

over this interval? Confirm your answer by graphing the function over the interval.

Solve each equation in x for exact solutions over the interval $[0, 2\pi)$ and each equation in θ for exact solutions over the interval $[0°, 360°)$. ***See Examples 1–4.***

7. $\cos 2x = \frac{\sqrt{3}}{2}$

8. $\cos 2x = -\frac{1}{2}$

9. $\sin 3\theta = -1$

10. $\sin 3\theta = 0$

11. $3 \tan 3x = \sqrt{3}$

12. $\cot 3x = \sqrt{3}$

13. $\sqrt{2} \cos 2\theta = -1$

14. $2\sqrt{3} \sin 2\theta = \sqrt{3}$

15. $\sin \frac{x}{2} = \sqrt{2} - \sin \frac{x}{2}$

16. $\tan 4x = 0$

17. $\sin x = \sin 2x$

18. $\cos 2x - \cos x = 0$

19. $8 \sec^2 \frac{x}{2} = 4$

20. $\sin^2 \frac{x}{2} - 2 = 0$

21. $\sin \frac{\theta}{2} = \csc \frac{\theta}{2}$

22. $\sec \frac{\theta}{2} = \cos \frac{\theta}{2}$

23. $\cos 2x + \cos x = 0$

24. $\sin x \cos x = \frac{1}{4}$

Solve each equation (x in radians and θ in degrees) for all exact solutions where appropriate. Round approximate answers in radians to four decimal places and approximate answers in degrees to the nearest tenth. Write answers using the least possible nonnegative angle measures. ***See Examples 1–4.***

25. $\sqrt{2} \sin 3x - 1 = 0$

26. $-2 \cos 2x = \sqrt{3}$

27. $\cos \frac{\theta}{2} = 1$

28. $\sin \frac{\theta}{2} = 1$

29. $2\sqrt{3} \sin \frac{x}{2} = 3$

30. $2\sqrt{3} \cos \frac{x}{2} = -3$

31. $2 \sin \theta = 2 \cos 2\theta$ **32.** $\cos \theta - 1 = \cos 2\theta$ **33.** $1 - \sin x = \cos 2x$

34. $\sin 2x = 2 \cos^2 x$ **35.** $3 \csc^2 \frac{x}{2} = 2 \sec x$ **36.** $\cos x = \sin^2 \frac{x}{2}$

37. $2 - \sin 2\theta = 4 \sin 2\theta$ **38.** $4 \cos 2\theta = 8 \sin \theta \cos \theta$

39. $2 \cos^2 2\theta = 1 - \cos 2\theta$ **40.** $\sin \theta - \sin 2\theta = 0$

Solve each equation for solutions over the interval $[0, 2\pi)$. Write solutions as exact values or to four decimal places, as appropriate. ***See Example 4.***

41. $\sin \frac{x}{2} - \cos \frac{x}{2} = 0$ **42.** $\sin \frac{x}{2} + \cos \frac{x}{2} = 1$

43. $\tan 2x + \sec 2x = 3$ **44.** $\tan 2x - \sec 2x = 2$

The following equations cannot be solved by algebraic methods. Use a graphing calculator to find all solutions over the interval $[0, 2\pi)$. Express solutions to four decimal places.

45. $2 \sin 2x - x^3 + 1 = 0$ **46.** $3 \cos \frac{x}{2} + \sqrt{x} - 2 = -\frac{1}{2}x + 2$

(Modeling) Solve each problem. ***See Example 5.***

47. ***Pressure of a Plucked String*** If a string with a fundamental frequency of 110 Hz is plucked in the middle, it will vibrate at the odd harmonics of 110, 330, 550, . . . Hz but not at the even harmonics of 220, 440, 660, . . . Hz. The resulting pressure P caused by the string can be modeled by the equation

$$P = 0.003 \sin 220\pi t + \frac{0.003}{3} \sin 660\pi t + \frac{0.003}{5} \sin 1100\pi t + \frac{0.003}{7} \sin 1540\pi t.$$

(*Source:* Benade, A., *Fundamentals of Musical Acoustics,* Dover Publications. Roederer, J., *Introduction to the Physics and Psychophysics of Music,* Second Edition, Springer-Verlag.)

(a) Graph P in the viewing window $[0, 0.03]$ by $[-0.005, 0.005]$.

(b) Use the graph to describe the shape of the sound wave that is produced.

(c) Refer to **Section 6.2, Exercise 61.** At lower frequencies, the inner ear will hear a tone only when the eardrum is moving outward. Determine the times over the interval $[0, 0.03]$ when this will occur.

48. ***Hearing Beats in Music*** Musicians sometimes tune instruments by playing the same tone on two different instruments and listening for a phenomenon known as **beats.** Beats occur when two tones vary in frequency by only a few hertz. When the two instruments are in tune, the beats disappear. The ear hears beats because the pressure slowly rises and falls as a result of this slight variation in the frequency. This phenomenon can be seen using a graphing calculator. (*Source:* Pierce, J., *The Science of Musical Sound,* Scientific American Books.)

(a) Consider the two tones with frequencies of 220 Hz and 223 Hz and pressures $P_1 = 0.005 \sin 440\pi t$ and $P_2 = 0.005 \sin 446\pi t$, respectively. Graph the pressure $P = P_1 + P_2$ felt by an eardrum over the 1-sec interval $[0.15, 1.15]$. How many beats are there in 1 sec?

(b) Repeat part (a) with frequencies of 220 and 216 Hz.

(c) Determine a simple way to find the number of beats per second if the frequency of each tone is given.

49. *Hearing Difference Tones* When a musical instrument creates a tone of 110 Hz, it also creates tones at 220, 330, 440, 550, 660, . . . Hz. A small speaker cannot reproduce the 110-Hz vibration but it can reproduce the higher frequencies, which are the **upper harmonics.** The low tones can still be heard because the speaker produces **difference tones** of the upper harmonics. The difference between consecutive frequencies is 110 Hz, and this difference tone will be heard by a listener. (*Source:* Benade, A., *Fundamentals of Musical Acoustics,* Dover Publications.)

(a) We can model this phenomenon using a graphing calculator. In the window $[0, 0.03]$ by $[-1, 1]$, graph the upper harmonics represented by the pressure

$$P = \frac{1}{2}\sin[2\pi(220)t] + \frac{1}{3}\sin[2\pi(330)t] + \frac{1}{4}\sin[2\pi(440)t].$$

(b) Estimate all t-coordinates where P is maximum.

(c) What does a person hear in addition to the frequencies of 220, 330, and 440 Hz?

(d) Graph the pressure produced by a speaker that can vibrate at 110 Hz and above.

50. *Daylight Hours in New Orleans* The seasonal variation in length of daylight can be modeled by a sine function. For example, the daily number of hours of daylight in New Orleans is given by

$$h = \frac{35}{3} + \frac{7}{3}\sin\frac{2\pi x}{365},$$

where x is the number of days after March 21 (disregarding leap year). (*Source:* Bushaw, D., et al., *A Sourcebook of Applications of School Mathematics*, Mathematical Association of America.)

(a) On what date will there be about 14 hr of daylight?

(b) What date has the least number of hours of daylight?

(c) When will there be about 10 hr of daylight?

51. *Average Monthly Temperature in Vancouver* The following function approximates average monthly temperature y (in °F) in Vancouver, Canada. Here x represents the month, where $x = 1$ corresponds to January, $x = 2$ corresponds to February, and so on. (*Source:* www.weather.com)

$$f(x) = 14\sin\left[\frac{\pi}{6}(x - 4)\right] + 50$$

When is the average monthly temperature **(a)** 64°F **(b)** 39°F?

52. *Average Monthly Temperature in Phoenix* The following function approximates average monthly temperature y (in °F) in Phoenix, Arizona. Here x represents the month, where $x = 1$ corresponds to January, $x = 2$ corresponds to February, and so on. (*Source:* www.weather.com)

$$f(x) = 19.5\cos\left[\frac{\pi}{6}(x - 7)\right] + 70.5$$

When is the average monthly temperature **(a)** 70.5°F **(b)** 55°F?

(Modeling) Alternating Electric Current *The study of alternating electric current requires the solutions of equations of the form*

$$i = I_{\text{max}} \sin 2\pi ft,$$

for time t in seconds, where i is instantaneous current in amperes, I_{max} is maximum current in amperes, and f is the number of cycles per second. (Source: Hannon, R. H., *Basic Technical Mathematics with Calculus,* W. B. Saunders Company.*) Find the least positive value of t, given the following data.*

53. $i = 40, I_{\text{max}} = 100, f = 60$

54. $i = 50, I_{\text{max}} = 100, f = 120$

55. $i = I_{\text{max}}, f = 60$

56. $i = \frac{1}{2} I_{\text{max}}, f = 60$

Chapter 6 Quiz (Sections 6.1–6.3)

1. Graph $y = \cos^{-1} x$, and indicate the coordinates of three points on the graph. Give the domain and range.

2. Find the exact value of each real number y.

(a) $y = \sin^{-1}\left(-\frac{\sqrt{2}}{2}\right)$ **(b)** $y = \tan^{-1} \sqrt{3}$ **(c)** $y = \sec^{-1}\left(-\frac{2\sqrt{3}}{3}\right)$

3. Use a calculator to give each value in decimal degrees.

(a) $\theta = \arccos 0.92341853$ **(b)** $\theta = \cot^{-1}(-1.0886767)$

4. Give the exact value of each expression without using a calculator.

(a) $\cos\left(\tan^{-1} \frac{4}{5}\right)$ **(b)** $\sin\left(\cos^{-1}\left(-\frac{1}{2}\right) + \tan^{-1}(-\sqrt{3})\right)$

Solve each equation for exact solutions over the interval $[0°, 360°)$.

5. $2 \sin \theta - \sqrt{3} = 0$ **6.** $\cos \theta + 1 = 2 \sin^2 \theta$

Solve each equation for solutions over the interval $[0, 2\pi)$.

7. $\tan^2 x - 5 \tan x + 3 = 0$ **8.** $3 \cot 2x - \sqrt{3} = 0$

9. Solve $\cos \frac{x}{2} + \sqrt{3} = -\cos \frac{x}{2}$, giving all solutions in radians.

10. *(Modeling) Electromotive Force* In an electric circuit, suppose that

$$V = \cos 2\pi t$$

models the electromotive force in volts at t seconds. Find the least value of t where $0 \le t \le \frac{1}{2}$ for each value of V.

(a) $V = 1$ **(b)** $V = 0.30$

6.4 Equations Involving Inverse Trigonometric Functions

- Solving for x in Terms of y Using Inverse Functions
- Solving Inverse Trigonometric Equations

Solving for x in Terms of y Using Inverse Functions

EXAMPLE 1 Solving an Equation for a Specified Variable

Solve $y = 3 \cos 2x$ for x, where x is restricted to the interval $\left[0, \frac{\pi}{2}\right]$.

SOLUTION We want $\cos 2x$ alone on one side of the equation so that we can solve for $2x$, and then for x.

$$y = 3 \cos 2x$$ Our goal is to isolate x.

$$\frac{y}{3} = \cos 2x$$ Divide by 3. (Appendix A)

$$2x = \arccos \frac{y}{3}$$ Definition of arccosine (Section 6.1)

$$x = \frac{1}{2} \arccos \frac{y}{3}$$ Multiply by $\frac{1}{2}$.

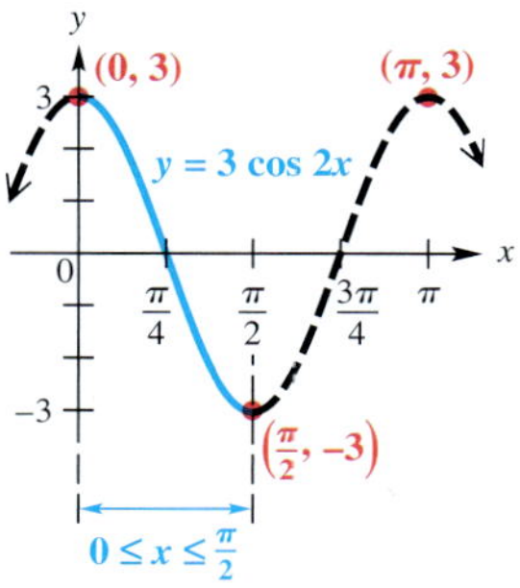

Figure 31

An equivalent form of this answer is $x = \frac{1}{2}\cos^{-1}\frac{y}{3}$.

Because the function $y = 3\cos 2x$ is periodic, with period π, there are infinitely many domain values (x-values) that will result in a given range value (y-value). For example, the x-values 0 and π both correspond to the y-value 3. See **Figure 31.** The restriction $0 \le x \le \frac{\pi}{2}$ given in the original problem ensures that this function is one-to-one, and, correspondingly, that

$$x = \frac{1}{2}\operatorname{arccos}\frac{y}{3}$$

has a one-to-one relationship. Thus, each y-value in $[-3, 3]$ substituted into this equation will lead to a single x-value.

✔ ***Now Try Exercise 9.***

Solving Inverse Trigonometric Equations

EXAMPLE 2 Solving an Equation Involving an Inverse Trigonometric Function

Solve $2 \arcsin x = \pi$.

SOLUTION First solve for arcsin x, and then for x.

$$2 \arcsin x = \pi \quad \text{Original equation}$$

$$\arcsin x = \frac{\pi}{2} \quad \text{Divide by 2.}$$

$$x = \sin\frac{\pi}{2} \quad \text{Definition of arcsine (Section 6.1)}$$

$$x = 1 \quad \text{(Section 3.3)}$$

CHECK

$$2 \arcsin x = \pi \quad \text{Original equation}$$

$$2 \arcsin 1 \stackrel{?}{=} \pi \quad \text{Let } x = 1.$$

$$2\left(\frac{\pi}{2}\right) \stackrel{?}{=} \pi \quad \text{Substitute the inverse value.}$$

$$\pi = \pi \ ✓ \quad \text{True}$$

The solution set is $\{1\}$.

✔ ***Now Try Exercise 25.***

EXAMPLE 3 Solving an Equation Involving Inverse Trigonometric Functions

Solve $\cos^{-1} x = \sin^{-1}\frac{1}{2}$.

SOLUTION Let $\sin^{-1}\frac{1}{2} = u$. Then $\sin u = \frac{1}{2}$, and for u in quadrant I we have the following.

$$\cos^{-1} x = \sin^{-1}\frac{1}{2} \quad \text{Original equation}$$

$$\cos^{-1} x = u \quad \text{Substitute.}$$

$$\cos u = x \quad \text{Alternative form (Section 6.1)}$$

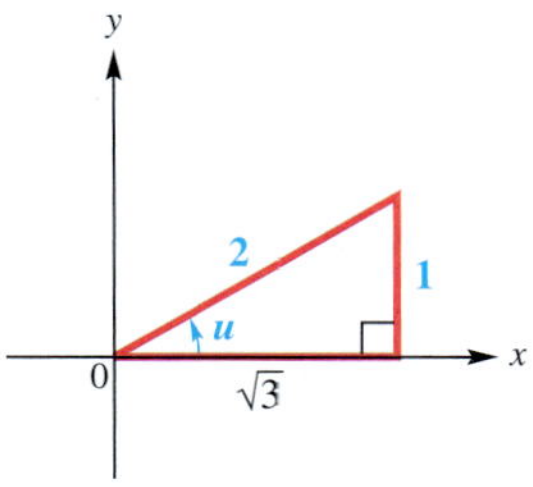

Figure 32

Sketch a triangle and label it using the facts that u is in quadrant I and $\sin u = \frac{1}{2}$. See **Figure 32.** Since $x = \cos u$, $x = \frac{\sqrt{3}}{2}$, and the solution set is $\left\{\frac{\sqrt{3}}{2}\right\}$. *Check.*

✔ ***Now Try Exercise 33.***

EXAMPLE 4 Solving an Inverse Trigonometric Equation Using an Identity

Solve $\arcsin x - \arccos x = \frac{\pi}{6}$.

SOLUTION Isolate one inverse function on one side of the equation.

$$\arcsin x - \arccos x = \frac{\pi}{6} \quad \text{Original equation}$$

$$\arcsin x = \arccos x + \frac{\pi}{6} \quad \text{Add } \arccos x. \quad (1)$$

$$x = \sin\left(\arccos x + \frac{\pi}{6}\right) \quad \text{Definition of arcsine}$$

Let $u = \arccos x$. The arccosine function yields angles in quadrants I and II, so $0 \leq u \leq \pi$ by definition.

$$x = \sin\left(u + \frac{\pi}{6}\right) \quad \text{Substitute.}$$

$$x = \sin u \cos \frac{\pi}{6} + \cos u \sin \frac{\pi}{6} \quad \text{Sine sum identity (Section 5.4)} \quad (2)$$

From equation (1) and by the definition of the arcsine function,

$$-\frac{\pi}{2} \leq \arccos x + \frac{\pi}{6} \leq \frac{\pi}{2} \quad \text{Range of arcsine is } \left[-\frac{\pi}{2}, \frac{\pi}{2}\right].$$

$$-\frac{2\pi}{3} \leq \arccos x \leq \frac{\pi}{3}. \quad \text{Subtract } \tfrac{\pi}{6} \text{ from each part. (Appendix A)}$$

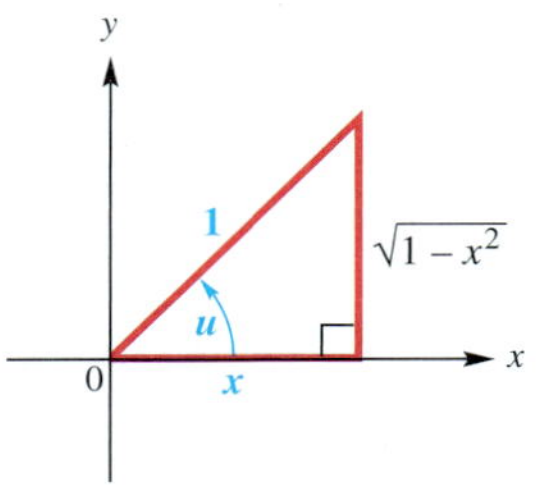

Figure 33

Since $0 \leq \arccos x \leq \pi$ **and** $-\frac{2\pi}{3} \leq \arccos x \leq \frac{\pi}{3}$, the intersection yields $0 \leq \arccos x \leq \frac{\pi}{3}$. This places u in quadrant I, and we can sketch the triangle in **Figure 33.** From this triangle we find that $\sin u = \sqrt{1 - x^2}$. Now substitute into equation (2) using $\sin u = \sqrt{1 - x^2}$, $\sin \frac{\pi}{6} = \frac{1}{2}$, $\cos \frac{\pi}{6} = \frac{\sqrt{3}}{2}$, and $\cos u = x$.

$$x = \sin u \cos \frac{\pi}{6} + \cos u \sin \frac{\pi}{6} \quad (2)$$

$$x = \left(\sqrt{1 - x^2}\right)\frac{\sqrt{3}}{2} + x \cdot \frac{1}{2} \quad \text{Substitute.}$$

$$2x = \left(\sqrt{1 - x^2}\right)\sqrt{3} + x \quad \text{Multiply by 2.}$$

$$x = \left(\sqrt{3}\right)\sqrt{1 - x^2} \quad \text{Subtract } x\text{; commutative property}$$

Square *each* factor.

$$x^2 = 3(1 - x^2) \quad \text{Square each side.}$$

$$x^2 = 3 - 3x^2 \quad \text{Distributive property}$$

$$4x^2 = 3 \quad \text{Add } 3x^2.$$

$$x^2 = \frac{3}{4} \quad \text{Divide by 4.}$$

Choose the positive square root, $x > 0$.

$$x = \sqrt{\frac{3}{4}} \quad \text{Take the square root on each side. (Appendix A)}$$

$$x = \frac{\sqrt{3}}{2} \quad \text{Quotient rule: } \sqrt[n]{\frac{a}{b}} = \frac{\sqrt[n]{a}}{\sqrt[n]{b}}$$

CHECK

$$\arcsin x - \arccos x = \frac{\pi}{6} \quad \text{Original equation}$$

$$\arcsin \frac{\sqrt{3}}{2} - \arccos \frac{\sqrt{3}}{2} \stackrel{?}{=} \frac{\pi}{6} \quad \text{Let } x = \tfrac{\sqrt{3}}{2}.$$

$$\frac{\pi}{3} - \frac{\pi}{6} \stackrel{?}{=} \frac{\pi}{6} \quad \text{Substitute inverse values.}$$

$$\frac{\pi}{6} = \frac{\pi}{6} \quad ✓ \text{ True}$$

The solution set is $\left\{\frac{\sqrt{3}}{2}\right\}$.

✔ *Now Try Exercise 35.*

6.4 Exercises

Concept Check *Answer each question.*

1. Which one of the following equations has solution 0?

A. $\arctan 1 = x$ **B.** $\arccos 0 = x$ **C.** $\arcsin 0 = x$

2. Which one of the following equations has solution $\frac{\pi}{4}$?

A. $\arcsin \frac{\sqrt{2}}{2} = x$ **B.** $\arccos\left(-\frac{\sqrt{2}}{2}\right) = x$ **C.** $\arctan \frac{\sqrt{3}}{3} = x$

3. Which one of the following equations has solution $\frac{3\pi}{4}$?

A. $\arctan 1 = x$ **B.** $\arcsin \frac{\sqrt{2}}{2} = x$ **C.** $\arccos\left(-\frac{\sqrt{2}}{2}\right) = x$

4. Which one of the following equations has solution $-\frac{\pi}{6}$?

A. $\arctan \frac{\sqrt{3}}{3} = x$ **B.** $\arccos\left(-\frac{1}{2}\right) = x$ **C.** $\arcsin\left(-\frac{1}{2}\right) = x$

Solve each equation for x, where x is restricted to the given interval. ***See Example 1.***

5. $y = 5 \cos x$, for x in $[0, \pi]$

6. $y = \frac{1}{4} \sin x$, for x in $\left[-\frac{\pi}{2}, \frac{\pi}{2}\right]$

7. $y = \frac{1}{2} \cot 3x$, for x in $\left(0, \frac{\pi}{3}\right)$

8. $y = \frac{1}{12} \sec x$, for x in $\left[0, \frac{\pi}{2}\right) \cup \left(\frac{\pi}{2}, \pi\right]$

9. $y = 3 \tan 2x$, for x in $\left(-\frac{\pi}{4}, \frac{\pi}{4}\right)$

10. $y = 3 \sin \frac{x}{2}$, for x in $[-\pi, \pi]$

11. $y = 6 \cos \frac{x}{4}$, for x in $[0, 4\pi]$

12. $y = -\sin \frac{x}{3}$, for x in $\left[-\frac{3\pi}{2}, \frac{3\pi}{2}\right]$

13. $y = -2 \cos 5x$, for x in $\left[0, \frac{\pi}{5}\right]$

14. $y = 3 \cot 5x$, for x in $\left(0, \frac{\pi}{5}\right)$

15. $y = \cos(x + 3)$, for x in $[-3, \pi - 3]$

16. $y = \tan(2x - 1)$, for x in $\left(\frac{1}{2} - \frac{\pi}{4}, \frac{1}{2} + \frac{\pi}{4}\right)$

17. $y = \sin x - 2$, for x in $\left[-\frac{\pi}{2}, \frac{\pi}{2}\right]$

18. $y = \cot x + 1$, for x in $(0, \pi)$

19. $y = -4 + 2\sin x$, for x in $\left[-\frac{\pi}{2}, \frac{\pi}{2}\right]$

20. $y = 4 + 3\cos x$, for x in $[0, \pi]$

21. $y = \sqrt{2} + 3\sec 2x$, for x in $\left[0, \frac{\pi}{4}\right) \cup \left(\frac{\pi}{4}, \frac{\pi}{2}\right]$

22. $y = -\sqrt{3} + 2\csc \frac{x}{2}$, for x in $[-\pi, 0) \cup (0, \pi]$

23. Refer to **Exercise 17.** A student attempting to solve this equation wrote as the first step $y = \sin(x - 2)$, inserting parentheses as shown. Explain why this is incorrect.

24. Explain why the equation $\sin^{-1} x = \cos^{-1} 2$ cannot have a solution. (No work is required.)

Solve each equation for exact solutions. ***See Examples 2 and 3.***

25. $-4 \arcsin x = \pi$

26. $6 \arccos x = 5\pi$

27. $\frac{4}{3}\cos^{-1}\frac{x}{4} = \pi$

28. $4\pi + 4\tan^{-1} x = \pi$

29. $2 \arccos\left(\frac{x - \pi}{3}\right) = 2\pi$

30. $\arccos\left(x - \frac{\pi}{3}\right) = \frac{\pi}{6}$

31. $\arcsin x = \arctan \frac{3}{4}$

32. $\arctan x = \arccos \frac{5}{13}$

33. $\cos^{-1} x = \sin^{-1} \frac{3}{5}$

34. $\cot^{-1} x = \tan^{-1} \frac{4}{3}$

Solve each equation for exact solutions. ***See Example 4.***

35. $\sin^{-1} x - \tan^{-1} 1 = -\frac{\pi}{4}$

36. $\sin^{-1} x + \tan^{-1} \sqrt{3} = \frac{2\pi}{3}$

37. $\arccos x + 2 \arcsin \frac{\sqrt{3}}{2} = \pi$

38. $\arccos x + 2 \arcsin \frac{\sqrt{3}}{2} = \frac{\pi}{3}$

39. $\arcsin 2x + \arccos x = \frac{\pi}{6}$

40. $\arcsin 2x + \arcsin x = \frac{\pi}{2}$

41. $\cos^{-1} x + \tan^{-1} x = \frac{\pi}{2}$

42. $\sin^{-1} x + \tan^{-1} x = 0$

43. Provide graphical support for the solution in **Example 4** by showing that the graph of

$$y = \arcsin x - \arccos x - \frac{\pi}{6} \quad \text{has } x\text{-intercept} \quad \frac{\sqrt{3}}{2} \approx 0.8660254.$$

44. Provide graphical support for the solution in **Example 4** by showing that the x-coordinate of the point of intersection of the graphs of

$$Y_1 = \arcsin X - \arccos X \quad \text{and} \quad Y_2 = \frac{\pi}{6} \quad \text{is} \quad \frac{\sqrt{3}}{2} \approx 0.8660254.$$

The following equations cannot be solved by algebraic methods. Use a graphing calculator to find all solutions over the interval $[0, 6]$. Express solutions to four decimal places.

45. $(\arctan x)^3 - x + 2 = 0$

46. $\pi \sin^{-1}(0.2x) - 3 = -\sqrt{x}$

(Modeling) Solve each problem.

47. ***Tone Heard by a Listener*** When two sources located at different positions produce the same pure tone, the human ear will often hear one sound that is equal to the sum of the individual tones. Since the sources are at different locations, they will have different phase angles ϕ. If two speakers located at different positions produce pure tones $P_1 = A_1 \sin(2\pi ft + \phi_1)$ and $P_2 = A_2 \sin(2\pi ft + \phi_2)$, where $-\frac{\pi}{4} \le \phi_1, \phi_2 \le \frac{\pi}{4}$, then the resulting tone heard by a listener can be written as $P = A \sin(2\pi ft + \phi)$, where

$$A = \sqrt{(A_1 \cos \phi_1 + A_2 \cos \phi_2)^2 + (A_1 \sin \phi_1 + A_2 \sin \phi_2)^2}$$

$$\text{and} \quad \phi = \arctan\left(\frac{A_1 \sin \phi_1 + A_2 \sin \phi_2}{A_1 \cos \phi_1 + A_2 \cos \phi_2}\right).$$

(*Source:* Fletcher, N. and T. Rossing, *The Physics of Musical Instruments,* Second Edition, Springer-Verlag.)

(a) Calculate A and ϕ if $A_1 = 0.0012$, $\phi_1 = 0.052$, $A_2 = 0.004$, and $\phi_2 = 0.61$. Also find an expression for $P = A \sin(2\pi ft + \phi)$ if $f = 220$.

(b) Graph $Y_1 = P$ and $Y_2 = P_1 + P_2$ on the same coordinate axes over the interval $[0, 0.01]$. Are the two graphs the same?

48. ***Tone Heard by a Listener*** Repeat **Exercise 47.** Use $A_1 = 0.0025$, $\phi_1 = \frac{\pi}{7}$, $A_2 = 0.001$; $\phi_2 = \frac{\pi}{6}$, and $f = 300$.

49. ***Depth of Field*** When a large-view camera is used to take a picture of an object that is not parallel to the film, the lens board should be tilted so that the planes containing the subject, the lens board, and the film intersect in a line. This gives the best "depth of field." See the figure. (*Source:* Bushaw, D., et al., *A Sourcebook of Applications of School Mathematics,* Mathematical Association of America.)

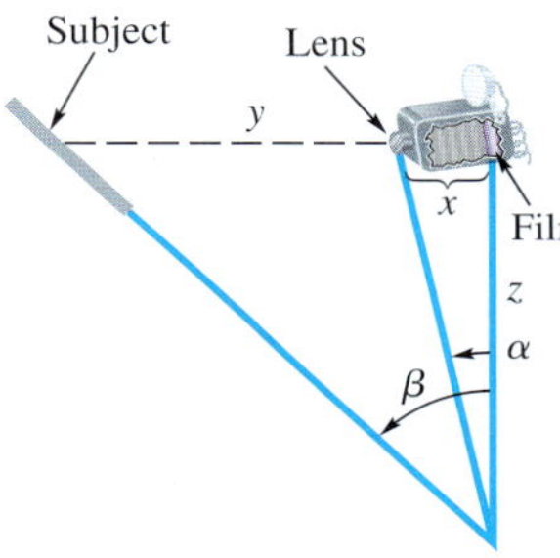

(a) Write two equations, one relating α, x, and z, and the other relating β, x, y, and z.

(b) Eliminate z from the equations in part (a) to get one equation relating α, β, x, and y.

(c) Solve the equation from part (b) for α.

(d) Solve the equation from part (b) for β.

50. ***Programming Language for Inverse Functions*** In Visual Basic, a widely used programming language for PCs, the only inverse trigonometric function available is arctangent. The other inverse trigonometric functions can be expressed in terms of arctangent as follows.

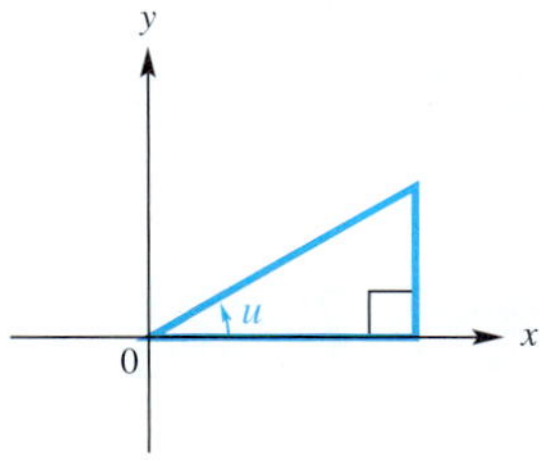

(a) Let $u = \arcsin x$. Solve the equation for x in terms of u.

(b) Use the result of part (a) to label the three sides of the triangle in the figure in terms of x.

(c) Use the triangle from part (b) to write an equation for $\tan u$ in terms of x.

(d) Solve the equation from part (c) for u.

51. ***Alternating Electric Current*** In the study of alternating electric current, instantaneous voltage is modeled by

$$E = E_{\text{max}} \sin 2\pi ft,$$

where f is the number of cycles per second, E_{max} is the maximum voltage, and t is time in seconds.

(a) Solve the equation for t.

(b) Find the least positive value of t if $E_{\text{max}} = 12$, $E = 5$, and $f = 100$. Use a calculator.

52. ***Viewing Angle of an Observer*** While visiting a museum, Marsha Langlois views a painting that is 3 ft high and hangs 6 ft above the ground. See the figure. Assume her eyes are 5 ft above the ground, and let x be the distance from the spot where she is standing to the wall displaying the painting.

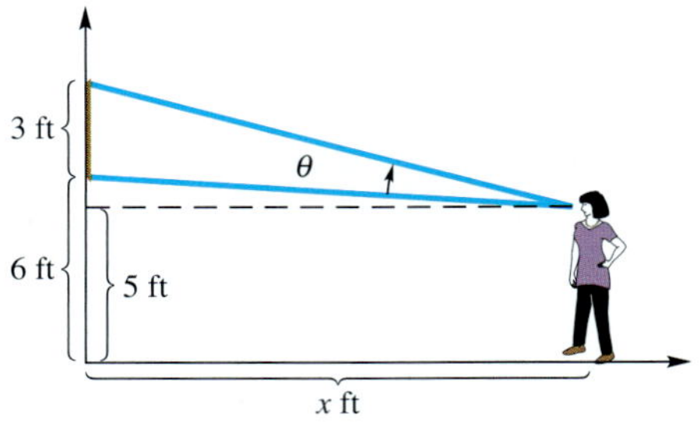

(a) Show that θ, the viewing angle subtended by the painting, is given by

$$\theta = \tan^{-1}\left(\frac{4}{x}\right) - \tan^{-1}\left(\frac{1}{x}\right).$$

(b) Find the value of x to the nearest hundredth for each value of θ.

(i) $\theta = \frac{\pi}{6}$ **(ii)** $\theta = \frac{\pi}{8}$

(c) Find the value of θ to the nearest hundredth for each value of x.

(i) $x = 4$ **(ii)** $x = 3$

53. ***Movement of an Arm*** In the equation below, t is time (in seconds) and y is the angle formed by a rhythmically moving arm.

$$y = \frac{1}{3}\sin\frac{4\pi t}{3}$$

(a) Solve the equation for t.

(b) At what time, to the nearest hundredth of a second, does the arm first form an angle of 0.3 radian?

54. The function $y = \sec^{-1} x$ is not found on graphing calculators. However, with some models it can be graphed as

$$y = \frac{\pi}{2} - ((x > 0) - (x < 0))\left(\frac{\pi}{2} - \tan^{-1}\left(\sqrt{(x^2 - 1)}\right)\right).$$

Use the formula to obtain the graph of $y = \sec^{-1} x$ in the window $[-4, 4]$ by $[0, \pi]$.

Chapter 6 Test Prep

Key Terms

6.1 one-to-one function
inverse function

New Symbols

f^{-1}	inverse of function f	$\cot^{-1} x$ ($\text{arccot}\, x$)	inverse cotangent of x
$\sin^{-1} x$ ($\arcsin x$)	inverse sine of x	$\sec^{-1} x$ ($\text{arcsec}\, x$)	inverse secant of x
$\cos^{-1} x$ ($\arccos x$)	inverse cosine of x	$\csc^{-1} x$ ($\text{arccsc}\, x$)	inverse cosecant of x
$\tan^{-1} x$ ($\arctan x$)	inverse tangent of x		

Quick Review

Concepts	Examples

6.1 Inverse Circular Functions

Inverse Function	Domain	Range: Interval	Range: Quadrants of the Unit Circle
$y = \sin^{-1} x$	$[-1, 1]$	$\left[-\frac{\pi}{2}, \frac{\pi}{2}\right]$	I and IV
$y = \cos^{-1} x$	$[-1, 1]$	$[0, \pi]$	I and II
$y = \tan^{-1} x$	$(-\infty, \infty)$	$\left(-\frac{\pi}{2}, \frac{\pi}{2}\right)$	I and IV
$y = \cot^{-1} x$	$(-\infty, \infty)$	$(0, \pi)$	I and II
$y = \sec^{-1} x$	$(-\infty, -1] \cup [1, \infty)$	$\left[0, \frac{\pi}{2}\right) \cup \left(\frac{\pi}{2}, \pi\right]$	I and II
$y = \csc^{-1} x$	$(-\infty, -1] \cup [1, \infty)$	$\left[-\frac{\pi}{2}, 0\right) \cup \left(0, \frac{\pi}{2}\right]$	I and IV

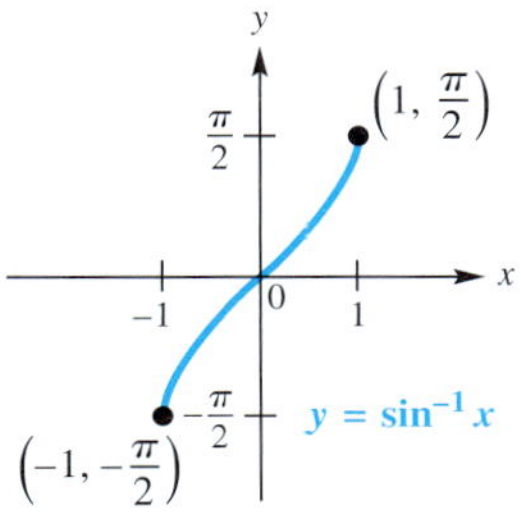

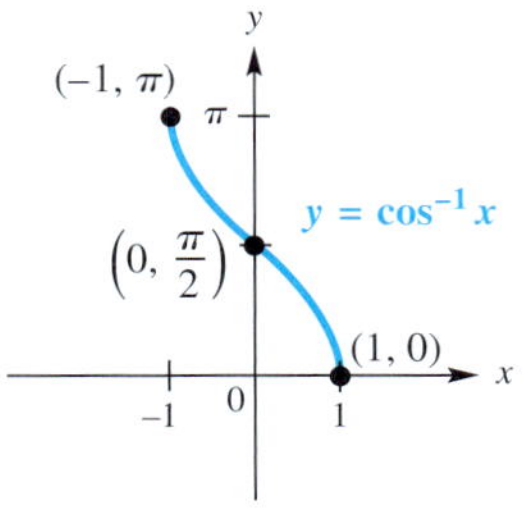

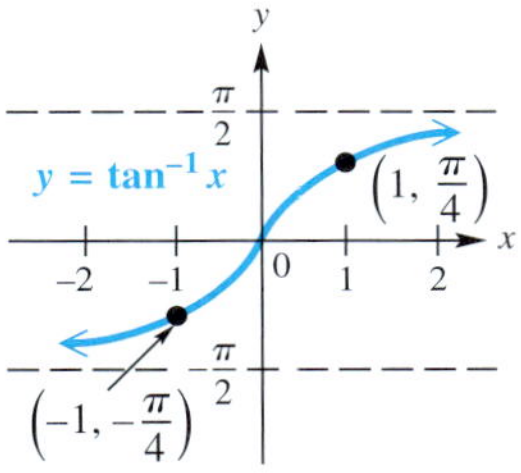

See **Section 6.1** for graphs of the other inverse circular (trigonometric) functions.

Evaluate $y = \cos^{-1} 0$.

Write $y = \cos^{-1} 0$ as $\cos y = 0$. Then

$$y = \frac{\pi}{2},$$

because $\cos \frac{\pi}{2} = 0$ and $\frac{\pi}{2}$ is in the range of $\cos^{-1} x$.

Use a calculator to find y in radians if $y = \sec^{-1}(-3)$. With the calculator in radian mode, enter $\sec^{-1}(-3)$ as $\cos^{-1}\left(\frac{1}{-3}\right)$ to get

$$y \approx 1.9106332.$$

Evaluate $\sin\left(\tan^{-1}\left(-\frac{3}{4}\right)\right)$.

Let $u = \tan^{-1}\left(-\frac{3}{4}\right)$. Then $\tan u = -\frac{3}{4}$. Since $\tan u$ is negative when u is in quadrant IV, sketch a triangle as shown.

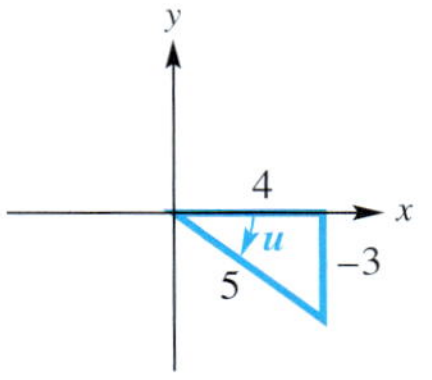

We want $\sin\left(\tan^{-1}\left(-\frac{3}{4}\right)\right) = \sin u$. From the triangle,

$$\sin u = -\frac{3}{5}.$$

6.2 Trigonometric Equations I

6.3 Trigonometric Equations II

Solving a Trigonometric Equation

1. Decide whether the equation is linear or quadratic in form, so that you can determine the solution method.
2. If only one trigonometric function is present, solve the equation for that function.
3. If more than one trigonometric function is present, rearrange the equation so that one side equals 0. Then try to factor and set each factor equal to 0 to solve.
4. If the equation is quadratic in form, but not factorable, use the quadratic formula. Check that solutions are in the desired interval.
5. Try using identities to change the form of the equation. It may be helpful to square each side of the equation first. In this case, check for extraneous solutions.

Solve $\tan \theta + \sqrt{3} = 2\sqrt{3}$ over the interval $[0°, 360°)$. Use a linear method.

$\tan \theta + \sqrt{3} = 2\sqrt{3}$	Original equation
$\tan \theta = \sqrt{3}$	Subtract $\sqrt{3}$.
$\theta = 60°$	Definition of inverse tangent

Another solution over $[0°, 360°)$ is

$$\theta = 60° + 180° = 240°.$$

The solution set is $\{60°, 240°\}$.

(continued)

Concepts	Examples
	Solve $2\cos^2 x = 1$ for all solutions, using a double-angle identity. $2\cos^2 x = 1$ Original equation $2\cos^2 x - 1 = 0$ Subtract 1. $\cos 2x = 0$ Cosine double-angle identity $2x = \frac{\pi}{2} + 2n\pi$ and $2x = \frac{3\pi}{2} + 2n\pi$ Add integer multiples of 2π. $x = \frac{\pi}{4} + n\pi$ and $x = \frac{3\pi}{4} + n\pi$ Divide by 2. All solutions are given by the following set, where π represents the period of $\cos 2x$. $\left\{\frac{\pi}{4} + n\pi, \frac{3\pi}{4} + n\pi, \text{ where } n \text{ is any integer}\right\}$
6.4 Equations Involving Inverse Trigonometric Functions	
We solve equations of the form $y = f(x)$, where $f(x)$ is a trigonometric function, using inverse trigonometric functions.	Solve $y = 2\sin 3x$ for x, where x is restricted to the interval $\left[-\frac{\pi}{6}, \frac{\pi}{6}\right]$. $y = 2\sin 3x$ Original equation $\frac{y}{2} = \sin 3x$ Divide by 2. $3x = \arcsin \frac{y}{2}$ Definition of arcsine $x = \frac{1}{3}\arcsin \frac{y}{2}$ Multiply by $\frac{1}{3}$.
Techniques introduced in this section also show how to solve equations that involve inverse functions.	Solve. $4\tan^{-1} x = \pi$ Original equation $\tan^{-1} x = \frac{\pi}{4}$ Divide by 4. $x = \tan \frac{\pi}{4} = 1$ Definition of arctangent The solution set is $\{1\}$.

Chapter 6 Review Exercises

1. Graph the inverse sine, cosine, and tangent functions, indicating three points on each graph. Give the domain and range for each.

Concept Check *Determine whether each statement is* true *or* false. *If false, tell why.*

2. The ranges of the inverse tangent and inverse cotangent functions are the same.
3. It is true that $\sin \frac{11\pi}{6} = -\frac{1}{2}$, and therefore $\arcsin\left(-\frac{1}{2}\right) = \frac{11\pi}{6}$.
4. For all x, $\tan(\tan^{-1} x) = x$.

Give the exact real number value of y. Do not use a calculator.

5. $y = \sin^{-1} \frac{\sqrt{2}}{2}$ **6.** $y = \arccos\left(-\frac{1}{2}\right)$ **7.** $y = \tan^{-1}(-\sqrt{3})$

8. $y = \arcsin(-1)$ **9.** $y = \cos^{-1}\left(-\frac{\sqrt{2}}{2}\right)$ **10.** $y = \arctan \frac{\sqrt{3}}{3}$

11. $y = \sec^{-1}(-2)$ **12.** $y = \text{arccsc}\, \frac{2\sqrt{3}}{3}$ **13.** $y = \text{arccot}(-1)$

Give the degree measure of θ. Do not use a calculator.

14. $\theta = \arccos \frac{1}{2}$ **15.** $\theta = \arcsin\left(-\frac{\sqrt{3}}{2}\right)$ **16.** $\theta = \tan^{-1} 0$

Use a calculator to give the degree measure of θ to the nearest hundredth.

17. $\theta = \arctan 1.7804675$ **18.** $\theta = \sin^{-1}(-0.66045320)$ **19.** $\theta = \cos^{-1} 0.80396577$

20. $\theta = \cot^{-1} 4.5046388$ **21.** $\theta = \text{arcsec}\, 3.4723155$ **22.** $\theta = \csc^{-1} 7.4890096$

Evaluate the following without using a calculator.

23. $\cos(\arccos(-1))$ **24.** $\sin\left(\arcsin\left(-\frac{\sqrt{3}}{2}\right)\right)$ **25.** $\arccos\left(\cos \frac{3\pi}{4}\right)$

26. $\text{arcsec}(\sec \pi)$ **27.** $\tan^{-1}\left(\tan \frac{\pi}{4}\right)$ **28.** $\cos^{-1}(\cos 0)$

29. $\sin\left(\arccos \frac{3}{4}\right)$ **30.** $\cos(\arctan 3)$ **31.** $\cos(\csc^{-1}(-2))$

32. $\sec\left(2 \sin^{-1}\left(-\frac{1}{3}\right)\right)$ **33.** $\tan\left(\arcsin \frac{3}{5} + \arccos \frac{5}{7}\right)$

Write each of the following as an algebraic (nontrigonometric) expression in u, $u > 0$.

34. $\cos\left(\arctan \frac{u}{\sqrt{1-u^2}}\right)$ **35.** $\tan\left(\text{arcsec}\, \frac{\sqrt{u^2+1}}{u}\right)$

Solve each equation for exact solutions over the interval $[0, 2\pi)$ where appropriate. Round approximate solutions to four decimal places.

36. $\sin^2 x = 1$ **37.** $2 \tan x - 1 = 0$

38. $3 \sin^2 x - 5 \sin x + 2 = 0$ **39.** $\tan x = \cot x$

40. $\sec^2 2x = 2$ **41.** $\tan^2 2x - 1 = 0$

Give all exact solutions, in radians, for each equation.

42. $\sec \frac{x}{2} = \cos \frac{x}{2}$ **43.** $\cos 2x + \cos x = 0$ **44.** $4 \sin x \cos x = \sqrt{3}$

Solve each equation for exact solutions over the interval $[0°, 360°)$ where appropriate. Round approximate solutions to the nearest tenth of a degree.

45. $\sin^2 \theta + 3 \sin \theta + 2 = 0$ **46.** $2 \tan^2 \theta = \tan \theta + 1$

47. $\sin 2\theta = \cos 2\theta + 1$ **48.** $2 \sin 2\theta = 1$

49. $3 \cos^2 \theta + 2 \cos \theta - 1 = 0$ **50.** $5 \cot^2 \theta - \cot \theta - 2 = 0$

Give all exact solutions, in degrees, for each equation.

51. $2\sqrt{3} \cos \frac{\theta}{2} = -3$ **52.** $\sin \theta - \cos 2\theta = 0$ **53.** $\tan \theta - \sec \theta = 1$

Solve each equation for x. In Exercises 58–61, x is restricted to the given interval.

54. $4\pi - 4\cot^{-1} x = \pi$

55. $\frac{4}{3} \arctan \frac{x}{2} = \pi$

56. $\arccos x = \arcsin \frac{2}{7}$

57. $\arccos x + \arctan 1 = \frac{11\pi}{12}$

58. $y = 3\cos \frac{x}{2}$, for x in $[0, 2\pi]$

59. $y = \frac{1}{2}\sin x$, for x in $\left[-\frac{\pi}{2}, \frac{\pi}{2}\right]$

60. $y = \frac{4}{5}\sin x - \frac{3}{5}$, for x in $\left[-\frac{\pi}{2}, \frac{\pi}{2}\right]$

61. $y = \frac{1}{2}\tan(3x + 2)$, for x in $\left(-\frac{2}{3} - \frac{\pi}{6}, -\frac{2}{3} + \frac{\pi}{6}\right)$

62. Solve $d = 550 + 450\cos\left(\frac{\pi}{50}t\right)$ for t, where t is in the interval $[0, 50]$.

(Modeling) Solve each problem.

63. *Viewing Angle of an Observer* A 10-ft-wide chalkboard is situated 5 ft from the left wall of a classroom. See the figure. A student sitting next to the wall x feet from the front of the classroom has a viewing angle of θ radians.

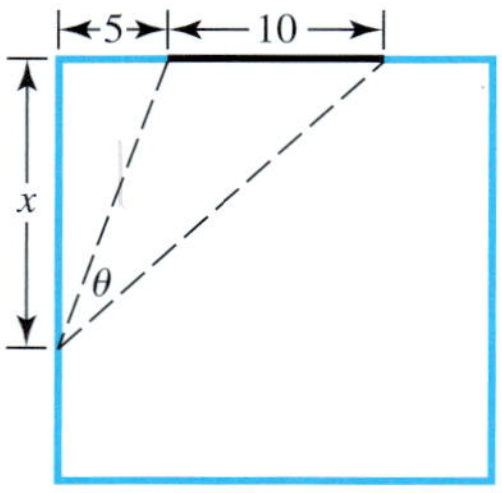

(a) Show that the value of θ is given by the function

$$f(x) = \arctan\left(\frac{15}{x}\right) - \arctan\left(\frac{5}{x}\right).$$

(b) Graph $f(x)$ with a graphing calculator to estimate the value of x that maximizes the viewing angle.

64. *Snell's Law* Recall Snell's law from **Exercises 69 and 70** of **Section 2.3:**

$$\frac{c_1}{c_2} = \frac{\sin \theta_1}{\sin \theta_2},$$

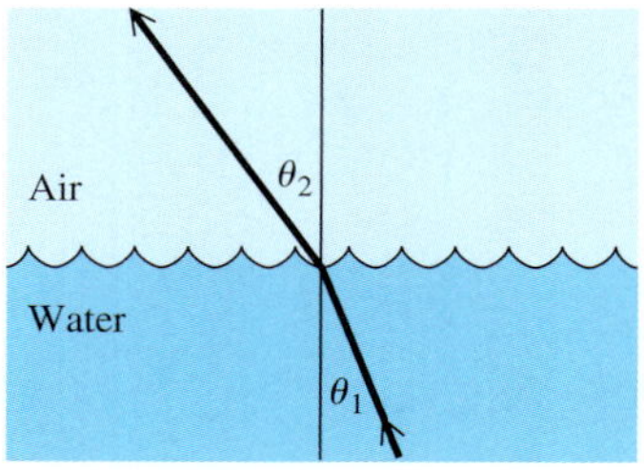

where c_1 is the speed of light in one medium, c_2 is the speed of light in a second medium, and θ_1 and θ_2 are the angles shown in the figure. Suppose a light is shining up through water into the air as in the figure. As θ_1 increases, θ_2 approaches 90°, at which point no light will emerge from the water. Assume the ratio $\frac{c_1}{c_2}$ in this case is 0.752. For what value of θ_1 does $\theta_2 = 90°$? This value of θ_1 is the **critical angle** for water.

65. *Snell's Law* Refer to **Exercise 64.** What happens when θ_1 is greater than the critical angle?

66. *British Nautical Mile* The British nautical mile is defined as the length of a minute of arc of a meridian. Since Earth is flat at its poles, the nautical mile, in feet, is given by

$$L = 6077 - 31\cos 2\theta,$$

where θ is the latitude in degrees. See the figure. (*Source:* Bushaw, D., et al., *A Sourcebook of Applications of School Mathematics,* Mathematical Association of America.)

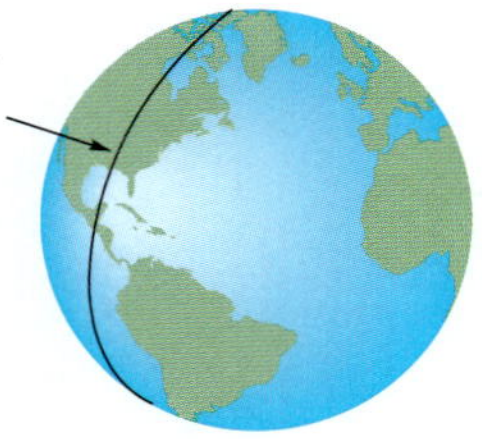

(a) Find the latitude between 0° and 90° at which the nautical mile is 6074 ft.

(b) At what latitude between 0° and 180° is the nautical mile 6108 ft?

(c) In the United States, the nautical mile is defined everywhere as 6080.2 ft. At what latitude between 0° and 90° does this agree with the British nautical mile?

67. The function $y = \csc^{-1} x$ is not found on graphing calculators. However, with some models it can be graphed as follows.

$$y = ((x > 0) - (x < 0))\left(\frac{\pi}{2} - \tan^{-1}\left(\sqrt{(x^2 - 1)}\right)\right)$$

Use the formula to obtain the graph of $y = \csc^{-1} x$ in the window $[-4, 4]$ by $\left[-\frac{\pi}{2}, \frac{\pi}{2}\right]$.

68. **(a)** Use the graph of $y = \sin^{-1} x$ to approximate $\sin^{-1} 0.4$.
(b) Use the inverse sine key of a graphing calculator to approximate $\sin^{-1} 0.4$.

Chapter 6 Test

1. Graph $y = \sin^{-1} x$, and indicate the coordinates of three points on the graph. Give the domain and range.

2. Find the exact value of y for each equation.

(a) $y = \arccos\left(-\frac{1}{2}\right)$ **(b)** $y = \sin^{-1}\left(-\frac{\sqrt{3}}{2}\right)$

(c) $y = \tan^{-1} 0$ **(d)** $y = \text{arcsec}(-2)$

3. Give the degree measure of θ.

(a) $\theta = \arccos \frac{\sqrt{3}}{2}$ **(b)** $\theta = \tan^{-1}(-1)$

(c) $\theta = \cot^{-1}(-1)$ **(d)** $\theta = \csc^{-1}\left(-\frac{2\sqrt{3}}{3}\right)$

4. Use a calculator to give each value in decimal degrees to the nearest hundredth.

(a) $\sin^{-1} 0.67610476$ **(b)** $\sec^{-1} 1.0840880$

(c) $\cot^{-1}(-0.7125586)$

5. Find each exact value.

(a) $\cos\left(\arcsin \frac{2}{3}\right)$ **(b)** $\sin\left(2 \cos^{-1} \frac{1}{3}\right)$

6. Explain why $\sin^{-1} 3$ is not defined.

7. Explain why $\arcsin\left(\sin \frac{5\pi}{6}\right) \neq \frac{5\pi}{6}$.

8. Write $\tan(\arcsin u)$ as an algebraic (nontrigonometric) expression in u, $u > 0$.

Solve each equation for exact solutions over the interval $[0°, 360°)$ where appropriate. Round approximate solutions to the nearest tenth of a degree.

9. $-3 \sec \theta + 2\sqrt{3} = 0$ **10.** $\sin^2 \theta = \cos^2 \theta + 1$ **11.** $\csc^2 \theta - 2 \cot \theta = 4$

Solve each equation for exact solutions over the interval $[0, 2\pi)$ where appropriate. Round approximate solutions to four decimal places.

12. $\cos x = \cos 2x$ **13.** $\sqrt{2} \cos 3x - 1 = 0$ **14.** $\sin x \cos x = \frac{1}{3}$

Solve each equation for all exact solutions in radians (for x) or in degrees (for θ). Write answers using the least possible nonnegative angle measures.

15. $\sin^2 \theta = -\cos 2\theta$ **16.** $2\sqrt{3} \sin \frac{x}{2} = 3$ **17.** $\csc x - \cot x = 1$

18. Solve each equation for x, where x is restricted to the given interval.

(a) $y = \cos 3x, \text{ for } x \text{ in } \left[0, \frac{\pi}{3}\right]$

(b) $y = 4 + 3 \cot x, \text{ for } x \text{ in } (0, \pi)$

19. Solve each equation for exact solutions.

(a) $\arcsin x = \arctan \frac{4}{3}$ **(b)** $\operatorname{arccot} x + 2 \arcsin \frac{\sqrt{3}}{2} = \pi$

20. *(Modeling) Movement of a Runner's Arm* A runner's arm swings rhythmically according to the model

$$y = \frac{\pi}{8} \cos\left[\pi\left(t - \frac{1}{3}\right)\right],$$

where y represents the angle between the actual position of the upper arm and the downward vertical position, and t represents time in seconds. At what times over the interval $[0, 3)$ is the angle y equal to 0?

7 Applications of Trigonometry and Vectors

Surveyors use a method known as *triangulation* to measure distances when direct measurements cannot be made due to obstructions in the line of sight.

7.1 Oblique Triangles and the Law of Sines

- Congruency and Oblique Triangles
- Derivation of the Law of Sines
- Solving SAA and ASA Triangles (Case 1)
- Area of a Triangle

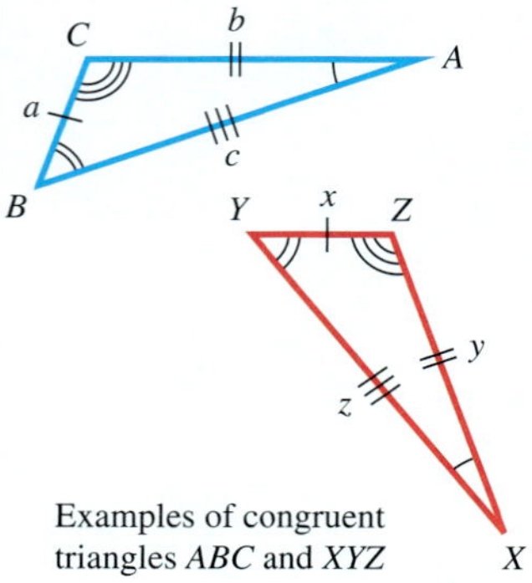

Examples of congruent triangles ABC and XYZ

Congruency and Oblique Triangles The concepts of solving triangles developed in **Chapter 2** can be extended to *all* triangles. The following axioms from geometry enable us to prove that two triangles are congruent (that is, their corresponding sides and angles are equal).

Congruence Axioms

Side-Angle-Side (SAS)	If two sides and the included angle of one triangle are equal, respectively, to two sides and the included angle of a second triangle, then the triangles are congruent.
Angle-Side-Angle (ASA)	If two angles and the included side of one triangle are equal, respectively, to two angles and the included side of a second triangle, then the triangles are congruent.
Side-Side-Side (SSS)	If three sides of one triangle are equal, respectively, to three sides of a second triangle, then the triangles are congruent.

If a side and *any* two angles are given **(SAA),** the third angle is easily determined by the angle sum formula $(A + B + C = 180°)$, and then the ASA axiom can be applied. Keep in mind that whenever SAS, ASA, or SSS is given, the triangle is unique.

A triangle that is not a right triangle is called an **oblique triangle.** ***The measures of the three sides and the three angles of a triangle can be found if at least one side and any other two measures are known.*** There are four possible cases.

Data Required for Solving Oblique Triangles

Case 1	One side and two angles are known (SAA or ASA).
Case 2	Two sides and one angle not included between the two sides are known (SSA). This case may lead to more than one triangle.
Case 3	Two sides and the angle included between the two sides are known (SAS).
Case 4	Three sides are known (SSS).

NOTE ***If we know three angles of a triangle, we cannot find unique side lengths since AAA assures us only of similarity, not congruence.*** For example, there are infinitely many triangles ABC of different sizes with $A = 35°$, $B = 65°$, and $C = 80°$.

Case 1, discussed in this section, and Case 2, discussed in **Section 7.2,** require the *law of sines.* Cases 3 and 4, discussed in **Section 7.3,** require the *law of cosines.*

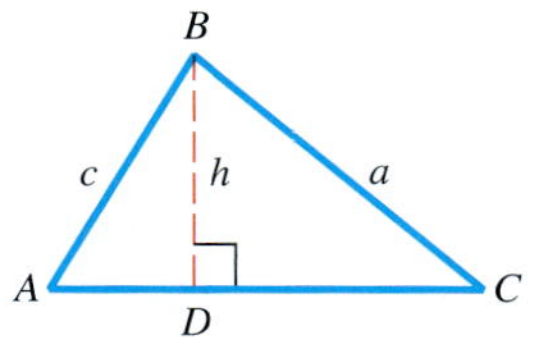

Acute triangle ABC

(a)

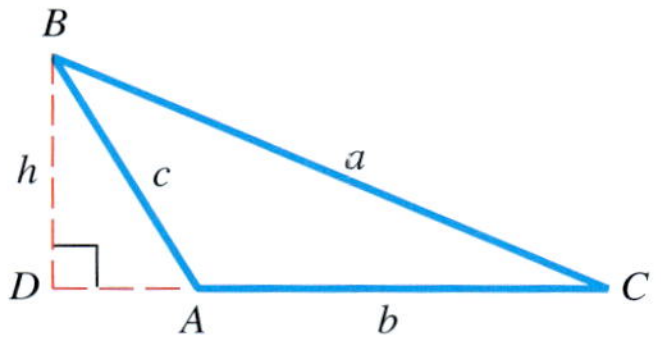

Obtuse triangle ABC

(b)

We label oblique triangles as we did right triangles: side a opposite angle A, side b opposite angle B, and side c opposite angle C.

Figure 1

Derivation of the Law of Sines To derive the law of sines, we start with an oblique triangle, such as the **acute triangle** in **Figure 1(a)** or the **obtuse triangle** in **Figure 1(b).** This discussion applies to both triangles. First, construct the perpendicular from B to side AC (or its extension). Let h be the length of this perpendicular. Then c is the hypotenuse of right triangle ADB, and a is the hypotenuse of right triangle BDC.

In triangle ADB, $\sin A = \frac{h}{c}$, or $h = c \sin A$.

(Section 2.1)

In triangle BDC, $\sin C = \frac{h}{a}$, or $h = a \sin C$.

Since $h = c \sin A$ and $h = a \sin C$, we set these two expressions equal.

$$a \sin C = c \sin A$$

$$\frac{a}{\sin A} = \frac{c}{\sin C} \quad \text{Divide each side by } \sin A \sin C.$$

In a similar way, by constructing perpendicular lines from the other vertices, we can show that these two equations are also true.

$$\frac{a}{\sin A} = \frac{b}{\sin B} \quad \text{and} \quad \frac{b}{\sin B} = \frac{c}{\sin C}$$

This discussion proves the following theorem.

Law of Sines

In any triangle ABC, with sides a, b, and c,

$$\frac{a}{\sin A} = \frac{b}{\sin B}, \quad \frac{a}{\sin A} = \frac{c}{\sin C}, \quad \text{and} \quad \frac{b}{\sin B} = \frac{c}{\sin C}.$$

This can be written in compact form as follows.

$$\frac{a}{\sin A} = \frac{b}{\sin B} = \frac{c}{\sin C}$$

That is, according to the law of sines, the lengths of the sides in a triangle are proportional to the sines of the measures of the angles opposite them.

In practice we can also use an alternative form of the law of sines.

$$\frac{\sin A}{a} = \frac{\sin B}{b} = \frac{\sin C}{c} \quad \text{Alternative form}$$

NOTE When using the law of sines, a good strategy is to select an equation so that the unknown variable is in the numerator and all other variables are known. This makes computation easier.

Solving SAA and ASA Triangles (Case 1) If two angles and one side of a triangle are known (Case 1, SAA or ASA), then the law of sines can be used to solve the triangle.

EXAMPLE 1 Applying the Law of Sines (SAA)

Solve triangle ABC if $A = 32.0°$, $B = 81.8°$, and $a = 42.9$ cm.

SOLUTION Start by drawing a triangle, roughly to scale, and labeling the given parts as in **Figure 2.** Since the values of A, B, and a are known, use the form of the law of sines that involves these variables, and then solve for b.

Be sure to label a sketch carefully to help set up the correct equation.

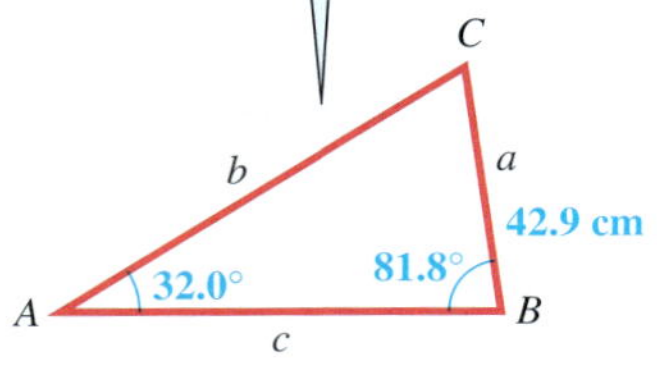

Figure 2

Choose a form that has the unknown variable in the numerator.

$$\frac{a}{\sin A} = \frac{b}{\sin B} \quad \text{Law of sines}$$

$$\frac{42.9}{\sin 32.0°} = \frac{b}{\sin 81.8°} \quad \text{Substitute the given values.}$$

$$b = \frac{42.9 \sin 81.8°}{\sin 32.0°} \quad \text{Multiply by sin 81.8° and rewrite.}$$

$$b \approx 80.1 \text{ cm} \quad \text{Approximate with a calculator.}$$

To find C, use the fact that the sum of the angles of any triangle is 180°.

$$A + B + C = 180° \quad \text{Angle sum formula (Section 1.2)}$$

$$C = 180° - A - B \quad \text{Solve for } C.$$

$$C = 180° - 32.0° - 81.8° \quad \text{Substitute.}$$

$$C = 66.2° \quad \text{Subtract.}$$

Now use the law of sines to find c. (The Pythagorean theorem does not apply because this is not a right triangle.)

$$\frac{a}{\sin A} = \frac{c}{\sin C} \quad \text{Law of sines}$$

$$\frac{42.9}{\sin 32.0°} = \frac{c}{\sin 66.2°} \quad \text{Substitute known values.}$$

$$c = \frac{42.9 \sin 66.2°}{\sin 32.0°} \quad \text{Multiply by sin 66.2° and rewrite.}$$

$$c \approx 74.1 \text{ cm} \quad \text{Approximate with a calculator.}$$

✔ *Now Try Exercise 9.*

CAUTION Whenever possible, use the given values in solving triangles, rather than values obtained in intermediate steps, to avoid rounding errors.

EXAMPLE 2 Applying the Law of Sines (ASA)

Kurt Daniels wishes to measure the distance across the Gasconade River. See **Figure 3.** He determines that $C = 112.90°$, $A = 31.10°$, and $b = 347.6$ ft. Find the distance a across the river.

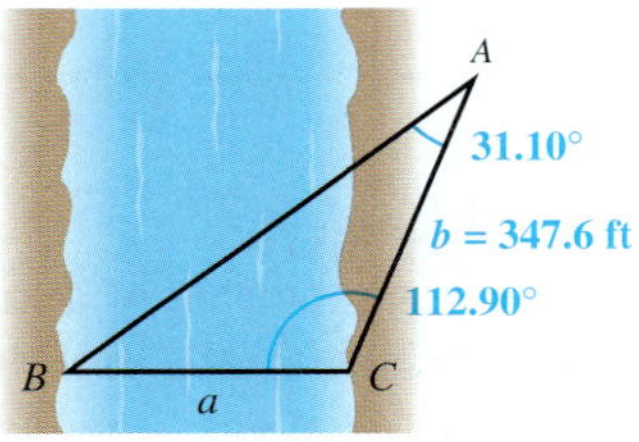

Figure 3

SOLUTION To use the law of sines, one side and the angle opposite it must be known. Since b is the only side whose length is given, angle B must be found before the law of sines can be used.

$$B = 180° - A - C \quad \text{Angle sum formula, solved for } B$$

$$B = 180° - 31.10° - 112.90° \quad \text{Substitute the given values.}$$

$$B = 36.00° \quad \text{Subtract.}$$

Now use the form of the law of sines involving A, B, and b to find a.

Solve for a.

$$\frac{a}{\sin A} = \frac{b}{\sin B} \quad \text{Law of sines}$$

$$\frac{a}{\sin 31.10^\circ} = \frac{347.6}{\sin 36.00^\circ} \quad \text{Substitute known values.}$$

$$a = \frac{347.6 \sin 31.10^\circ}{\sin 36.00^\circ} \quad \text{Multiply by } \sin 31.10^\circ.$$

$$a \approx 305.5 \text{ ft} \quad \text{Use a calculator.}$$

✔ ***Now Try Exercise 25.***

The next example involves the concept of bearing, first discussed in **Chapter 2.**

EXAMPLE 3 Applying the Law of Sines (ASA)

Two ranger stations are on an east-west line 110 mi apart. A forest fire is located on a bearing of N 42° E from the western station at A and a bearing of N 15° E from the eastern station at B. To the nearest ten miles, how far is the fire from the western station?

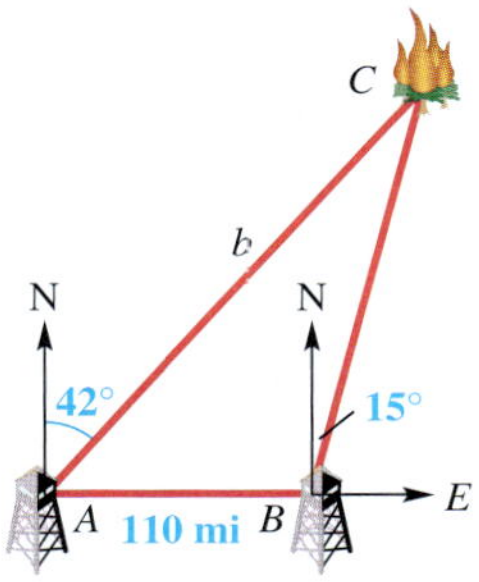

Figure 4

SOLUTION **Figure 4** shows the two stations at points A and B and the fire at point C. Angle $BAC = 90^\circ - 42^\circ = 48^\circ$, the obtuse angle at B measures $90^\circ + 15^\circ = 105^\circ$, and the third angle, C, measures $180^\circ - 105^\circ - 48^\circ = 27^\circ$. We use the law of sines to find side b.

Solve for b.

$$\frac{b}{\sin B} = \frac{c}{\sin C} \quad \text{Law of sines}$$

$$\frac{b}{\sin 105^\circ} = \frac{110}{\sin 27^\circ} \quad \text{Substitute known values.}$$

$$b = \frac{110 \sin 105^\circ}{\sin 27^\circ} \quad \text{Multiply by } \sin 105^\circ.$$

$$b \approx 230 \text{ mi} \quad \text{Use a calculator and give two significant digits.}$$

✔ ***Now Try Exercise 27.***

NOTE There is another method for describing bearing and it was first introduced in Method 1 in **Chapter 2.** It involves measuring clockwise from due north, using a single degree measure between 0° and 360°.

Area of a Triangle The method used to derive the law of sines can also be used to derive a formula to find the area of a triangle. A familiar formula for the area of a triangle is

$$\mathcal{A} = \frac{1}{2}bh, \quad \text{where } \mathcal{A} \text{ represents area, } b \text{ base, and } h \text{ height.}$$

This formula cannot always be used easily because in practice, h is often unknown. To find another formula, refer to acute triangle ABC in **Figure 5(a)** or obtuse triangle ABC in **Figure 5(b),** shown on the next page.

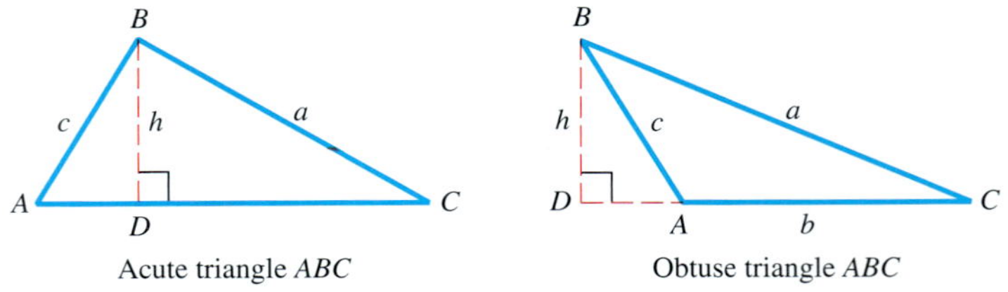

Figure 5

A perpendicular has been drawn from B to the base of the triangle (or the extension of the base). Consider right triangle ADB in either figure.

$$\sin A = \frac{h}{c}, \quad \text{or} \quad h = c \sin A$$

Substitute into the formula for the area of a triangle.

$$\mathcal{A} = \frac{1}{2}bh = \frac{1}{2}bc \sin A$$

Any other pair of sides and the angle between them could have been used.

Area of a Triangle (SAS)

In any triangle ABC, the area $\mathcal{A}$ is given by the following formulas.

$$\mathcal{A} = \frac{1}{2}bc \sin A, \quad \mathcal{A} = \frac{1}{2}ab \sin C, \quad \text{and} \quad \mathcal{A} = \frac{1}{2}ac \sin B$$

That is, the area is half the product of the lengths of two sides and the sine of the angle included between them.

NOTE If the included angle measures 90°, its sine is 1 and the formula becomes the familiar $\mathcal{A} = \frac{1}{2}bh$.

EXAMPLE 4 Finding the Area of a Triangle (SAS)

Find the area of triangle ABC in **Figure 6.**

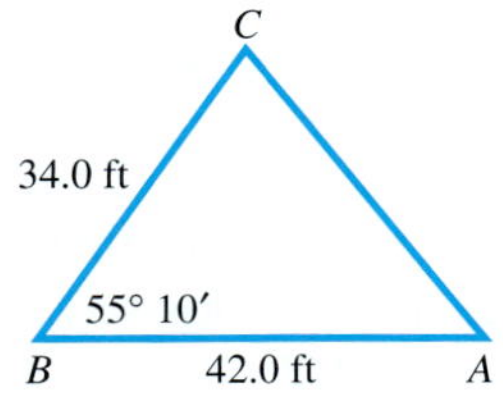

Figure 6

SOLUTION Substitute $B = 55° \, 10'$, $a = 34.0$ ft, and $c = 42.0$ ft into the area formula.

$$\mathcal{A} = \frac{1}{2}ac \sin B = \frac{1}{2}(34.0)(42.0) \sin 55° \, 10' \approx 586 \text{ ft}^2$$

✔ *Now Try Exercise 43.*

EXAMPLE 5 Finding the Area of a Triangle (ASA)

Find the area of triangle ABC in **Figure 7.**

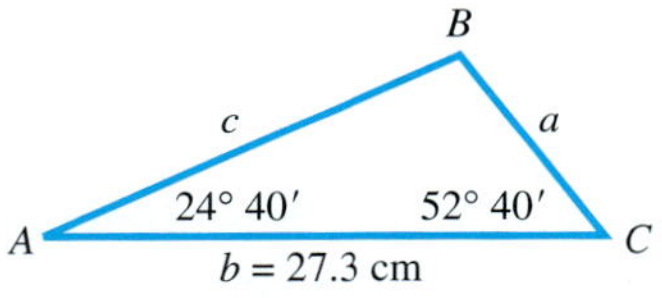

Figure 7

SOLUTION Before the area formula can be used, we must find either a or c. Begin by using the fact that the sum of the measures of the angles of any triangle is 180°.

$180° = A + B + C$	Angle sum formula
$B = 180° - 24° \, 40' - 52° \, 40'$	Substitute and solve for B.
$B = 102° \, 40'$	Subtract.

Next use the law of sines to find a.

Solve for a.

$$\frac{a}{\sin A} = \frac{b}{\sin B} \qquad \text{Law of sines}$$

$$\frac{a}{\sin 24^\circ 40'} = \frac{27.3}{\sin 102^\circ 40'} \qquad \text{Substitute known values.}$$

$$a = \frac{27.3 \sin 24^\circ 40'}{\sin 102^\circ 40'} \qquad \text{Multiply by } \sin 24^\circ 40'.$$

$$a \approx 11.7 \text{ cm} \qquad \text{Use a calculator.}$$

Now that we know two sides, a and b, and their included angle C, we find the area.

11.7 is only an approximation. In practice, use your calculator value.

$$\mathcal{A} = \frac{1}{2}ab \sin C \approx \frac{1}{2}(11.7)(27.3) \sin 52^\circ 40' \approx 127 \text{ cm}^2$$

✔ **Now Try Exercise 49.**

7.1 Exercises

1. *Concept Check* Consider the oblique triangle ABC. Which one of the following proportions is *not* valid?

A. $\dfrac{a}{b} = \dfrac{\sin A}{\sin B}$ **B.** $\dfrac{a}{\sin A} = \dfrac{b}{\sin B}$

C. $\dfrac{\sin A}{a} = \dfrac{b}{\sin B}$ **D.** $\dfrac{\sin A}{a} = \dfrac{\sin B}{b}$

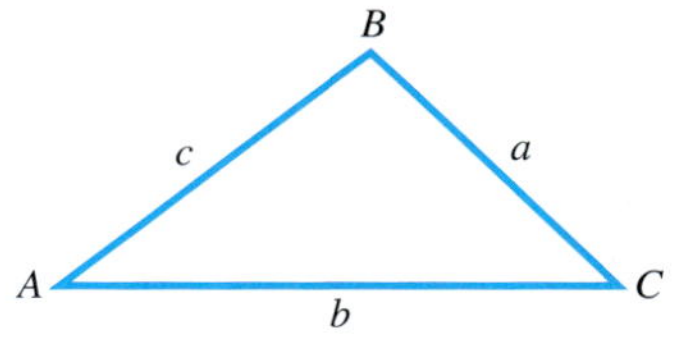

2. *Concept Check* Which two of the following situations do *not* provide sufficient information for solving a triangle by the law of sines?

A. We are given two angles and the side included between them.

B. We are given two angles and a side opposite one of them.

C. We are given two sides and the angle included between them.

D. We are given three sides.

Find the length of each side a. Do not use a calculator.

3.

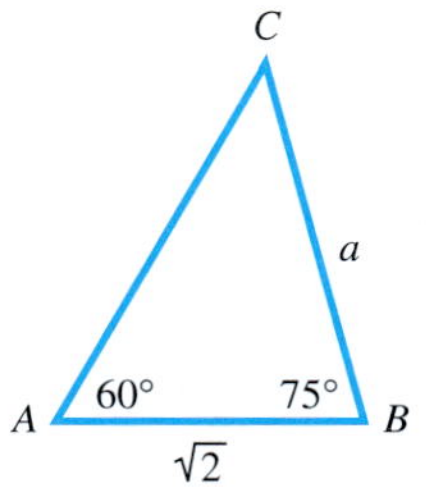

4.

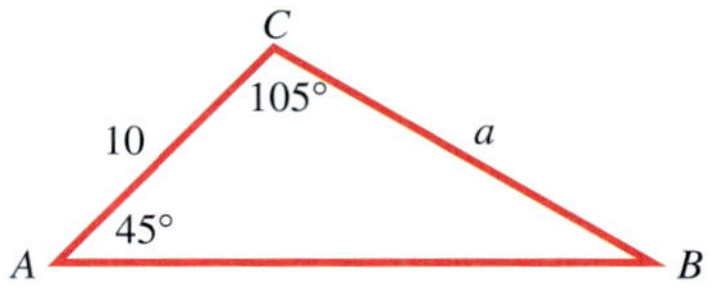

Determine the remaining sides and angles of each triangle ABC. **See Example 1.**

5.

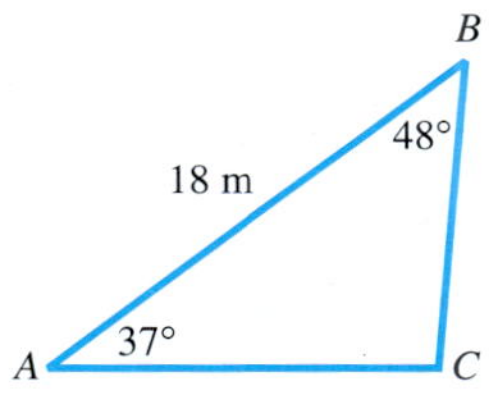

6.

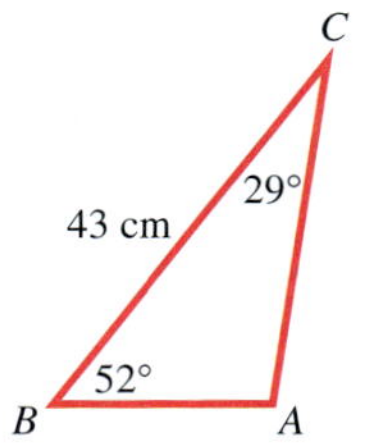

7. 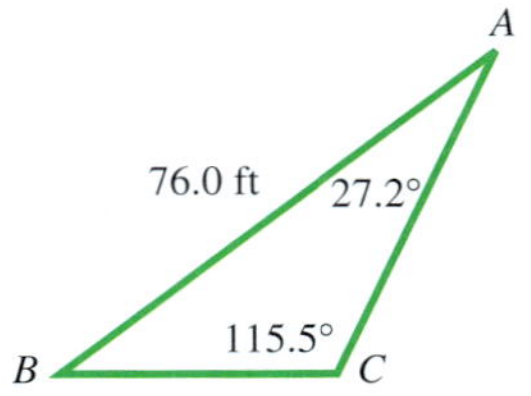

8. 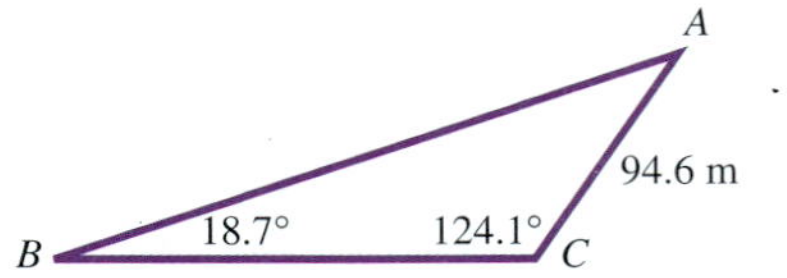

9. $A = 68.41°$, $B = 54.23°$, $a = 12.75$ ft

10. $C = 74.08°$, $B = 69.38°$, $c = 45.38$ m

11. $A = 87.2°$, $b = 75.9$ yd, $C = 74.3°$

12. $B = 38° 40'$, $a = 19.7$ cm, $C = 91° 40'$

13. $B = 20° 50'$, $C = 103° 10'$, $AC = 132$ ft

14. $A = 35.3°$, $B = 52.8°$, $AC = 675$ ft

15. $A = 39.70°$, $C = 30.35°$, $b = 39.74$ m

16. $C = 71.83°$, $B = 42.57°$, $a = 2.614$ cm

17. $B = 42.88°$, $C = 102.40°$, $b = 3974$ ft

18. $C = 50.15°$, $A = 106.1°$, $c = 3726$ yd

19. $A = 39° 54'$, $a = 268.7$ m, $B = 42° 32'$

20. $C = 79° 18'$, $c = 39.81$ mm, $A = 32° 57'$

21. Explain why we cannot use the law of sines to solve a triangle if we are given only the lengths of the three sides of the triangle.

22. Suppose that we are solving **Example 1** and we begin (as seen there) by solving for b and C. Explain why it is a better idea to solve for c using a and sin A rather than using b and sin B.

23. Eli Maor, a perceptive trigonometry student, makes this statement: "If we know *any* two angles and one side of a triangle, then the triangle is uniquely determined." Explain why this is true, referring to the congruence axioms given in this section.

24. *Concept Check* If a is twice as long as b, is A necessarily twice as large as B?

Solve each problem. ***See Examples 2 and 3.***

25. *Distance across a River* To find the distance AB across a river, a surveyor laid off a distance $BC = 354$ m on one side of the river. It is found that $B = 112° 10'$ and $C = 15° 20'$. Find AB. See the figure.

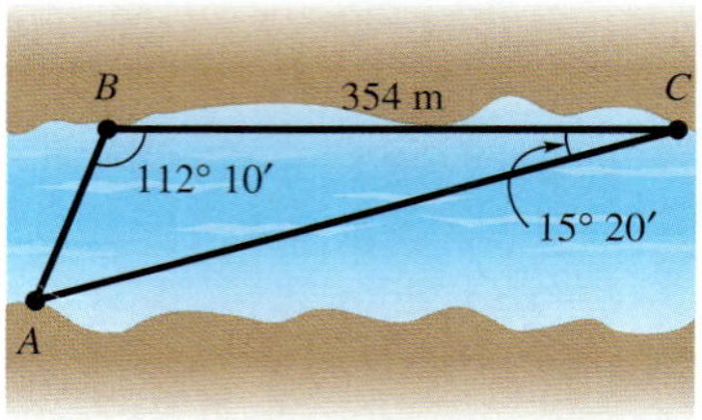

26. *Distance across a Canyon* To determine the distance RS across a deep canyon, Rhonda lays off a distance $TR = 582$ yd. She then finds that $T = 32° 50'$ and $R = 102° 20'$. Find RS.

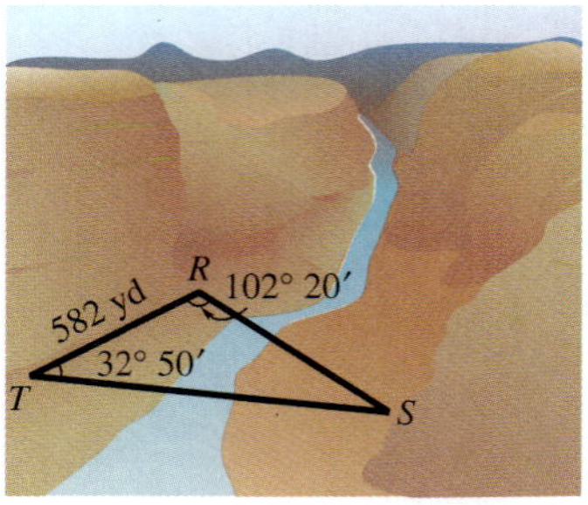

27. *Distance a Ship Travels* A ship is sailing due north. At a certain point the bearing of a lighthouse 12.5 km away is N 38.8° E. Later on, the captain notices that the bearing of the lighthouse has become S 44.2° E. How far did the ship travel between the two observations of the lighthouse?

28. *Distance between Radio Direction Finders* Radio direction finders are placed at points A and B, which are 3.46 mi apart on an east-west line, with A west of B. From A the bearing of a certain radio transmitter is 47.7°, and from B the bearing is 302.5°. Find the distance of the transmitter from A.

29. *Distance between a Ship and a Lighthouse* The bearing of a lighthouse from a ship was found to be N 37° E. After the ship sailed 2.5 mi due south, the new bearing was N 25° E. Find the distance between the ship and the lighthouse at each location.

30. *Distance across a River* Standing on one bank of a river flowing north, Mark notices a tree on the opposite bank at a bearing of 115.45°. Lisa is on the same bank as Mark, but 428.3 m away. She notices that the bearing of the tree is 45.47°. The two banks are parallel. What is the distance across the river?

31. *Height of a Balloon* A balloonist is directly above a straight road 1.5 mi long that joins two villages. She finds that the town closer to her is at an angle of depression of 35°, and the farther town is at an angle of depression of 31°. How high above the ground is the balloon?

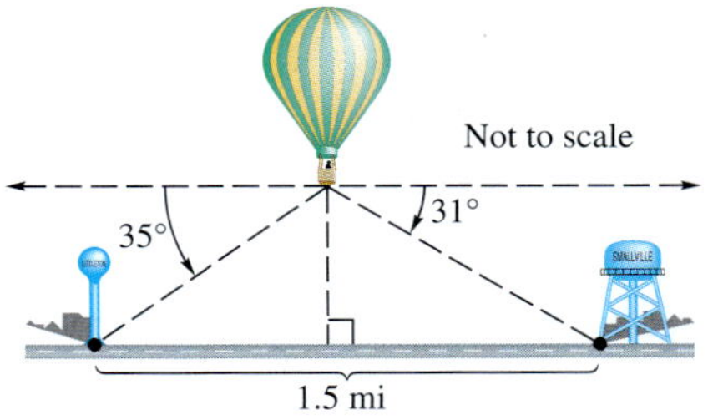

32. *Measurement of a Folding Chair* A folding chair is to have a seat 12.0 in. deep with angles as shown in the figure. How far down from the seat should the crossing legs be joined? (Find length x in the figure.)

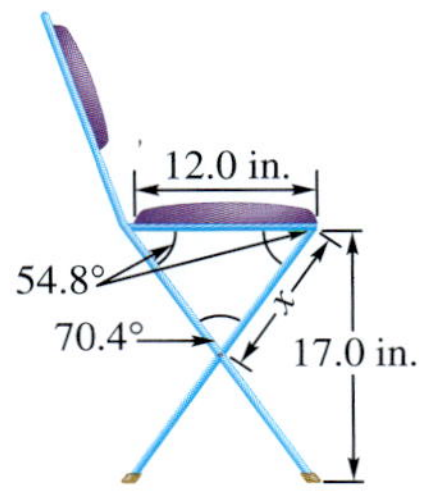

33. *Angle Formed by Radii of Gears* Three gears are arranged as shown in the figure. Find angle θ.

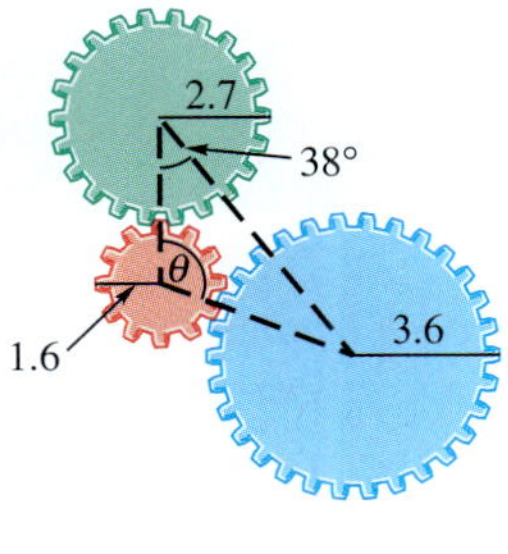

34. *Distance between Atoms* Three atoms with atomic radii of 2.0, 3.0, and 4.5 are arranged as in the figure. Find the distance between the centers of atoms A and C.

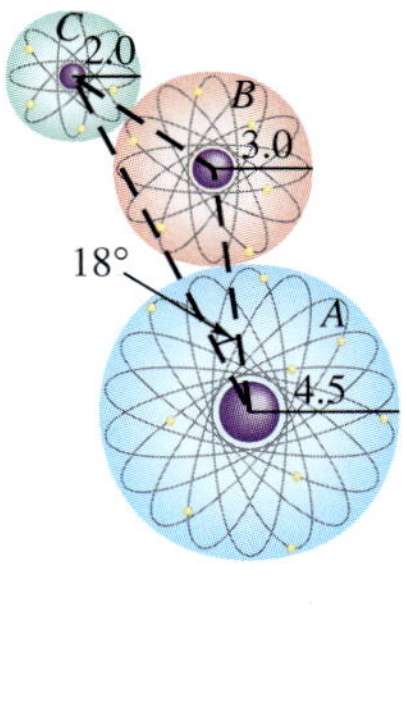

35. *Distance to the Moon* Since the moon is a relatively close celestial object, its distance can be measured directly by taking two different photographs at precisely the same time from two different locations. The moon will have a different angle of elevation at each location. On April 29, 1976, at 11:35 A.M., the lunar angles of elevation during a partial solar eclipse at Bochum in upper Germany and at Donaueschingen in lower Germany were measured as 52.6997° and 52.7430°, respectively. The two cities are 398 km apart. Calculate the distance to the moon from Bochum on this day, and compare it with the actual value of 406,000 km. Disregard the curvature of Earth in this calculation. (*Source:* Scholosser, W., T. Schmidt-Kaler, and E. Milone, *Challenges of Astronomy,* Springer-Verlag.)

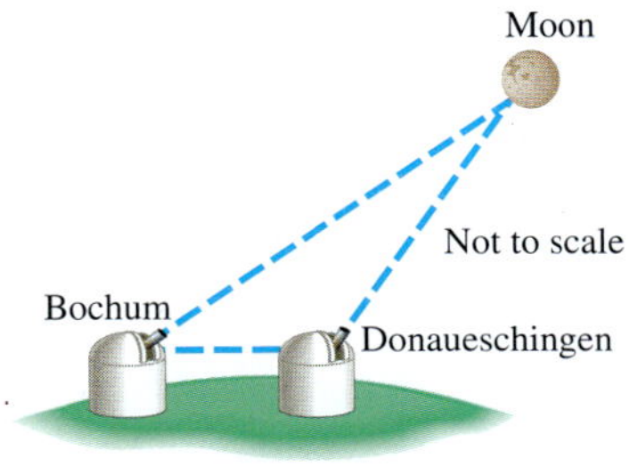

36. ***Ground Distances Measured by Aerial Photography*** The distance covered by an aerial photograph is determined by both the focal length of the camera and the tilt of the camera from the perpendicular to the ground. A camera lens with a 12-in. focal length will have an angular coverage of 60°. If an aerial photograph is taken with this camera tilted $\theta = 35°$ at an altitude of 5000 ft, calculate to the nearest foot the ground distance d that will be shown in this photograph. (*Source:* Brooks, R. and D. Johannes, *Phytoarchaeology,* Dioscorides Press.)

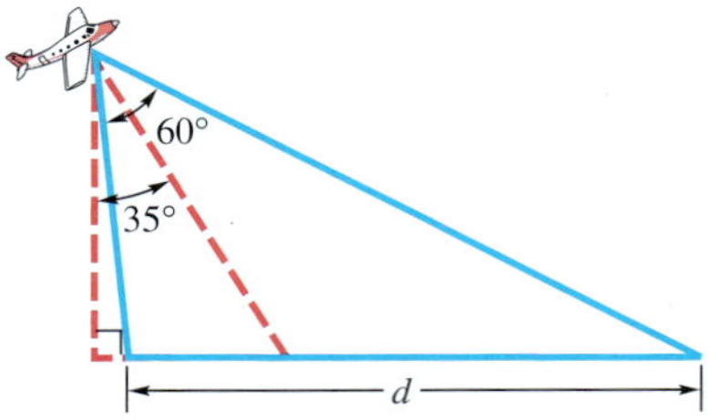

37. ***Ground Distances Measured by Aerial Photography*** Refer to **Exercise 36.** A camera lens with a 6-in. focal length has an angular coverage of 86°. Suppose an aerial photograph is taken vertically with no tilt at an altitude of 3500 ft over ground with an increasing slope of 5°, as shown in the figure. Calculate the ground distance CB that will appear in the resulting photograph. (*Source:* Moffitt, F. and E. Mikhail, *Photogrammetry,* Third Edition, Harper & Row.)

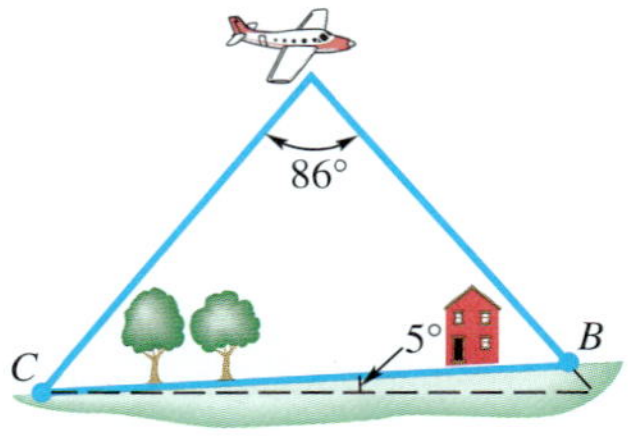

38. ***Ground Distances Measured by Aerial Photography*** Repeat **Exercise 37** if the camera lens has an 8.25-in. focal length with an angular coverage of 72°.

Find the area of each triangle using the formula $\mathcal{A} = \frac{1}{2}bh$*, and then verify that the formula* $\mathcal{A} = \frac{1}{2}ab \sin C$ *gives the same result.*

39.

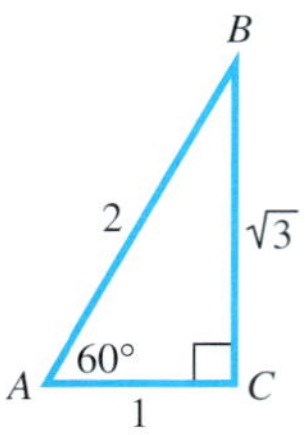

40.

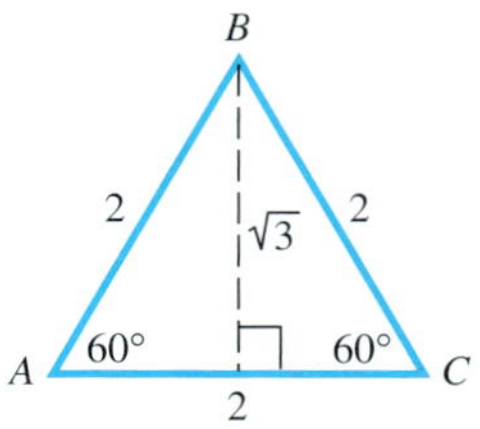

41.

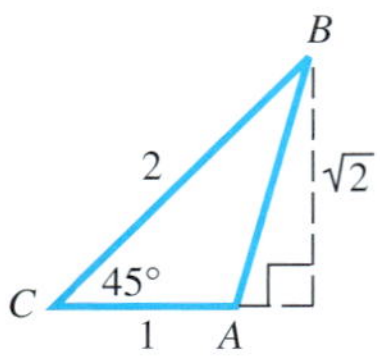

42.

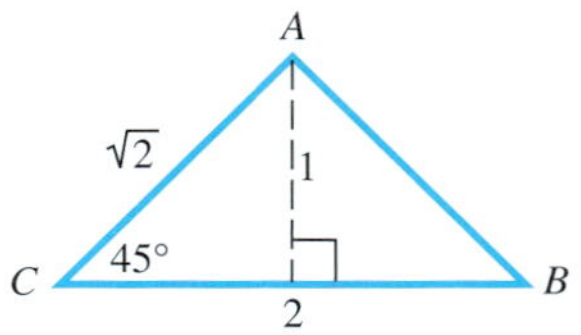

Find the area of each triangle ABC. ***See Examples 4 and 5.***

43. $A = 42.5°$, $b = 13.6$ m, $c = 10.1$ m

44. $C = 72.2°$, $b = 43.8$ ft, $a = 35.1$ ft

45. $B = 124.5°$, $a = 30.4$ cm, $c = 28.4$ cm

46. $C = 142.7°$, $a = 21.9$ km, $b = 24.6$ km

47. $A = 56.80°$, $b = 32.67$ in., $c = 52.89$ in. **48.** $A = 34.97°$, $b = 35.29$ m, $c = 28.67$ m

49. $A = 30.50°$, $b = 13.00$ cm, $C = 112.60°$ **50.** $A = 59.80°$, $b = 15.00$ m, $C = 53.10°$

Solve each problem.

51. *Area of a Metal Plate* A painter is going to apply a special coating to a triangular metal plate on a new building. Two sides measure 16.1 m and 15.2 m. She knows that the angle between these sides is 125°. What is the area of the surface she plans to cover with the coating?

52. *Area of a Triangular Lot* A real estate agent wants to find the area of a triangular lot. A surveyor takes measurements and finds that two sides are 52.1 m and 21.3 m, and the angle between them is 42.2°. What is the area of the triangular lot?

53. *Triangle Inscribed in a Circle* For a triangle inscribed in a circle of radius r, the law of sines ratios $\frac{a}{\sin A}$, $\frac{b}{\sin B}$, and $\frac{c}{\sin C}$ have value $2r$. The circle in the figure has diameter 1. What are the values of a, b, and c? (*Note:* This result provides an alternative way to define the sine function for angles between 0° and 180°. It was used nearly 2000 years ago by the mathematician Ptolemy to construct one of the earliest trigonometric tables.)

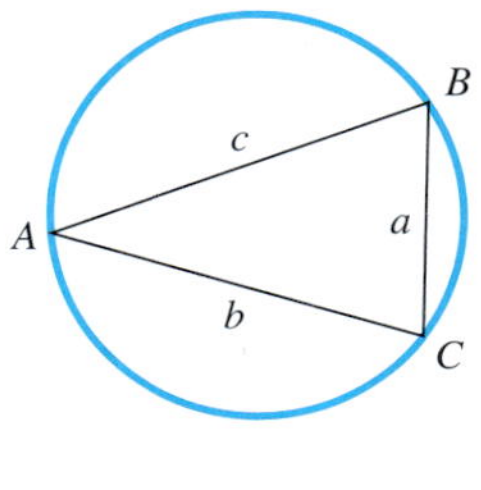

54. *Theorem of Ptolemy* The following theorem is also attributed to Ptolemy: *In a quadrilateral inscribed in a circle, the product of the diagonals is equal to the sum of the products of the opposite sides.* (*Source:* Eves, H., *An Introduction to the History of Mathematics,* Sixth Edition, Saunders College Publishing.) The circle in the figure has diameter 1. Explain why the lengths of the line segments are as shown, and then apply Ptolemy's theorem to derive the formula for the sine of the sum of two angles.

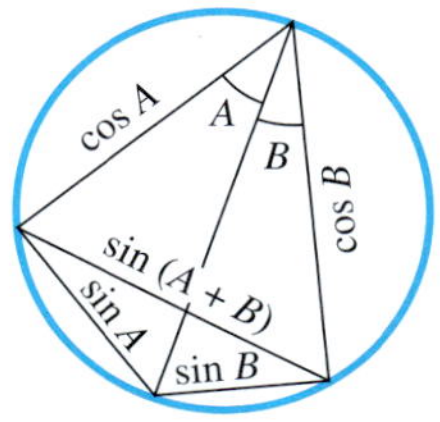

55. *Law of Sines* Several of the exercises on right triangle applications involved a figure similar to the one shown here, in which angles α and β and the length of line segment AB are known, and the length of side CD is to be determined. Use the law of sines to obtain x in terms of α, β, and d.

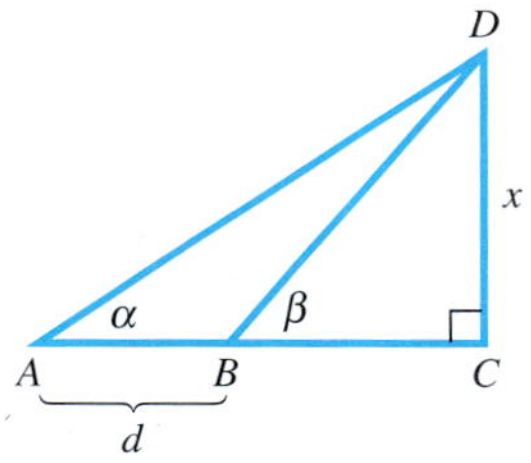

56. *Aerial Photography* Aerial photographs can be used to provide coordinates of ordered pairs to determine distances on the ground. Suppose we assign coordinates as shown in the figure. If an object's photographic coordinates are (x, y), then its ground coordinates (X, Y) in feet can be computed using the following formulas.

$$X = \frac{(a - h)x}{f \sec \theta - y \sin \theta}, \quad Y = \frac{(a - h)y \cos \theta}{f \sec \theta - y \sin \theta}$$

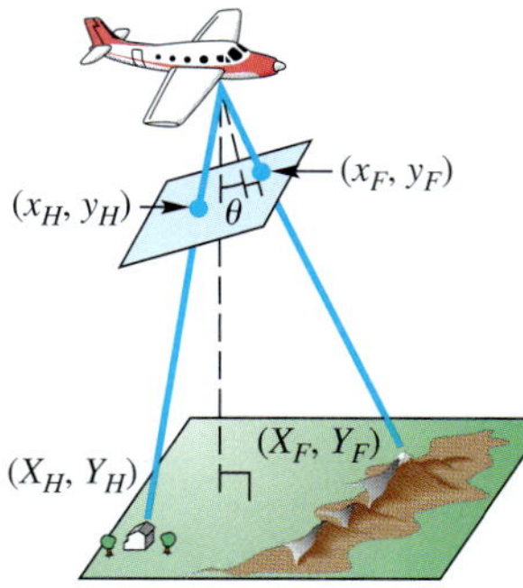

Here, f is focal length of the camera in inches, a is altitude in feet of the airplane, and h is elevation in feet of the object. Suppose that a house has photographic coordinates $(x_H, y_H) = (0.9, 3.5)$ with elevation 150 ft, and a nearby forest fire has photographic coordinates $(x_F, y_F) = (2.1, -2.4)$ and is at elevation 690 ft. Also suppose the photograph was taken at 7400 ft by a camera with focal length 6 in. and tilt angle $\theta = 4.1°$. (*Source:* Moffitt, F. and E. Mikhail, *Photogrammetry,* Third Edition, Harper & Row.)

(a) Use the formulas to find the ground coordinates of the house and the fire to the nearest tenth of a foot.

(b) Use the distance formula given in **Appendix B** to find the distance on the ground between the house and the fire to the nearest tenth of a foot.

7.2 The Ambiguous Case of the Law of Sines

- Description of the Ambiguous Case
- Solving SSA Triangles (Case 2)
- Analyzing Data for Possible Number of Triangles

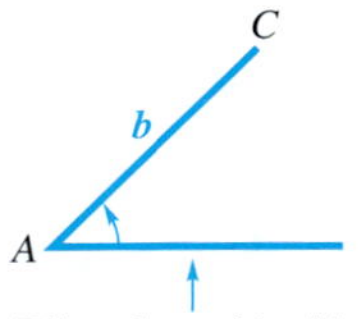

B lies along this side if a triangle exists.

Figure 8

Description of the Ambiguous Case We used the law of sines to solve triangles involving Case 1, SAA or ASA, in **Section 7.1.** If we are given the lengths of two sides and the angle opposite one of them (Case 2, SSA), then zero, one, or two such triangles may exist. (There is no SSA congruence axiom.)

Suppose we know the measure of acute angle A of triangle ABC, the length of side a, and the length of side b, as shown in **Figure 8.** Now we must draw the side of length a opposite angle A. The table shows possible outcomes. This situation (SSA) is called the **ambiguous case** of the law of sines.

As shown in the table, if angle A is acute, there are four possible outcomes. If A is obtuse, there are two possible outcomes.

Angle A is	Possible Number of Triangles	Sketch	Applying Law of Sines Leads to
Acute	0	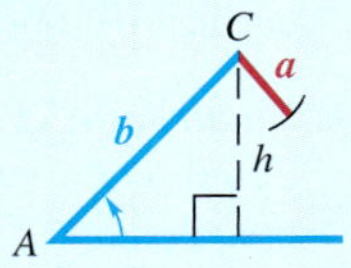	$\sin B > 1,\ a < h < b$
Acute	1	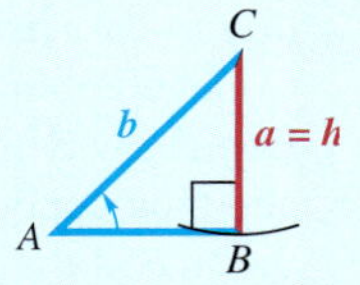	$\sin B = 1,\ a = h < b$
Acute	1	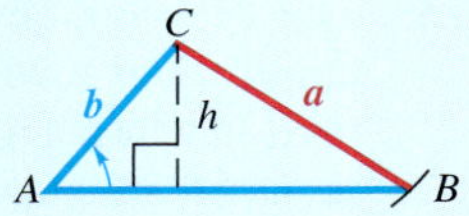	$0 < \sin B < 1,\ a \geq b$
Acute	2	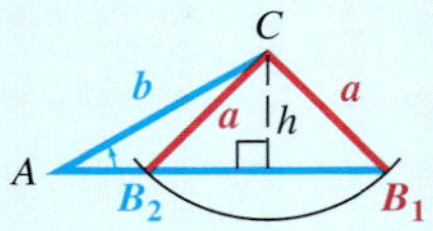	$0 < \sin B_1 < 1,\ h < a < b,$ $A + B_2 < 180°$
Obtuse	0	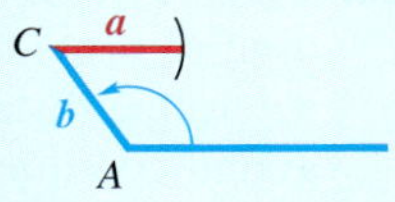	$\sin B \geq 1,\ a \leq b$
Obtuse	1	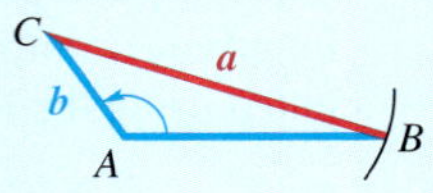	$0 < \sin B < 1,\ a > b$

The following basic facts help determine which situation applies.

Applying the Law of Sines

1. For any angle θ of a triangle, $0 < \sin\theta \leq 1$. If $\sin\theta = 1$, then $\theta = 90°$ and the triangle is a right triangle.
2. $\sin\theta = \sin(180° - \theta)$ (Supplementary angles have the same sine value.)
3. The smallest angle is opposite the shortest side, the largest angle is opposite the longest side, and the middle-valued angle is opposite the intermediate side (assuming the triangle has sides that are all of different lengths).

Solving SSA Triangles (Case 2)

EXAMPLE 1 Solving the Ambiguous Case (No Such Triangle)

Solve triangle ABC if $B = 55° \, 40'$, $b = 8.94$ m, and $a = 25.1$ m.

SOLUTION We are given B, b, and a, so we can use the law of sines to find A.

Choose a form that has the unknown variable in the numerator.

$$\frac{\sin A}{a} = \frac{\sin B}{b} \qquad \text{Law of sines (alternative form) (Section 7.1)}$$

$$\frac{\sin A}{25.1} = \frac{\sin 55° \, 40'}{8.94} \qquad \text{Substitute the given values.}$$

$$\sin A = \frac{25.1 \sin 55° \, 40'}{8.94} \qquad \text{Multiply by 25.1.}$$

$$\sin A \approx 2.3184379 \qquad \text{Use a calculator.}$$

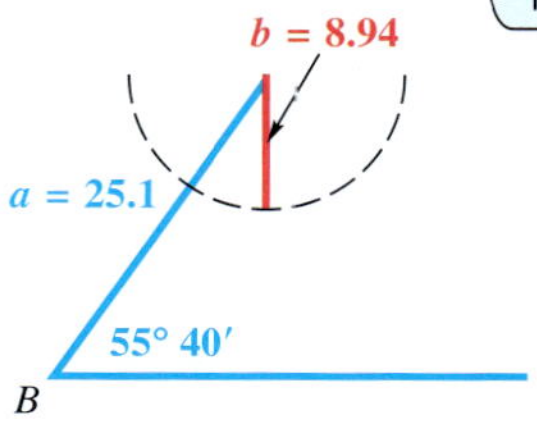

Figure 9

Because $\sin A$ cannot be greater than 1, there can be no such angle A—and thus no triangle with the given information. An attempt to sketch such a triangle leads to the situation shown in **Figure 9.**

✔ *Now Try Exercise 17.*

NOTE In the ambiguous case, we are given two sides and an angle opposite one of the sides (SSA). For example, suppose b, c, and angle C are given. This situation represents the ambiguous case because angle C is opposite side c.

EXAMPLE 2 Solving the Ambiguous Case (Two Triangles)

Solve triangle ABC if $A = 55.3°$, $a = 22.8$ ft, and $b = 24.9$ ft.

SOLUTION To begin, use the law of sines to find angle B.

$$\frac{\sin A}{a} = \frac{\sin B}{b} \qquad \text{Solve for } \sin B.$$

$$\frac{\sin 55.3°}{22.8} = \frac{\sin B}{24.9} \qquad \text{Substitute the given values.}$$

$$\sin B = \frac{24.9 \sin 55.3°}{22.8} \qquad \text{Multiply by 24.9 and rewrite.}$$

$$\sin B \approx 0.8978678 \qquad \text{Use a calculator.}$$

There are two angles B between 0° and 180° that satisfy this condition. Since $\sin B \approx 0.8978678$, to the nearest tenth one value of B is

$$B_1 = 63.9°. \qquad \text{Use the inverse sine function. (Section 6.1)}$$

Supplementary angles have the same sine value, so another *possible* value of B is

$$B_2 = 180° - 63.9° = 116.1°. \qquad \text{(Section 1.1)}$$

To see whether $B_2 = 116.1°$ is a valid possibility, add 116.1° to the measure of A, 55.3°. Since $116.1° + 55.3° = 171.4°$, and this sum is less than 180°, it is a valid angle measure for this triangle.

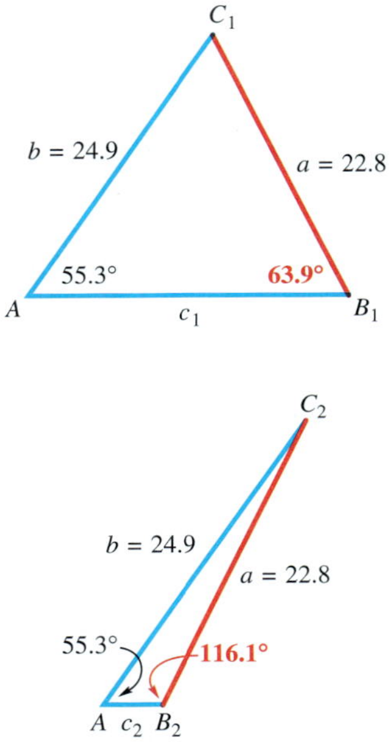

Figure 10

Now separately solve triangles AB_1C_1 and AB_2C_2 shown in **Figure 10.** Begin with AB_1C_1. Find C_1 first.

$$C_1 = 180° - A - B_1 \quad \text{Angle sum formula (Section 1.2)}$$

$$C_1 = 180° - 55.3° - 63.9° \quad \text{Substitute.}$$

$$C_1 = 60.8° \quad \text{Subtract.}$$

Now, use the law of sines to find c_1.

$$\frac{a}{\sin A} = \frac{c_1}{\sin C_1} \quad \text{Solve for } c_1.$$

$$\frac{22.8}{\sin 55.3°} = \frac{c_1}{\sin 60.8°} \quad \text{Substitute.}$$

$$c_1 = \frac{22.8 \sin 60.8°}{\sin 55.3°} \quad \text{Multiply by } \sin 60.8°.$$

$$c_1 \approx 24.2 \text{ ft} \quad \text{Use a calculator.}$$

To solve triangle AB_2C_2, first find C_2.

$$C_2 = 180° - A - B_2 \quad \text{Angle sum formula}$$

$$C_2 = 180° - 55.3° - 116.1° \quad \text{Substitute.}$$

$$C_2 = 8.6° \quad \text{Subtract.}$$

Use the law of sines to find c_2.

$$\frac{a}{\sin A} = \frac{c_2}{\sin C_2} \quad \text{Solve for } c_2.$$

$$\frac{22.8}{\sin 55.3°} = \frac{c_2}{\sin 8.6°} \quad \text{Substitute.}$$

$$c_2 = \frac{22.8 \sin 8.6°}{\sin 55.3°} \quad \text{Multiply by } \sin 8.6°.$$

$$c_2 \approx 4.15 \text{ ft} \quad \text{Use a calculator.}$$

✔ ***Now Try Exercise 25.***

The ambiguous case results in zero, one, or two triangles. The following guidelines can be used to determine how many triangles there are.

Number of Triangles Satisfying the Ambiguous Case (SSA)

Let sides a and b and angle A be given in triangle ABC. (The law of sines can be used to calculate the value of $\sin B$.)

1. If applying the law of sines results in an equation having $\sin B > 1$, then *no triangle* satisfies the given conditions.

2. If $\sin B = 1$, then *one triangle* satisfies the given conditions and $B = 90°$.

3. If $0 < \sin B < 1$, then either *one or two triangles* satisfy the given conditions.

(a) If $\sin B = k$, then let $B_1 = \sin^{-1} k$ and use B_1 for B in the first triangle.

(b) Let $B_2 = 180° - B_1$. If $A + B_2 < 180°$, then a second triangle exists. In this case, use B_2 for B in the second triangle.

EXAMPLE 3 **Solving the Ambiguous Case (One Triangle)**

Solve triangle ABC, given $A = 43.5°$, $a = 10.7$ in., and $c = 7.2$ in.

SOLUTION To find angle C, use an alternative form of the law of sines.

$$\frac{\sin C}{c} = \frac{\sin A}{a} \quad \text{Law of sines}$$

$$\frac{\sin C}{7.2} = \frac{\sin 43.5°}{10.7} \quad \text{Substitute the given values.}$$

$$\sin C = \frac{7.2 \sin 43.5°}{10.7} \quad \text{Multiply by 7.2.}$$

$$\sin C \approx 0.46319186 \quad \text{Use a calculator.}$$

$$C \approx 27.6° \quad \text{Use the inverse sine function.}$$

There is another angle C that has sine value 0.46319186. It is

$$C = 180° - 27.6° = 152.4°.$$

However, notice in the given information that $c < a$, meaning that in the triangle, angle C must have measure *less than* angle A. Notice also that when we add this obtuse value to the given angle $A = 43.5°$, we obtain

$$152.4° + 43.5° = 195.9°,$$

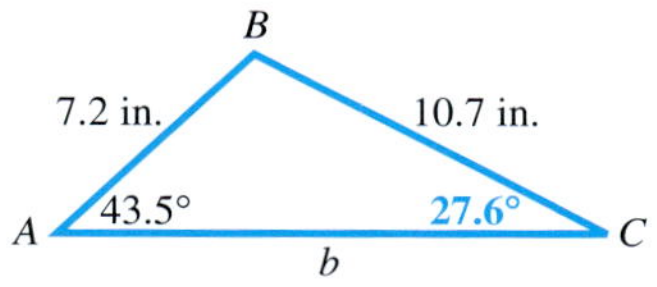

Figure 11

which is greater than 180°. Thus either of these approaches shows that there can be only one triangle. See **Figure 11.** Then

$$B = 180° - 27.6° - 43.5° \quad \text{Substitute.}$$

$$B = 108.9°, \quad \text{Subtract.}$$

and we can find side b with the law of sines.

$$\frac{b}{\sin B} = \frac{a}{\sin A} \quad \text{Law of sines}$$

$$\frac{b}{\sin 108.9°} = \frac{10.7}{\sin 43.5°} \quad \text{Substitute known values.}$$

$$b = \frac{10.7 \sin 108.9°}{\sin 43.5°} \quad \text{Multiply by } \sin 108.9°.$$

$$b \approx 14.7 \text{ in.} \quad \text{Use a calculator.}$$

✔ ***Now Try Exercise 21.***

Analyzing Data for Possible Number of Triangles

EXAMPLE 4 **Analyzing Data Involving an Obtuse Angle**

Without using the law of sines, explain why $A = 104°$, $a = 26.8$ m, and $b = 31.3$ m cannot be valid for a triangle ABC.

SOLUTION Since A is an obtuse angle, it is the largest angle, and so the longest side of the triangle must be a. However, we are given $b > a$.

Thus, $B > A$, which is impossible if A is obtuse.

Therefore, no such triangle ABC exists.

✔ ***Now Try Exercise 33.***

7.2 Exercises

1. *Concept Check* Which one of the following sets of data does *not* determine a unique triangle?

A. $A = 40°, B = 60°, C = 80°$ **B.** $a = 5, b = 12, c = 13$

C. $a = 3, b = 7, C = 50°$ **D.** $a = 2, b = 2, c = 2$

2. *Concept Check* Which one of the following sets of data determines a unique triangle?

A. $A = 50°, B = 50°, C = 80°$ **B.** $a = 3, b = 5, c = 20$

C. $A = 40°, B = 20°, C = 30°$ **D.** $a = 7, b = 24, c = 25$

Concept Check In each figure, a line segment of length L is to be drawn from the given point to the positive x-axis in order to form a triangle. For what value(s) of L can you draw the following?

(a) *two triangles* **(b)** *exactly one triangle* **(c)** *no triangle*

3.

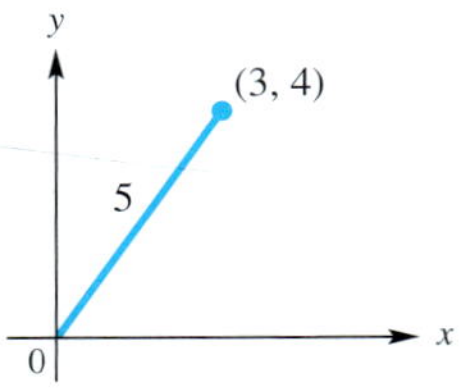

4.

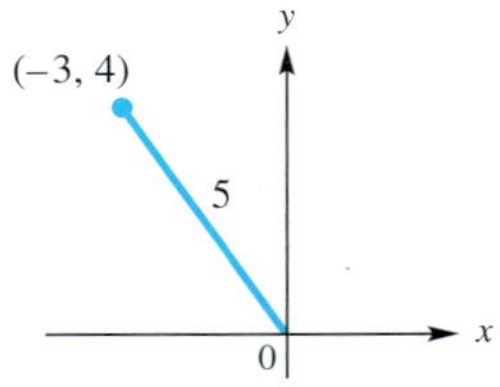

Determine the number of triangles ABC possible with the given parts. **See Examples 1–4.**

5. $a = 50, b = 26, A = 95°$

6. $a = 35, b = 30, A = 40°$

7. $a = 31, b = 26, B = 48°$

8. $B = 54°, c = 28, b = 23$

9. $a = 50, b = 61, A = 58°$

10. $b = 60, a = 82, B = 100°$

Find each angle B. Do not use a calculator.

11.

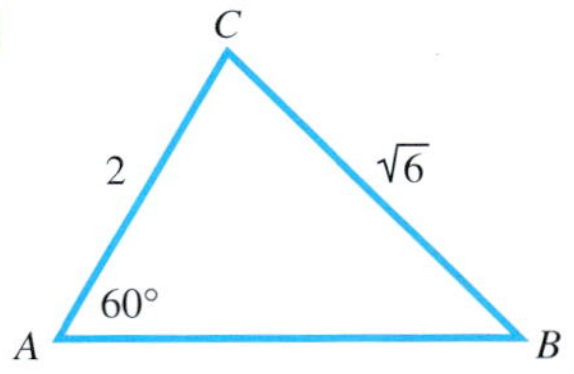

12.

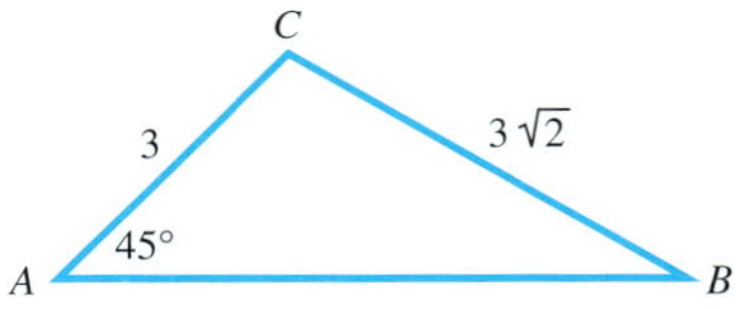

Find the unknown angles in triangle ABC for each triangle that exists. **See Examples 1–3.**

13. $A = 29.7°, b = 41.5$ ft, $a = 27.2$ ft

14. $B = 48.2°, a = 890$ cm, $b = 697$ cm

15. $C = 41° 20', b = 25.9$ m, $c = 38.4$ m

16. $B = 48° 50', a = 3850$ in., $b = 4730$ in.

17. $B = 74.3°, a = 859$ m, $b = 783$ m

18. $C = 82.2°, a = 10.9$ km, $c = 7.62$ km

19. $A = 142.13°, b = 5.432$ ft, $a = 7.297$ ft

20. $B = 113.72°, a = 189.6$ yd, $b = 243.8$ yd

Solve each triangle ABC that exists. **See Examples 1–3.**

21. $A = 42.5°,\ a = 15.6$ ft, $b = 8.14$ ft

22. $C = 52.3°,\ a = 32.5$ yd, $c = 59.8$ yd

23. $B = 72.2°,\ b = 78.3$ m, $c = 145$ m

24. $C = 68.5°,\ c = 258$ cm, $b = 386$ cm

25. $A = 38° 40',\ a = 9.72$ m, $b = 11.8$ m

26. $C = 29° 50',\ a = 8.61$ m, $c = 5.21$ m

27. $A = 96.80°,\ b = 3.589$ ft, $a = 5.818$ ft

28. $C = 88.70°,\ b = 56.87$ m, $c = 112.4$ m

29. $B = 39.68°,\ a = 29.81$ m, $b = 23.76$ m

30. $A = 51.20°,\ c = 7986$ cm, $a = 7208$ cm

31. Apply the law of sines to the following: $a = \sqrt{5},\ c = 2\sqrt{5},\ A = 30°$. What is the value of $\sin C$? What is the measure of C? Based on its angle measures, what kind of triangle is triangle ABC?

32. Explain the condition that must exist to determine that there is no triangle satisfying the given values of a, b, and B, once the value of $\sin A$ is found.

33. Without using the law of sines, explain why no triangle ABC exists satisfying $A = 103° 20',\ a = 14.6$ ft, $b = 20.4$ ft.

34. Apply the law of sines to the data given in **Example 4.** Describe what happens when you try to find the measure of angle B using a calculator.

Use the law of sines to solve each problem.

35. ***Distance between Inaccessible Points*** To find the distance between a point X and an inaccessible point Z, a line segment XY is constructed. It is found that $XY = 960$ m, angle $XYZ = 43° 30'$, and angle $YZX = 95° 30'$. Find the distance between X and Z to the nearest meter.

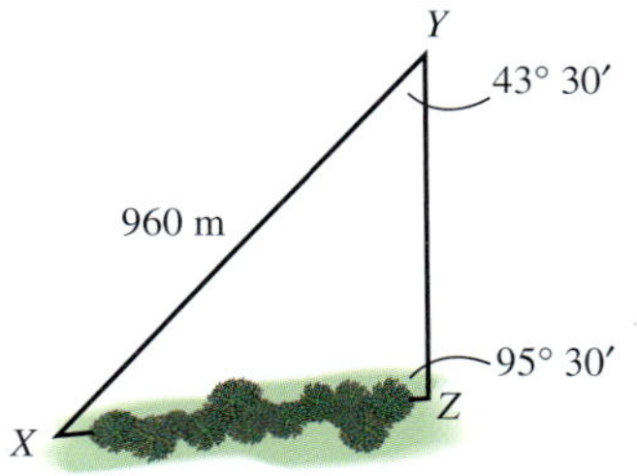

36. ***Height of an Antenna Tower*** The angle of elevation from the top of a building 45.0 ft high to the top of a nearby antenna tower is 15° 20′. From the base of the building, the angle of elevation of the tower is 29° 30′. Find the height of the tower.

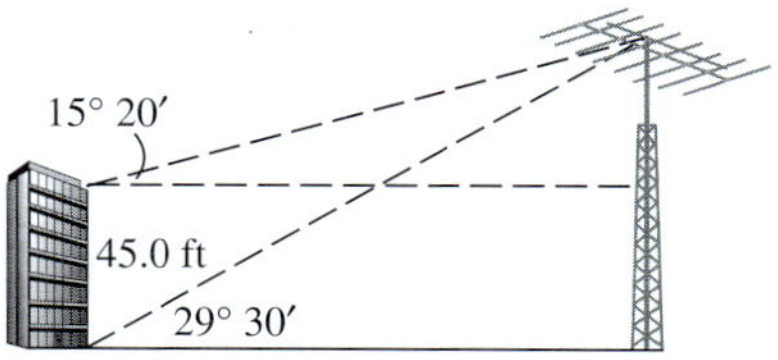

37. ***Height of a Building*** A flagpole 95.0 ft tall is on the top of a building. From a point on level ground, the angle of elevation of the top of the flagpole is 35.0°, and the angle of elevation of the bottom of the flagpole is 26.0°. Find the height of the building.

38. ***Flight Path of a Plane*** A pilot flies her plane on a heading of 35° 00′ from point X to point Y, which is 400 mi from X. Then she turns and flies on a heading of 145° 00′ to point Z, which is 400 mi from her starting point X. What is the heading of Z from X, and what is the distance YZ?

Use the law of sines to prove that each statement is true for any triangle ABC, with corresponding sides a, b, and c.

39. $\dfrac{a + b}{b} = \dfrac{\sin A + \sin B}{\sin B}$

40. $\dfrac{a - b}{a + b} = \dfrac{\sin A - \sin B}{\sin A + \sin B}$

Relating Concepts

For individual or collaborative investigation. *(Exercises 41–44)*

Colors of the U.S. Flag *The flag of the United States includes the colors red, white, and blue. Which color is predominant? Clearly the answer is either red or white. (It can be shown that only* 18.73% *of the total area is blue.)* (Source: *Banks, R.,* Slicing Pizzas, Racing Turtles, and Further Adventures in Applied Mathematics, *Princeton University Press.) Work* **Exercises 41–44 in order** *to determine the answer to this question.*

41. Let R denote the radius of the circumscribing circle of a five-pointed star appearing on the American flag. The star can be decomposed into ten congruent triangles. In the figure, r is the radius of the circumscribing circle of the pentagon in the interior of the star. Show that the area of a star is

$$\mathcal{A} = \left[5\frac{\sin A \sin B}{\sin(A + B)}\right]R^2.$$

(*Hint:* $\sin C = \sin[180° - (A + B)] = \sin(A + B)$.)

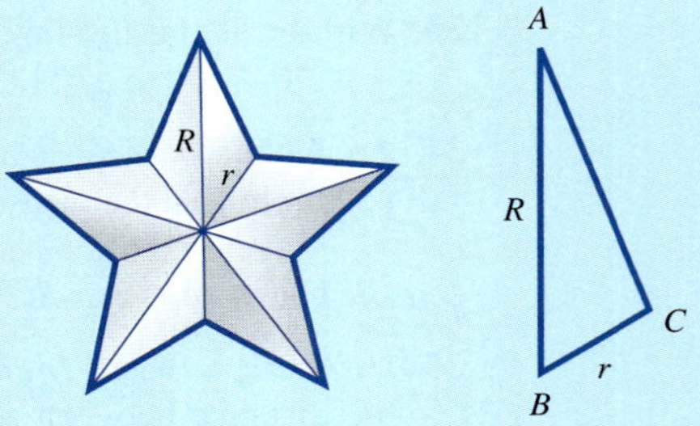

42. Angles A and B have values 18° and 36°, respectively. Express the area $\mathcal{A}$ of a star in terms of its radius, R.

43. To determine whether red or white is predominant, we must know the measurements of the flag. Consider a flag of width 10 in., length 19 in., length of each upper stripe 11.4 in., and radius R of the circumscribing circle of each star 0.308 in. The thirteen stripes consist of six matching pairs of red and white stripes and one additional red, upper stripe. Therefore, we must compare the area of a red, upper stripe with the total area of the 50 white stars.

(a) Compute the area of the red, upper stripe.

(b) Compute the total area of the 50 white stars.

44. Which color occupies the greatest area on the flag?

7.3 The Law of Cosines

- Derivation of the Law of Cosines
- Solving SAS and SSS Triangles (Cases 3 and 4)
- Heron's Formula for the Area of a Triangle
- Derivation of Heron's Formula

As mentioned in **Section 7.1,** if we are given two sides and the included angle (Case 3) or three sides (Case 4) of a triangle, then a unique triangle is determined. These are the SAS and SSS cases, respectively. Both cases require using the *law of cosines.*

The property of triangles given at the top of the next page is important when applying the law of cosines to solve a triangle.

Triangle Side Length Restriction

In any triangle, the sum of the lengths of any two sides must be greater than the length of the remaining side.

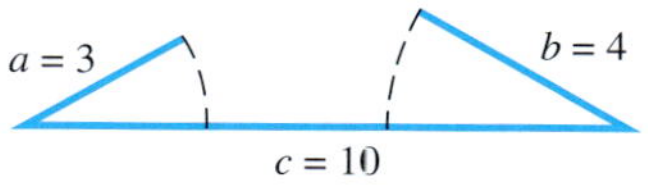

Figure 12

For example, it would be impossible to construct a triangle with sides of lengths 3, 4, and 10. See **Figure 12.**

Derivation of the Law of Cosines To derive the law of cosines, let ABC be any oblique triangle. Choose a coordinate system so that vertex B is at the origin and side BC is along the positive x-axis. See **Figure 13.**

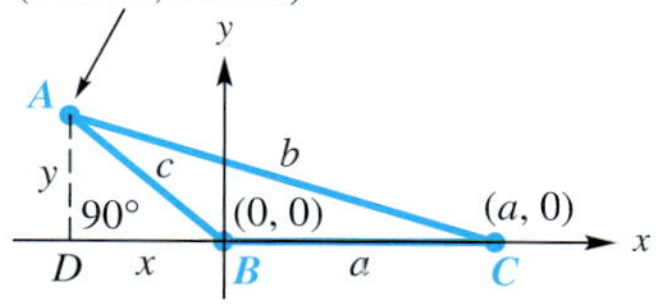

Figure 13

Let (x, y) be the coordinates of vertex A of the triangle. Then the following are true for angle B, whether obtuse or acute.

$$\sin B = \frac{y}{c} \quad \text{and} \quad \cos B = \frac{x}{c} \qquad \text{(Section 2.2)}$$

$$y = c \sin B \quad \text{and} \quad x = c \cos B \qquad \text{Here } x \text{ is negative when } B \text{ is obtuse.}$$

Thus, the coordinates of point A become $(c \cos B, c \sin B)$.

Point C in **Figure 13** has coordinates $(a, 0)$, AC has length b, and point A has coordinates $(c \cos B, c \sin B)$. We can use the distance formula to write an equation.

$$b = \sqrt{(c \cos B - a)^2 + (c \sin B - 0)^2} \qquad \text{(Appendix B)}$$

$$b^2 = (c \cos B - a)^2 + (c \sin B)^2 \qquad \text{Square each side.}$$

$$= (c^2 \cos^2 B - 2ac \cos B + a^2) + c^2 \sin^2 B \qquad \text{Multiply; } (x - y)^2 = x^2 - 2xy + y^2$$

Remember the middle term.

$$= a^2 + c^2(\cos^2 B + \sin^2 B) - 2ac \cos B \qquad \text{Properties of real numbers}$$

$$= a^2 + c^2(1) - 2ac \cos B \qquad \text{Fundamental identity (Section 5.1)}$$

$$b^2 = a^2 + c^2 - 2ac \cos B$$

This result is one form of the law of cosines. In our work, we could just as easily have placed A or C at the origin. This would have given the same result, but with the variables rearranged.

Law of Cosines

In any triangle ABC, with sides a, b, and c, the following hold.

$$a^2 = b^2 + c^2 - 2bc \cos A,$$

$$b^2 = a^2 + c^2 - 2ac \cos B,$$

$$c^2 = a^2 + b^2 - 2ab \cos C$$

That is, according to the law of cosines, the square of a side of a triangle is equal to the sum of the squares of the other two sides, minus twice the product of those two sides and the cosine of the angle included between them.

NOTE If we let $C = 90°$ in the third form of the law of cosines, then $\cos C = \cos 90° = 0$, and the formula becomes $c^2 = a^2 + b^2$, the Pythagorean theorem **(Appendix B).** The Pythagorean theorem is a special case of the law of cosines.

Solving SAS and SSS Triangles (Cases 3 and 4)

EXAMPLE 1 Applying the Law of Cosines (SAS)

A surveyor wishes to find the distance between two inaccessible points A and B on opposite sides of a lake. While standing at point C, she finds that $b = 259$ m, $a = 423$ m, and angle ACB measures $132°\ 40'$. Find the distance c. See **Figure 14.**

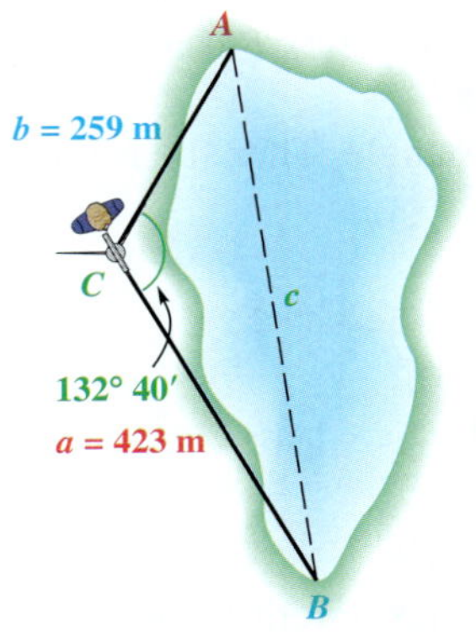

Figure 14

SOLUTION We can use the law of cosines here because we know the lengths of two sides of the triangle and the measure of the included angle.

$c^2 = a^2 + b^2 - 2ab \cos C$ Law of cosines

$c^2 = 423^2 + 259^2 - 2(423)(259) \cos 132°\ 40'$ Substitute.

$c^2 \approx 394{,}510.6$ Use a calculator.

$c \approx 628$ Take the square root of each side. Choose the positive root.

The distance between the points is approximately 628 m.

✔ *Now Try Exercise 39.*

EXAMPLE 2 Applying the Law of Cosines (SAS)

Solve triangle ABC if $A = 42.3°$, $b = 12.9$ m, and $c = 15.4$ m.

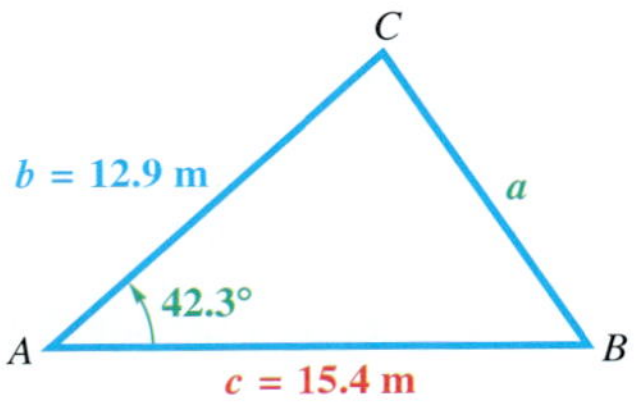

Figure 15

SOLUTION See **Figure 15.** We start by finding a with the law of cosines.

$a^2 = b^2 + c^2 - 2bc \cos A$ Law of cosines

$a^2 = 12.9^2 + 15.4^2 - 2(12.9)(15.4) \cos 42.3°$ Substitute.

$a^2 \approx 109.7$ Use a calculator.

$a \approx 10.47$ m Take square roots and choose the positive root.

Of the two remaining angles B and C, B must be the smaller since it is opposite the shorter of the two sides b and c. Therefore, B cannot be obtuse.

$\frac{\sin A}{a} = \frac{\sin B}{b}$ Law of sines (alternative form) **(Section 7.1)**

$\frac{\sin 42.3°}{10.47} = \frac{\sin B}{12.9}$ Substitute.

$\sin B = \frac{12.9 \sin 42.3°}{10.47}$ Multiply by 12.9 and rewrite.

$B \approx 56.0°$ Use the inverse sine function. **(Section 6.1)**

The easiest way to find C is to subtract the measures of A and B from 180°.

$C = 180° - A - B$ Angle sum formula **(Section 1.2)**, solved for C

$C \approx 180° - 42.3° - 56.0°$ Substitute.

$C \approx 81.7°$ Subtract.

✔ *Now Try Exercise 19.*

CAUTION Had we used the law of sines to find C rather than B in **Example 2,** we would not have known whether C was equal to $81.7°$ or its supplement, $98.3°$.

EXAMPLE 3 Applying the Law of Cosines (SSS)

Solve triangle ABC if $a = 9.47$ ft, $b = 15.9$ ft, and $c = 21.1$ ft.

SOLUTION We can use the law of cosines to solve for any angle of the triangle. We solve for C, the largest angle. We will know that C is obtuse if $\cos C < 0$.

$$c^2 = a^2 + b^2 - 2ab \cos C \quad \text{Law of cosines}$$

$$\cos C = \frac{a^2 + b^2 - c^2}{2ab} \quad \text{Solve for } \cos C.$$

$$\cos C = \frac{9.47^2 + 15.9^2 - 21.1^2}{2(9.47)(15.9)} \quad \text{Substitute.}$$

$$\cos C \approx -0.34109402 \quad \text{Use a calculator.}$$

$$C \approx 109.9° \quad \text{Use the inverse cosine function. (Section 6.1)}$$

Now use the law of sines to find B.

$$\frac{\sin C}{c} = \frac{\sin B}{b} \quad \text{Law of sines (alternative form)}$$

$$\frac{\sin 109.9°}{21.1} = \frac{\sin B}{15.9} \quad \text{Substitute.}$$

$$\sin B = \frac{15.9 \sin 109.9°}{21.1} \quad \text{Multiply by 15.9 and rewrite.}$$

$$B \approx 45.1° \quad \text{Use the inverse sine function.}$$

Since $A = 180° - B - C$, we have $A \approx 180° - 45.1° - 109.9° \approx 25.0°$.

✔ *Now Try Exercise 23.*

Figure 16

Trusses are frequently used to support roofs on buildings, as illustrated in **Figure 16.** The simplest type of roof truss is a triangle, as shown in **Figure 17.** (*Source:* Riley, W., L. Sturges, and D. Morris, *Statics and Mechanics of Materials,* John Wiley and Sons.)

EXAMPLE 4 Designing a Roof Truss (SSS)

Find angle B to the nearest degree for the truss shown in **Figure 17.**

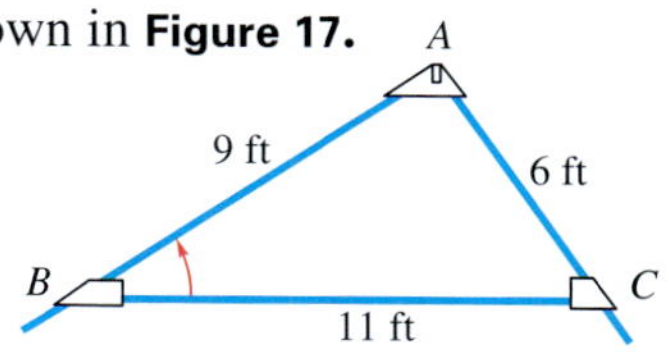

Figure 17

SOLUTION

$$b^2 = a^2 + c^2 - 2ac \cos B \quad \text{Law of cosines}$$

$$\cos B = \frac{a^2 + c^2 - b^2}{2ac} \quad \text{Solve for } \cos B.$$

$$\cos B = \frac{11^2 + 9^2 - 6^2}{2(11)(9)} \quad \text{Let } a = 11,\ b = 6, \text{ and } c = 9.$$

$$\cos B \approx 0.83838384 \quad \text{Use a calculator.}$$

$$B \approx 33° \quad \text{Use the inverse cosine function.}$$

✔ *Now Try Exercise 49.*

Four possible cases can occur when we solve an oblique triangle. They are summarized in the following table. In all four cases, it is assumed that the given information actually produces a triangle.

Oblique Triangle	Suggested Procedure for Solving
Case 1: One side and two angles are known. **(SAA or ASA)**	***Step 1*** Find the remaining angle using the angle sum formula $(A + B + C = 180°)$. ***Step 2*** Find the remaining sides using the law of sines.
Case 2: Two sides and one angle (not included between the two sides) are known. **(SSA)**	*This is the ambiguous case. There may be no triangle, one triangle, or two triangles.* ***Step 1*** Find an angle using the law of sines. ***Step 2*** Find the remaining angle using the angle sum formula. ***Step 3*** Find the remaining side using the law of sines. *If two triangles exist, repeat Steps 2 and 3.*
Case 3: Two sides and the included angle are known. **(SAS)**	***Step 1*** Find the third side using the law of cosines. ***Step 2*** Find the smaller of the two remaining angles using the law of sines. ***Step 3*** Find the remaining angle using the angle sum formula.
Case 4: Three sides are known. **(SSS)**	***Step 1*** Find the largest angle using the law of cosines. ***Step 2*** Find either remaining angle using the law of sines. ***Step 3*** Find the remaining angle using the angle sum formula.

Heron's Formula for the Area of a Triangle A formula for finding the area of a triangle given the lengths of the three sides, known as **Heron's formula,** is named after the Greek mathematician Heron of Alexandria, who lived around A.D. 75. It is found in his work *Metrica*. Heron's formula can be used for the case SSS.

Heron's Area Formula (SSS)

If a triangle has sides of lengths a, b, and c, with **semiperimeter**

$$s = \frac{1}{2}(a + b + c),$$

then the area $\mathcal{A}$ of the triangle is given by the following formula.

$$\mathcal{A} = \sqrt{s(s - a)(s - b)(s - c)}$$

That is, according to Heron's formula, the area of a triangle is the square root of the product of four factors: (1) the semiperimeter, (2) the semiperimeter minus the first side, (3) the semiperimeter minus the second side, and (4) the semiperimeter minus the third side.

A derivation of Heron's formula is given at the end of this section.

EXAMPLE 5 Using Heron's Formula to Find an Area (SSS)

The distance "as the crow flies" from Los Angeles to New York is 2451 mi, from New York to Montreal is 331 mi, and from Montreal to Los Angeles is 2427 mi. What is the area of the triangular region having these three cities as vertices? (Ignore the curvature of Earth.)

SOLUTION In **Figure 18,** we let $a = 2451$, $b = 331$, and $c = 2427$.

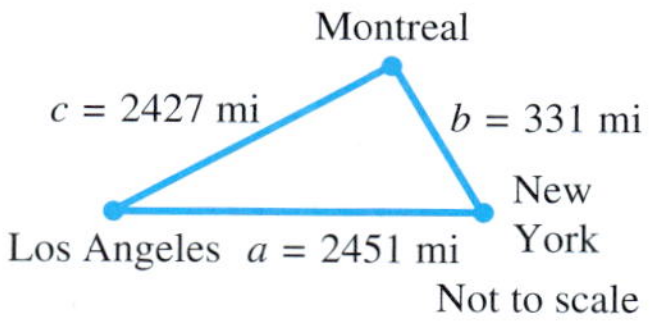

Figure 18

Here,

$$s = \frac{1}{2}(2451 + 331 + 2427) \quad \text{Semiperimeter}$$

$$s = 2604.5. \quad \text{Add, and then multiply.}$$

Now use Heron's formula to find the area $\mathcal{A}$.

$$\mathcal{A} = \sqrt{s(s-a)(s-b)(s-c)}$$

Don't forget the factor s.

$$\mathcal{A} = \sqrt{2604.5(2604.5 - 2451)(2604.5 - 331)(2604.5 - 2427)}$$

$$\mathcal{A} \approx 401{,}700 \text{ mi}^2 \quad \text{Use a calculator.}$$

✔ *Now Try Exercise 73.*

Derivation of Heron's Formula A trigonometric derivation of Heron's formula illustrates some ingenious manipulation involving the law of cosines, algebraic techniques, double-angle identities, and the area formula $\mathcal{A} = \frac{1}{2}bc \sin A$.

Let triangle ABC have sides of lengths a, b, and c. Apply the law of cosines.

$$a^2 = b^2 + c^2 - 2bc \cos A \quad \text{Law of cosines}$$

$$\cos A = \frac{b^2 + c^2 - a^2}{2bc} \quad \text{Solve for } \cos A. \quad (1)$$

The perimeter of the triangle is $a + b + c$, so half of the perimeter (the semi-perimeter) is given by the formula in equation (2) below.

$$s = \frac{1}{2}(a + b + c) \quad (2)$$

$$2s = a + b + c \quad \text{Multiply by 2.} \quad (3)$$

$$b + c - a = 2s - 2a \quad \text{Subtract } 2a \text{ from each side and rewrite.}$$

$$b + c - a = 2(s - a) \quad \text{Factor.} \quad (4)$$

Subtract $2b$ and $2c$ in a similar way in equation (3) to obtain equations (5) and (6).

$$a - b + c = 2(s - b) \quad (5)$$

$$a + b - c = 2(s - c) \quad (6)$$

Now we obtain an expression for $1 - \cos A$.

$$1 - \cos A = 1 - \underbrace{\frac{b^2 + c^2 - a^2}{2bc}}_{\cos A,\ \text{from (1)}}$$

$$= \frac{2bc + a^2 - b^2 - c^2}{2bc} \qquad \text{Find a common denominator, and distribute the } - \text{ sign.}$$

$$= \frac{a^2 - (b^2 - 2bc + c^2)}{2bc} \qquad \text{Regroup.}$$

Pay attention to signs.

$$= \frac{a^2 - (b - c)^2}{2bc} \qquad \text{Factor the perfect square trinomial.}$$

$$= \frac{[a - (b - c)][a + (b - c)]}{2bc} \qquad \text{Factor the difference of squares.}$$

$$= \frac{(a - b + c)(a + b - c)}{2bc} \qquad \text{Distributive property}$$

$$= \frac{2(s - b) \cdot 2(s - c)}{2bc} \qquad \text{From (5) and (6)}$$

$$1 - \cos A = \frac{2(s - b)(s - c)}{bc} \qquad \text{Lowest terms} \quad (7)$$

Similarly, it can be shown that

$$1 + \cos A = \frac{2s(s - a)}{bc}. \qquad (8)$$

Recall the double-angle identities for $\cos 2\theta$ from **Section 5.5.**

$$\cos 2\theta = 2\cos^2\theta - 1$$

$$\cos A = 2\cos^2\left(\frac{A}{2}\right) - 1 \qquad \text{Let } \theta = \tfrac{A}{2}.$$

$$1 + \cos A = 2\cos^2\left(\frac{A}{2}\right) \qquad \text{Add 1.}$$

$$\underbrace{\frac{2s(s - a)}{bc}}_{\text{From (8)}} = 2\cos^2\left(\frac{A}{2}\right) \qquad \text{Substitute.}$$

$$\frac{s(s - a)}{bc} = \cos^2\left(\frac{A}{2}\right) \qquad \text{Divide by 2.}$$

$$\cos\left(\frac{A}{2}\right) = \sqrt{\frac{s(s - a)}{bc}} \qquad (9)$$

$$\cos 2\theta = 1 - 2\sin^2\theta$$

$$\cos A = 1 - 2\sin^2\left(\frac{A}{2}\right) \qquad \text{Let } \theta = \tfrac{A}{2}.$$

$$1 - \cos A = 2\sin^2\left(\frac{A}{2}\right) \qquad \text{Subtract 1. Multiply by } -1.$$

$$\underbrace{\frac{2(s - b)(s - c)}{bc}}_{\text{From (7)}} = 2\sin^2\left(\frac{A}{2}\right) \qquad \text{Substitute.}$$

$$\frac{(s - b)(s - c)}{bc} = \sin^2\left(\frac{A}{2}\right) \qquad \text{Divide by 2.}$$

$$\sin\left(\frac{A}{2}\right) = \sqrt{\frac{(s - b)(s - c)}{bc}} \qquad (10)$$

The area of triangle ABC can be expressed as follows.

$$\mathcal{A} = \frac{1}{2} bc \sin A \qquad \text{(Section 7.2)}$$

$$2\mathcal{A} = bc \sin A \qquad \text{Multiply by 2.}$$

$$\frac{2\mathcal{A}}{bc} = \sin A \qquad \text{Divide by } bc. \quad (11)$$

Recall the double-angle identity for sin 2θ.

$$\sin 2\theta = 2\sin\theta\cos\theta \qquad \text{(Section 5.5)}$$

$$\sin A = 2\sin\left(\frac{A}{2}\right)\cos\left(\frac{A}{2}\right) \qquad \text{Let } \theta = \tfrac{A}{2}.$$

$$\frac{2\mathscr{A}}{bc} = 2\sin\left(\frac{A}{2}\right)\cos\left(\frac{A}{2}\right) \qquad \text{Use equation (11).}$$

$$\frac{2\mathscr{A}}{bc} = 2\sqrt{\frac{(s-b)(s-c)}{bc}}\cdot\sqrt{\frac{s(s-a)}{bc}} \qquad \text{Use equations (9) and (10).}$$

$$\frac{2\mathscr{A}}{bc} = 2\sqrt{\frac{s(s-a)(s-b)(s-c)}{b^2c^2}} \qquad \text{Multiply.}$$

$$\frac{2\mathscr{A}}{bc} = \frac{2\sqrt{s(s-a)(s-b)(s-c)}}{bc} \qquad \text{Simplify the denominator.}$$

$$\mathscr{A} = \sqrt{s(s-a)(s-b)(s-c)} \qquad \text{Heron's formula}$$

7.3 Exercises

Concept Check *Assume a triangle ABC has standard labeling.*

(a) Determine whether SAA, ASA, SSA, SAS, *or* SSS *is given.*

(b) Decide whether the law of sines or the law of cosines should be used to begin solving the triangle.

1. a, b, and C **2.** A, C, and c **3.** a, b, and A **4.** a, B, and C

5. A, B, and c **6.** a, c, and A **7.** a, b, and c **8.** b, c, and A

Find the length of the remaining side of each triangle. Do not use a calculator.

9.

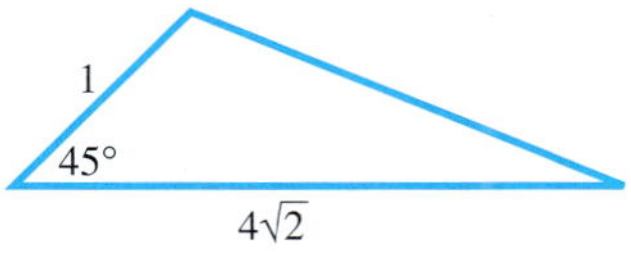

10.

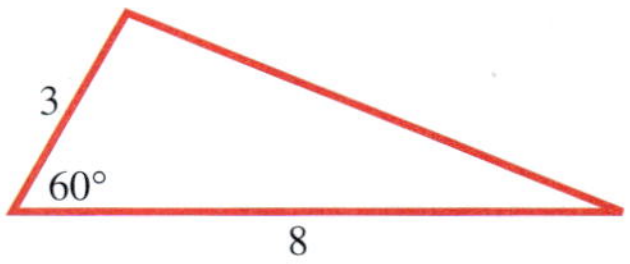

Find the measure of θ in each triangle. Do not use a calculator.

11.

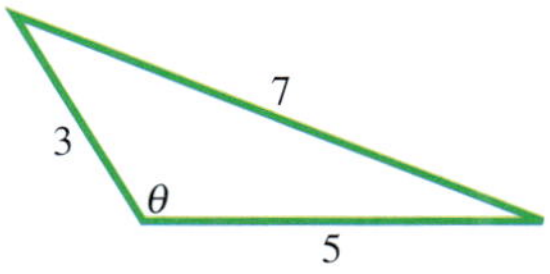

12.

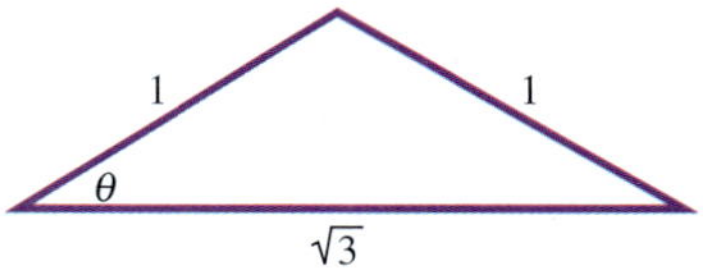

Solve each triangle. Approximate values to the nearest tenth.

13.

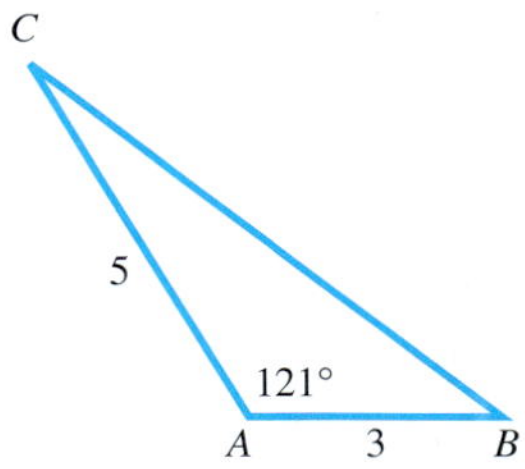

14.

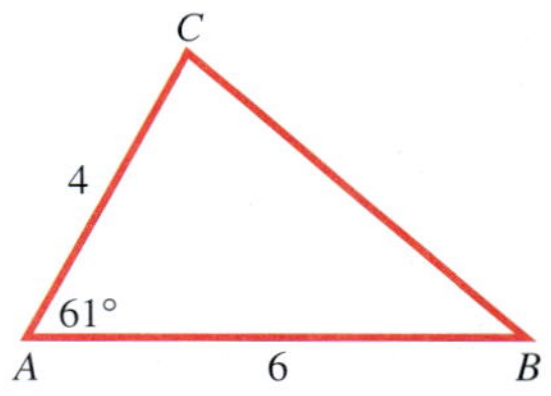

15.

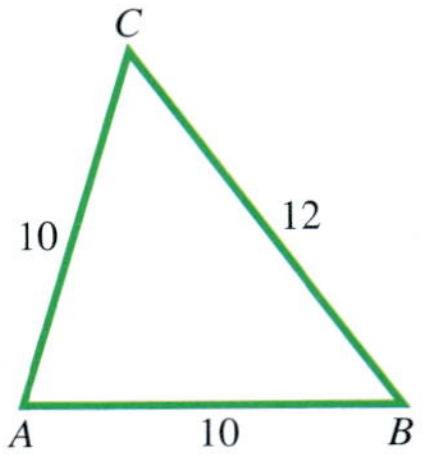

16.

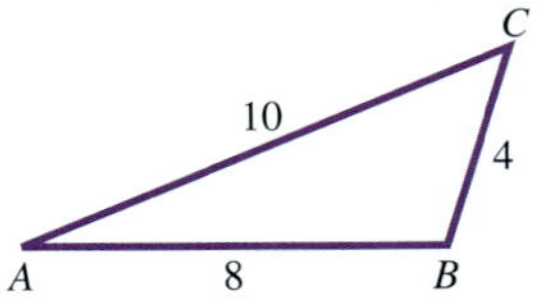

17.

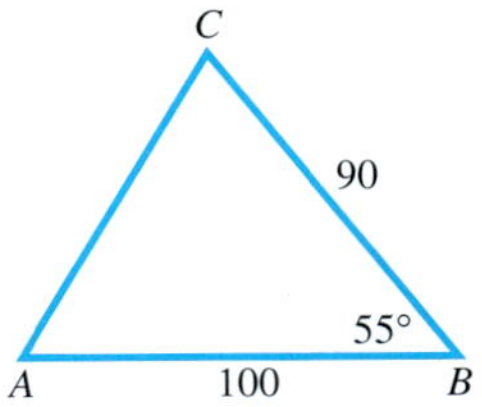

18.

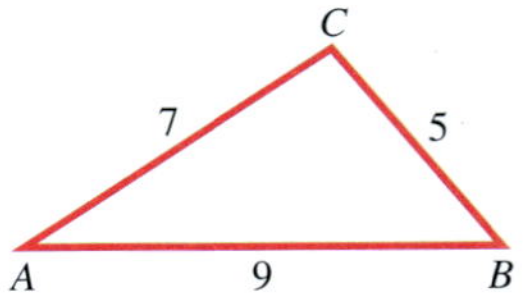

Solve each triangle. **See Examples 2 and 3.**

19. $A = 41.4°$, $b = 2.78$ yd, $c = 3.92$ yd

20. $C = 28.3°$, $b = 5.71$ in., $a = 4.21$ in.

21. $C = 45.6°$, $b = 8.94$ m, $a = 7.23$ m

22. $A = 67.3°$, $b = 37.9$ km, $c = 40.8$ km

23. $a = 9.3$ cm, $b = 5.7$ cm, $c = 8.2$ cm

24. $a = 28$ ft, $b = 47$ ft, $c = 58$ ft

25. $a = 42.9$ m, $b = 37.6$ m, $c = 62.7$ m

26. $a = 189$ yd, $b = 214$ yd, $c = 325$ yd

27. $a = 965$ ft, $b = 876$ ft, $c = 1240$ ft

28. $a = 324$ m, $b = 421$ m, $c = 298$ m

29. $A = 80° \, 40'$, $b = 143$ cm, $c = 89.6$ cm

30. $C = 72° \, 40'$, $a = 327$ ft, $b = 251$ ft

31. $B = 74.8°$, $a = 8.92$ in., $c = 6.43$ in.

32. $C = 59.7°$, $a = 3.73$ mi, $b = 4.70$ mi

33. $A = 112.8°$, $b = 6.28$ m, $c = 12.2$ m

34. $B = 168.2°$, $a = 15.1$ cm, $c = 19.2$ cm

35. $a = 3.0$ ft, $b = 5.0$ ft, $c = 6.0$ ft

36. $a = 4.0$ ft, $b = 5.0$ ft, $c = 8.0$ ft

37. Refer to **Figure 12.** If you attempt to find any angle of a triangle with the values $a = 3$, $b = 4$, and $c = 10$ by using the law of cosines, what happens?

38. "The shortest distance between two points is a straight line." Explain how this is related to the geometric property that states that the sum of the lengths of any two sides of a triangle must be greater than the length of the remaining side.

Solve each problem. **See Examples 1–4.**

39. ***Distance across a River*** Points A and B are on opposite sides of False River. From a third point, C, the angle between the lines of sight to A and B is 46.3°. If AC is 350 m long and BC is 286 m long, find AB.

40. *Distance across a Ravine* Points X and Y are on opposite sides of a ravine. From a third point Z, the angle between the lines of sight to X and Y is $37.7°$. If XZ is 153 m long and YZ is 103 m long, find XY.

41. *Angle in a Parallelogram* A parallelogram has sides of length 25.9 cm and 32.5 cm. The longer diagonal has length 57.8 cm. Find the measure of the angle opposite the longer diagonal.

42. *Diagonals of a Parallelogram* The sides of a parallelogram are 4.0 cm and 6.0 cm. One angle is $58°$ while another is $122°$. Find the lengths of the diagonals of the parallelogram.

43. *Flight Distance* Airports A and B are 450 km apart, on an east-west line. Tom flies in a northeast direction from A to airport C. From C he flies 359 km on a bearing of $128° 40'$ to B. How far is C from A?

44. *Distance Traveled by a Plane* An airplane flies 180 mi from point X at a bearing of $125°$, and then turns and flies at a bearing of $230°$ for 100 mi. How far is the plane from point X?

45. *Distance between Ends of the Vietnam Memorial* The Vietnam Veterans Memorial in Washington, D.C., is V-shaped with equal sides of length 246.75 ft. The angle between these sides measures $125° 12'$. Find the distance between the ends of the two sides. (*Source:* Pamphlet obtained at Vietnam Veterans Memorial.)

46. *Distance between Two Ships* Two ships leave a harbor together, traveling on courses that have an angle of $135° 40'$ between them. If each travels 402 mi, how far apart are they?

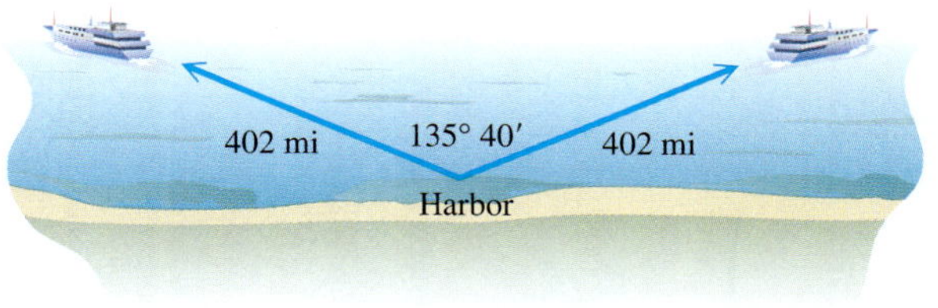

47. *Distance between a Ship and a Rock* A ship is sailing east. At one point, the bearing of a submerged rock is $45° 20'$. After the ship has sailed 15.2 mi, the bearing of the rock has become $308° 40'$. Find the distance of the ship from the rock at the latter point.

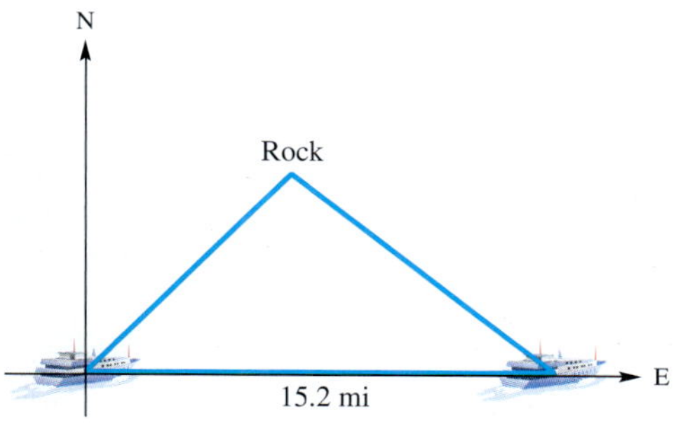

48. *Distance between a Ship and a Submarine* From an airplane flying over the ocean, the angle of depression to a submarine lying under the surface is $24° 10'$. At the same moment, the angle of depression from the airplane to a battleship is $17° 30'$. See the figure. The distance from the airplane to the battleship is 5120 ft. Find the distance between the battleship and the submarine. (Assume the airplane, submarine, and battleship are in a vertical plane.)

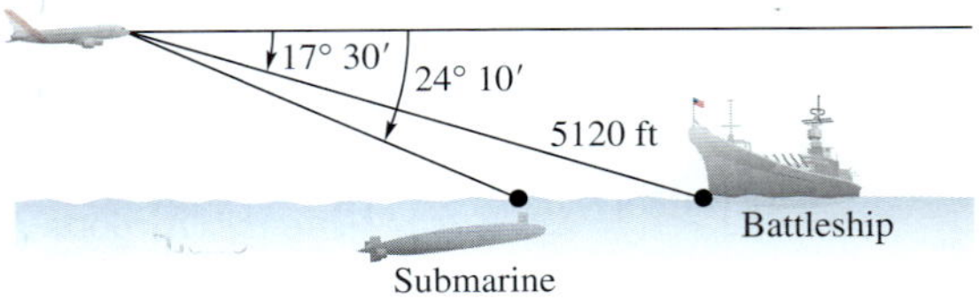

49. *Truss Construction* A triangular truss is shown in the figure. Find angle θ.

50. *Truss Construction* Find angle β in the truss shown in the figure.

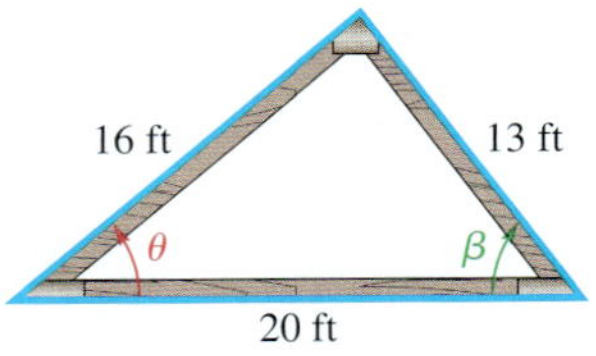

51. *Distance between a Beam and Cables* A weight is supported by cables attached to both ends of a balance beam, as shown in the figure. What angles are formed between the beam and the cables?

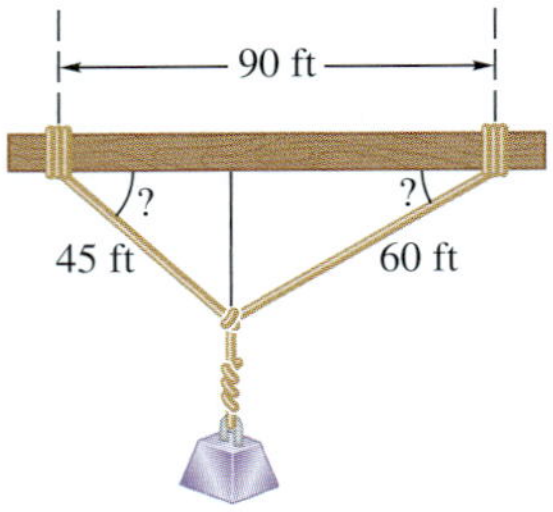

52. *Distance between Points on a Crane* A crane with a counterweight is shown in the figure. Find the horizontal distance between points A and B to the nearest foot.

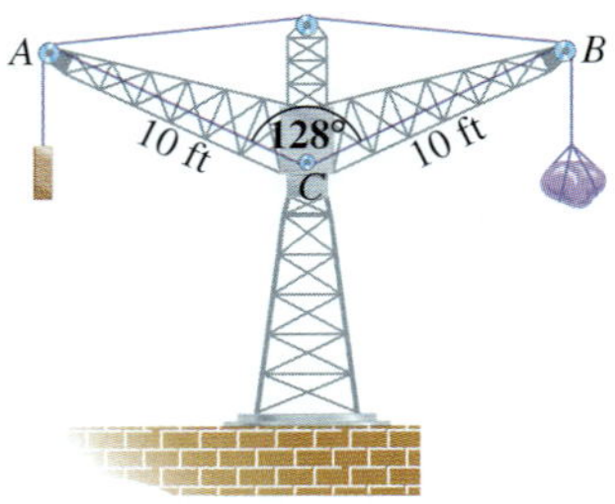

53. *Distance on a Baseball Diamond* A baseball diamond is a square, 90.0 ft on a side, with home plate and the three bases as vertices. The pitcher's position is 60.5 ft from home plate. Find the distance from the pitcher's position to each of the bases.

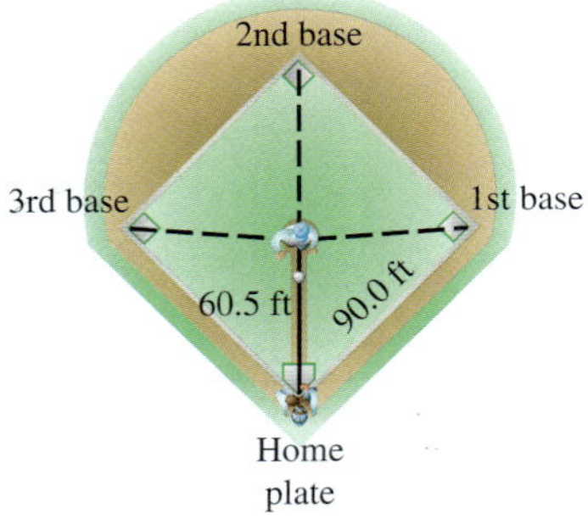

54. *Distance on a Softball Diamond* A softball diamond is a square, 60.0 ft on a side, with home plate and the three bases as vertices. The pitcher's position is 46.0 ft from home plate. Find the distance from the pitcher's position to each of the bases.

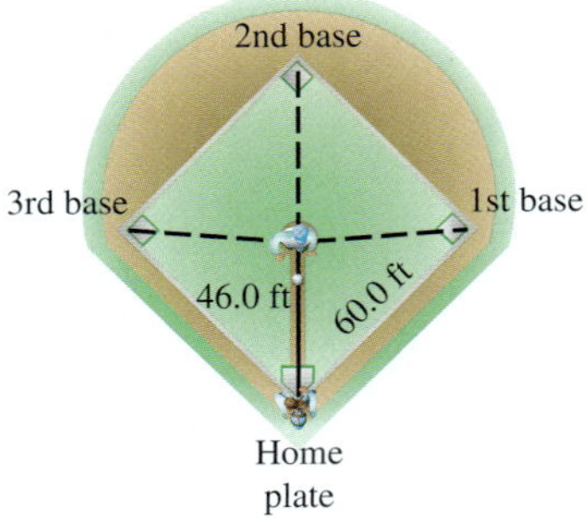

55. *Distance between a Ship and a Point* Starting at point A, a ship sails 18.5 km on a bearing of 189°, then turns and sails 47.8 km on a bearing of 317°. Find the distance of the ship from point A.

56. *Distance between Two Factories* Two factories blow their whistles at exactly 5:00. A man hears the two blasts at 3 sec and 6 sec after 5:00, respectively. The angle between his lines of sight to the two factories is 42.2°. If sound travels 344 m per sec, how far apart are the factories?

57. *Measurement Using Triangulation* Surveyors are often confronted with obstacles, such as trees, when measuring the boundary of a lot. One technique used to obtain an accurate measurement is the so-called **triangulation method.** In this technique, a triangle is constructed around the obstacle and one angle and two sides of the triangle are measured. Use this technique to find the length of the property line (the straight line between the two markers) in the figure. (*Source:* Kavanagh, B., *Surveying Principles and Applications,* Sixth Edition, Prentice-Hall.)

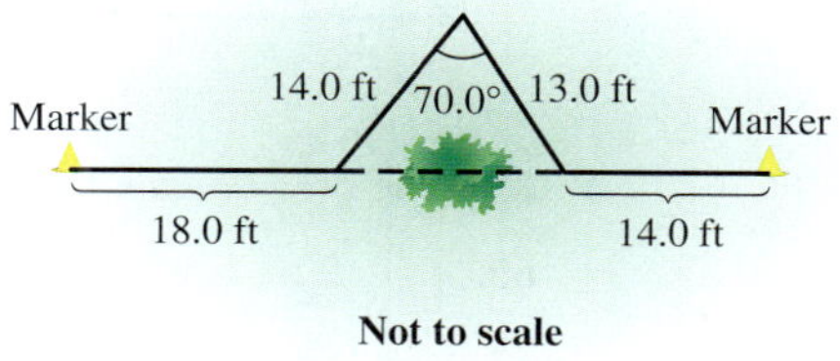

58. *Path of a Ship* A ship sailing due east in the North Atlantic has been warned to change course to avoid icebergs. The captain turns and sails on a bearing of 62°, then changes course again to a bearing of 115° until the ship reaches its original course. See the figure. How much farther did the ship have to travel to avoid the icebergs?

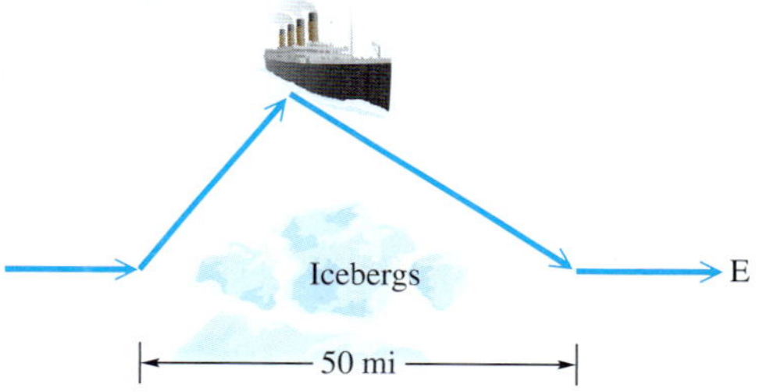

59. *Length of a Tunnel* To measure the distance through a mountain for a proposed tunnel, a point C is chosen that can be reached from each end of the tunnel. See the figure. If $AC = 3800$ m, $BC = 2900$ m, and angle $C = 110°$, find the length of the tunnel.

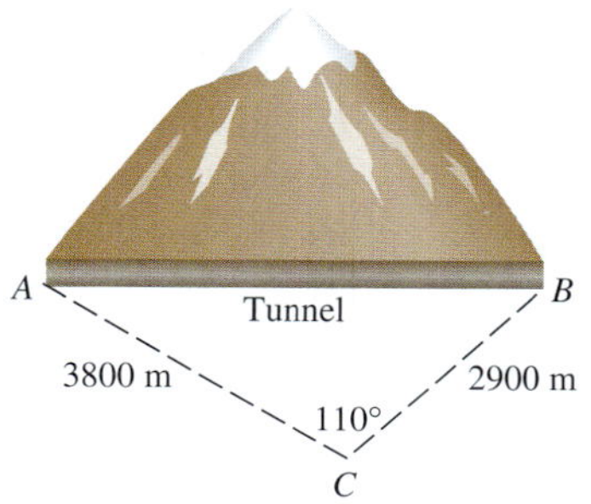

60. *Distance between an Airplane and a Mountain* A person in a plane flying straight north observes a mountain at a bearing of 24.1°. At that time, the plane is 7.92 km from the mountain. A short time later, the bearing to the mountain becomes 32.7°. How far is the airplane from the mountain when the second bearing is taken?

Find the measure of each angle θ to two decimal places.

61.

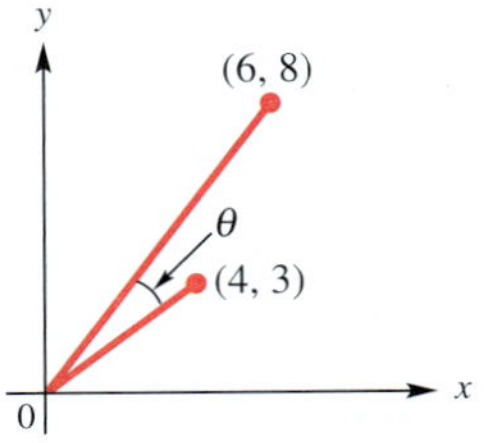

62.

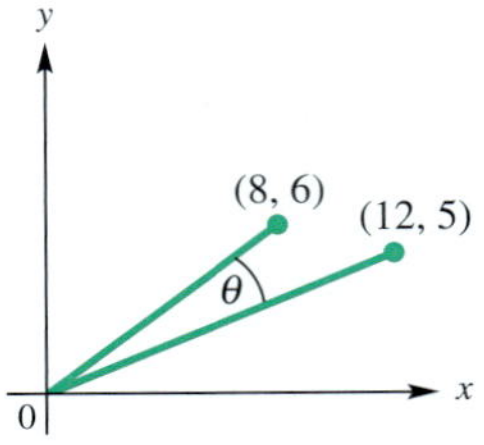

Find the exact area of each triangle using the formula $\mathcal{A} = \frac{1}{2}bh$, and then verify that Heron's formula gives the same result.

63.

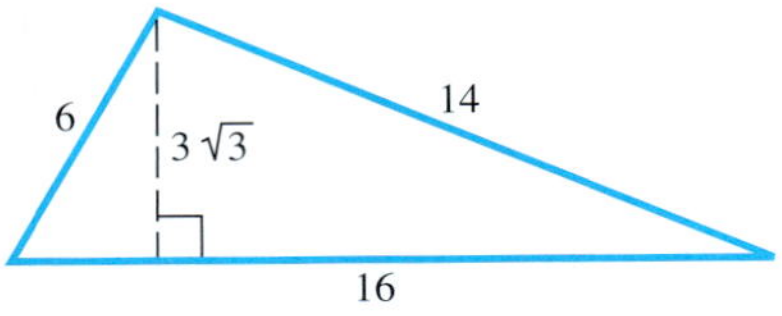

64.

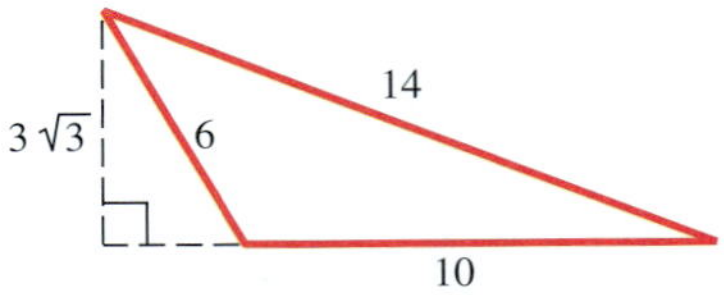

Find the area of each triangle ABC. **See Example 5.**

65. $a = 12$ m, $b = 16$ m, $c = 25$ m

66. $a = 22$ in., $b = 45$ in., $c = 31$ in.

67. $a = 154$ cm, $b = 179$ cm, $c = 183$ cm

68. $a = 25.4$ yd, $b = 38.2$ yd, $c = 19.8$ yd

69. $a = 76.3$ ft, $b = 109$ ft, $c = 98.8$ ft

70. $a = 15.89$ m, $b = 21.74$ m, $c = 10.92$ m

Solve each problem. **See Example 5.**

71. *Perfect Triangles* A **perfect triangle** is a triangle whose sides have whole number lengths and whose area is numerically equal to its perimeter. Show that the triangle with sides of length 9, 10, and 17 is perfect.

72. *Heron Triangles* A **Heron triangle** is a triangle having integer sides and area. Show that each of the following is a Heron triangle.

(a) $a = 11, b = 13, c = 20$ **(b)** $a = 13, b = 14, c = 15$
(c) $a = 7, b = 15, c = 20$ **(d)** $a = 9, b = 10, c = 17$

73. *Area of the Bermuda Triangle* Find the area of the Bermuda Triangle if the sides of the triangle have approximate lengths 850 mi, 925 mi, and 1300 mi.

74. *Required Amount of Paint* A painter needs to cover a triangular region 75 m by 68 m by 85 m. A can of paint covers 75 m^2 of area. How many cans (to the next higher number of cans) will be needed?

75. Consider triangle ABC shown here.

(a) Use the law of sines to find candidates for the value of angle C. Round angle measures to the nearest tenth of a degree.
(b) Rework part (a) using the law of cosines.
(c) Why is the law of cosines a better method in this case?

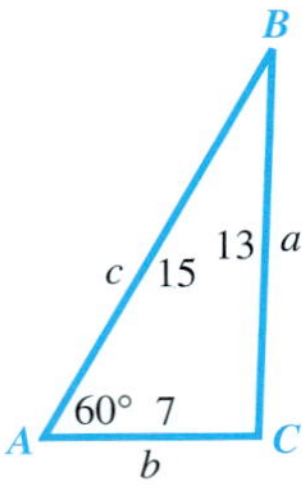

76. Show that the measure of angle A is twice the measure of angle B. (*Hint:* Use the law of cosines to find $\cos A$ and $\cos B$, and then show that $\cos A = 2\cos^2 B - 1$.)

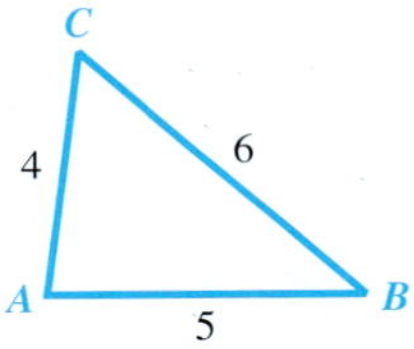

Relating Concepts

For individual or collaborative investigation *(Exercises 77–80)*

We have introduced two new formulas for the area of a triangle in this chapter. You should now be able to find the area $\mathcal{A}$ *of a triangle using one of three formulas.*

(a) $\mathcal{A} = \frac{1}{2}bh$

(b) $\mathcal{A} = \frac{1}{2}ab \sin C \quad \left(\text{or } \mathcal{A} = \frac{1}{2}ac \sin B \text{ or } \mathcal{A} = \frac{1}{2}bc \sin A\right)$

(c) $\mathcal{A} = \sqrt{s(s-a)(s-b)(s-c)}$ (Heron's formula)

Another area formula can be used when the coordinates of the vertices of a triangle are given. If the vertices are the ordered pairs (x_1, y_1), (x_2, y_2), *and* (x_3, y_3), *then the following is valid.*

(d) $\mathcal{A} = \frac{1}{2}\left|(x_1y_2 - y_1x_2 + x_2y_3 - y_2x_3 + x_3y_1 - y_3x_1)\right|$

Work Exercises 77–80 in order, *showing that the various formulas all lead to the same area.*

77. Draw a triangle with vertices $A(2, 5)$, $B(-1, 3)$, and $C(4, 0)$, and use the distance formula to find the lengths of the sides a, b, and c.

78. Find the area of triangle ABC using formula (b). (First use the law of cosines to find the measure of an angle.)

79. Find the area of triangle ABC using formula (c)—that is, Heron's formula.

80. Find the area of triangle ABC using new formula (d).

Chapter 7 Quiz (Sections 7.1–7.3)

Find the indicated part of each triangle ABC.

1. Find A if $B = 30.6°$, $b = 7.42$ in., and $c = 4.54$ in.

2. Find a if $A = 144°$, $c = 135$ m, and $b = 75.0$ m.

3. Find C if $a = 28.4$ ft, $b = 16.9$ ft, and $c = 21.2$ ft.

4. Find the area of the triangle shown here.

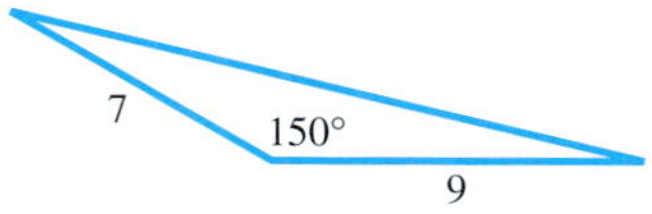

5. Find the area of triangle ABC if $a = 19.5$ km, $b = 21.0$ km, and $c = 22.5$ km.

6. For triangle ABC with $c = 345$, $a = 534$, and $C = 25.4°$, there are two possible values for angle A. What are they?

7. Solve triangle ABC if $c = 326$, $A = 111°$, and $B = 41.0°$.

8. ***Height of a Balloon*** The angles of elevation of a hot air balloon from two observation points X and Y on level ground are 42° 10′ and 23° 30′, respectively. As shown in the figure, points X, Y, and Z are in the same vertical plane and points X and Y are 12.2 mi apart. Approximate the height of the balloon to the nearest tenth of a mile.

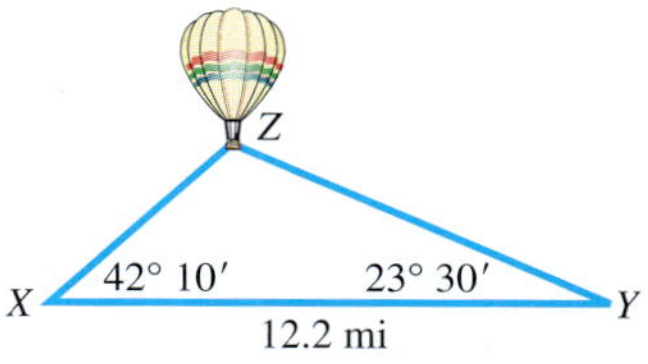

9. *Volcano Movement* To help predict eruptions from the volcano Mauna Loa on the island of Hawaii, scientists keep track of the volcano's movement by using a "super triangle" with vertices on the three volcanoes shown on the map at the right. Find BC given that $AB = 22.47928$ mi, $AC = 28.14276$ mi, and $A = 58.56989°$.

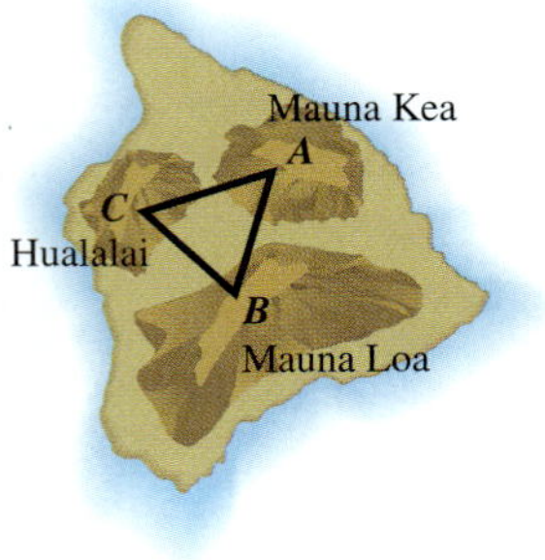

10. *Distance between Two Towns* To find the distance between two small towns, an electronic distance measuring (EDM) instrument is placed on a hill from which both towns are visible. The distance to each town from the EDM and the angle between the two lines of sight are measured. See the figure. Find the distance between the towns.

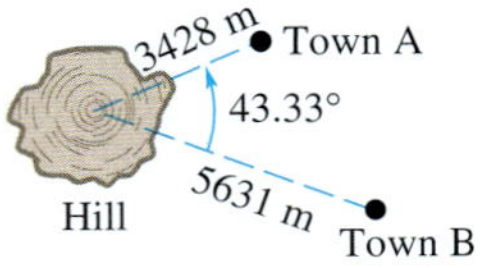

7.4 Vectors, Operations, and the Dot Product

- Basic Terminology
- Algebraic Interpretation of Vectors
- Operations with Vectors
- Dot Product and the Angle between Vectors

Basic Terminology Quantities that involve magnitudes, such as 45 lb or 60 mph, can be represented by real numbers called **scalars.** Other quantities, called **vector quantities,** involve both magnitude *and* direction. Typical vector quantities are velocity, acceleration, and force. For example, traveling 50 mph *east* represents a vector quantity.

A vector quantity can be represented with a directed line segment (a segment that uses an arrowhead to indicate direction) called a **vector.** The *length* of the vector represents the **magnitude** of the vector quantity. The *direction* of the vector, indicated by the arrowhead, represents the direction of the quantity. See **Figure 19.**

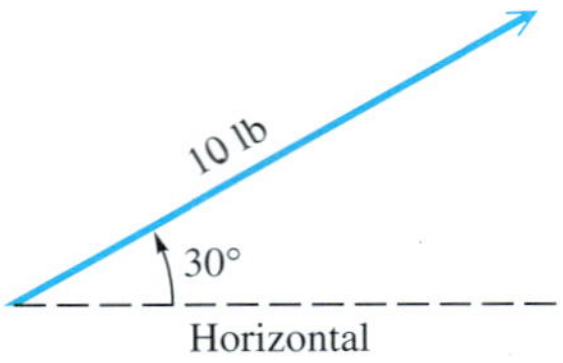

This vector represents a force of 10 lb applied at an angle 30° above the horizontal.

Figure 19

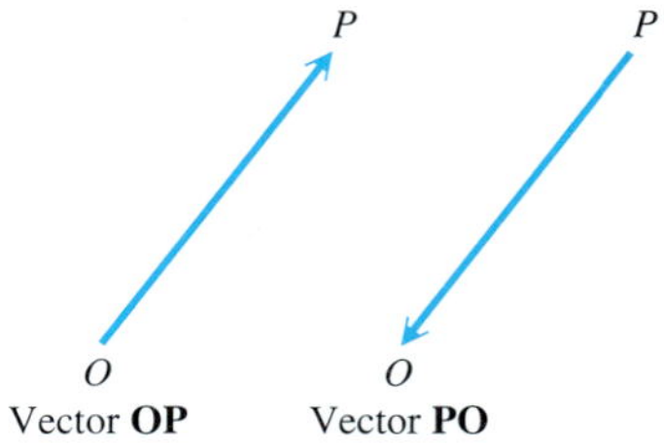

Vectors may be named with one lowercase or uppercase letter or with two uppercase letters.

Figure 20

When we indicate vectors in print, it is customary to use boldface type or an arrow over the letter or letters. Thus, **OP** and $\overrightarrow{OP}$ both represent the vector **OP**. When two letters name a vector, the first indicates the **initial point** and the second indicates the **terminal point** of the vector. Knowing these points gives the direction of the vector. For example, vectors **OP** and **PO** in **Figure 20** are not the same vector. They have the same magnitude but *opposite* directions. The magnitude of vector **OP** is written $|\mathbf{OP}|$.

Two vectors are equal if and only if they have the same direction and the same magnitude. In **Figure 21,** vectors **A** and **B** are equal, as are vectors **C** and **D**. As **Figure 21** shows, equal vectors need not coincide, but they must be parallel and in the same direction. Vectors **A** and **E** are unequal because they do not have the same direction, while $\mathbf{A} \neq \mathbf{F}$ because they have different magnitudes.

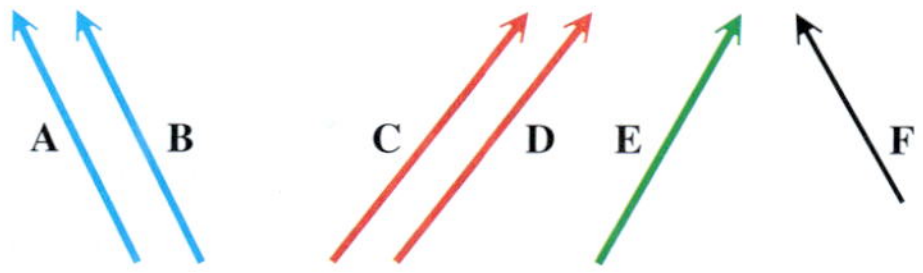

Figure 21

The sum of two vectors is also a vector. There are two ways to find the sum of two vectors **A** and **B** geometrically.

1. Place the initial point of vector **B** at the terminal point of vector **A**, as shown in **Figure 22(a).** The vector with the same initial point as **A** and the same terminal point as **B** is the sum $\mathbf{A} + \mathbf{B}$.

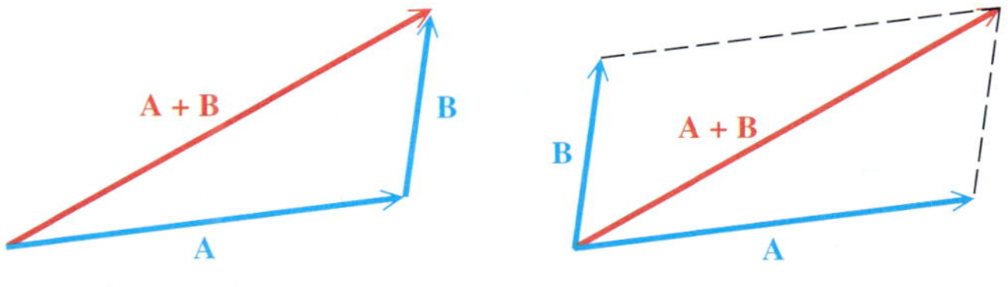

Figure 22

2. Use the **parallelogram rule.** Place vectors **A** and **B** so that their initial points coincide, as in **Figure 22(b).** Then, complete a parallelogram that has **A** and **B** as two sides. The diagonal of the parallelogram with the same initial point as **A** and **B** is the sum $\mathbf{A} + \mathbf{B}$.

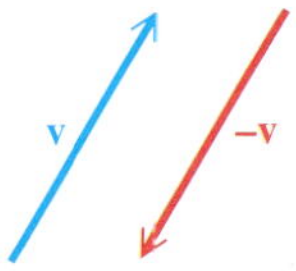

Vectors **v** and **–v** are opposites.

Figure 23

Parallelograms can be used to show that vector $\mathbf{B} + \mathbf{A}$ is the same as vector $\mathbf{A} + \mathbf{B}$, or that $\mathbf{A} + \mathbf{B} = \mathbf{B} + \mathbf{A}$, so ***vector addition is commutative.*** The vector sum $\mathbf{A} + \mathbf{B}$ is the **resultant** of vectors **A** and **B**.

For every vector **v** there is a vector $-\mathbf{v}$ that has the same magnitude as **v** but opposite direction. Vector $-\mathbf{v}$ is the **opposite** of **v**. See **Figure 23.** The sum of **v** and $-\mathbf{v}$ has magnitude 0 and is the **zero vector.** As with real numbers, to subtract vector **B** from vector **A**, find the vector sum $\mathbf{A} + (-\mathbf{B})$. See **Figure 24.**

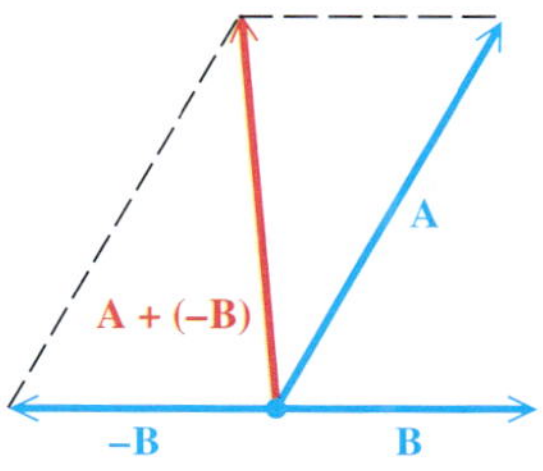

Figure 24

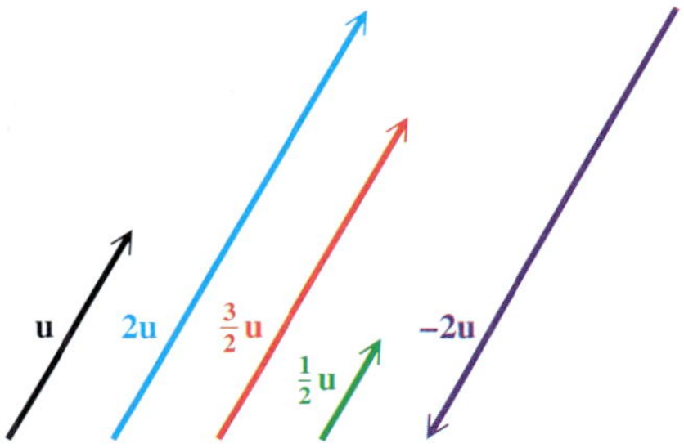

Figure 25

The product of a real number (or scalar) k and a vector **u** is the vector $k \cdot \mathbf{u}$, which has magnitude $|k|$ times the magnitude of **u**. As suggested by **Figure 25,** the vector $k \cdot \mathbf{u}$ has the same direction as **u** if $k > 0$, and has the opposite direction if $k < 0$.

LOOKING AHEAD TO CALCULUS

In addition to two-dimensional vectors in a plane, calculus courses introduce three-dimensional vectors in space. The magnitude of the two-dimensional vector $\langle a, b \rangle$ is given by $\sqrt{a^2 + b^2}$. If we extend this to the three-dimensional vector $\langle a, b, c \rangle$, the expression becomes $\sqrt{a^2 + b^2 + c^2}$. Similar extensions are made for other concepts.

Algebraic Interpretation of Vectors A vector with its initial point at the origin in a rectangular coordinate system is called a **position vector.** A position vector **u** with its endpoint at the point (a, b) is written $\langle a, b \rangle$, so

$$\mathbf{u} = \langle a, b \rangle.$$

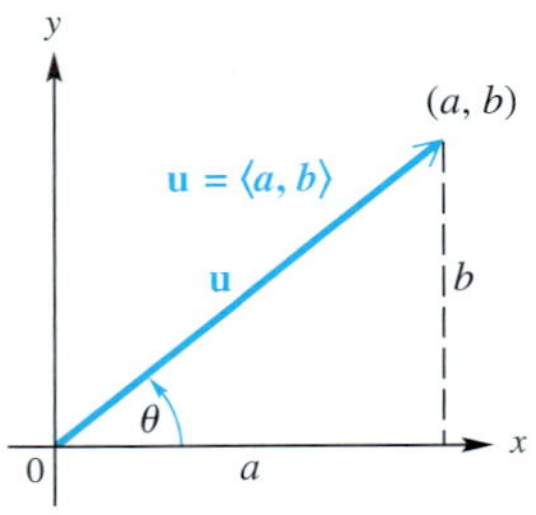

Figure 26

This means that every vector in the real plane corresponds to an ordered pair of real numbers. ***Thus, geometrically a vector is a directed line segment while algebraically it is an ordered pair.*** The numbers a and b are the **horizontal component** and the **vertical component,** respectively, of vector **u**.

Figure 26 shows the vector $\mathbf{u} = \langle a, b \rangle$. The positive angle between the x-axis and a position vector is the **direction angle** for the vector. In **Figure 26,** θ is the direction angle for vector **u**.

From **Figure 26,** we can see that the magnitude and direction of a vector are related to its horizontal and vertical components.

Magnitude and Direction Angle of a Vector $\langle a, b \rangle$

The magnitude (length) of vector $\mathbf{u} = \langle a, b \rangle$ is given by the following.

$$|\mathbf{u}| = \sqrt{a^2 + b^2}$$

The direction angle θ satisfies $\tan \theta = \frac{b}{a}$, where $a \neq 0$.

EXAMPLE 1 Finding Magnitude and Direction Angle

Find the magnitude and direction angle for $\mathbf{u} = \langle 3, -2 \rangle$.

ALGEBRAIC SOLUTION

The magnitude is $|\mathbf{u}| = \sqrt{3^2 + (-2)^2} = \sqrt{13}$. To find the direction angle θ, start with $\tan \theta = \frac{b}{a} = \frac{-2}{3} = -\frac{2}{3}$. Vector **u** has a positive horizontal component and a negative vertical component, placing the position vector in quadrant IV. A calculator gives $\tan^{-1}\left(-\frac{2}{3}\right) \approx -33.7°$. Adding 360° yields the direction angle $\theta \approx 326.3°$. See **Figure 27.**

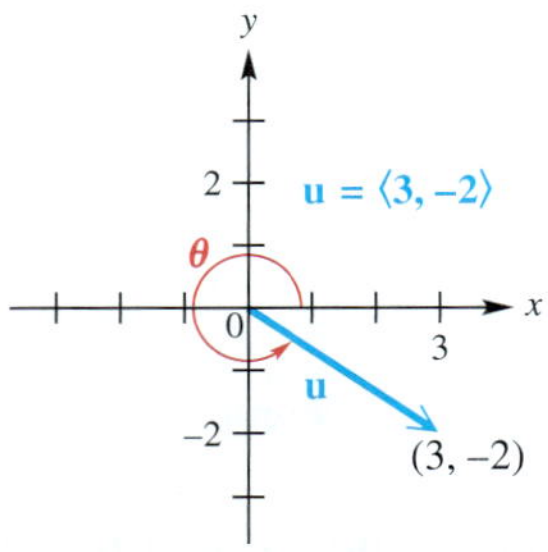

Figure 27

GRAPHING CALCULATOR SOLUTION

A calculator returns the magnitude and direction angle, given the horizontal and vertical components. An approximation for $\sqrt{13}$ is given, and the direction angle has a measure with least possible absolute value. We must add 360° to the value of θ to obtain the positive direction angle. See **Figure 28.**

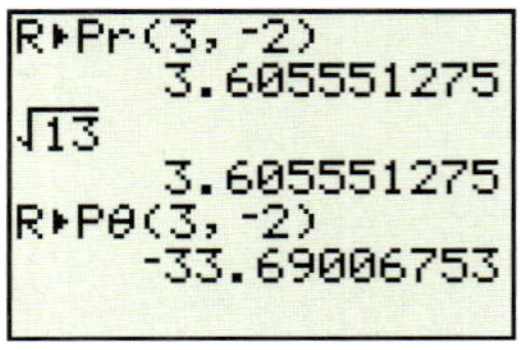

Figure 28

For more information, see your owner's manual or the graphing calculator manual that accompanies this text.

✔ *Now Try Exercise 33.*

Horizontal and Vertical Components

The horizontal and vertical components, respectively, of a vector **u** having magnitude $|\mathbf{u}|$ and direction angle θ are the following.

$$a = |\mathbf{u}| \cos \theta \quad \text{and} \quad b = |\mathbf{u}| \sin \theta$$

That is, $\mathbf{u} = \langle a, b \rangle = \langle |\mathbf{u}| \cos \theta, |\mathbf{u}| \sin \theta \rangle$.

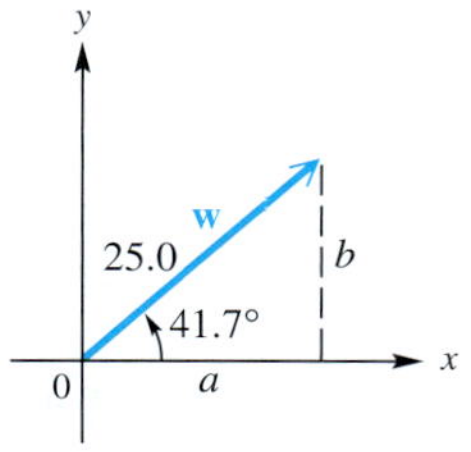

Figure 29

EXAMPLE 2 Finding Horizontal and Vertical Components

Vector **w** in **Figure 29** has magnitude 25.0 and direction angle 41.7°. Find the horizontal and vertical components.

ALGEBRAIC SOLUTION

Use the two formulas in the box, with $|\mathbf{w}| = 25.0$ and $\theta = 41.7°$.

$$a = 25.0 \cos 41.7° \qquad b = 25.0 \sin 41.7°$$

$$a \approx 18.7 \qquad b \approx 16.6$$

Therefore, $\mathbf{w} = \langle 18.7, 16.6 \rangle$. The horizontal component is 18.7, and the vertical component is 16.6 (rounded to the nearest tenth).

GRAPHING CALCULATOR SOLUTION

See **Figure 30.** The results support the algebraic solution.

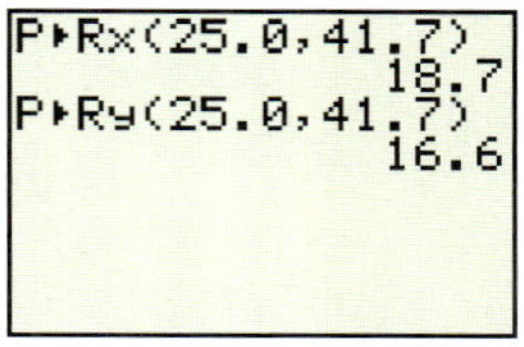

Figure 30

✔ *Now Try Exercise 37.*

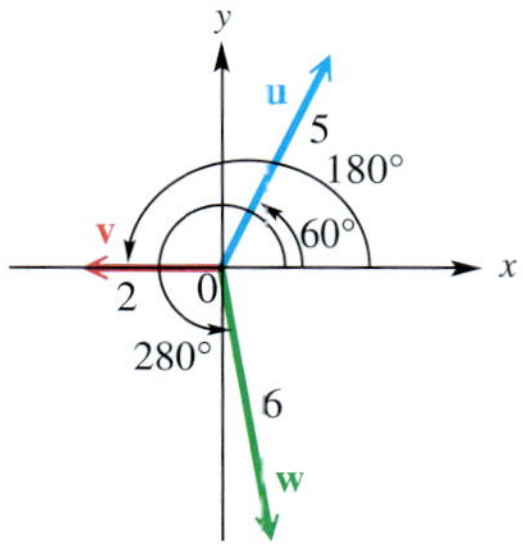

Figure 31

EXAMPLE 3 Writing Vectors in the Form $\langle a, b \rangle$

Write each vector in **Figure 31** in the form $\langle a, b \rangle$.

SOLUTION

$$\mathbf{u} = \langle 5 \cos 60°, 5 \sin 60° \rangle = \left\langle 5 \cdot \frac{1}{2}, 5 \cdot \frac{\sqrt{3}}{2} \right\rangle = \left\langle \frac{5}{2}, \frac{5\sqrt{3}}{2} \right\rangle$$

$$\mathbf{v} = \langle 2 \cos 180°, 2 \sin 180° \rangle = \langle 2(-1), 2(0) \rangle = \langle -2, 0 \rangle$$

$$\mathbf{w} = \langle 6 \cos 280°, 6 \sin 280° \rangle \approx \langle 1.0419, -5.9088 \rangle \quad \text{Use a calculator.}$$

✔ *Now Try Exercises 43 and 45.*

The following geometric properties of parallelograms are helpful when studying applications of vectors.

Properties of Parallelograms

1. A parallelogram is a quadrilateral whose opposite sides are parallel.
2. The opposite sides and opposite angles of a parallelogram are equal, and adjacent angles of a parallelogram are supplementary.
3. The diagonals of a parallelogram bisect each other, but they do not necessarily bisect the angles of the parallelogram.

EXAMPLE 4 Finding the Magnitude of a Resultant

Two forces of 15 and 22 newtons act on a point in the plane. (A **newton** is a unit of force that equals 0.225 lb.) If the angle between the forces is 100°, find the magnitude of the resultant force.

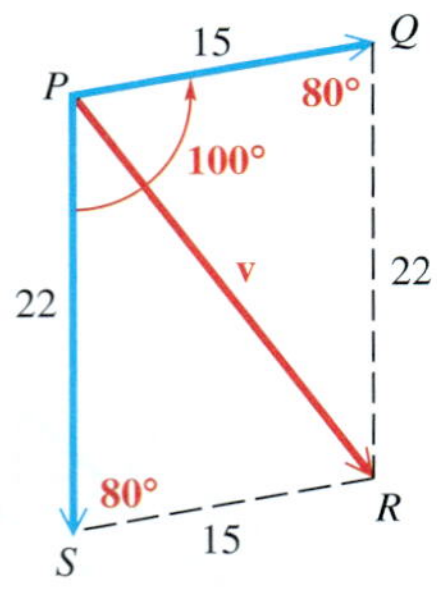

Figure 32

SOLUTION As shown in **Figure 32,** a parallelogram that has the forces as adjacent sides can be formed. The angles of the parallelogram adjacent to angle P measure 80°, since adjacent angles of a parallelogram are supplementary. Opposite sides of the parallelogram are equal in length. The resultant force divides the parallelogram into two triangles. Use the law of cosines with either triangle.

$$|\mathbf{v}|^2 = 15^2 + 22^2 - 2(15)(22)\cos 80^\circ$$ Law of cosines (Section 7.3)

$$\approx 225 + 484 - 115$$ Evaluate powers and multiply.

$$|\mathbf{v}|^2 \approx 594$$ Add and subtract.

$$|\mathbf{v}| \approx 24$$ Take the positive square root. (Appendix A)

To the nearest unit, the magnitude of the resultant force is 24 newtons.

Now Try Exercise 49.

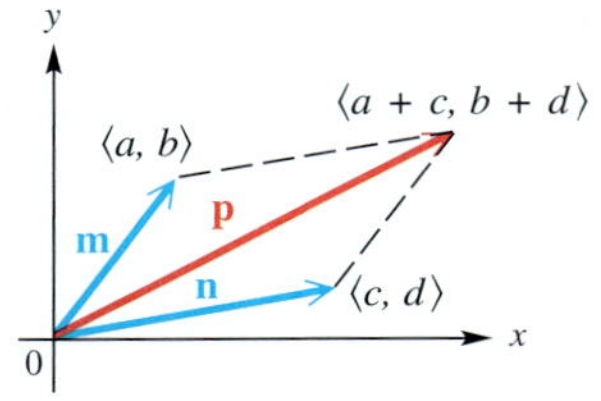

Figure 33

Operations with Vectors As shown in **Figure 33,** $\mathbf{m} = \langle a, b \rangle$, $\mathbf{n} = \langle c, d \rangle$, and $\mathbf{p} = \langle a + c, b + d \rangle$. Using geometry, we can show that the endpoints of the three vectors and the origin form a parallelogram. Since a diagonal of this parallelogram gives the resultant of **m** and **n**, we have $\mathbf{p} = \mathbf{m} + \mathbf{n}$ or

$$\langle a + c, b + d \rangle = \langle a, b \rangle + \langle c, d \rangle.$$

Similarly, we can verify the following vector operations.

Vector Operations

Let a, b, c, d, and k represent real numbers.

$$\langle a, b \rangle + \langle c, d \rangle = \langle a + c, b + d \rangle$$

$$k \cdot \langle a, b \rangle = \langle ka, kb \rangle$$

If $\mathbf{u} = \langle a_1, a_2 \rangle$, then $-\mathbf{u} = \langle -a_1, -a_2 \rangle$.

$$\langle a, b \rangle - \langle c, d \rangle = \langle a, b \rangle + (-\langle c, d \rangle) = \langle a - c, b - d \rangle$$

EXAMPLE 5 Performing Vector Operations

Let $\mathbf{u} = \langle -2, 1 \rangle$ and $\mathbf{v} = \langle 4, 3 \rangle$. See **Figure 34.** Find and illustrate each of the following.

(a) $\mathbf{u} + \mathbf{v}$

(b) $-2\mathbf{u}$

(c) $3\mathbf{u} - 2\mathbf{v}$

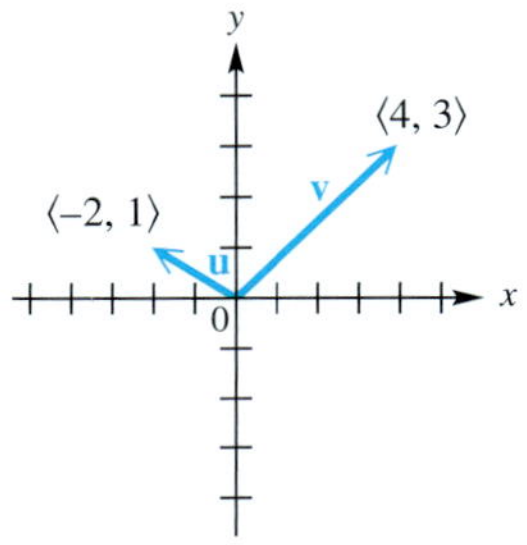

Figure 34

SOLUTION See **Figure 35.**

(a) $\mathbf{u} + \mathbf{v} = \langle -2, 1 \rangle + \langle 4, 3 \rangle$
$= \langle -2 + 4, 1 + 3 \rangle$
$= \langle 2, 4 \rangle$

(b) $-2\mathbf{u} = -2 \cdot \langle -2, 1 \rangle$
$= \langle -2(-2), -2(1) \rangle$
$= \langle 4, -2 \rangle$

(c) $3\mathbf{u} - 2\mathbf{v} = 3 \cdot \langle -2, 1 \rangle - 2 \cdot \langle 4, 3 \rangle$
$= \langle -6, 3 \rangle - \langle 8, 6 \rangle$
$= \langle -6 - 8, 3 - 6 \rangle$
$= \langle -14, -3 \rangle$

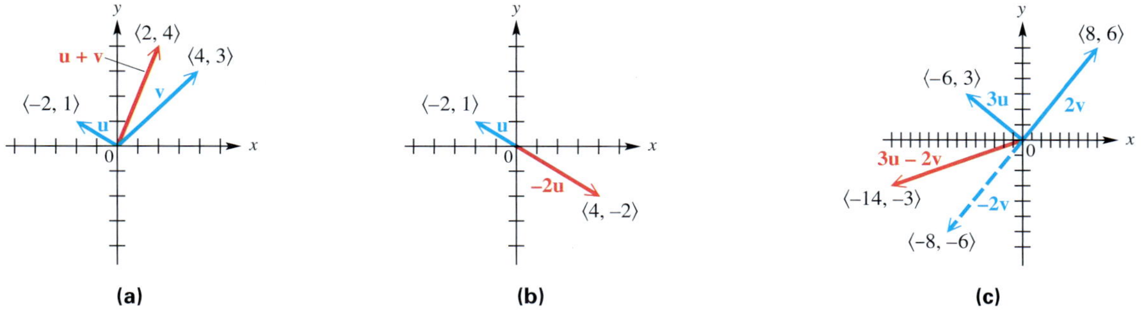

Figure 35

✔ *Now Try Exercises 59, 61, and 63.*

A **unit vector** is a vector that has magnitude 1. Two very useful unit vectors are defined as follows and shown in **Figure 36(a).**

$$\mathbf{i} = \langle 1, 0 \rangle \qquad \mathbf{j} = \langle 0, 1 \rangle$$

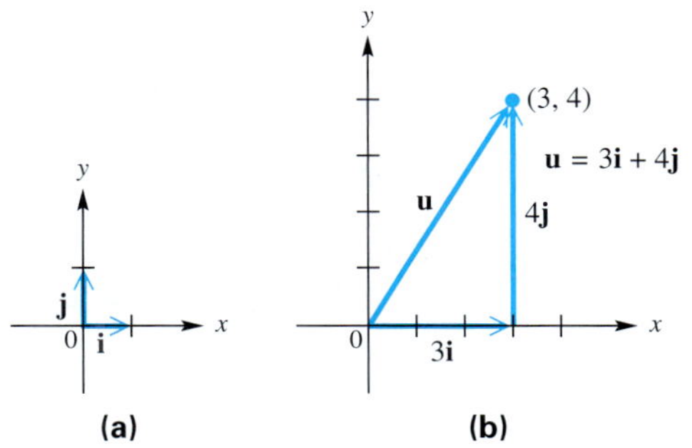

Figure 36

With the unit vectors **i** and **j**, we can express any other vector $\langle a, b \rangle$ in the form $a\mathbf{i} + b\mathbf{j}$, as shown in **Figure 36(b)**, where $\langle 3, 4 \rangle = 3\mathbf{i} + 4\mathbf{j}$. The vector operations previously given can be restated, using $a\mathbf{i} + b\mathbf{j}$ notation.

i, j Form for Vectors

If $\mathbf{v} = \langle a, b \rangle$, then

$$\mathbf{v} = a\mathbf{i} + b\mathbf{j}, \quad \text{where } \mathbf{i} = \langle 1, 0 \rangle \text{ and } \mathbf{j} = \langle 0, 1 \rangle.$$

Dot Product and the Angle between Vectors ***The* dot product *of two vectors is a real number, not a vector.*** It is also known as the *inner product*. Dot products are used to determine the angle between two vectors, to derive geometric theorems, and to solve physics problems.

Dot Product

The **dot product** of the two vectors $\mathbf{u} = \langle a, b \rangle$ and $\mathbf{v} = \langle c, d \rangle$ is denoted $\mathbf{u} \cdot \mathbf{v}$, read "**u** dot **v**," and given by the following.

$$\mathbf{u} \cdot \mathbf{v} = ac + bd$$

That is, the dot product of two vectors is the sum of the product of their first components and the product of their second components.

EXAMPLE 6 Finding Dot Products

Find each dot product.

(a) $\langle 2, 3 \rangle \cdot \langle 4, -1 \rangle$ **(b)** $\langle 6, 4 \rangle \cdot \langle -2, 3 \rangle$

SOLUTION

(a) $\langle 2, 3 \rangle \cdot \langle 4, -1 \rangle = 2(4) + 3(-1)$
$= 5$

(b) $\langle 6, 4 \rangle \cdot \langle -2, 3 \rangle = 6(-2) + 4(3)$
$= 0$

✔ *Now Try Exercises 71 and 73.*

The following properties of dot products can be verified by using the definitions presented so far.

Properties of the Dot Product

For all vectors **u**, **v**, and **w** and real numbers k, the following hold.

(a) $\mathbf{u} \cdot \mathbf{v} = \mathbf{v} \cdot \mathbf{u}$ **(b)** $\mathbf{u} \cdot (\mathbf{v} + \mathbf{w}) = \mathbf{u} \cdot \mathbf{v} + \mathbf{u} \cdot \mathbf{w}$

(c) $(\mathbf{u} + \mathbf{v}) \cdot \mathbf{w} = \mathbf{u} \cdot \mathbf{w} + \mathbf{v} \cdot \mathbf{w}$ **(d)** $(k\mathbf{u}) \cdot \mathbf{v} = k(\mathbf{u} \cdot \mathbf{v}) = \mathbf{u} \cdot (k\mathbf{v})$

(e) $\mathbf{0} \cdot \mathbf{u} = 0$ **(f)** $\mathbf{u} \cdot \mathbf{u} = |\mathbf{u}|^2$

For example, to prove the first part of (d), we let $\mathbf{u} = \langle a, b \rangle$ and $\mathbf{v} = \langle c, d \rangle$.

$$\begin{aligned}
(k\mathbf{u}) \cdot \mathbf{v} &= \big(k\langle a, b \rangle\big) \cdot \langle c, d \rangle && \text{Substitute.} \\
&= \langle ka, kb \rangle \cdot \langle c, d \rangle && \text{Multiply by scalar } k. \\
&= kac + kbd && \text{Dot product} \\
&= k(ac + bd) && \text{Distributive property} \\
&= k\big(\langle a, b \rangle \cdot \langle c, d \rangle\big) && \text{Dot product} \\
&= k(\mathbf{u} \cdot \mathbf{v}) && \text{Substitute.}
\end{aligned}$$

The proofs of the remaining properties are similar.

The dot product of two vectors can be positive, 0, or negative. A geometric interpretation of the dot product explains when each of these cases occurs. This interpretation involves the angle between the two vectors. Consider the vectors $\mathbf{u} = \langle a_1, a_2 \rangle$ and $\mathbf{v} = \langle b_1, b_2 \rangle$, as shown in **Figure 37.** The **angle θ between u and v** is defined to be the angle having the two vectors as its sides for which $0° \le \theta \le 180°$.

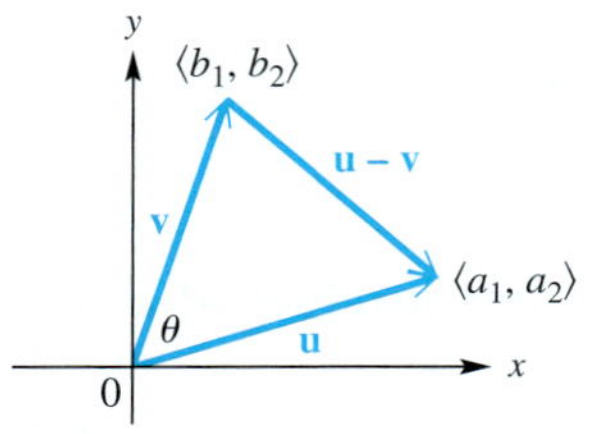

Figure 37

We can use the law of cosines to develop a formula to find angle θ in **Figure 37.**

$$|\mathbf{u} - \mathbf{v}|^2 = |\mathbf{u}|^2 + |\mathbf{v}|^2 - 2|\mathbf{u}||\mathbf{v}| \cos\theta$$ Law of cosines applied to **Figure 37** (Section 7.3)

$$\left(\sqrt{(a_1 - b_1)^2 + (a_2 - b_2)^2}\right)^2 = \left(\sqrt{a_1{}^2 + a_2{}^2}\right)^2 + \left(\sqrt{b_1{}^2 + b_2{}^2}\right)^2 - 2|\mathbf{u}||\mathbf{v}| \cos\theta$$ Magnitude of a vector

$$a_1{}^2 - 2a_1b_1 + b_1{}^2 + a_2{}^2 - 2a_2b_2 + b_2{}^2 = a_1{}^2 + a_2{}^2 + b_1{}^2 + b_2{}^2 - 2|\mathbf{u}||\mathbf{v}| \cos\theta$$ Square.

$$-2a_1b_1 - 2a_2b_2 = -2|\mathbf{u}||\mathbf{v}| \cos\theta$$ Subtract like terms from each side.

$$a_1b_1 + a_2b_2 = |\mathbf{u}||\mathbf{v}| \cos\theta$$ Divide by -2.

$$\mathbf{u} \cdot \mathbf{v} = |\mathbf{u}||\mathbf{v}| \cos\theta$$ Definition of dot product

$$\cos\theta = \frac{\mathbf{u} \cdot \mathbf{v}}{|\mathbf{u}||\mathbf{v}|}$$ Divide by $|\mathbf{u}||\mathbf{v}|$ and rewrite.

Geometric Interpretation of Dot Product

If θ is the angle between the two nonzero vectors **u** and **v**, where $0° \le \theta \le 180°$, then the following holds.

$$\cos\theta = \frac{\mathbf{u} \cdot \mathbf{v}}{|\mathbf{u}||\mathbf{v}|}$$

EXAMPLE 7 Finding the Angle between Two Vectors

Find the angle θ between the two vectors.

(a) $\mathbf{u} = \langle 3, 4\rangle$ and $\mathbf{v} = \langle 2, 1\rangle$ **(b)** $\mathbf{u} = \langle 2, -6\rangle$ and $\mathbf{v} = \langle 6, 2\rangle$

SOLUTION

(a)
$$\cos\theta = \frac{\mathbf{u} \cdot \mathbf{v}}{|\mathbf{u}||\mathbf{v}|} = \frac{\langle 3, 4\rangle \cdot \langle 2, 1\rangle}{|\langle 3, 4\rangle|\,|\langle 2, 1\rangle|}$$ Substitute values.

$$= \frac{3(2) + 4(1)}{\sqrt{9 + 16} \cdot \sqrt{4 + 1}}$$ Use the definitions.

$$= \frac{10}{5\sqrt{5}} \approx 0.894427191$$ Use a calculator.

Therefore, $\theta \approx \cos^{-1} 0.894427191 \approx 26.57°$. Use the inverse cosine function. (Section 6.1)

(b)
$$\cos\theta = \frac{\mathbf{u} \cdot \mathbf{v}}{|\mathbf{u}||\mathbf{v}|} = \frac{\langle 2, -6\rangle \cdot \langle 6, 2\rangle}{|\langle 2, -6\rangle|\,|\langle 6, 2\rangle|}$$ Substitute values.

$$= \frac{2(6) + (-6)(2)}{\sqrt{4 + 36} \cdot \sqrt{36 + 4}}$$ Use the definitions.

$$= \frac{0}{40} = 0$$ Evaluate.

$$\theta = \cos^{-1} 0 = 90°$$ $\cos^{-1} 0 = 90°$

✔ ***Now Try Exercises 77 and 79.***

For angles θ between 0° and 180°, cos θ is positive, 0, or negative when θ is less than, equal to, or greater than 90°, respectively. Therefore, the dot product of nonzero vectors is positive, 0, or negative according to this table.

Dot Product	Angle between Vectors
Positive	Acute
0	Right
Negative	Obtuse

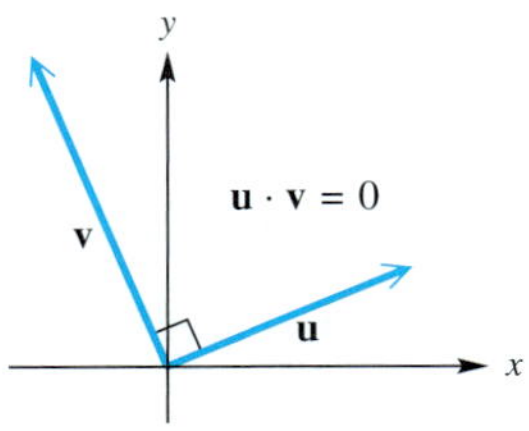

Orthogonal vectors

Figure 38

Thus, in **Example 7** on the preceding page, the vectors in part (a) form an acute angle, and those in part (b) form a right angle. If $\mathbf{u} \cdot \mathbf{v} = 0$ for two nonzero vectors **u** and **v**, then $\cos \theta = 0$ and $\theta = 90°$. Thus, **u** and **v** are perpendicular vectors, also called **orthogonal vectors.** See **Figure 38.**

7.4 Exercises

Concept Check Exercises 1–4 refer to the vectors **m** *through* **t** *at the right.*

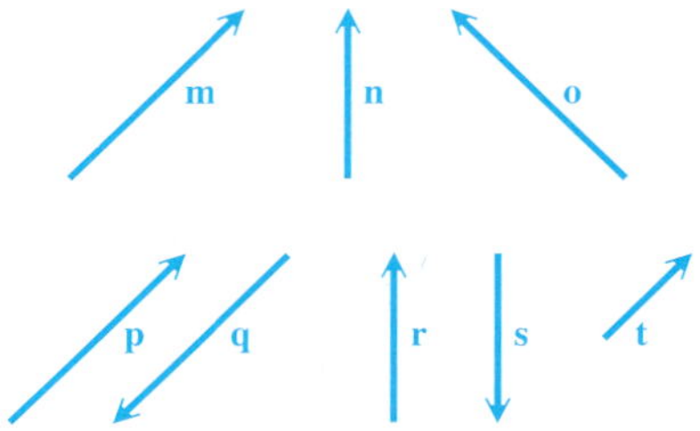

1. Name all pairs of vectors that appear to be equal.

2. Name all pairs of vectors that are opposites.

3. Name all pairs of vectors where the first is a scalar multiple of the other, with the scalar positive.

4. Name all pairs of vectors where the first is a scalar multiple of the other, with the scalar negative.

Concept Check Refer to vectors **a** *through* **h** *below. Make a copy or a sketch of each vector, and then draw a sketch to represent each vector in Exercises 5–16. For example, find* **a** + **e** *by placing* **a** *and* **e** *so that their initial points coincide. Then use the parallelogram rule to find the resultant, as shown in the figure on the right.*

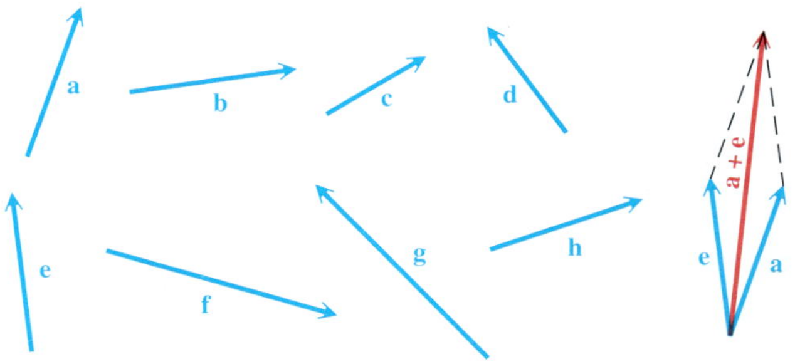

5. $-\mathbf{b}$	**6.** $-\mathbf{g}$	**7.** $2\mathbf{c}$	**8.** $2\mathbf{h}$
9. $\mathbf{a} + \mathbf{b}$	**10.** $\mathbf{h} + \mathbf{g}$	**11.** $\mathbf{a} - \mathbf{c}$	**12.** $\mathbf{d} - \mathbf{e}$
13. $\mathbf{a} + (\mathbf{b} + \mathbf{c})$	**14.** $(\mathbf{a} + \mathbf{b}) + \mathbf{c}$	**15.** $\mathbf{c} + \mathbf{d}$	**16.** $\mathbf{d} + \mathbf{c}$

17. From the results of **Exercises 13 and 14,** does it appear that vector addition is associative?

18. From the results of **Exercises 15 and 16,** does it appear that vector addition is commutative?

In Exercises 19–24, use the figure to find each vector: ***(a)*** $\mathbf{u} + \mathbf{v}$ ***(b)*** $\mathbf{u} - \mathbf{v}$ ***(c)*** $-\mathbf{u}$. *Use vector notation as in* ***Example 3.***

19.
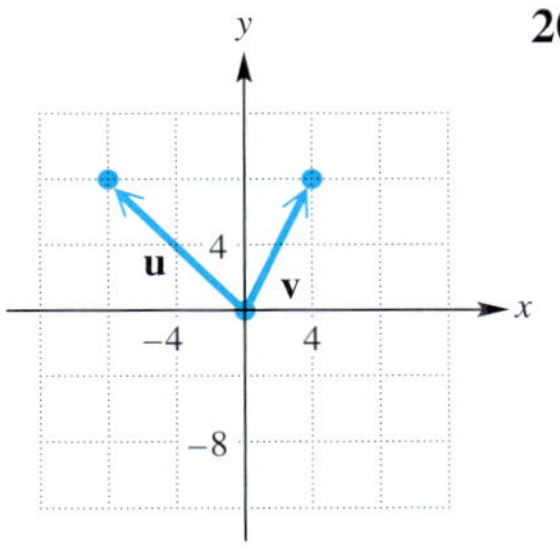

20.
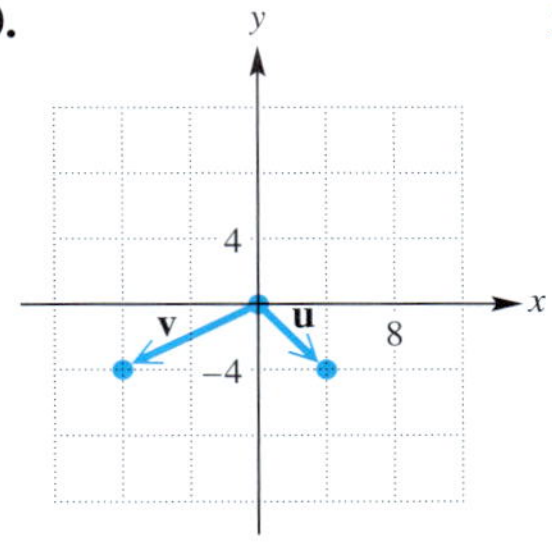

21.
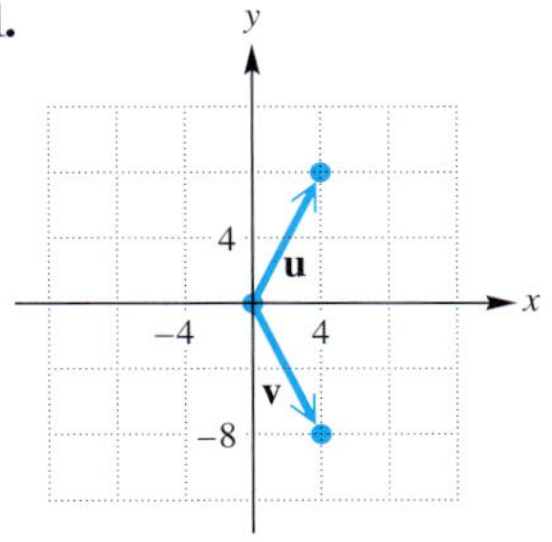

22.
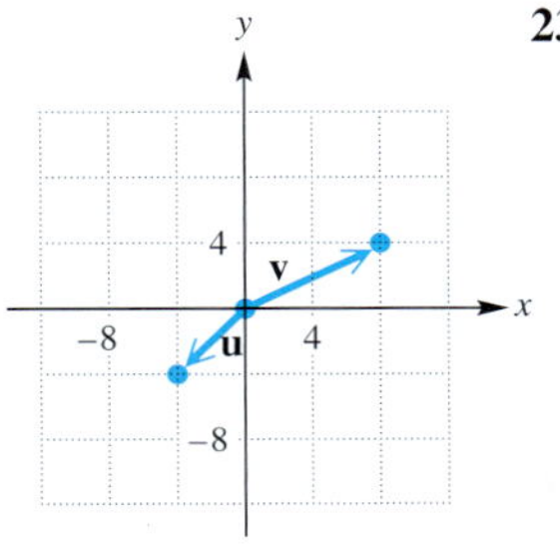

23.
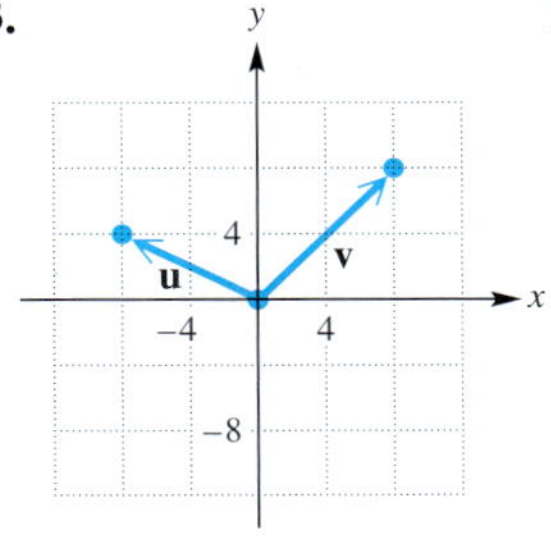

24.
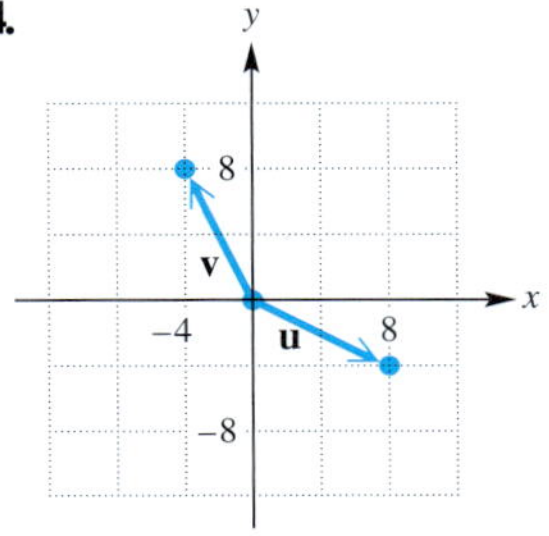

Given vectors **u** *and* **v**, *find:* ***(a)*** $2\mathbf{u}$ ***(b)*** $2\mathbf{u} + 3\mathbf{v}$ ***(c)*** $\mathbf{v} - 3\mathbf{u}$.

25. $\mathbf{u} = 2\mathbf{i}$, $\mathbf{v} = \mathbf{i} + \mathbf{j}$

26. $\mathbf{u} = -\mathbf{i} + 2\mathbf{j}$, $\mathbf{v} = \mathbf{i} - \mathbf{j}$

27. $\mathbf{u} = \langle -1, 2 \rangle$, $\mathbf{v} = \langle 3, 0 \rangle$

28. $\mathbf{u} = \langle -2, -1 \rangle$, $\mathbf{v} = \langle -3, 2 \rangle$

For each pair of vectors **u** *and* **v** *with angle* θ *between them, sketch the resultant.*

29. $|\mathbf{u}| = 12$, $|\mathbf{v}| = 20$, $\theta = 27°$

30. $|\mathbf{u}| = 8$, $|\mathbf{v}| = 12$, $\theta = 20°$

31. $|\mathbf{u}| = 20$, $|\mathbf{v}| = 30$, $\theta = 30°$

32. $|\mathbf{u}| = 50$, $|\mathbf{v}| = 70$, $\theta = 40°$

Find the magnitude and direction angle for each vector. ***See Example 1.***

33. $\langle 15, -8 \rangle$

34. $\langle -7, 24 \rangle$

35. $\langle -4, 4\sqrt{3} \rangle$

36. $\langle 8\sqrt{2}, -8\sqrt{2} \rangle$

For each of the following, vector **v** *has the given direction and magnitude. Find the magnitudes of the horizontal and vertical components of* **v**, *if* θ *is the direction angle of* **v** *from the horizontal.* ***See Example 2.***

37. $\theta = 20°$, $|\mathbf{v}| = 50$

38. $\theta = 50°$, $|\mathbf{v}| = 26$

39. $\theta = 35°\ 50'$, $|\mathbf{v}| = 47.8$

40. $\theta = 27°\ 30'$, $|\mathbf{v}| = 15.4$

41. $\theta = 128.5°$, $|\mathbf{v}| = 198$

42. $\theta = 146.3°$, $|\mathbf{v}| = 238$

Write each vector in the form $\langle a, b \rangle$. **See Example 3.**

43.

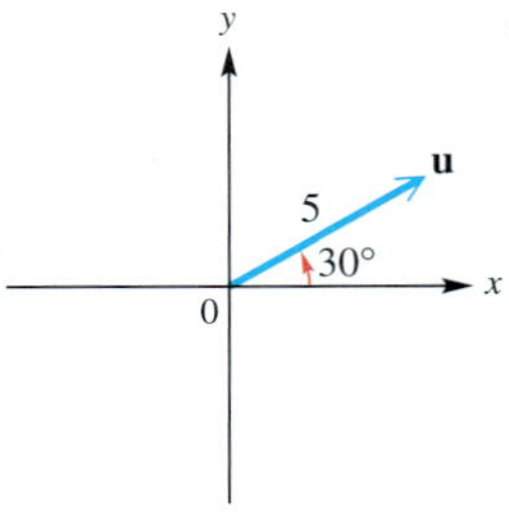

44.

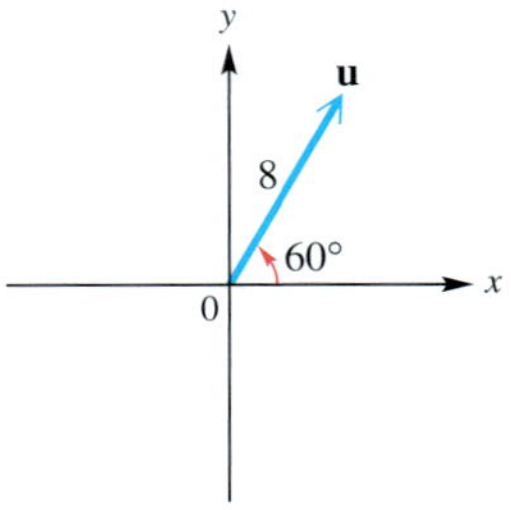

45.

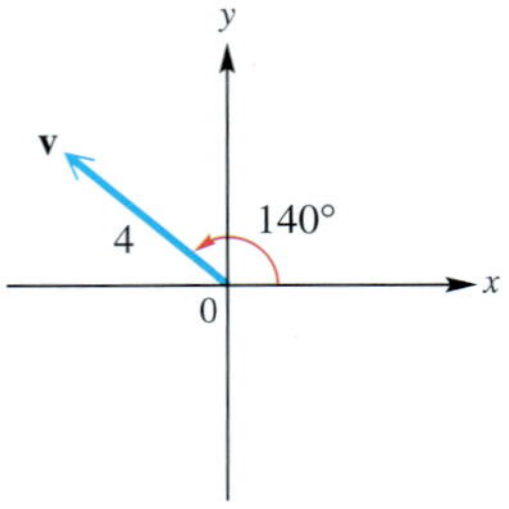

46.

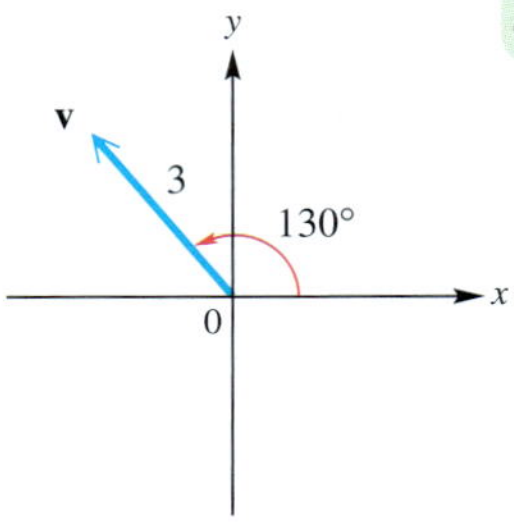

47.

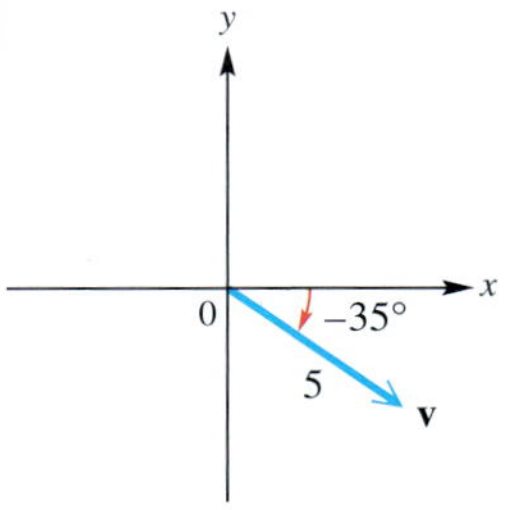

48.

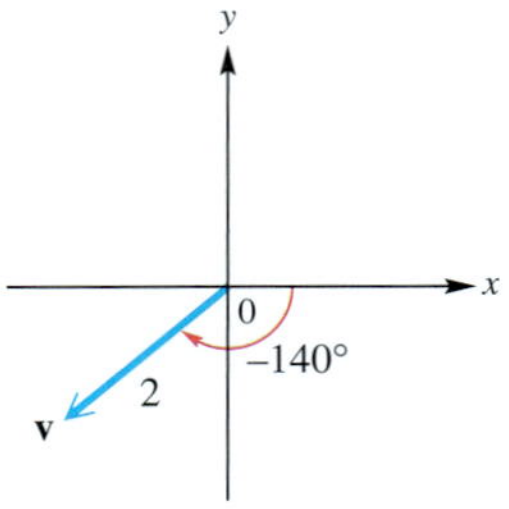

Two forces act at a point in the plane. The angle between the two forces is given. Find the magnitude of the resultant force. **See Example 4.**

49. forces of 250 and 450 newtons, forming an angle of 85°

50. forces of 19 and 32 newtons, forming an angle of 118°

51. forces of 116 and 139 lb, forming an angle of 140° 50′

52. forces of 37.8 and 53.7 lb, forming an angle of 68.5°

Use the parallelogram rule to find the magnitude of the resultant force for the two forces shown in each figure. Round answers to the nearest tenth.

53.

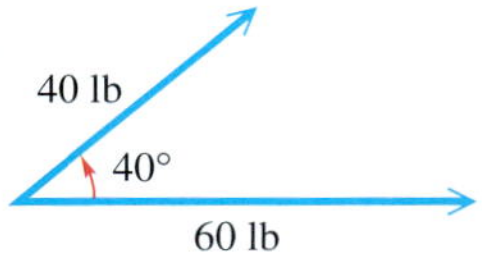

54.

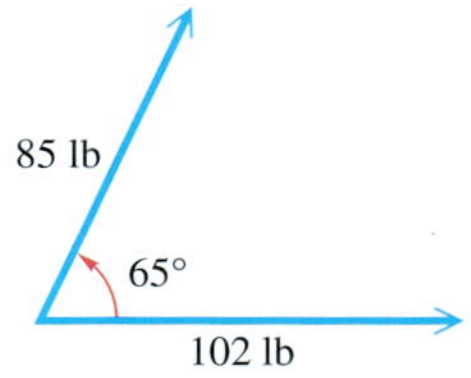

55.

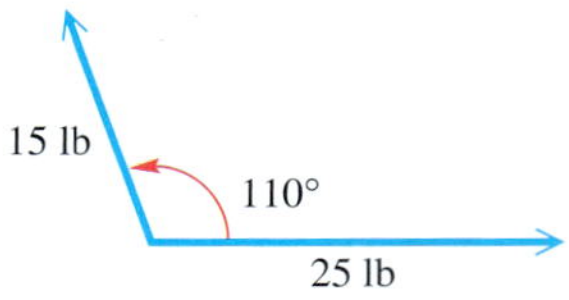

56.

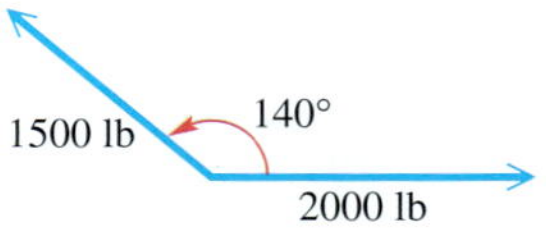

57. **Concept Check** If $\mathbf{u} = \langle a, b \rangle$ and $\mathbf{v} = \langle c, d \rangle$, what is the vector notation for $\mathbf{u} + \mathbf{v}$?

58. Explain how to add vectors.

Given $\mathbf{u} = \langle -2, 5 \rangle$ *and* $\mathbf{v} = \langle 4, 3 \rangle$, *find each of the following.* **See Example 5.**

59. $\mathbf{u} - \mathbf{v}$ **60.** $\mathbf{v} - \mathbf{u}$ **61.** $-4\mathbf{u}$ **62.** $-5\mathbf{v}$

63. $3\mathbf{u} - 6\mathbf{v}$ **64.** $-2\mathbf{u} + 4\mathbf{v}$ **65.** $\mathbf{u} + \mathbf{v} - 3\mathbf{u}$ **66.** $2\mathbf{u} + \mathbf{v} - 6\mathbf{v}$

Write each vector in the form $a\mathbf{i} + b\mathbf{j}$. **See Figure 36(b).**

67. $\langle -5, 8 \rangle$ **68.** $\langle 6, -3 \rangle$ **69.** $\langle 2, 0 \rangle$ **70.** $\langle 0, -4 \rangle$

Find the dot product for each pair of vectors. ***See Example 6.***

71. $\langle 6, -1 \rangle, \langle 2, 5 \rangle$ **72.** $\langle -3, 8 \rangle, \langle 7, -5 \rangle$ **73.** $\langle 5, 2 \rangle, \langle -4, 10 \rangle$

74. $\langle 7, -2 \rangle, \langle 4, 14 \rangle$ **75.** $4\mathbf{i}, 5\mathbf{i} - 9\mathbf{j}$ **76.** $2\mathbf{i} + 4\mathbf{j}, -\mathbf{j}$

Find the angle between each pair of vectors. ***See Example 7.***

77. $\langle 2, 1 \rangle, \langle -3, 1 \rangle$ **78.** $\langle 1, 7 \rangle, \langle 1, 1 \rangle$ **79.** $\langle 1, 2 \rangle, \langle -6, 3 \rangle$

80. $\langle 4, 0 \rangle, \langle 2, 2 \rangle$ **81.** $3\mathbf{i} + 4\mathbf{j}, \mathbf{j}$ **82.** $-5\mathbf{i} + 12\mathbf{j}, 3\mathbf{i} + 2\mathbf{j}$

Let $\mathbf{u} = \langle -2, 1 \rangle$, $\mathbf{v} = \langle 3, 4 \rangle$, *and* $\mathbf{w} = \langle -5, 12 \rangle$. *Evaluate each expression.*

83. $(3\mathbf{u}) \cdot \mathbf{v}$ **84.** $\mathbf{u} \cdot (3\mathbf{v})$ **85.** $\mathbf{u} \cdot \mathbf{v} - \mathbf{u} \cdot \mathbf{w}$ **86.** $\mathbf{u} \cdot (\mathbf{v} - \mathbf{w})$

Determine whether each pair of vectors is orthogonal. ***See Example 7(b).***

87. $\langle 1, 2 \rangle, \langle -6, 3 \rangle$ **88.** $\langle 1, 1 \rangle, \langle 1, -1 \rangle$

89. $\langle 1, 0 \rangle, \langle \sqrt{2}, 0 \rangle$ **90.** $\langle 3, 4 \rangle, \langle 6, 8 \rangle$

91. $\sqrt{5}\,\mathbf{i} - 2\mathbf{j}, -5\mathbf{i} + 2\sqrt{5}\,\mathbf{j}$ **92.** $-4\mathbf{i} + 3\mathbf{j}, 8\mathbf{i} - 6\mathbf{j}$

Relating Concepts

For individual or collaborative investigation *(Exercises 93–98)*

Consider the two vectors **u** *and* **v** *shown. Assume all values are exact.* ***Work Exercises 93–98 in order.***

93. Use trigonometry alone (without using vector notation) to find the magnitude and direction angle of $\mathbf{u} + \mathbf{v}$. Use the law of cosines and the law of sines in your work.

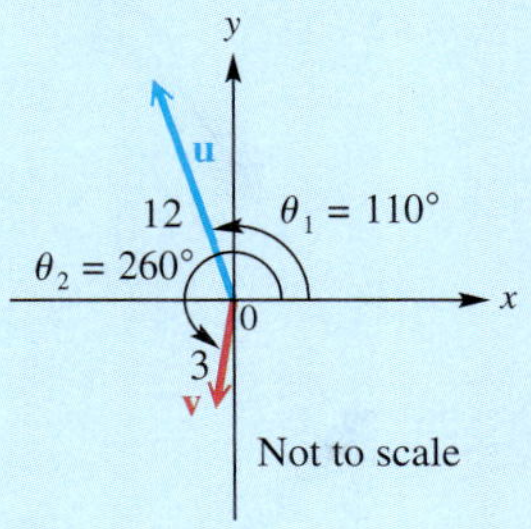

94. Find the horizontal and vertical components of **u**, using your calculator.

95. Find the horizontal and vertical components of **v**, using your calculator.

96. Find the horizontal and vertical components of $\mathbf{u} + \mathbf{v}$ by adding the results you obtained in **Exercises 94 and 95.**

97. Use your calculator to find the magnitude and direction angle of the vector $\mathbf{u} + \mathbf{v}$.

98. Compare your answers in **Exercises 93 and 97.** What do you notice? Which method of solution do you prefer?

7.5 Applications of Vectors

- The Equilibrant
- Incline Applications
- Navigation Applications

The Equilibrant The previous section covered methods for finding the resultant of two vectors. Sometimes it is necessary to find a vector that will counterbalance the resultant. This opposite vector is called the **equilibrant.** That is, the equilibrant of vector **u** is the vector $-\mathbf{u}$.

EXAMPLE 1 Finding the Magnitude and Direction of an Equilibrant

Find the magnitude of the equilibrant of forces of 48 newtons and 60 newtons acting on a point A, if the angle between the forces is 50°. Then find the angle between the equilibrant and the 48-newton force.

SOLUTION

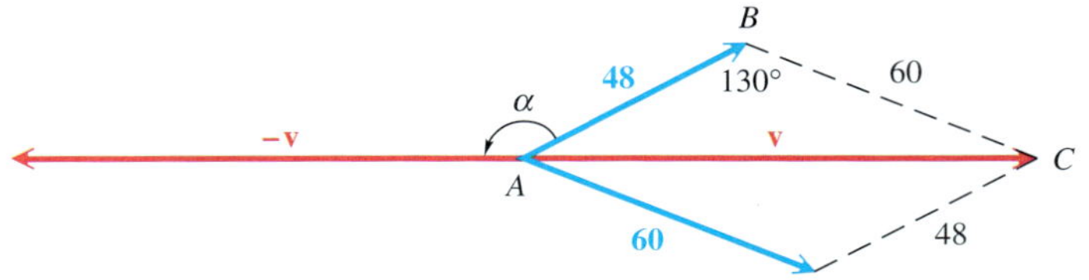

Figure 39

In **Figure 39,** the equilibrant is $-\mathbf{v}$. The magnitude of $\mathbf{v}$, and hence of $-\mathbf{v}$, is found by using triangle ABC and the law of cosines.

$$|\mathbf{v}|^2 = 48^2 + 60^2 - 2(48)(60)\cos 130° \quad \text{Law of cosines (Section 7.3)}$$

$$|\mathbf{v}|^2 \approx 9606.5 \quad \text{Use a calculator.}$$

$$|\mathbf{v}| \approx 98 \text{ newtons} \quad \text{Two significant digits (Section 2.4)}$$

The required angle, labeled α in **Figure 39,** can be found by subtracting angle CAB from 180°. Use the law of sines to find angle CAB.

$$\frac{\sin CAB}{60} = \frac{\sin 130°}{98} \quad \text{Law of sines (alternative form) (Section 7.1)}$$

$$\sin CAB \approx 0.46900680 \quad \text{Multiply by 60 and use a calculator.}$$

$$CAB \approx 28° \quad \text{Use the inverse sine function. (Section 6.1)}$$

Finally, $\alpha \approx 180° - 28° = 152°$.

✔ *Now Try Exercise 1.*

Incline Applications We can use vectors to solve incline problems.

EXAMPLE 2 Finding a Required Force

Find the force required to keep a 50-lb wagon from sliding down a ramp inclined at 20° to the horizontal. (Assume there is no friction.)

SOLUTION In **Figure 40,** the vertical 50-lb force **BA** represents the force of gravity. It is the sum of vectors **BC** and $-$**AC**. The vector **BC** represents the force with which the weight pushes against the ramp. The vector **BF** represents the force that would pull the weight up the ramp. Since vectors **BF** and **AC** are equal, $|\mathbf{AC}|$ gives the magnitude of the required force.

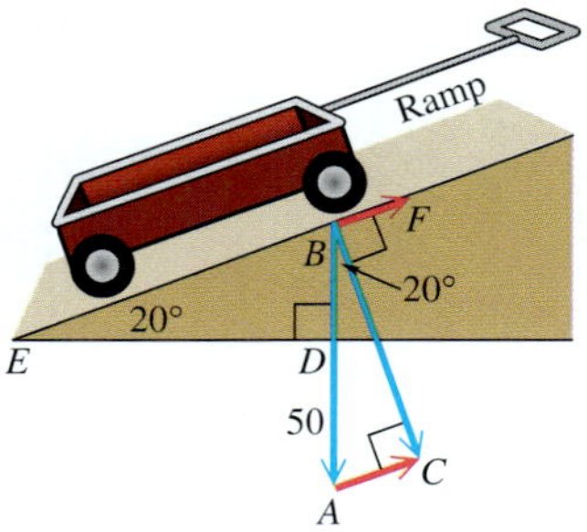

Figure 40

Vectors **BF** and **AC** are parallel, so angle EBD equals angle A. Since angle BDE and angle C are right angles, triangles CBA and DEB have two corresponding angles equal and, thus, are similar triangles. Therefore, angle ABC equals angle E, which is 20°. From right triangle ABC, we have the following.

$$\sin 20° = \frac{|\mathbf{AC}|}{50} \quad \text{(Section 2.1)}$$

$$|\mathbf{AC}| = 50 \sin 20° \quad \text{Multiply by 50 and rewrite.}$$

$$|\mathbf{AC}| \approx 17 \quad \text{Use a calculator.}$$

A force of approximately 17 lb will keep the wagon from sliding down the ramp.

✔ *Now Try Exercise 9.*

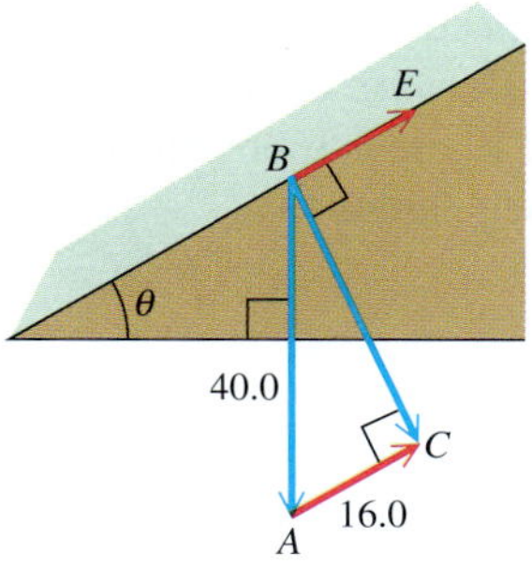

Figure 41

EXAMPLE 3 **Finding an Incline Angle**

A force of 16.0 lb is required to hold a 40.0-lb lawn mower on an incline. What angle does the incline make with the horizontal?

SOLUTION **Figure 41** illustrates the situation. Consider right triangle ABC. Angle B equals angle θ, the magnitude of vector **BA** represents the weight of the mower, and vector **AC** equals vector **BE**, which represents the force required to hold the mower on the incline.

$$\sin B = \frac{16.0}{40.0} \qquad \sin B = \frac{\text{side opposite } B}{\text{hypotenuse}} \text{ (Section 2.1)}$$

$$\sin B = 0.4 \qquad \text{Simplify.}$$

$$B \approx 23.6^\circ \qquad \text{Use the inverse sine function.}$$

Therefore, the hill makes an angle of about 23.6° with the horizontal.

✔ *Now Try Exercise 11.*

Navigation Applications Problems involving bearing (defined in **Section 2.5**) can also be worked with vectors.

EXAMPLE 4 **Applying Vectors to a Navigation Problem**

A ship leaves port on a bearing of 28.0° and travels 8.20 mi. The ship then turns due east and travels 4.30 mi. How far is the ship from port? What is its bearing from port?

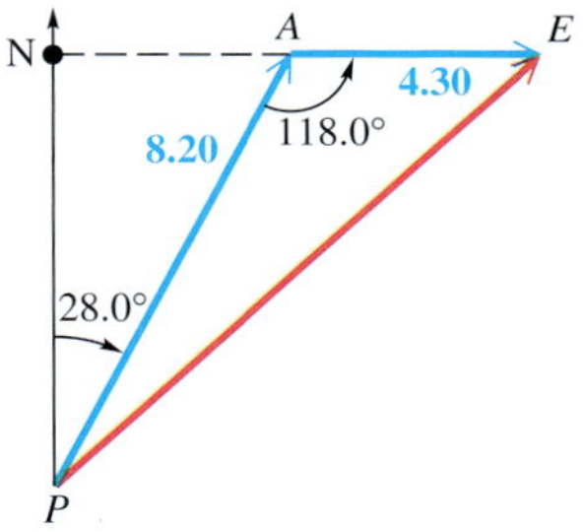

Figure 42

SOLUTION In **Figure 42,** vectors **PA** and **AE** represent the ship's path. The magnitude and bearing of the resultant **PE** can be found as follows. Triangle PNA is a right triangle, so

$$\text{angle } NAP = 90^\circ - 28.0^\circ = 62.0^\circ,$$

and $$\text{angle } PAE = 180^\circ - 62.0^\circ = 118.0^\circ.$$

Use the law of cosines to find $|\mathbf{PE}|$, the magnitude of vector **PE**.

$$|\mathbf{PE}|^2 = 8.20^2 + 4.30^2 - 2(8.20)(4.30)\cos 118.0^\circ \qquad \text{Law of cosines}$$

$$|\mathbf{PE}|^2 \approx 118.84 \qquad \text{Evaluate.}$$

$$|\mathbf{PE}| \approx 10.9 \qquad \text{Square root property (Appendix A)}$$

The ship is about 10.9 mi from port.

To find the bearing of the ship from port, first find angle APE. Use the law of sines.

$$\frac{\sin APE}{4.30} = \frac{\sin 118.0^\circ}{10.9} \qquad \text{Law of sines}$$

$$\sin APE = \frac{4.30 \sin 118.0^\circ}{10.9} \qquad \text{Multiply by 4.30.}$$

$$APE \approx 20.4^\circ \qquad \text{Use the inverse sine function.}$$

Now add 20.4° to 28.0° to find that the bearing is 48.4°.

✔ *Now Try Exercise 15.*

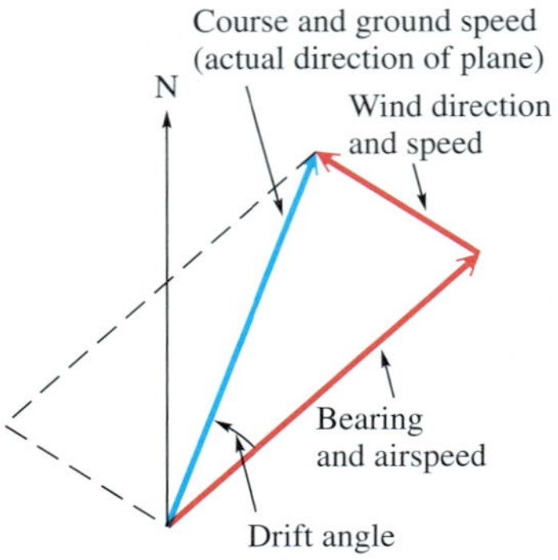

Figure 43

In air navigation, the **airspeed** of a plane is its speed relative to the air, and the **ground speed** is its speed relative to the ground. Because of wind, these two speeds are usually different. The ground speed of the plane is represented by the vector sum of the airspeed and windspeed vectors. See **Figure 43.**

EXAMPLE 5 Applying Vectors to a Navigation Problem

An airplane that is following a bearing of 239° at an airspeed of 425 mph encounters a wind blowing at 36.0 mph from a direction of 115°. Find the resulting bearing and ground speed of the plane.

SOLUTION An accurate sketch is essential to the solution of this problem. We have included two sets of geographical axes, which enable us to determine measures of necessary angles. Analyze **Figure 44** carefully.

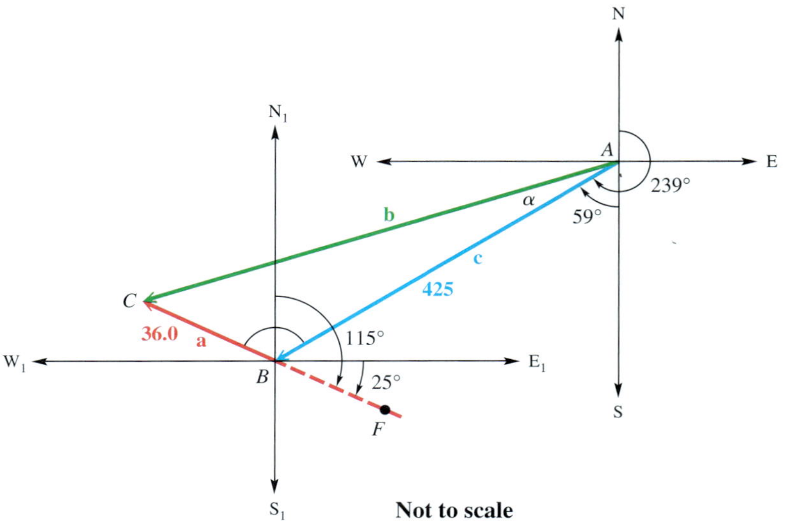

Figure 44

Vector **c** represents the airspeed and bearing of the plane, and vector **a** represents the speed and direction of the wind. Angle ABC has as its measure the sum of angle ABN_1 and angle N_1BC.

- Angle SAB measures $239° - 180° = 59°$. Because angle ABN_1 is an alternate interior angle to it, $ABN_1 = 59°$.
- Angle E_1BF measures $115° - 90° = 25°$. Thus, angle CBW_1 also measures 25° because it is a vertical angle. Angle N_1BC is the complement of 25°, which is

$$90° - 25° = 65°.$$

By these results,

$$\text{angle } ABC = 59° + 65° = 124°.$$

To find $|\mathbf{b}|$, we use the law of cosines.

$$|\mathbf{b}|^2 = |\mathbf{a}|^2 + |\mathbf{c}|^2 - 2|\mathbf{a}||\mathbf{c}| \cos ABC \qquad \text{Law of cosines}$$

$$|\mathbf{b}|^2 = 36.0^2 + 425^2 - 2(36.0)(425) \cos 124° \qquad \text{Substitute.}$$

$$|\mathbf{b}|^2 \approx 199{,}032 \qquad \text{Use a calculator.}$$

$$|\mathbf{b}| \approx 446 \qquad \text{Square root property}$$

The ground speed is approximately 446 mph.

To find the resulting bearing of **b**, we must find the measure of angle α in **Figure 44** and then add it to 239°. To find α, we use the law of sines.

$$\frac{\sin \alpha}{36.0} = \frac{\sin 124^\circ}{446}$$

To maintain accuracy, use all the significant digits that your calculator allows.

$$\sin \alpha = \frac{36.0 \sin 124^\circ}{446} \quad \text{Multiply by 36.0.}$$

$$\alpha = \sin^{-1}\left(\frac{36.0 \sin 124^\circ}{446}\right) \quad \text{Use the inverse sine function.}$$

$$\alpha \approx 4^\circ \quad \text{Use a calculator.}$$

Add 4° to 239° to find the resulting bearing of 243°.

✔ **Now Try Exercise 21.**

7.5 Exercises

Solve each problem. ***See Examples 1–3.***

1. *Direction and Magnitude of an Equilibrant* Two tugboats are pulling a disabled speedboat into port with forces of 1240 lb and 1480 lb. The angle between these forces is 28.2°. Find the direction and magnitude of the equilibrant.

2. *Direction and Magnitude of an Equilibrant* Two rescue vessels are pulling a broken-down motorboat toward a boathouse with forces of 840 lb and 960 lb. The angle between these forces is 24.5°. Find the direction and magnitude of the equilibrant.

3. *Angle between Forces* Two forces of 692 newtons and 423 newtons act at a point. The resultant force is 786 newtons. Find the angle between the forces.

4. *Angle between Forces* Two forces of 128 lb and 253 lb act at a point. The resultant force is 320 lb. Find the angle between the forces.

5. *Magnitudes of Forces* A force of 176 lb makes an angle of 78° 50′ with a second force. The resultant of the two forces makes an angle of 41° 10′ with the first force. Find the magnitudes of the second force and of the resultant.

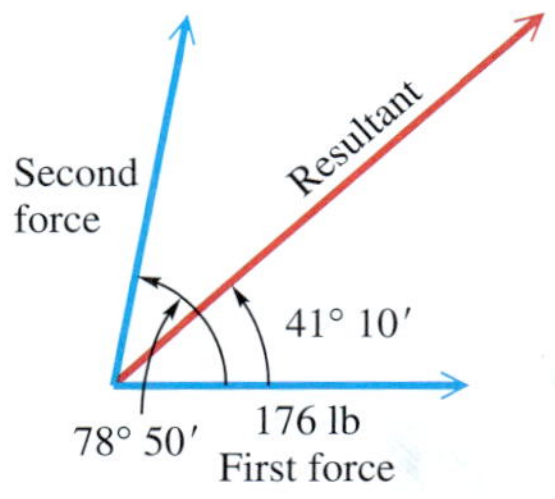

6. *Magnitudes of Forces* A force of 28.7 lb makes an angle of 42° 10′ with a second force. The resultant of the two forces makes an angle of 32° 40′ with the first force. Find the magnitudes of the second force and of the resultant.

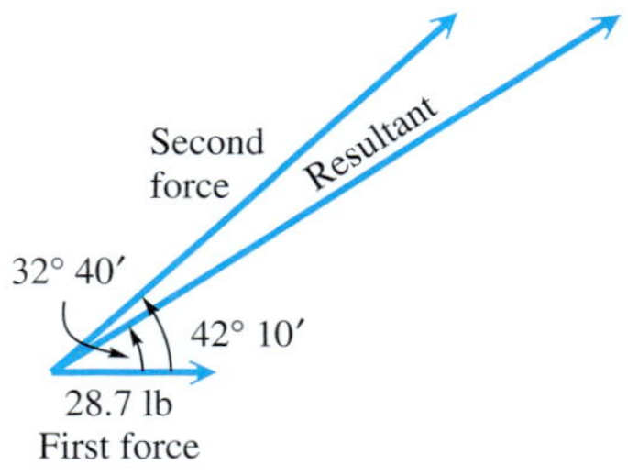

7. *Angle of a Hill Slope* A force of 25 lb is required to hold an 80-lb crate on a hill. What angle does the hill make with the horizontal?

8. *Force Needed to Keep a Car Parked* Find the force required to keep a 3000-lb car parked on a hill that makes an angle of 15° with the horizontal.

9. ***Force Needed for a Monolith*** To build the pyramids in Egypt, it is believed that giant causeways were constructed to transport the building materials to the site. One such causeway is said to have been 3000 ft long, with a slope of about 2.3°. How much force would be required to hold a 60-ton monolith on this causeway?

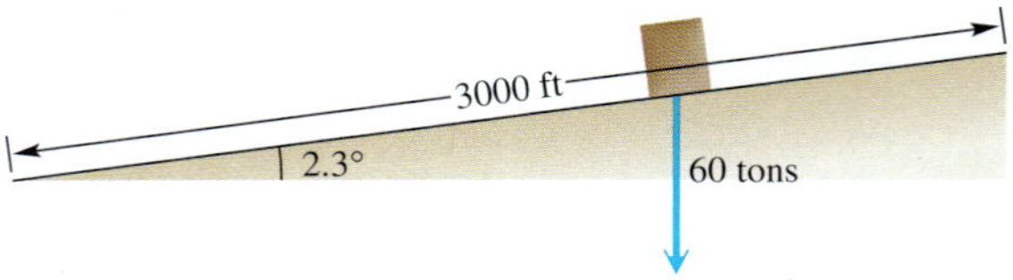

10. ***Force Needed for a Monolith*** If the causeway in **Exercise 9** were 500 ft longer and the monolith weighed 10 tons more, how much force would be required?

11. ***Incline Angle*** A force of 18.0 lb is required to hold a 60.0-lb stump grinder on an incline. What angle does the incline make with the horizontal?

12. ***Incline Angle*** A force of 30.0 lb is required to hold an 80.0-lb pressure washer on an incline. What angle does the incline make with the horizontal?

13. ***Weight of a Box*** Two people are carrying a box. One person exerts a force of 150 lb at an angle of 62.4° with the horizontal. The other person exerts a force of 114 lb at an angle of 54.9°. Find the weight of the box.

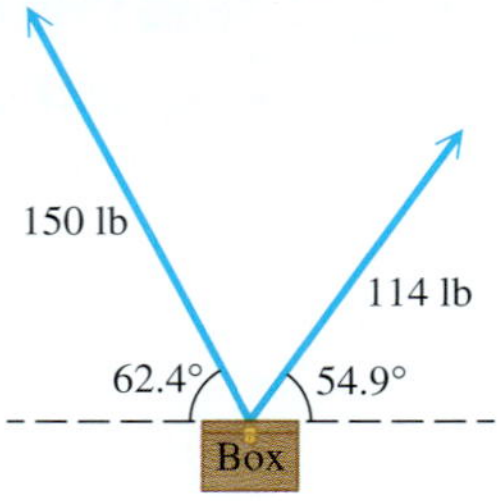

14. ***Weight of a Crate and Tension of a Rope*** A crate is supported by two ropes. One rope makes an angle of 46° 20′ with the horizontal and has a tension of 89.6 lb on it. The other rope is horizontal. Find the weight of the crate and the tension in the horizontal rope.

Solve each problem. ***See Examples 4 and 5.***

15. ***Distance and Bearing of a Ship*** A ship leaves port on a bearing of 34.0° and travels 10.4 mi. The ship then turns due east and travels 4.6 mi. How far is the ship from port, and what is its bearing from port?

16. ***Distance and Bearing of a Luxury Liner*** A luxury liner leaves port on a bearing of 110.0° and travels 8.8 mi. It then turns due west and travels 2.4 mi. How far is the liner from port, and what is its bearing from port?

17. ***Distance of a Ship from Its Starting Point*** Starting at point A, a ship sails 18.5 km on a bearing of 189°, then turns and sails 47.8 km on a bearing of 317°. Find the distance of the ship from point A.

18. ***Distance of a Ship from Its Starting Point*** Starting at point X, a ship sails 15.5 km on a bearing of 200°, then turns and sails 2.4 km on a bearing of 320°. Find the distance of the ship from point X.

19. ***Distance and Direction of a Motorboat*** A motorboat sets out in the direction N 80° 00′ E. The speed of the boat in still water is 20.0 mph. If the current is flowing directly south, and the actual direction of the motorboat is due east, find the speed of the current and the actual speed of the motorboat.

20. *Movement of a Motorboat* Suppose you would like to cross a 132-ft-wide river in a motorboat. Assume that the motorboat can travel at 7.0 mph relative to the water and that the current is flowing west at the rate of 3.0 mph. The bearing θ is chosen so that the motorboat will land at a point exactly across from the starting point.

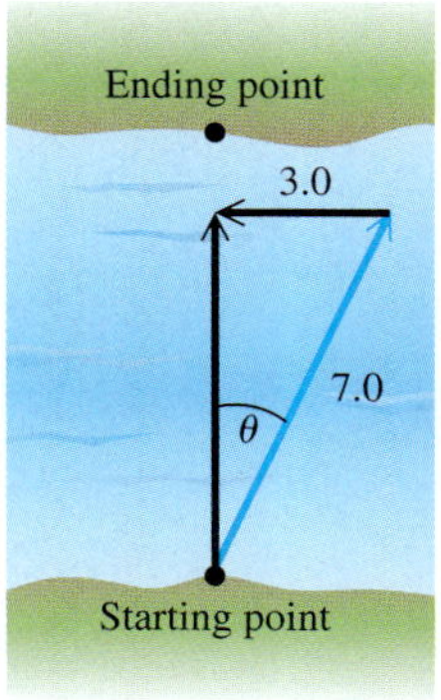

(a) At what speed will the motorboat be traveling relative to the banks?

(b) How long will it take for the motorboat to make the crossing?

(c) What is the measure of angle θ?

21. *Bearing and Ground Speed of a Plane* An airline route from San Francisco to Honolulu is on a bearing of 233.0°. A jet flying at 450 mph on that bearing encounters a wind blowing at 39.0 mph from a direction of 114.0°. Find the resulting bearing and ground speed of the plane.

22. *Path Traveled by a Plane* The aircraft carrier *Tallahassee* is traveling at sea on a steady course with a bearing of 30° at 32 mph. Patrol planes on the carrier have enough fuel for 2.6 hr of flight when traveling at a speed of 520 mph. One of the pilots takes off on a bearing of 338° and then turns and heads in a straight line, so as to be able to catch the carrier and land on the deck at the exact instant that his fuel runs out. If the pilot left at 2 P.M., at what time did he turn to head for the carrier?

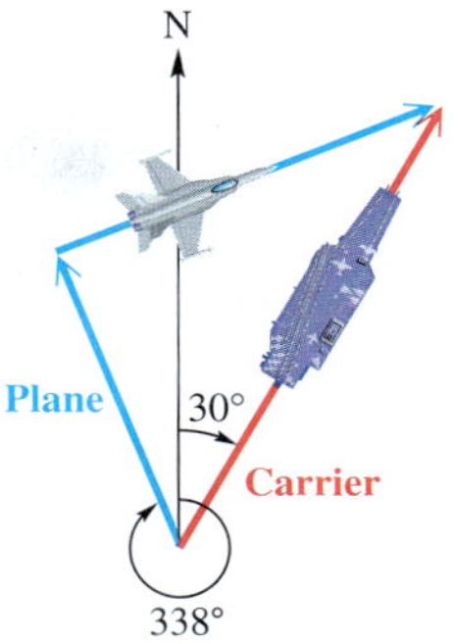

23. *Airspeed and Ground Speed* A pilot wants to fly on a bearing of 74.9°. By flying due east, he finds that a 42.0-mph wind, blowing from the south, puts him on course. Find the airspeed and the ground speed.

24. *Bearing of a Plane* A plane flies 650 mph on a bearing of 175.3°. A 25-mph wind, from a direction of 266.6°, blows against the plane. Find the resulting bearing of the plane.

25. *Bearing and Ground Speed of a Plane* A pilot is flying at 190.0 mph. He wants his flight path to be on a bearing of 64° 30′. A wind is blowing from the south at 35.0 mph. Find the bearing he should fly, and find the plane's ground speed.

26. *Bearing and Ground Speed of a Plane* A pilot is flying at 168 mph. She wants her flight path to be on a bearing of 57° 40′. A wind is blowing from the south at 27.1 mph. Find the bearing the pilot should fly, and find the plane's ground speed.

27. *Bearing and Airspeed of a Plane* What bearing and airspeed are required for a plane to fly 400 mi due north in 2.5 hr if the wind is blowing from a direction of 328° at 11 mph?

28. *Ground Speed and Bearing of a Plane* A plane is headed due south with an airspeed of 192 mph. A wind from a direction of 78.0° is blowing at 23.0 mph. Find the ground speed and resulting bearing of the plane.

29. *Ground Speed and Bearing of a Plane* An airplane is headed on a bearing of 174° at an airspeed of 240 km per hr. A 30-km-per-hr wind is blowing from a direction of 245°. Find the ground speed and resulting bearing of the plane.

30. ***Velocity of a Star*** The space velocity **v** of a star relative to the sun can be expressed as the resultant vector of two perpendicular vectors—the radial velocity $\mathbf{v}_r$ and the tangential velocity $\mathbf{v}_t$, where $\mathbf{v} = \mathbf{v}_r + \mathbf{v}_t$. If a star is located near the sun and its space velocity is large, then its motion across the sky will also be large. Barnard's Star is a relatively close star with a distance of 35 trillion mi from the sun. It moves across the sky through an angle of 10.34″ per year, which is the largest motion of any known star. Its radial velocity is $\mathbf{v}_r = 67$ mi per sec toward the sun. (*Sources:* Zeilik, M., S. Gregory, and E. Smith, *Introductory Astronomy and Astrophysics,* Second Edition, Saunders College Publishing; Acker, A. and C. Jaschek, *Astronomical Methods and Calculations,* John Wiley and Sons.)

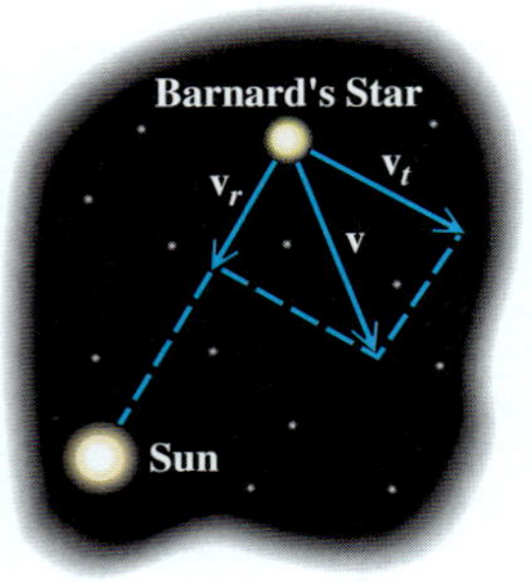

Not to scale

(a) Approximate the tangential velocity $\mathbf{v}_t$ of Barnard's Star. (*Hint:* Use the arc length formula $s = r\theta$ from **Section 3.2.**)

(b) Compute the magnitude of **v**.

31. ***(Modeling) Measuring Rainfall*** Suppose that vector **R** models the amount of rainfall in inches and the direction it falls, and vector **A** models the area in square inches and the orientation of the opening of a rain gauge, as illustrated in the figure. The total volume V of water collected in the rain gauge is given by $V = |\mathbf{R} \cdot \mathbf{A}|$. This formula calculates the volume of water collected even if the wind is blowing the rain in a slanted direction or the rain gauge is not exactly vertical. Let $\mathbf{R} = \mathbf{i} - 2\mathbf{j}$ and $\mathbf{A} = 0.5\mathbf{i} + \mathbf{j}$.

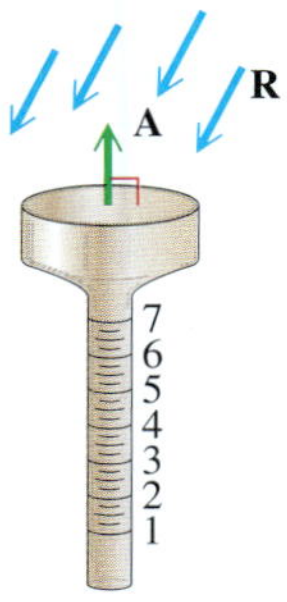

(a) Find $|\mathbf{R}|$ and $|\mathbf{A}|$. Interpret your results.

(b) Calculate V and interpret this result.

32. ***Concept Check*** In **Exercise 31,** for the rain gauge to collect the maximum amount of water, what should be true about vectors **R** and **A**?

Summary Exercises on Applications of Trigonometry and Vectors

These summary exercises provide practice with applications that involve solving triangles and using vectors.

1. ***Wires Supporting a Flagpole*** A flagpole stands vertically on a hillside that makes an angle of 20° with the horizontal. Two supporting wires are attached as shown in the figure. What are the lengths of the supporting wires?

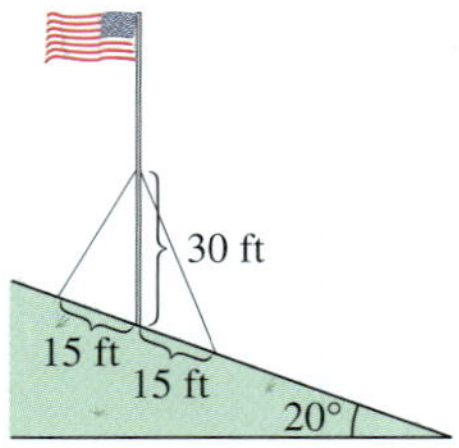

2. ***Distance between a Pin and a Rod*** A slider crank mechanism is shown in the figure. Find the distance between the wrist pin W and the connecting rod center C.

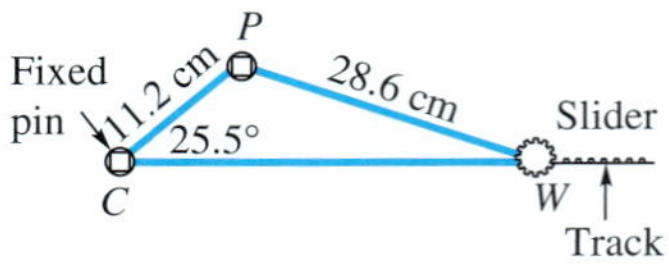

3. ***Distance between Two Lighthouses*** Two lighthouses are located on a north-south line. From lighthouse A, the bearing of a ship 3742 m away is 129° 43′. From lighthouse B, the bearing of a ship is 39° 43′. Find the distance between the lighthouses.

4. *Hot-Air Balloon* A hot-air balloon is rising straight up at the speed of 15.0 ft per sec. Then a wind starts blowing horizontally at 5.00 ft per sec. What will the new speed of the balloon be and what angle with the horizontal will the balloon's path make?

5. *Playing on a Swing* Mary is playing with her daughter Brittany on a swing. Starting from rest, Mary pulls the swing through an angle of 40° and holds it briefly before releasing the swing. If Brittany weighs 50 lb, what horizontal force, to the nearest pound, must Mary apply while holding the swing?

6. *Height of an Airplane* Two observation points A and B are 950 ft apart. From these points the angles of elevation of an airplane are 52° and 57°. See the figure. Find the height of the airplane.

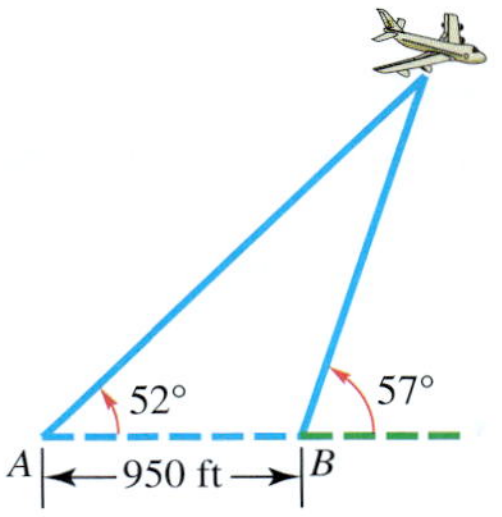

7. *Wind and Vectors* A wind can be described by $\mathbf{v} = 6\mathbf{i} + 8\mathbf{j}$, where vector $\mathbf{j}$ points north and represents a south wind of 1 mph.

(a) What is the speed of the wind?
(b) Find $3\mathbf{v}$ and interpret the result.
(c) Interpret the direction and speed of the wind if it changes to $\mathbf{u} = -8\mathbf{i} + 8\mathbf{j}$.

8. *Ground Speed and Bearing* A plane with an airspeed of 355 mph is on a bearing of 62°. A wind is blowing from west to east at 28.5 mph. Find the ground speed and the actual bearing of the plane.

9. *Property Survey* A surveyor reported the following data about a piece of property: "The property is triangular in shape, with dimensions as shown in the figure." Use the law of sines to see whether such a piece of property could exist.

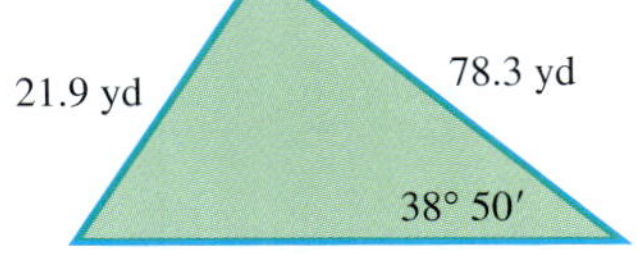

Can such a triangle exist?

10. *Property Survey* A triangular piece of property has the dimensions shown. It turns out that the surveyor did not consider every possible case. Use the law of sines to show why.

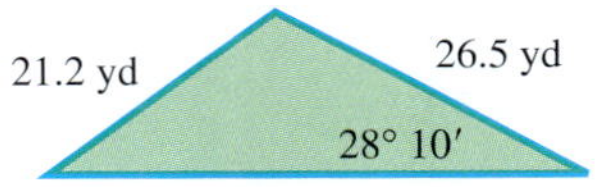

Chapter 7 Test Prep

Key Terms

7.1 Side-Angle-Side (SAS), Angle-Side-Angle (ASA), Side-Side-Side (SSS), oblique triangle, Side-Angle-Angle (SAA)

7.2 ambiguous case

7.3 semiperimeter

7.4 scalar, vector quantity, vector, magnitude, initial point, terminal point, parallelogram rule, resultant, opposite (of a vector), zero vector, position vector, horizontal component, vertical component, direction angle, unit vector, dot product, inner product, angle between two vectors, orthogonal vectors

7.5 equilibrant, airspeed, ground speed

New Symbols

OP or $\overrightarrow{OP}$	vector **OP**
$\lvert \mathbf{OP} \rvert$	magnitude of vector **OP**
$\langle a, b \rangle$	position vector
i, j	unit vectors

Quick Review

Concepts / **Examples**

7.1 Oblique Triangles and the Law of Sines

Law of Sines

In any triangle ABC, with sides a, b, and c, the following holds.

$$\frac{a}{\sin A} = \frac{b}{\sin B} = \frac{c}{\sin C}$$

In triangle ABC, find c, to the nearest hundredth, if $A = 44°$, $C = 62°$, and $a = 12.00$ units. Then find its area.

$$\frac{a}{\sin A} = \frac{c}{\sin C} \quad \text{Law of sines}$$

$$\frac{12.00}{\sin 44°} = \frac{c}{\sin 62°} \quad \text{Substitute.}$$

$$c = \frac{12.00 \sin 62°}{\sin 44°} \quad \text{Multiply by } \sin 62° \text{ and rewrite.}$$

$$c \approx 15.25 \text{ units} \quad \text{Use a calculator.}$$

Area of a Triangle

In any triangle ABC, the area is half the product of the lengths of two sides and the sine of the angle between them.

$$\mathcal{A} = \frac{1}{2}bc \sin A, \quad \mathcal{A} = \frac{1}{2}ab \sin C, \quad \mathcal{A} = \frac{1}{2}ac \sin B$$

For triangle ABC above, apply the appropriate area formula.

$$\mathcal{A} = \frac{1}{2}ac \sin B \quad \text{Area formula}$$

$$= \frac{1}{2}(12.00)(15.25) \sin 74° \quad B = 180° - 44° - 62°$$

$$\approx 87.96 \text{ sq units} \quad \text{Use a calculator.}$$

Concepts	Examples

7.2 The Ambiguous Case of the Law of Sines

Ambiguous Case

If we are given the lengths of two sides and the angle opposite one of them (for example, A, a, and b in triangle ABC), then it is possible that zero, one, or two such triangles exist. If A is acute, h is the altitude from C, and

- $a < h < b$, then there is no triangle.
- $a = h$ and $h < b$, then there is one triangle (a right triangle).
- $a \geq b$, then there is one triangle.
- $h < a < b$, then there are two triangles.

If A is obtuse and

- $a \leq b$, then there is no triangle.
- $a > b$, then there is one triangle.

See the guidelines in **Section 7.2** that illustrate the possible outcomes.

Solve triangle ABC, given $A = 44.5°$, $a = 11.0$ in., and $c = 7.0$ in.

Find angle C.

$$\frac{\sin C}{7.0} = \frac{\sin 44.5°}{11.0} \quad \text{Law of sines}$$

$$\sin C \approx 0.4460 \quad \text{Solve for } \sin C.$$

$$C \approx 26.5° \quad \text{Use inverse sine.}$$

Another angle with this sine value is

$$180° - 26.5° \approx 153.5°.$$

However, $153.5° + 44.5° > 180°$, so there is only one triangle.

$$B \approx 180° - 44.5° - 26.5° \quad \text{Angle sum formula}$$

$$B \approx 109° \quad \text{Subtract.}$$

Use the law of sines again to solve for b.

$$b \approx 14.8 \text{ in.}$$

7.3 The Law of Cosines

Law of Cosines

In any triangle ABC, with sides a, b, and c, the following hold.

$$\mathbf{a^2 = b^2 + c^2 - 2bc \cos A}$$

$$\mathbf{b^2 = a^2 + c^2 - 2ac \cos B}$$

$$\mathbf{c^2 = a^2 + b^2 - 2ab \cos C}$$

In triangle ABC, find C if $a = 11$ units, $b = 13$ units, and $c = 20$ units. Then find its area.

$$c^2 = a^2 + b^2 - 2ab \cos C \quad \text{Law of cosines}$$

$$20^2 = 11^2 + 13^2 - 2(11)(13) \cos C \quad \text{Substitute.}$$

$$400 = 121 + 169 - 286 \cos C \quad \text{Square and multiply.}$$

$$\frac{400 - 121 - 169}{-286} = \cos C \quad \text{Solve for } \cos C.$$

$$C = \cos^{-1}\left(\frac{400 - 121 - 169}{-286}\right)$$

$$C \approx 113°$$

Heron's Area Formula

If a triangle has sides of lengths a, b, and c, with semiperimeter

$$\mathbf{s = \frac{1}{2}(a + b + c),}$$

then the area $\mathcal{A}$ of the triangle is given by the following.

$$\mathbf{\mathcal{A} = \sqrt{s(s - a)(s - b)(s - c)}}$$

The semiperimeter s is

$$s = \frac{1}{2}(11 + 13 + 20) = 22,$$

so

$$\mathcal{A} = \sqrt{22(22 - 11)(22 - 13)(22 - 20)} = 66 \text{ sq units.}$$

(continued)

Concepts	Examples

7.4 Vectors, Operations, and the Dot Product

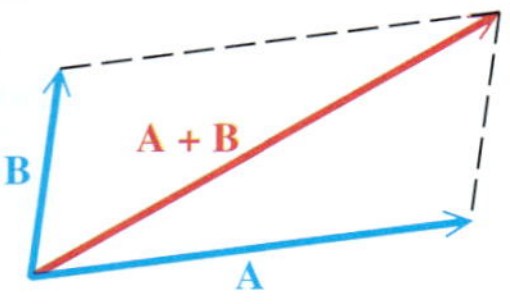

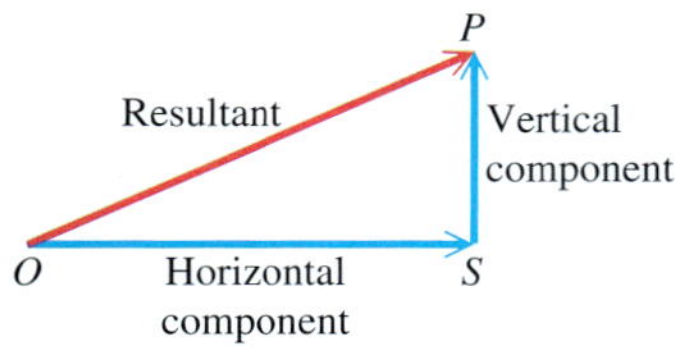

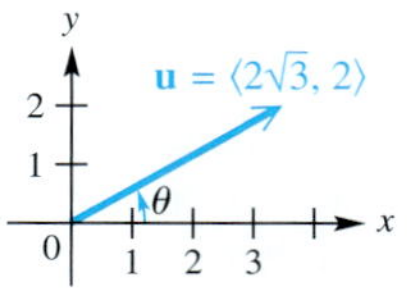

Magnitude and Direction Angle of a Vector

The magnitude (length) of vector $\mathbf{u} = \langle a, b \rangle$ is given by the following.

$$|\mathbf{u}| = \sqrt{a^2 + b^2}$$

The direction angle θ satisfies $\tan\theta = \frac{b}{a}$, where $a \neq 0$.

Example:

$$|\mathbf{u}| = \sqrt{(2\sqrt{3})^2 + 2^2} = \sqrt{16} = 4$$

Since $\tan\theta = \frac{2}{2\sqrt{3}} = \frac{1}{\sqrt{3}} \cdot \frac{\sqrt{3}}{\sqrt{3}} = \frac{\sqrt{3}}{3}$, it follows that $\theta = 30°$.

Vector Operations

Let a, b, c, d, and k represent real numbers.

$$\langle a, b\rangle + \langle c, d\rangle = \langle a + c, b + d\rangle$$

$$k \cdot \langle a, b\rangle = \langle ka, kb\rangle$$

If $\mathbf{u} = \langle a_1, a_2\rangle$, then $-\mathbf{u} = \langle -a_1, -a_2\rangle$.

$$\langle a, b\rangle - \langle c, d\rangle = \langle a, b\rangle + (-\langle c, d\rangle) = \langle a - c, b - d\rangle$$

If $\mathbf{u} = \langle a, b\rangle$ has direction angle θ, then

$$\mathbf{u} = \langle |\mathbf{u}|\cos\theta, |\mathbf{u}|\sin\theta\rangle.$$

Examples:

$$\langle 4, 6\rangle + \langle -8, 3\rangle = \langle -4, 9\rangle$$

$$5\langle -2, 1\rangle = \langle -10, 5\rangle$$

$$-\langle -9, 6\rangle = \langle 9, -6\rangle$$

$$\langle 4, 6\rangle - \langle -8, 3\rangle = \langle 12, 3\rangle$$

For $\mathbf{u}$ defined at the top of the column,

$$\mathbf{u} = \langle 4\cos 30°, 4\sin 30°\rangle$$

$$= \langle 2\sqrt{3}, 2\rangle \qquad \cos 30° = \tfrac{\sqrt{3}}{2}, \quad \sin 30° = \tfrac{1}{2}$$

i, j Form for Vectors

If $\mathbf{v} = \langle a, b\rangle$, then

$$\mathbf{v} = a\mathbf{i} + b\mathbf{j}, \quad \text{where } \mathbf{i} = \langle 1, 0\rangle \text{ and } \mathbf{j} = \langle 0, 1\rangle.$$

and $\quad \mathbf{u} = 2\sqrt{3}\,\mathbf{i} + 2\mathbf{j}.$

Dot Product

The dot product of the two vectors $\mathbf{u} = \langle a, b\rangle$ and $\mathbf{v} = \langle c, d\rangle$, denoted $\mathbf{u} \cdot \mathbf{v}$, is given by the following.

$$\mathbf{u} \cdot \mathbf{v} = ac + bd$$

If θ is the angle between $\mathbf{u}$ and $\mathbf{v}$, where $0° \le \theta \le 180°$, then the following holds.

$$\cos\theta = \frac{\mathbf{u} \cdot \mathbf{v}}{|\mathbf{u}||\mathbf{v}|}$$

Examples:

$$\langle 2, 1\rangle \cdot \langle 5, -2\rangle = 2 \cdot 5 + 1(-2) = 8$$

Find the angle θ between $\mathbf{u} = \langle 3, 1\rangle$ and $\mathbf{v} = \langle 2, -3\rangle$.

$$\cos\theta = \frac{\langle 3, 1\rangle \cdot \langle 2, -3\rangle}{\sqrt{3^2 + 1^2} \cdot \sqrt{2^2 + (-3)^2}}$$

$$\cos\theta = \frac{6 + (-3)}{\sqrt{10} \cdot \sqrt{13}}$$

$$\cos\theta = \frac{3}{\sqrt{130}}$$

$$\theta = \cos^{-1}\frac{3}{\sqrt{130}}$$

$$\theta \approx 74.7°$$

Chapter 7 Review Exercises

Use the law of sines to find the indicated part of each triangle ABC.

1. Find b if $C = 74.2°$, $c = 96.3$ m, $B = 39.5°$.

2. Find B if $A = 129.7°$, $a = 127$ ft, $b = 69.8$ ft.

3. Find B if $C = 51.3°$, $c = 68.3$ m, $b = 58.2$ m.

4. Find b if $a = 165$ m, $A = 100.2°$, $B = 25.0°$.

5. Find A if $B = 39° 50'$, $b = 268$ m, $a = 340$ m.

6. Find A if $C = 79° 20'$, $c = 97.4$ mm, $a = 75.3$ mm.

7. If we are given a, A, and C in a triangle ABC, does the possibility of the ambiguous case exist? If not, explain why.

8. Can triangle ABC exist if $a = 4.7$, $b = 2.3$, and $c = 7.0$? If not, explain why. Answer this question without using trigonometry.

9. Given $a = 10$ and $B = 30°$, determine the values of b for which A has
   (a) exactly one value **(b)** two possible values **(c)** no value.

10. Explain why there can be no triangle ABC satisfying $A = 140°$, $a = 5$, and $b = 7$.

Use the law of cosines to find the indicated part of each triangle ABC.

11. Find A if $a = 86.14$ in., $b = 253.2$ in., $c = 241.9$ in.

12. Find b if $B = 120.7°$, $a = 127$ ft, $c = 69.8$ ft.

13. Find a if $A = 51° 20'$, $c = 68.3$ m, $b = 58.2$ m.

14. Find B if $a = 14.8$ m, $b = 19.7$ m, $c = 31.8$ m.

15. Find a if $A = 60°$, $b = 5.0$ cm, $c = 21$ cm.

16. Find A if $a = 13$ ft, $b = 17$ ft, $c = 8$ ft.

Solve each triangle ABC having the given information.

17. $A = 25.2°$, $a = 6.92$ yd, $b = 4.82$ yd

18. $A = 61.7°$, $a = 78.9$ m, $b = 86.4$ m

19. $a = 27.6$ cm, $b = 19.8$ cm, $C = 42° 30'$

20. $a = 94.6$ yd, $b = 123$ yd, $c = 109$ yd

Find the area of each triangle ABC with the given information.

21. $b = 840.6$ m, $c = 715.9$ m, $A = 149.3°$

22. $a = 6.90$ ft, $b = 10.2$ ft, $C = 35° 10'$

23. $a = 0.913$ km, $b = 0.816$ km, $c = 0.582$ km

24. $a = 43$ m, $b = 32$ m, $c = 51$ m

Solve each problem.

25. **Distance across a Canyon** To measure the distance AB across a canyon for a power line, a surveyor measures angles B and C and the distance BC, as shown in the figure. What is the distance from A to B?

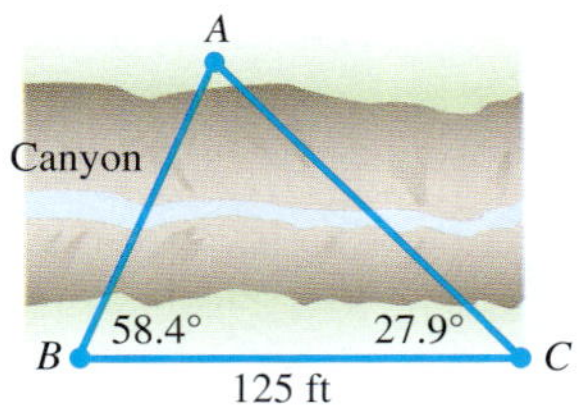

26. *Length of a Brace* A banner on an 8.0-ft pole is to be mounted on a building at an angle of 115°, as shown in the figure. Find the length of the brace.

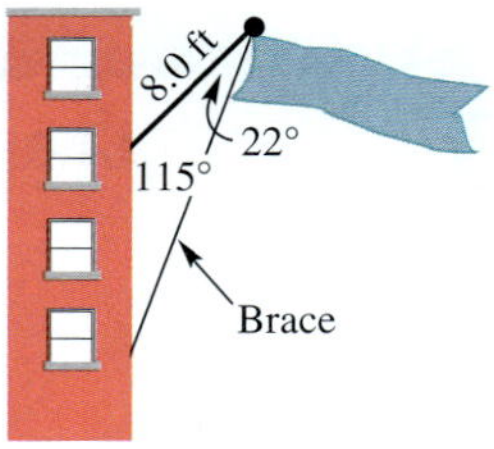

27. *Height of a Tree* A tree leans at an angle of 8.0° from the vertical. From a point 7.0 m from the bottom of the tree, the angle of elevation to the top of the tree is 68°. How tall is the leaning tree?

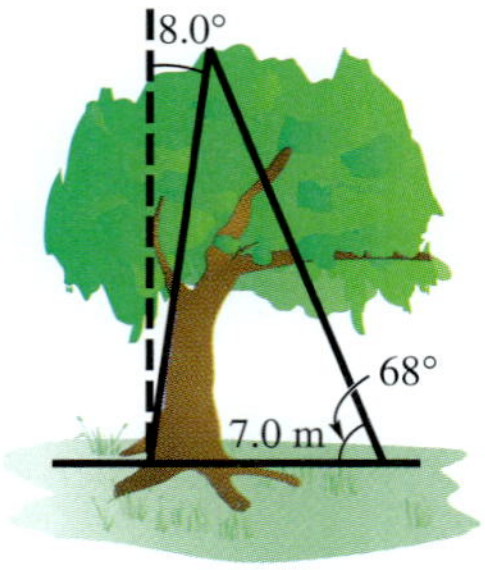

28. *Hanging Sculpture* A hanging sculpture is to be hung in an art gallery with two wires of lengths 15.0 ft and 12.2 ft so that the angle between them is 70.3°. How far apart should the ends of the wire be placed on the ceiling?

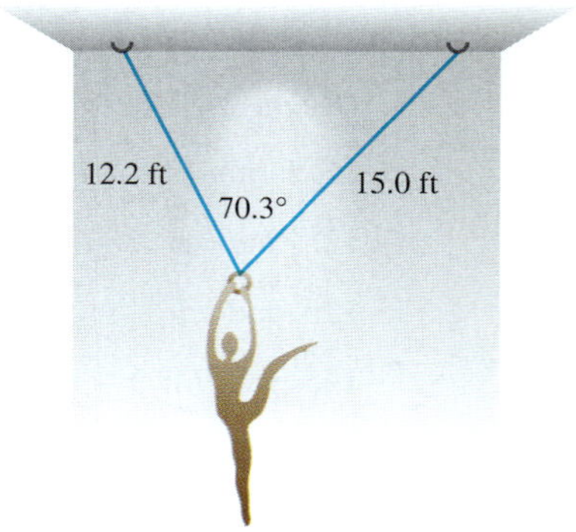

29. *Height of a Tree* A hill makes an angle of 14.3° with the horizontal. From the base of the hill, the angle of elevation to the top of a tree on top of the hill is 27.2°. The distance along the hill from the base to the tree is 212 ft. Find the height of the tree.

30. *Pipeline Position* A pipeline is to run between points A and B, which are separated by a protected wetlands area. To avoid the wetlands, the pipe will run from point A to C and then to B. The distances involved are $AB = 150$ km, $AC = 102$ km, and $BC = 135$ km. What angle should be used at point C?

31. *Distance between Two Boats* Two boats leave a dock together. Each travels in a straight line. The angle between their courses measures 54° 10′. One boat travels 36.2 km per hr, and the other travels 45.6 km per hr. How far apart will they be after 3 hr?

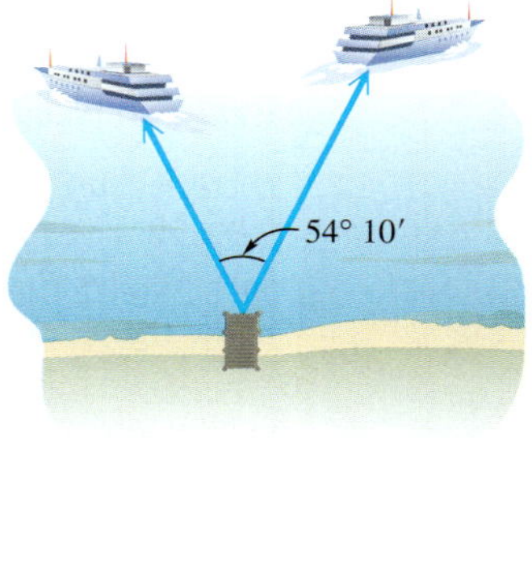

32. *Distance from a Ship to a Lighthouse* A ship sailing parallel to shore sights a lighthouse at an angle of 30° from its direction of travel. After the ship travels 2.0 mi farther, the angle has increased to 55°. At that time, how far is the ship from the lighthouse?

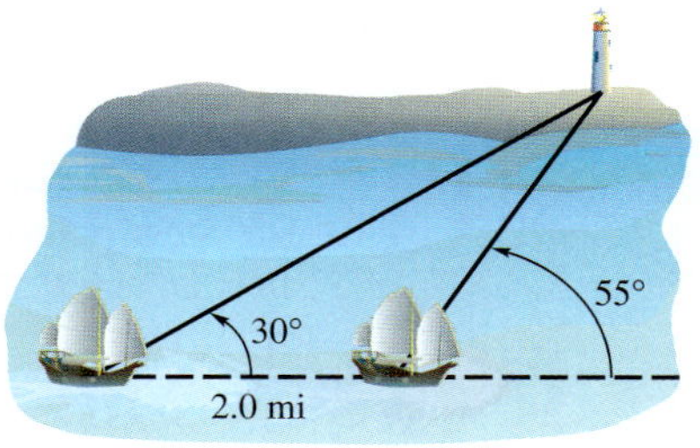

33. *Area of a Triangle* Find the area of the triangle shown in the figure using Heron's area formula.

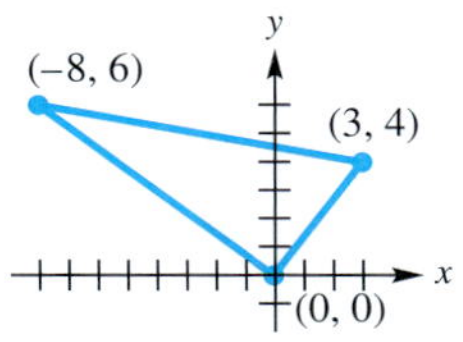

34. Show that the triangle in **Exercise 33** is a right triangle. Then use the formula $\mathcal{A} = \frac{1}{2}ac \sin B$, with $B = 90°$, to find the area.

Use the given vectors to sketch each of the following.

35. $\mathbf{a} - \mathbf{b}$

36. $\mathbf{a} + 3\mathbf{c}$

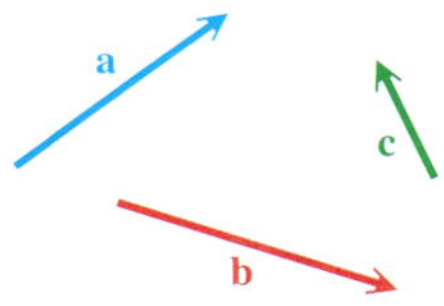

Given two forces and the angle between them, find the magnitude of the resultant force.

37.

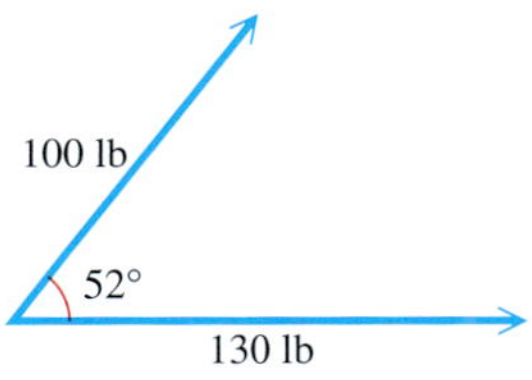

38. two forces of 142 and 215 newtons, forming an angle of 112°

Vector **v** *has the given magnitude and direction angle. Find the magnitudes of the horizontal and vertical components of* **v**.

39. $|\mathbf{v}| = 964$, $\theta = 154° 20'$

40. $|\mathbf{v}| = 50$, $\theta = 45°$ (Give exact values.)

Find the magnitude and direction angle for **u** *rounded to the nearest tenth.*

41. $\mathbf{u} = \langle -9, 12 \rangle$

42. $\mathbf{u} = \langle 21, -20 \rangle$

43. Let $\mathbf{v} = 2\mathbf{i} - \mathbf{j}$ and $\mathbf{u} = -3\mathbf{i} + 2\mathbf{j}$. Express each in terms of **i** and **j**.

(a) $2\mathbf{v} + \mathbf{u}$ **(b)** $2\mathbf{v}$ **(c)** $\mathbf{v} - 3\mathbf{u}$

Find the angle between the vectors. Round to the nearest tenth of a degree. If the vectors are orthogonal, say so.

44. $\langle 3, -2 \rangle, \langle -1, 3 \rangle$ **45.** $\langle 5, -3 \rangle, \langle 3, 5 \rangle$ **46.** $\langle 0, 4 \rangle, \langle -4, 4 \rangle$

Solve each problem.

47. *Weight of a Sled and Passenger* Paula and Steve are pulling their daughter Jessie on a sled. Steve pulls with a force of 18 lb at an angle of 10°. Paula pulls with a force of 12 lb at an angle of 15°. Find the magnitude of the resultant force on Jessie and the sled.

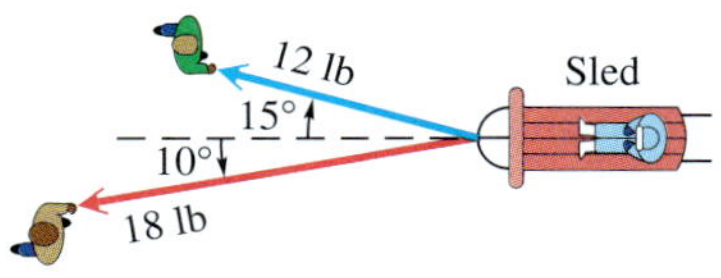

48. *Force Placed on a Barge* One boat pulls a barge with a force of 100 newtons. Another boat pulls the barge at an angle of 45° to the first force, with a force of 200 newtons. Find the resultant force acting on the barge, to the nearest unit, and the angle between the resultant and the first boat, to the nearest tenth.

49. *Direction and Speed of a Plane* A plane has an airspeed of 520 mph. The pilot wishes to fly on a bearing of 310°. A wind of 37 mph is blowing from a bearing of 212°. In what direction should the pilot fly, and what will be her ground speed?

50. *Angle of a Hill* A 186-lb force is required to hold a 2800-lb car on a hill. What angle does the hill make with the horizontal?

51. *Incline Force* Find the force required to keep a 75-lb sled from sliding down an incline that makes an angle of 27° with the horizontal. (Assume there is no friction.)

52. *Speed and Direction of a Boat* A boat travels 15 km per hr in still water. The boat is traveling across a large river, on a bearing of 130°. The current in the river, coming from the west, has a speed of 7 km per hr. Find the resulting speed of the boat and its resulting direction of travel.

Other Formulas from Trigonometry The following identities involve all six parts of a triangle ABC and are thus useful for checking answers.

$$\frac{a+b}{c} = \frac{\cos \frac{1}{2}(A-B)}{\sin \frac{1}{2}C} \quad \text{Newton's formula}$$

$$\frac{a-b}{c} = \frac{\sin \frac{1}{2}(A-B)}{\cos \frac{1}{2}C} \quad \text{Mollweide's formula}$$

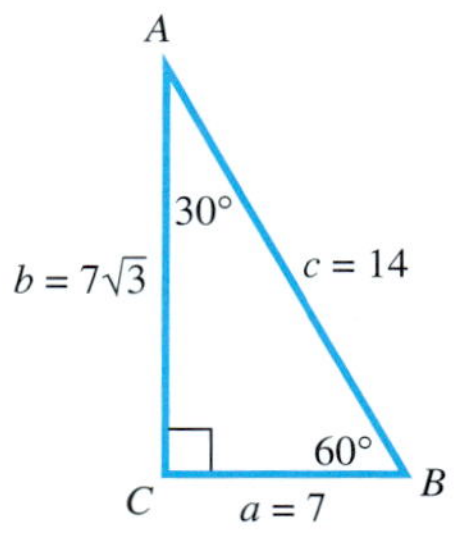

53. Apply Newton's formula to the given triangle to verify the accuracy of the information.

54. Apply Mollweide's formula to the given triangle to verify the accuracy of the information.

55. *Law of Tangents* In addition to the law of sines and the law of cosines, there is a **law of tangents.** In any triangle *ABC*,

$$\frac{\tan \frac{1}{2}(A-B)}{\tan \frac{1}{2}(A+B)} = \frac{a-b}{a+b}.$$

Verify this law for the triangle *ABC* with $a = 2$, $b = 2\sqrt{3}$, $A = 30°$, and $B = 60°$.

Chapter 7 Test

Find the indicated part of each triangle ABC.

1. Find C if $A = 25.2°$, $a = 6.92$ yd, and $b = 4.82$ yd.

2. Find c if $C = 118°$, $a = 75.0$ km, and $b = 131$ km.

3. Find B if $a = 17.3$ ft, $b = 22.6$ ft, $c = 29.8$ ft.

4. Find the area of triangle ABC if $a = 14$, $b = 30$, and $c = 40$.

5. Find the area of triangle XYZ shown here.

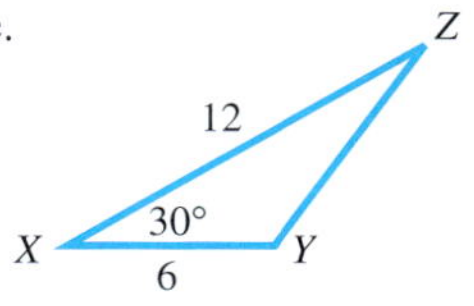

6. Given $a = 10$ and $B = 150°$ in triangle ABC, determine the values of b for which A has

(a) exactly one value **(b)** two possible values **(c)** no value.

Solve each triangle ABC.

7. $A = 60°$, $b = 30$ m, $c = 45$ m

8. $b = 1075$ in., $c = 785$ in., $C = 38° 30'$

9. Find the magnitude and the direction angle for the vector shown in the figure.

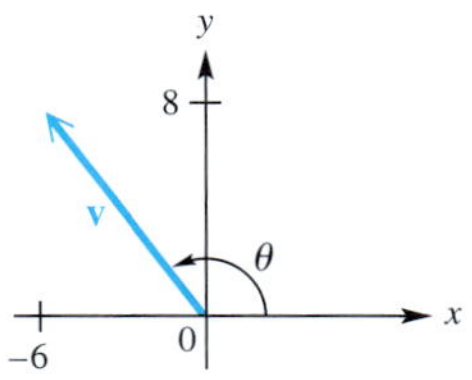

10. Use the given vectors to sketch $\mathbf{a} + \mathbf{b}$.

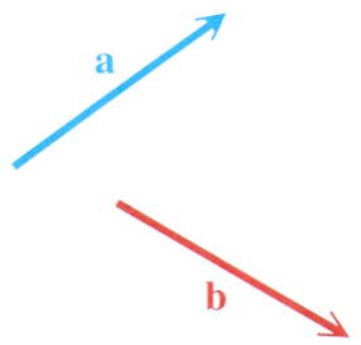

11. For the vectors $\mathbf{u} = \langle -1, 3 \rangle$ and $\mathbf{v} = \langle 2, -6 \rangle$, find each of the following.

(a) $\mathbf{u} + \mathbf{v}$ **(b)** $-3\mathbf{v}$ **(c)** $\mathbf{u} \cdot \mathbf{v}$ **(d)** $|\mathbf{u}|$

12. Find the measure of the angle θ between $\mathbf{u} = \langle 4, 3 \rangle$ and $\mathbf{v} = \langle 1, 5 \rangle$.

13. Show that the vectors $\mathbf{u} = \langle -4, 7 \rangle$ and $\mathbf{v} = \langle -14, -8 \rangle$ are orthogonal vectors.

Solve each problem.

14. ***Height of a Balloon*** The angles of elevation of a balloon from two points A and B on level ground are $24° 50'$ and $47° 20'$, respectively. As shown in the figure, points A, B, and C are in the same vertical plane and points A and B are 8.4 mi apart. Approximate the height of the balloon above the ground to the nearest tenth of a mile.

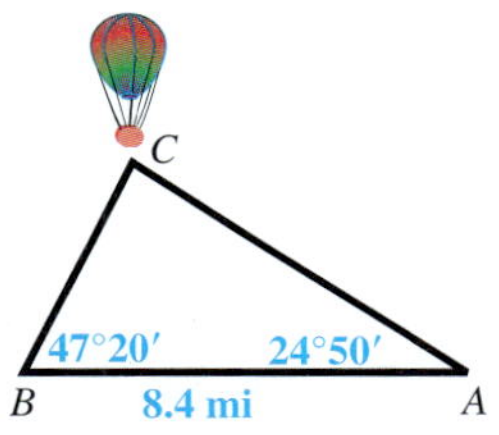

15. ***Horizontal and Vertical Components*** Find the horizontal and vertical components of the vector with magnitude 569 and direction angle 127.5° from the horizontal. Give your answer in the form $\langle a, b \rangle$.

16. ***Radio Direction Finders*** Radio direction finders are placed at points A and B, which are 3.46 mi apart on an east-west line, with A west of B. From A, the bearing of a certain illegal pirate radio transmitter is 48°, and from B the bearing is 302°. Find the distance between the transmitter and A to the nearest hundredth of a mile.

17. ***Height of a Tree*** A tree leans at an angle of 8.0° from the vertical, as shown in the figure. From a point 8.0 m from the bottom of the tree, the angle of elevation to the top of the tree is 66°. Find the height of the leaning tree.

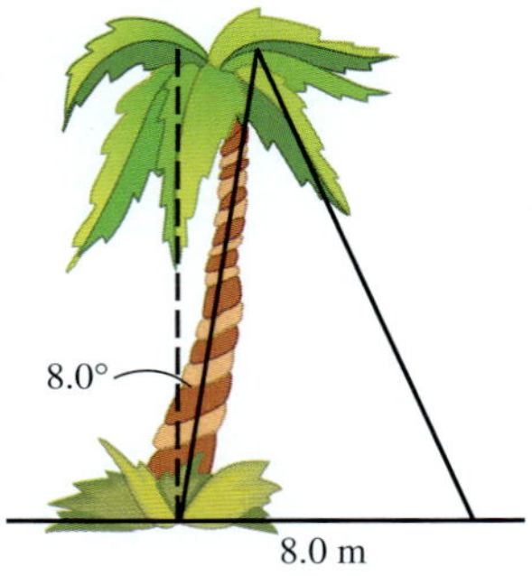

18. ***Walking Dogs on Leashes*** While Michael is walking his two dogs, Duke and Prince, they reach a corner and must wait for a WALK sign. Michael is holding the two leashes in the same hand, and the dogs are pulling on their leashes at the angles and forces shown in the figure. Find the magnitude of the equilibrant force (to the nearest tenth of a pound) that Michael must apply to restrain the dogs.

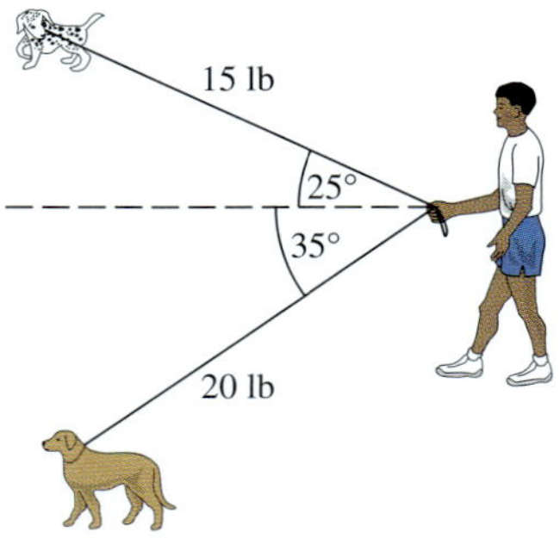

19. ***Bearing and Airspeed*** Find the bearing and airspeed required for a plane to fly 630 mi due north in 3.0 hr if the wind is blowing from a direction of 318° at 15 mph. Approximate the bearing to the nearest degree and the airspeed to the nearest 10 mph.

20. ***Incline Angle*** A force of 16.0 lb is required to hold a 50.0-lb wheelbarrow on an incline. What angle does the incline make with the horizontal?

8 Complex Numbers, Polar Equations, and Parametric Equations

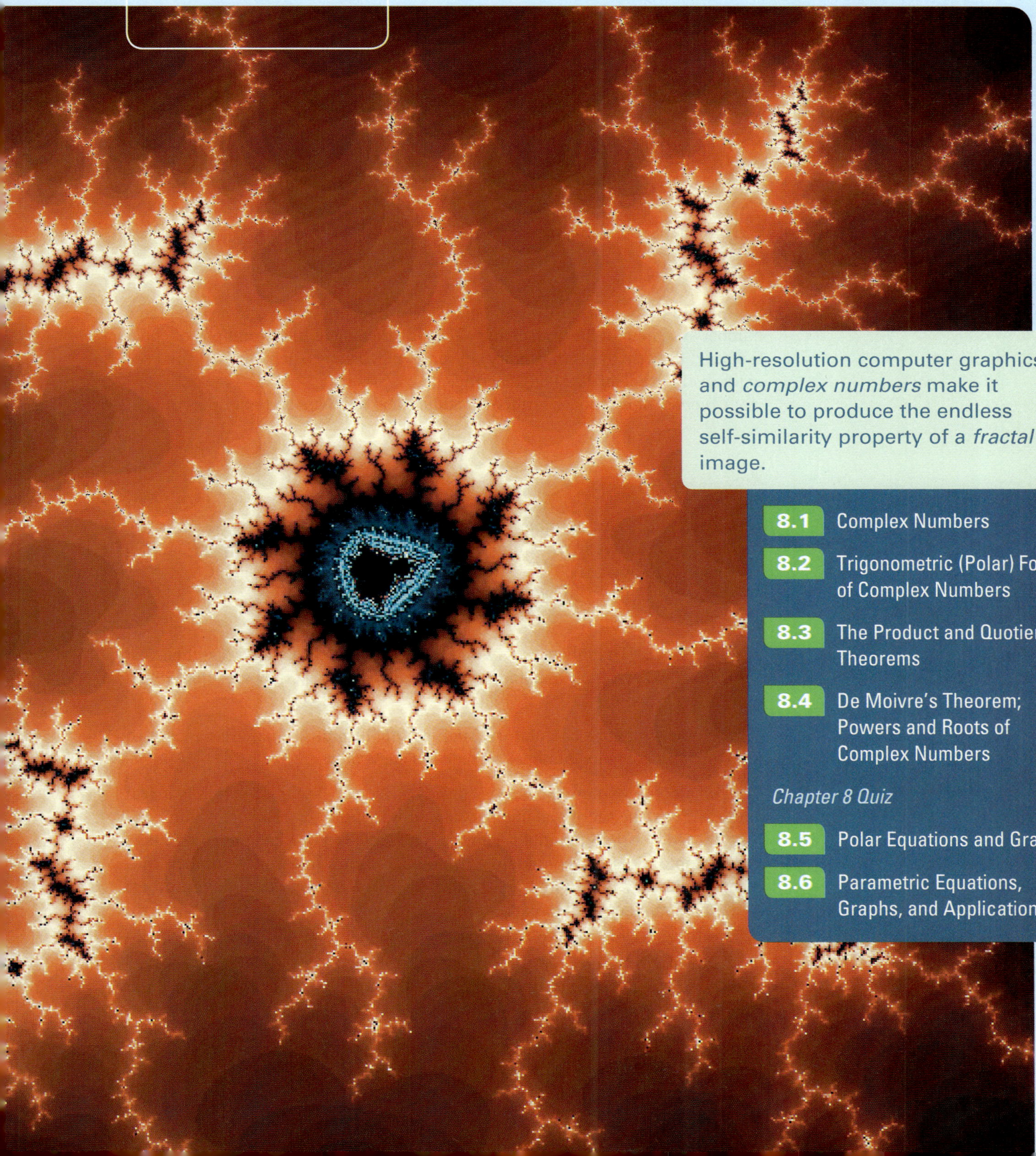

High-resolution computer graphics and *complex numbers* make it possible to produce the endless self-similarity property of a *fractal* image.

8.1 Complex Numbers

- Basic Concepts of Complex Numbers
- Complex Solutions of Equations
- Operations on Complex Numbers

Basic Concepts of Complex Numbers

The set of real numbers does not include all the numbers needed in algebra. For example, there is no real number solution of the equation

$$x^2 = -1,$$

since no real number, when squared, gives -1. To extend the real number system to include solutions of equations of this type, the number i is defined to have the following property.

LOOKING AHEAD TO CALCULUS

The letters j and k are also used to represent $\sqrt{-1}$ in calculus and some applications (electronics, for example).

The Imaginary Unit i

$$i = \sqrt{-1}, \quad \text{and therefore} \quad i^2 = -1.$$

(Note that $-i$ is also a square root of -1.)

Square roots of negative numbers were not incorporated into an integrated number system until the 16th century. They were then used as solutions of equations and later (in the 18th century) in surveying. Today, such numbers are used extensively in science and engineering.

Complex numbers are formed by adding real numbers and multiples of i.

Complex Number

If a and b are real numbers, then any number of the form $a + bi$ is a **complex number.** In the complex number $a + bi$, a is the **real part** and b is the **imaginary part.***

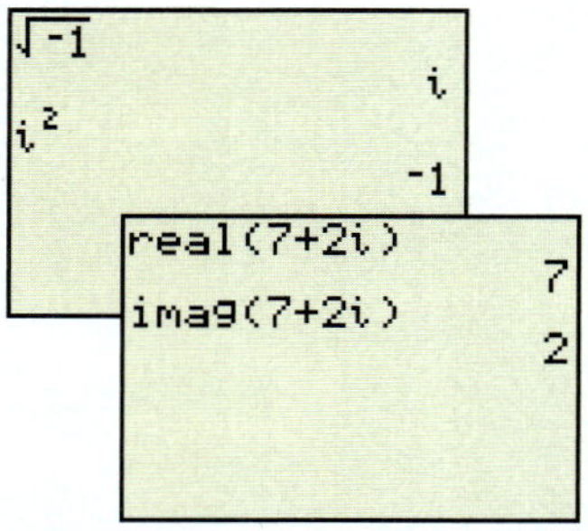

The calculator is in complex number mode. The top screen supports the definition of i. The bottom screen shows how the calculator returns the real and imaginary parts of the complex number $7 + 2i$.

Figure 1

Two complex numbers $a + bi$ and $c + di$ are equal provided that their real parts are equal and their imaginary parts are equal; that is, they are equal if and only if $a = c$ and $b = d$.

Some graphing calculators, such as the TI-83/84 Plus, are capable of working with complex numbers, as seen in **Figure 1.** ■

For a complex number $a + bi$, if $b = 0$, then $a + bi = a$, which is a real number. Thus, the set of real numbers is a subset of the set of complex numbers. If $a = 0$ and $b \neq 0$, the complex number is said to be a **pure imaginary number.** For example, $3i$ is a pure imaginary number. A pure imaginary number, or a number such as $7 + 2i$ with $a \neq 0$ and $b \neq 0$, is a **nonreal complex number.** A complex number written in the form $a + bi$ (or $a + ib$) is in **standard form.** (The form $a + ib$ is used to write expressions such as $i\sqrt{5}$, since $\sqrt{5}i$ could be mistaken for $\sqrt{5i}$.)

The relationships among the subsets of the complex numbers are shown in **Figure 2** on the next page.

*In some texts, the term bi is defined to be the imaginary part.

Complex Numbers $a + bi$, for a and b Real

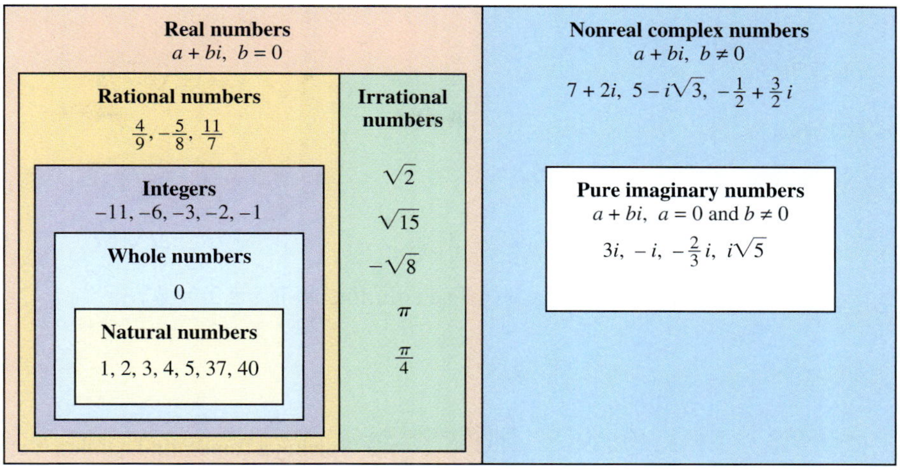

Figure 2

For a positive real number a, the expression $\sqrt{-a}$ is defined as follows.

The Expression $\sqrt{-a}$

If $a > 0$, then $\qquad \sqrt{-a} = i\sqrt{a}.$

EXAMPLE 1 Writing $\sqrt{-a}$ as $i\sqrt{a}$

Write as the product of a real number and i, using the definition of $\sqrt{-a}$.

(a) $\sqrt{-16}$ **(b)** $\sqrt{-70}$ **(c)** $\sqrt{-48}$

SOLUTION

(a) $\sqrt{-16} = i\sqrt{16} = 4i$ **(b)** $\sqrt{-70} = i\sqrt{70}$

(c) $\sqrt{-48} = i\sqrt{48} = i\sqrt{16 \cdot 3} = 4i\sqrt{3}$ Product rule for radicals: $\sqrt[n]{ab} = \sqrt[n]{a} \cdot \sqrt[n]{b}$

✔ ***Now Try Exercises 17, 19, and 21.***

Complex Solutions of Equations

EXAMPLE 2 Solving Quadratic Equations for Complex Solutions

Solve each equation.

(a) $x^2 = -9$ **(b)** $x^2 + 24 = 0$

SOLUTION

(a) Take the square root on each side, remembering that we must find both roots, which is indicated by the $\pm$ sign.

$$x^2 = -9$$

Take *both* square roots.

$$x = \pm\sqrt{-9} \quad \text{Square root property (Appendix A)}$$

$$x = \pm i\sqrt{9} \quad \sqrt{-a} = i\sqrt{a}$$

$$x = \pm 3i \quad \sqrt{9} = 3$$

The solution set is $\{\pm 3i\}$.

(b)

$$x^2 + 24 = 0$$

$$x^2 = -24 \quad \text{Subtract 24.}$$

$$x = \pm\sqrt{-24} \quad \text{Square root property}$$

$$x = \pm i\sqrt{24} \quad \sqrt{-a} = i\sqrt{a}$$

$$x = \pm i\sqrt{4}\cdot\sqrt{6} \quad \text{Product rule for radicals}$$

$$x = \pm 2i\sqrt{6} \quad \sqrt{4} = 2$$

The solution set is $\{\pm 2i\sqrt{6}\}$.

✔ **Now Try Exercises 25 and 27.**

EXAMPLE 3 Solving a Quadratic Equation (Complex Solutions)

Solve $9x^2 + 5 = 6x$.

SOLUTION Write the equation in standard form. Then use the quadratic formula.

$$9x^2 - 6x + 5 = 0 \quad \text{Standard form (Appendix A)}$$

$$x = \frac{-b \pm \sqrt{b^2 - 4ac}}{2a} \quad \text{Quadratic formula (Appendix A)}$$

The fraction bar extends under $-b$.

$$= \frac{-(-6) \pm \sqrt{(-6)^2 - 4(9)(5)}}{2(9)} \quad a = 9,\ b = -6,\ c = 5$$

$$= \frac{6 \pm \sqrt{-144}}{18} \quad \text{Simplify.}$$

$$= \frac{6 \pm 12i}{18} \quad \sqrt{-144} = 12i$$

Factor first, then divide out the common factor.

$$= \frac{6(1 \pm 2i)}{6\cdot 3} \quad \text{Factor.}$$

$$x = \frac{1 \pm 2i}{3} \quad \text{Write in lowest terms.}$$

Written in standard form, the solution set is $\left\{\frac{1}{3} \pm \frac{2}{3}i\right\}$.

✔ **Now Try Exercise 29.**

Operations on Complex Numbers Products or quotients with negative radicands are simplified by first rewriting $\sqrt{-a}$ as $i\sqrt{a}$ for a positive number a. Then the properties of real numbers and the fact that $i^2 = -1$ are applied.

CAUTION ***When working with negative radicands, use the definition $\sqrt{-a} = i\sqrt{a}$ before using any of the other rules for radicals.*** In particular, the rule $\sqrt{c}\cdot\sqrt{d} = \sqrt{cd}$ is valid only when c and d are *not* both negative. For example,

$$\sqrt{-4}\cdot\sqrt{-9} = 2i\cdot 3i = 6i^2 = -6 \quad \text{is correct,}$$

whereas $\sqrt{-4}\cdot\sqrt{-9} = \sqrt{(-4)(-9)} = \sqrt{36} = 6$ is incorrect.

EXAMPLE 4 **Finding Products and Quotients Involving $\sqrt{-a}$**

Multiply or divide, as indicated. Simplify each answer.

(a) $\sqrt{-7} \cdot \sqrt{-7}$ **(b)** $\sqrt{-6} \cdot \sqrt{-10}$ **(c)** $\dfrac{\sqrt{-20}}{\sqrt{-2}}$ **(d)** $\dfrac{\sqrt{-48}}{\sqrt{24}}$

SOLUTION

(a)
$$\begin{aligned}\sqrt{-7} \cdot \sqrt{-7} &= i\sqrt{7} \cdot i\sqrt{7}\\ &= i^2 \cdot (\sqrt{7})^2\\ &= -1 \cdot 7 \quad i^2 = -1\\ &= -7\end{aligned}$$

First write all square roots in terms of i.

(b)
$$\begin{aligned}\sqrt{-6} \cdot \sqrt{-10} &= i\sqrt{6} \cdot i\sqrt{10}\\ &= i^2 \cdot \sqrt{60}\\ &= -1\sqrt{4 \cdot 15}\\ &= -1 \cdot 2\sqrt{15}\\ &= -2\sqrt{15}\end{aligned}$$

(c) $\dfrac{\sqrt{-20}}{\sqrt{-2}} = \dfrac{i\sqrt{20}}{i\sqrt{2}} = \sqrt{\dfrac{20}{2}} = \sqrt{10}$ Quotient rule for radicals: $\dfrac{\sqrt[n]{a}}{\sqrt[n]{b}} = \sqrt[n]{\dfrac{a}{b}}$

(d) $\dfrac{\sqrt{-48}}{\sqrt{24}} = \dfrac{i\sqrt{48}}{\sqrt{24}} = i\sqrt{\dfrac{48}{24}} = i\sqrt{2}$ Quotient rule for radicals

✔ ***Now Try Exercises 37, 39, 41, and 43.***

EXAMPLE 5 **Simplifying a Quotient Involving $\sqrt{-a}$**

Write $\dfrac{-8 + \sqrt{-128}}{4}$ in standard form $a + bi$.

SOLUTION

$$\begin{aligned}\frac{-8 + \sqrt{-128}}{4} &= \frac{-8 + \sqrt{-64 \cdot 2}}{4}\\ &= \frac{-8 + 8i\sqrt{2}}{4} && \sqrt{-64} = 8i\\ &= \frac{4(-2 + 2i\sqrt{2})}{4} && \text{Factor.}\\ &= -2 + 2i\sqrt{2} && \text{Lowest terms}\end{aligned}$$

Be sure to factor before simplifying.

✔ ***Now Try Exercise 49.***

With the definitions $i^2 = -1$ and $\sqrt{-a} = i\sqrt{a}$ for $a > 0$, all properties of real numbers are extended to complex numbers.

Addition and Subtraction of Complex Numbers

For complex numbers $a + bi$ and $c + di$,

$$(a + bi) + (c + di) = (a + c) + (b + d)i$$

and

$$(a + bi) - (c + di) = (a - c) + (b - d)i.$$

That is, to add or subtract complex numbers, add or subtract the real parts, and add or subtract the imaginary parts.

EXAMPLE 6 **Adding and Subtracting Complex Numbers**

Find each sum or difference.

(a) $(3 - 4i) + (-2 + 6i)$ **(b)** $(-4 + 3i) - (6 - 7i)$

SOLUTION

(a) $(3 - 4i) + (-2 + 6i) = \underbrace{[3 + (-2)]}_{\text{Add real parts.}} + \underbrace{[-4 + 6]}_{\text{Add imaginary parts.}}i$ Commutative, associative and distributive properties

$= 1 + 2i$

(b) $(-4 + 3i) - (6 - 7i) = (-4 - 6) + [3 - (-7)]i$

$= -10 + 10i$

✔ **Now Try Exercises 55, 57, and 59.**

The product of two complex numbers is found by multiplying as though the numbers were binomials and using the fact that $i^2 = -1$, as follows.

$(a + bi)(c + di) = ac + adi + bic + bidi$ FOIL (Multiply **F**irst, **O**uter, **I**nner, **L**ast terms.)

$= ac + adi + bci + bdi^2$ Associative property

$= ac + (ad + bc)i + bd(-1)$ Distributive property; $i^2 = -1$

$= (ac - bd) + (ad + bc)i$ Group like terms.

Multiplication of Complex Numbers

For complex numbers $a + bi$ and $c + di$,

$$(a + bi)(c + di) = (ac - bd) + (ad + bc)i.$$

This definition is not practical in routine calculations. To find a given product, it is easier just to multiply as with binomials.

EXAMPLE 7 **Multiplying Complex Numbers**

Find each product.

(a) $(2 - 3i)(3 + 4i)$ **(b)** $(4 + 3i)^2$ **(c)** $(6 + 5i)(6 - 5i)$

SOLUTION

(a) $(2 - 3i)(3 + 4i) = 2(3) + 2(4i) - 3i(3) - 3i(4i)$ FOIL

$= 6 + 8i - 9i - 12i^2$ Multiply.

$= 6 - i - 12(-1)$ Combine like terms; $i^2 = -1$.

$= 18 - i$ Standard form

(b) $(4+3i)^2 = 4^2 + 2(4)(3i) + (3i)^2$ — Square of a binomial

Remember to add twice the product of the two terms.

$= 16 + 24i + 9i^2$ — Multiply.

$= 16 + 24i + 9(-1)$ — $i^2 = -1$

$= 7 + 24i$ — Standard form

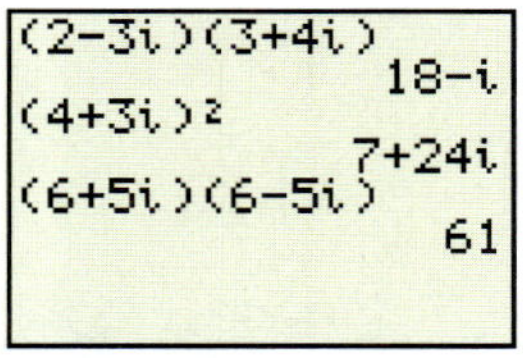

This screen shows how the TI-83/84 Plus displays the results found in **Example 7.**

(c) $(6+5i)(6-5i) = 6^2 - (5i)^2$ — Product of the sum and difference of two terms: $(x+y)(x-y) = x^2 - y^2$

$= 36 - 25(-1)$ — Square 6 and 5; $i^2 = -1$.

$= 36 + 25$ — Multiply.

$= 61$, or $61 + 0i$ — Standard form

✔ *Now Try Exercises 63, 67, and 71.*

Example 7(c) showed that $(6+5i)(6-5i) = 61$. The numbers $6+5i$ and $6-5i$ differ only in the sign of their imaginary parts and are called **complex conjugates.** ***The product of a complex number and its conjugate is always a real number.*** This product is the sum of the squares of the real and imaginary parts.

Property of Complex Conjugates

For real numbers a and b,

$$(a + bi)(a - bi) = a^2 + b^2.$$

To find the quotient of two complex numbers in standard form, we multiply both the numerator and the denominator by the complex conjugate of the denominator.

EXAMPLE 8 Dividing Complex Numbers

Write each quotient in standard form $a + bi$.

(a) $\dfrac{3+2i}{5-i}$ **(b)** $\dfrac{3}{i}$

SOLUTION

(a) $\dfrac{3+2i}{5-i} = \dfrac{(3+2i)(5+i)}{(5-i)(5+i)}$ — Multiply by the complex conjugate of the denominator in both the numerator and the denominator.

$= \dfrac{15 + 3i + 10i + 2i^2}{25 - i^2}$ — Multiply.

$= \dfrac{13 + 13i}{26}$ — Combine like terms; $i^2 = -1$.

$= \dfrac{13}{26} + \dfrac{13i}{26}$ — $\dfrac{a+bi}{c} = \dfrac{a}{c} + \dfrac{bi}{c}$

$= \dfrac{1}{2} + \dfrac{1}{2}i$ — Write in lowest terms and standard form.

CHECK $\left(\dfrac{1}{2} + \dfrac{1}{2}i\right)(5 - i) = 3 + 2i$ ✓ Quotient × Divisor = Dividend

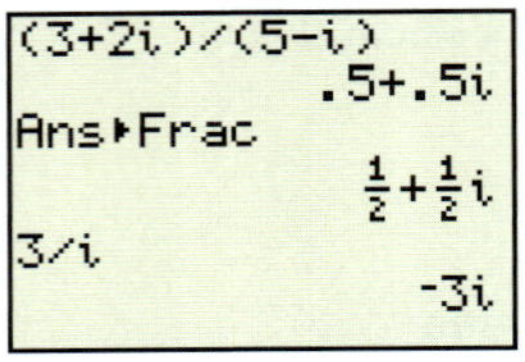

This screen supports the results in **Example 8.**

(b) $\frac{3}{i} = \frac{3(-i)}{i(-i)}$ $\quad -i$ is the conjugate of i.

$= \frac{-3i}{-i^2}$ $\quad$ Multiply.

$= \frac{-3i}{1}$ $\quad -i^2 = -(-1) = 1$

$= -3i$, or $0 - 3i$ $\quad$ Standard form

✔ *Now Try Exercises 81 and 87.*

Powers of i can be simplified using the facts

$$i^2 = -1 \quad \text{and} \quad i^4 = (i^2)^2 = (-1)^2 = 1.$$

Consider the following powers of i.

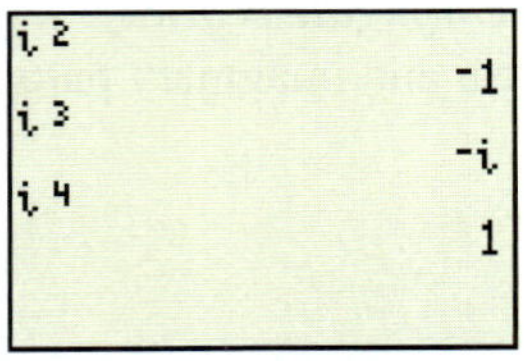

Powers of i can be found on the TI-83/84 Plus calculator.

$i^1 = i$ $\qquad\qquad$ $i^5 = i^4 \cdot i = 1 \cdot i = i$

$i^2 = -1$ $\qquad\qquad$ $i^6 = i^4 \cdot i^2 = 1(-1) = -1$

$i^3 = i^2 \cdot i = (-1) \cdot i = -i$ $\qquad\qquad$ $i^7 = i^4 \cdot i^3 = 1 \cdot (-i) = -i$

$i^4 = i^2 \cdot i^2 = (-1)(-1) = 1$ $\qquad\qquad$ $i^8 = i^4 \cdot i^4 = 1 \cdot 1 = 1$ and so on.

Powers of i cycle through the same four outcomes (i, −1, −i, and 1) since i^4 has the same multiplicative property as 1. Also, any power of i with an exponent that is a multiple of 4 has value 1.

EXAMPLE 9 Simplifying Powers of *i*

Simplify each power of i.

(a) i^{15} $\qquad\qquad$ **(b)** i^{-3}

SOLUTION

(a) Since $i^4 = 1$, write the given power as a product involving i^4.

$$i^{15} = i^{12} \cdot i^3 = (i^4)^3 \cdot i^3 = 1^3(-i) = -i$$

(b) Multiply i^{-3} by 1 in the form of i^4 to create the least positive exponent for i.

$$i^{-3} = i^{-3} \cdot 1 = i^{-3} \cdot i^4 = i \qquad i^4 = 1$$

✔ *Now Try Exercises 93 and 101.*

8.1 Exercises

Concept Check *Determine whether each statement is* true *or* false. *If it is false, tell why.*

1. Every real number is a complex number.

2. No real number is a pure imaginary number.

3. Every pure imaginary number is a complex number.

4. A number can be both real and complex.

5. There is no real number that is a complex number.

6. A complex number might not be a pure imaginary number.

Identify each number as real, complex, pure imaginary, *or* nonreal complex. *(More than one of these descriptions will apply.)*

7. -4 **8.** 0 **9.** $13i$ **10.** $-7i$ **11.** $5 + i$

12. $-6 - 2i$ **13.** π **14.** $\sqrt{24}$ **15.** $\sqrt{-25}$ **16.** $\sqrt{-36}$

Write each number as the product of a real number and i. ***See Example 1.***

17. $\sqrt{-25}$ **18.** $\sqrt{-36}$ **19.** $\sqrt{-10}$ **20.** $\sqrt{-15}$

21. $\sqrt{-288}$ **22.** $\sqrt{-500}$ **23.** $-\sqrt{-18}$ **24.** $-\sqrt{-80}$

Solve each quadratic equation and express all nonreal complex solutions in terms of i. ***See Examples 2 and 3.***

25. $x^2 = -16$ **26.** $x^2 = -36$

27. $x^2 + 12 = 0$ **28.** $x^2 + 48 = 0$

29. $3x^2 + 2 = -4x$ **30.** $2x^2 + 3x = -2$

31. $x^2 - 6x + 14 = 0$ **32.** $x^2 + 4x + 11 = 0$

33. $4(x^2 - x) = -7$ **34.** $3(3x^2 - 2x) = -7$

35. $x^2 + 1 = -x$ **36.** $x^2 + 2 = 2x$

Multiply or divide, as indicated. Simplify each answer. ***See Example 4.***

37. $\sqrt{-13} \cdot \sqrt{-13}$ **38.** $\sqrt{-17} \cdot \sqrt{-17}$ **39.** $\sqrt{-3} \cdot \sqrt{-8}$

40. $\sqrt{-5} \cdot \sqrt{-15}$ **41.** $\dfrac{\sqrt{-30}}{\sqrt{-10}}$ **42.** $\dfrac{\sqrt{-70}}{\sqrt{-7}}$

43. $\dfrac{\sqrt{-24}}{\sqrt{8}}$ **44.** $\dfrac{\sqrt{-54}}{\sqrt{27}}$ **45.** $\dfrac{\sqrt{-10}}{\sqrt{-40}}$

46. $\dfrac{\sqrt{-8}}{\sqrt{-72}}$ **47.** $\dfrac{\sqrt{-6} \cdot \sqrt{-2}}{\sqrt{3}}$ **48.** $\dfrac{\sqrt{-12} \cdot \sqrt{-6}}{\sqrt{8}}$

Write each number in standard form a + bi. ***See Example 5.***

49. $\dfrac{-6 - \sqrt{-24}}{2}$ **50.** $\dfrac{-9 - \sqrt{-18}}{3}$ **51.** $\dfrac{10 + \sqrt{-200}}{5}$

52. $\dfrac{20 + \sqrt{-8}}{2}$ **53.** $\dfrac{-3 + \sqrt{-18}}{24}$ **54.** $\dfrac{-5 + \sqrt{-50}}{10}$

Find each sum or difference. Write the answer in standard form. ***See Example 6.***

55. $(3 + 2i) + (9 - 3i)$ **56.** $(4 - i) + (8 + 5i)$

57. $(-2 + 4i) - (-4 + 4i)$ **58.** $(-3 + 2i) - (-4 + 2i)$

59. $(2 - 5i) - (3 + 4i) - (-1 - 9i)$ **60.** $(-4 - i) - (2 + 3i) + (6 + 4i)$

61. $-i\sqrt{2} - 2 - (6 - 4i\sqrt{2}) - (5 - i\sqrt{2})$

62. $3\sqrt{7} - (4\sqrt{7} - i) - 4i + (-2\sqrt{7} + 5i)$

Find each product. Write the answer in standard form. **See Example 7.**

63. $(2+i)(3-2i)$ **64.** $(-2+3i)(4-2i)$ **65.** $(2+4i)(-1+3i)$

66. $(1+3i)(2-5i)$ **67.** $(3-2i)^2$ **68.** $(2+i)^2$

69. $(3+i)(3-i)$ **70.** $(5+i)(5-i)$ **71.** $(-2-3i)(-2+3i)$

72. $(6-4i)(6+4i)$ **73.** $(\sqrt{6}+i)(\sqrt{6}-i)$ **74.** $(\sqrt{2}-4i)(\sqrt{2}+4i)$

75. $i(3-4i)(3+4i)$ **76.** $i(2+7i)(2-7i)$ **77.** $3i(2-i)^2$

78. $-5i(4-3i)^2$ **79.** $(2+i)(2-i)(4+3i)$ **80.** $(3-i)(3+i)(2-6i)$

Find each quotient. Write the answer in standard form a + bi. **See Example 8.**

81. $\frac{6+2i}{1+2i}$ **82.** $\frac{14+5i}{3+2i}$ **83.** $\frac{2-i}{2+i}$

84. $\frac{4-3i}{4+3i}$ **85.** $\frac{1-3i}{1+i}$ **86.** $\frac{-3+4i}{2-i}$

87. $\frac{-5}{i}$ **88.** $\frac{-6}{i}$ **89.** $\frac{8}{-i}$

90. $\frac{12}{-i}$ **91.** $\frac{2}{3i}$ **92.** $\frac{5}{9i}$

Simplify each power of i. **See Example 9.**

93. i^{25} **94.** i^{29} **95.** i^{22}

96. i^{26} **97.** i^{23} **98.** i^{27}

99. i^{32} **100.** i^{40} **101.** i^{-13}

102. i^{-14} **103.** $\frac{1}{i^{-11}}$ **104.** $\frac{1}{i^{-12}}$

105. Suppose that your friend, Kathy Strautz, tells you that she has discovered a method of simplifying a positive power of i. "Just divide the exponent by 2. Your answer is then the simplified form of i^2 raised to the quotient times i raised to the remainder." Explain why her method works.

106. Explain why the following method of simplifying i^{-42} works.

$$i^{-42}=\frac{1}{i^{42}}=\frac{1}{(i^2)^{21}}=\frac{1}{(-1)^{21}}=\frac{1}{-1}=-1$$

107. Show that $\frac{\sqrt{2}}{2}+\frac{\sqrt{2}}{2}i$ is a square root of i.

108. Show that $\frac{\sqrt{3}}{2}+\frac{1}{2}i$ is a cube root of i.

109. Show that $-2+i$ is a solution of the equation $x^2+4x+5=0$.

110. Show that $-3+4i$ is a solution of the equation $x^2+6x+25=0$.

(Modeling) Alternating Current *Complex numbers are used to describe current, I, voltage, E, and impedance, Z (the opposition to current). These three quantities are related by the equation E = IZ. Thus, if any two of these quantities are known, the third can be found. In each exercise, solve the equation E = IZ for the missing variable.*

111. $I=8+6i,\quad Z=6+3i$ **112.** $I=10+6i,\quad Z=8+5i$

113. $I=7+5i,\quad E=28+54i$ **114.** $E=35+55i,\quad Z=6+4i$

(Modeling) Impedance ***Impedance*** *is a measure of the opposition to the flow of alternating electrical current found in common electrical outlets. It consists of two parts,* ***resistance*** *and* ***reactance.*** *Resistance occurs when a light bulb is turned on, while reactance is produced when electricity passes through a coil of wire like that found in electric motors. Impedance Z in* ohms (Ω) *can be expressed as a complex number, where the real part represents resistance and the imaginary part represents reactance.*

For example, if the resistive part is 3 ohms *and the reactive part is* 4 ohms, *then the impedance could be described by the complex number* $Z = 3 + 4i$. *In the series circuit shown in the figure, the total impedance will be the sum of the individual impedances.* (*Source:* Wilcox, G. and C. Hesselberth, *Electricity for Engineering Technology,* Allyn & Bacon.)

115. The circuit contains two light bulbs and two electric motors. Assuming that the light bulbs are pure resistive and the motors are pure reactive, find the total impedance in this circuit and express it in the form $Z = a + bi$.

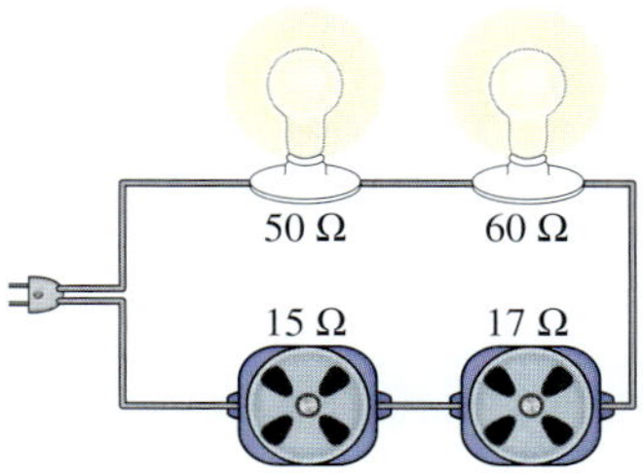

116. The phase angle θ measures the phase difference between the voltage and the current in an electrical circuit. θ (in degrees) can be determined by the equation $\tan \theta = \frac{b}{a}$. Find θ for this circuit.

8.2 Trigonometric (Polar) Form of Complex Numbers

- The Complex Plane and Vector Representation
- Trigonometric (Polar) Form
- Converting between Rectangular and Trigonometric (Polar) Forms
- An Application of Complex Numbers to Fractals

The Complex Plane and Vector Representation Unlike real numbers, complex numbers cannot be ordered. One way to organize and illustrate them is by using a graph.

To graph a complex number such as $2 - 3i$, we modify the familiar coordinate system by calling the horizontal axis the **real axis** and the vertical axis the **imaginary axis.** Then complex numbers can be graphed in this **complex plane,** as shown in **Figure 3.** ***Each complex number*** $a + bi$ ***determines a unique position vector with initial point*** $(0, 0)$ ***and terminal point*** (a, b).

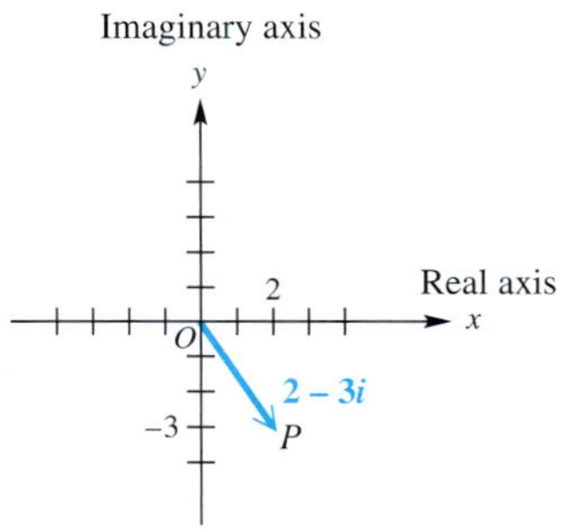

Figure 3

NOTE This geometric representation is the reason that $a + bi$ is called the **rectangular form** of a complex number. (*Rectangular form* is also called *standard form.*)

Recall that the sum of the two complex numbers $4 + i$ and $1 + 3i$ is

$$(4 + i) + (1 + 3i) = 5 + 4i. \quad \text{(Section 8.1)}$$

Graphically, the sum of two complex numbers is represented by the vector that is the resultant of the vectors corresponding to the two numbers, as shown in **Figure 4.**

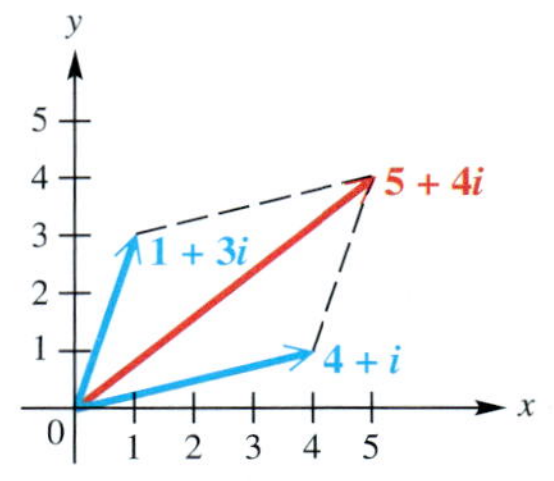

Figure 4

EXAMPLE 1 Expressing the Sum of Complex Numbers Graphically

Find the sum of $6 - 2i$ and $-4 - 3i$. Graph both complex numbers and their resultant.

SOLUTION The sum is found by adding the two numbers.

$$(6 - 2i) + (-4 - 3i) = 2 - 5i \quad \text{Add real parts, and add imaginary parts.}$$

The graphs are shown in **Figure 5.**

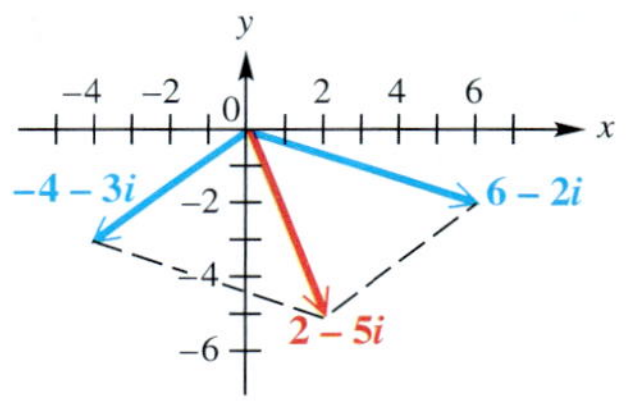

Figure 5

✔ *Now Try Exercise 13.*

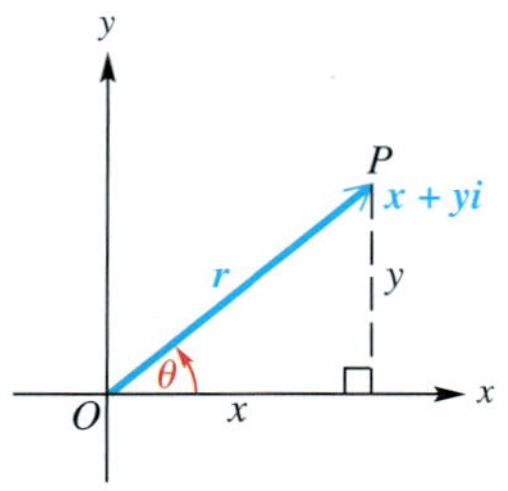

Figure 6

Trigonometric (Polar) Form **Figure 6** shows the complex number $x + yi$ that corresponds to a vector **OP** with direction angle θ and magnitude r. The following relationships among x, y, r, and θ can be verified from **Figure 6.**

Relationships among x, y, r, and θ

$$x = r\cos\theta \qquad y = r\sin\theta$$

$$r = \sqrt{x^2 + y^2} \qquad \tan\theta = \frac{y}{x}, \quad \text{if } x \neq 0$$

Substituting $x = r\cos\theta$ and $y = r\sin\theta$ into $x + yi$ gives the following.

$$\begin{aligned} x + yi &= r\cos\theta + (r\sin\theta)i && \text{Substitute.} \\ &= r(\cos\theta + i\sin\theta) && \text{Factor out } r. \end{aligned}$$

Trigonometric (Polar) Form of a Complex Number

The expression

$$r(\cos\theta + i\sin\theta)$$

is the **trigonometric form** (or **polar form**) of the complex number $x + yi$. The expression $\cos\theta + i\sin\theta$ is sometimes abbreviated **cis** $\boldsymbol{\theta}$**.** Using this notation, $\boldsymbol{r(\cos\theta + i\sin\theta)}$ is written $\boldsymbol{r}$ **cis** $\boldsymbol{\theta}$**.**

The number r is the **absolute value** (or **modulus**) of $x + yi$, and θ is the **argument** of $x + yi$. In this section, we choose the value of θ in the interval $[0°, 360°)$. However, any angle coterminal with θ also could serve as the argument.

EXAMPLE 2 Converting from Trigonometric Form to Rectangular Form

Express $2(\cos 300° + i\sin 300°)$ in rectangular form.

ALGEBRAIC SOLUTION

$2(\cos 300° + i\sin 300°)$

$= 2\left(\frac{1}{2} - i\frac{\sqrt{3}}{2}\right)$ $\quad\cos 300° = \frac{1}{2}$; $\sin 300° = -\frac{\sqrt{3}}{2}$ (Section 2.2)

$= 1 - i\sqrt{3}$ $\quad$ Distributive property

Note that the real part is positive and the imaginary part is negative. This is consistent with 300° being a quadrant IV angle.

GRAPHING CALCULATOR SOLUTION

We use a calculator in degree mode to confirm the algebraic solution. See **Figure 7.**

The imaginary part is an approximation for $-\sqrt{3}$.

Figure 7

✔ *Now Try Exercise 29.*

Converting between Rectangular and Trigonometric (Polar) Forms To convert from rectangular form to trigonometric form, we use the following procedure.

Converting from Rectangular to Trigonometric Form

Step 1 Sketch a graph of the number $x + yi$ in the complex plane.

Step 2 Find r by using the equation $r = \sqrt{x^2 + y^2}$.

Step 3 Find θ by using the equation $\tan\theta = \frac{y}{x}$, where $x \neq 0$, choosing the quadrant indicated in Step 1.

CAUTION Errors often occur in Step 3. ***Be sure to choose the correct quadrant for*** $\boldsymbol{\theta}$ ***by referring to the graph sketched in Step 1.***

EXAMPLE 3 Converting from Rectangular to Trigonometric Form

Write each complex number in trigonometric form.

(a) $-\sqrt{3}+i$ (Use radian measure.) **(b)** $-3i$ (Use degree measure.)

SOLUTION

(a) We start by sketching the graph of $-\sqrt{3}+i$ in the complex plane, as shown in **Figure 8.** Next, we use $x=-\sqrt{3}$ and $y=1$ to find r and θ.

$$r=\sqrt{x^2+y^2}=\sqrt{(-\sqrt{3})^2+1^2}=\sqrt{3+1}=2$$

and $$\tan\theta=\frac{y}{x}=\frac{1}{-\sqrt{3}}=-\frac{1}{\sqrt{3}}\cdot\frac{\sqrt{3}}{\sqrt{3}}=-\frac{\sqrt{3}}{3}$$

Rationalize the denominator.

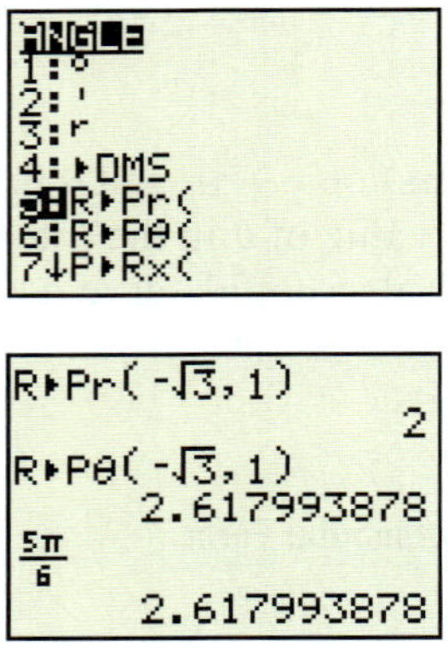

Choices 5 and 6 in the top screen show how to convert from rectangular (x, y) form to trigonometric form. The calculator is in radian mode. The results agree with our algebraic results in **Example 3(a).**

Since $\tan\theta=-\frac{\sqrt{3}}{3}$, the reference angle for θ in radians is $\frac{\pi}{6}$. From the graph, we see that θ is in quadrant II, so $\theta=\pi-\frac{\pi}{6}=\frac{5\pi}{6}$. Therefore,

Be sure to choose the correct quadrant.

$$-\sqrt{3}+i=2\left(\cos\frac{5\pi}{6}+i\sin\frac{5\pi}{6}\right),\quad\text{or}\quad 2\text{ cis }\frac{5\pi}{6}.$$

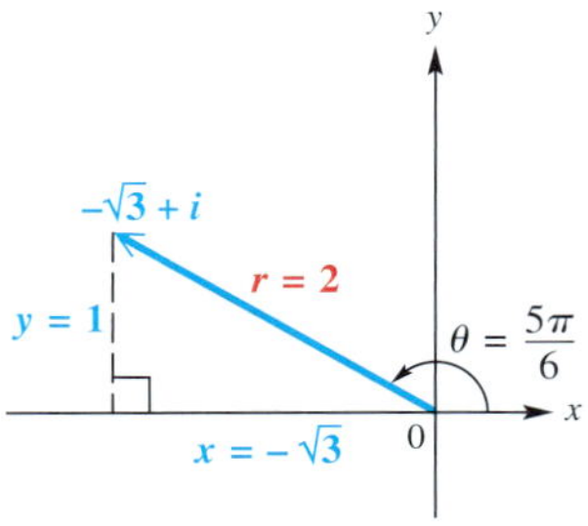

Figure 8

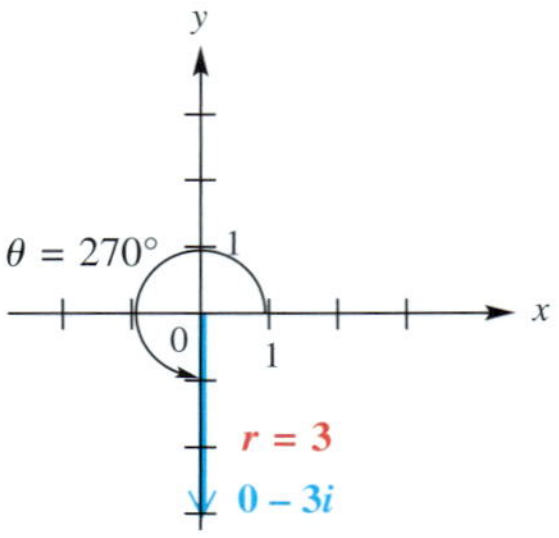

Figure 9

(b) See **Figure 9.** Since $-3i=0-3i$, we have $x=0$ and $y=-3$.

$$r=\sqrt{0^2+(-3)^2}=\sqrt{0+9}=\sqrt{9}=3 \quad \text{Substitute.}$$

We cannot find θ by using $\tan\theta=\frac{y}{x}$, because $x=0$. However, the graph suggests that the value for θ is 270°.

$$-3i=3(\cos 270^\circ+i\sin 270^\circ),\quad\text{or}\quad 3\text{ cis }270^\circ \quad \text{Trigonometric form}$$

✔ ***Now Try Exercises 41 and 47.***

EXAMPLE 4 Converting between Trigonometric and Rectangular Forms Using Calculator Approximations

Write each complex number in its alternative form, using calculator approximations as necessary.

(a) $6(\cos 115^\circ+i\sin 115^\circ)$ **(b)** $5-4i$

SOLUTION

(a) Since 115° does not have a special angle as a reference angle, we cannot find exact values for cos 115° and sin 115°. Use a calculator set in degree mode.

$$\begin{aligned}&6(\cos 115^\circ+i\sin 115^\circ)\\&\approx 6(-0.4226182617+0.906307787i) && \text{Use a calculator.}\\&\approx -2.5357+5.4378i && \text{Four decimal places}\end{aligned}$$

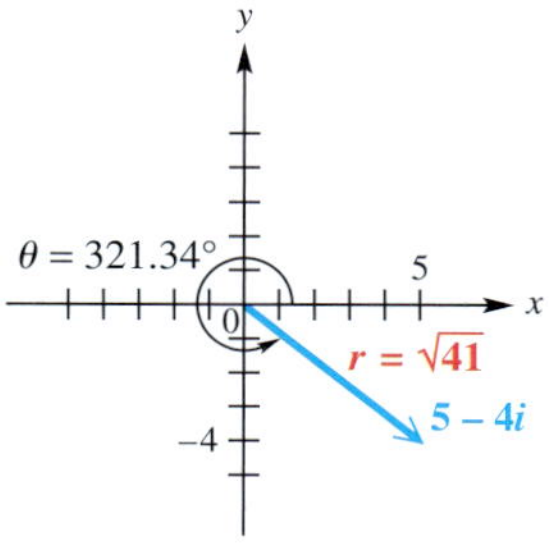

Figure 10

(b) A sketch of $5 - 4i$ shows that θ must be in quadrant IV. See **Figure 10.**

$$r = \sqrt{5^2 + (-4)^2} = \sqrt{41} \quad \text{and} \quad \tan \theta = -\frac{4}{5}$$

Use a calculator to find that one measure of θ is $-38.66°$. In order to express θ in the interval $[0, 360°)$, we find $\theta = 360° - 38.66° = 321.34°$.

$$5 - 4i = \sqrt{41} \text{ cis } 321.34°$$

✔ *Now Try Exercises 53 and 57.*

An Application of Complex Numbers to Fractals At its basic level, a **fractal** is a unique, enchanting geometric figure with an endless self-similarity property. A fractal image repeats itself infinitely with ever-decreasing dimensions. If we look at smaller and smaller portions of a fractal image, we will continue to see the whole—it is much like looking into two parallel mirrors that are facing each other.

EXAMPLE 5 Deciding Whether a Complex Number Is in the Julia Set

The fractal called the **Julia set** is shown in **Figure 11.** To determine whether a complex number $z = a + bi$ is in this Julia set, perform the following sequence of calculations.

$$z^2 - 1, \quad (z^2 - 1)^2 - 1, \quad [(z^2 - 1)^2 - 1]^2 - 1, \quad \ldots$$

If the absolute values of any of the resulting complex numbers exceed 2, then the complex number z is not in the Julia set. Otherwise z is part of this set and the point (a, b) should be shaded in the graph.

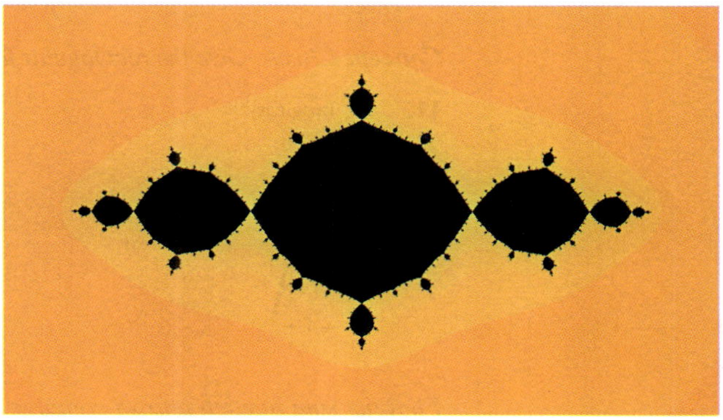

Figure 11

Determine whether each number belongs to the Julia set.

(a) $z = 0 + 0i$ **(b)** $z = 1 + 1i$

SOLUTION

(a) Here

$$z = 0 + 0i = 0,$$
$$z^2 - 1 = 0^2 - 1 = -1,$$
$$(z^2 - 1)^2 - 1 = (-1)^2 - 1 = 0,$$
$$[(z^2 - 1)^2 - 1]^2 - 1 = 0^2 - 1 = -1, \quad \text{and so on.}$$

We see that the calculations repeat as $0, -1, 0, -1$, and so on. The absolute values are either 0 or 1, which do not exceed 2, so $0 + 0i$ is in the Julia set and the point $(0, 0)$ is part of the graph.

(b) For $z = 1 + 1i$, we have the following.

$$z^2 - 1 = (1 + i)^2 - 1 \quad \text{Substitute for } z\text{; } 1 + 1i = 1 + i.$$

$$= (1 + 2i + i^2) - 1 \quad \text{Square the binomial; } (x+y)^2 = x^2 + 2xy + y^2.$$

$$= -1 + 2i \quad i^2 = -1$$

The absolute value is

$$\sqrt{(-1)^2 + 2^2} = \sqrt{5}.$$

Since $\sqrt{5}$ is greater than 2, the number $1 + 1i$ is not in the Julia set and $(1, 1)$ is not part of the graph.

✔ *Now Try Exercise 63.*

8.2 Exercises

1. *Concept Check* The absolute value (or modulus) of a complex number represents the ______________ of the vector representing it in the complex plane.

2. *Concept Check* What is the geometric interpretation of the argument of a complex number?

Graph each complex number. ***See Example 1.***

3. $-3 + 2i$ **4.** $6 - 5i$ **5.** $\sqrt{2} + \sqrt{2}i$ **6.** $2 - 2i\sqrt{3}$

7. $-4i$ **8.** $3i$ **9.** -8 **10.** 2

Concept Check Give the rectangular form of the complex number shown.

11.

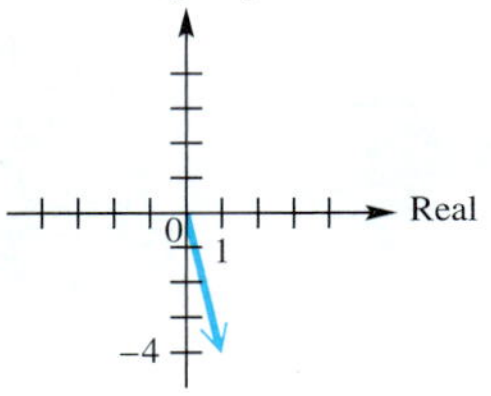

12.

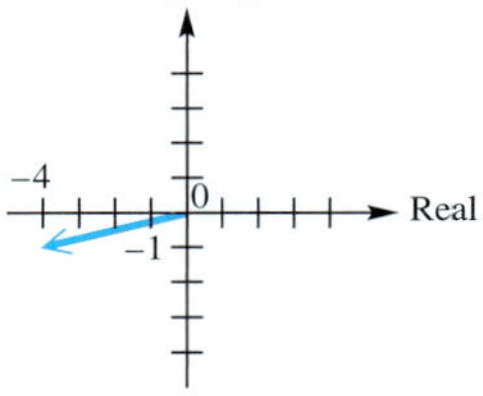

Find the sum of each pair of complex numbers. In Exercises 13–16, graph both complex numbers and their resultant. ***See Example 1.***

13. $4 - 3i, -1 + 2i$ **14.** $2 + 3i, -4 - i$ **15.** $5 - 6i, -5 + 3i$

16. $7 - 3i, -4 + 3i$ **17.** $-3, 3i$ **18.** $6, -2i$

19. $-5 - 8i, -1$ **20.** $4 - 2i, 5$ **21.** $7 + 6i, 3i$

22. $2 + 6i, -2i$ **23.** $\frac{1}{2} + \frac{2}{3}i, \frac{2}{3} + \frac{1}{2}i$ **24.** $-\frac{1}{5} + \frac{2}{7}i, \frac{3}{7} - \frac{3}{4}i$

Write each complex number in rectangular form. ***See Example 2.***

25. $2(\cos 45° + i \sin 45°)$ **26.** $4(\cos 60° + i \sin 60°)$

27. $10(\cos 90° + i \sin 90°)$ **28.** $8(\cos 270° + i \sin 270°)$

29. $4(\cos 240° + i \sin 240°)$ **30.** $2(\cos 330° + i \sin 330°)$

31. $3 \text{ cis } 150°$ **32.** $2 \text{ cis } 30°$

33. $5 \text{ cis } 300°$

34. $6 \text{ cis } 135°$

35. $\sqrt{2} \text{ cis } 225°$

36. $\sqrt{3} \text{ cis } 315°$

37. $4(\cos(-30°) + i \sin(-30°))$

38. $\sqrt{2}(\cos(-60°) + i \sin(-60°))$

Write each complex number in trigonometric form $r(\cos \theta + i \sin \theta)$, with θ in the interval $[0°, 360°)$. ***See Example 3.***

39. $-3 - 3i\sqrt{3}$

40. $1 + i\sqrt{3}$

41. $\sqrt{3} - i$

42. $4\sqrt{3} + 4i$

43. $-5 - 5i$

44. $-2 + 2i$

45. $2 + 2i$

46. $4 + 4i$

47. $5i$

48. $-2i$

49. -4

50. 7

Perform each conversion, using a calculator to approximate answers as necessary. ***See Example 4.***

	Rectangular Form	Trigonometric Form
51.	$2 + 3i$	________
52.	________	$\cos 35° + i \sin 35°$
53.	________	$3(\cos 250° + i \sin 250°)$
54.	$-4 + i$	________
55.	$12i$	________
56.	________	$3 \text{ cis } 180°$
57.	$3 + 5i$	________
58.	________	$\text{cis } 110.5°$

Concept Check The complex number z, where $z = x + yi$, can be graphed in the plane as (x, y). Describe the graphs of all complex numbers z satisfying the conditions in Exercises 59–62.

59. The absolute value of z is 1.

60. The real and imaginary parts of z are equal.

61. The real part of z is 1.

62. The imaginary part of z is 1.

Julia Set Refer to ***Example 5*** *to solve Exercises 63 and 64.*

63. Is $z = -0.2i$ in the Julia set?

64. The graph of the Julia set in **Figure 11** appears to be symmetric with respect to both the x-axis and the y-axis. Complete the following to show that this is true.

(a) Show that complex conjugates have the same absolute value.

(b) Compute $z_1^2 - 1$ and $z_2^2 - 1$, where $z_1 = a + bi$ and $z_2 = a - bi$.

(c) Discuss why if (a, b) is in the Julia set, then so is $(a, -b)$.

(d) Conclude that the graph of the Julia set must be symmetric with respect to the x-axis.

(e) Using a similar argument, show that the Julia set must also be symmetric with respect to the y-axis.

In Exercises 65 and 66, suppose $z = r(\cos \theta + i \sin \theta)$.

65. Use vectors to show that the conjugate of z is

$$r[\cos(360° - \theta) + i \sin(360° - \theta)], \quad \text{or} \quad r(\cos \theta - i \sin \theta).$$

66. Use vectors to show that

$$-z = r[\cos(\theta + \pi) + i \sin(\theta + \pi)].$$

Concept Check *In Exercises 67–69, identify the geometric condition (A, B, or C) that implies the situation.*

A. *The corresponding vectors have opposite directions.*

B. *The terminal points of the vectors corresponding to* $a + bi$ *and* $c + di$ *lie on a horizontal line.*

C. *The corresponding vectors have the same direction.*

67. The difference between two nonreal complex numbers $a + bi$ and $c + di$ is a real number.

68. The absolute value of the sum of two complex numbers $a + bi$ and $c + di$ is equal to the sum of their absolute values.

69. The absolute value of the difference of two complex numbers $a + bi$ and $c + di$ is equal to the sum of their absolute values.

70. Show that z and iz have the same absolute value. How are the graphs of these two numbers related?

8.3 The Product and Quotient Theorems

- Products of Complex Numbers in Trigonometric Form
- Quotients of Complex Numbers in Trigonometric Form

Products of Complex Numbers in Trigonometric Form Using the FOIL method to multiply complex numbers in rectangular form, we find the product of $1 + i\sqrt{3}$ and $-2\sqrt{3} + 2i$ as follows.

$$\left(1 + i\sqrt{3}\right)\left(-2\sqrt{3} + 2i\right)$$

$$= -2\sqrt{3} + 2i - 2i(3) + 2i^2\sqrt{3} \quad \text{FOIL}$$

$$= -2\sqrt{3} + 2i - 6i - 2\sqrt{3} \quad i^2 = -1 \text{ (Section 8.1)}$$

$$= -4\sqrt{3} - 4i \quad \text{Combine like terms.}$$

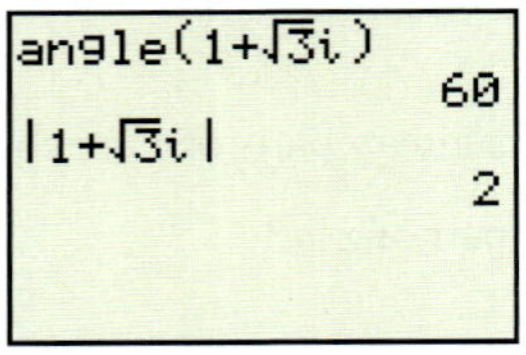

With the TI-83/84 Plus calculator in complex and degree modes, the MATH menu can be used to find the angle and the magnitude (absolute value) of the vector that corresponds to a given complex number.

We can also find this same product by first converting the complex numbers $1 + i\sqrt{3}$ and $-2\sqrt{3} + 2i$ to trigonometric form.

$$1 + i\sqrt{3} = 2(\cos 60^\circ + i \sin 60^\circ)$$

and

$$-2\sqrt{3} + 2i = 4(\cos 150^\circ + i \sin 150^\circ) \quad \text{(Section 8.2)}$$

If we multiply the trigonometric forms and use identities for the cosine and the sine of the sum of two angles, then the result is as follows.

$$[2(\cos 60^\circ + i \sin 60^\circ)][4(\cos 150^\circ + i \sin 150^\circ)]$$

$$= 2 \cdot 4(\cos 60^\circ \cdot \cos 150^\circ + i \sin 60^\circ \cdot \cos 150^\circ + i \cos 60^\circ \cdot \sin 150^\circ + i^2 \sin 60^\circ \cdot \sin 150^\circ) \quad \text{Multiply the absolute values and FOIL.}$$

$$= 8[(\cos 60^\circ \cdot \cos 150^\circ - \sin 60^\circ \cdot \sin 150^\circ) + i(\sin 60^\circ \cdot \cos 150^\circ + \cos 60^\circ \cdot \sin 150^\circ)] \quad i^2 = -1; \text{ Factor out } i.$$

$$= 8[\cos(60^\circ + 150^\circ) + i \sin(60^\circ + 150^\circ)] \quad \text{Use identities for } \cos(A + B) \text{ and } \sin(A + B). \text{ (Section 5.3)}$$

$$= 8(\cos 210^\circ + i \sin 210^\circ) \quad \text{Add.}$$

The absolute value of the product, 8, is equal to the product of the absolute values of the factors, $2 \cdot 4$, and the argument of the product, 210°, is equal to the sum of the arguments of the factors, $60^\circ + 150^\circ$.

The product obtained when multiplying by the first method is the rectangular form of the product obtained when multiplying by the second method.

$$8(\cos 210° + i \sin 210°)$$

$$= 8\left(-\frac{\sqrt{3}}{2} - \frac{1}{2}i\right) \qquad \cos 210° = -\tfrac{\sqrt{3}}{2};\ \sin 210° = -\tfrac{1}{2} \text{ (Section 2.2)}$$

$$= -4\sqrt{3} - 4i \qquad \text{Rectangular form}$$

We can generalize this work in the *product theorem.*

Product Theorem

If $r_1(\cos \theta_1 + i \sin \theta_1)$ and $r_2(\cos \theta_2 + i \sin \theta_2)$ are any two complex numbers, then the following holds.

$$[r_1(\cos \theta_1 + i \sin \theta_1)] \cdot [r_2(\cos \theta_2 + i \sin \theta_2)]$$
$$= r_1 r_2[\cos(\theta_1 + \theta_2) + i \sin(\theta_1 + \theta_2)]$$

In compact form, this is written

$$(r_1 \text{ cis } \theta_1)(r_2 \text{ cis } \theta_2) = r_1 r_2 \text{ cis}(\theta_1 + \theta_2).$$

That is, to multiply complex numbers in trigonometric form, multiply their absolute values and add their arguments.

EXAMPLE 1 Using the Product Theorem

Find the product of $3(\cos 45° + i \sin 45°)$ and $2(\cos 135° + i \sin 135°)$. Write the result in rectangular form.

SOLUTION

$$[3(\cos 45° + i \sin 45°)][2(\cos 135° + i \sin 135°)]$$

$$= 3 \cdot 2[\cos(45° + 135°) + i \sin(45° + 135°)] \qquad \text{Product theorem}$$

$$= 6(\cos 180° + i \sin 180°) \qquad \text{Multiply and add.}$$

$$= 6(-1 + i \cdot 0) \qquad \cos 180° = -1;\ \sin 180° = 0 \text{ (Section 2.2)}$$

$$= 6(-1), \quad \text{or} \quad -6 \qquad \text{Rectangular form}$$

✔ **Now Try Exercise 7.**

Quotients of Complex Numbers in Trigonometric Form The rectangular form of the quotient of $1 + i\sqrt{3}$ and $-2\sqrt{3} + 2i$ is found as follows.

$$\frac{1 + i\sqrt{3}}{-2\sqrt{3} + 2i}$$

$$= \frac{(1 + i\sqrt{3})(-2\sqrt{3} - 2i)}{(-2\sqrt{3} + 2i)(-2\sqrt{3} - 2i)} \qquad \text{Multiply both numerator and denominator by the conjugate of the denominator. (Section 8.1)}$$

$$= \frac{-2\sqrt{3} - 2i - 6i - 2i^2\sqrt{3}}{12 - 4i^2} \qquad \text{FOIL; } (x + y)(x - y) = x^2 - y^2$$

$$= \frac{-8i}{16}, \quad \text{or} \quad -\frac{1}{2}i \qquad \text{Simplify.}$$

Writing $1 + i\sqrt{3}$, $-2\sqrt{3} + 2i$, and $-\frac{1}{2}i$ in trigonometric form gives

$$1 + i\sqrt{3} = 2(\cos 60^\circ + i \sin 60^\circ),$$

$$-2\sqrt{3} + 2i = 4(\cos 150^\circ + i \sin 150^\circ),$$

Use $r = \sqrt{x^2 + y^2}$ and $\tan \theta = \frac{y}{x}$. (Section 8.2)

and
$$-\frac{1}{2}i = \frac{1}{2}[\cos(-90^\circ) + i \sin(-90^\circ)].$$

Here, the absolute value of the quotient, $\frac{1}{2}$, is the quotient of the two absolute values, $\frac{2}{4} = \frac{1}{2}$. The argument of the quotient, -90°, is the difference of the two arguments,

$$60^\circ - 150^\circ = -90^\circ.$$

Generalizing this work leads to the *quotient theorem.*

Quotient Theorem

If $r_1(\cos \theta_1 + i \sin \theta_1)$ and $r_2(\cos \theta_2 + i \sin \theta_2)$ are any two complex numbers, where $r_2(\cos \theta_2 + i \sin \theta_2) \neq 0$, then the following holds.

$$\frac{r_1(\cos \theta_1 + i \sin \theta_1)}{r_2(\cos \theta_2 + i \sin \theta_2)} = \frac{r_1}{r_2}[\cos(\theta_1 - \theta_2) + i \sin(\theta_1 - \theta_2)]$$

In compact form, this is written

$$\frac{r_1 \operatorname{cis} \theta_1}{r_2 \operatorname{cis} \theta_2} = \frac{r_1}{r_2} \operatorname{cis}(\theta_1 - \theta_2).$$

That is, to divide complex numbers in trigonometric form, divide their absolute values and subtract their arguments.

EXAMPLE 2 Using the Quotient Theorem

Find the quotient $\frac{10 \operatorname{cis}(-60^\circ)}{5 \operatorname{cis} 150^\circ}$. Write the result in rectangular form.

SOLUTION

$$\frac{10 \operatorname{cis}(-60^\circ)}{5 \operatorname{cis} 150^\circ}$$

$= \frac{10}{5} \operatorname{cis}(-60^\circ - 150^\circ)$ — Quotient theorem

$= 2 \operatorname{cis}(-210^\circ)$ — Divide and subtract.

$= 2[\cos(-210^\circ) + i \sin(-210^\circ)]$ — Rewrite.

$= 2\left[-\frac{\sqrt{3}}{2} + i\left(\frac{1}{2}\right)\right]$ — $\cos(-210^\circ) = -\frac{\sqrt{3}}{2}$; $\sin(-210^\circ) = \frac{1}{2}$ (Section 2.2)

$= -\sqrt{3} + i$ — Rectangular form

✔ ***Now Try Exercise 17.***

EXAMPLE 3 Using the Product and Quotient Theorems with a Calculator

Use a calculator to find the following. Write the results in rectangular form.

(a) $(9.3 \text{ cis } 125.2°)(2.7 \text{ cis } 49.8°)$ **(b)** $\dfrac{10.42\left(\cos \frac{3\pi}{4} + i \sin \frac{3\pi}{4}\right)}{5.21\left(\cos \frac{\pi}{5} + i \sin \frac{\pi}{5}\right)}$

SOLUTION

(a)

$$(9.3 \text{ cis } 125.2°)(2.7 \text{ cis } 49.8°)$$

$$= 9.3(2.7) \text{ cis}(125.2° + 49.8°) \quad \text{Product theorem}$$

$$= 25.11 \text{ cis } 175°$$

Multiply the absolute values and add the arguments.

$$= 25.11(\cos 175° + i \sin 175°) \quad \text{Equivalent form}$$

$$\approx 25.11[-0.99619470 + i(0.08715574)] \quad \text{Use a calculator.}$$

$$\approx -25.0144 + 2.1885i \quad \text{Rectangular form}$$

(b)

$$\frac{10.42\left(\cos \frac{3\pi}{4} + i \sin \frac{3\pi}{4}\right)}{5.21\left(\cos \frac{\pi}{5} + i \sin \frac{\pi}{5}\right)}$$

$$= \frac{10.42}{5.21}\left[\cos\left(\frac{3\pi}{4} - \frac{\pi}{5}\right) + i \sin\left(\frac{3\pi}{4} - \frac{\pi}{5}\right)\right] \quad \text{Quotient theorem}$$

$$= 2\left(\cos \frac{11\pi}{20} + i \sin \frac{11\pi}{20}\right) \quad \frac{3\pi}{4} = \frac{15\pi}{20};\ \frac{\pi}{5} = \frac{4\pi}{20}$$

Divide the absolute values and subtract the arguments.

$$\approx -0.3129 + 1.9754i \quad \text{Rectangular form}$$

Now Try Exercises 27 and 29.

8.3 Exercises

Concept Check *Fill in the blanks with the correct responses.*

1. When multiplying two complex numbers in trigonometric form, we ________ their absolute values and ________ their arguments.

2. When dividing two complex numbers in trigonometric form, we ________ their absolute values and ________ their arguments.

Find each product and write it in rectangular form. ***See Example 1.***

3. $[3(\cos 60° + i \sin 60°)][2(\cos 90° + i \sin 90°)]$

4. $[4(\cos 30° + i \sin 30°)][5(\cos 120° + i \sin 120°)]$

5. $[4(\cos 60° + i \sin 60°)][6(\cos 330° + i \sin 330°)]$

6. $[8(\cos 300° + i \sin 300°)][5(\cos 120° + i \sin 120°)]$

7. $[2(\cos 135° + i \sin 135°)][2(\cos 225° + i \sin 225°)]$

8. $[8(\cos 210° + i \sin 210°)][2(\cos 330° + i \sin 330°)]$

9. $(\sqrt{3} \text{ cis } 45°)(\sqrt{3} \text{ cis } 225°)$

10. $(\sqrt{6} \text{ cis } 120°)[\sqrt{6} \text{ cis}(-30°)]$

11. $(5 \text{ cis } 90°)(3 \text{ cis } 45°)$

12. $(3 \text{ cis } 300°)(7 \text{ cis } 270°)$

Find each quotient and write it in rectangular form. In Exercises 19–24, first convert the numerator and the denominator to trigonometric form. **See Example 2.**

13. $\dfrac{4(\cos 150° + i \sin 150°)}{2(\cos 120° + i \sin 120°)}$

14. $\dfrac{24(\cos 150° + i \sin 150°)}{2(\cos 30° + i \sin 30°)}$

15. $\dfrac{10(\cos 50° + i \sin 50°)}{5(\cos 230° + i \sin 230°)}$

16. $\dfrac{12(\cos 23° + i \sin 23°)}{6(\cos 293° + i \sin 293°)}$

17. $\dfrac{3 \text{ cis } 305°}{9 \text{ cis } 65°}$

18. $\dfrac{16 \text{ cis } 310°}{8 \text{ cis } 70°}$

19. $\dfrac{8}{\sqrt{3} + i}$

20. $\dfrac{2i}{-1 - i\sqrt{3}}$

21. $\dfrac{-i}{1 + i}$

22. $\dfrac{1}{2 - 2i}$

23. $\dfrac{2\sqrt{6} - 2i\sqrt{2}}{\sqrt{2} - i\sqrt{6}}$

24. $\dfrac{-3\sqrt{2} + 3i\sqrt{6}}{\sqrt{6} + i\sqrt{2}}$

Use a calculator to perform the indicated operations. Give answers in rectangular form, expressing real and imaginary parts to four decimal places. **See Example 3.**

25. $[2.5(\cos 35° + i \sin 35°)][3.0(\cos 50° + i \sin 50°)]$

26. $[4.6(\cos 12° + i \sin 12°)][2.0(\cos 13° + i \sin 13°)]$

27. $(12 \text{ cis } 18.5°)(3 \text{ cis } 12.5°)$

28. $(4 \text{ cis } 19.25°)(7 \text{ cis } 41.75°)$

29. $\dfrac{45\left(\cos \frac{2\pi}{3} + i \sin \frac{2\pi}{3}\right)}{22.5\left(\cos \frac{3\pi}{5} + i \sin \frac{3\pi}{5}\right)}$

30. $\dfrac{30\left(\cos \frac{2\pi}{5} + i \sin \frac{2\pi}{5}\right)}{10\left(\cos \frac{\pi}{7} + i \sin \frac{\pi}{7}\right)}$

31. $\left[2 \text{ cis } \dfrac{5\pi}{9}\right]^2$

32. $\left[24.3 \text{ cis } \dfrac{7\pi}{12}\right]^2$

Relating Concepts

For individual or collaborative investigation *(Exercises 33–39)*

Consider the following complex numbers, and **work Exercises 33–39 in order.**

$$w = -1 + i \quad \text{and} \quad z = -1 - i$$

33. Multiply w and z using their rectangular forms and the FOIL method from **Section 8.1.** Leave the product in rectangular form.

34. Find the trigonometric forms of w and z.

35. Multiply w and z using their trigonometric forms and the method described in this section.

36. Use the result of **Exercise 35** to find the rectangular form of wz. How does this compare to your result in **Exercise 33?**

37. Find the quotient $\frac{w}{z}$ using their rectangular forms and multiplying both the numerator and the denominator by the conjugate of the denominator. Leave the quotient in rectangular form.

38. Use the trigonometric forms of w and z, found in **Exercise 34,** to divide w by z using the method described in this section.

39. Use the result of **Exercise 38** to find the rectangular form of $\frac{w}{z}$. How does this compare to your result in **Exercise 37?**

40. Note that $(r \text{ cis } \theta)^2 = (r \text{ cis } \theta)(r \text{ cis } \theta) = r^2 \text{ cis}(\theta + \theta) = r^2 \text{ cis } 2\theta$. Explain how we can square a complex number in trigonometric form. (In the next section, we will develop this idea more fully.)

41. Without actually performing the operations, state why the following products are the same.

$$[2(\cos 45° + i \sin 45°)] \cdot [5(\cos 90° + i \sin 90°)]$$

and $$[2[\cos(-315°) + i \sin(-315°)]] \cdot [5[\cos(-270°) + i \sin(-270°)]]$$

42. Show that $\frac{1}{z} = \frac{1}{r}(\cos \theta - i \sin \theta)$, where $z = r(\cos \theta + i \sin \theta)$.

(Modeling) Solve each problem.

43. ***Electrical Current*** The alternating current in an electric inductor is $I = \frac{E}{Z}$ amperes, where E is voltage and $Z = R + X_L i$ is impedance. If $E = 8(\cos 20° + i \sin 20°)$, $R = 6$, and $X_L = 3$, find the current. Give the answer in rectangular form, with real and imaginary parts to the nearest hundredth.

44. ***Electrical Current*** The current I in a circuit with voltage E, resistance R, capacitive reactance X_c, and inductive reactance X_L is

$$I = \frac{E}{R + (X_L - X_c)i}.$$

Find I if $E = 12(\cos 25° + i \sin 25°)$, $R = 3$, $X_L = 4$, and $X_c = 6$. Give the answer in rectangular form, with real and imaginary parts to the nearest tenth.

(Modeling) Impedance *In the parallel electrical circuit shown in the figure, the impedance Z can be calculated using the equation*

$$Z = \frac{1}{\frac{1}{Z_1} + \frac{1}{Z_2}},$$

where Z_1 and Z_2 are the impedances for the branches of the circuit.

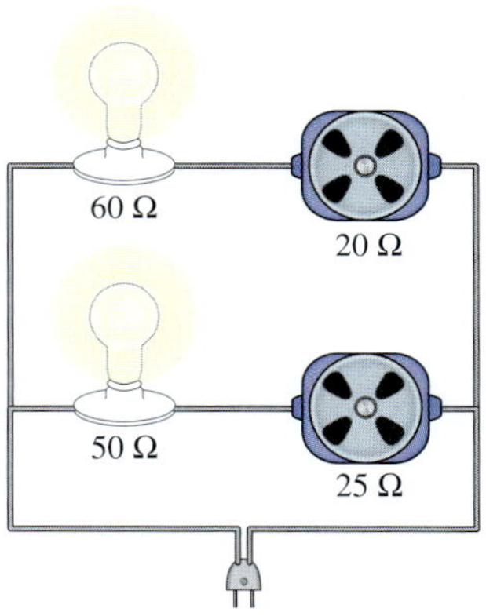

45. If $Z_1 = 50 + 25i$ and $Z_2 = 60 + 20i$, calculate Z.

46. Determine the angle θ for the value of Z found in **Exercise 45.**

8.4 De Moivre's Theorem; Powers and Roots of Complex Numbers

- Powers of Complex Numbers (De Moivre's Theorem)
- Roots of Complex Numbers

Powers of Complex Numbers (De Moivre's Theorem) Because raising a number to a positive integer power is a repeated application of the product rule, it would seem likely that a theorem for finding powers of complex numbers exists. Consider the following.

$$[r(\cos \theta + i \sin \theta)]^2$$

$$= [r(\cos \theta + i \sin \theta)][r(\cos \theta + i \sin \theta)] \qquad a^2 = a \cdot a$$

$$= r \cdot r[\cos(\theta + \theta) + i \sin(\theta + \theta)] \qquad \text{Product theorem (Section 8.3)}$$

$$= r^2(\cos 2\theta + i \sin 2\theta) \qquad \text{Multiply and add.}$$

Abraham De Moivre (1667–1754)

Named after this French expatriate friend of Isaac Newton, De Moivre's theorem relates complex numbers and trigonometry.

In the same way,

$$[r(\cos\theta + i\sin\theta)]^3 \quad \text{is equivalent to} \quad r^3(\cos 3\theta + i\sin 3\theta).$$

These results suggest the following theorem for positive integer values of n. Although the theorem is stated and can be proved for all n, we use it only for positive integer values of n and their reciprocals.

De Moivre's Theorem

If $r(\cos\theta + i\sin\theta)$ is a complex number, and if n is any real number, then the following holds.

$$[r(\cos\theta + i\sin\theta)]^n = r^n(\cos n\theta + i\sin n\theta)$$

In compact form, this is written

$$[r \text{ cis } \theta]^n = r^n(\text{cis } n\theta).$$

EXAMPLE 1 Finding a Power of a Complex Number

Find $(1 + i\sqrt{3})^8$ and express the result in rectangular form.

SOLUTION First write $1 + i\sqrt{3}$ in trigonometric form as

$$2(\cos 60° + i\sin 60°). \quad \text{(Section 8.2)}$$

Now, apply De Moivre's theorem.

$(1 + i\sqrt{3})^8$

$= [2(\cos 60° + i\sin 60°)]^8$

$= 2^8[\cos(8 \cdot 60°) + i\sin(8 \cdot 60°)]$ De Moivre's theorem

$= 256(\cos 480° + i\sin 480°)$ Apply the exponent and multiply.

$= 256(\cos 120° + i\sin 120°)$ 480° and 120° are coterminal. (Section 1.1)

$= 256\left(-\frac{1}{2} + i\frac{\sqrt{3}}{2}\right)$ $\cos 120° = -\frac{1}{2}$; $\sin 120° = \frac{\sqrt{3}}{2}$ (Section 2.2)

$= -128 + 128i\sqrt{3}$ Rectangular form

✔ *Now Try Exercise 7.*

Roots of Complex Numbers Every nonzero complex number has exactly n distinct complex nth roots. De Moivre's theorem can be extended to find all nth roots of a complex number.

*n*th Root

For a positive integer n, the complex number $a + bi$ is an ***n*th root** of the complex number $x + yi$ if

$$(a + bi)^n = x + yi.$$

To find the three complex cube roots of $8(\cos 135° + i \sin 135°)$, for example, look for a complex number, say $r(\cos \alpha + i \sin \alpha)$, that will satisfy

$$[r(\cos \alpha + i \sin \alpha)]^3 = 8(\cos 135° + i \sin 135°).$$

By De Moivre's theorem, this equation becomes

$$r^3(\cos 3\alpha + i \sin 3\alpha) = 8(\cos 135° + i \sin 135°).$$

Set $r^3 = 8$ and $\cos 3\alpha + i \sin 3\alpha = \cos 135° + i \sin 135°$, to satisfy this equation. The first of these conditions implies that $r = 2$, and the second implies that

$$\cos 3\alpha = \cos 135° \quad \text{and} \quad \sin 3\alpha = \sin 135°.$$

For these equations to be satisfied, 3α must represent an angle that is coterminal with 135°. Therefore, we must have

$$3\alpha = 135° + 360° \cdot k, \quad k \text{ any integer}$$

or

$$\alpha = \frac{135° + 360° \cdot k}{3}, \quad k \text{ any integer.}$$

Now, let k take on the integer values 0, 1, and 2.

$$\text{If } k = 0, \text{ then} \quad \alpha = \frac{135° + 360° \cdot 0}{3} = 45°.$$

$$\text{If } k = 1, \text{ then} \quad \alpha = \frac{135° + 360° \cdot 1}{3} = \frac{495°}{3} = 165°.$$

$$\text{If } k = 2, \text{ then} \quad \alpha = \frac{135° + 360° \cdot 2}{3} = \frac{855°}{3} = 285°.$$

In the same way, $\alpha = 405°$ when $k = 3$. But note that $405° = 45° + 360°$, so $\sin 405° = \sin 45°$ and $\cos 405° = \cos 45°$. Similarly, if $k = 4$, then $\alpha = 525°$, which has the same sine and cosine values as 165°. Continuing with larger values of k would repeat solutions already found. Therefore, all of the cube roots (three of them) can be found by letting $k = 0$, 1, and 2, respectively.

When $k = 0$, the root is $2(\cos 45° + i \sin 45°)$.

When $k = 1$, the root is $2(\cos 165° + i \sin 165°)$.

When $k = 2$, the root is $2(\cos 285° + i \sin 285°)$.

In summary, we see that $2(\cos 45° + i \sin 45°)$, $2(\cos 165° + i \sin 165°)$, and $2(\cos 285° + i \sin 285°)$ are the three cube roots of $8(\cos 135° + i \sin 135°)$.

*n*th Root Theorem

If n is any positive integer, r is a positive real number, and θ is in degrees, then the nonzero complex number $r(\cos \theta + i \sin \theta)$ has exactly n distinct nth roots, given by the following.

$$\sqrt[n]{r}(\cos \alpha + i \sin \alpha) \quad \text{or} \quad \sqrt[n]{r} \operatorname{cis} \alpha,$$

where

$$\alpha = \frac{\theta + 360° \cdot k}{n}, \quad \text{or} \quad \alpha = \frac{\theta}{n} + \frac{360° \cdot k}{n}, \quad k = 0, 1, 2, \ldots, n - 1$$

If θ is in radians, then

$$\alpha = \frac{\theta + 2\pi k}{n}, \quad \text{or} \quad \alpha = \frac{\theta}{n} + \frac{2\pi k}{n}, \quad k = 0, 1, 2, \ldots, n - 1.$$

EXAMPLE 2 Finding Complex Roots

Find the two square roots of $4i$. Write the roots in rectangular form.

SOLUTION First write $4i$ in trigonometric form.

$$4\left(\cos \frac{\pi}{2} + i \sin \frac{\pi}{2}\right) \quad \text{Trigonometric form}$$

Here $r = 4$ and $\theta = \frac{\pi}{2}$. The square roots have absolute value $\sqrt{4} = 2$ and arguments as follows.

$$\alpha = \frac{\frac{\pi}{2}}{2} + \frac{2\pi k}{2} = \frac{\pi}{4} + \pi k$$

Be careful simplifying here.

Since there are two square roots, let $k = 0$ and 1.

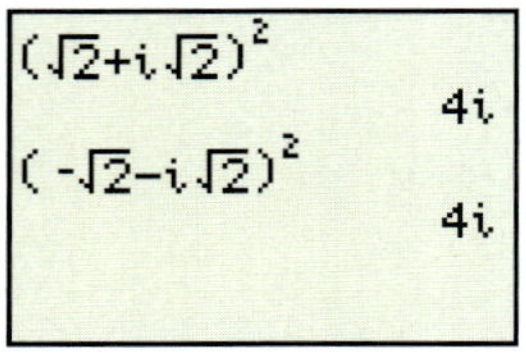

This screen confirms the result of **Example 2.**

$$\text{If } k = 0, \text{ then } \quad \alpha = \frac{\pi}{4} + \pi \cdot 0 = \frac{\pi}{4}.$$

$$\text{If } k = 1, \text{ then } \quad \alpha = \frac{\pi}{4} + \pi \cdot 1 = \frac{5\pi}{4}.$$

Using these values for α, the square roots are $2 \text{ cis } \frac{\pi}{4}$ and $2 \text{ cis } \frac{5\pi}{4}$, which can be written in rectangular form as

$$\sqrt{2} + i\sqrt{2} \quad \text{and} \quad -\sqrt{2} - i\sqrt{2}.$$

✔ *Now Try Exercise 17(a).*

EXAMPLE 3 Finding Complex Roots

Find all fourth roots of $-8 + 8i\sqrt{3}$. Write the roots in rectangular form.

SOLUTION $-8 + 8i\sqrt{3} = 16 \text{ cis } 120°$ Write in trigonometric form.

Here $r = 16$ and $\theta = 120°$. The fourth roots of this number have absolute value $\sqrt[4]{16} = 2$ and arguments as follows.

$$\alpha = \frac{120°}{4} + \frac{360° \cdot k}{4} = 30° + 90° \cdot k$$

Since there are four fourth roots, let $k = 0$, 1, 2, and 3.

$$\text{If } k = 0, \text{ then } \quad \alpha = 30° + 90° \cdot 0 = 30°.$$

$$\text{If } k = 1, \text{ then } \quad \alpha = 30° + 90° \cdot 1 = 120°.$$

$$\text{If } k = 2, \text{ then } \quad \alpha = 30° + 90° \cdot 2 = 210°.$$

$$\text{If } k = 3, \text{ then } \quad \alpha = 30° + 90° \cdot 3 = 300°.$$

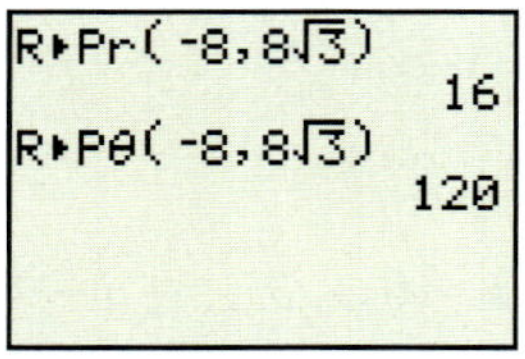

Degree mode

This screen shows how a calculator finds r and θ for the number in **Example 3.**

Using these angles, the fourth roots are

$$2 \text{ cis } 30°, \quad 2 \text{ cis } 120°, \quad 2 \text{ cis } 210°, \quad \text{and} \quad 2 \text{ cis } 300°.$$

These four roots can be written in rectangular form as

$$\sqrt{3} + i, \quad -1 + i\sqrt{3}, \quad -\sqrt{3} - i, \quad \text{and} \quad 1 - i\sqrt{3}.$$

The graphs of these roots lie on a circle with center at the origin and radius 2. See **Figure 12.** The roots are equally spaced about the circle, 90° apart.

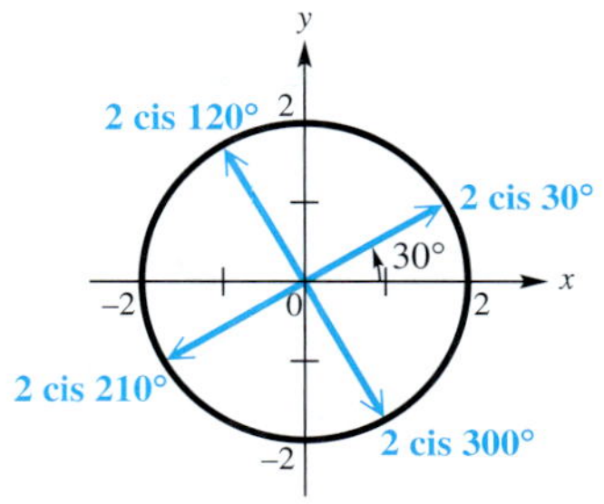

Figure 12

✔ ***Now Try Exercises 23(a) and (b).***

EXAMPLE 4 Solving an Equation (Complex Roots)

Find all complex number solutions of $x^5 - 1 = 0$. Graph them as vectors in the complex plane.

SOLUTION Write the equation as

$$x^5 - 1 = 0, \quad \text{or} \quad x^5 = 1.$$

There is only one real number solution, 1, but there are five complex number solutions. To find these solutions, first write 1 in trigonometric form.

$$1 = 1 + 0i = 1(\cos 0° + i \sin 0°) \quad \text{Trigonometric form}$$

The absolute value of the fifth roots is $\sqrt[5]{1} = 1$. The arguments are given by

$$0° + 72° \cdot k, \quad k = 0, 1, 2, 3, \text{ and } 4.$$

By using these arguments, we find that the fifth roots are as follows.

$$1(\cos 0° + i \sin 0°), \quad k = 0$$

$$1(\cos 72° + i \sin 72°), \quad k = 1$$

$$1(\cos 144° + i \sin 144°), \quad k = 2$$

$$1(\cos 216° + i \sin 216°), \quad k = 3$$

$$1(\cos 288° + i \sin 288°) \quad k = 4$$

The solution set of the equation can be written as

$$\{\text{cis } 0°, \text{cis } 72°, \text{cis } 144°, \text{cis } 216°, \text{cis } 288°\}.$$

The first of these roots equals 1. The others cannot easily be expressed in rectangular form but can be approximated with a calculator.

The tips of the arrows representing the five fifth roots all lie on a unit circle and are equally spaced around it every 72°, as shown in **Figure 13** on the next page.

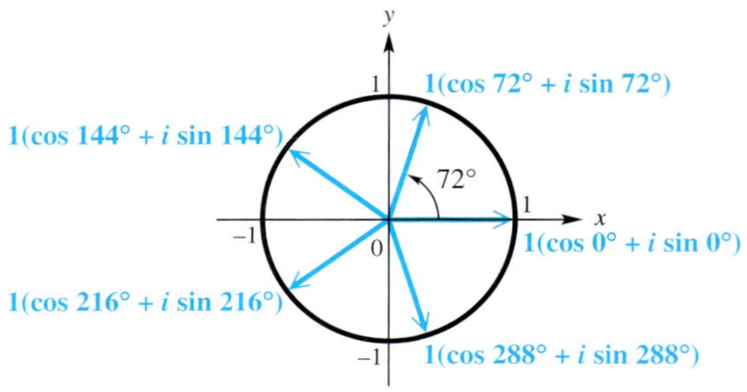

Figure 13

✔ *Now Try Exercise 35.*

8.4 Exercises

Find each power. Write each answer in rectangular form. ***See Example 1.***

1. $[3(\cos 30° + i \sin 30°)]^3$

2. $[2(\cos 135° + i \sin 135°)]^4$

3. $(\cos 45° + i \sin 45°)^8$

4. $[2(\cos 120° + i \sin 120°)]^3$

5. $[3 \text{ cis } 100°]^3$

6. $[3 \text{ cis } 40°]^3$

7. $(\sqrt{3} + i)^5$

8. $(2 - 2i\sqrt{3})^4$

9. $(2\sqrt{2} - 2i\sqrt{2})^6$

10. $\left(\frac{\sqrt{2}}{2} - \frac{\sqrt{2}}{2}i\right)^8$

11. $(-2 - 2i)^5$

12. $(-1 + i)^7$

*In Exercises 13–24, **(a)** find all cube roots of each complex number. Leave answers in trigonometric form. **(b)** Graph each cube root as a vector in the complex plane.* ***See Examples 2 and 3.***

13. $\cos 0° + i \sin 0°$

14. $\cos 90° + i \sin 90°$

15. $8 \text{ cis } 60°$

16. $27 \text{ cis } 300°$

17. $-8i$

18. $27i$

19. -64

20. 27

21. $1 + i\sqrt{3}$

22. $2 - 2i\sqrt{3}$

23. $-2\sqrt{3} + 2i$

24. $\sqrt{3} - i$

Find and graph all specified roots of 1.

25. second (square)

26. fourth

27. sixth

Find and graph all specified roots of i.

28. second (square)

29. third (cube)

30. fourth

Find all complex number solutions of each equation. Leave answers in trigonometric form. ***See Example 4.***

31. $x^3 - 1 = 0$

32. $x^3 + 1 = 0$

33. $x^3 + i = 0$

34. $x^4 + i = 0$

35. $x^3 - 8 = 0$

36. $x^3 + 27 = 0$

37. $x^4 + 1 = 0$

38. $x^4 + 16 = 0$

39. $x^4 - i = 0$

40. $x^5 - i = 0$

41. $x^3 - (4 + 4i\sqrt{3}) = 0$

42. $x^4 - (8 + 8i\sqrt{3}) = 0$

43. Solve the cubic equation

$$x^3 - 1 = 0$$

by factoring the left side as the difference of two cubes and setting each factor equal to 0. Apply the quadratic formula as needed. Then compare your solutions to those of **Exercise 31.**

44. Solve the cubic equation

$$x^3 + 27 = 0$$

by factoring the left side as the sum of two cubes and setting each factor equal to 0. Apply the quadratic formula as needed. Then compare your solutions to those of **Exercise 36.**

Relating Concepts

For individual or collaborative investigation *(Exercises 45–48)*

*In **Chapter 5** we derived identities, or formulas, for* $\cos 2\theta$ *and* $\sin 2\theta$. *These identities can also be derived using De Moivre's theorem.* ***Work Exercises 45–48 in order,*** *to see how this is done.*

45. De Moivre's theorem states that $(\cos\theta + i\sin\theta)^2 =$ ______________.

46. Expand the left side of the equation in **Exercise 45** as a binomial and collect terms to write the left side in the form $a + bi$.

47. Use the result of **Exercise 46** to obtain the double-angle formula for cosine.

48. Repeat **Exercise 47,** but find the double-angle formula for sine.

Solve each problem.

49. *Mandelbrot Set* The fractal known as the **Mandelbrot set** is shown in the figure. To determine if a complex number $z = a + bi$ is in this set, perform the following sequence of calculations. Repeatedly compute

$$z, \quad z^2 + z, \quad (z^2 + z)^2 + z,$$
$$\left[(z^2 + z)^2 + z\right]^2 + z, \ldots .$$

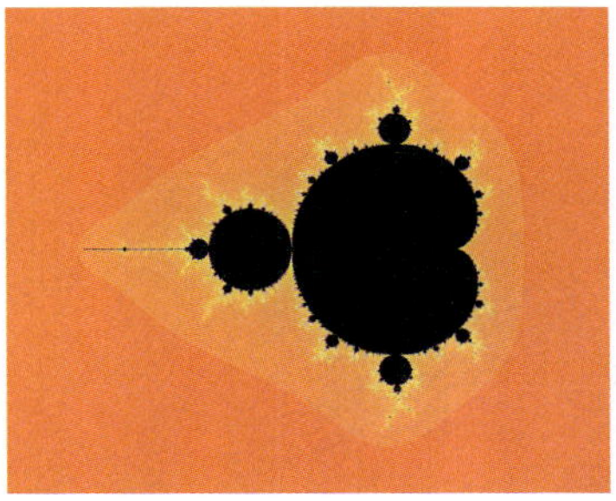

In a manner analogous to the Julia set, the complex number z does not belong to the Mandelbrot set if any of the resulting absolute values exceeds 2. Otherwise z is in the set and the point (a, b) should be shaded in the graph. Determine whether or not the following numbers belong to the Mandelbrot set. (*Source:* Lauwerier, H., *Fractals,* Princeton University Press.)

(a) $z = 0 + 0i$ **(b)** $z = 1 - 1i$ **(c)** $z = -0.5i$

50. *Basins of Attraction* The fractal shown in the figure is the solution to Cayley's problem of determining the basins of attraction for the cube roots of unity. The three cube roots of unity are

$$w_1 = 1, \quad w_2 = -\frac{1}{2} + \frac{\sqrt{3}}{2}i,$$

and $$w_3 = -\frac{1}{2} - \frac{\sqrt{3}}{2}i.$$

This fractal can be generated by repeatedly evaluating the function

$$f(z) = \frac{2z^3 + 1}{3z^2},$$

where z is a complex number. One begins by picking $z_1 = a + bi$ and then successively computing $z_2 = f(z_1)$, $z_3 = f(z_2)$, $z_4 = f(z_3)$, ... If the resulting values of $f(z)$ approach w_1, color the pixel at (a, b) red. If they approach w_2, color it blue, and if they approach w_3, color it yellow. If this process continues for a large number of different z_1, the fractal in the figure will appear. Determine the appropriate color of the pixel for each value of z_1. (*Source:* Crownover, R., *Introduction to Fractals and Chaos,* Jones and Bartlett Publishers.)

(a) $z_1 = i$ **(b)** $z_1 = 2 + i$ **(c)** $z_1 = -1 - i$

51. The screens here illustrate how a pentagon can be graphed using a graphing calculator. Note that a pentagon has five sides, and the T-step is $\frac{360}{5} = 72$. The display at the bottom of the graph screen indicates that one fifth root of 1 is $1 + 0i = 1$. Use this technique to find all fifth roots of 1, and express the real and imaginary parts in decimal form.

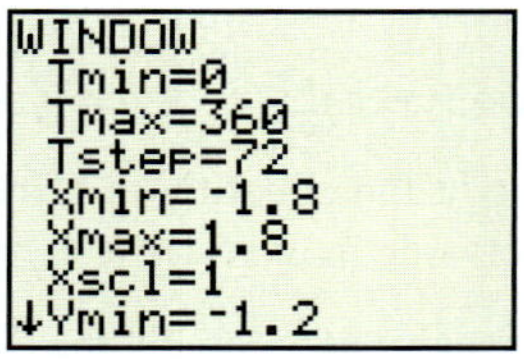

This is a continuation of the previous screen.

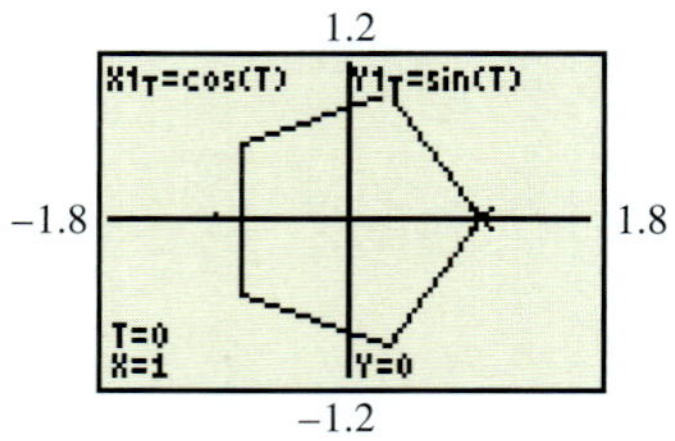

The calculator is in parametric, degree, and connected graph modes.

52. Use the method of **Exercise 51** to find the first three of the ten 10th roots of 1.

53. One of the three cube roots of a complex number is $2 + 2i\sqrt{3}$. Determine the rectangular form of its other two cube roots.

Use a calculator to find all solutions of each equation in rectangular form.

54. $x^2 + 2 - i = 0$

55. $x^2 - 3 + 2i = 0$

56. $x^3 + 4 - 5i = 0$

57. $x^5 + 2 + 3i = 0$

58. *Concept Check* How many complex 64th roots does 1 have? How many are real? How many are not?

59. *Concept Check* *True* or *false*: Every real number must have two distinct real square roots.

60. *Concept Check* *True* or *false*: Some real numbers have three real cube roots.

61. Show that if z is an nth root of 1, then so is $\frac{1}{z}$.

62. Explain why a real number can have only one real cube root.

63. Explain why the n nth roots of 1 are equally spaced around the unit circle.

64. Refer to **Figure 13.** A regular pentagon can be created by joining the tips of the arrows. Explain how you can use this principle to create a regular octagon.

Chapter 8 Quiz (Sections 8.1–8.4)

1. Multiply or divide as indicated. Simplify each answer.

(a) $\sqrt{-24} \cdot \sqrt{-3}$ **(b)** $\dfrac{\sqrt{-8}}{\sqrt{72}}$

2. Write each of the following in rectangular form for the complex numbers

$$w = 3 + 5i \quad \text{and} \quad z = -4 + i.$$

(a) $w + z$ (and give a geometric representation)

(b) $w - z$ **(c)** wz **(d)** $\dfrac{w}{z}$

3. Express each of the following in rectangular form.

(a) $(1 - i)^3$ **(b)** i^{33}

4. Solve $3x^2 - x + 4 = 0$ over the complex number system.

5. Write each complex number in trigonometric (polar) form, where $0° \le \theta < 360°$.

(a) $-4i$ **(b)** $1 - i\sqrt{3}$ **(c)** $-3 - i$

6. Write each complex number in rectangular form.

(a) $4(\cos 60° + i \sin 60°)$ **(b)** $5 \text{ cis } 130°$

(c) $7(\cos 270° + i \sin 270°)$ **(d)** $2 \text{ cis } 0°$

7. Write each of the following in the form specified for the complex numbers

$$w = 12(\cos 80° + i \sin 80°) \quad \text{and} \quad z = 3(\cos 50° + i \sin 50°).$$

(a) wz (trigonometric form) **(b)** $\dfrac{w}{z}$ (rectangular form)

(c) z^3 (rectangular form) **(d)** w^3 (rectangular form)

8. Find the four complex fourth roots of -16. Express them in both trigonometric and rectangular forms.

8.5 Polar Equations and Graphs

- Polar Coordinate System
- Graphs of Polar Equations
- Converting from Polar to Rectangular Equations
- Classifying Polar Equations

Polar Coordinate System

Previously we have used the rectangular coordinate system to graph points and equations. In the rectangular coordinate system, each point in the plane is specified by giving two numbers (x, y). These represent the directed distances from a pair of perpendicular axes, the x-axis and the y-axis.

Now we consider the **polar coordinate system** which is based on a point, called the **pole,** and a ray, called the **polar axis.** The polar axis is usually drawn in the direction of the positive x-axis, as shown in **Figure 14.**

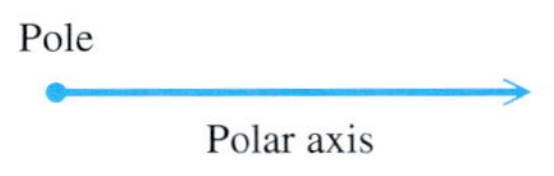

Figure 14

In **Figure 15** the pole has been placed at the origin of a rectangular coordinate system so that the polar axis coincides with the positive x-axis. Point P has rectangular coordinates (x, y). Point P can also be located by giving the directed angle θ from the positive x-axis to ray OP and the *directed* distance r from the pole to point P. The ordered pair (r, θ) gives the **polar coordinates** of point P. If $r > 0$ then point P lies on the terminal side of θ, and if $r < 0$ then point P lies on the ray pointing in the opposite direction of the terminal side of θ, a distance $|r|$ from the pole. **Figure 16** shows rectangular axes superimposed on a polar coordinate grid.

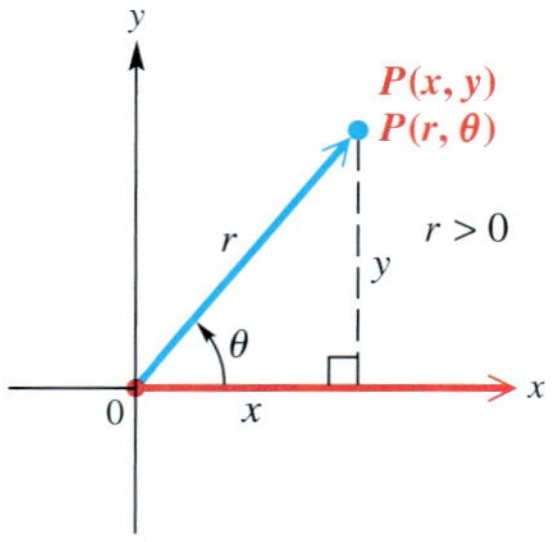

Figure 15

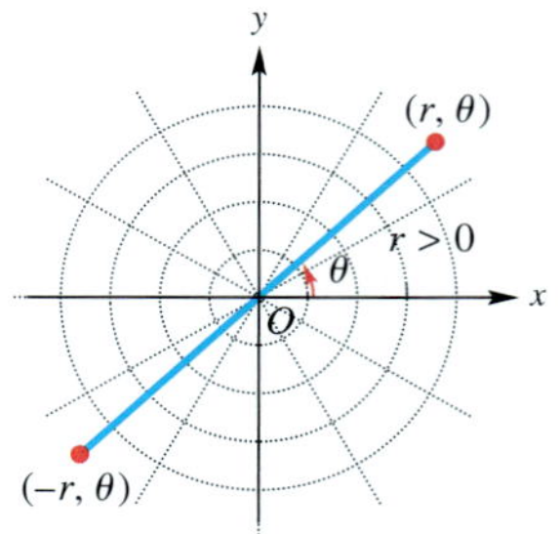

Figure 16

Rectangular and Polar Coordinates

If a point has rectangular coordinates (x, y) and polar coordinates (r, θ), then these coordinates are related as follows.

$$x = r\cos\theta \qquad y = r\sin\theta$$

$$r^2 = x^2 + y^2 \qquad \tan\theta = \frac{y}{x}, \quad \text{if } x \neq 0$$

EXAMPLE 1 Plotting Points with Polar Coordinates

Plot each point by hand in the polar coordinate system. Then determine the rectangular coordinates of each point.

(a) $P(2, 30°)$ **(b)** $Q\left(-4, \frac{2\pi}{3}\right)$ **(c)** $R\left(5, -\frac{\pi}{4}\right)$

SOLUTION

(a) In the point $P(2, 30°)$, $r = 2$ and $\theta = 30°$, so P is located 2 units from the origin in the positive direction on a ray making a 30° angle with the polar axis, as shown in **Figure 17.**

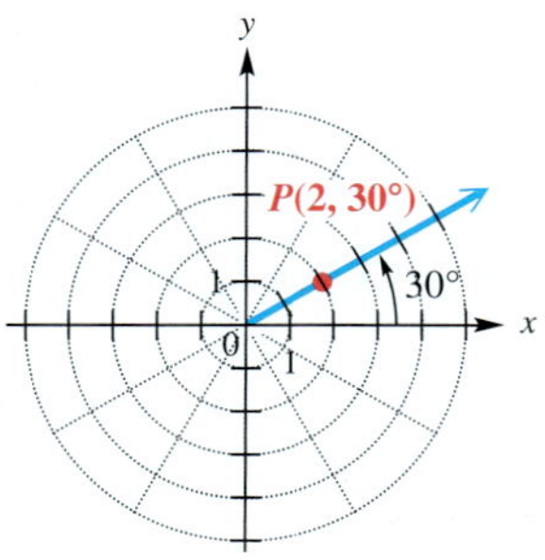

Figure 17

We find the rectangular coordinates as follows.

$x = r\cos\theta$	$y = r\sin\theta$	Conversion equations
$x = 2\cos 30°$	$y = 2\sin 30°$	Substitute.
$x = 2\left(\frac{\sqrt{3}}{2}\right)$	$y = 2\left(\frac{1}{2}\right)$	(Section 2.1)
$x = \sqrt{3}$	$y = 1$	Multiply.

The rectangular coordinates are $(\sqrt{3}, 1)$.

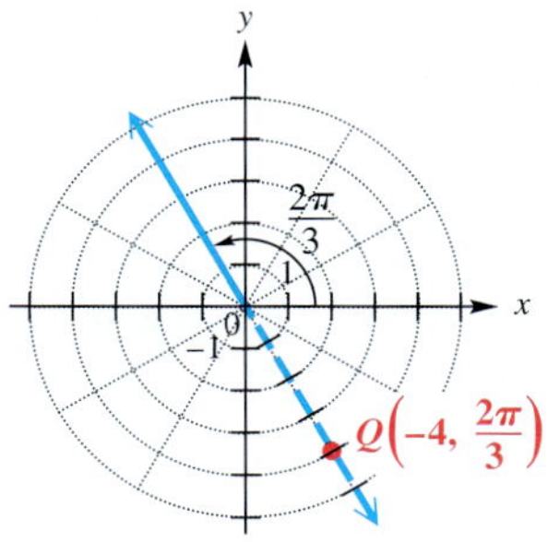

Figure 18

(b) In the point $Q\left(-4, \frac{2\pi}{3}\right)$, r is *negative*, so Q is 4 units in the *opposite* direction from the pole on an extension of the $\frac{2\pi}{3}$ ray. See **Figure 18.** The rectangular coordinates are

$$x = -4 \cos \frac{2\pi}{3} = -4\left(-\frac{1}{2}\right) = 2$$

and
$$y = -4 \sin \frac{2\pi}{3} = -4\left(\frac{\sqrt{3}}{2}\right) = -2\sqrt{3}.$$

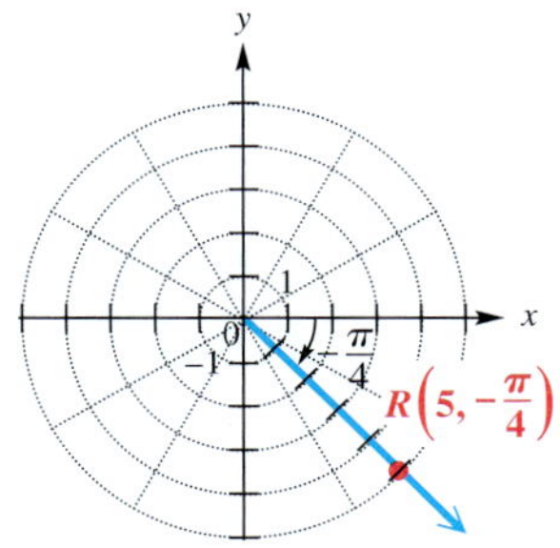

Figure 19

(c) Point $R\left(5, -\frac{\pi}{4}\right)$ is shown in **Figure 19.** Since θ is negative, the angle is measured in the clockwise direction.

$$x = 5 \cos\left(-\frac{\pi}{4}\right) = \frac{5\sqrt{2}}{2} \quad \text{and} \quad y = 5 \sin\left(-\frac{\pi}{4}\right) = -\frac{5\sqrt{2}}{2}$$

✔ *Now Try Exercises 3(a), (c), 5(a), (c), and 11(a), (c).*

While a given point in the plane can have only one pair of rectangular coordinates, this same point can have an infinite number of pairs of polar coordinates. For example, (2, 30°) locates the same point as

$$(2, 390°), \quad (2, -330°), \quad \text{and} \quad (-2, 210°).$$

EXAMPLE 2 Giving Alternative Forms for Coordinates of a Point

(a) Give three other pairs of polar coordinates for the point $P(3, 140°)$.

(b) Determine two pairs of polar coordinates for the point with rectangular coordinates $(-1, 1)$.

SOLUTION

(a) Three pairs that could be used for the point are $(3, -220°)$, $(-3, 320°)$, and $(-3, -40°)$. See **Figure 20.**

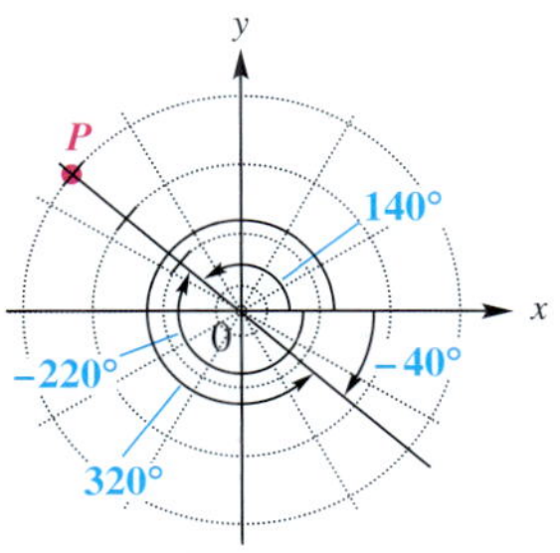

Figure 20

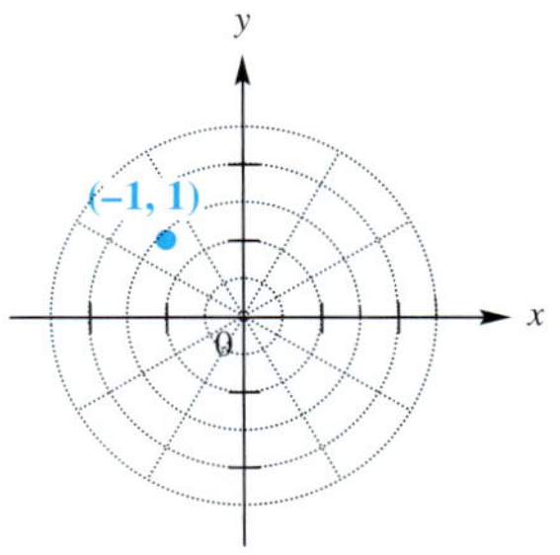

Figure 21

(b) As shown in **Figure 21,** the point $(-1, 1)$ lies in the second quadrant. Since $\tan \theta = \frac{1}{-1} = -1$, one possible value for θ is 135°. Also,

$$r = \sqrt{x^2 + y^2} = \sqrt{(-1)^2 + 1^2} = \sqrt{2}.$$

Two pairs of polar coordinates are $\left(\sqrt{2}, 135°\right)$ and $\left(-\sqrt{2}, 315°\right)$.

✔ *Now Try Exercises 3(b), 5(b), 11(b), and 15.*

LOOKING AHEAD TO CALCULUS

Techniques studied in calculus associated with derivatives and integrals provide methods of finding slopes of tangent lines to polar curves, areas bounded by such curves, and lengths of their arcs.

Graphs of Polar Equations

Equations in x and y are **rectangular** (or **Cartesian**) **equations.** An equation in which r and θ are the variables instead of x and y is a **polar equation.**

$$r = 3 \sin \theta, \quad r = 2 + \cos \theta, \quad r = \theta \qquad \text{Polar equations}$$

Although the rectangular forms of lines and circles are the ones most often encountered, they can also be defined in terms of polar coordinates. The polar equation of the line $ax + by = c$ can be derived as follows.

Line:

$$ax + by = c \qquad \text{Rectangular equation of a line}$$

$$a(r \cos \theta) + b(r \sin \theta) = c \qquad \text{Convert to polar coordinates.}$$

$$r(a \cos \theta + b \sin \theta) = c \qquad \text{Factor out } r.$$

This is the polar equation of $ax + by = c$.

$$\boldsymbol{r = \frac{c}{a \cos \theta + b \sin \theta}} \qquad \text{Polar equation of a line}$$

For the circle $x^2 + y^2 = a^2$, the polar equation can be found in a similar manner.

Circle:

$$x^2 + y^2 = a^2 \qquad \text{Rectangular equation of a circle}$$

$$r^2 = a^2 \qquad x^2 + y^2 = r^2$$

These are polar equations of $x^2 + y^2 = a^2$.

$$\boldsymbol{r = \pm a} \qquad \text{Polar equation of a circle; } r \text{ can be negative in polar coordinates.}$$

We use these forms in the next example.

EXAMPLE 3 Finding Polar Equations of Lines and Circles

For each rectangular equation, give the equivalent polar equation and sketch its graph.

(a) $y = x - 3$ **(b)** $x^2 + y^2 = 4$

SOLUTION

(a) This is the equation of a line.

$$y = x - 3$$

$$x - y = 3 \qquad \text{Write in standard form, } ax + by = c.$$

$$r \cos \theta - r \sin \theta = 3 \qquad \text{Substitute for } x \text{ and } y.$$

$$r(\cos \theta - \sin \theta) = 3 \qquad \text{Factor out } r.$$

$$r = \frac{3}{\cos \theta - \sin \theta} \qquad \text{Divide by } \cos \theta - \sin \theta.$$

A traditional graph is shown in **Figure 22(a),** and a calculator graph is shown in **Figure 22(b).**

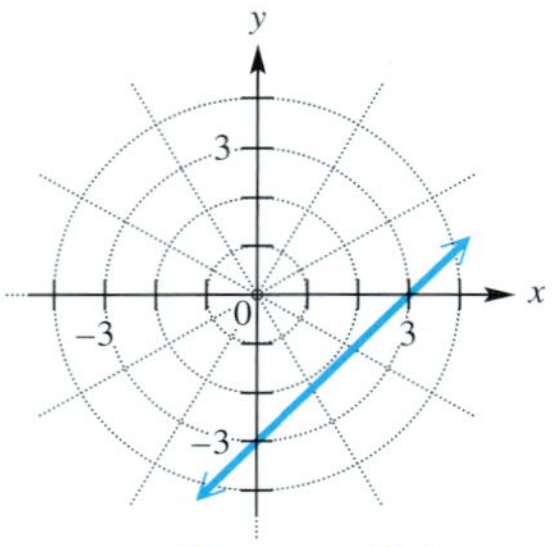

$y = x - 3$ (rectangular)

$r = \dfrac{3}{\cos \theta - \sin \theta}$ (polar)

(a)

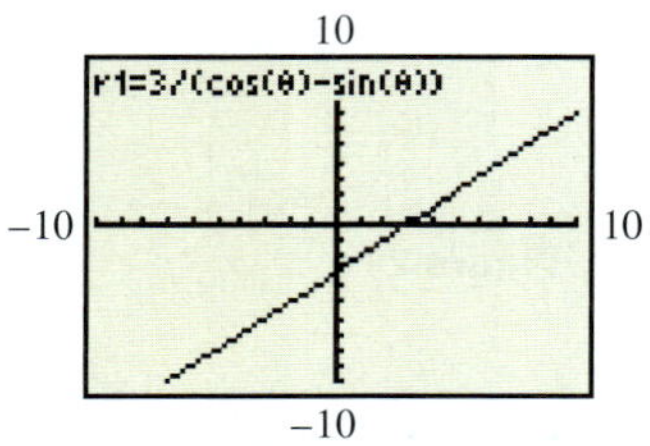

Polar graphing mode

(b)

Figure 22

(b) The graph of $x^2 + y^2 = 4$ is a circle with center at the origin and radius 2.

$$x^2 + y^2 = 4 \qquad \text{(Appendix B)}$$

$$r^2 = 4 \qquad x^2 + y^2 = r^2$$

$$r = 2 \quad \text{or} \quad r = -2$$

In polar coordinates, we may have $r < 0$.

The graphs of $r = 2$ and $r = -2$ coincide. See **Figure 23** on the next page.

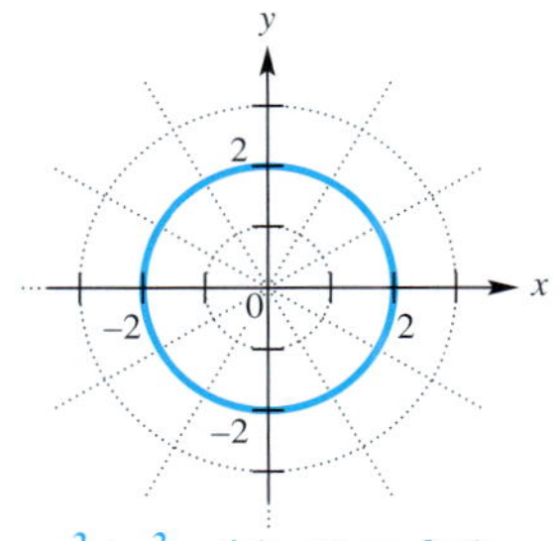

$x^2 + y^2 = 4$ (rectangular)
$r = 2$ (polar)

(a)

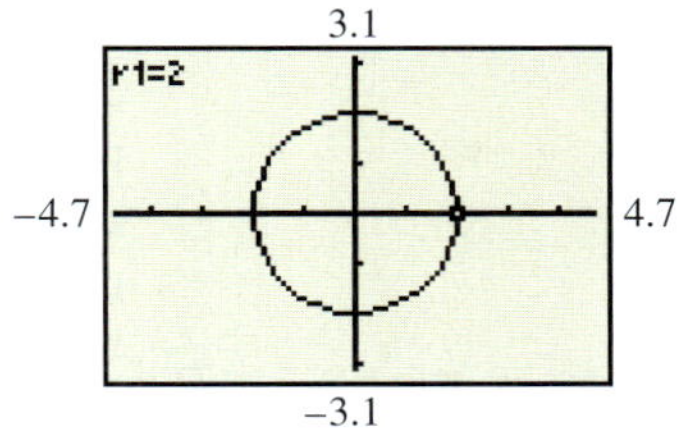

Polar graphing mode

(b)

Figure 23

✔ *Now Try Exercises 27 and 29.*

To graph polar equations, evaluate r for various values of θ until a pattern appears, and then join the points with a smooth curve.

EXAMPLE 4 Graphing a Polar Equation (Cardioid)

Graph $r = 1 + \cos \theta$.

ALGEBRAIC SOLUTION

To graph this equation, find some ordered pairs (as in the table). Once the pattern of values of r becomes clear, it is not necessary to find more ordered pairs. The table includes approximated values for $\cos \theta$ and r.

θ	$\cos \theta$	$r = 1 + \cos \theta$	θ	$\cos \theta$	$r = 1 + \cos \theta$
0°	1	2	135°	−0.7	0.3
30°	0.9	1.9	150°	−0.9	0.1
45°	0.7	1.7	180°	−1	0
60°	0.5	1.5	270°	0	1
90°	0	1	315°	0.7	1.7
120°	−0.5	0.5	330°	0.9	1.9

Connect the points in order—from $(2, 0°)$ to $(1.9, 30°)$ to $(1.7, 45°)$ and so on. See **Figure 24.** This curve is called a **cardioid** because of its heart shape. The curve has been graphed on a **polar grid.**

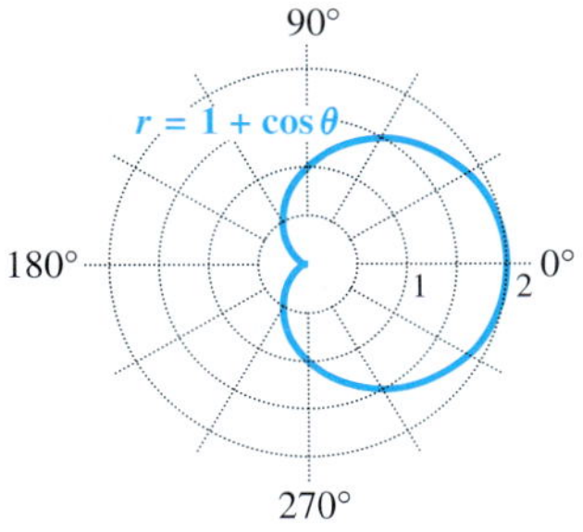

Figure 24

GRAPHING CALCULATOR SOLUTION

We choose degree mode and graph values of θ in the interval $[0°, 360°]$. The screens in **Figure 25(a)** show the choices needed to generate the graph in **Figure 25(b).**

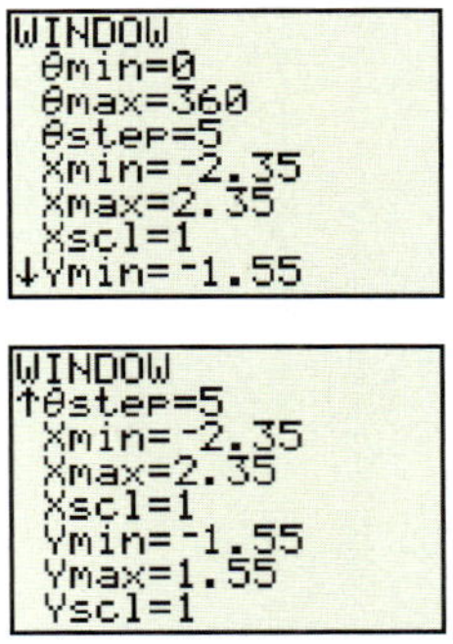

This is a continuation of the previous screen.

(a)

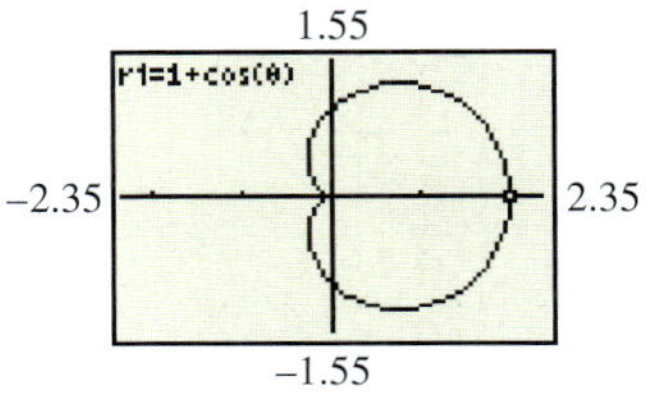

(b)

Figure 25

✔ *Now Try Exercise 45.*

EXAMPLE 5 Graphing a Polar Equation (Rose)

Graph $r = 3 \cos 2\theta$.

SOLUTION Because the argument is 2θ, the graph requires a greater number of points than when the argument is just θ. We complete the table using selected angle measures through 360° in order to see the pattern of the graph. Approximate values in the table have been rounded to the nearest tenth.

θ	2θ	$\cos 2\theta$	$r = 3\cos 2\theta$	θ	2θ	$\cos 2\theta$	$r = 3\cos 2\theta$
0°	0°	1	3	120°	240°	−0.5	−1.5
15°	30°	0.9	2.6	135°	270°	0	0
30°	60°	0.5	1.5	180°	360°	1	3
45°	90°	0	0	225°	450°	0	0
60°	120°	−0.5	−1.5	270°	540°	−1	−3
75°	150°	−0.9	−2.6	315°	630°	0	0
90°	180°	−1	−3	360°	720°	1	3

Plotting the points from the table in order gives the graph of a **four-leaved rose.** Note in **Figure 26(a)** how the graph is developed with a continuous curve, beginning with the upper half of the right horizontal leaf and ending with the lower half of that leaf. As the graph is traced, the curve goes through the pole four times. This can actually be seen as a calculator graphs the curve. See **Figure 26(b).**

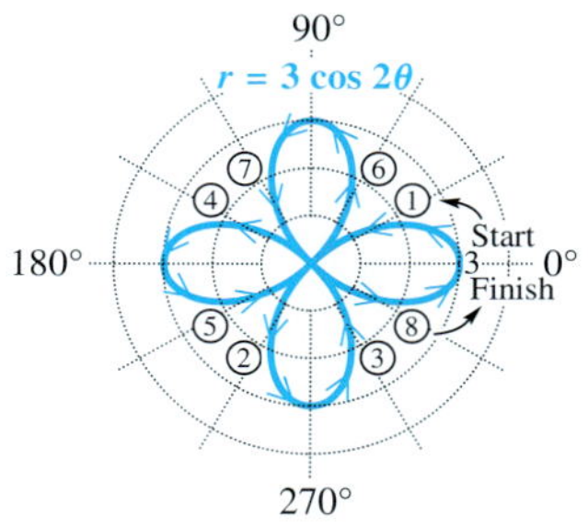

(a)

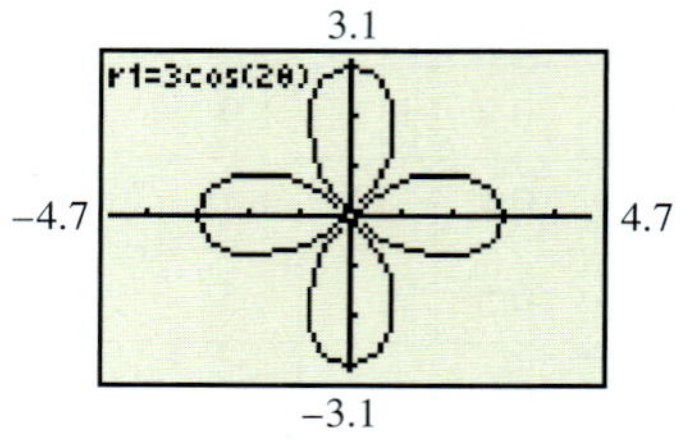

Polar graphing mode

(b)

Figure 26

✔ *Now Try Exercise 49.*

NOTE To sketch the graph of $r = 3 \cos 2\theta$ in polar coordinates, it may be helpful to first sketch the graph of $y = 3 \cos 2x$ in rectangular coordinates. The minimum and maximum values of this function may be used to determine the location of the tips of the rose petals, and the x-intercepts of this function may be used to determine where the polar graph passes through the pole.

The equation $r = 3 \cos 2\theta$ in **Example 5** has a graph that belongs to a family of curves called **roses.** The graphs of

$$r = a \sin n\theta \quad \text{and} \quad r = a \cos n\theta$$

are roses, with n leaves if n is odd, and $2n$ leaves if n is even. The absolute value of a determines the length of the leaves.

EXAMPLE 6 Graphing a Polar Equation (Lemniscate)

Graph $r^2 = \cos 2\theta$.

ALGEBRAIC SOLUTION

Complete a table of ordered pairs, and sketch the graph, as in **Figure 27.** The point $(-1, 0°)$, with r negative, may be plotted as $(1, 180°)$. Also, $(-0.7, 30°)$ may be plotted as $(0.7, 210°)$, and so on.

Values of θ for $45° < \theta < 135°$ are not included in the table because the corresponding values of $\cos 2\theta$ are negative (quadrants II and III) and so do not have real square roots. Values of θ larger than 180° give 2θ larger than 360° and would repeat the points already found. This curve is called a **lemniscate.**

θ	0°	30°	45°	135°	150°	180°
2θ	0°	60°	90°	270°	300°	360°
$\cos 2\theta$	1	0.5	0	0	0.5	1
$r = \pm\sqrt{\cos 2\theta}$	± 1	± 0.7	0	0	± 0.7	± 1

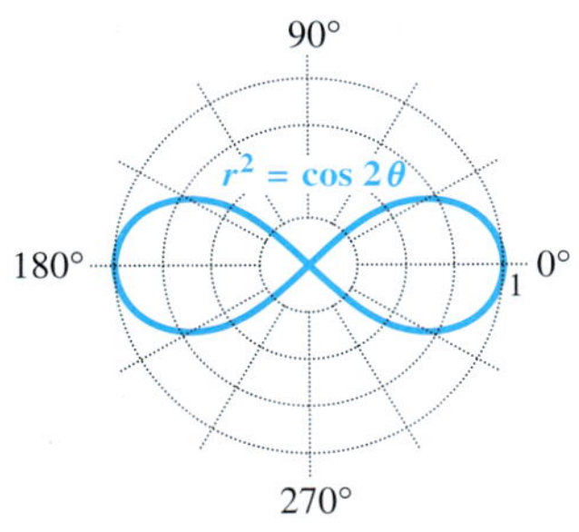

Figure 27

GRAPHING CALCULATOR SOLUTION

To graph $r^2 = \cos 2\theta$ with a graphing calculator, first solve for r by considering both square roots. Enter the two polar equations as

$$r_1 = \sqrt{\cos 2\theta}$$

and $$r_2 = -\sqrt{\cos 2\theta}.$$

See **Figures 28(a) and (b).**

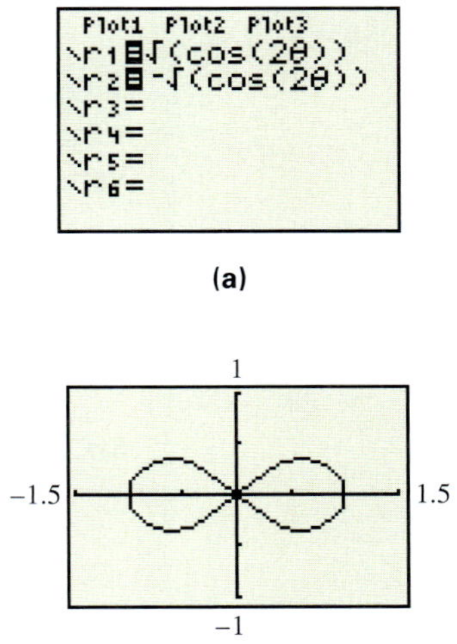

(a)

(b)

Figure 28

✔ *Now Try Exercise 51.*

EXAMPLE 7 Graphing a Polar Equation (Spiral of Archimedes)

Graph $r = 2\theta$ (with θ measured in radians).

SOLUTION Some ordered pairs are shown in the table. Since $r = 2\theta$ rather than a trigonometric function of θ, we must also consider negative values of θ. Radian measures have been rounded. The graph in **Figure 29(a)** on the next page is a **spiral of Archimedes. Figure 29(b)** shows a calculator graph of this spiral.

θ (radians)	$r = 2\theta$	θ (radians)	$r = 2\theta$
$-\pi$	-6.3	$\frac{\pi}{3}$	2.1
$-\frac{\pi}{2}$	-3.1	$\frac{\pi}{2}$	3.1
$-\frac{\pi}{4}$	-1.6	π	6.3
0	0	$\frac{3\pi}{2}$	9.4
$\frac{\pi}{6}$	1	2π	12.6

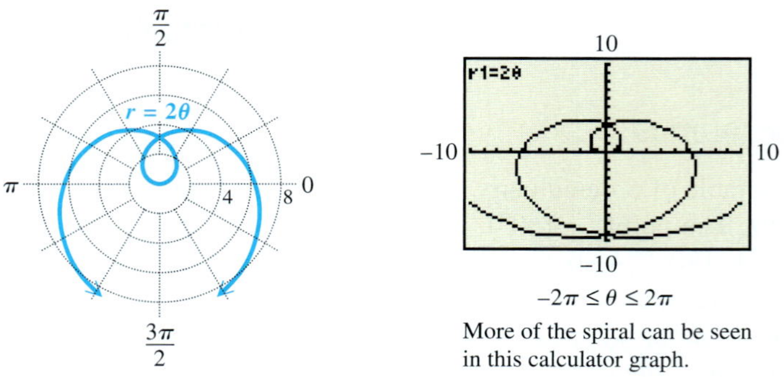

(a)

$-2\pi \le \theta \le 2\pi$

More of the spiral can be seen in this calculator graph.

(b)

Figure 29

✔ *Now Try Exercise 67.*

Converting from Polar to Rectangular Equations In **Example 3** we converted rectangular equations to polar equations. We conclude with an example that converts a polar equation to a rectangular one.

EXAMPLE 8 Converting a Polar Equation to a Rectangular Equation

Convert the equation $r = \frac{4}{1 + \sin\theta}$ to rectangular coordinates, and graph.

SOLUTION

$$r = \frac{4}{1 + \sin\theta} \quad \text{Polar equation}$$

$$r(1 + \sin\theta) = 4 \quad \text{Multiply by } 1 + \sin\theta.$$

$$r + r\sin\theta = 4 \quad \text{Distributive property}$$

$$\sqrt{x^2 + y^2} + y = 4 \quad \text{Let } r = \sqrt{x^2 + y^2} \text{ and } r\sin\theta = y.$$

$$\sqrt{x^2 + y^2} = 4 - y \quad \text{Subtract } y.$$

$$x^2 + y^2 = (4 - y)^2 \quad \text{Square each side.}$$

$$x^2 + y^2 = 16 - 8y + y^2 \quad \text{Expand the right side.}$$

$$x^2 = -8y + 16 \quad \text{Subtract } y^2.$$

$$x^2 = -8(y - 2) \quad \text{Rectangular equation}$$

The final equation represents a parabola and is graphed in **Figure 30.**

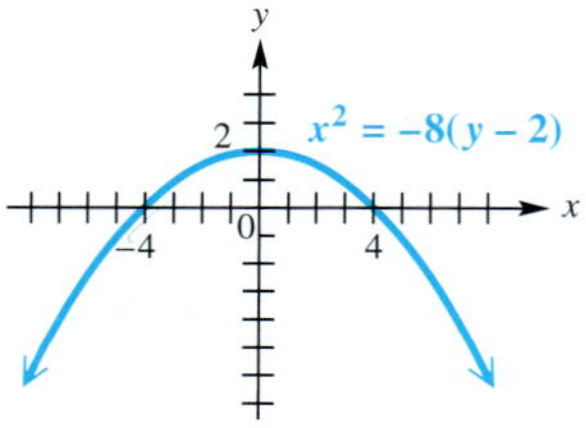

Figure 30

✔ *Now Try Exercise 59.*

The conversion in **Example 8** is not necessary when one is using a graphing calculator. **Figure 31** shows the graph of $r = \frac{4}{1 + \sin\theta}$, graphed directly with the calculator in polar mode. ■

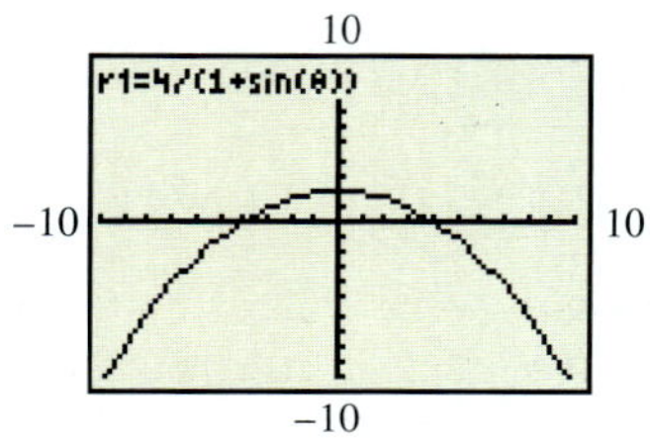

$0° \le \theta \le 360°$

Figure 31

Classifying Polar Equations The table on the next page summarizes common polar graphs and forms of their equations. (In addition to circles, lemniscates, and roses, we include **limaçons.** Cardioids are a special case of limaçons, where $\left|\frac{a}{b}\right| = 1$.)

Circles and Lemniscates

Circles | **Lemniscates**

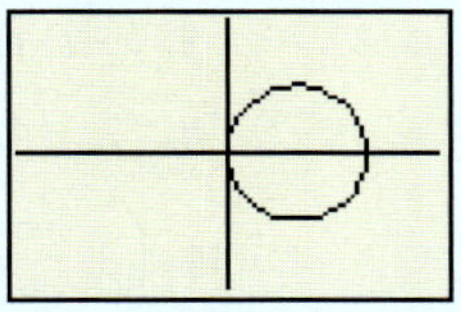

$r = a \cos \theta$

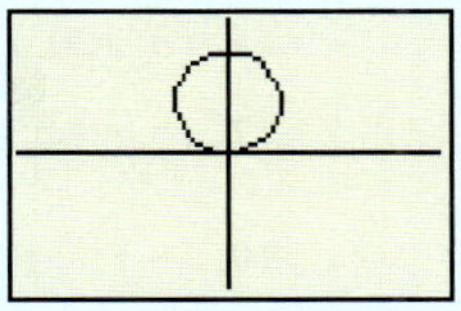

$r = a \sin \theta$

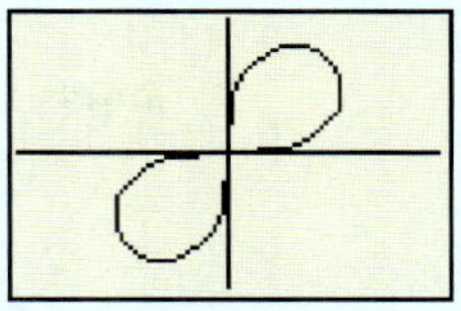

$r^2 = a^2 \sin 2\theta$

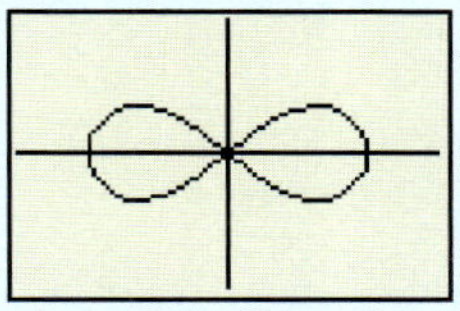

$r^2 = a^2 \cos 2\theta$

Limaçons

$r = a \pm b \sin \theta$ or $r = a \pm b \cos \theta$

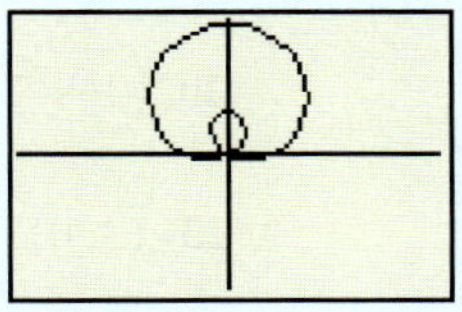

$\frac{a}{b} < 1$

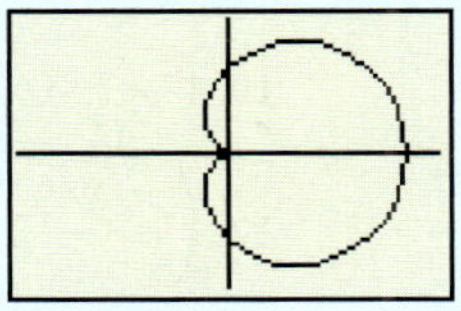

$\frac{a}{b} = 1$

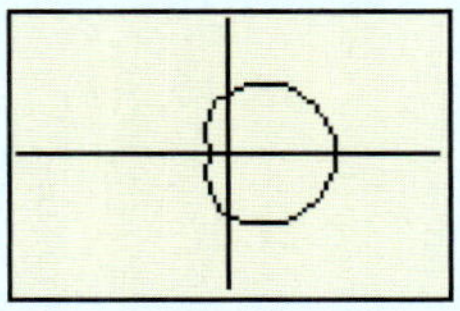

$1 < \frac{a}{b} < 2$

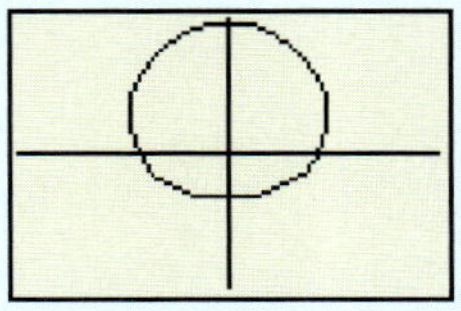

$\frac{a}{b} \geq 2$

Rose Curves

$2n$ leaves if n is even, $n \geq 2$ | n leaves if n is odd

$n = 2$

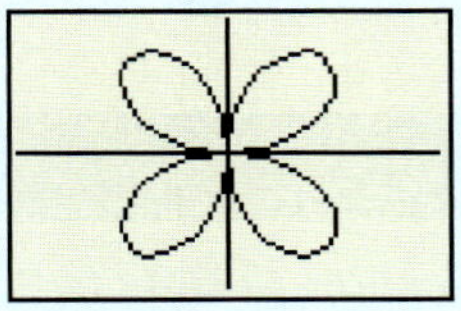

$r = a \sin n\theta$

$n = 4$

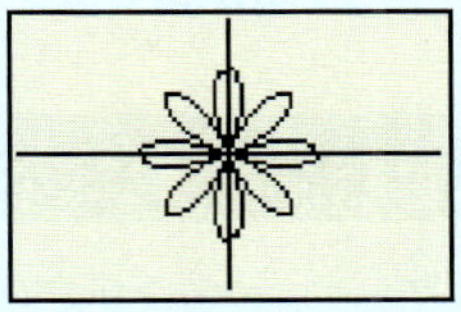

$r = a \cos n\theta$

$n = 3$

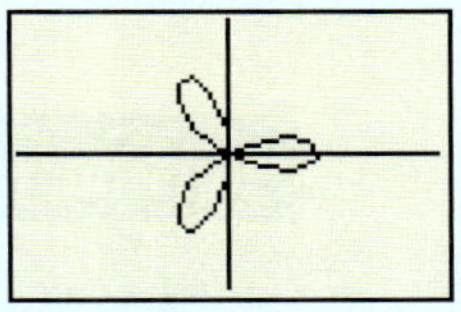

$r = a \cos n\theta$

$n = 5$

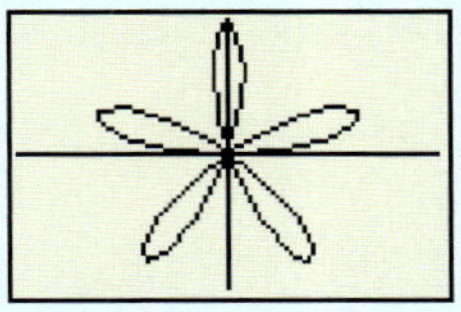

$r = a \sin n\theta$

NOTE Some other polar curves are the **cissoid, kappa curve, conchoid, trisectrix, cruciform, strophoid,** and **lituus.** Refer to older textbooks on analytic geometry or the Internet to investigate them.

8.5 Exercises

1. *Concept Check* For each point given in polar coordinates, state the quadrant in which the point lies if it is graphed in a rectangular coordinate system.

(a) $(5, 135°)$ **(b)** $(2, 60°)$ **(c)** $(6, -30°)$ **(d)** $(4.6, 213°)$

2. *Concept Check* For each point given in polar coordinates, state the axis on which the point lies if it is graphed in a rectangular coordinate system. Also state whether it is on the positive portion or the negative portion of the axis. (For example, $(5, 0°)$ lies on the positive x-axis.)

(a) $(7, 360°)$ **(b)** $(4, 180°)$ **(c)** $(2, -90°)$ **(d)** $(8, 450°)$

For each pair of polar coordinates, (a) plot the point, (b) give two other pairs of polar coordinates for the point, and (c) give the rectangular coordinates for the point. ***See Examples 1 and 2.***

3. $(1, 45°)$ **4.** $(3, 120°)$ **5.** $(-2, 135°)$

6. $(-4, 30°)$ **7.** $(5, -60°)$ **8.** $(2, -45°)$

9. $(-3, -210°)$ **10.** $(-1, -120°)$ **11.** $\left(3, \frac{5\pi}{3}\right)$

12. $\left(4, \frac{3\pi}{2}\right)$ **13.** $\left(-2, \frac{\pi}{3}\right)$ **14.** $\left(-5, \frac{5\pi}{6}\right)$

For each pair of rectangular coordinates, (a) plot the point and (b) give two pairs of polar coordinates for the point, where $0° \le \theta < 360°$. ***See Example 2(b).***

15. $(1, -1)$ **16.** $(1, 1)$ **17.** $(0, 3)$

18. $(0, -3)$ **19.** $(\sqrt{2}, \sqrt{2})$ **20.** $(-\sqrt{2}, \sqrt{2})$

21. $\left(\frac{\sqrt{3}}{2}, \frac{3}{2}\right)$ **22.** $\left(-\frac{\sqrt{3}}{2}, -\frac{1}{2}\right)$ **23.** $(3, 0)$

24. $(-2, 0)$ **25.** $\left(-\frac{3}{2}, -\frac{3\sqrt{3}}{2}\right)$ **26.** $\left(\frac{1}{2}, -\frac{\sqrt{3}}{2}\right)$

For each rectangular equation, give its equivalent polar equation and sketch its graph. ***See Example 3.***

27. $x - y = 4$ **28.** $x + y = -7$ **29.** $x^2 + y^2 = 16$

30. $x^2 + y^2 = 9$ **31.** $2x + y = 5$ **32.** $3x - 2y = 6$

Relating Concepts

For individual or collaborative investigation *(Exercises 33–40)*

In rectangular coordinates, the graph of

$$ax + by = c$$

is a horizontal line if $a = 0$ *or a vertical line if* $b = 0$. ***Work Exercises 33–40 in order,*** *to determine the general forms of polar equations for horizontal and vertical lines.*

33. Begin with the equation $y = k$, whose graph is a horizontal line. Make a trigonometric substitution for y using r and θ.

34. Solve the equation in **Exercise 33** for r.

35. Rewrite the equation in **Exercise 34** using the appropriate reciprocal function.

36. Sketch the graph of the equation

$$r = 3 \csc \theta.$$

What is the corresponding rectangular equation?

37. Begin with the equation $x = k$, whose graph is a vertical line. Make a trigonometric substitution for x using r and θ.

38. Solve the equation in **Exercise 37** for r.

39. Rewrite the equation in **Exercise 38** using the appropriate reciprocal function.

40. Sketch the graph of $r = 3 \sec \theta$. What is the corresponding rectangular equation?

Concept Check In Exercises 41–44, match each equation with its polar graph from choices A–D.

41. $r = 3$ **42.** $r = \cos 3\theta$ **43.** $r = \cos 2\theta$ **44.** $r = \dfrac{2}{\cos\theta + \sin\theta}$

A.

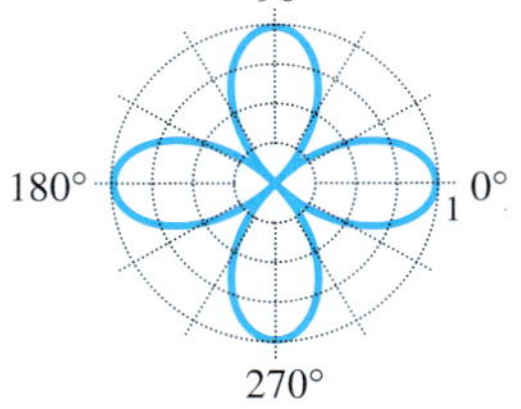

B.

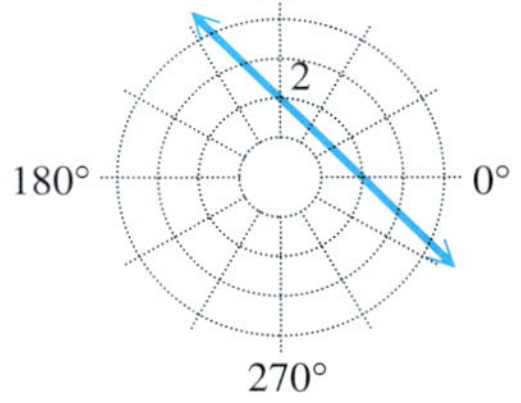

C.

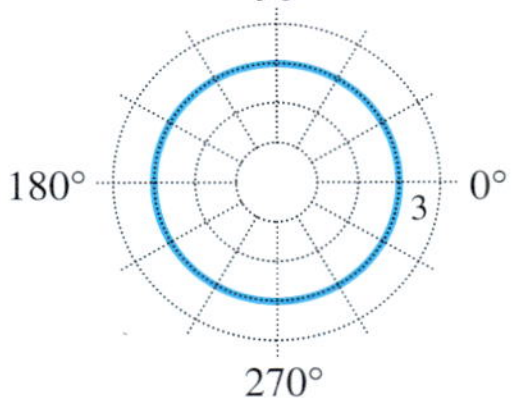

D.

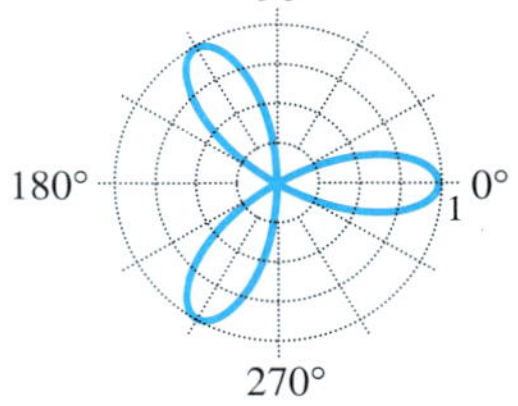

Give a complete graph of each polar equation. In Exercises 45–54, also identify the type of polar graph. ***See Examples 4–6.***

45. $r = 2 + 2\cos\theta$

46. $r = 8 + 6\cos\theta$

47. $r = 3 + \cos\theta$

48. $r = 2 - \cos\theta$

49. $r = 4\cos 2\theta$

50. $r = 3\cos 5\theta$

51. $r^2 = 4\cos 2\theta$

52. $r^2 = 4\sin 2\theta$

53. $r = 4 - 4\cos\theta$

54. $r = 6 - 3\cos\theta$

55. $r = 2\sin\theta\tan\theta$
(This is a **cissoid.**)

56. $r = \dfrac{\cos 2\theta}{\cos\theta}$
(This is a **cissoid with a loop.**)

For each equation, find an equivalent equation in rectangular coordinates, and graph. ***See Example 8.***

57. $r = 2\sin\theta$

58. $r = 2\cos\theta$

59. $r = \dfrac{2}{1 - \cos\theta}$

60. $r = \dfrac{3}{1 - \sin\theta}$

61. $r = -2\cos\theta - 2\sin\theta$

62. $r = \dfrac{3}{4\cos\theta - \sin\theta}$

63. $r = 2\sec\theta$

64. $r = -5\csc\theta$

65. $r = \dfrac{2}{\cos\theta + \sin\theta}$

66. $r = \dfrac{2}{2\cos\theta + \sin\theta}$

67. Graph $r = \theta$, a spiral of Archimedes. (**See Example 7.**) Use both positive and nonpositive values for θ.

68. Use a graphing calculator window of $[-1250, 1250]$ by $[-1250, 1250]$, in degree mode, to graph more of

$$r = 2\theta \text{ (a spiral of Archimedes)}$$

than what is shown in **Figure 29.** Use $-1250° \leq \theta \leq 1250°$.

69. Find the polar equation of the line that passes through the points $(1, 0°)$ and $(2, 90°)$.

70. Explain how to plot a point (r, θ) in polar coordinates, if $r < 0$.

Concept Check The polar graphs in this section exhibit symmetry. (**See Appendix D.**) *Visualize an xy-plane superimposed on the polar coordinate system, with the pole at the origin and the polar axis on the positive x-axis. Then a polar graph may be symmetric with respect to the x-axis (the polar axis), the y-axis* $\left(\text{the line } \theta = \frac{\pi}{2}\right)$, *or the origin (the pole). Use this information to work Exercises 71 and 72.*

71. Complete the missing ordered pairs in the graphs below.

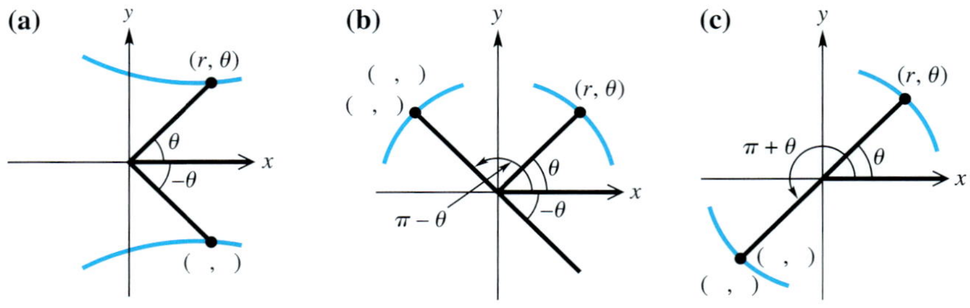

72. Based on your results in **Exercise 71,** fill in the blanks with the correct responses.

(a) The graph of $r = f(\theta)$ is symmetric with respect to the polar axis if substitution of ____________ for θ leads to an equivalent equation.

(b) The graph of $r = f(\theta)$ is symmetric with respect to the vertical line $\theta = \frac{\pi}{2}$ if substitution of ____________ for θ leads to an equivalent equation.

(c) Alternatively, the graph of $r = f(\theta)$ is symmetric with respect to the vertical line $\theta = \frac{\pi}{2}$ if substitution of ____________ for r and ____________ for θ leads to an equivalent equation.

(d) The graph of $r = f(\theta)$ is symmetric with respect to the pole if substitution of ____________ for r leads to an equivalent equation.

(e) Alternatively, the graph of $r = f(\theta)$ is symmetric with respect to the pole if substitution of ____________ for θ leads to an equivalent equation.

(f) In general, the completed statements in parts (a)–(e) mean that the graphs of polar equations of the form $r = a \pm b \cos \theta$ (where a may be 0) are symmetric with respect to ____________.

(g) In general, the completed statements in parts (a)–(e) mean that the graphs of polar equations of the form $r = a \pm b \sin \theta$ (where a may be 0) are symmetric with respect to ____________.

The graph of $r = a\theta$ *in polar coordinates is an example of the spiral of Archimedes. With your calculator set to radian mode, use the given value of a and interval of* θ *to graph the spiral in the window specified.*

73. $a = 1,\ 0 \le \theta \le 4\pi,\ [-15, 15]$ by $[-15, 15]$

74. $a = 2,\ -4\pi \le \theta \le 4\pi,\ [-30, 30]$ by $[-30, 30]$

75. $a = 1.5,\ -4\pi \le \theta \le 4\pi,\ [-20, 20]$ by $[-20, 20]$

76. $a = -1,\ 0 \le \theta \le 12\pi,\ [-40, 40]$ by $[-40, 40]$

Find the polar coordinates of the points of intersection of the given curves for the specified interval of θ.

77. $r = 4 \sin \theta,\ r = 1 + 2 \sin \theta;\ 0 \le \theta < 2\pi$

78. $r = 3,\ r = 2 + 2 \cos \theta;\ 0° \le \theta < 360°$

79. $r = 2 + \sin \theta,\ r = 2 + \cos \theta;\ 0 \le \theta < 2\pi$

80. $r = \sin 2\theta,\ r = \sqrt{2} \cos \theta;\ 0 \le \theta < \pi$

(Modeling) Solve each problem.

81. ***Orbits of Satellites*** The polar equation

$$r = \frac{a(1 - e^2)}{1 + e \cos \theta}$$

can be used to graph the orbits of the satellites of our sun, where a is the average distance in astronomical units from the sun and e is a constant called the **eccentricity**. The sun will be located at the pole. The table lists the values of a and e.

Satellite	a	e
Mercury	0.39	0.206
Venus	0.78	0.007
Earth	1.00	0.017
Mars	1.52	0.093
Jupiter	5.20	0.048
Saturn	9.54	0.056
Uranus	19.20	0.047
Neptune	30.10	0.009
Pluto	39.40	0.249

Source: Karttunen, H., P. Kröger, H. Oja, M. Putannen, and K. Donners (Editors), *Fundamental Astronomy, 4th edition,* Springer-Verlag. Zeilik, M., S. Gregory, and E. Smith, *Introductory Astronomy and Astrophysics,* Saunders College Publishers.

(a) Graph the orbits of the four closest satellites on the same polar grid. Choose a viewing window that results in a graph with nearly circular orbits.

(b) Plot the orbits of Earth, Jupiter, Uranus, and Pluto on the same polar grid. How does Earth's distance from the sun compare to the others' distances from the sun?

(c) Use graphing to determine whether or not Pluto is always farthest from the sun.

82. ***Radio Towers and Broadcasting Patterns*** Many times radio stations do not broadcast in all directions with the same intensity. To avoid interference with an existing station to the north, a new station may be licensed to broadcast only east and west. To create an east-west signal, two radio towers are sometimes used, as illustrated in the figure. Locations where the radio signal is received correspond to the interior of the curve

$$r^2 = 40{,}000 \cos 2\theta,$$

where the polar axis (or positive x-axis) points east.

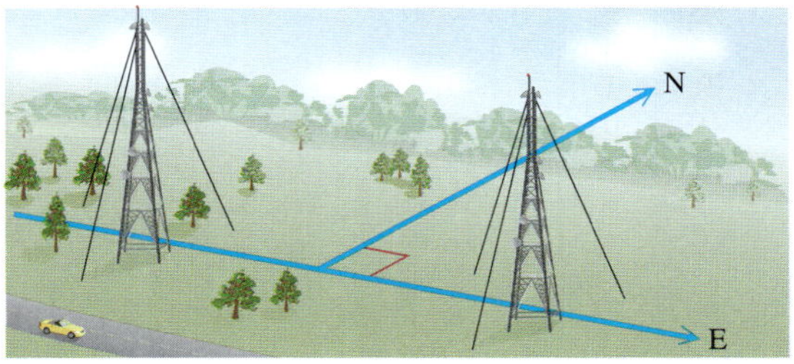

(a) Graph $r^2 = 40{,}000 \cos 2\theta$ for $0° \le \theta \le 360°$, where distances are in miles. Assuming the radio towers are located near the pole, use the graph to describe the regions where the signal can be received and where the signal cannot be received.

(b) Suppose a radio signal pattern is given by

$$r^2 = 22{,}500 \sin 2\theta.$$

Graph this pattern and interpret the results.

8.6 Parametric Equations, Graphs, and Applications

- Basic Concepts
- Parametric Graphs and Their Rectangular Equivalents
- The Cycloid
- Applications of Parametric Equations

Basic Concepts

Throughout this text, we have graphed sets of ordered pairs of real numbers that correspond to a function of the form $y = f(x)$ or $r = g(\theta)$. Another way to determine a set of ordered pairs involves two functions f and g defined by $x = f(t)$ and $y = g(t)$, where t is a real number in some interval I. Each value of t leads to a corresponding x-value and a corresponding y-value, and thus to an ordered pair (x, y).

Parametric Equations of a Plane Curve

A **plane curve** is a set of points (x, y) such that $x = f(t)$, $y = g(t)$, and f and g are both defined on an interval I. The equations $x = f(t)$ and $y = g(t)$ are **parametric equations** with **parameter** $\boldsymbol{t}$.

Graphing calculators are capable of graphing plane curves defined by parametric equations. The calculator must be set in parametric mode, and the window requires intervals for the parameter t, as well as for x and y. ■

Parametric Graphs and Their Rectangular Equivalents

EXAMPLE 1 Graphing a Plane Curve Defined Parametrically

Let $x = t^2$ and $y = 2t + 3$, for t in $[-3, 3]$. Graph the set of ordered pairs (x, y).

ALGEBRAIC SOLUTION

Make a table of corresponding values of t, x, and y over the domain of t. Plot the points as shown in **Figure 32.** The graph is a portion of a parabola with horizontal axis $y = 3$. The arrowheads indicate the direction the curve traces as t increases.

t	x	y
−3	9	−3
−2	4	−1
−1	1	1
0	0	3
1	1	5
2	4	7
3	9	9

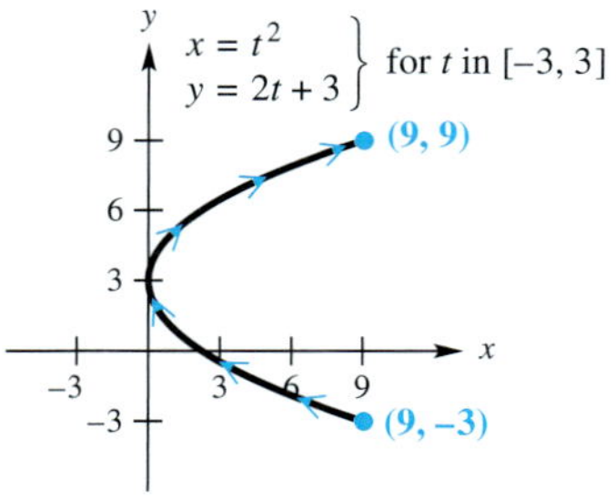

Figure 32

GRAPHING CALCULATOR SOLUTION

We set the parameters of the TI-83/84 Plus as shown in the top two screens to obtain the bottom screen in **Figure 33.**

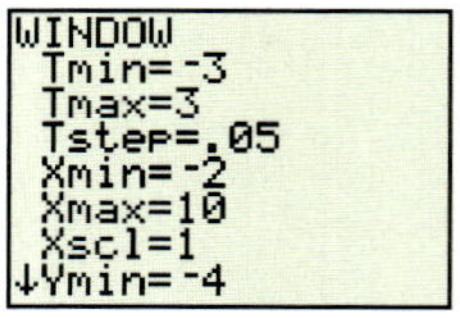

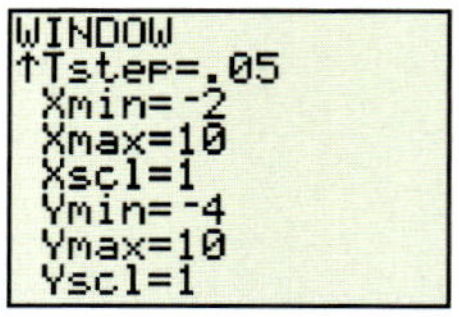

This is a continuation of the previous screen.

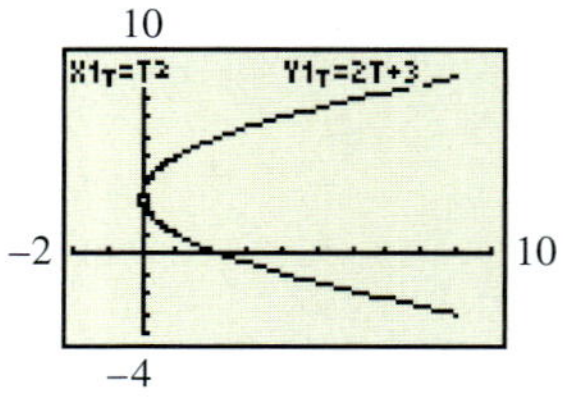

Figure 33

✔ *Now Try Exercise 5(a).*

EXAMPLE 2 Finding an Equivalent Rectangular Equation

Find a rectangular equation for the plane curve of **Example 1** defined as follows:

$$x = t^2, \quad y = 2t + 3, \qquad \text{for } t \text{ in } [-3, 3].$$

SOLUTION To eliminate the parameter t, first solve either equation for t. Here, only the second equation, $y = 2t + 3$, leads to a unique solution for t, so we choose it.

$$y = 2t + 3 \quad \text{Choose the simpler equation.}$$

$$2t = y - 3 \quad \text{Subtract 3 and rewrite. (Appendix A)}$$

$$t = \frac{y - 3}{2} \quad \text{Divide by 2.}$$

Now substitute this result into the first equation to eliminate the parameter t.

$$x = t^2$$

$$x = \left(\frac{y - 3}{2}\right)^2 \quad \text{Substitute for } t.$$

$$x = \frac{(y - 3)^2}{4} \quad \left(\frac{a}{b}\right)^2 = \frac{a^2}{b^2}$$

$$4x = (y - 3)^2 \quad \text{Multiply by 4.}$$

This is the equation of a horizontal parabola opening to the right, which agrees with the graph given in **Figure 32.** Because t is in $[-3, 3]$, x is in $[0, 9]$ and y is in $[-3, 9]$. The rectangular equation must be given with its restricted domain as

$$4x = (y - 3)^2, \qquad \text{for } x \text{ in } [0, 9].$$

✔ ***Now Try Exercise 5(b).***

EXAMPLE 3 Graphing a Plane Curve Defined Parametrically

Graph the plane curve defined by $x = 2 \sin t$, $y = 3 \cos t$, for t in $[0, 2\pi]$.

SOLUTION To convert to a rectangular equation, it is not productive here to solve either equation for t. Instead, we use the fact that $\sin^2 t + \cos^2 t = 1$ to apply another approach. Square both sides of each equation; solve one for $\sin^2 t$, the other for $\cos^2 t$.

$x = 2 \sin t$	$y = 3 \cos t$	Given equations
$x^2 = 4 \sin^2 t$	$y^2 = 9 \cos^2 t$	Square each side.
$\frac{x^2}{4} = \sin^2 t$	$\frac{y^2}{9} = \cos^2 t$	Divide.

Now add corresponding sides of the two equations.

$$\frac{x^2}{4} + \frac{y^2}{9} = \sin^2 t + \cos^2 t$$

$$\frac{x^2}{4} + \frac{y^2}{9} = 1 \qquad \sin^2 t + \cos^2 t = 1 \text{ (Section 5.1)}$$

This is an equation of an **ellipse.** See **Figure 34** on the next page for traditional and calculator graphs. (Ellipses are covered in more detail in college algebra courses.)

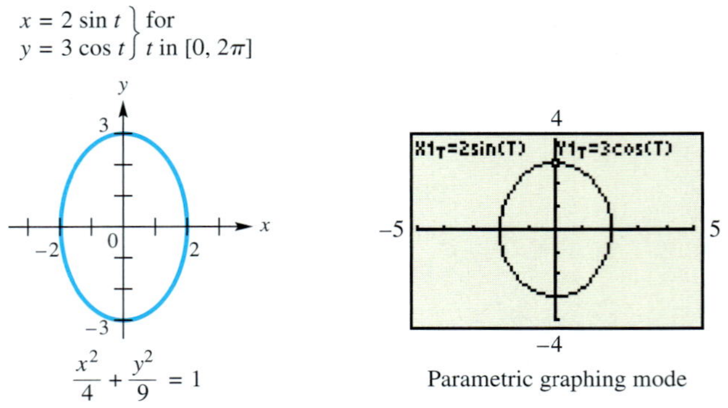

Figure 34

✔ *Now Try Exercise 27.*

Parametric representations of a curve are not unique. In fact, there are infinitely many parametric representations of a given curve. If the curve can be described by a rectangular equation $y = f(x)$, with domain X, then one simple parametric representation is

$$x = t, \quad y = f(t), \qquad \text{for } t \text{ in } X.$$

EXAMPLE 4 Finding Alternative Parametric Equation Forms

Give two parametric representations for the equation of the parabola.

$$y = (x - 2)^2 + 1$$

SOLUTION The simplest choice is to let

$$x = t, \quad y = (t - 2)^2 + 1, \qquad \text{for } t \text{ in } (-\infty, \infty).$$

Another choice, which leads to a simpler equation for y, is

$$x = t + 2, \quad y = t^2 + 1, \qquad \text{for } t \text{ in } (-\infty, \infty).$$

✔ *Now Try Exercise 29.*

NOTE Sometimes trigonometric functions are desirable. One choice in **Example 4** might be

$$x = 2 + \tan t, \quad y = \sec^2 t, \qquad \text{for } t \text{ in } \left(-\frac{\pi}{2}, \frac{\pi}{2}\right).$$

The Cycloid The *cycloid* is a special case of the **trochoid**—a curve traced out by a point at a given distance from the center of a circle as the circle rolls along a straight line. If the given point is on the *circumference* of the circle, then the path traced as the circle rolls along a straight line is a **cycloid,** which is defined parametrically as follows.

$$\mathbf{x = at - a \sin t, \quad y = a - a \cos t, \qquad \text{for } t \text{ in } (-\infty, \infty)}$$

Other curves related to trochoids are **hypotrochoids** and **epitrochoids,** which are traced out by a point that is a given distance from the center of a circle that rolls not on a straight line, but on the inside or outside, respectively, of another circle. The classic Spirograph toy can be used to draw these curves.

EXAMPLE 5 Graphing a Cycloid

Graph the cycloid.

$$x = t - \sin t, \quad y = 1 - \cos t, \quad \text{for } t \text{ in } [0, 2\pi]$$

ALGEBRAIC SOLUTION

There is no simple way to find a rectangular equation for the cycloid from its parametric equations. Instead, begin with a table using selected values for t in $[0, 2\pi]$. Approximate values have been rounded as necessary.

t	0	$\frac{\pi}{4}$	$\frac{\pi}{2}$	π	$\frac{3\pi}{2}$	2π
x	0	0.08	0.6	π	5.7	2π
y	0	0.3	1	2	1	0

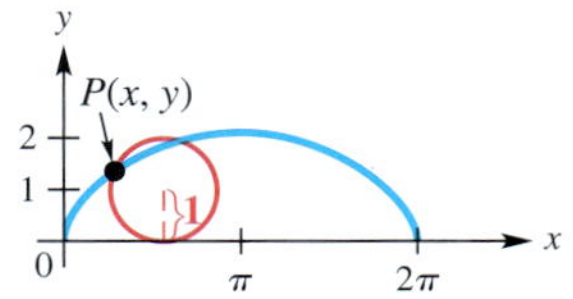

Figure 35

Plotting the ordered pairs (x, y) from the table of values leads to the portion of the graph in **Figure 35** from 0 to 2π.

GRAPHING CALCULATOR SOLUTION

It is easier to graph a cycloid with a graphing calculator in parametric mode than with traditional methods. See **Figure 36.**

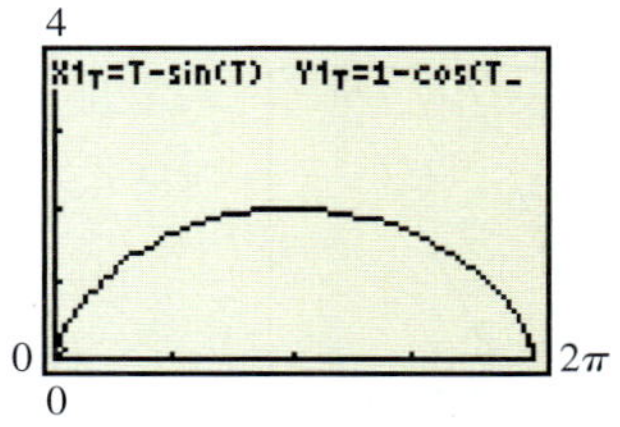

Figure 36

Using a larger interval for t would show that the cycloid repeats the pattern shown here every 2π units.

Now Try Exercise 33.

Figure 37

The cycloid has an interesting physical property. If a flexible cord or wire goes through points P and Q as in **Figure 37,** and a bead is allowed to slide due to the force of gravity without friction along this path from P to Q, the path that requires the shortest time takes the shape of the graph of an inverted cycloid.

LOOKING AHEAD TO CALCULUS

At any time t, the velocity of an object is given by the vector $\mathbf{v} = \langle \sim'(t), g'(t) \rangle$. The object's speed at time t is

$$|\mathbf{v}| = \sqrt{(\sim'(t))^2 + (g'(t))^2}.$$

Applications of Parametric Equations Parametric equations are used to simulate motion. If a ball is thrown with a velocity of v feet per second at an angle θ with the horizontal, its flight can be modeled by the parametric equations

$$\mathbf{x = (v \cos \theta)t \quad \text{and} \quad y = (v \sin \theta)t - 16t^2 + h,}$$

where t is in seconds and h is the ball's initial height in feet above the ground. Here, x gives the horizontal position information and y gives the vertical position information. The term $-16t^2$ occurs because gravity is pulling downward. See **Figure 38.** These equations ignore air resistance.

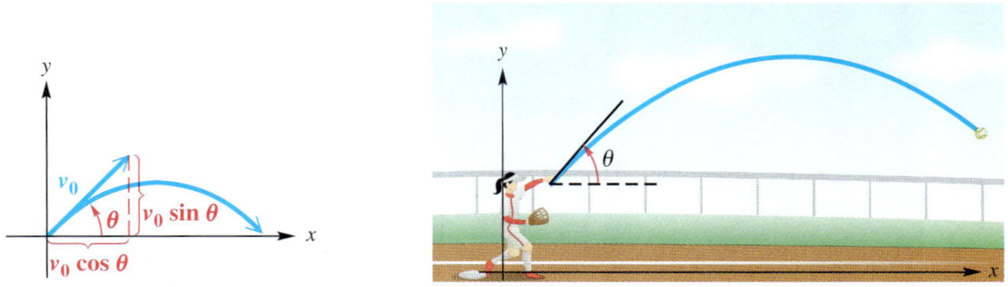

Figure 38

EXAMPLE 6 Simulating Motion with Parametric Equations

Three golf balls are hit simultaneously into the air at 132 ft per sec (90 mph) at angles of 30°, 50°, and 70° with the horizontal.

(a) Assuming the ground is level, determine graphically which ball travels the greatest distance. Estimate this distance.

(b) Which ball reaches the greatest height? Estimate this height.

SOLUTION

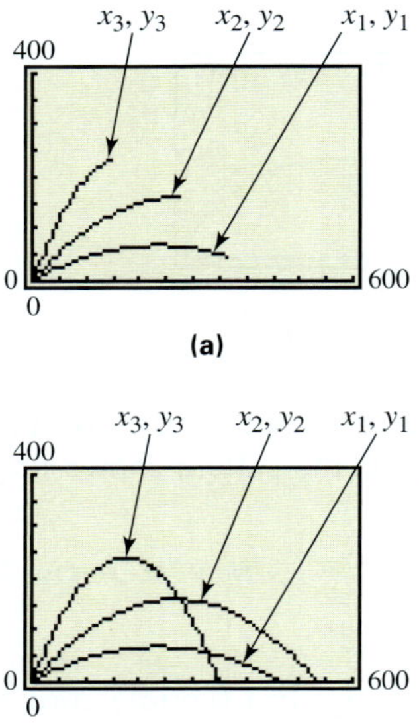

(a)

(b)

Figure 39

(a) Use the following parametric equations to model the flight of the golf balls.

$$x = (v \cos \theta)t \quad \text{and} \quad y = (v \sin \theta)t - 16t^2 + h$$

Substitute $h = 0$, $v = 132$ ft per sec, and $\theta = 30°, 50°$, and $70°$ to write three sets of parametric equations.

$$x_1 = (132 \cos 30°)t, \quad y_1 = (132 \sin 30°)t - 16t^2$$

$$x_2 = (132 \cos 50°)t, \quad y_2 = (132 \sin 50°)t - 16t^2$$

$$x_3 = (132 \cos 70°)t, \quad y_3 = (132 \sin 70°)t - 16t^2$$

The graphs of the three sets of parametric equations are shown in **Figure 39(a),** where $0 \le t \le 9$. From the graph in **Figure 39(b),** we can see that the ball hit at 50° travels the greatest distance. Using the TRACE feature of the TI-83/84 Plus, we estimate this distance to be about 540 ft.

(b) Again, use the TRACE feature to find that the ball hit at 70° reaches the greatest height, about 240 ft.

✔ *Now Try Exercise 39.*

NOTE The TI-83/84 Plus graphing calculator allows the user to view the graphing of more than one equation either *sequentially* or *simultaneously.* By choosing the latter, one can view the three golf balls in **Figure 39** in flight at the same time.

EXAMPLE 7 Examining Parametric Equations of Flight

Jack Lukas launches a small rocket from a table that is 3.36 ft above the ground. Its initial velocity is 64 ft per sec, and it is launched at an angle of 30° with respect to the ground. Find the rectangular equation that models its path. What type of path does the rocket follow?

SOLUTION The path of the rocket is defined by the parametric equations

$$x = (64 \cos 30°)t \quad \text{and} \quad y = (64 \sin 30°)t - 16t^2 + 3.36$$

or, equivalently,

$$x = 32\sqrt{3}t \quad \text{and} \quad y = -16t^2 + 32t + 3.36.$$

From $x = 32\sqrt{3}t$, we solve for t to obtain

$$t = \frac{x}{32\sqrt{3}}. \quad \text{Divide by } 32\sqrt{3}.$$

Substituting for t in the other parametric equation yields the following.

$$y = -16t^2 + 32t + 3.36$$

$$y = -16\left(\frac{x}{32\sqrt{3}}\right)^2 + 32\left(\frac{x}{32\sqrt{3}}\right) + 3.36 \quad \text{Let } t = \frac{x}{32\sqrt{3}}.$$

$$y = -\frac{1}{192}x^2 + \frac{\sqrt{3}}{3}x + 3.36 \quad \text{Simplify.}$$

Because this equation defines a parabola, we can conclude that the rocket follows a parabolic path.

✔ *Now Try Exercise 43(a).*

EXAMPLE 8 Analyzing the Path of a Projectile

Determine the total flight time and the horizontal distance traveled by the rocket in **Example 7.**

ALGEBRAIC SOLUTION

The equation $y = -16t^2 + 32t + 3.36$ tells the vertical position of the rocket at time t. We need to determine those values of t for which $y = 0$ since these values correspond to the rocket at ground level. This yields

$$0 = -16t^2 + 32t + 3.36.$$

Using the quadratic formula, the solutions are $t = -0.1$ or $t = 2.1$. Since t represents time, $t = -0.1$ is an unacceptable answer. Therefore, the flight time is 2.1 sec.

The rocket was in the air for 2.1 sec, so we can use $t = 2.1$ and the parametric equation that models the horizontal position, $x = 32\sqrt{3}t$, to obtain

$$x = 32\sqrt{3}(2.1) \approx 116.4 \text{ ft.}$$

GRAPHING CALCULATOR SOLUTION

Figure 40 shows that when T = 2.1, the horizontal distance X covered is approximately 116.4 ft, which agrees with the algebraic solution.

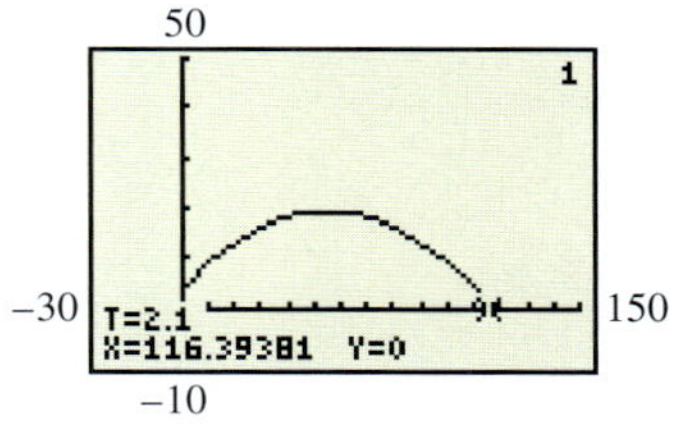

Figure 40

✔ *Now Try Exercise 43(b).*

8.6 Exercises

Concept Check Match the ordered pair from Column II with the pair of parametric equations in Column I on whose graph the point lies. In each case, consider the given value of t.

I	II
1. $x = 3t + 6,\ y = -2t + 4;\quad t = 2$	**A.** $(5, 25)$
2. $x = \cos t,\ y = \sin t;\quad t = \frac{\pi}{4}$	**B.** $(7, 2)$
3. $x = t,\ y = t^2;\quad t = 5$	**C.** $(12, 0)$
4. $x = t^2 + 3,\ y = t^2 - 2;\quad t = 2$	**D.** $\left(\frac{\sqrt{2}}{2}, \frac{\sqrt{2}}{2}\right)$

*For each plane curve, **(a)** graph the curve, and **(b)** find a rectangular equation for the curve.* ***See Examples 1 and 2.***

5. $x = t + 2,\ y = t^2,$ for t in $[-1, 1]$

6. $x = 2t,\ y = t + 1,$ for t in $[-2, 3]$

7. $x = \sqrt{t},\ y = 3t - 4,$ for t in $[0, 4]$

8. $x = t^2,\ y = \sqrt{t},$ for t in $[0, 4]$

9. $x = t^3 + 1,\ y = t^3 - 1,$ for t in $(-\infty, \infty)$

10. $x = 2t - 1,\ y = t^2 + 2,$ for t in $(-\infty, \infty)$

11. $x = 2 \sin t,\ y = 2 \cos t,$ for t in $[0, 2\pi]$

12. $x = \sqrt{5} \sin t,\ y = \sqrt{3} \cos t,$ for t in $[0, 2\pi]$

13. $x = 3 \tan t,\ y = 2 \sec t,$ for t in $\left(-\frac{\pi}{2}, \frac{\pi}{2}\right)$

14. $x = \cot t,\ y = \csc t,$ for t in $(0, \pi)$

15. $x = \sin t,\ y = \csc t,$ for t in $(0, \pi)$

16. $x = \tan t,\ y = \cot t,$ for t in $\left(0, \frac{\pi}{2}\right)$

17. $x = t,\ y = \sqrt{t^2 + 2},$ for t in $(-\infty, \infty)$

18. $x = \sqrt{t},\ y = t^2 - 1,$ for t in $[0, \infty)$

19. $x = 2 + \sin t,\ y = 1 + \cos t,$ for t in $[0, 2\pi]$

20. $x = 1 + 2 \sin t,\ y = 2 + 3 \cos t,$ for t in $[0, 2\pi]$

21. $x = t + 2,\ y = \dfrac{1}{t + 2},$ for $t \neq -2$

22. $x = t - 3,\ y = \dfrac{2}{t - 3},$ for $t \neq 3$

23. $x = t + 2,\ y = t - 4,$ for t in $(-\infty, \infty)$

24. $x = t^2 + 2,\ y = t^2 - 4,$ for t in $(-\infty, \infty)$

Graph each plane curve defined by the parametric equations for t in $[0, 2\pi]$. Then find a rectangular equation for the plane curve. ***See Example 3.***

25. $x = 3 \cos t,\ y = 3 \sin t$

26. $x = 2 \cos t,\ y = 2 \sin t$

27. $x = 3 \sin t,\ y = 2 \cos t$

28. $x = 4 \sin t,\ y = 3 \cos t$

Give two parametric representations for the equation of each parabola. ***See Example 4.***

29. $y = (x + 3)^2 - 1$

30. $y = (x + 4)^2 + 2$

31. $y = x^2 - 2x + 3$

32. $y = x^2 - 4x + 6$

Graph each cycloid defined by the given equations for t in the specified interval. ***See Example 5.***

33. $x = 2t - 2 \sin t,\ y = 2 - 2 \cos t,$ for t in $[0, 4\pi]$

34. $x = t - \sin t,\ y = 1 - \cos t,$ for t in $[0, 4\pi]$

Lissajous Figures *The screen shown here is an example of a* ***Lissajous figure.*** *Lissajous figures occur in electronics and may be used to find the frequency of an unknown voltage. Graph each Lissajous figure for t in $[0, 6.5]$ in the window $[-6, 6]$ by $[-4, 4]$.*

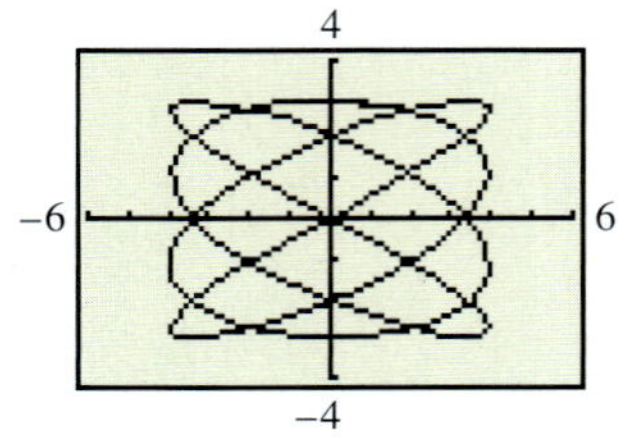

35. $x = 2 \cos t,\ y = 3 \sin 2t$

36. $x = 3 \cos 2t,\ y = 3 \sin 3t$

37. $x = 3 \sin 4t,\ y = 3 \cos 3t$

38. $x = 4 \sin 4t,\ y = 3 \sin 5t$

(Modeling) In Exercises 39–42, do the following. ***See Examples 6–8.***

(a) *Determine the parametric equations that model the path of the projectile.*
(b) *Determine the rectangular equation that models the path of the projectile.*
(c) *Determine approximately how long the projectile is in flight and the horizontal distance covered.*

39. *Flight of a Model Rocket* A model rocket is launched from the ground with velocity 48 ft per sec at an angle of 60° with respect to the ground.

40. *Flight of a Golf Ball* Tyler is playing golf. He hits a golf ball from the ground at an angle of 60° with respect to the ground at velocity 150 ft per sec.

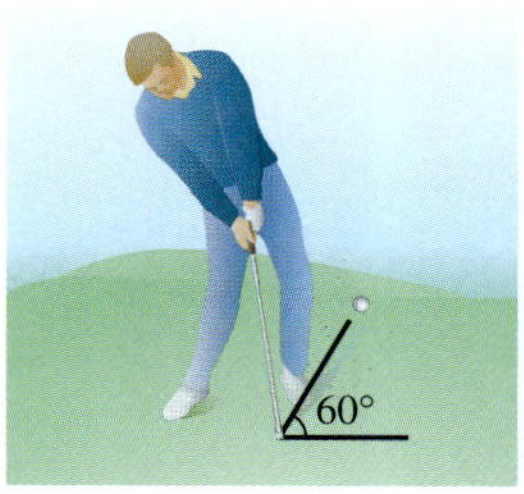

41. *Flight of a Softball* Sally hits a softball when it is 2 ft above the ground. The ball leaves her bat at an angle of 20° with respect to the ground at velocity 88 ft per sec.

42. *Flight of a Baseball* Carlos hits a baseball when it is 2.5 ft above the ground. The ball leaves his bat at an angle of 29° from the horizontal with velocity 136 ft per sec.

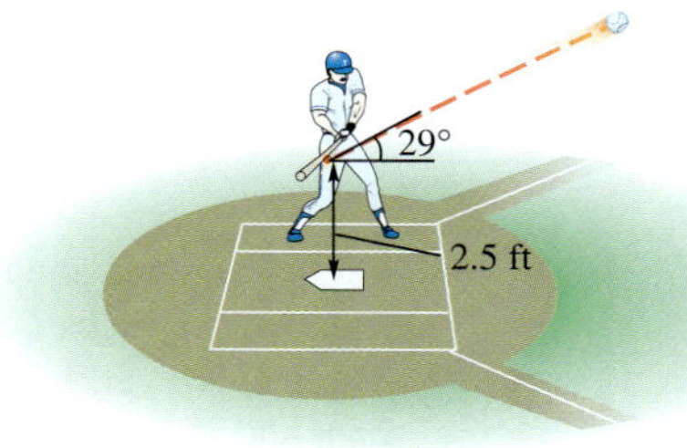

(Modeling) Solve each problem. ***See Examples 7 and 8.***

43. *Path of a Rocket* A rocket is launched from the top of an 8-ft ladder. Its initial velocity is 128 ft per sec, and it is launched at an angle of 60° with respect to the ground.

(a) Find the rectangular equation that models its path. What type of path does the rocket follow?

(b) Determine the total flight time and the horizontal distance the rocket travels.

44. *Simulating Gravity on the Moon* If an object is thrown on the moon, then the parametric equations of flight are

$$x = (v \cos \theta)t \quad \text{and} \quad y = (v \sin \theta)t - 2.66t^2 + h.$$

Estimate the distance that a golf ball hit at 88 ft per sec (60 mph) at an angle of 45° with the horizontal travels on the moon if the moon's surface is level.

45. *Flight of a Baseball* A baseball is hit from a height of 3 ft at a 60° angle above the horizontal. Its initial velocity is 64 ft per sec.

(a) Write parametric equations that model the flight of the baseball.

(b) Determine the horizontal distance traveled by the ball in the air. Assume that the ground is level.

(c) What is the maximum height of the baseball? At that time, how far has the ball traveled horizontally?

(d) Would the ball clear a 5-ft-high fence that is 100 ft from the batter?

(Modeling) Path of a Projectile *In Exercises 46 and 47, a projectile has been launched from the ground with initial velocity 88 ft per sec. The parametric equations modeling the path of the projectile are supplied.*

(a) *Graph the parametric equations.*
(b) *Approximate θ, the angle the projectile makes with the horizontal at launch, to the nearest tenth of a degree.*
(c) *Based on your answer to part (b), write parametric equations for the projectile using the cosine and sine functions.*

46. $x = 82.69295063t, \ y = -16t^2 + 30.09777261t$

47. $x = 56.56530965t, \ y = -16t^2 + 67.41191099t$

48. Give two parametric representations of the line through the point (x_1, y_1) with slope m.

49. Give two parametric representations of the parabola $y = a(x - h)^2 + k$.

50. Give a parametric representation of the rectangular equation $\frac{x^2}{a^2} - \frac{y^2}{b^2} = 1$.

51. Give a parametric representation of the rectangular equation $\frac{x^2}{a^2} + \frac{y^2}{b^2} = 1$.

52. The spiral of Archimedes has polar equation $r = a\theta$, where $r^2 = x^2 + y^2$. Show that a parametric representation of the spiral of Archimedes is

$$x = a\theta \cos \theta, \quad y = a\theta \sin \theta, \qquad \text{for } \theta \text{ in } (-\infty, \infty).$$

53. Show that the **hyperbolic spiral** $r\theta = a$, where $r^2 = x^2 + y^2$, is given parametrically by

$$x = \frac{a \cos \theta}{\theta}, \quad y = \frac{a \sin \theta}{\theta}, \qquad \text{for } \theta \text{ in } (-\infty, 0) \cup (0, \infty).$$

54. The parametric equations $x = \cos t$, $y = \sin t$, for t in $[0, 2\pi]$ and the parametric equations $x = \cos t$, $y = -\sin t$, for t in $[0, 2\pi]$ both have the unit circle as their graph. However, in one case the circle is traced out clockwise (as t moves from 0 to 2π), and in the other case the circle is traced out counterclockwise. For which pair of equations is the circle traced out in the clockwise direction?

Concept Check *Consider the parametric equations $x = f(t)$, $y = g(t)$, for t in $[a, b]$, with $c > 0$, $d > 0$.*

55. How is the graph affected if the equation $x = f(t)$ is replaced by $x = c + f(t)$?

56. How is the graph affected if the equation $y = g(t)$ is replaced by $y = d + g(t)$?

Chapter 8 Test Prep

Key Terms

8.1 imaginary unit, complex number, real part, imaginary part, pure imaginary number, nonreal complex number, standard form, complex conjugates

8.2 real axis, imaginary axis, complex plane, rectangular form of a complex number, trigonometric (polar) form of a complex number, absolute value (modulus), argument

8.4 *n*th root of a complex number

8.5 polar coordinate system, pole, polar axis, polar coordinates, rectangular (Cartesian) equation, polar equation, cardioid, polar grid, rose curve, lemniscate, spiral of Archimedes, limaçon

8.6 plane curve, parametric equations of a plane curve, parameter, cycloid

New Symbols

i	imaginary unit	$a + bi$	complex number

Quick Review

Concepts	Examples

8.1 Complex Numbers

Definition of i

$$i = \sqrt{-1} \quad \text{and} \quad i^2 = -1$$

Definition of Complex Number

$$a + bi$$

(a: Real part; bi: Imaginary part)

In the complex number $3 - 4i$, the real part is 3 and the imaginary part is -4.

Definition of $\sqrt{-a}$

For $a > 0$, $\quad \sqrt{-a} = i\sqrt{a}.$

Simplify.

$$\sqrt{-4} = 2i$$

$$\sqrt{-12} = i\sqrt{12} = 2i\sqrt{3}$$

Adding and Subtracting Complex Numbers
Add or subtract the real parts, and add or subtract the imaginary parts.

$$\begin{aligned}(2 + 3i) + (3 + i) - (2 - i) &= (2 + 3 - 2) + (3 + 1 + 1)i \\ &= 3 + 5i\end{aligned}$$

Multiplying and Dividing Complex Numbers
Multiply complex numbers as with binomials, and use the fact that $i^2 = -1$.

$$\begin{aligned}(6 + i)(3 - 2i) &= 18 - 12i + 3i - 2i^2 && \text{FOIL} \\ &= (18 + 2) + (-12 + 3)i && i^2 = -1 \\ &= 20 - 9i\end{aligned}$$

Divide complex numbers by multiplying the numerator and denominator by the complex conjugate of the denominator.

$$\frac{3 + i}{1 + i} = \frac{(3 + i)(1 - i)}{(1 + i)(1 - i)} = \frac{3 - 3i + i - i^2}{1 - i^2}$$

$$= \frac{4 - 2i}{2} = \frac{2(2 - i)}{2} = 2 - i$$

(continued)

Concepts	Examples

8.2 Trigonometric (Polar) Form of Complex Numbers

Trigonometric (Polar) Form of Complex Numbers

Let the complex number $x + yi$ correspond to the vector with direction angle θ and magnitude r.

$$x = r \cos \theta \qquad y = r \sin \theta$$

$$r = \sqrt{x^2 + y^2} \qquad \tan \theta = \frac{y}{x}, \quad \text{if } x \neq 0$$

The expression

$$r(\cos \theta + i \sin \theta) \quad \text{or} \quad r \text{ cis } \theta$$

is the trigonometric form (or polar form) of $x + yi$.

Write $2(\cos 60° + i \sin 60°)$ in rectangular form.

$$2(\cos 60° + i \sin 60°) = 2\left(\frac{1}{2} + i \cdot \frac{\sqrt{3}}{2}\right)$$
$$= 1 + i\sqrt{3}$$

Write $-\sqrt{2} + i\sqrt{2}$ in trigonometric form.

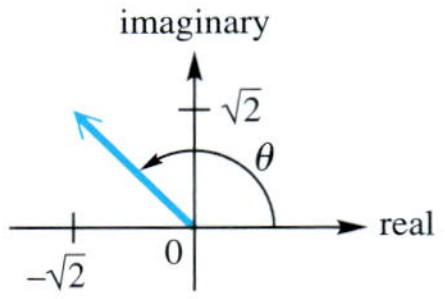

$$r = \sqrt{(-\sqrt{2})^2 + (\sqrt{2})^2} = 2$$

$\tan \theta = -1$ and θ is in quadrant II, so $\theta = 180° - 45° = 135°$. Therefore,

$$-\sqrt{2} + i\sqrt{2} = 2 \text{ cis } 135°.$$

8.3 The Product and Quotient Theorems

Product and Quotient Theorems

For any two complex numbers $r_1(\cos \theta_1 + i \sin \theta_1)$ and $r_2(\cos \theta_2 + i \sin \theta_2)$, the following hold.

$$[r_1(\cos \theta_1 + i \sin \theta_1)] \cdot [r_2(\cos \theta_2 + i \sin \theta_2)]$$
$$= r_1 r_2[\cos(\theta_1 + \theta_2) + i \sin(\theta_1 + \theta_2)]$$

and

$$\frac{r_1(\cos \theta_1 + i \sin \theta_1)}{r_2(\cos \theta_2 + i \sin \theta_2)}$$
$$= \frac{r_1}{r_2}[\cos(\theta_1 - \theta_2) + i \sin(\theta_1 - \theta_2)],$$

where $r_2 \text{ cis } \theta_2 \neq 0$

Let $\quad z_1 = 4(\cos 135° + i \sin 135°)$

and $\quad z_2 = 2(\cos 45° + i \sin 45°).$

$$z_1 z_2 = 8(\cos 180° + i \sin 180°)$$
$$= 8(-1 + i \cdot 0)$$
$$= -8$$

$$\frac{z_1}{z_2} = 2(\cos 90° + i \sin 90°)$$
$$= 2(0 + i \cdot 1)$$
$$= 2i$$

8.4 De Moivre's Theorem; Powers and Roots of Complex Numbers

De Moivre's Theorem

$$[r(\cos \theta + i \sin \theta)]^n = r^n(\cos n\theta + i \sin n\theta)$$

***n*th Root Theorem**

If n is any positive integer, r is a positive real number, and θ is in degrees, then the nonzero complex number $r(\cos \theta + i \sin \theta)$ has exactly n distinct nth roots, given by the following.

$$\sqrt[n]{r}(\cos \alpha + i \sin \alpha), \quad \text{or} \quad \sqrt[n]{r} \text{ cis } \alpha,$$

where

$$\alpha = \frac{\theta + 360° \cdot k}{n}, \quad k = 0, 1, 2, \ldots, n - 1$$

If θ is in radians, then

$$\alpha = \frac{\theta + 2\pi k}{n}, \quad k = 0, 1, 2, \ldots, n - 1.$$

Let $z = 4(\cos 180° + i \sin 180°)$. Find z^3 and the square roots of z.

$$z^3 = 4^3(\cos 3 \cdot 180° + i \sin 3 \cdot 180°)$$
$$= 64(\cos 540° + i \sin 540°)$$
$$= 64(-1 + i \cdot 0)$$
$$= -64$$

For the given z, $r = 4$ and $\theta = 180°$. Its square roots are

$$\sqrt{4}\left(\cos \frac{180°}{2} + i \sin \frac{180°}{2}\right) = 2(0 + i \cdot 1)$$
$$= 2i$$

and $\sqrt{4}\left(\cos \frac{180° + 360°}{2} + i \sin \frac{180° + 360°}{2}\right)$

$$= 2(0 + i(-1))$$
$$= -2i.$$

Concepts	Examples

8.5 Polar Equations and Graphs

Rectangular and Polar Coordinates

The following relationships hold between the point (x, y) in the rectangular coordinate plane and the same point (r, θ) in the polar coordinate plane.

$$x = r\cos\theta \qquad y = r\sin\theta$$

$$r^2 = x^2 + y^2 \qquad \tan\theta = \frac{y}{x}, \quad \text{if } x \neq 0$$

Find the rectangular coordinates for the point $(5, 60°)$ in polar coordinates.

$$x = 5\cos 60° = 5\left(\frac{1}{2}\right) = \frac{5}{2}$$

$$y = 5\sin 60° = 5\left(\frac{\sqrt{3}}{2}\right) = \frac{5\sqrt{3}}{2}$$

The rectangular coordinates are $\left(\frac{5}{2}, \frac{5\sqrt{3}}{2}\right)$.

Find polar coordinates for $(-1, -1)$ in rectangular coordinates.

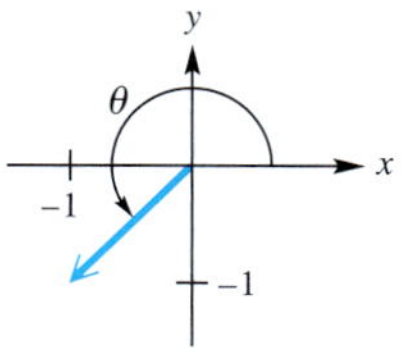

$$r = \sqrt{(-1)^2 + (-1)^2} = \sqrt{2}$$

$\tan\theta = 1$ and θ is in quadrant III, so $\theta = 225°$.

One pair of polar coordinates for $(-1, -1)$ is $(\sqrt{2}, 225°)$.

Polar Equations and Graphs

$r = a\cos\theta$, $r = a\sin\theta$ — Circles

$r^2 = a^2\sin 2\theta$, $r^2 = a^2\cos 2\theta$ — Lemniscates

$r = a \pm b\sin\theta$, $r = a \pm b\cos\theta$ — Limaçons

$r = a\sin n\theta$, $r = a\cos n\theta$ — Rose curves

Graph $r = 4\cos 2\theta$.

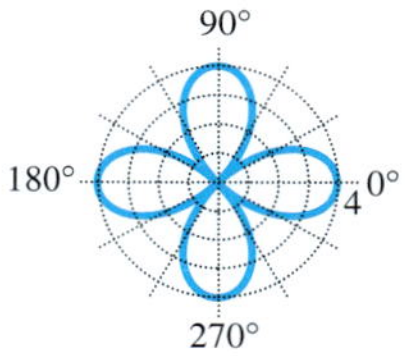

$r = 4\cos 2\theta$

8.6 Parametric Equations, Graphs, and Applications

Plane Curve

A **plane curve** is a set of points (x, y) such that $x = f(t)$, $y = g(t)$, and f and g are both defined on an interval I. The equations

$$x = f(t) \quad \text{and} \quad y = g(t)$$

are **parametric equations** with **parameter *t*.**

Graph $x = 2 - \sin t$, $y = \cos t - 1$, for $0 \le t \le 2\pi$.

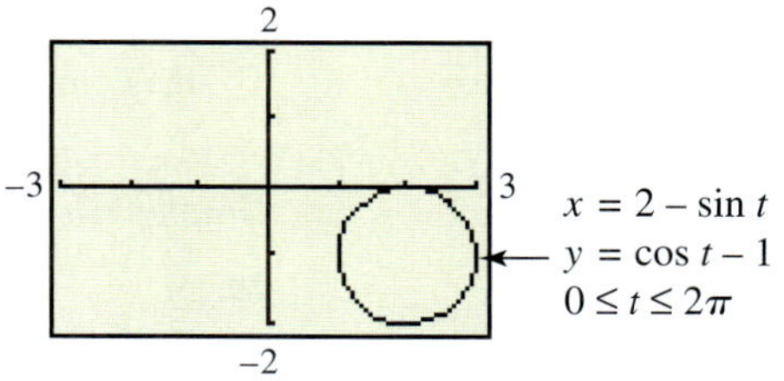

Flight of an Object

If an object has initial velocity v, has initial height h, and travels such that its initial angle of elevation is θ, then its flight after t seconds is modeled by the following parametric equations.

$$x = (v\cos\theta)t \quad \text{and} \quad y = (v\sin\theta)t - 16t^2 + h$$

Joe kicks a football from the ground at an angle of 45° with a velocity of 48 ft per sec. Give the parametric equations that model the path of the football and the distance it travels before hitting the ground.

$$x = (48\cos 45°)t = 24\sqrt{2}t$$

$$y = (48\sin 45°)t - 16t^2 = 24\sqrt{2}t - 16t^2$$

When the ball hits the ground, $y = 0$.

$24\sqrt{2}t - 16t^2 = 0$ — Substitute $y = 0$.

$8t(3\sqrt{2} - 2t) = 0$ — Factor.

$t = 0$ (Reject) or $t = \frac{3\sqrt{2}}{2}$ — Zero-factor property

The distance it travels is $x = 24\sqrt{2}\left(\frac{3\sqrt{2}}{2}\right) = 72$ ft.

Chapter 8 Review Exercises

Write each number as the product of a real number and i.

1. $\sqrt{-9}$

2. $\sqrt{-12}$

Solve each quadratic equation over the set of complex numbers.

3. $x^2 = -81$

4. $x(2x + 3) = -4$

Perform each operation. Write answers in rectangular form.

5. $(1 - i) - (3 + 4i) + 2i$

6. $(2 - 5i) + (9 - 10i) - 3$

7. $(6 - 5i) + (2 + 7i) - (3 - 2i)$

8. $(4 - 2i) - (6 + 5i) - (3 - i)$

9. $(3 + 5i)(8 - i)$

10. $(4 - i)(5 + 2i)$

11. $(2 + 6i)^2$

12. $(6 - 3i)^2$

13. $(1 - i)^3$

14. $(2 + i)^3$

15. $\dfrac{25 - 19i}{5 + 3i}$

16. $\dfrac{2 - 5i}{1 + i}$

17. $\dfrac{2 + i}{1 - 5i}$

18. $\dfrac{3 + 2i}{i}$

19. i^{53}

20. i^{-41}

Perform each operation. Write answers in rectangular form.

21. $[5(\cos 90° + i \sin 90°)][6(\cos 180° + i \sin 180°)]$

22. $[3 \text{ cis } 135°][2 \text{ cis } 105°]$

23. $\dfrac{2(\cos 60° + i \sin 60°)}{8(\cos 300° + i \sin 300°)}$

24. $\dfrac{4 \text{ cis } 270°}{2 \text{ cis } 90°}$

25. $(\sqrt{3} + i)^3$

26. $(2 - 2i)^5$

27. $(\cos 100° + i \sin 100°)^6$

28. *Concept Check* The vector representing a real number will lie on the ____-axis in the complex plane.

Graph each complex number as a vector.

29. $5i$

30. $-4 + 2i$

31. $3 - 3i\sqrt{3}$

32. Find the sum of $7 + 3i$ and $-2 + i$. Graph both complex numbers and their resultant.

Perform each conversion, using a calculator to approximate answers as necessary.

	Rectangular Form	Trigonometric Form
33.	$-2 + 2i$	________
34.	________	$3(\cos 90° + i \sin 90°)$
35.	________	$2(\cos 225° + i \sin 225°)$
36.	$-4 + 4i\sqrt{3}$	________
37.	$1 - i$	________
38.	________	$4 \text{ cis } 240°$
39.	$-4i$	________
40.	________	$7 \text{ cis } 310°$

Concept Check The complex number z, where $z = x + yi$, can be graphed in the plane as (x, y). Describe the graph of all complex numbers z satisfying the given conditions.

41. The imaginary part of z is the negative of the real part of z.

42. The absolute value of z is 2.

Find all roots as indicated. Express them in trigonometric form.

43. the cube roots of $1 - i$

44. the fifth roots of $-2 + 2i$

45. *Concept Check* How many real sixth roots does -64 have?

46. *Concept Check* How many real fifth roots does -32 have?

Solve each equation. Leave answers in trigonometric form.

47. $x^4 + 16 = 0$

48. $x^3 + 125 = 0$

49. $x^2 + i = 0$

50. Convert $(5, 315°)$ to rectangular coordinates.

51. Convert $(-1, \sqrt{3})$ to polar coordinates, with $0° \le \theta < 360°$ and $r > 0$.

52. *Concept Check* What will the graph of $r = k$ be, for $k > 0$?

Identify and graph each polar equation for θ in $[0°, 360°)$.

53. $r = 4 \cos \theta$

54. $r = -1 + \cos \theta$

55. $r = 2 \sin 4\theta$

56. $r = \dfrac{2}{2 \cos \theta - \sin \theta}$

Find an equivalent equation in rectangular coordinates.

57. $r = \dfrac{3}{1 + \cos \theta}$

58. $r = \sin \theta + \cos \theta$

59. $r = 2$

Find an equivalent equation in polar coordinates.

60. $y = x$

61. $y = x^2$

62. $x^2 + y^2 = 25$

In Exercises 63–66, identify the geometric symmetry (A, B, or C) that the graph will possess.

A. *symmetry with respect to the origin*

B. *symmetry with respect to the y-axis*

C. *symmetry with respect to the x-axis*

63. Whenever (r, θ) is on the graph, so is $(-r, -\theta)$.

64. Whenever (r, θ) is on the graph, so is $(-r, \theta)$.

65. Whenever (r, θ) is on the graph, so is $(r, -\theta)$.

66. Whenever (r, θ) is on the graph, so is $(r, \pi - \theta)$.

In Exercises 67–70, find a polar equation having the given graph.

67.

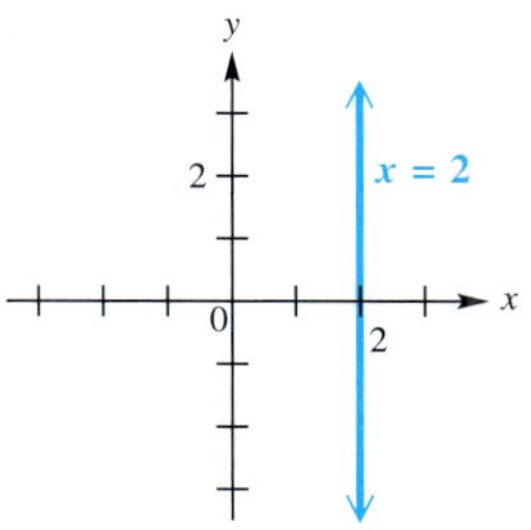

68.

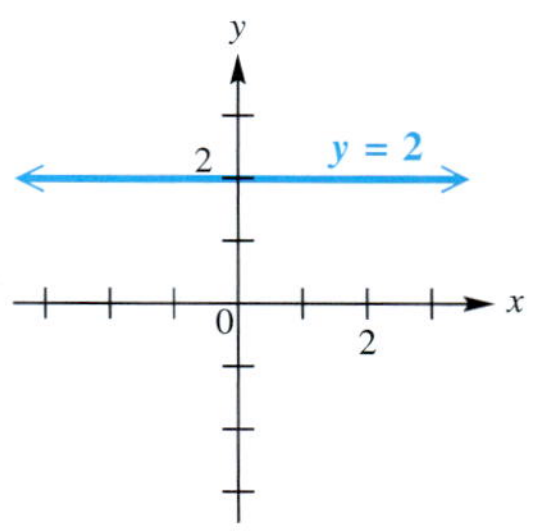

69.

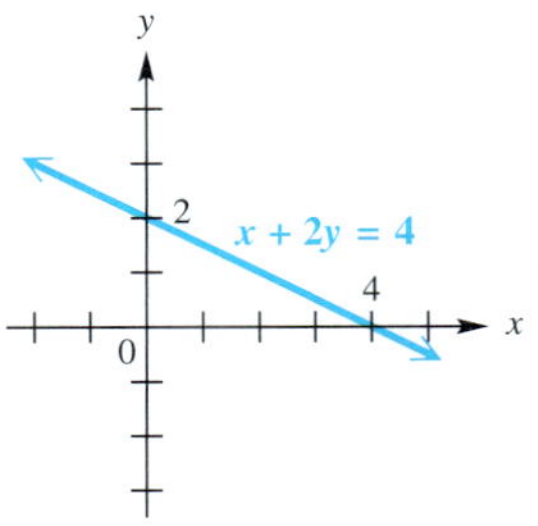

70.

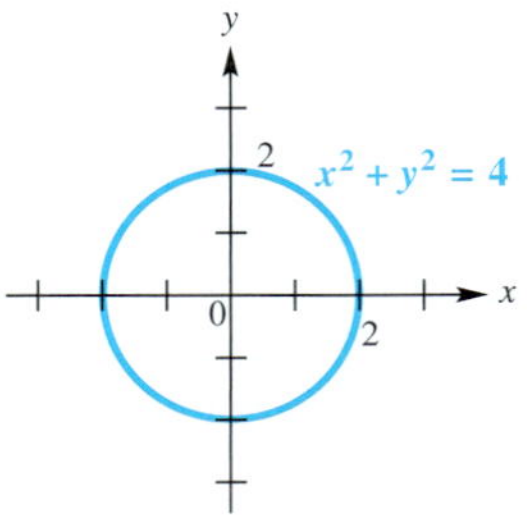

71. Graph the plane curve defined by the parametric equations $x = t + \cos t$, $y = \sin t$, for t in $[0, 2\pi]$.

72. Show that the distance between (r_1, θ_1) and (r_2, θ_2) in polar coordinates is given by

$$d = \sqrt{r_1{}^2 + r_2{}^2 - 2r_1r_2\cos(\theta_1 - \theta_2)}.$$

Find a rectangular equation for each plane curve with the given parametric equations.

73. $x = \sqrt{t-1}$, $y = \sqrt{t}$, for t in $[1, \infty)$

74. $x = 3t + 2$, $y = t - 1$, for t in $[-5, 5]$

75. $x = 5\tan t$, $y = 3\sec t$, for t in $\left(-\frac{\pi}{2}, \frac{\pi}{2}\right)$

76. $x = t^2 + 5$, $y = \frac{1}{t^2+1}$, for t in $(-\infty, \infty)$

77. $x = \cos 2t$, $y = \sin t$, for t in $(-\pi, \pi)$

78. Find a pair of parametric equations whose graph is the circle having center $(3, 4)$ and passing through the origin.

79. *Flight of a Baseball* A batter hits a baseball when it is 3.2 ft above the ground. It leaves the bat with velocity 118 ft per sec at an angle of 27° with respect to the ground.

(a) Determine the parametric equations that model the path of the baseball.
(b) Determine the rectangular equation that models the path of the baseball.
(c) Determine approximately how long the projectile is in flight and the horizontal distance it covers.

80. *Mandelbrot Set* Consider the complex number $z = 1 + i$. Compute the value of $z^2 + z$, and show that its absolute value exceeds 2, indicating that $1 + i$ is not in the Mandelbrot set.

Chapter 8 Test

1. Multiply or divide as indicated. Simplify each answer.

(a) $\sqrt{-8} \cdot \sqrt{-6}$ **(b)** $\frac{\sqrt{-2}}{\sqrt{8}}$ **(c)** $\frac{\sqrt{-20}}{\sqrt{-180}}$

2. For the complex numbers $w = 2 - 4i$ and $z = 5 + i$, find each of the following in rectangular form.

(a) $w + z$ (and give a geometric representation) **(b)** $w - z$ **(c)** wz **(d)** $\frac{w}{z}$

3. Express each of the following in rectangular form.

(a) i^{15} (b) $(1 + i)^2$

4. Solve $2x^2 - x + 4 = 0$ over the set of complex numbers.

5. Write each complex number in trigonometric (polar) form, where $0° \le \theta < 360°$.

(a) $3i$ (b) $1 + 2i$ (c) $-1 - i\sqrt{3}$

6. Write each complex number in rectangular form.

(a) $3(\cos 30° + i \sin 30°)$ (b) $4 \text{ cis } 40°$ (c) $3(\cos 90° + i \sin 90°)$

7. For the complex numbers $w = 8(\cos 40° + i \sin 40°)$ and $z = 2(\cos 10° + i \sin 10°)$, find each of the following in the form specified.

(a) wz (trigonometric form) (b) $\frac{w}{z}$ (rectangular form) (c) z^3 (rectangular form)

8. Find the four complex fourth roots of $-16i$. Express them in trigonometric form.

9. Convert the given rectangular coordinates to polar coordinates. Give two pairs of polar coordinates for each point.

(a) $(0, 5)$ (b) $(-2, -2)$

10. Convert the given polar coordinates to rectangular coordinates.

(a) $(3, 315°)$ (b) $(-4, 90°)$

Identify and graph each polar equation for θ in $[0°, 360°)$.

11. $r = 1 - \cos \theta$

12. $r = 3 \cos 3\theta$

13. Convert each polar equation to a rectangular equation, and sketch its graph.

(a) $r = \dfrac{4}{2 \sin \theta - \cos \theta}$ (b) $r = 6$

Graph each pair of parametric equations.

14. $x = 4t - 3,\ y = t^2,$ for t in $[-3, 4]$

15. $x = 2 \cos 2t,\ y = 2 \sin 2t,$ for t in $[0, 2\pi]$

16. *Julia Set* Consider the complex number $z = -1 + i$. Compute the value of $z^2 - 1$, and show that its absolute value exceeds 2, indicating that $-1 + i$ is not in the Julia set.

Appendices

A Equations and Inequalities

- Equations
- Solving Linear Equations
- Solving Quadratic Equations
- Inequalities
- Solving Linear Inequalities and Using Interval Notation
- Solving Three-Part Inequalities

Equations

An **equation** is a statement that two expressions are equal.

$$x + 2 = 9, \quad 11x = 5x + 6x, \quad x^2 - 2x - 1 = 0 \quad \text{Equations}$$

To *solve* an equation means to find all numbers that make the equation a true statement. These numbers are the **solutions,** or **roots,** of the equation. A number that is a solution of an equation is said to *satisfy* the equation, and the solutions of an equation make up its **solution set.** Equations with the same solution set are **equivalent equations.** For example,

$$x = 4, \quad x + 1 = 5, \quad \text{and} \quad 6x + 3 = 27 \quad \text{are equivalent equations}$$

because they have the same solution set, $\{4\}$. However, the equations

$$x^2 = 9 \quad \text{and} \quad x = 3 \quad \text{are } \textit{not} \text{ equivalent,}$$

since the first has solution set $\{-3, 3\}$ while the solution set of the second is $\{3\}$.

One way to solve an equation is to rewrite it as a series of simpler equivalent equations using the **addition and multiplication properties of equality.** Let a, b, and c represent real numbers.

$$\textbf{If } a = b, \textbf{ then } a + c = b + c.$$

$$\textbf{If } a = b \textbf{ and } c \neq 0, \textbf{ then } ac = bc.$$

These properties can be extended: The same number may be subtracted from each side of an equation, and each side may be divided by the same nonzero number, without changing the solution set.

Solving Linear Equations

We use the properties of equality to solve *linear equations.*

Linear Equation in One Variable

A **linear equation in one variable** is an equation that can be written in the form

$$ax + b = 0,$$

where a and b are real numbers with $a \neq 0$.

A linear equation is a **first-degree equation** since the greatest degree of the variable is 1.

$$3x + \sqrt{2} = 0, \quad \frac{3}{4}x = 12, \quad 0.5(x + 3) = 2x - 6 \quad \text{Linear equations}$$

$$\sqrt{x} + 2 = 5, \quad \frac{1}{x} = -8, \quad x^2 + 3x + 0.2 = 0 \quad \text{Nonlinear equations}$$

EXAMPLE 1 Solving a Linear Equation

Solve $3(2x - 4) = 7 - (x + 5)$.

SOLUTION

$3(2x - 4) = 7 - (x + 5)$ — Be careful with signs.

$6x - 12 = 7 - x - 5$ — Distributive property

$6x - 12 = 2 - x$ — Combine like terms.

$6x - 12 + x = 2 - x + x$ — Add x to each side.

$7x - 12 = 2$ — Combine like terms.

$7x - 12 + 12 = 2 + 12$ — Add 12 to each side.

$7x = 14$ — Combine like terms.

$\frac{7x}{7} = \frac{14}{7}$ — Divide each side by 7.

$x = 2$

CHECK (A check of the solution is recommended.)

$3(2x - 4) = 7 - (x + 5)$ — Original equation

$3(2 \cdot 2 - 4) \stackrel{?}{=} 7 - (2 + 5)$ — Let $x = 2$.

$3(4 - 4) \stackrel{?}{=} 7 - (7)$ — Work inside the parentheses.

$0 = 0$ ✓ — True

Since replacing x with 2 results in a true statement, 2 is a solution of the given equation. The solution set is $\{2\}$.

✔ *Now Try Exercise 9.*

EXAMPLE 2 Solving a Linear Equation with Fractions

Solve $\frac{2x + 4}{3} + \frac{1}{2}x = \frac{1}{4}x - \frac{7}{3}$.

SOLUTION

$$\frac{2x + 4}{3} + \frac{1}{2}x = \frac{1}{4}x - \frac{7}{3}$$

Distribute to *all* terms within the parentheses.

$12\left(\frac{2x + 4}{3} + \frac{1}{2}x\right) = 12\left(\frac{1}{4}x - \frac{7}{3}\right)$ — Multiply by 12, the LCD of the fractions.

$12\left(\frac{2x + 4}{3}\right) + 12\left(\frac{1}{2}x\right) = 12\left(\frac{1}{4}x\right) - 12\left(\frac{7}{3}\right)$ — Distributive property

$4(2x + 4) + 6x = 3x - 28$ — Multiply.

$8x + 16 + 6x = 3x - 28$ — Distributive property

$14x + 16 = 3x - 28$ — Combine like terms.

$11x = -44$ — Subtract $3x$. Subtract 16.

$x = -4$ — Divide each side by 11.

CHECK

$\frac{2x + 4}{3} + \frac{1}{2}x = \frac{1}{4}x - \frac{7}{3}$ — Original equation

$\frac{2(-4) + 4}{3} + \frac{1}{2}(-4) \stackrel{?}{=} \frac{1}{4}(-4) - \frac{7}{3}$ — Let $x = -4$.

$$\frac{-4}{3} + (-2) \stackrel{?}{=} -1 - \frac{7}{3} \quad \text{Simplify.}$$

$$-\frac{10}{3} = -\frac{10}{3} \checkmark \quad \text{True}$$

The solution set is $\{-4\}$.

✔ **Now Try Exercise 11.**

An equation satisfied by every number that is a meaningful replacement for the variable is an **identity.**

$$3(x + 1) = 3x + 3 \quad \text{Identity}$$

An equation that is satisfied by some numbers but not others is a **conditional equation.** The equations in **Examples 1 and 2** are conditional equations.

$$2x = 4 \quad \text{Conditional equation}$$

An equation that has no solution is a **contradiction.**

$$x = x + 1 \quad \text{Contradiction}$$

EXAMPLE 3 Identifying Types of Equations

Determine whether each equation is an *identity,* a *conditional equation,* or a *contradiction.* Give the solution set.

(a) $-2(x + 4) + 3x = x - 8$ **(b)** $5x - 4 = 11$ **(c)** $3(3x - 1) = 9x + 7$

SOLUTION

(a) $-2(x + 4) + 3x = x - 8$

$$-2x - 8 + 3x = x - 8 \quad \text{Distributive property}$$

$$x - 8 = x - 8 \quad \text{Combine like terms.}$$

$$0 = 0 \quad \text{Subtract } x. \text{ Add 8.}$$

When a *true* statement such as $0 = 0$ results, the equation is an identity, and the solution set is **{all real numbers}.**

(b) $5x - 4 = 11$

$$5x = 15 \quad \text{Add 4 to each side.}$$

$$x = 3 \quad \text{Divide each side by 5.}$$

This is a conditional equation, and its solution set is $\{3\}$.

(c) $3(3x - 1) = 9x + 7$

$$9x - 3 = 9x + 7 \quad \text{Distributive property}$$

$$-3 = 7 \quad \text{Subtract } 9x.$$

When a *false* statement such as $-3 = 7$ results, the equation is a contradiction, and the solution set is the **empty set,** or **null set,** symbolized $\emptyset$.

✔ **Now Try Exercises 23, 25, and 27.**

Solving Quadratic Equations A *quadratic equation* is defined as follows.

Quadratic Equation in One Variable

An equation that can be written in the form

$$ax^2 + bx + c = 0,$$

where a, b, and c are real numbers with $a \neq 0$, is a **quadratic equation.** The given form is called **standard form.**

A quadratic equation is a **second-degree equation**—that is, an equation with a squared variable term and no terms of greater degree.

$$x^2 = 25, \quad 4x^2 + 4x - 5 = 0, \quad 3x^2 = 4x - 8 \quad \text{Quadratic equations}$$

Factoring, the simplest method of solving a quadratic equation, depends on the **zero-factor property.**

If a and b are complex numbers with $ab = 0$, then $a = 0$ or $b = 0$ or both equal zero.

EXAMPLE 4 Using the Zero-Factor Property

Solve $6x^2 + 7x = 3$.

SOLUTION

Don't factor out x here.

$$6x^2 + 7x = 3$$

$$6x^2 + 7x - 3 = 0 \quad \text{Standard form}$$

$$(3x - 1)(2x + 3) = 0 \quad \text{Factor.}$$

$$3x - 1 = 0 \quad \text{or} \quad 2x + 3 = 0 \quad \text{Zero-factor property}$$

$$3x = 1 \quad \text{or} \quad 2x = -3 \quad \text{Solve each equation.}$$

$$x = \frac{1}{3} \quad \text{or} \quad x = -\frac{3}{2}$$

CHECK $6x^2 + 7x = 3$ Original equation

$$6\left(\frac{1}{3}\right)^2 + 7\left(\frac{1}{3}\right) \stackrel{?}{=} 3 \quad \text{Let } x = \tfrac{1}{3}. \qquad 6\left(-\frac{3}{2}\right)^2 + 7\left(-\frac{3}{2}\right) \stackrel{?}{=} 3 \quad \text{Let } x = -\tfrac{3}{2}.$$

$$\frac{6}{9} + \frac{7}{3} \stackrel{?}{=} 3 \qquad \frac{54}{4} - \frac{21}{2} \stackrel{?}{=} 3$$

$$3 = 3 \checkmark \text{ True} \qquad 3 = 3 \checkmark \text{ True}$$

Both values check, since true statements result. The solution set is $\left\{\frac{1}{3}, -\frac{3}{2}\right\}$.

✔ *Now Try Exercise 35.*

A quadratic equation of the form $x^2 = k$ can be solved by the **square root property.**

$$\textbf{If } x^2 = k, \quad \textbf{then} \quad x = \sqrt{k} \quad \textbf{or} \quad x = -\sqrt{k}.$$

That is, the solution set of $x^2 = k$ is $\left\{\sqrt{k}, -\sqrt{k}\right\}$, which may be abbreviated $\left\{\pm\sqrt{k}\right\}$.

EXAMPLE 5 Using the Square Root Property

Solve each quadratic equation.

(a) $x^2 = 17$ **(b)** $(x-4)^2 = 12$

SOLUTION

(a) By the square root property, the solution set of $x^2 = 17$ is $\{\pm\sqrt{17}\}$.

(b)

$$(x-4)^2 = 12$$

$$x - 4 = \pm\sqrt{12} \quad \text{Generalized square root property}$$

$$x = 4 \pm \sqrt{12} \quad \text{Add 4.}$$

$$x = 4 \pm 2\sqrt{3} \quad \sqrt{12} = \sqrt{4 \cdot 3} = 2\sqrt{3}$$

CHECK $(x-4)^2 = 12$ Original equation

Let $x = 4 + 2\sqrt{3}$.

$$(4 + 2\sqrt{3} - 4)^2 \stackrel{?}{=} 12$$

$$(2\sqrt{3})^2 \stackrel{?}{=} 12$$

$$2^2 \cdot (\sqrt{3})^2 \stackrel{?}{=} 12$$

$$12 = 12 \ ✓ \ \text{True}$$

Let $x = 4 - 2\sqrt{3}$.

$$(4 - 2\sqrt{3} - 4)^2 \stackrel{?}{=} 12$$

$$(-2\sqrt{3})^2 \stackrel{?}{=} 12$$

$$(-2)^2 \cdot (\sqrt{3})^2 \stackrel{?}{=} 12$$

$$12 = 12 \ ✓ \ \text{True}$$

The solution set is $\{4 \pm 2\sqrt{3}\}$.

✓ ***Now Try Exercises 45 and 49.***

Any quadratic equation can be solved by the **quadratic formula,** which says that the solutions of the quadratic equation $ax^2 + bx + c = 0$, where $a \neq 0$, are given by

$$x = \frac{-b \pm \sqrt{b^2 - 4ac}}{2a}.$$

This formula is derived in algebra courses.

EXAMPLE 6 Using the Quadratic Formula

Solve $x^2 - 4x = -2$ using the quadratic formula.

SOLUTION

$$x^2 - 4x + 2 = 0 \quad \text{Write in standard form. Here } a = 1, b = -4, \text{ and } c = 2.$$

$$x = \frac{-b \pm \sqrt{b^2 - 4ac}}{2a} \quad \text{Quadratic formula}$$

$$x = \frac{-(-4) \pm \sqrt{(-4)^2 - 4(1)(2)}}{2(1)} \quad \text{Substitute } a = 1, b = -4, \text{ and } c = 2.$$

The fraction bar extends *under* $-b$.

$$x = \frac{4 \pm \sqrt{16 - 8}}{2} \quad \text{Simplify.}$$

$$x = \frac{4 \pm 2\sqrt{2}}{2} \quad \sqrt{16 - 8} = \sqrt{8} = \sqrt{4 \cdot 2} = 2\sqrt{2}$$

$$x = \frac{2(2 \pm \sqrt{2})}{2} \quad \text{Factor out 2 in the numerator.}$$

Factor first, then divide.

$$x = 2 \pm \sqrt{2} \quad \text{Lowest terms}$$

The solution set is $\{2 \pm \sqrt{2}\}$.

✓ ***Now Try Exercise 55.***

Inequalities An **inequality** says that one expression is greater than, greater than or equal to, less than, or less than or equal to another. As with equations, a value of the variable for which the inequality is true is a solution of the inequality, and the set of all solutions is the solution set of the inequality. Two inequalities with the same solution set are equivalent.

Inequalities are solved with the **properties of inequality.** For real numbers a, b, and c:

1. **If $a < b$, then $a + c < b + c$.**
2. **If $a < b$ and if $c > 0$, then $ac < bc$.**
3. **If $a < b$ and if $c < 0$, then $ac > bc$.**

Replacing $<$ with $>$, $\leq$, or $\geq$ results in similar properties. (Restrictions on c remain the same.) Multiplication may be replaced by division in Properties 2 and 3. ***Always remember to reverse the direction of the inequality symbol when multiplying or dividing by a negative number.***

Solving Linear Inequalities and Using Interval Notation The definition of a *linear inequality* is similar to the definition of a linear equation.

Linear Inequality in One Variable

A **linear inequality in one variable** is an inequality that can be written in the form

$$ax + b > 0,$$

where a and b are real numbers, with $a \neq 0$. (Any of the symbols $\geq$, $<$, and $\leq$ may also be used.)

EXAMPLE 7 Solving a Linear Inequality

Solve $-3x + 5 > -7$.

SOLUTION

$$-3x + 5 > -7$$

$$-3x + 5 - 5 > -7 - 5 \quad \text{Subtract 5.}$$

$$-3x > -12 \quad \text{Combine like terms.}$$

$$\frac{-3x}{-3} < \frac{-12}{-3} \quad \text{Divide by } -3. \text{ Reverse the direction of the inequality symbol when multiplying or dividing by a negative number.}$$

Don't forget to reverse the inequality symbol here.

$$x < 4$$

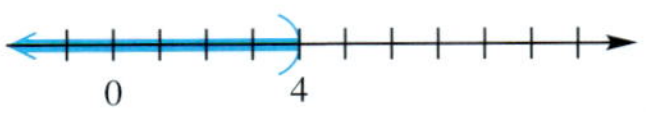

Figure 1

Thus, the original inequality $-3x + 5 > -7$ is satisfied by any real number less than 4. The solution set can be written using **set-builder notation** as $\{x \mid x < 4\}$, which is read "the set of all x such that x is less than 4." A graph of the solution set is shown in **Figure 1**, where the parenthesis is used to show that 4 itself does not belong to the solution set.

The solution set $\{x \mid x < 4\}$ is an example of an **interval.** We can use **interval notation** to write intervals. With this notation, we write the interval as

$$(-\infty, 4). \quad \text{Interval notation}$$

The symbol $-\infty$ does not represent an actual number. Rather it is used to show that the interval includes all real numbers less than 4. The interval $(-\infty, 4)$ is an example of an **open interval,** since the endpoint, 4, is not part of the interval.

A **closed interval** includes both endpoints. A square bracket is used to show that a number *is* part of the graph, and a parenthesis is used to indicate that a number *is not* part of the graph.

✔ ***Now Try Exercise 79.***

In the table that follows, we assume that $a < b$.

Type of Interval	Set	Interval Notation	Graph
Open interval	$\{x \mid x > a\}$	(a, ∞)	
	$\{x \mid a < x < b\}$	(a, b)	
	$\{x \mid x < b\}$	$(-\infty, b)$	
Other intervals	$\{x \mid x \geq a\}$	$[a, \infty)$	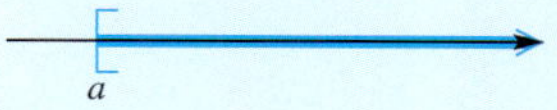
	$\{x \mid a < x \leq b\}$	$(a, b]$	
	$\{x \mid a \leq x < b\}$	$[a, b)$	
	$\{x \mid x \leq b\}$	$(-\infty, b]$	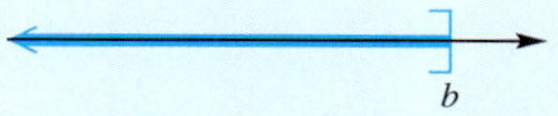
Closed interval	$\{x \mid a \leq x \leq b\}$	$[a, b]$	
Disjoint interval	$\{x \mid x < a \text{ or } x > b\}$	$(-\infty, a) \cup (b, \infty)$	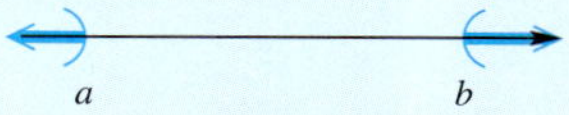
All real numbers	$\{x \mid x \text{ is a real number}\}$	$(-\infty, \infty)$	

Solving Three-Part Inequalities The inequality $-2 < 5 + 3x < 20$ says that $5 + 3x$ is *between* -2 and 20. This inequality is solved using an extension of the properties of inequality given earlier, working with all three expressions at the same time.

EXAMPLE 8 Solving a Three-Part Inequality

Solve $-2 < 5 + 3x < 20$.

SOLUTION

$$-2 < 5 + 3x < 20$$

$$-2 - 5 < 5 + 3x - 5 < 20 - 5 \quad \text{Subtract 5 from each part.}$$

$$-7 < 3x < 15 \quad \text{Combine like terms in each part.}$$

$$\frac{-7}{3} < \frac{3x}{3} < \frac{15}{3} \quad \text{Divide each part by 3.}$$

$$-\frac{7}{3} < x < 5 \quad \text{Simplify.}$$

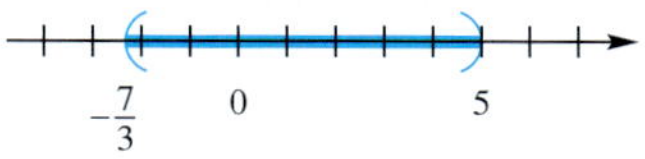

Figure 2

The solution set, graphed in **Figure 2,** is the interval $\left(-\frac{7}{3}, 5\right)$.

✔ ***Now Try Exercise 91.***

Appendix A Exercises

Concept Check In Exercises 1–4, decide whether each statement is true or false.

1. The solution set of $2x + 5 = x - 3$ is $\{-8\}$.

2. The equation $5(x - 8) = 5x - 40$ is an example of an identity.

3. The equations $x^2 = 4$ and $x + 2 = 4$ are equivalent equations.

4. It is possible for a linear equation to have exactly two solutions.

5. *Concept Check* Which one is not a linear equation?

A. $5x + 7(x - 1) = -3x$ **B.** $9x^2 - 4x + 3 = 0$

C. $7x + 8x = 13x$ **D.** $0.04x - 0.08x = 0.40$

6. In solving the equation $3(2x - 8) = 6x - 24$, a student obtains the result $0 = 0$ and gives the solution set $\{0\}$. Is this correct? Explain.

Solve each equation. ***See Examples 1 and 2.***

7. $5x + 4 = 3x - 4$

8. $9x + 11 = 7x + 1$

9. $6(3x - 1) = 8 - (10x - 14)$

10. $4(-2x + 1) = 6 - (2x - 4)$

11. $\frac{5}{6}x - 2x + \frac{4}{3} = \frac{5}{3}$

12. $\frac{7}{4} + \frac{1}{5}x - \frac{3}{2} = \frac{4}{5}x$

13. $3x + 5 - 5(x + 1) = 6x + 7$

14. $5(x + 3) + 4x - 3 = -(2x - 4) + 2$

15. $2[x - (4 + 2x) + 3] = 2x + 2$

16. $4[2x - (3 - x) + 5] = -6x - 28$

17. $\frac{1}{14}(3x - 2) = \frac{x + 10}{10}$

18. $\frac{1}{15}(2x + 5) = \frac{x + 2}{9}$

19. $0.2x - 0.5 = 0.1x + 7$

20. $0.01x + 3.1 = 2.03x - 2.96$

21. $-4(2x - 6) + 8x = 5x + 24 + x$

22. $-8(3x + 4) + 6x = 4(x - 8) + 4x$

Determine whether each equation is an identity, *a* conditional equation, *or a* contradiction. *Give the solution set.* ***See Example 3.***

23. $4(2x + 7) = 2x + 22 + 3(2x + 2)$

24. $\frac{1}{2}(6x + 20) = x + 4 + 2(x + 3)$

25. $2(x - 8) = 3x - 16$

26. $-8(x + 5) = -8x - 5(x + 8)$

27. $4(x + 7) = 2(x + 12) + 2(x + 1)$

28. $-6(2x + 1) - 3(x - 4) = -15x + 1$

Concept Check Use choices A–D to answer each question in Exercises 29–32.

A. $3x^2 - 17x - 6 = 0$ **B.** $(2x + 5)^2 = 7$

C. $x^2 + x = 12$ **D.** $(3x - 1)(x - 7) = 0$

29. Which equation is set up for direct use of the zero-factor property? Solve it.

30. Which equation is set up for direct use of the square root property? Solve it.

31. Which one or more of these equations can be solved by the quadratic formula?

32. Only one of the equations is set up so that the values of a, b, and c for the quadratic formula can be determined immediately. Which one is it? Solve it.

Solve each equation by the zero-factor property. ***See Example 4.***

33. $x^2 - 5x + 6 = 0$ **34.** $x^2 + 2x - 8 = 0$ **35.** $5x^2 - 3x - 2 = 0$

36. $2x^2 - x - 15 = 0$ **37.** $-4x^2 + x = -3$ **38.** $-6x^2 + 7x = -10$

39. $x^2 - 100 = 0$ **40.** $x^2 - 64 = 0$ **41.** $4x^2 - 4x + 1 = 0$

42. $9x^2 - 12x + 4 = 0$ **43.** $25x^2 + 30x + 9 = 0$ **44.** $36x^2 + 60x + 25 = 0$

Solve each equation by the square root property. ***See Example 5.***

45. $x^2 = 16$ **46.** $x^2 = 121$ **47.** $27 - x^2 = 0$

48. $48 - x^2 = 0$ **49.** $(3x - 1)^2 = 12$ **50.** $(4x + 1)^2 = 20$

Solve each equation by the quadratic formula. ***See Example 6.***

51. $x^2 - 4x + 3 = 0$ **52.** $x^2 - 7x + 12 = 0$ **53.** $2x^2 - x - 28 = 0$

54. $4x^2 - 3x - 10 = 0$ **55.** $x^2 - 2x - 2 = 0$ **56.** $x^2 - 10x + 18 = 0$

57. $2x^2 + x = 10$ **58.** $3x^2 + 2x = 5$ **59.** $-2x^2 + 4x + 3 = 0$

60. $-3x^2 + 6x + 5 = 0$ **61.** $\frac{1}{2}x^2 + \frac{1}{4}x - 3 = 0$ **62.** $\frac{2}{3}x^2 + \frac{1}{4}x = 3$

63. $0.2x^2 + 0.4x - 0.3 = 0$ **64.** $0.1x^2 - 0.1x = 0.3$

65. $(4x - 1)(x + 2) = 4x$ **66.** $(3x + 2)(x - 1) = 3x$

Concept Check *Match the inequality in each exercise in Column I with its equivalent interval notation in Column II.*

I

67. $x < -6$

68. $x \le 6$

69. $-2 < x \le 6$

70. $x^2 \ge 0$

71. $x \ge -6$

72. $6 \le x$

73. 

74. 

75. 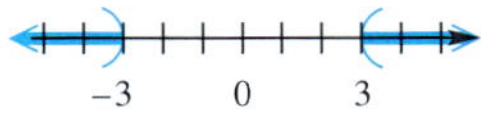

76. 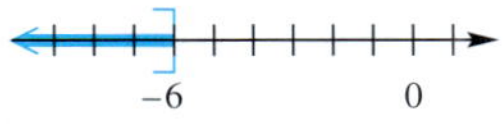

II

A. $(-2, 6]$

B. $[-2, 6)$

C. $(-\infty, -6]$

D. $[6, \infty)$

E. $(-\infty, -3) \cup (3, \infty)$

F. $(-\infty, -6)$

G. $(0, 8)$

H. $(-\infty, \infty)$

I. $[-6, \infty)$

J. $(-\infty, 6]$

77. Explain how to determine whether to use a parenthesis or a square bracket when graphing the solution set of a linear inequality.

78. **Concept Check** The three-part inequality $a < x < b$ means "a is less than x and x is less than b." Which one of the following inequalities is not satisfied by some real number x?

A. $-3 < x < 10$ **B.** $0 < x < 6$

C. $-3 < x < -1$ **D.** $-8 < x < -10$

Solve each inequality. Write each solution set in interval notation. **See Example 7.**

79. $-2x + 8 \le 16$

80. $-3x - 8 \le 7$

81. $-2x - 2 \le 1 + x$

82. $-4x + 3 \ge -2 + x$

83. $3(x + 5) + 1 \ge 5 + 3x$

84. $6x - (2x + 3) \ge 4x - 5$

85. $8x - 3x + 2 < 2(x + 7)$

86. $2 - 4x + 5(x - 1) < -6(x - 2)$

87. $\dfrac{4x + 7}{-3} \le 2x + 5$

88. $\dfrac{2x - 5}{-8} \le 1 - x$

89. $\dfrac{1}{3}x + \dfrac{2}{5}x - \dfrac{1}{2}(x + 3) \le \dfrac{1}{10}$

90. $-\dfrac{2}{3}x - \dfrac{1}{6}x + \dfrac{2}{3}(x + 1) \le \dfrac{4}{3}$

Solve each inequality. Write each solution set in interval notation. **See Example 8.**

91. $-5 < 5 + 2x < 11$

92. $-7 < 2 + 3x < 5$

93. $10 \le 2x + 4 \le 16$

94. $-6 \le 6x + 3 \le 21$

95. $-11 > -3x + 1 > -17$

96. $2 > -6x + 3 > -3$

97. $-4 \le \dfrac{x + 1}{2} \le 5$

98. $-5 \le \dfrac{x - 3}{3} \le 1$

99. $-3 \le \dfrac{x - 4}{-5} < 4$

100. $1 \le \dfrac{4x - 5}{-2} < 9$

B Graphs of Equations

- The Rectangular Coordinate System
- The Pythagorean Theorem and the Distance Formula
- The Midpoint Formula
- Graphing Equations
- Circles

The Rectangular Coordinate System Each real number corresponds to a point on a number line. This idea is extended to **ordered pairs** of real numbers by using two perpendicular number lines, one horizontal and one vertical, that intersect at their zero-points. This point of intersection is the **origin.** The horizontal line is the ***x*-axis,** and the vertical line is the ***y*-axis.** See **Figure1.**

The x-axis and y-axis together make up a **rectangular coordinate system,** or **Cartesian coordinate system** (named for one of its coinventors, René Descartes. The other coinventor was Pierre de Fermat). The plane into which the coordinate system is introduced is the **coordinate plane,** or ***xy*-plane.** See **Figure 1.** The x-axis and y-axis divide the plane into four regions, or **quadrants,** labeled as shown. The points on the x-axis and y-axis belong to no quadrant.

Each point P in the xy-plane corresponds to a unique ordered pair (a, b) of real numbers. The point P corresponding to the ordered pair (a, b) often is written $P(a, b)$ as in **Figure 1** and referred to as "the point (a, b)." The numbers a and b are the **coordinates** of point P. To locate on the xy-plane the point corresponding to the ordered pair $(3, 4)$, for example, start at the origin, move 3 units in the positive x-direction, and then move 4 units in the positive y-direction. See **Figure 2.** Point A corresponds to the ordered pair $(3, 4)$.

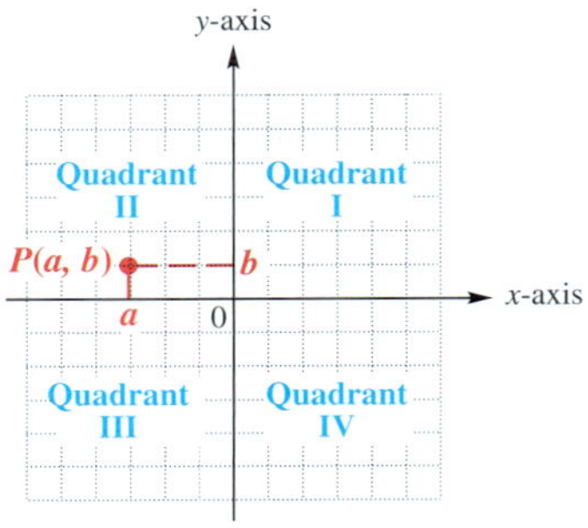

Figure 1

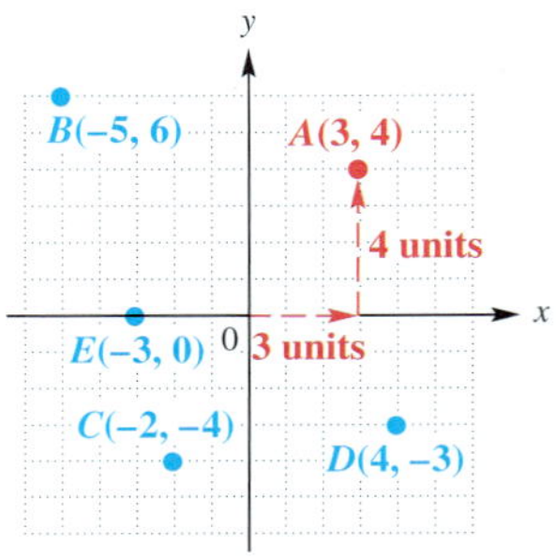

Figure 2

The Pythagorean Theorem and the Distance Formula The distance between any two points in a plane can be found by using a formula derived from the **Pythagorean theorem.**

Pythagorean Theorem

In a right triangle, the sum of the squares of the lengths of the legs is equal to the square of the length of the hypotenuse.

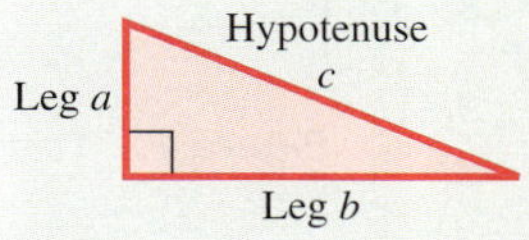

$$a^2 + b^2 = c^2$$

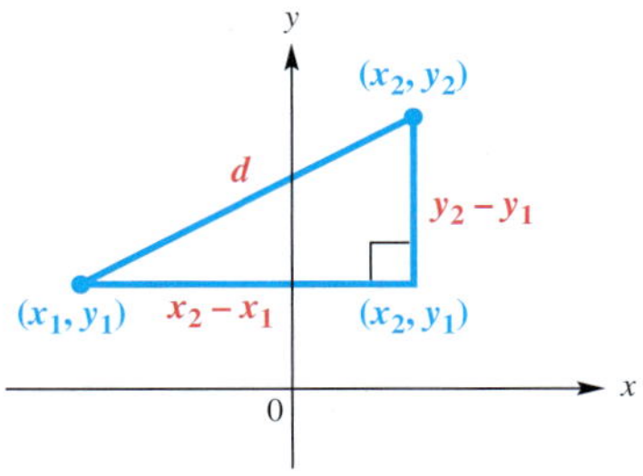

Figure 3

To find the distance between two points (x_1, y_1) and (x_2, y_2), draw the line segment connecting the points, as shown in **Figure 3.** Complete a right triangle by drawing a line through (x_1, y_1) parallel to the x-axis and a line through (x_2, y_2) parallel to the y-axis. The ordered pair at the right angle of this triangle is (x_2, y_1).

The horizontal side of the right triangle in **Figure 3** has length $x_2 - x_1$, while the vertical side has length $y_2 - y_1$. If d represents the distance between the two original points, then by the Pythagorean theorem,

$$d^2 = (x_2 - x_1)^2 + (y_2 - y_1)^2.$$

Solving for d, we obtain the **distance formula.**

Distance Formula

Suppose that $P(x_1, y_1)$ and $R(x_2, y_2)$ are two points in a coordinate plane. The distance between P and R, written $d(P, R)$, is given by the following formula.

$$d(P, R) = \sqrt{(x_2 - x_1)^2 + (y_2 - y_1)^2}$$

That is, the distance between two points in a coordinate plane is the square root of the sum of the square of the difference between their x-coordinates and the square of the difference between their y-coordinates.

EXAMPLE 1 Using the Distance Formula

Find the distance between $P(-8, 4)$ and $Q(3, -2)$.

SOLUTION Use the distance formula.

$$\begin{aligned} d(P, Q) &= \sqrt{(x_2 - x_1)^2 + (y_2 - y_1)^2} && \text{Distance formula} \\ &= \sqrt{[3 - (-8)]^2 + (-2 - 4)^2} && x_1 = -8,\ y_1 = 4,\ x_2 = 3,\ y_2 = -2 \\ &= \sqrt{11^2 + (-6)^2} \\ &= \sqrt{121 + 36} \\ &= \sqrt{157} \end{aligned}$$

Be careful when subtracting a negative number.

✔ *Now Try Exercise 9(a).*

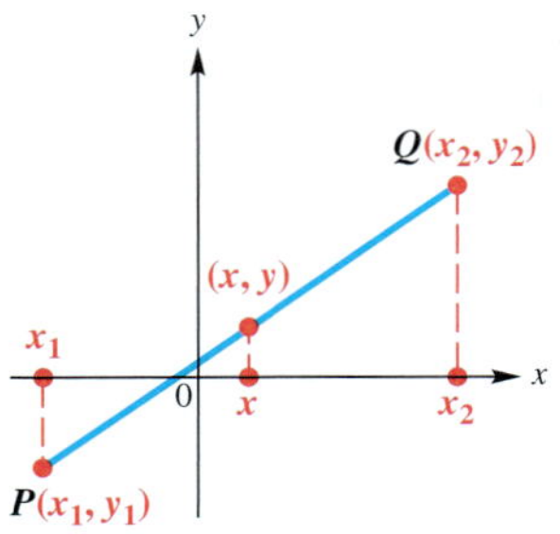

Figure 4

The Midpoint Formula The midpoint of a line segment is equidistant from the endpoints of the segment. The **midpoint formula** is used to find the coordinates of the midpoint of a line segment. To develop the midpoint formula, let $P(x_1, y_1)$ and $Q(x_2, y_2)$ be any two distinct points in a plane. (Although **Figure 4** shows $x_1 < x_2$, no particular order is required.) Let (x, y) be the midpoint of the segment joining P and Q. Draw vertical lines from each of the three points to the x-axis, as shown in **Figure 4.**

Since (x, y) is the midpoint of the line segment joining P and Q, the distance between x and x_1 equals the distance between x and x_2.

$$x_2 - x = x - x_1$$

$$x_2 + x_1 = 2x \qquad \text{Add } x \text{ and } x_1. \text{ (Appendix A)}$$

$$x = \frac{x_1 + x_2}{2} \qquad \text{Divide by 2 and rewrite.}$$

Similarly, the y-coordinate is $\frac{y_1 + y_2}{2}$, yielding the following formula.

Midpoint Formula

The midpoint M of the line segment with endpoints $P(x_1, y_1)$ and $Q(x_2, y_2)$ has the following coordinates.

$$\left(\frac{x_1 + x_2}{2}, \frac{y_1 + y_2}{2}\right)$$

That is, the x-coordinate of the midpoint of a line segment is the* average *of the x-coordinates of the segment's endpoints, and the y-coordinate is the* average *of the y-coordinates of the segment's endpoints.

EXAMPLE 2 Using the Midpoint Formula

Find the coordinates of the midpoint M of the segment with endpoints $(8, -4)$ and $(-6, 1)$.

SOLUTION The coordinates of M are found using the midpoint formula.

$$\left(\frac{8 + (-6)}{2}, \frac{-4 + 1}{2}\right) = \left(1, -\frac{3}{2}\right) \qquad \text{Substitute in } \left(\frac{x_1 + x_2}{2}, \frac{y_1 + y_2}{2}\right).$$

The coordinates of midpoint M are $\left(1, -\frac{3}{2}\right)$.

✔ ***Now Try Exercise 9(b).***

Graphing Equations Ordered pairs are used to express the solutions of equations in two variables. When an ordered pair represents the solution of an equation with the variables x and y, the x-value is written first. For example, we say that $(1, 2)$ is a solution of $2x - y = 0$, since substituting 1 for x and 2 for y in the equation gives a true statement.

$$2x - y = 0$$

$$2(1) - 2 \stackrel{?}{=} 0 \qquad \text{Let } x = 1 \text{ and } y = 2.$$

$$0 = 0 \ ✓ \qquad \text{True}$$

EXAMPLE 3 Finding Ordered-Pair Solutions of Equations

For each equation, find at least three ordered pairs that are solutions.

(a) $y = 4x - 1$ **(b)** $x = \sqrt{y - 1}$ **(c)** $y = x^2 - 4$

SOLUTION

(a) Choose any real number for x or y and substitute in the equation to get the corresponding value of the other variable. For example, let $x = -2$ and then let $y = 3$.

$y = 4x - 1$		$y = 4x - 1$	
$y = 4(-2) - 1$	Let $x = -2$.	$3 = 4x - 1$	Let $y = 3$.
$y = -8 - 1$	Multiply.	$4 = 4x$	Add 1.
$y = -9$	Subtract.	$1 = x$	Divide by 4.

This gives the ordered pairs $(-2, -9)$ and $(1, 3)$. Verify that the ordered pair $(0, -1)$ is also a solution.

(b)

$x = \sqrt{y - 1}$	Given equation
$1 = \sqrt{y - 1}$	Let $x = 1$.
$1 = y - 1$	Square each side.
$2 = y$	Add 1.

One ordered pair is $(1, 2)$. Verify that the ordered pairs $(0, 1)$ and $(2, 5)$ are also solutions of the equation.

(c) A table provides an organized method for determining ordered pairs. Here, we let x equal -2, -1, 0, 1, and 2 in

$$y = x^2 - 4$$

and determine the corresponding y-values.

x	y	
-2	0	$(-2)^2 - 4 = 4 - 4 = 0$
-1	-3	$(-1)^2 - 4 = 1 - 4 = -3$
0	-4	$0^2 - 4 = -4$
1	-3	$1^2 - 4 = -3$
2	0	$2^2 - 4 = 0$

Five ordered pairs are $(-2, 0)$, $(-1, -3)$, $(0, -4)$, $(1, -3)$, and $(2, 0)$.

✔ ***Now Try Exercises 17(a), 21(a), and 23(a).***

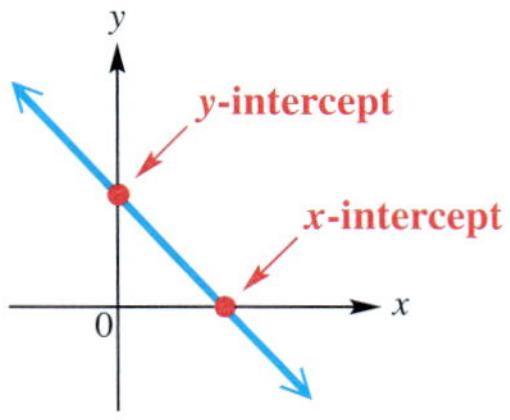

The **graph** of an equation is found by plotting ordered pairs that are solutions of the equation. The **intercepts** of the graph are good points to plot first. An ***x*-intercept** is an x-value where the graph intersects the x-axis. A ***y*-intercept** is a y-value where the graph intersects the y-axis. In other words, the x-intercept is the x-coordinate of an ordered pair where $y = 0$, and the y-intercept is the y-coordinate of an ordered pair where $x = 0$.

A general algebraic approach for graphing an equation using intercepts and point-plotting follows on the next page.

Graphing an Equation by Point Plotting

Step 1 Find the intercepts.

Step 2 Find as many additional ordered pairs as needed.

Step 3 Plot the ordered pairs from Steps 1 and 2.

Step 4 Join the points from Step 3 with a smooth line or curve.

EXAMPLE 4 Graphing Equations

Graph each of the equations here, from **Example 3.**

(a) $y = 4x - 1$ **(b)** $x = \sqrt{y - 1}$ **(c)** $y = x^2 - 4$

SOLUTION

(a) ***Step 1*** Let $y = 0$ to find the x-intercept, and let $x = 0$ to find the y-intercept.

$y = 4x - 1$		$y = 4x - 1$	
$0 = 4x - 1$	Let $y = 0$.	$y = 4(0) - 1$	Let $x = 0$.
$1 = 4x$		$y = 0 - 1$	
$\frac{1}{4} = x$	x-intercept*	$y = -1$	y-intercept*

These intercepts lead to the ordered pairs $\left(\frac{1}{4}, 0\right)$ and $(0, -1)$. The y-intercept yields one of the ordered pairs we found in **Example 3(a).**

Step 2 We use the other ordered pairs from **Example 3(a):** $(-2, -9)$ and $(1, 3)$.

Step 3 Plot the four ordered pairs from Steps 1 and 2 as shown in **Figure 5.**

Step 4 Join the points plotted in Step 3 with a straight line. This line, shown in **Figure 5,** is the graph of the equation $y = 4x - 1$.

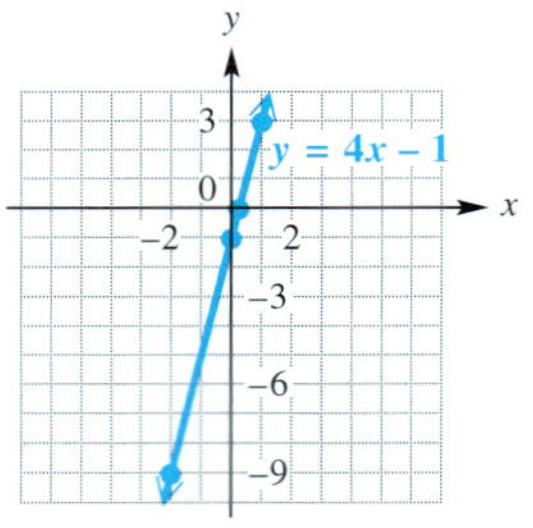

Figure 5

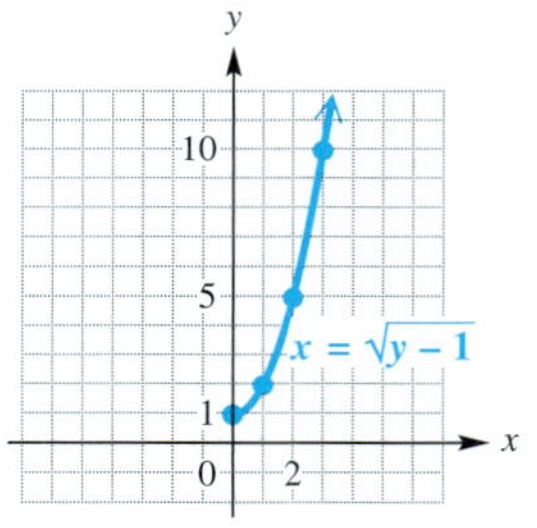

Figure 6

(b) For $x = \sqrt{y - 1}$, the y-intercept 1 was found in **Example 3(b).** Solve

$$x = \sqrt{0 - 1} \quad \text{Let } y = 0.$$

for the x-intercept. Since the quantity under the radical is negative, there is no x-intercept. In fact, $y - 1$ must be greater than or equal to 0, so y must be greater than or equal to 1.

We start by plotting the ordered pairs from **Example 3(b)** and then join the points with a smooth curve as in **Figure 6.** To confirm the direction the curve will take as x increases, we find another solution, $(3, 10)$.

---

*The intercepts are sometimes defined as ordered pairs, such as $\left(\frac{1}{4}, 0\right)$ and $(0, -1)$ instead of numbers, such as x-intercept $\frac{1}{4}$ and y-intercept -1. In this text, we define them as numbers.

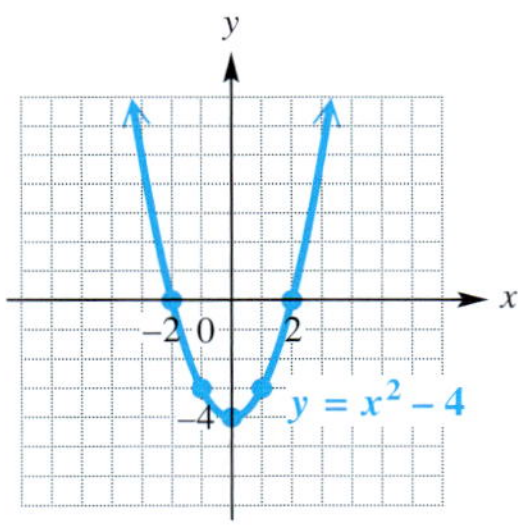

Figure 7

(c) In **Example 3(c),** we made a table of five ordered pairs that satisfy the equation $y = x^2 - 4$.

$$(-2, 0), \quad (-1, -3), \quad (0, -4), \quad (1, -3), \quad (2, 0)$$

$(-2, 0)$: x-intercept; $(0, -4)$: y-intercept; $(2, 0)$: x-intercept

Plotting the points and joining them with a smooth curve gives the graph in **Figure 7.** This curve is called a **parabola.**

✔ ***Now Try Exercises 17(b), 21(b), and 23(b).***

Circles By definition, a **circle** is the set of all points in a plane that lie a given distance from a given point. The given distance is the **radius** of the circle, and the given point is the **center.**

We can find the equation of a circle from its definition by using the distance formula. Suppose that the point (h, k) is the center and the circle has radius r, where $r > 0$. Let (x, y) represent any point on the circle. See **Figure 8.**

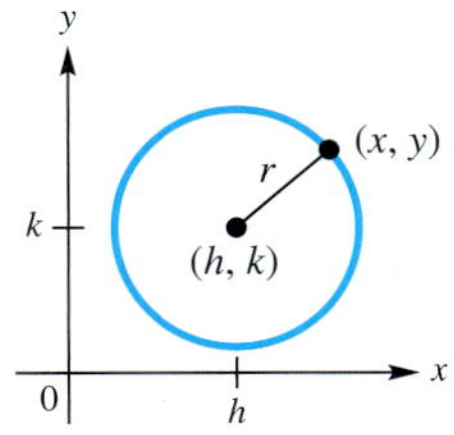

Figure 8

$$\sqrt{(x_2 - x_1)^2 + (y_2 - y_1)^2} = r \qquad \text{Distance formula}$$

$$\sqrt{(x - h)^2 + (y - k)^2} = r \qquad (h, k) = (x_1, y_1) \text{ and } (x, y) = (x_2, y_2)$$

$$(x - h)^2 + (y - k)^2 = r^2 \qquad \text{Square each side.}$$

Center-Radius Form of the Equation of a Circle

A circle with center (h, k) and radius r has equation

$$(x - h)^2 + (y - k)^2 = r^2,$$

which is the **center-radius form** of the equation of the circle. As a special case, a circle with center $(0, 0)$ and radius r has the following equation.

$$x^2 + y^2 = r^2$$

EXAMPLE 5 Finding the Center-Radius Form

Find the center-radius form of the equation of each circle described.

(a) center at $(-3, 4)$, radius 6 **(b)** center at $(0, 0)$, radius 3

SOLUTION

(a)

$$(x - h)^2 + (y - k)^2 = r^2 \qquad \text{Center-radius form}$$

$$[x - (-3)]^2 + (y - 4)^2 = 6^2 \qquad \text{Substitute. Let } (h, k) = (-3, 4) \text{ and } r = 6.$$

Watch signs here.

$$(x + 3)^2 + (y - 4)^2 = 36 \qquad \text{Simplify.}$$

(b) The center is the origin and $r = 3$.

$$x^2 + y^2 = r^2 \qquad \text{Special case of the center-radius form}$$

$$x^2 + y^2 = 3^2 \qquad \text{Let } r = 3.$$

$$x^2 + y^2 = 9 \qquad \text{Apply the exponent.}$$

✔ ***Now Try Exercises 29(a) and 35(a).***

EXAMPLE 6 Graphing Circles

Graph each circle discussed in **Example 5.**

(a) $(x + 3)^2 + (y - 4)^2 = 36$ **(b)** $x^2 + y^2 = 9$

SOLUTION

(a) Writing the given equation in center-radius form

$$[x - (-3)]^2 + (y - 4)^2 = 6^2$$

gives $(-3, 4)$ as the center and 6 as the radius. See **Figure 9.**

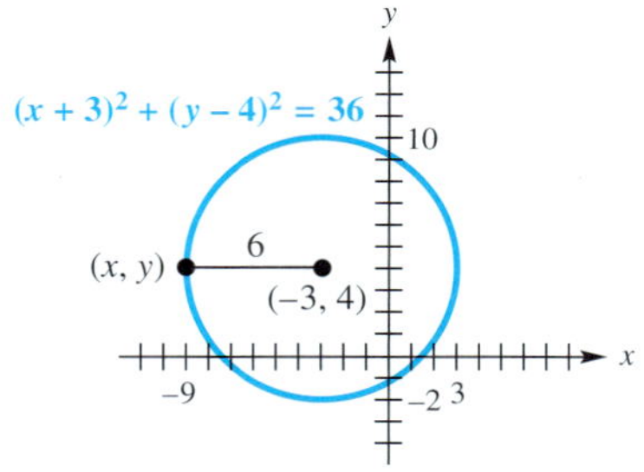

Figure 9

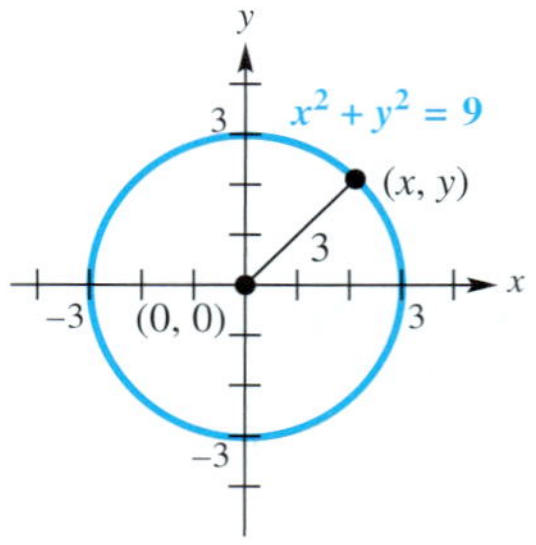

Figure 10

(b) The graph with center $(0, 0)$ and radius 3 is shown in **Figure 10.**

✔ *Now Try Exercises 29(b) and 35(b).*

Appendix B Exercises

Graph the points on a coordinate system and identify the quadrant or axis for each point.

1. $(3, 2)$ **2.** $(-7, 6)$ **3.** $(-7, -4)$ **4.** $(8, -5)$

5. $(0, 5)$ **6.** $(-8, 0)$ **7.** $(4.5, 7)$ **8.** $(-7.5, 8)$

*For the points P and Q, find **(a)** the distance $d(P, Q)$ and **(b)** the coordinates of the midpoint of the segment PQ.* ***See Examples 1 and 2.***

9. $P(8, 2), Q(3, 5)$ **10.** $P(-8, 4), Q(3, -5)$

11. $P(-6, -5), Q(6, 10)$ **12.** $P(6, -2), Q(4, 6)$

13. $P(3\sqrt{2}, 4\sqrt{5}), Q(\sqrt{2}, -\sqrt{5})$ **14.** $P(-\sqrt{7}, 8\sqrt{3}), Q(5\sqrt{7}, -\sqrt{3})$

Solve each problem.

15. *Bachelor's Degree Attainment* The graph shows a straight line that approximates the percentage of Americans 25 years and older who had earned bachelor's degrees or higher for the years 1990–2008. Use the midpoint formula and the two given points to estimate the percent in 1999. Compare your answer with the actual percent of 25.2.

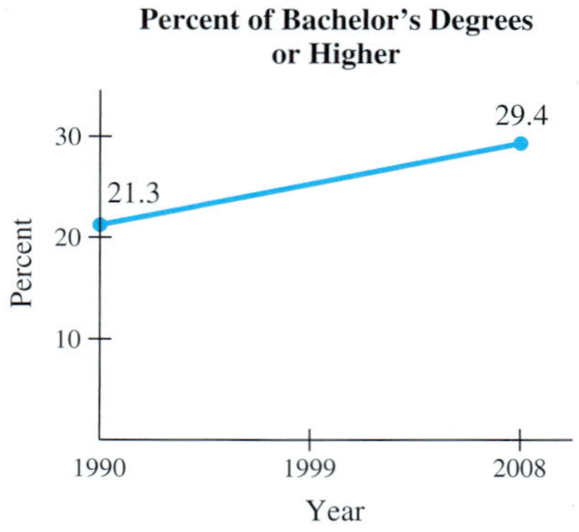

Source: U.S. Census Bureau.

16. ***Poverty Level Income Cutoffs*** The table lists how poverty level income cutoffs (in dollars) for a family of four have changed over time. Use the midpoint formula to approximate the poverty level cutoff in 2006 to the nearest dollar.

Year	Income (in dollars)
1980	8414
1990	13,359
2000	17,604
2004	19,307
2008	22,025

Source: U.S. Census Bureau.

For each equation, ***(a)*** *give a table with at least three ordered pairs that are solutions, and* ***(b)*** *graph the equation.* ***See Examples 3 and 4.***

17. $y = \frac{1}{2}x - 2$

18. $y = -x + 3$

19. $2x + 3y = 5$

20. $3x - 2y = 6$

21. $y = x^2$

22. $y = x^2 + 2$

23. $y = \sqrt{x - 3}$

24. $y = \sqrt{x} - 3$

25. $y = |x - 2|$

26. $y = -|x + 4|$

27. $y = x^3$

28. $y = -x^3$

In Exercises 29–40, ***(a)*** *find the center-radius form of the equation of each circle, and* ***(b)*** *graph it.* ***See Examples 5 and 6.***

29. center $(0, 0)$, radius 6

30. center $(0, 0)$, radius 9

31. center $(2, 0)$, radius 6

32. center $(3, 0)$, radius 3

33. center $(0, 4)$, radius 4

34. center $(0, -3)$, radius 7

35. center $(-2, 5)$, radius 4

36. center $(4, 3)$, radius 5

37. center $(5, -4)$, radius 7

38. center $(-3, -2)$, radius 6

39. center $(\sqrt{2}, \sqrt{2})$, radius $\sqrt{2}$

40. center $(-\sqrt{3}, -\sqrt{3})$, radius $\sqrt{3}$

Connecting Graphs with Equations *In Exercises 41–44, use each graph to determine the equation of the circle in center-radius form.*

41.

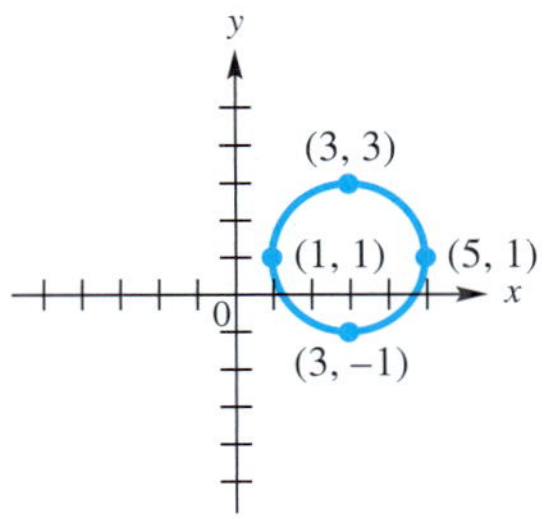

42.

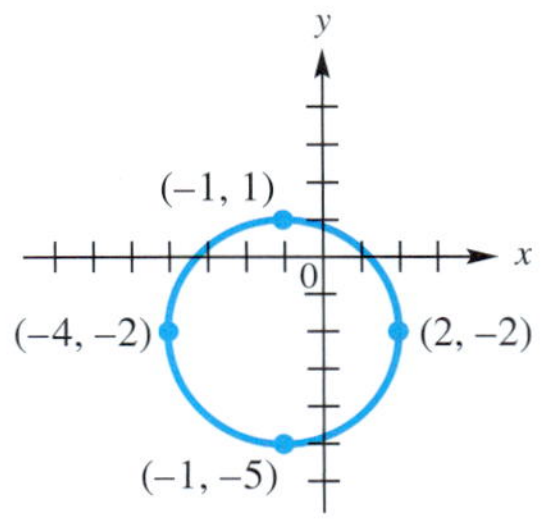

43.

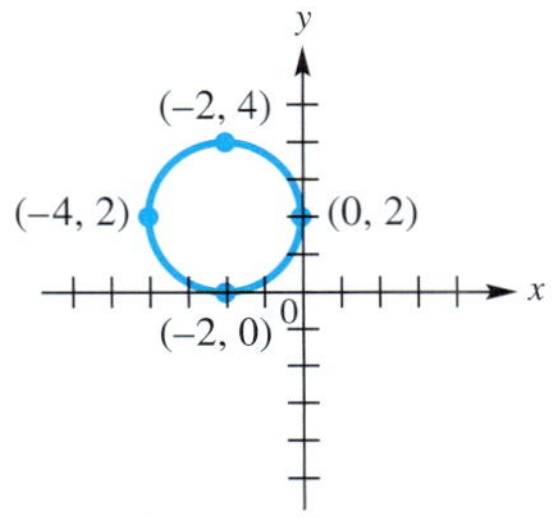

44.

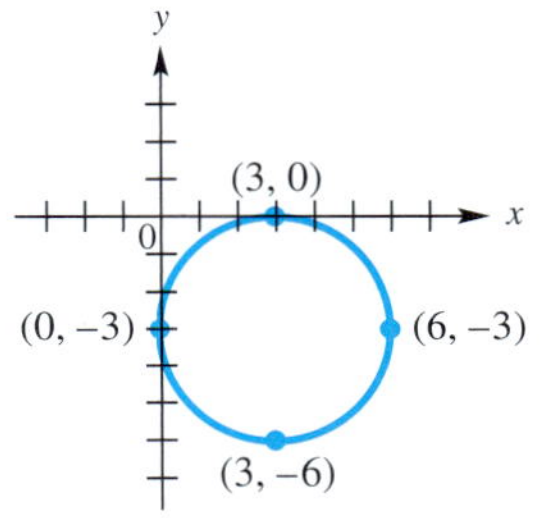

C Functions

- Relations and Functions
- Domain and Range
- Determining Whether Relations Are Functions
- Function Notation
- Increasing, Decreasing, and Constant Functions

Relations and Functions

In algebra, we use ordered pairs to represent related quantities. For example, $(3, \$10.50)$ might indicate that you pay \$10.50 for 3 gallons of gas. Since the amount you pay *depends* on the number of gallons pumped, the amount (in dollars) is called the *dependent variable,* and the number of gallons pumped is called the *independent variable.*

Generalizing, if the value of the second component y depends on the value of the first component x, then y is the **dependent variable** and x is the **independent variable.**

Independent variable → x; Dependent variable → y

$$(x, y)$$

A set of ordered pairs such as $\{(3, 10.50), (8, 28.00), (10, 35.00)\}$ is a *relation.* A special kind of relation called a *function* is very important in mathematics and its applications.

Relation and Function

A **relation** is a set of ordered pairs. A **function** is a relation in which, for each distinct value of the first component of the ordered pairs, there is *exactly one* value of the second component.

EXAMPLE 1 Deciding Whether Relations Define Functions

Decide whether each relation defines a function.

$$F = \{(1, 2), (-2, 4), (3, 4)\}$$
$$G = \{(1, 1), (1, 2), (1, 3), (2, 3)\}$$
$$H = \{(-4, 1), (-2, 1), (-2, 0)\}$$

SOLUTION Relation F is a function, because for each distinct x-value there is exactly one y-value. We can show this correspondence as follows.

$$\{1, -2, 3\} \quad x\text{-values of } F$$
$$\downarrow \quad \downarrow \quad \downarrow$$
$$\{2, 4, 4\} \quad y\text{-values of } F$$

As the correspondence below shows, relation G is not a function because one first component corresponds to *more than one* second component.

$$\{1, 2\} \quad x\text{-values of } G$$
$$\{1, 2, 3\} \quad y\text{-values of } G$$

In relation H the last two ordered pairs have the same x-value paired with two different y-values (-2 is paired with both 1 and 0), so H is a relation but not a function. ***In a function, no two ordered pairs can have the same first component and different second components.***

Different y-values

$$H = \{(-4, 1), (-2, 1), (-2, 0)\} \quad \text{Not a function}$$

Same x-value

✔ *Now Try Exercises 1 and 3.*

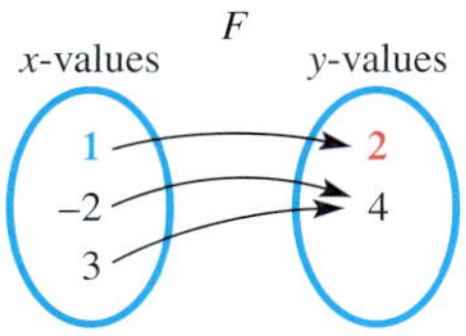

F is a function.

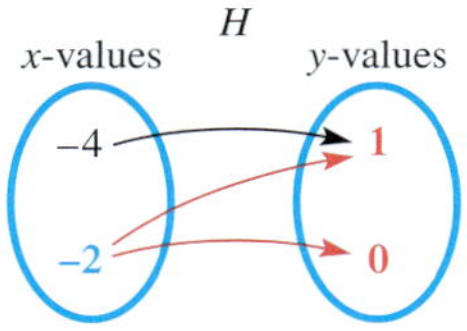

H is not a function.

Figure 1

Relations and functions can also be expressed as a correspondence or *mapping* from one set to another, as shown in **Figure 1** for function F and relation H from **Example 1.** The arrow from 1 to 2 indicates that the ordered pair $(1, 2)$ belongs to F—each first component is paired with exactly one second component. In the mapping for relation H, which is not a function, the first component -2 is paired with two different second components, 1 and 0.

Since relations and functions are sets of ordered pairs, we can represent them using tables and graphs. A table and graph for function F are shown in **Figure 2.**

x	y
1	2
−2	4
3	4

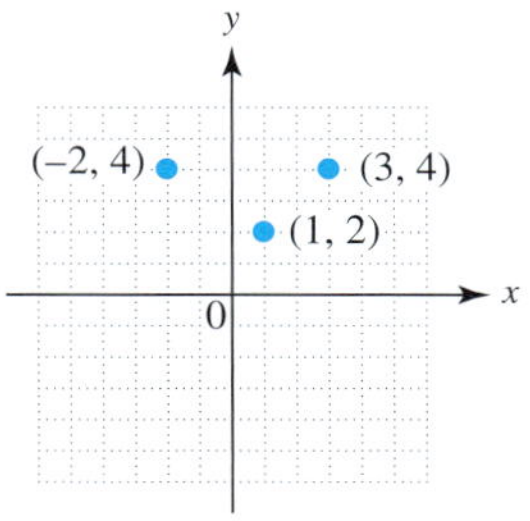

Graph of F

Figure 2

We can describe a relation or function using a rule that tells how to determine the value of the dependent variable for a specific value of the independent variable. The rule may be given in words: for instance, "the dependent variable is twice the independent variable." Usually the rule is an equation, such as the one below.

Dependent variable $\longrightarrow y = 2x \longleftarrow$ Independent variable

In a function, there is exactly one value of the dependent variable, the second component, for each value of the independent variable, the first component.

On this particular day, an *input* of pumping 7.870 gallons of gasoline led to an *output* of \$29.58 from the purchaser's wallet. This is an example of a function whose domain consists of numbers of gallons pumped, and whose range consists of amounts from the purchaser's wallet. Dividing the dollar amount by the number of gallons pumped gives the exact price of gasoline that day. Use your calculator to check this. Was this pump fair?

Domain and Range

For every relation there are two important sets of elements called the *domain* and *range*.

Domain and Range

In a relation consisting of ordered pairs (x, y), the set of all values of the independent variable (x) is the **domain.** The set of all values of the dependent variable (y) is the **range.**

EXAMPLE 2 Finding Domains and Ranges of Relations

Give the domain and range of each relation. Tell whether the relation defines a function.

(a) $\{(3, -1), (4, 2), (4, 5), (6, 8)\}$

(b)

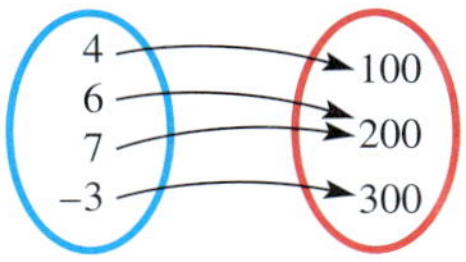

(c)

x	y
−5	2
0	2
5	2

SOLUTION

(a) The domain is the set of x-values, $\{3, 4, 6\}$. The range is the set of y-values, $\{-1, 2, 5, 8\}$. This relation is not a function because the same x-value, 4, is paired with two different y-values, 2 and 5.

(b) The domain is $\{4, 6, 7, -3\}$ and the range is $\{100, 200, 300\}$. This mapping defines a function. Each x-value corresponds to exactly one y-value.

(c) This relation is a set of ordered pairs, so the domain is the set of x-values $\{-5, 0, 5\}$ and the range is the set of y-values $\{2\}$. The table defines a function because each distinct x-value corresponds to exactly one y-value (even though it is the same y-value).

✔ ***Now Try Exercises 9, 11, and 13.***

EXAMPLE 3 Finding Domains and Ranges from Graphs

Give the domain and range of each relation.

(a)

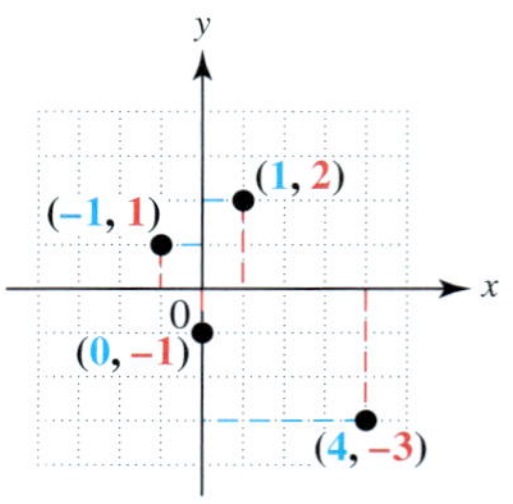

(b)

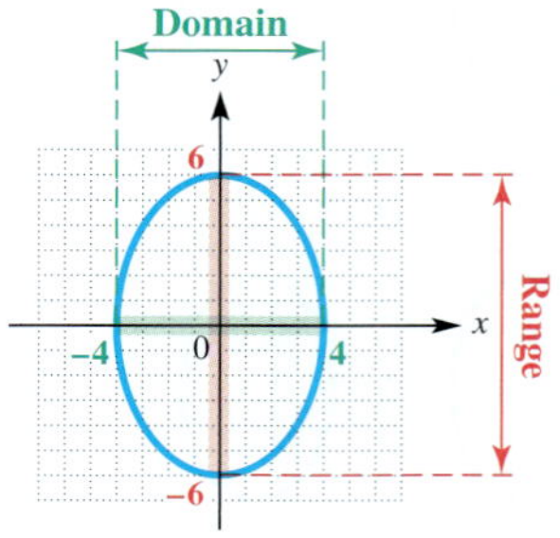

(c)

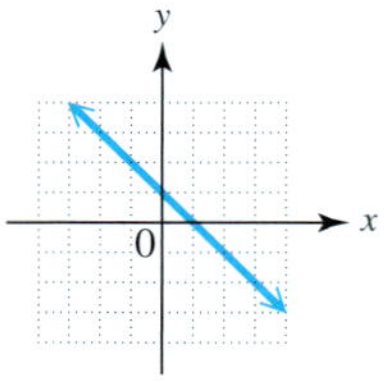

(d)

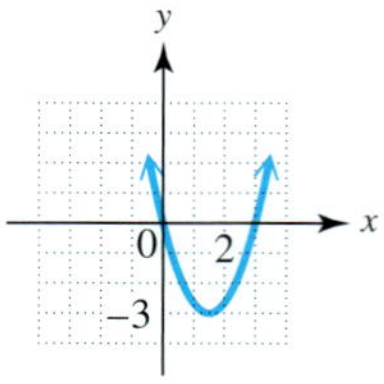

SOLUTION

(a) The domain is the set of x-values,

$$\{-1, 0, 1, 4\}.$$

The range is the set of y-values,

$$\{-3, -1, 1, 2\}.$$

(b) The x-values of the points on the graph include all numbers between -4 and 4, inclusive. The y-values include all numbers between -6 and 6, inclusive.

The domain is $[-4, 4]$. Use interval notation.

The range is $[-6, 6]$. (Appendix A)

(c) The arrowheads indicate that the line extends indefinitely left and right, as well as up and down. Therefore, both the domain and the range include all real numbers, which is written

$$(-\infty, \infty).$$ Interval notation for the set of real numbers

(d) The arrowheads indicate that the graph extends indefinitely left and right, as well as upward. The domain is $(-\infty, \infty)$. Because there is a least y-value, -3, the range includes all numbers greater than or equal to -3, written $[-3, \infty)$.

✔ *Now Try Exercises 15 and 17.*

Determining Whether Relations Are Functions Since each value of x leads to only one value of y in a function, any vertical line must intersect the graph in at most one point. This is the **vertical line test** for a function.

Vertical Line Test

If every vertical line intersects the graph of a relation in no more than one point, then the relation is a function.

The graph in **Figure 3(a)** represents a function because each vertical line intersects the graph in no more than one point. The graph in **Figure 3(b)** is not the graph of a function since a vertical line intersects the graph in more than one point.

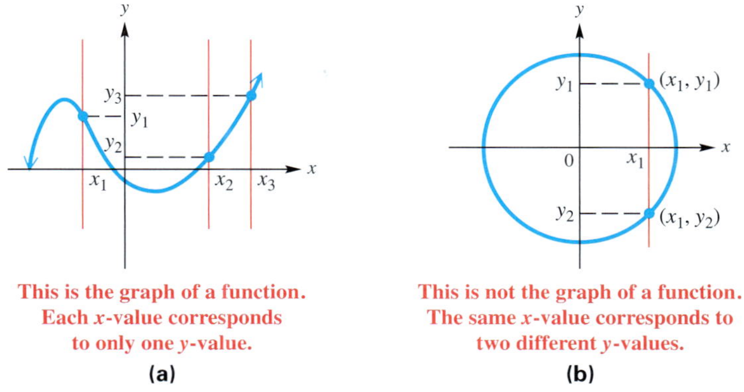

Figure 3

EXAMPLE 4 Using the Vertical Line Test

Use the vertical line test to determine whether each relation graphed in **Example 3** is a function.

SOLUTION We repeat each graph from **Example 3,** this time with vertical lines drawn through the graphs.

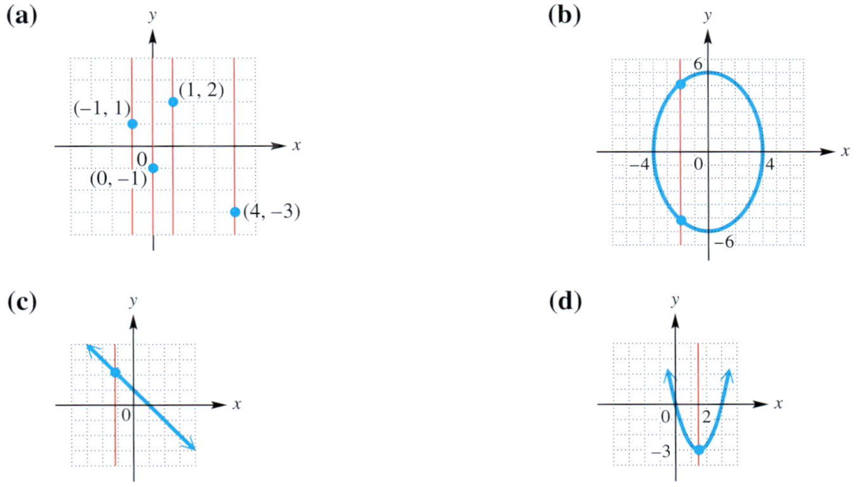

- The graphs of the relations in parts (a), (c), and (d) pass the vertical line test, since every vertical line intersects each graph no more than once. Thus, these graphs represent functions.
- The graph of the relation in part (b) fails the vertical line test, since the same x-value corresponds to two different y-values. Therefore, it is not the graph of a function.

✔ ***Now Try Exercises 15 and 17.***

The vertical line test is a simple method for identifying a function defined by a graph. Deciding whether a relation defined by an equation or an inequality is a function, as well as determining the domain and range, is more difficult. The next example gives some hints that may help.

EXAMPLE 5 Identifying Functions, Domains, and Ranges

Decide whether each relation defines a function and give the domain and range.

(a) $y = x + 4$ **(b)** $y = \sqrt{2x - 1}$ **(c)** $y^2 = x$ **(d)** $y = \dfrac{5}{x - 1}$

SOLUTION

(a) In the defining equation (or rule), $y = x + 4$, y is always found by adding 4 to x. Thus, each value of x corresponds to just one value of y, and the relation defines a function. The variable x can represent any real number, so the domain is

$$\{x \mid x \text{ is a real number}\}, \quad \text{or} \quad (-\infty, \infty).$$

Since y is always 4 more than x, y also may be any real number, and so the range is $(-\infty, \infty)$.

(b) For any choice of x in the domain of $y = \sqrt{2x - 1}$, there is exactly one corresponding value for y (the radical is a nonnegative number), so this equation defines a function. Since the equation involves a square root, the quantity under the radical sign cannot be negative.

$$2x - 1 \geq 0 \quad \text{Solve the inequality. (Appendix A)}$$

$$2x \geq 1 \quad \text{Add 1.}$$

$$x \geq \frac{1}{2} \quad \text{Divide by 2.}$$

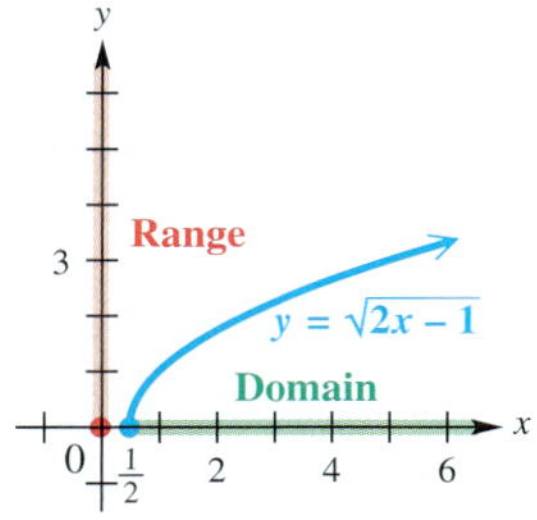

Figure 4

The domain of the function is $\left[\frac{1}{2}, \infty\right)$. Because the radical must represent a nonnegative number, as x takes values greater than or equal to $\frac{1}{2}$, the range is $\{y \mid y \geq 0\}$, or $[0, \infty)$. See **Figure 4.**

(c) The ordered pairs $(16, 4)$ and $(16, -4)$ both satisfy the equation $y^2 = x$. Since one value of x, 16, corresponds to two values of y, 4 and -4, this equation does not define a function.

Because x is equal to the square of y, the values of x must always be nonnegative. The domain of the relation is $[0, \infty)$. Any real number can be squared, so the range of the relation is $(-\infty, \infty)$. See **Figure 5.**

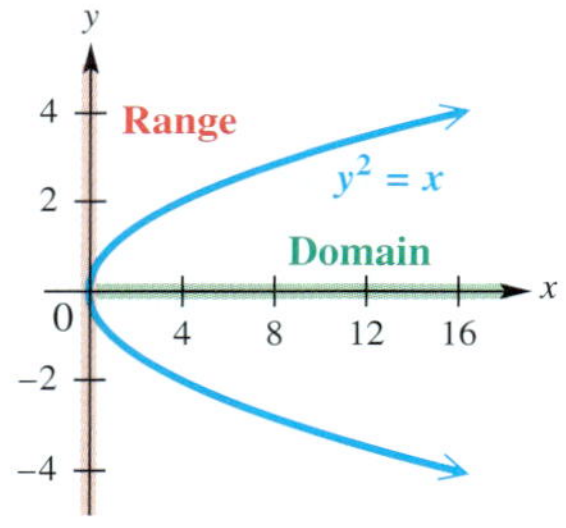

Figure 5

(d) Given any value of x in the domain of

$$y = \frac{5}{x - 1},$$

we find y by subtracting 1 from x, and then dividing the result into 5. This process produces exactly one value of y for each value in the domain, so this equation defines a function.

The domain of $y = \frac{5}{x - 1}$ includes all real numbers except those that make the denominator 0. We find these numbers by setting the denominator equal to 0 and solving for x.

$$x - 1 = 0$$

$$x = 1 \quad \text{Add 1. (Appendix A)}$$

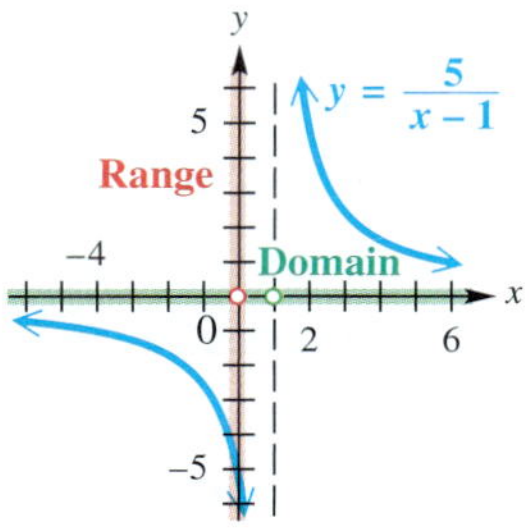

Figure 6

Thus, the domain includes all real numbers except 1, written as the interval $(-\infty, 1) \cup (1, \infty)$. Values of y can be positive or negative, but never 0, because a fraction cannot equal 0 unless its numerator is 0. Therefore, the range is the interval $(-\infty, 0) \cup (0, \infty)$, as shown in **Figure 6.**

✔ *Now Try Exercises 23, 25, and 31.*

Function Notation When a function f is defined with a rule or an equation using x and y for the independent and dependent variables, we say, "y is a function of x" to emphasize that y *depends on* x. We use the notation

$$y = f(x),$$

called **function notation,** to express this and read $f(x)$ as **"f of x."** The letter f is the name given to this function.

For example, if $y = 3x - 5$, we can name the function f and write

$$f(x) = 3x - 5.$$

Note that $f(x)$ is just another name for the dependent variable y. For example, if $y = f(x) = 3x - 5$ and $x = 2$, then we find y, or $f(2)$, by replacing x with 2.

$f(2) = 3 \cdot 2 - 5$ Let $x = 2$.

$f(2) = 1$ Multiply, and then subtract.

The statement "In the function f, if $x = 2$, then $y = 1$" represents the ordered pair $(2, 1)$ and is abbreviated with function notation as follows.

$$f(2) = 1$$

The symbol $f(2)$ is read "f of 2" or "f at 2."

These ideas can be illustrated as follows.

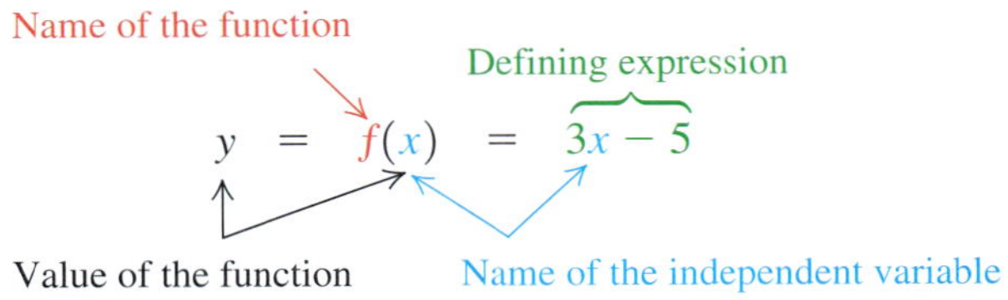

EXAMPLE 6 Using Function Notation

Let $f(x) = -x^2 + 5x - 3$ and $g(x) = 2x + 3$. Find and simplify each of the following.

(a) $f(2)$ **(b)** $f(q)$ **(c)** $g(a + 1)$

SOLUTION

(a) $f(x) = -x^2 + 5x - 3$

$f(2) = -2^2 + 5 \cdot 2 - 3$ Replace x with 2.

$= -4 + 10 - 3$ Apply the exponent and multiply.

$= 3$ Add and subtract.

Thus, $f(2) = 3$, and the ordered pair $(2, 3)$ belongs to f.

(b) $f(x) = -x^2 + 5x - 3$

$f(q) = -q^2 + 5q - 3$ Replace x with q.

(c) $g(x) = 2x + 3$

$g(a + 1) = 2(a + 1) + 3$ Replace x with $a + 1$.

$= 2a + 2 + 3$ Distributive property

$= 2a + 5$ Add.

✔ ***Now Try Exercises 35, 43, and 49.***

Functions can be evaluated in a variety of ways, as shown in **Example 7.**

EXAMPLE 7 Using Function Notation

For each function, find $f(3)$.

(a) $f(x) = 3x - 7$

(b) $f = \{(-3, 5), (0, 3), (3, 1), (6, -1)\}$

(c)

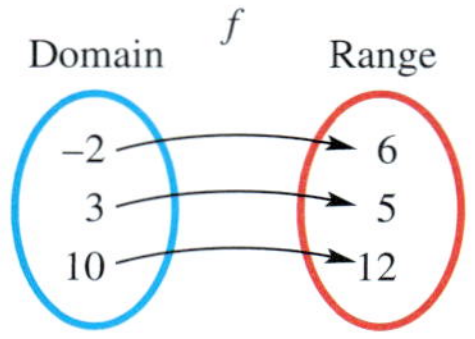

(d)

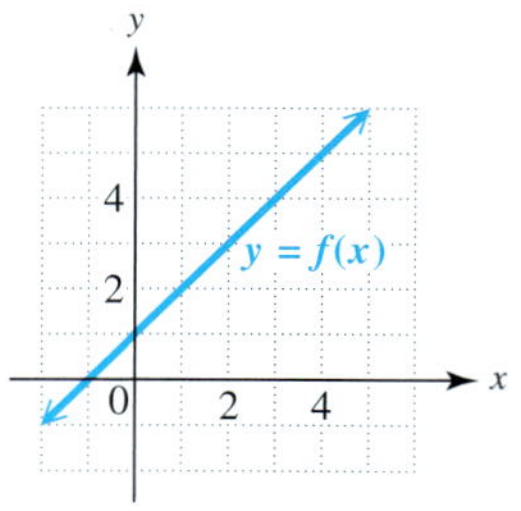

SOLUTION

(a) $f(x) = 3x - 7$

$f(3) = 3(3) - 7$ Replace x with 3.

$f(3) = 2$ Simplify.

(b) For $f = \{(-3, 5), (0, 3), (3, 1), (6, -1)\}$, we want $f(3)$, the y-value of the ordered pair where $x = 3$. As indicated by the ordered pair $(3, 1)$, when $x = 3$, $y = 1$, so $f(3) = 1$.

(c) In the mapping, the domain element 3 is paired with 5 in the range, so $f(3) = 5$.

(d) To evaluate $f(3)$ using the graph, find 3 on the x-axis. See **Figure 7.** Then move up until the graph of f is reached. Moving horizontally to the y-axis gives 4 for the corresponding y-value. Thus, $f(3) = 4$.

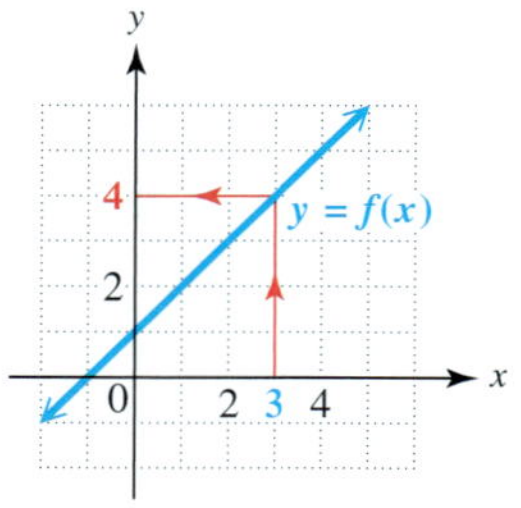

Figure 7

✔ *Now Try Exercises 51, 53, and 55.*

Increasing, Decreasing, and Constant Functions Informally speaking, a function *increases* on an interval of its domain if its graph rises from left to right on the interval. It *decreases* on an interval of its domain if its graph falls from left to right on the interval. It is *constant* on an interval of its domain if its graph is horizontal on the interval.

For example, consider **Figure 8.** The function increases on the interval $[-2, 1]$ because the y-values continue to get larger for x-values in that interval. Similarly, the function is constant on the interval $[1, 4]$ because the y-values are always 5 for all x-values there. Finally, the function decreases on the interval $[4, 6]$ because there the y-values continuously get smaller. ***The intervals refer to the x-values where the y-values either increase, decrease, or are constant.***

The formal definitions of these concepts follow.

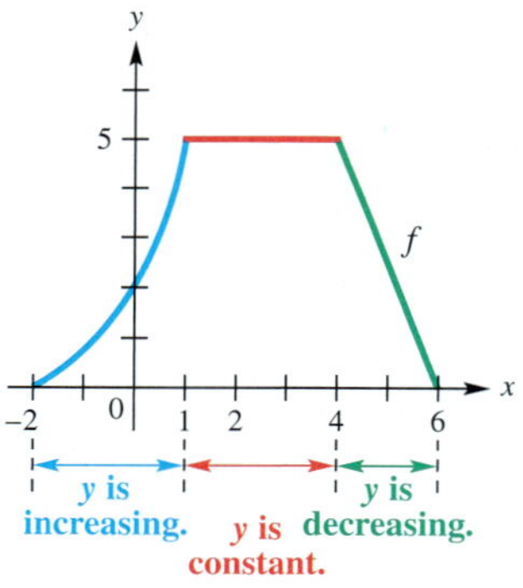

Figure 8

Increasing, Decreasing, and Constant Functions

Suppose that a function f is defined over an interval I and x_1 and x_2 are in I.

(a) f **increases** on I if, whenever $x_1 < x_2$, $f(x_1) < f(x_2)$.

(b) f **decreases** on I if, whenever $x_1 < x_2$, $f(x_1) > f(x_2)$.

(c) f is **constant** on I if, for every x_1 and x_2, $f(x_1) = f(x_2)$.

Figure 9 illustrates these ideas.

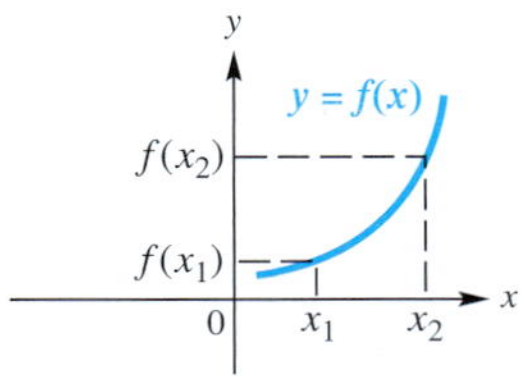

Whenever $x_1 < x_2$, and $f(x_1) < f(x_2)$, f is increasing.

(a)

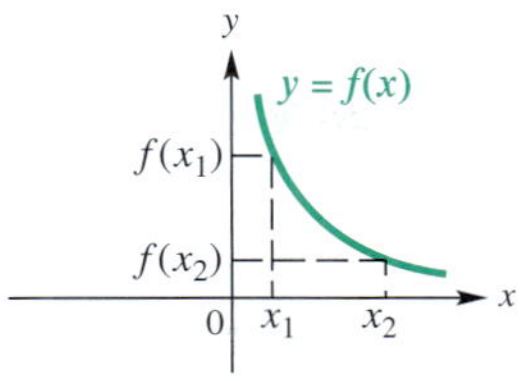

Whenever $x_1 < x_2$, and $f(x_1) > f(x_2)$, f is decreasing.

(b)

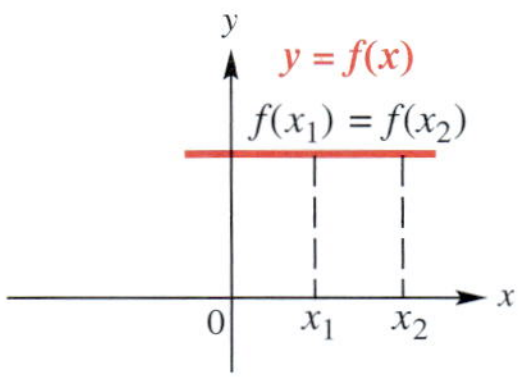

For every x_1 and x_2, if $f(x_1) = f(x_2)$, then f is constant.

(c)

Figure 9

NOTE To decide whether a function is increasing, decreasing, or constant on an interval, ask yourself, ***"What does y do as x goes from left to right?"***

There can be confusion regarding whether endpoints of an interval should be included when determining intervals over which a function is increasing or decreasing. For example, consider the graph of $y = f(x) = x^2 + 4$, shown in **Figure 10.**

Is f increasing on $[0, \infty)$ or just on $(0, \infty)$?

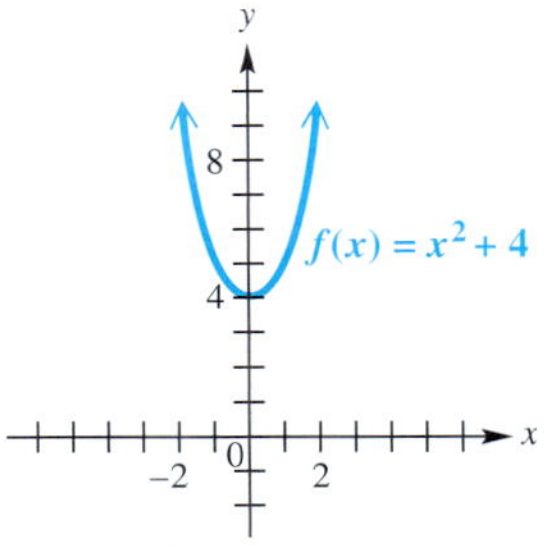

$f(x) = x^2 + 4$ increases on $[0, \infty)$ and decreases on $(-\infty, 0]$.

Figure 10

The definition of increasing and decreasing allows us to include 0 as a part of the interval I over which this function is increasing, because if we let $x_1 = 0$, then $f(0) < f(x_2)$ whenever $0 < x_2$. Thus, $f(x) = x^2 + 4$ is increasing on $[0, \infty)$. A similar discussion can be used to show that this function is decreasing on $(-\infty, 0]$. Do not confuse these concepts by saying that f both increases and decreases at the point $(0, 0)$.

The concepts of increasing and decreasing functions apply to intervals of the domain, not to individual points.

It is not incorrect to say that $f(x) = x^2 + 4$ is increasing on $(0, \infty)$—there are infinitely many intervals over which it increases. However, we generally give the largest possible interval when determining where a function increases or decreases. (*Source:* Stewart J., *Calculus,* Fourth Edition, Brooks/Cole Publishing Company, p. 21.)

EXAMPLE 8 **Determining Intervals over Which a Function Is Increasing, Decreasing, or Constant**

Figure 11 shows the graph of a function. Determine the intervals over which the function is increasing, decreasing, or constant.

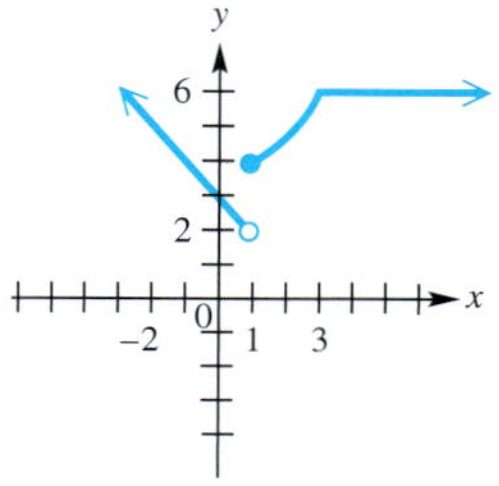

Figure 11

SOLUTION We should ask, "What is happening to the y-values as the x-values are getting larger?" Moving from left to right on the graph, we see the following:

- On the interval $(-\infty, 1)$, the y-values are *decreasing*.
- On the interval $[1, 3]$, the y-values are *increasing*.
- On the interval $[3, \infty)$, the y-values are *constant* (and equal to 6).

Therefore, the function is decreasing on $(-\infty, 1)$, increasing on $[1, 3]$, and constant on $[3, \infty)$.

✔ *Now Try Exercise 61.*

Appendix C Exercises

Decide whether each relation defines a function. ***See Example 1.***

1. $\{(5, 1), (3, 2), (4, 9), (7, 8)\}$

2. $\{(8, 0), (5, 7), (9, 3), (3, 8)\}$

3. $\{(2, 4), (0, 2), (2, 6)\}$

4. $\{(9, -2), (-3, 5), (9, 1)\}$

5. $\{(-3, 1), (4, 1), (-2, 7)\}$

6. $\{(-12, 5), (-10, 3), (8, 3)\}$

7.

x	y
3	−4
7	−4
10	−4

8.

x	y
−4	$\sqrt{2}$
0	$\sqrt{2}$
4	$\sqrt{2}$

Decide whether each relation defines a function and give the domain and range. ***See Examples 1–4.***

9. $\{(1, 1), (1, -1), (0, 0), (2, 4), (2, -4)\}$

10. $\{(2, 5), (3, 7), (3, 9), (5, 11)\}$

11.

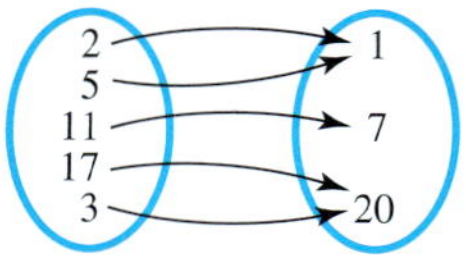

12.

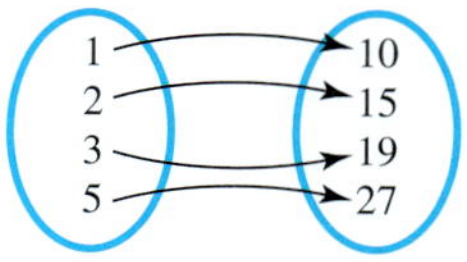

13.

x	y
0	0
−1	1
−2	2

14.

x	y
0	0
1	−1
2	−2

15.

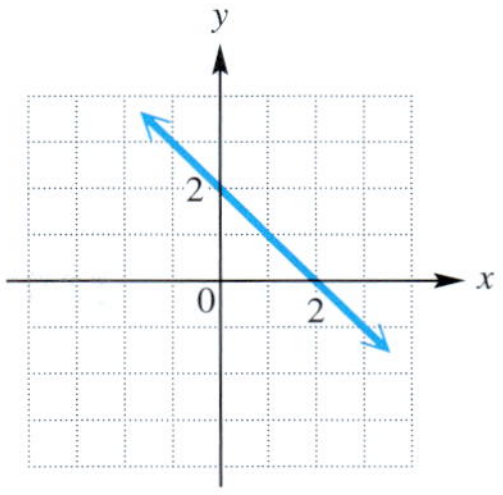

16.

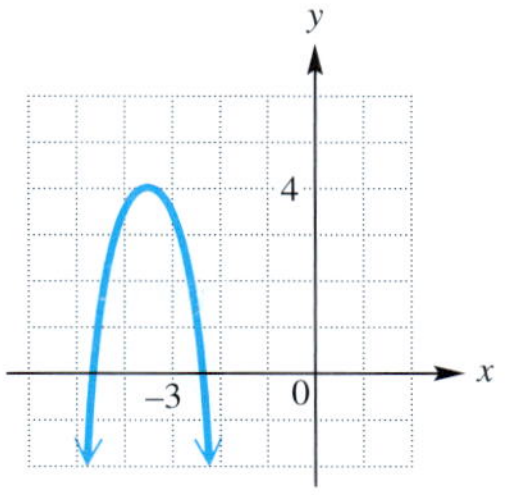

17.

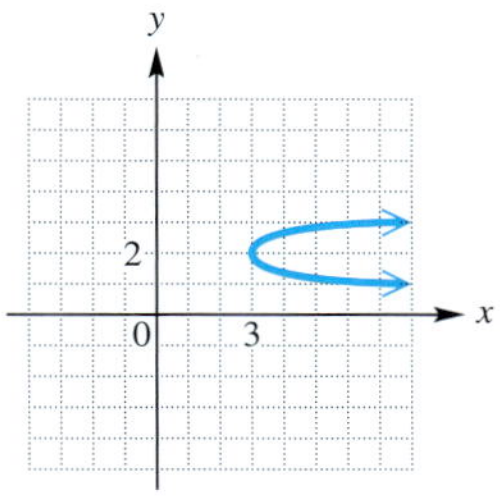

18.

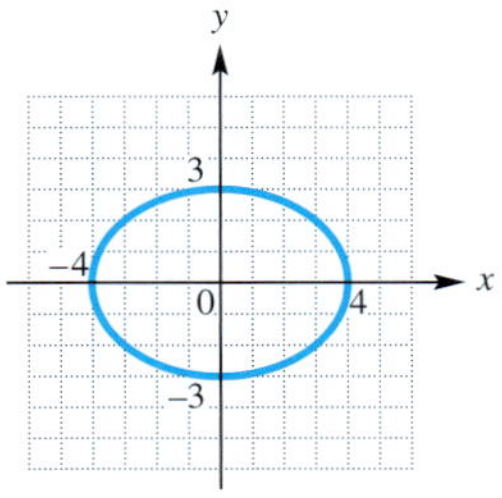

19.

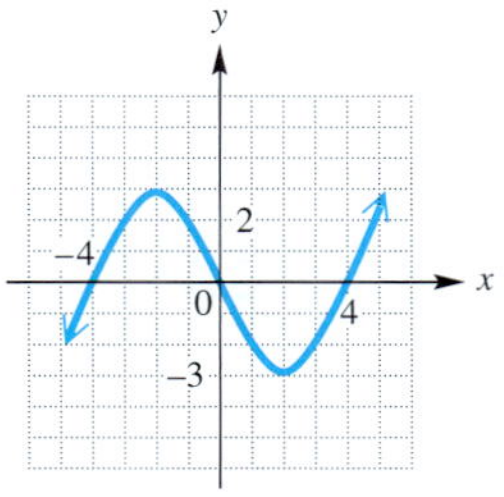

20.

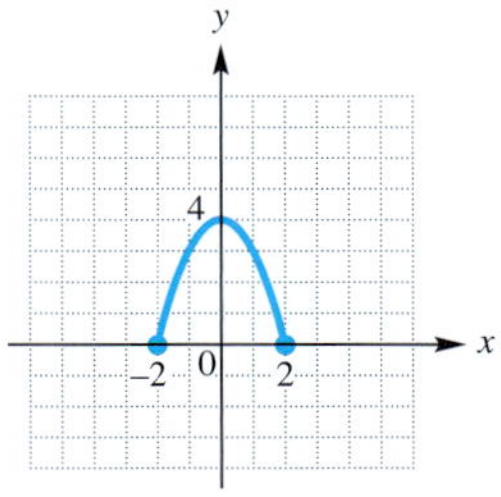

Decide whether each relation defines y as a function of x. Give the domain and range. ***See Example 5.***

21. $y = x^2$ **22.** $y = x^3$ **23.** $x = y^6$ **24.** $x = y^4$

25. $y = 2x - 5$ **26.** $y = -6x + 4$ **27.** $y = \sqrt{x}$ **28.** $y = -\sqrt{x}$

29. $xy = 2$ **30.** $xy = -6$ **31.** $y = \sqrt{4x + 1}$

32. $y = \sqrt{7 - 2x}$ **33.** $y = \dfrac{2}{x - 3}$ **34.** $y = \dfrac{-7}{x - 5}$

Let $f(x) = -3x + 4$ *and* $g(x) = -x^2 + 4x + 1$. *Find and simplify each of the following.* ***See Example 6.***

35. $f(0)$ **36.** $f(-3)$ **37.** $g(-2)$ **38.** $g(10)$

39. $f\left(\frac{1}{3}\right)$ **40.** $f\left(-\frac{7}{3}\right)$ **41.** $g\left(\frac{1}{2}\right)$ **42.** $g\left(-\frac{1}{4}\right)$

43. $f(p)$ **44.** $g(k)$ **45.** $f(-x)$ **46.** $g(-x)$

47. $f(x + 2)$ **48.** $f(a + 4)$ **49.** $f(2m - 3)$ **50.** $f(3t - 2)$

For each function, find ***(a)*** $f(2)$ *and* ***(b)*** $f(-1)$. ***See Example 7.***

51. $f = \{(-1, 3), (4, 7), (0, 6), (2, 2)\}$ **52.** $f = \{(2, 5), (3, 9), (-1, 11), (5, 3)\}$

53.

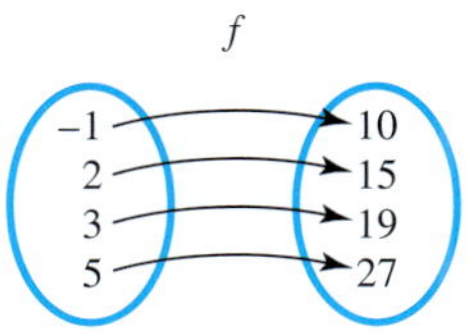

54.

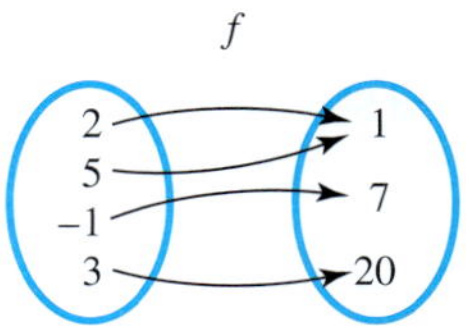

55.

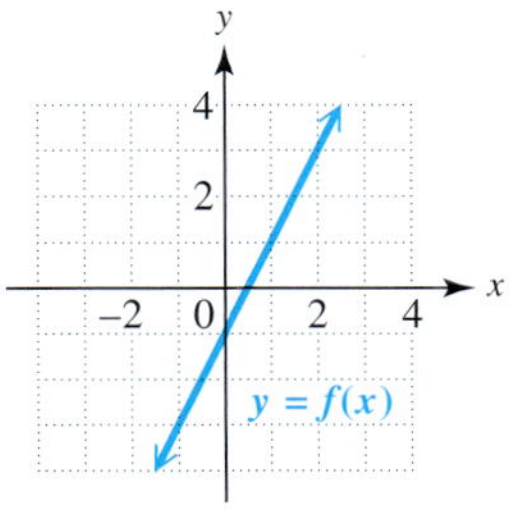

56.

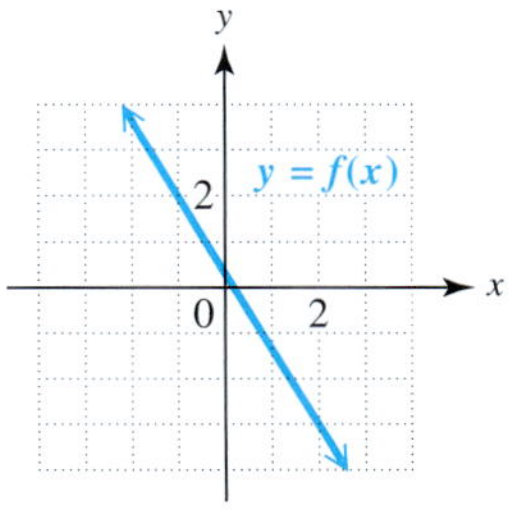

In Exercises 57–60, use the graph of $y = f(x)$ to find each function value: ***(a)*** $f(-2)$, ***(b)*** $f(0)$, ***(c)*** $f(1)$, *and* ***(d)*** $f(4)$. ***See Example 7(d).***

57.

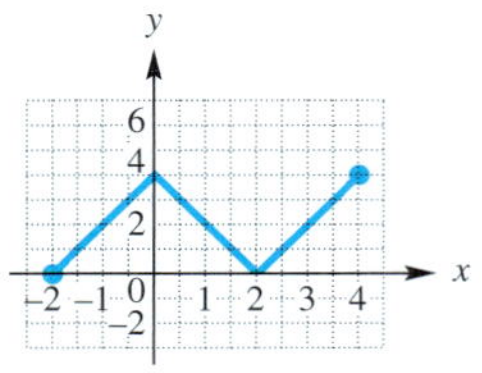

58.

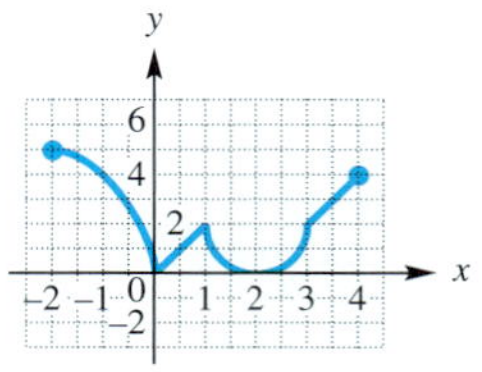

59.

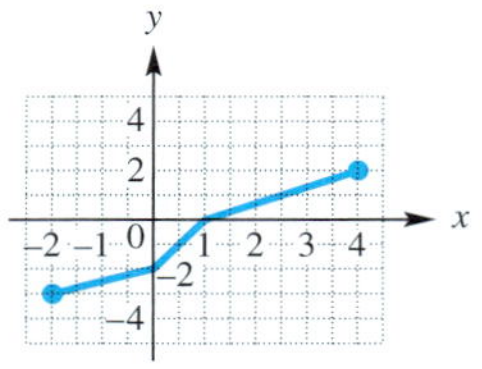

60.

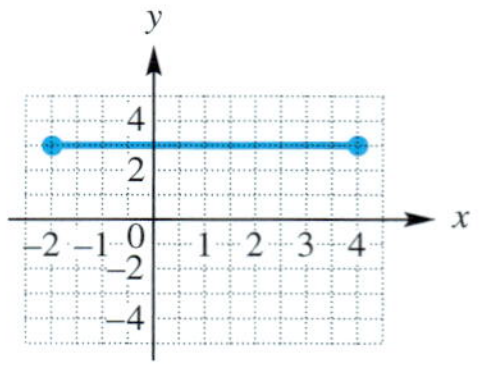

Determine the intervals of the domain for which each function is ***(a)*** *increasing,* ***(b)*** *decreasing, and* ***(c)*** *constant.* ***See Example 8.***

61.

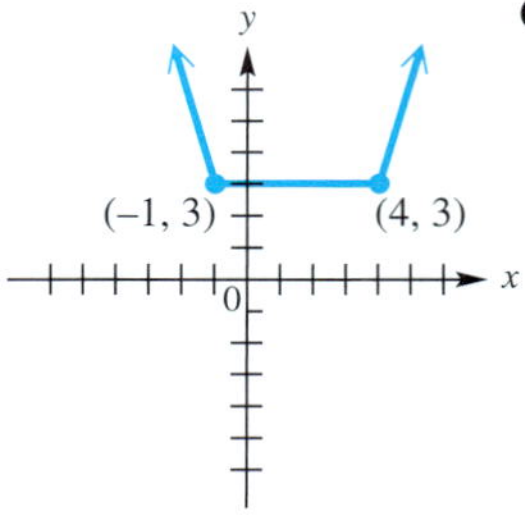

62.

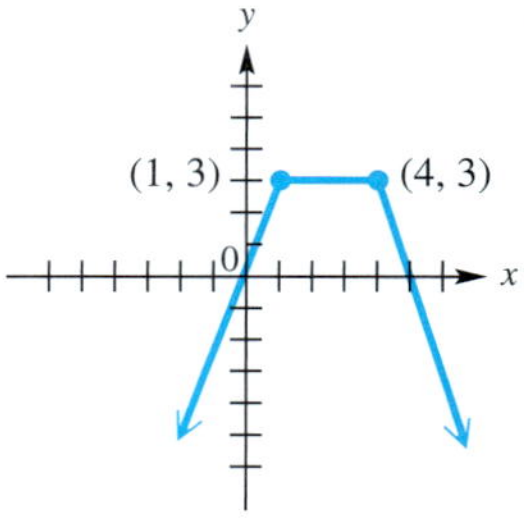

63.

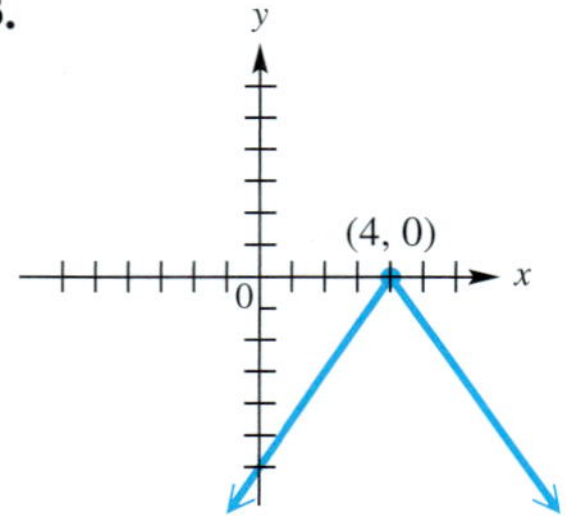

64.

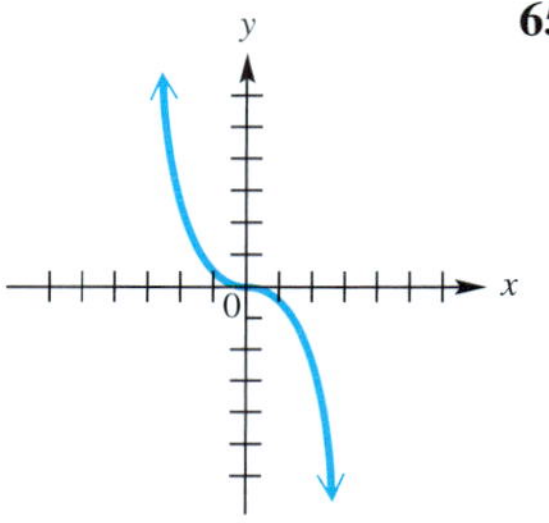

65.

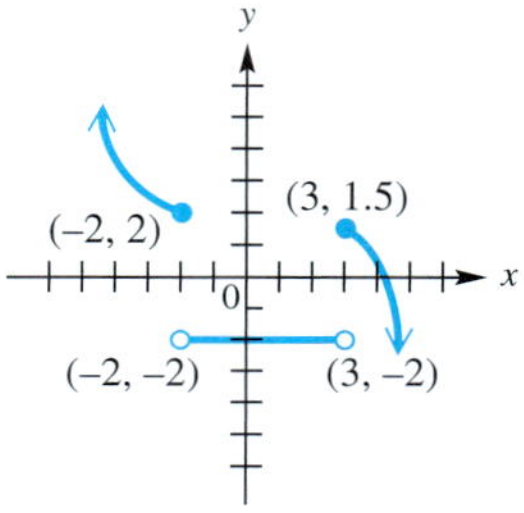

66.

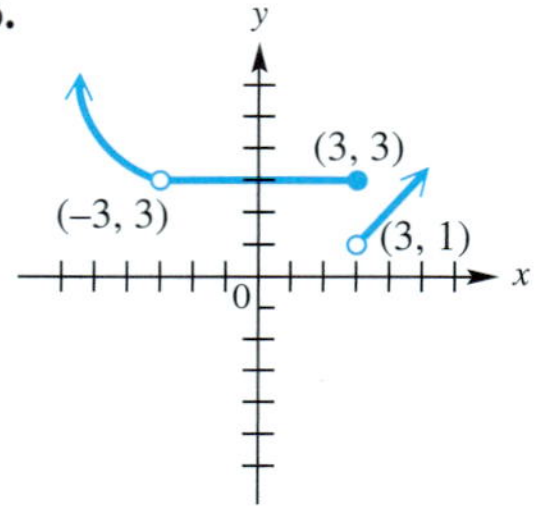

D Graphing Techniques

- Stretching and Shrinking
- Reflecting
- Symmetry
- Translations

Graphing techniques presented in this section show how to graph functions that are defined by altering the equation of a basic function.

NOTE Recall from algebra that $|a|$ is the absolute value of a number a.

$$|a| = \begin{cases} a \text{ if } a \text{ is positive or } 0 \\ -a \text{ if } a \text{ is negative} \end{cases}$$

Thus, $|2| = |2|$ and $|-2| = |2|$.

We use absolute value functions to illustrate many of the graphing techniques in this section.

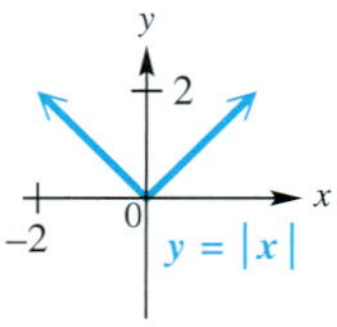

Graph of the absolute value function

Stretching and Shrinking We begin by considering how the graphs of $y = af(x)$ and $y = f(ax)$ compare to the graph of $y = f(x)$, where $a > 0$.

EXAMPLE 1 Stretching or Shrinking a Graph

Graph each function.

(a) $g(x) = 2|x|$ **(b)** $h(x) = \frac{1}{2}|x|$ **(c)** $k(x) = |2x|$

SOLUTION

(a) Comparing the tables of values for $f(x) = |x|$ and $g(x) = 2|x|$ in **Figure 1,** we see that for corresponding x-values, the y-values of g are each twice those of f. The graph of $f(x) = |x|$ is *vertically stretched.* The graph of $g(x)$, shown in blue in **Figure 1,** is narrower than that of $f(x)$, shown in red for comparison.

x	$f(x) = \lvert x \rvert$	$g(x) = 2\lvert x \rvert$
-2	2	4
-1	1	2
0	0	0
1	1	2
2	2	4

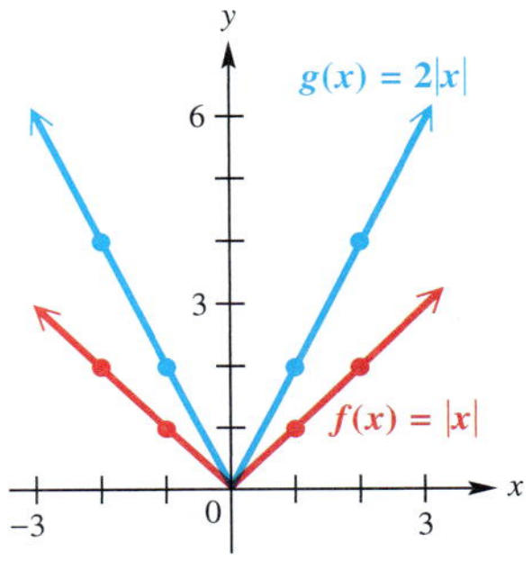

Figure 1

(b) The graph of $h(x) = \frac{1}{2}|x|$ is also the same general shape as that of $f(x)$, but here the coefficient $\frac{1}{2}$ is between 0 and 1 and causes a *vertical shrink.* The graph of $h(x)$ is wider than the graph of $f(x)$, as we see by comparing the tables of values. See **Figure 2.**

x	$f(x) = \lvert x \rvert$	$h(x) = \frac{1}{2}\lvert x \rvert$
-2	2	1
-1	1	$\frac{1}{2}$
0	0	0
1	1	$\frac{1}{2}$
2	2	1

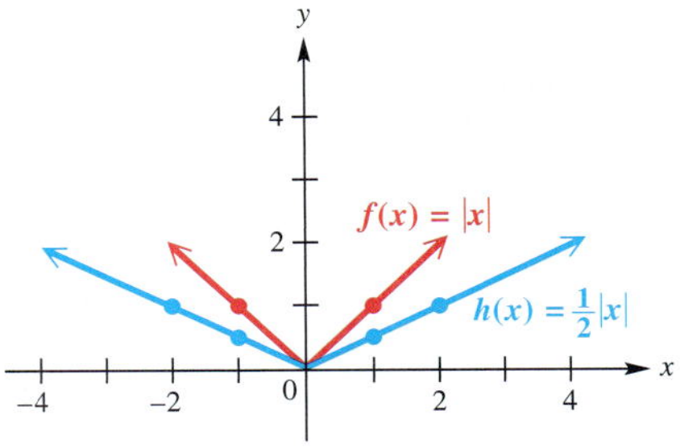

Figure 2

(c) Use the property of absolute value $|ab| = |a| \cdot |b|$ to rewrite $|2x|$.

$$k(x) = |2x| = |2| \cdot |x| = 2|x|$$

Therefore, the graph of $k(x) = |2x|$ is the same as the graph of $g(x) = 2|x|$ in part (a). This is a *horizontal shrink* of the graph of $f(x) = |x|$. See **Figure 2**.

✔ *Now Try Exercises 7 and 9.*

Vertical Stretching or Shrinking of the Graph of a Function

Suppose that $a > 0$. If a point (x, y) lies on the graph of $y = f(x)$, then the point (x, ay) lies on the graph of $\mathbf{y = af(x)}$.

(a) If $a > 1$, then the graph of $y = af(x)$ is a **vertical stretching** of the graph of $y = f(x)$.

(b) If $0 < a < 1$, then the graph of $y = af(x)$ is a **vertical shrinking** of the graph of $y = f(x)$.

Figure 3 shows graphical interpretations of vertical stretching and shrinking. In both cases, the x-intercepts remain the same but the y-intercepts *are* affected.

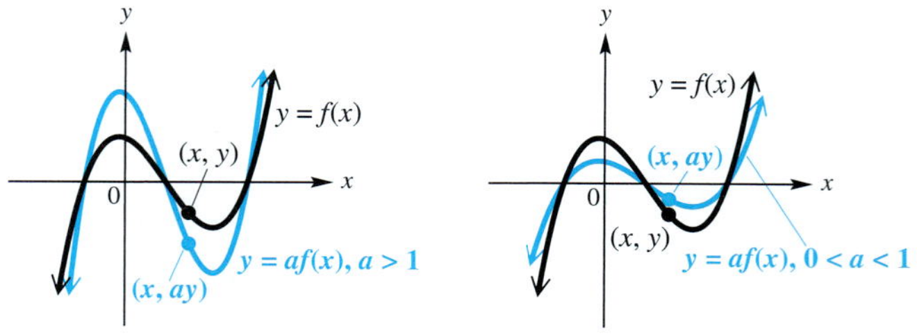

Vertical stretching
$a > 1$

Vertical shrinking
$0 < a < 1$

Figure 3

Graphs of functions can also be stretched and shrunk horizontally.

Horizontal Stretching or Shrinking of the Graph of a Function

Suppose that $a > 0$. If a point (x, y) lies on the graph of $y = f(x)$, then the point $\left(\frac{x}{a}, y\right)$ lies on the graph of $\mathbf{y = f(ax)}$.

(a) If $0 < a < 1$, then the graph of $y = f(ax)$ is a **horizontal stretching** of the graph of $y = f(x)$.

(b) If $a > 1$, then the graph of $y = f(ax)$ is a **horizontal shrinking** of the graph of $y = f(x)$.

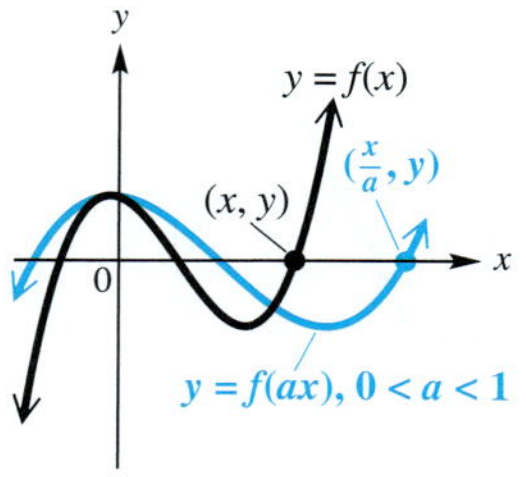

Horizontal stretching
$0 < a < 1$

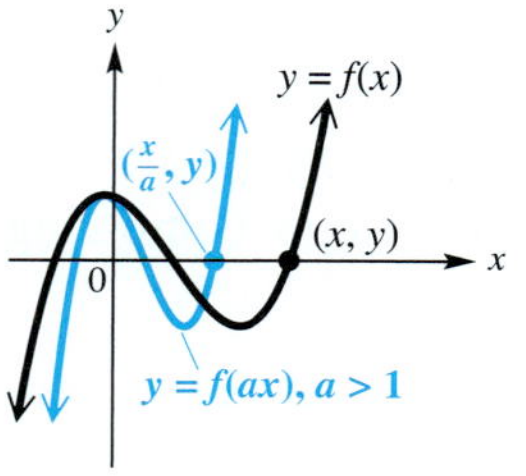

Horizontal shrinking
$a > 1$

Figure 4

See **Figure 4** for graphical interpretations of horizontal stretching and shrinking. In both cases, the y-intercept remains the same but the x-intercepts *are* affected.

Reflecting Forming the mirror image of a graph across a line is called **reflecting the graph across the line.**

EXAMPLE 2 Reflecting a Graph across an Axis

Graph each function.

(a) $g(x) = -\sqrt{x}$ **(b)** $h(x) = \sqrt{-x}$

SOLUTION

(a) The tables of values for $g(x) = -\sqrt{x}$ and $f(x) = \sqrt{x}$ are shown with their graphs in **Figure 5** on the next page. As the tables suggest, every y-value of the graph of $g(x) = -\sqrt{x}$ is the negative of the corresponding y-value of $f(x) = \sqrt{x}$. This has the effect of reflecting the graph across the x-axis.

x	$f(x) = \sqrt{x}$	$g(x) = -\sqrt{x}$
0	0	0
1	1	-1
4	2	-2

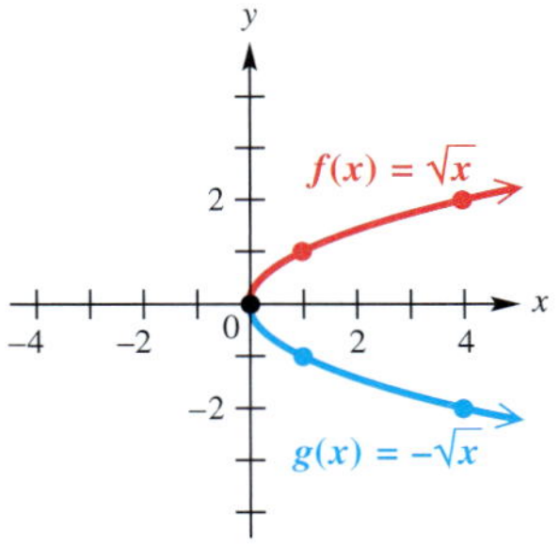

Figure 5

(b) The domain of $h(x) = \sqrt{-x}$ is $(-\infty, 0]$, while the domain of $f(x) = \sqrt{x}$ is $[0, \infty)$. Choosing x-values for $h(x)$ that are negatives of those used for $f(x)$, we see that corresponding y-values are the same. The graph of h is a reflection of the graph of f across the y-axis. See **Figure 6.**

x	$f(x) = \sqrt{x}$	$h(x) = \sqrt{-x}$
-4	undefined	2
-1	undefined	1
0	0	0
1	1	undefined
4	2	undefined

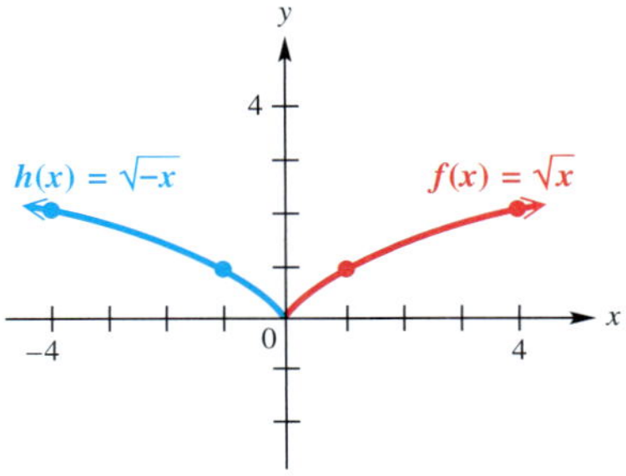

Figure 6

✔ ***Now Try Exercises 17 and 23.***

The graphs in **Example 2** suggest the following generalizations.

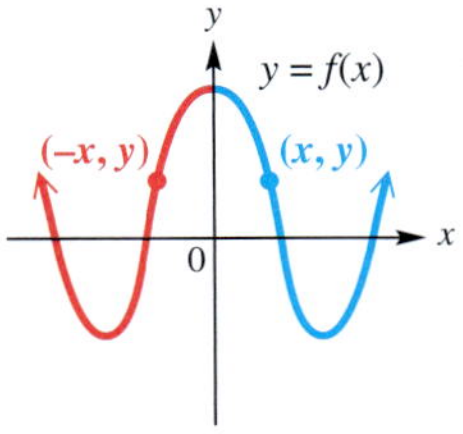

***y*-axis symmetry**

(a)

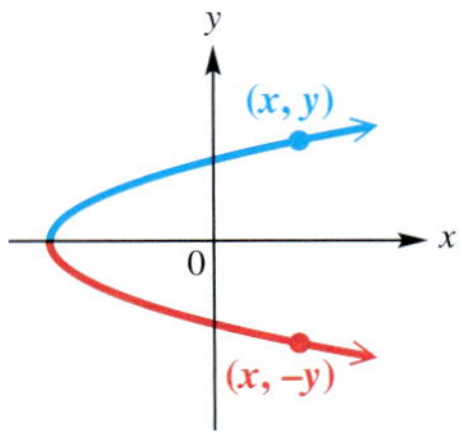

***x*-axis symmetry**

(b)

Figure 7

Reflecting across an Axis

The graph of $\boldsymbol{y = -f(x)}$ is the same as the graph of $y = f(x)$ reflected across the x-axis. (If a point (x, y) lies on the graph of $y = f(x)$, then $(x, -y)$ lies on this reflection.)

The graph of $\boldsymbol{y = f(-x)}$ is the same as the graph of $y = f(x)$ reflected across the y-axis. (If a point (x, y) lies on the graph of $y = f(x)$, then $(-x, y)$ lies on this reflection.)

Symmetry The graph of f shown in **Figure 7(a)** is cut in half by the y-axis with each half the mirror image of the other half. Such a graph is *symmetric with respect to the y-axis.* ***The point $(-x, y)$ is on the graph whenever the point (x, y) is on the graph.***

Similarly, if the graph in **Figure 7(b)** were folded in half along the x-axis, the portion at the top would exactly match the portion at the bottom. Such a graph is *symmetric with respect to the x-axis.* ***The point $(x, -y)$ is on the graph whenever the point (x, y) is on the graph.***

Symmetry with Respect to an Axis

The graph of an equation is **symmetric with respect to the y-axis** if the replacement of x with $-x$ results in an equivalent equation.

The graph of an equation is **symmetric with respect to the x-axis** if the replacement of y with $-y$ results in an equivalent equation.

EXAMPLE 3 Testing for Symmetry with Respect to an Axis

Test for symmetry with respect to the x-axis and the y-axis.

(a) $y = x^2 + 4$ **(b)** $x = y^2 - 3$ **(c)** $x^2 + y^2 = 16$ **(d)** $2x + y = 4$

SOLUTION

(a) In $y = x^2 + 4$, replace x with $-x$.

Use parentheses around $-x$.

$$y = x^2 + 4$$
$$y = (-x)^2 + 4 \quad \text{Equivalent}$$
$$y = x^2 + 4$$

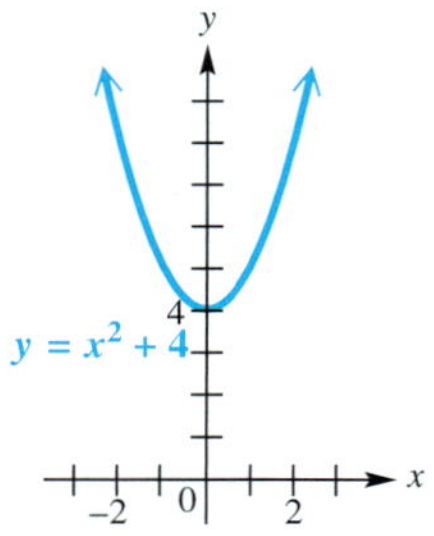

y-axis symmetry

Figure 8

The result is the same as the original equation, so the graph, shown in **Figure 8,** is symmetric with respect to the y-axis. Substituting $-y$ for y does not result in an equivalent equation, and thus the graph is *not* symmetric with respect to the x-axis.

(b) In $x = y^2 - 3$, replace y with $-y$.

$$x = (-y)^2 - 3 = y^2 - 3 \quad \text{Same as the original equation}$$

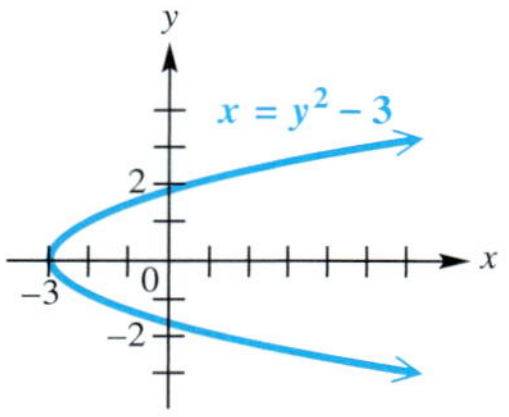

x-axis symmetry

Figure 9

The graph is symmetric with respect to the x-axis, as shown in **Figure 9.** It is *not* symmetric with respect to the y-axis.

(c) Substitute $-x$ for x and then $-y$ for y in $x^2 + y^2 = 16$.

$$(-x)^2 + y^2 = 16 \quad \text{and} \quad x^2 + (-y)^2 = 16$$

Both simplify to the original equation, $x^2 + y^2 = 16$. The graph, a circle of radius 4 centered at the origin, is symmetric with respect to *both* axes. See **Figure 10.**

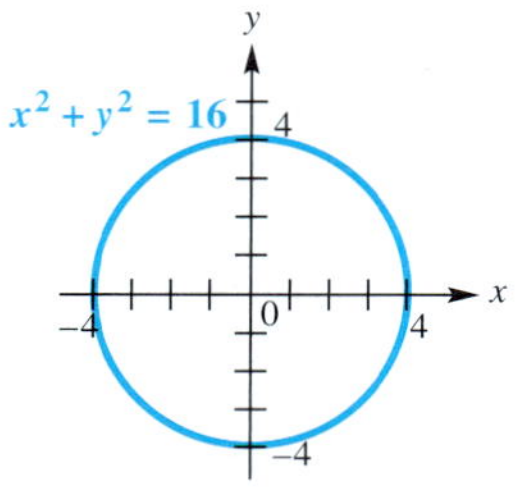

x-axis and y-axis symmetry

Figure 10

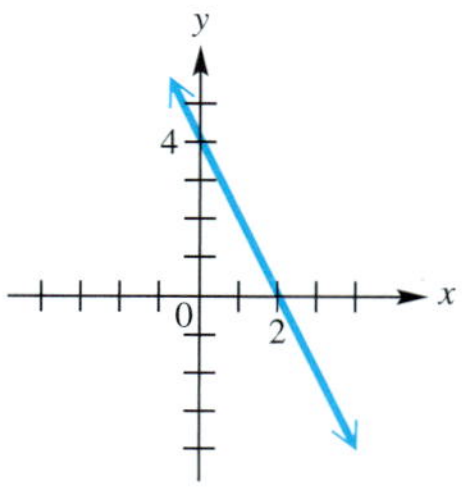

No x-axis or y-axis symmetry

Figure 11

(d) In $2x + y = 4$, replace x with $-x$ to get $-2x + y = 4$. Then replace y with $-y$ in the original equation to get $2x - y = 4$. Neither case produces an equivalent equation, so this graph is not symmetric with respect to either axis. See **Figure 11.**

✔ ***Now Try Exercise 33.***

Another kind of symmetry occurs when a graph can be rotated 180° about the origin, with the result coinciding exactly with the original graph. Symmetry of this type is called *symmetry with respect to the origin.* ***The point*** $(-x, -y)$ ***is on the graph whenever the point*** (x, y) ***is on the graph.*** See **Figure 12.**

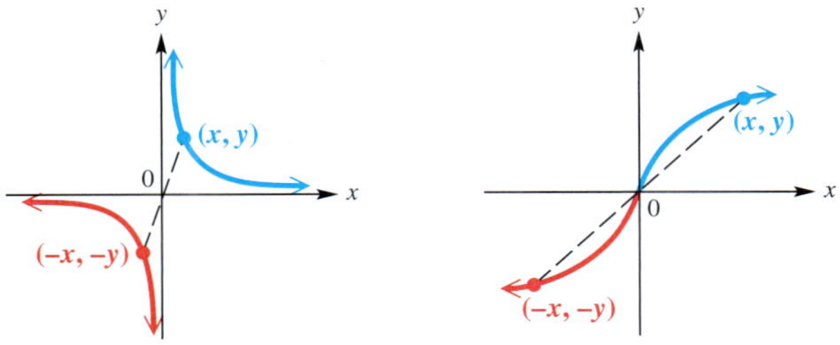

Origin symmetry

Figure 12

It is true that, for functions with origin symmetry, the origin becomes a midpoint of every line segment passing through the origin that connects two points on the graph of the function.

Symmetry with Respect to the Origin

The graph of an equation is **symmetric with respect to the origin** if the replacement of both x with $-x$ and y with $-y$ at the same time results in an equivalent equation.

EXAMPLE 4 Testing for Symmetry with Respect to the Origin

Are the following graphs symmetric with respect to the origin?

(a) $x^2 + y^2 = 16$ **(b)** $y = x^3$

SOLUTION

(a) Replace x with $-x$ and y with $-y$.

$$x^2 + y^2 = 16$$

Use parentheses around $-x$ and $-y$.

$$(-x)^2 + (-y)^2 = 16$$

$$x^2 + y^2 = 16$$

Equivalent

The graph, which is the circle shown in **Figure 10** in **Example 3(c),** is symmetric with respect to the origin.

(b) Replace x with $-x$ and y with $-y$.

$$y = x^3$$

$$-y = (-x)^3$$

$$-y = -x^3$$

$$y = x^3$$

Equivalent

The graph, which is that of the cubing function, is symmetric with respect to the origin and is shown in **Figure 13.**

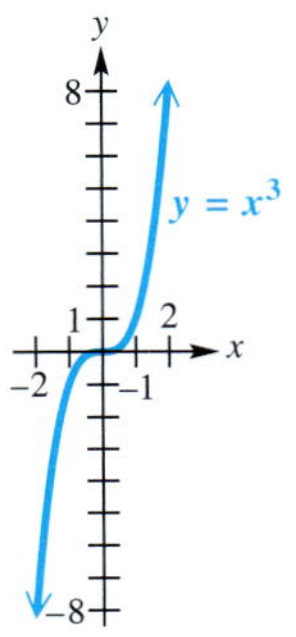

Origin symmetry

Figure 13

✔ *Now Try Exercise 37.*

Notice the following important concepts regarding symmetry:

- A graph symmetric with respect to both the x- and y-axes is automatically symmetric with respect to the origin. (See **Figure 10.**)
- A graph symmetric with respect to the origin need *not* be symmetric with respect to either axis. (See **Figure 13.**)
- Of the three types of symmetry—with respect to the x-axis, with respect to the y-axis, and with respect to the origin—a graph possessing any two types must also exhibit the third type of symmetry.

Translations The next examples show the results of horizontal and vertical shifts, or **translations,** of the graph of $f(x) = |x|$.

EXAMPLE 5 Translating a Graph Vertically

Graph $g(x) = |x| - 4$.

SOLUTION By comparing the table of values for $g(x) = |x| - 4$ and $f(x) = |x|$ shown with **Figure 14,** we see that for corresponding x-values, the y-values of g are each 4 *less* than those for f. Thus, the graph of $g(x) = |x| - 4$ is the same as that of $f(x) = |x|$, but translated 4 units down. See **Figure 14.** The lowest point is at $(0, -4)$. The graph is symmetric with respect to the y-axis and is therefore the graph of an even function.

x	$f(x) = \|x\|$	$g(x) = \|x\| - 4$
-4	4	0
-1	1	-3
0	0	-4
1	1	-3
4	4	0

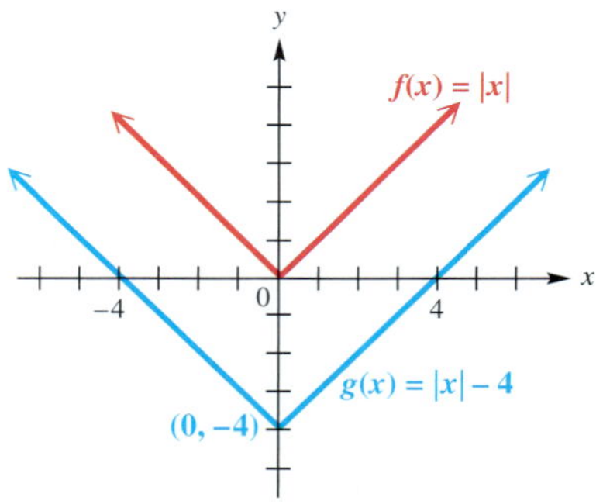

Figure 14

✔ *Now Try Exercise 49.*

The graphs in **Example 5** suggest the following generalization.

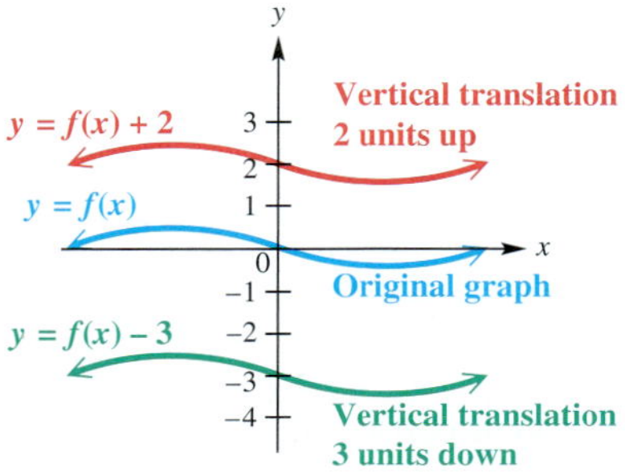

Figure 15

Vertical Translations

If a function g is defined by $\boldsymbol{g(x) = f(x) + c}$, where c is a real number, then for every point (x, y) on the graph of f, there will be a corresponding point $(x, y + c)$ on the graph of g.

The graph of g will be the same as the graph of f, but translated c units up if c is positive or $|c|$ units down if c is negative. The graph of g is called a **vertical translation** of the graph of f. See **Figure 15.**

EXAMPLE 6 Translating a Graph Horizontally

Graph $g(x) = |x - 4|$.

SOLUTION Comparing the tables of values given with **Figure 16** shows that for corresponding y-values, the x-values of g are each 4 *more* than those for f. The graph of $g(x) = |x - 4|$ is the same as that of $f(x) = |x|$, but translated 4 units to the right. The lowest point is at $(4, 0)$. As suggested by the graphs in **Figure 16,** this graph is symmetric with respect to the line $x = 4$.

x	$f(x) = \|x\|$	$g(x) = \|x - 4\|$
−2	2	6
0	0	4
2	2	2
4	4	0
6	6	2

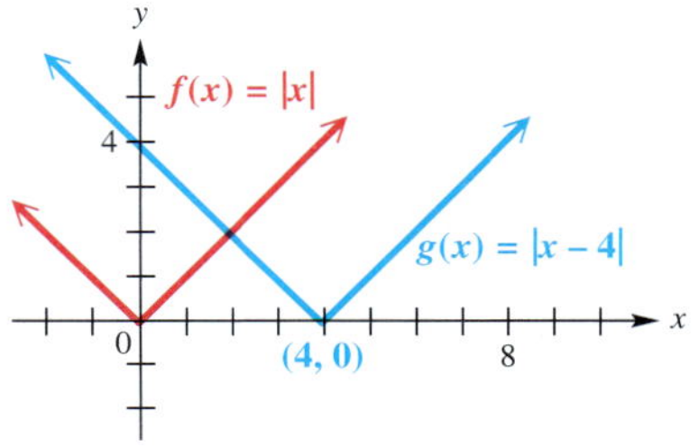

Figure 16

✔ *Now Try Exercise 47.*

The graphs in **Example 6** suggest the following generalization.

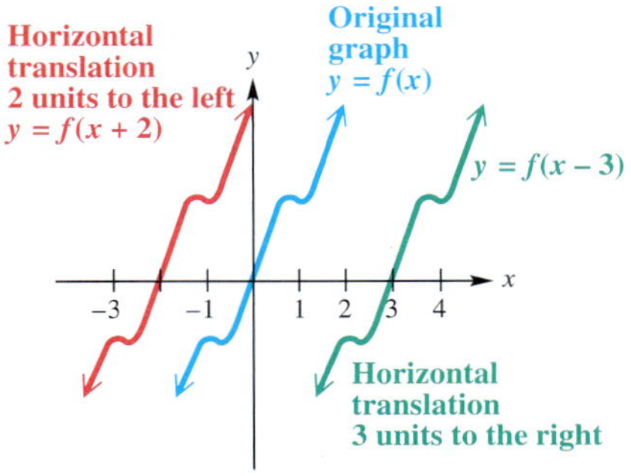

Figure 17

Horizontal Translations

If a function g is defined by $\mathbf{g(x) = f(x - c)}$, where c is a real number, then for every point (x, y) on the graph of f, there will be a corresponding point $(x + c, y)$ on the graph of g.

The graph of g will be the same as the graph of f, but translated c units to the right if c is positive or $|c|$ units to the left if c is negative. The graph of g is called a **horizontal translation** of the graph of f. See **Figure 17.**

$(c > 0)$ To Graph:	Shift the Graph of $y = f(x)$ by c Units:
$y = f(x) + c$	up
$y = f(x) - c$	down
$y = f(x + c)$	left
$y = f(x - c)$	right

Vertical and horizontal translations are summarized in the table in the margin, where f is a function, and c is a positive number.

EXAMPLE 7 Using More Than One Transformation

Graph each function.

(a) $f(x) = -|x + 3| + 1$ **(b)** $h(x) = |2x - 4|$ **(c)** $g(x) = -\frac{1}{2}x^2 + 4$

SOLUTION

(a) To graph $f(x) = -|x + 3| + 1$, the *lowest* point on the graph of $y = |x|$ is translated 3 units to the left and 1 unit up. The graph opens down because of the negative sign in front of the absolute value expression, making the lowest point now the highest point on the graph, as shown in **Figure 18**. The graph is symmetric with respect to the line $x = -3$.

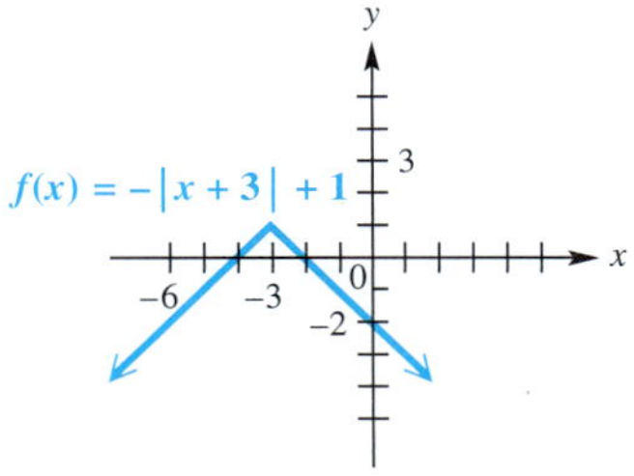

Figure 18

(b) To determine the horizontal translation, factor out 2.

$$\begin{aligned} h(x) &= |2x - 4| \\ &= |2(x-2)| && \text{Factor out 2.} \\ &= |2| \cdot |x-2| && |ab| = |a| \cdot |b| \\ &= 2|x-2| && |2| = 2 \end{aligned}$$

The graph of h is the graph of $y = |x|$ translated 2 units to the right, and vertically stretched by a factor of 2. Horizontal shrinking gives the same appearance as vertical stretching for this function. See **Figure 19.**

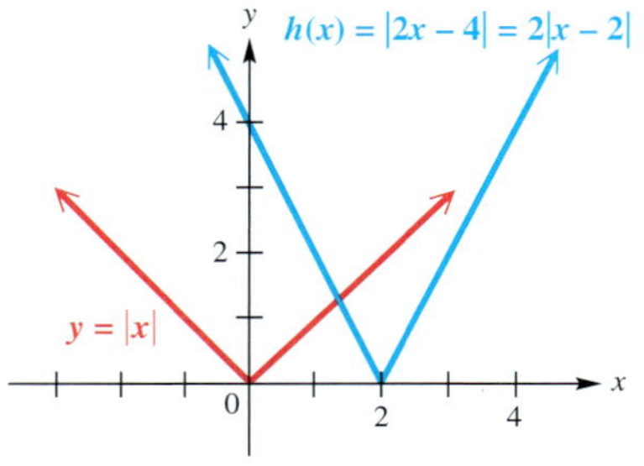

Figure 19

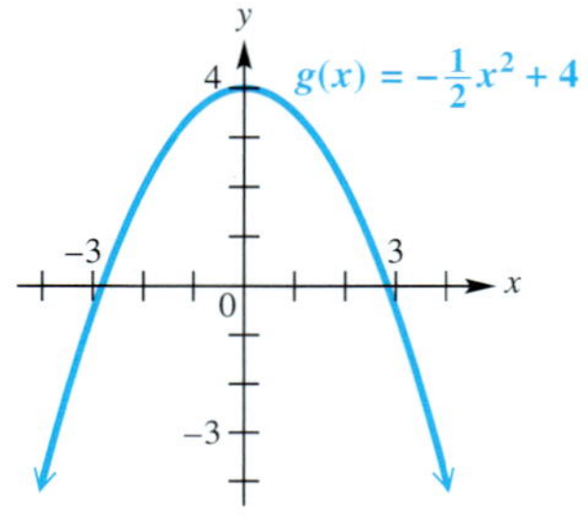

Figure 20

(c) The graph of $g(x) = -\frac{1}{2}x^2 + 4$ has the same shape as that of $y = x^2$, but it is wider (that is, shrunken vertically), reflected across the x-axis because the coefficient $-\frac{1}{2}$ is negative, and then translated 4 units up. See **Figure 20.**

✔ *Now Try Exercises 53, 55, and 63.*

Appendix D Exercises

1. *Concept Check* Match each equation in Column I with a description of its graph from Column II as it relates to the graph of $y = x^2$.

I	II
(a) $y = (x-7)^2$	**A.** a translation 7 units to the left
(b) $y = x^2 - 7$	**B.** a translation 7 units to the right
(c) $y = 7x^2$	**C.** a translation 7 units up
(d) $y = (x+7)^2$	**D.** a translation 7 units down
(e) $y = x^2 + 7$	**E.** a vertical stretching by a factor of 7

2. *Concept Check* Match each equation in Column I with a description of its graph from Column II as it relates to the graph of $y = \sqrt[3]{x}$.

I	II
(a) $y = 4\sqrt[3]{x}$	**A.** a translation 4 units to the right
(b) $y = -\sqrt[3]{x}$	**B.** a translation 4 units down
(c) $y = \sqrt[3]{-x}$	**C.** a reflection across the x-axis
(d) $y = \sqrt[3]{x-4}$	**D.** a reflection across the y-axis
(e) $y = \sqrt[3]{x} - 4$	**E.** a vertical stretching by a factor of 4

3. *Concept Check* Match each equation in parts (a)–(i) with the sketch of its graph.

(a) $y = x^2 + 2$ **(b)** $y = x^2 - 2$ **(c)** $y = (x + 2)^2$

(d) $y = (x - 2)^2$ **(e)** $y = 2x^2$ **(f)** $y = -x^2$

(g) $y = (x - 2)^2 + 1$ **(h)** $y = (x + 2)^2 + 1$ **(i)** $y = (x + 2)^2 - 1$

A.

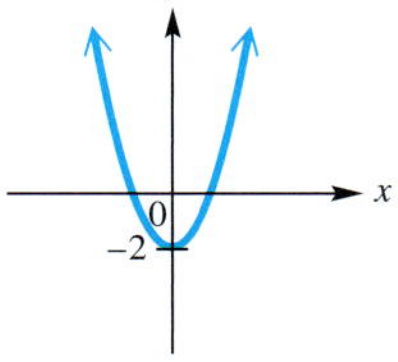

B.

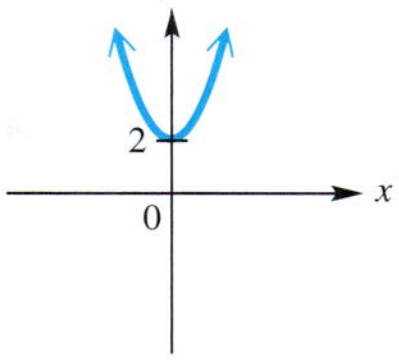

C.

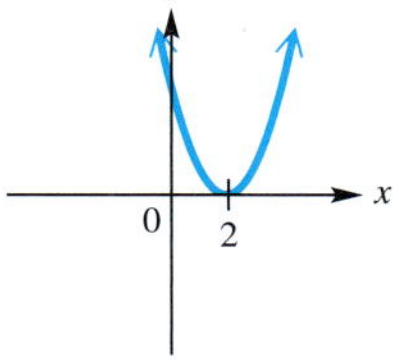

D.

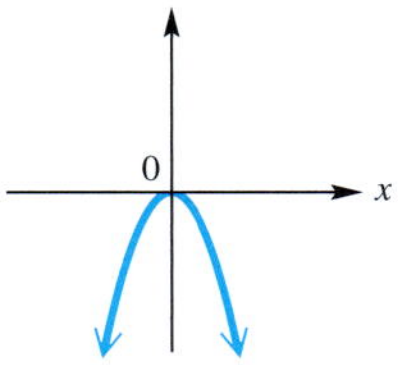

E.

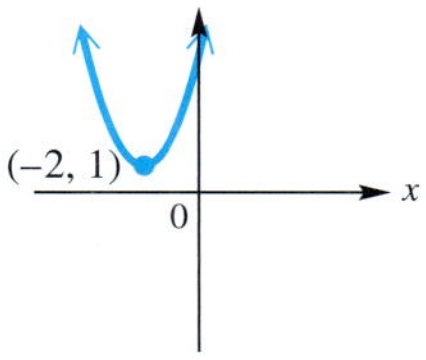

F.

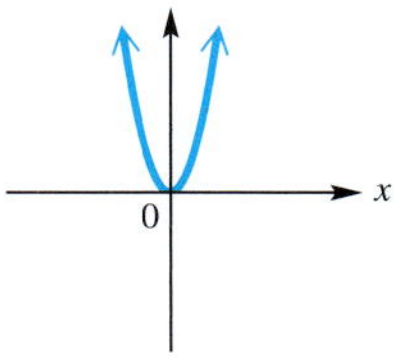

G.

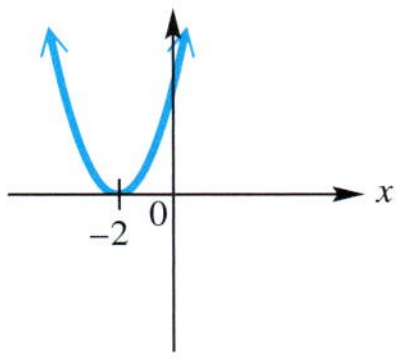

H.

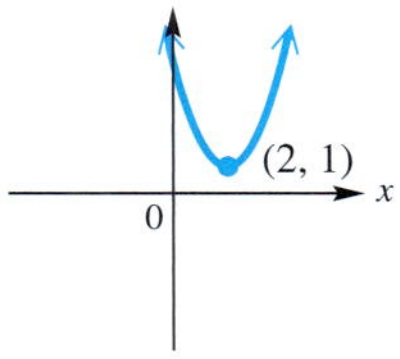

I.

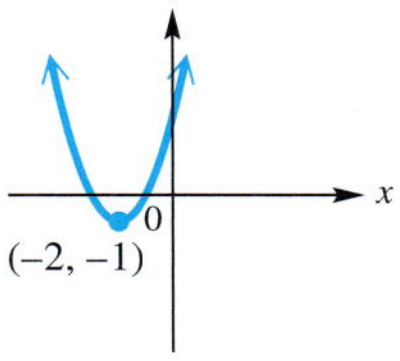

4. *Concept Check* Match each equation in parts (a)–(i) with the sketch of its graph.

(a) $y = \sqrt{x + 3}$ **(b)** $y = \sqrt{x} - 3$ **(c)** $y = \sqrt{x} + 3$

(d) $y = 3\sqrt{x}$ **(e)** $y = -\sqrt{x}$ **(f)** $y = \sqrt{x - 3}$

(g) $y = \sqrt{x - 3} + 2$ **(h)** $y = \sqrt{x + 3} + 2$ **(i)** $y = \sqrt{x - 3} - 2$

A.

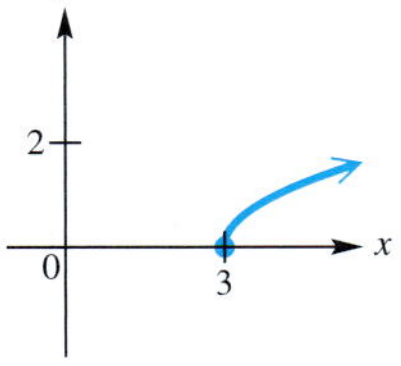

B.

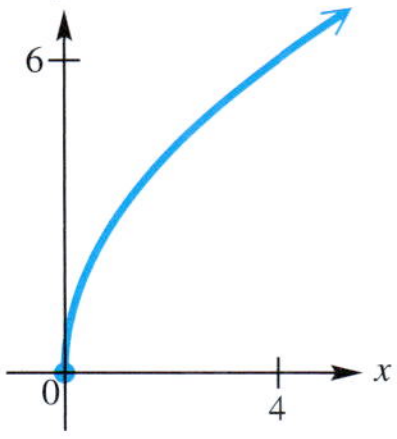

C.

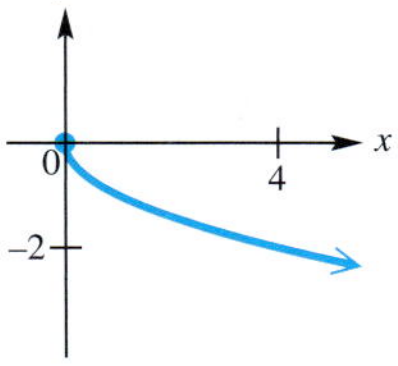

D.

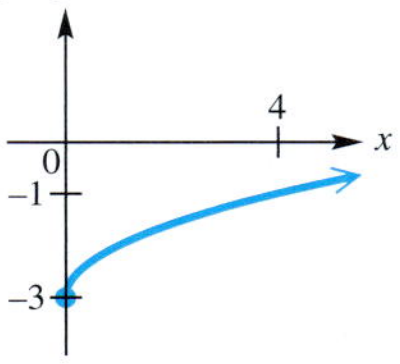

E.

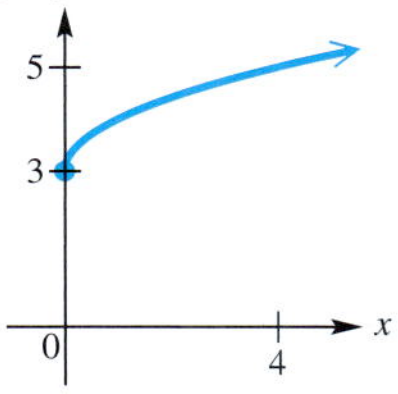

F.

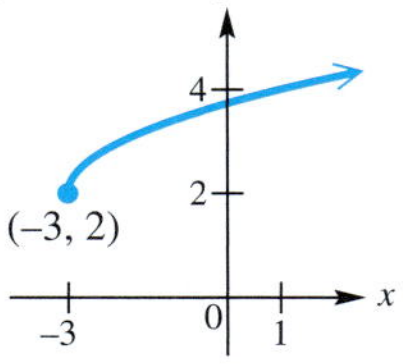

G.

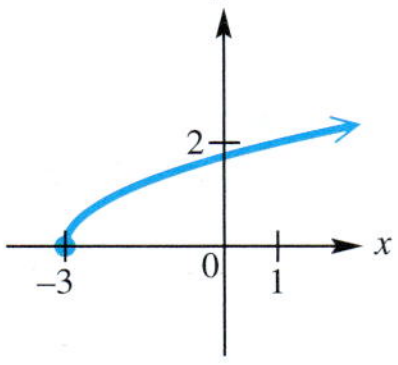

H.

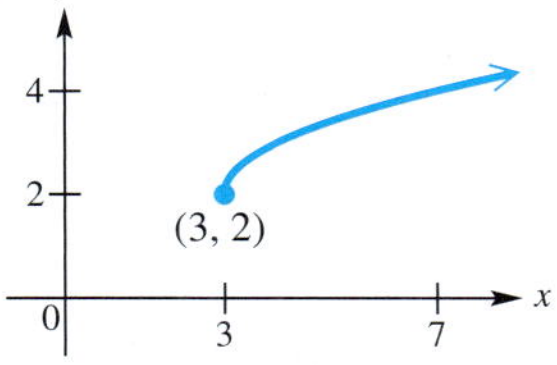

I.

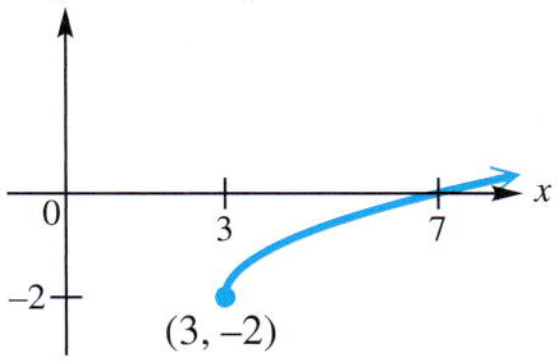

5. *Concept Check* Match each equation in parts (a)–(i) with the sketch of its graph.

(a) $y = |x - 2|$ **(b)** $y = |x| - 2$ **(c)** $y = |x| + 2$

(d) $y = 2|x|$ **(e)** $y = -|x|$ **(f)** $y = |-x|$

(g) $y = -2|x|$ **(h)** $y = |x - 2| + 2$ **(i)** $y = |x + 2| - 2$

A.
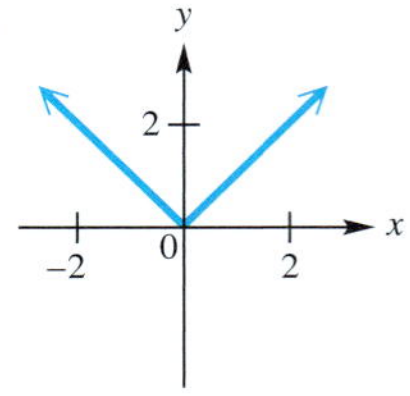

B.
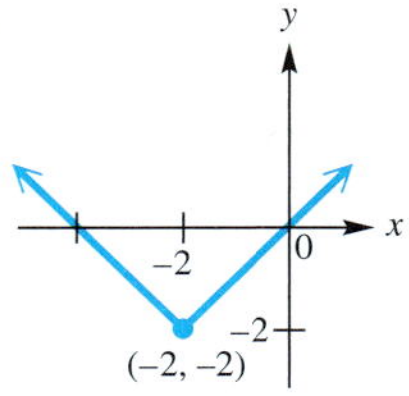

C.
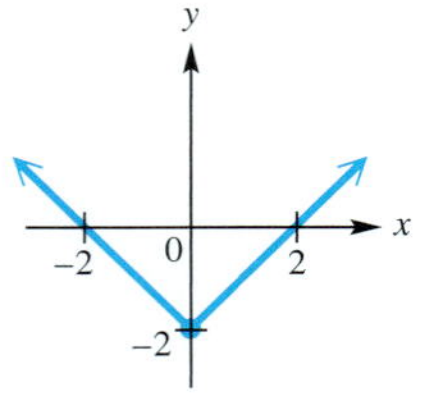

D.
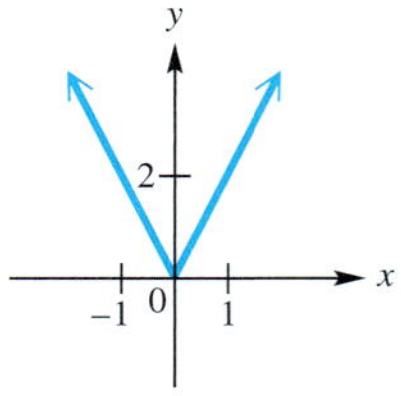

E.
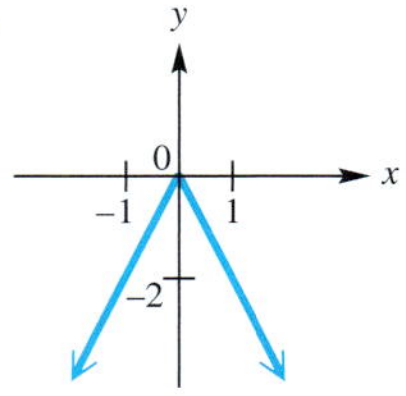

F.
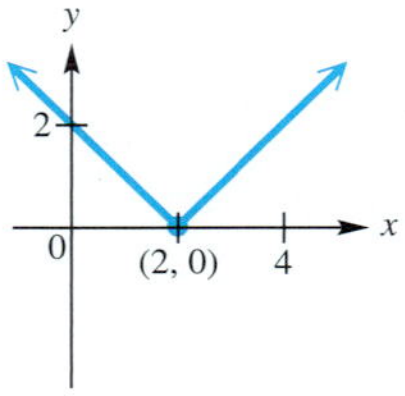

G.
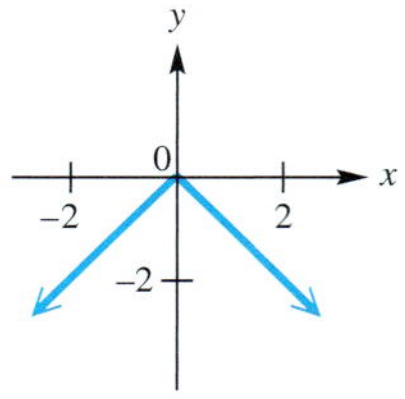

H.
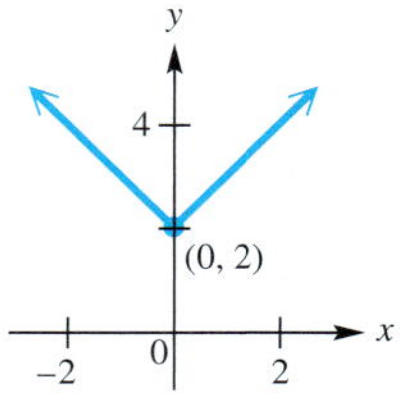

I.
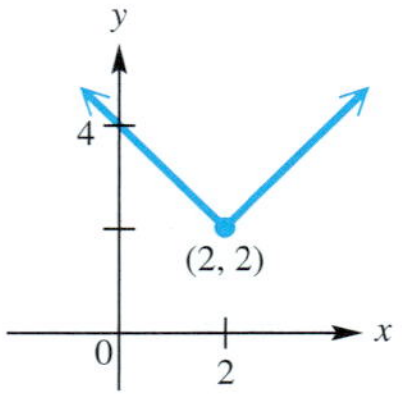

6. (a) Suppose the equation $y = F(x)$ is changed to $y = c \cdot F(x)$, for some constant c. What is the effect on the graph of $y = F(x)$? Discuss the effect depending on whether $c > 0$ or $c < 0$, and on whether $|c| > 1$ or $|c| < 1$.

(b) Suppose $y = F(x)$ is changed to $y = F(x + h)$. How are the graphs of these equations related? Is the graph of $y = F(x) + h$ the same as the graph of $y = F(x + h)$? If not, how do they differ?

Graph each function. ***See Examples 1 and 2.***

7. $y = 3|x|$ **8.** $y = 4|x|$ **9.** $y = \frac{2}{3}|x|$ **10.** $y = \frac{3}{4}|x|$

11. $y = 2x^2$ **12.** $y = 3x^2$ **13.** $y = \frac{1}{2}x^2$ **14.** $y = \frac{1}{3}x^2$

15. $y = -\frac{1}{2}x^2$ **16.** $y = -\frac{1}{3}x^2$ **17.** $y = -3|x|$ **18.** $y = -2|x|$

19. $y = \left|-\frac{1}{2}x\right|$ **20.** $y = \left|-\frac{1}{3}x\right|$ **21.** $y = \sqrt{4x}$

22. $y = \sqrt{9x}$ **23.** $y = -\sqrt{-x}$ **24.** $y = -|-x|$

Concept Check In Exercises 25–28, suppose the point $(8, 12)$ is on the graph of $y = f(x)$. Find a point on the graph of each function.

25. (a) $y = f(x + 4)$ **26. (a)** $y = \frac{1}{4}f(x)$ **27. (a)** $y = f(4x)$

(b) $y = f(x) + 4$ **(b)** $y = 4f(x)$ **(b)** $y = f\left(\frac{1}{4}x\right)$

28. **(a)** the reflection of the graph of $y = f(x)$ across the x-axis

(b) the reflection of the graph of $y = f(x)$ across the y-axis

*Concept Check Plot each point, and then plot the points that are symmetric to the given point with respect to the **(a)** x-axis, **(b)** y-axis, and **(c)** origin.*

29. $(5, -3)$ **30.** $(-6, 1)$ **31.** $(-4, -2)$ **32.** $(-8, 0)$

Without graphing, determine whether each equation has a graph that is symmetric with respect to the x-axis, *the* y-axis, *the* origin, *or* none of these. ***See Examples 3 and 4.***

33. $y = x^2 + 5$ **34.** $y = 2x^4 - 3$ **35.** $x^2 + y^2 = 12$ **36.** $y^2 - x^2 = -6$

37. $y = -4x^3 + x$ **38.** $y = x^3 - x$ **39.** $y = x^2 - x + 8$ **40.** $y = x + 15$

Graph each function. ***See Examples 5–7.***

41. $y = x^2 - 1$ **42.** $y = x^2 - 2$ **43.** $y = x^2 + 2$

44. $y = x^2 + 3$ **45.** $y = (x - 4)^2$ **46.** $y = (x - 2)^2$

47. $y = (x + 2)^2$ **48.** $y = (x + 3)^2$ **49.** $y = |x| - 1$

50. $y = |x + 3| + 2$ **51.** $y = -(x + 1)^3$ **52.** $y = -(x - 1)^3$

53. $y = 2x^2 - 1$ **54.** $y = 3x^2 - 2$ **55.** $f(x) = 2(x - 2)^2 - 4$

56. $f(x) = -3(x - 2)^2 + 1$ **57.** $f(x) = \sqrt{x + 2}$ **58.** $f(x) = \sqrt{x - 3}$

59. $f(x) = -\sqrt{x}$ **60.** $f(x) = \sqrt{x} - 2$ **61.** $f(x) = 2\sqrt{x} + 1$

62. $y = 3\sqrt{x} - 2$ **63.** $y = \frac{1}{2}x^3 - 4$ **64.** $y = \frac{1}{2}x^3 + 2$

Connecting Graphs with Equations Each of the following graphs is obtained from the graph of $f(x) = |x|$ or $g(x) = \sqrt{x}$ by applying several of the transformations discussed in this section. Describe the transformations and give the equation for the graph.

65.

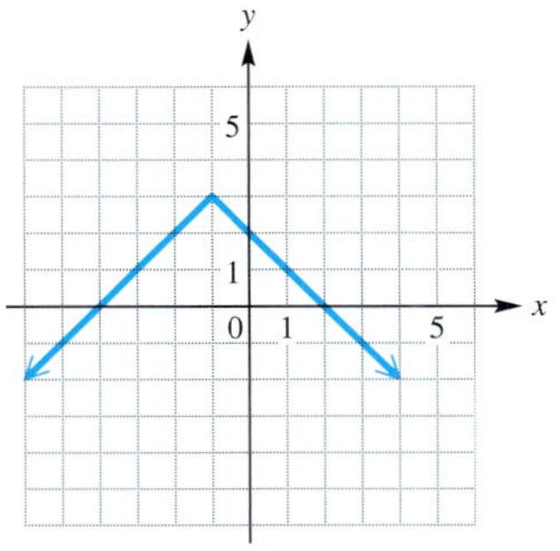

66.

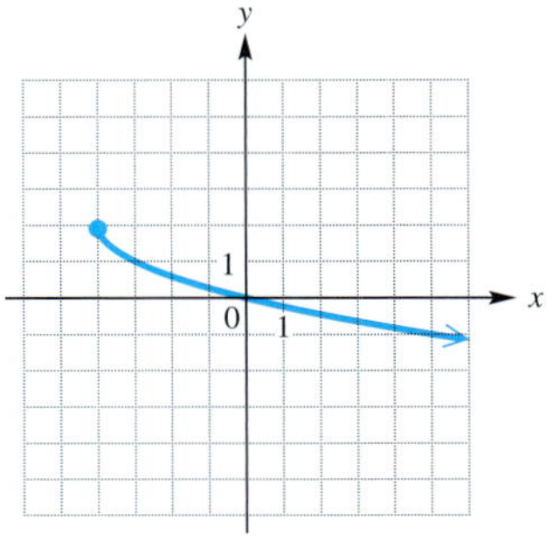

67.

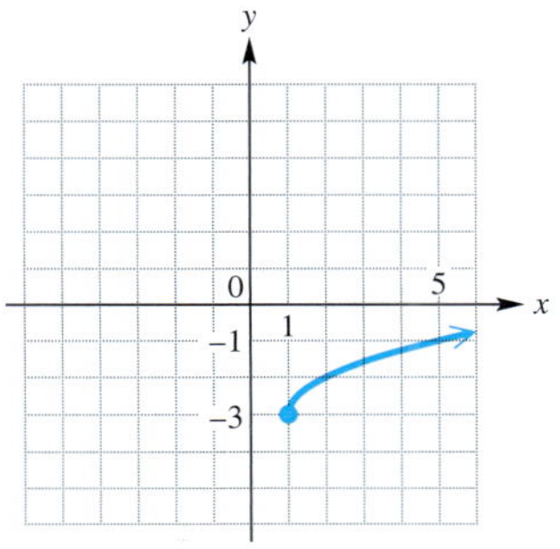

68.

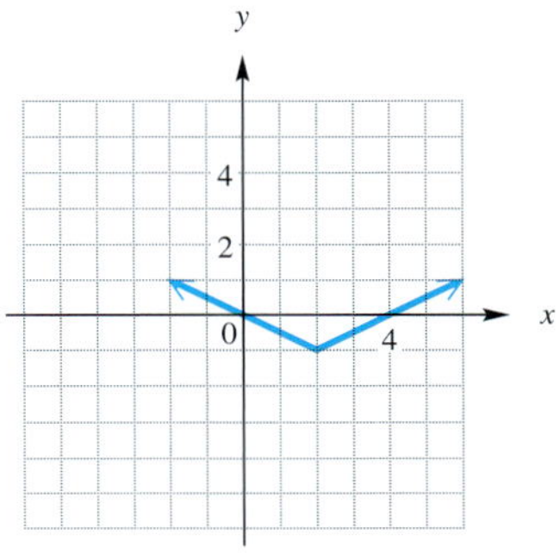

69.

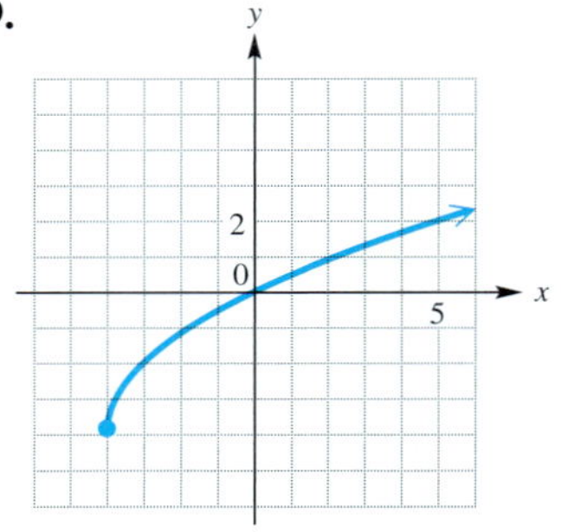

70.

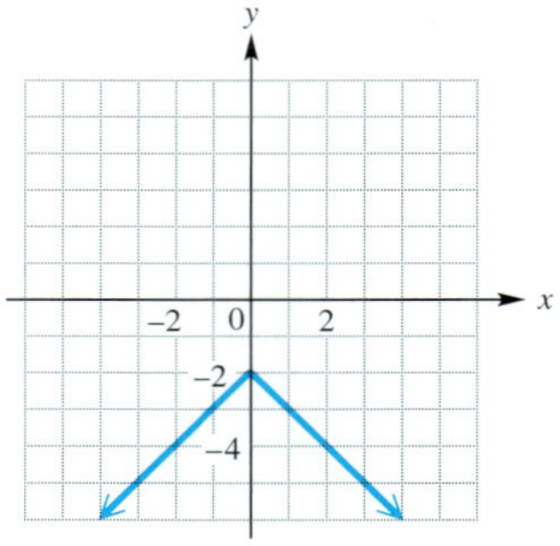

Concept Check Suppose that for a function f, $f(3) = 6$. For the given assumptions in Exercises 71–76, find another function value.

71. The graph of $y = f(x)$ is symmetric with respect to the origin.

72. The graph of $y = f(x)$ is symmetric with respect to the y-axis.

73. The graph of $y = f(x)$ is symmetric with respect to the line $x = 6$.

74. For all x, $f(-x) = f(x)$.

75. For all x, $f(-x) = -f(x)$.

76. f is an odd function.

Glossary

For a more complete discussion, see the section(s) in parentheses.

A

abscissa The x-value of a point may be called its abscissa. (Section 4.4)

absolute value (modulus) of a complex number When a complex number is written in trigonometric (or polar) form as $r(\cos\theta + i\sin\theta)$, the number r is the absolute value (or modulus) of the complex number. (Section 8.2)

acute angle An acute angle is an angle measuring between 0° and 90°. (Section 1.1)

addition of ordinates Addition of ordinates is a method for graphing a function that is the sum of two other functions, using addition of the y-values of the two functions at selected x-values. (Section 4.4)

airspeed In air navigation, the airspeed of a plane is its speed relative to the air. (Section 7.5)

ambiguous case The situation in which the lengths of two sides of a triangle and the measure of the angle opposite one of them are given (SSA) is the ambiguous case of the law of sines. Depending on the given measurements, this combination of given parts may result in 0, 1, or 2 possible triangles. (Section 7.2)

amplitude The amplitude of a periodic function is half the difference between the maximum and minimum values of the function. (Section 4.1)

angle An angle is formed by rotating a ray around its endpoint. (Section 1.1)

angle of depression The angle of depression from point X to point Y (below X) is the acute angle formed by ray XY and a horizontal ray with endpoint at X. (Section 2.4)

angle of elevation The angle of elevation from point X to point Y (above X) is the acute angle formed by ray XY and a horizontal ray with endpoint at X. (Section 2.4)

Angle-Side-Angle (ASA) The Angle-Side-Angle (ASA) congruence axiom states that if two angles and the included side of one triangle are equal, respectively, to two angles and the included side of a second triangle, then the triangles are congruent. (Section 7.1)

angle in standard position An angle is in standard position if its vertex is at the origin and its initial side is along the positive x-axis. (Section 1.1)

angle between two vectors The angle between two vectors is defined to be the angle θ, for $0° \le \theta \le 180°$, having the two vectors as its sides. (Section 7.4)

angular speed ω Angular speed ω (omega) measures the speed of rotation and is defined by $\omega = \frac{\theta}{t}$, where θ is the angle of rotation in radians and t is time. (Section 3.4)

argument of a complex number When a complex number is written in trigonometric (or polar) form as $r(\cos\theta + i\sin\theta)$, the angle θ is the argument of the complex number. (Section 8.2)

argument of a function The argument of a function is the expression containing the independent variable of the function. For example, in the function $y = f(x - d)$, the expression $x - d$ is the argument. (Section 4.2)

B

bearing Bearing is used to identify angles in navigation. One method for expressing bearing uses a single angle, with bearing measured in a clockwise direction from due north. A second method for expressing bearing starts with a north-south line and uses an acute angle to show the direction, either east or west, from this line. (Sections 2.5, 7.5)

C

cardioid A cardioid is a heart-shaped curve that is the graph of a polar equation of the form $r = a \pm b\sin\theta$ or $r = a \pm b\cos\theta$, where $\left|\frac{a}{b}\right| = 1$. (Section 8.5)

center of a circle The center of a circle is the given point that is a given distance from all points on the circle. (Appendix B)

circle A circle is the set of all points in a plane that lie a given distance from a given point. (Appendix B)

circular functions The trigonometric functions of arc lengths, or real numbers, are the circular functions. (Section 3.3)

closed interval A closed interval is an interval that includes both of its endpoints. (Appendix A)

cofunctions The function pairs sine and cosine, tangent and cotangent, and secant and cosecant are cofunctions. (Section 2.1)

complementary angles (complements) Two positive angles are complementary angles (or complements) if the sum of their measures is 90°. (Section 1.1)

complex conjugates The complex conjugate of $a + bi$ is $a - bi$. (Section 8.1)

complex number A complex number is a number of the form $a + bi$, where a and b are real numbers and $i = \sqrt{-1}$. (Section 8.1)

complex plane The complex plane is a two-dimensional representation of the complex numbers in which the horizontal axis is the real axis and the vertical axis is the imaginary axis. (Section 8.2)

conditional equation An equation that is satisfied by some numbers but not by others is a conditional equation. (Section 6.2, Appendix A)

congruent triangles Triangles that are both the same size and the same shape are congruent triangles. (Section 1.2)

constant function A function f is constant on an interval I if, for every x_1 and x_2 in I, $f(x_1) = f(x_2)$. (Appendix C)

contradiction An equation that has no solution is a contradiction. (Appendix A)

coordinate plane (*xy*-plane) The plane into which the rectangular coordinate system is introduced is the coordinate plane (or xy-plane). (Appendix B)

coordinates (in the *xy*-plane) The coordinates of a point in the xy-plane are the numbers in the ordered pair that correspond to that point. (Appendix B)

cosecant Let $P(x, y)$ be a point other than the origin on the terminal side of an angle θ in standard position. Let $r = \sqrt{x^2 + y^2}$ represent the distance from the origin to P. Then the cosecant function is defined by $\csc\theta = \frac{r}{y}\ (y \neq 0)$. (Section 1.3)

cosine Let $P(x, y)$ be a point other than the origin on the terminal side of an angle θ in standard position. Let $r = \sqrt{x^2 + y^2}$ represent the distance from the origin to P. Then the cosine function is defined by $\cos\theta = \frac{x}{r}$. (Section 1.3)

cotangent Let $P(x, y)$ be a point other than the origin on the terminal side of an angle θ in standard position. Let $r = \sqrt{x^2 + y^2}$ represent the distance from the origin to P. Then the cotangent function is defined by $\cot\theta = \frac{x}{y}\ (y \neq 0)$. (Section 1.3)

coterminal angles Two angles that have the same initial side and the same terminal side, but different measures of rotation, are coterminal angles. The measures of coterminal angles differ by a multiple of 360°. (Section 1.1)

cycloid A cycloid is a curve that represents the path traced by a fixed point on the circumference of a circle rolling along a line. (Section 8.6)

D

damped oscillatory motion Damped oscillatory motion is oscillatory motion that has been slowed down (damped) by the force of friction. Friction causes the amplitude of the motion to diminish gradually until the weight comes to rest. (Section 4.5)

decreasing function A function f is decreasing on an interval I if, whenever $x_1 < x_2$ in I, $f(x_1) > f(x_2)$. (Appendix C)

degree The degree is a unit of measure for angles. One degree, written 1°, represents $\frac{1}{360}$ of a rotation. (Section 1.1)

dependent variable If the value of the variable y depends on the value of the variable x, then y is the dependent variable. (Appendix C)

direction angle The positive angle between the x-axis and a position vector is the direction angle for the vector. (Section 7.4)

domain In a relation, the set of all values of the independent variable (x) is the domain. (Appendix C)

dot product The dot product of two vectors is the sum of the product of their first components and the product of their second components. The dot product of the two vectors $\mathbf{u} = \langle a, b \rangle$ and $\mathbf{v} = \langle c, d \rangle$ is denoted $\mathbf{u} \cdot \mathbf{v}$ and given by $\mathbf{u} \cdot \mathbf{v} = ac + bd$. (Section 7.4)

E

empty set (null set) The empty set (or null set), written $\emptyset$ or $\{ \}$, is the set containing no elements. (Appendix A)

endpoint of a ray In a given ray AB, point A is the endpoint of the ray. (Section 1.1)

equation An equation is a statement that two expressions are equal. (Appendix A)

equilibrant The opposite vector of the resultant of two vectors is called the equilibrant. (Section 7.5)

even function A function f is an even function if for all x in the domain of f, $f(-x) = f(x)$. The graph of an even function is symmetric with respect to the y-axis. (Section 4.1)

exact number A number that represents the result of counting, or a number that results from theoretical work and is not the result of a measurement, is an exact number. (Section 2.4)

F

four-leaved rose A four-leaved rose is a curve that is the graph of a polar equation of the form $r = a \sin 2\theta$ or $r = a \cos 2\theta$. (Section 8.5)

frequency In simple harmonic motion, the frequency is the number of cycles per unit of time, or the reciprocal of the period. (Section 4.5)

function A function is a relation (set of ordered pairs) in which, for each value of the first component of the ordered pairs, there is *exactly one* value of the second component. (Appendix C)

function notation Function notation $f(x)$ (read "f of x") represents the y-value of the function f for the indicated x-value. (Appendix C)

G

graph of an equation The graph of an equation is the set of all points that correspond to all of the ordered pairs that satisfy the equation. (Appendix B)

ground speed In air navigation, the ground speed of a plane is its speed relative to the ground. (Section 7.5)

H

horizontal component When a vector $\mathbf{u}$ is expressed as an ordered pair in the form $\mathbf{u} = \langle a, b \rangle$, the number a is the horizontal component of the vector. (Section 7.4)

I

identity An equation satisfied by every number that is a meaningful replacement for the variable is an identity. (Section 5.1, Appendix A)

imaginary axis In the complex plane, the vertical axis is the imaginary axis. (Section 8.2)

imaginary part In the complex number $a + bi$, b is the imaginary part. (Section 8.1)

imaginary unit The number i, defined by $i = \sqrt{-1}$ (and thus $i^2 = -1$), is the imaginary unit. (Section 8.1)

increasing function A function f is increasing on an interval I if, whenever $x_1 < x_2$ in I, $f(x_1) < f(x_2)$. (Appendix C)

independent variable If the value of the variable y depends on the value of the variable x, then x is the independent variable. (Appendix C)

inequality An inequality says that one expression is greater than, greater than or equal to, less than, or less than or equal to another. (Appendix A)

initial point When two letters are used to name a vector, the first letter indicates the initial (starting) point of the vector. (Section 7.4)

initial side When a ray is rotated around its endpoint to form an angle, the ray in its starting position is the initial side of the angle. (Section 1.1)

interval An interval is a portion of the real number line, which may or may not include its endpoint(s). (Appendix A)

interval notation Interval notation is a simplified notation for writing intervals. It uses parentheses and brackets to show whether the endpoints are included. (Appendix A)

inverse function The inverse function of the one-to-one function f is defined as $\{(y, x) \mid (x, y) \text{ belongs to } f\}$. (Section 6.1)

L

latitude Latitude gives the measure of a central angle with vertex at Earth's center whose initial side goes through the equator and whose terminal side goes through the given location. (Section 3.2)

lemniscate A lemniscate is a figure-eight-shaped curve that is the graph of a polar equation of the form $r^2 = a^2 \sin 2\theta$ or $r^2 = a^2 \cos 2\theta$. (Section 8.5)

limaçon A limaçon is the graph of a polar equation of the form $r = a \pm b \sin \theta$ or $r = a \pm b \cos \theta$. If $\left|\frac{a}{b}\right| = 1$, the limaçon is a cardioid. (Section 8.5)

line Two distinct points A and B determine the line AB. (Section 1.1)

line segment (segment) Line segment AB is the portion of line AB between A and B, including the endpoints A and B. (Section 1.1)

linear equation (first-degree equation) in one variable A linear equation in one variable is an equation that can be written in the form $ax + b = 0$, where a and b are real numbers with $a \neq 0$. (Appendix A)

linear inequality in one variable A linear inequality in one variable is an inequality that can be written in the form $ax + b > 0$, where a and b are real numbers with $a \neq 0$. (Any of the symbols $<$, $\geq$, and $\leq$ may also be used.) (Appendix A)

linear speed v Linear speed v measures the distance traveled per unit of time. (Section 3.4)

M

magnitude The length of a vector represents the magnitude of the vector quantity. (Section 7.4)

minute One minute, written $1'$, is $\frac{1}{60}$ of a degree. (Section 1.1)

N

negative angle A negative angle is an angle that is formed by clockwise rotation around its endpoint. (Section 1.1)

nonreal complex number A complex number $a + bi$ with $b \neq 0$ is a nonreal complex number. (Section 8.1)

***n*th root of a complex number** For a positive integer n, the complex number $a + bi$ is an nth root of the complex number $x + yi$ if $(a + bi)^n = x + yi$. (Section 8.4)

O

oblique triangle A triangle that is not a right triangle is an oblique triangle. (Section 7.1)

obtuse angle An obtuse angle is an angle measuring more than 90° but less than 180°. (Section 1.1)

odd function A function f is an odd function if for all x in the domain of f, $f(-x) = -f(x)$. The graph of an odd function is symmetric with respect to the origin. (Section 4.1)

one-to-one function If a function is defined so that each range element is used only once, then it is a one-to-one function. (Section 6.1)

open interval An open interval is an interval that does not include its endpoint(s). (Appendix A)

opposite of a vector The opposite of a vector $\mathbf{v}$ is a vector $-\mathbf{v}$ that has the same magnitude as $\mathbf{v}$ but opposite direction. (Section 7.4)

ordered pair An ordered pair consists of two components, written inside parentheses. Ordered pairs are used to identify points in the rectangular coordinate plane. (Appendix B)

ordinate The y-value of a point may be called its ordinate. (Section 4.4)

origin The point of intersection of the x-axis and the y-axis of a rectangular coordinate system is the origin. (Appendix B)

orthogonal vectors Orthogonal vectors are vectors that are perpendicular, meaning that the angle between the two vectors is 90°. (Section 7.4)

P

parallel lines Parallel lines are lines that lie in the same plane and do not intersect. (Section 1.2)

parallelogram rule The parallelogram rule is a geometric interpretation of the sum of two vectors. If the two vectors are placed so that their initial points coincide and a parallelogram is completed that has these two vectors as two of its sides, then the diagonal vector of the parallelogram that has the same initial point as the two vectors is their sum. (Section 7.4)

parameter A parameter is a variable in terms of which two or more other variables are expressed. In a pair of parametric equations $x = f(t)$ and $y = g(t)$, the variable t is the parameter. (Section 8.6)

parametric equations of a plane curve A pair of equations $x = f(t)$ and $y = g(t)$ are parametric equations of a plane curve. (Section 8.6)

period For a periodic function such that $f(x) = f(x + np)$, the least possible positive value of p is the period of the function. (Section 4.1)

periodic function A periodic function is a function f such that $f(x) = f(x + np)$, for every real number x in the domain of f, every integer n, and some positive real number p. (Section 4.1)

phase shift For periodic functions, a horizontal translation is a phase shift. (Section 4.2)

plane curve A plane curve is a set of points (x, y) such that $x = f(t)$ and $y = g(t)$, and f and g are both defined on an interval I. (Section 8.6)

polar axis The polar axis is a specific ray in the polar coordinate system that has the pole as its endpoint. The polar axis is usually drawn in the direction of the positive x-axis.(Section 8.5)

polar coordinates In the polar coordinate system, the ordered pair (r, θ) gives polar coordinates of point P, where r is the directed distance from the pole to P and θ is the directed angle from the positive x-axis to ray OP. (Section 8.5)

polar coordinate system The polar coordinate system is a coordinate system based on a point (the pole) and a ray (the polar axis). (Section 8.5)

polar equation A polar equation is an equation that uses polar coordinates. The variables are r and θ. (Section 8.5)

pole The pole is the single fixed point in the polar coordinate system that is the endpoint of the polar axis. The pole is usually placed at the origin of a rectangular coordinate system. (Section 8.5)

position vector A vector with its initial point at the origin is a position vector. (Section 7.4)

positive angle A positive angle is an angle that is formed by counterclockwise rotation around its endpoint. (Section 1.1)

pure imaginary number A complex number $a + bi$ in which $a = 0$ and $b \neq 0$ is a pure imaginary number. (Section 8.1)

Pythagorean theorem The Pythagorean theorem states that in a right triangle, the sum of the squares of the lengths of the legs is equal to the square of the length of the hypotenuse. (Appendix B)

Q

quadrantal angle A quadrantal angle is an angle that, when placed in standard position, has its terminal side along the x-axis or the y-axis. (Section 1.1)

quadrants The quadrants are the four regions into which the x-axis and y-axis divide the coordinate plane. (Appendix B)

quadratic equation (second-degree equation) An equation that can be written in the form $ax^2 + bx + c = 0$, where a, b, and c are real numbers with $a \neq 0$, is a quadratic equation. (Appendix A)

quadratic formula The quadratic formula $x = \frac{-b \pm \sqrt{b^2 - 4ac}}{2a}$ is a general formula that can be used to solve a quadratic equation of the form $ax^2 + bx + c = 0$. (Appendix A)

R

radian A radian is a unit of measure for angles. An angle with its vertex at the center of a circle that intercepts an arc on the circle equal in length to the radius of the circle has a measure of 1 radian. (Section 3.1)

radius The radius of a circle is the distance between the center and any point on the circle. (Appendix B)

range In a relation, the set of all values of the dependent variable (y) is the range. (Appendix C)

ray The portion of line AB that starts at A and continues through B, and on past B, is ray AB. (Section 1.1)

real axis In the complex plane, the horizontal axis is the real axis. (Section 8.2)

real part In the complex number $a + bi$, a is the real part. (Section 8.1)

reciprocal The reciprocal of a nonzero number x is $\frac{1}{x}$. (Section 1.4)

rectangular (Cartesian) coordinate system The x-axis and y-axis together make up a rectangular (or Cartesian) coordinate system. (Appendix B)

rectangular (Cartesian) equation A rectangular (or Cartesian) equation is an equation that uses rectangular coordinates. If it is an equation in two variables, the variables are x and y. (Section 8.5)

rectangular form (standard form) of a complex number The rectangular form (or standard form) of a complex number is $a + bi$, where a and b are real numbers. (Section 8.2)

reference angle The reference angle for an angle θ, written θ', is the positive acute angle made by the terminal side of angle θ and the x-axis. (Section 2.2)

reference arc The reference arc for a point on the unit circle is the shortest arc from the point itself to the nearest point on the x-axis. (Section 3.3)

relation A relation is a set of ordered pairs. (Appendix C)

resultant If **A** and **B** are vectors, then the vector sum **A** + **B** is the resultant of vectors **A** and **B**. (Section 7.4)

right angle A right angle is an angle measuring exactly 90°. (Section 1.1)

rose curve A rose curve is a member of a family of curves that resemble flowers. It is the graph of a polar equation of the form $r = a \sin n\theta$ or $r = a \cos n\theta$. (Section 8.5)

S

scalar A scalar is a quantity that involves a magnitude and can be represented by a real number. (Section 7.4)

scalar product The scalar product of a real number (or scalar) k and a vector **u** is the vector $k \cdot \mathbf{u}$, which has magnitude $|k|$ times the magnitude of **u**. (Section 7.4)

secant Let $P(x, y)$ be a point other than the origin on the terminal side of an angle θ in standard position. Let $r = \sqrt{x^2 + y^2}$ represent the distance from the origin to P. Then the secant function is defined by $\sec \theta = \frac{r}{x}$ $(x \neq 0)$. (Section 1.3)

second One second, written 1″, is $\frac{1}{60}$ of a minute. (Section 1.1)

sector of a circle A sector of a circle is the portion of the interior of a circle intercepted by a central angle. (Section 3.2)

semiperimeter The semiperimeter of a triangle is half the sum of the lengths of the three sides. (Section 7.3)

side of an angle One of the two rays (or line segments) with a common endpoint that form an angle is a side of the angle. (Section 1.1)

Side-Angle-Side (SAS) The Side-Angle-Side (SAS) congruence axiom states that if two sides and the included angle of one triangle are equal, respectively, to two sides and the included angle of a second triangle, then the triangles are congruent. (Section 7.1)

Side-Side-Side (SSS) The Side-Side-Side (SSS) congruence axiom states that if three sides of one triangle are equal, respectively, to three sides of a second triangle, then the triangles are congruent. (Section 7.1)

significant digit A significant digit is a digit obtained by actual measurement. (Section 2.4)

similar triangles Triangles that are the same shape, but not necessarily the same size, are similar triangles. (Section 1.2)

simple harmonic motion Simple harmonic motion is oscillatory motion about an equilibrium position. If friction is neglected, then this motion can be described by a sinusoid. (Section 4.5)

sine Let $P(x, y)$ be a point other than the origin on the terminal side of an angle θ in standard position. Let $r = \sqrt{x^2 + y^2}$ represent the distance from the origin to P. Then the sine function is defined by $\sin \theta = \frac{y}{r}$. (Section 1.3)

sine wave (sinusoid) The graph of a sine function is called a sine wave (or sinusoid). (Section 4.1)

solution (root) A solution (or root) of an equation is a number that makes the equation a true statement. (Appendix A)

solution set The solution set of an equation is the set of all numbers that satisfy the equation. (Appendix A)

spiral of Archimedes A spiral of Archimedes is an infinite curve that is the graph of a polar equation of the form $r = n\theta$. (Section 8.5)

standard form of a complex number A complex number written in the form $a + bi$ (or $a + ib$) is in standard form. (Section 8.1)

straight angle A straight angle is an angle measuring exactly 180°. (Section 1.1)

supplementary angles (supplements) Two positive angles are supplementary angles (or supplements) if the sum of their measures is 180°. (Section 1.1)

T

tangent Let $P(x, y)$ be a point other than the origin on the terminal side of an angle θ in standard position. Let $r = \sqrt{x^2 + y^2}$ represent the distance from the origin to P. Then the tangent function is defined by $\tan \theta = \frac{y}{x}$ $(x \neq 0)$. (Section 1.3)

terminal point When two letters are used to name a vector, the second letter indicates the terminal (ending) point of the vector. (Section 7.4)

terminal side When a ray is rotated around its endpoint to form an angle, the ray in its location after rotation is the terminal side of the angle. (Section 1.1)

translation A translation is a horizontal or vertical shift of a graph. (Appendix D)

transversal A line that intersects two or more other lines, which may be parallel, is a transversal. (Section 1.2)

trigonometric (polar) form of a complex number The expression $r(\cos \theta + i \sin \theta)$ is the trigonometric form (or polar form) of the complex number $x + yi$. The expression $\cos \theta + i \sin \theta$ is sometimes abbreviated as cis θ. (Section 8.2)

U

unit circle The unit circle is the circle with center at the origin and radius 1. (Section 3.3)

unit vector A unit vector is a vector that has magnitude 1. Two important unit vectors are $\mathbf{i} = \langle 1, 0 \rangle$ and $\mathbf{j} = \langle 0, 1 \rangle$. (Section 7.4)

V

vector A vector is a directed line segment that represents a vector quantity with direction and magnitude. (Section 7.4)

vector quantities Quantities that involve both magnitude and direction are vector quantities. (Section 7.4)

vertex of an angle The vertex of an angle is the endpoint of the ray that is rotated to form the angle. (Section 1.1)

vertical angles Vertical angles are opposite angles formed by intersecting lines. (Section 1.2)

vertical asymptote A vertical line that a graph approaches, but never touches or intersects, is a vertical asymptote. The line $x = a$ is a vertical asymptote if $|f(x)|$ increases without bound as x approaches a. (Section 4.3)

vertical component When a vector **u** is expressed as an ordered pair in the form $\mathbf{u} = \langle a, b \rangle$, the number b is the vertical component of the vector. (Section 7.4)

X

***x*-axis** The horizontal number line in a rectangular coordinate system is the x-axis. (Appendix B)

***x*-intercept** An x-intercept is the x-value of a point where the graph of an equation intersects the x-axis. (Appendix B)

Y

***y*-axis** The vertical number line in a rectangular coordinate system is the *y*-axis. (Appendix B)

***y*-intercept** A *y*-intercept is the *y*-value of a point where the graph of an equation intersects the *y*-axis. (Appendix B)

Z

zero-factor property The zero-factor property states that if the product of two (or more) complex numbers is 0, then at least one of the numbers must be 0. (Appendix A)

zero vector The zero vector is the vector with magnitude 0. (Section 7.4)

Solutions to Selected Exercises

Chapter 1 Trigonometric Functions

1.1 Exercises *(pages 7–10)*

47. $90° - 72° 58' 11''$

$$\begin{array}{rl} 89° 59' 60'' & \text{Write } 90° \text{ as } 89° 59' 60''. \\ \underline{-72° 58' 11''} & \\ 17° 01' 49'' & \end{array}$$

Thus, $90° - 72° 58' 11'' = 17° 01' 49''$.

133. 600 rotations per min

$$\begin{aligned} &= \frac{600}{60} \text{ rotations per sec} \\ &= 10 \text{ rotations per sec} \\ &= 5 \text{ rotations per } \tfrac{1}{2} \text{ sec} \\ &= 5(360°) \text{ per } \tfrac{1}{2} \text{ sec} \\ &= 1800° \text{ per } \tfrac{1}{2} \text{ sec} \end{aligned}$$

A point on the edge of the tire will move 1800° in $\frac{1}{2}$ sec.

1.2 Exercises *(pages 15–20)*

1. Angle 1 and the 55° angle are vertical angles, which are equal, so angle $1 = 55°$. Angle 5 and the 120° angle are interior angles on the same side of the transversal, which means they are supplements, so

$$\begin{aligned} \text{angle } 5 + 120° &= 180° \\ \text{angle } 5 &= 60°. \quad \text{Subtract } 120°. \end{aligned}$$

Since angles 3 and 5 are vertical angles, angle $3 = 60°$.

$$\begin{aligned} \text{angle } 1 + \text{angle } 2 + \text{angle } 3 &= 180° \\ 55° + \text{angle } 2 + 60° &= 180° \\ \text{angle } 2 &= 65° \quad \text{Subtract } 115°. \end{aligned}$$

Since angles 2 and 4 are vertical angles, angle $4 = 65°$. Angle 6 and the 120° angle are vertical angles, so angle $6 = 120°$. Angles 6 and 8 are supplements.

$$\begin{aligned} \text{angle } 6 + \text{angle } 8 &= 180° \\ 120° + \text{angle } 8 &= 180° \\ \text{angle } 8 &= 60° \quad \text{Subtract } 120°. \end{aligned}$$

Since angles 7 and 8 are vertical angles, angle $7 = 60°$.

$$\text{angle } 7 + \text{angle } 4 + \text{angle } 10 = 180°$$

The sum of the measures of the angles in a triangle is 180°.

$$\begin{aligned} 60° + 65° + \text{angle } 10 &= 180° \\ \text{angle } 10 &= 55° \quad \text{Subtract } 125°. \end{aligned}$$

Since angles 9 and 10 are vertical angles, angle $9 = 55°$. Thus, the measures of the angles are 1: 55°; 2: 65°; 3: 60°; 4: 65°; 5: 60°; 6: 120°; 7: 60°; 8: 60°; 9: 55°; 10: 55°.

33. The triangle is obtuse because it has an angle of 96°, which is between 90° and 180°. It is a scalene triangle because no two sides are equal.

1.3 Exercises *(pages 26–28)*

83. Evaluate $\tan 360° + 4 \sin 180° + 5 \cos^2 180°$.

$$\tan 360° = \tan 0° = \frac{y}{x} = \frac{0}{1} = 0$$

$$\sin 180° = \frac{y}{r} = \frac{0}{1} = 0$$

$$\cos 180° = \frac{x}{r} = \frac{-1}{1} = -1$$

$$\tan 360° + 4 \sin 180° + 5 \cos^2 180° = 0 + 4(0) + 5(-1)^2$$

Substitute; $\cos^2 x = (\cos x)^2$.

$$= 5$$

1.4 Exercises *(pages 36–38)*

69. We are given $\tan \theta = -\frac{15}{8}$, with θ in quadrant II. Draw θ in standard position in quadrant II. Because $\tan \theta = \frac{y}{x}$ and θ is in quadrant II, we can use the values $y = 15$ and $x = -8$ for a point on its terminal side.

$$\begin{aligned} r &= \sqrt{x^2 + y^2} = \sqrt{(-8)^2 + 15^2} = \sqrt{64 + 225} \\ &= \sqrt{289} = 17 \end{aligned}$$

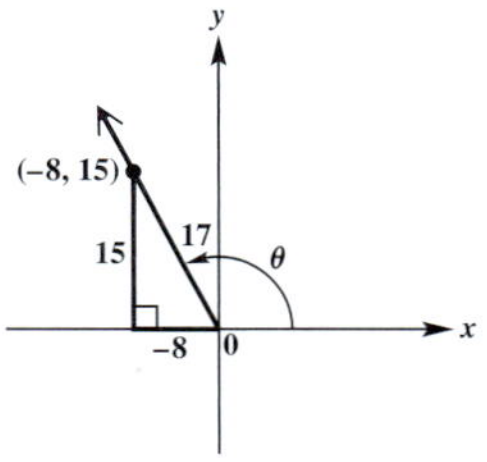

Use the values of x, y, and r and the definitions of the trigonometric functions to find the six trigonometric function values for θ.

$$\sin \theta = \frac{y}{r} = \frac{15}{17} \qquad \csc \theta = \frac{r}{y} = \frac{17}{15}$$

$$\cos \theta = \frac{x}{r} = \frac{-8}{17} = -\frac{8}{17} \qquad \sec \theta = \frac{r}{x} = \frac{17}{-8} = -\frac{17}{8}$$

$$\tan \theta = \frac{y}{x} = \frac{15}{-8} = -\frac{15}{8} \qquad \cot \theta = \frac{x}{y} = \frac{-8}{15} = -\frac{8}{15}$$

85. Multiply the compound inequality $90° < \theta < 180°$ by 2 to find that $180° < 2\theta < 360°$. Thus, 2θ must lie in quadrant III or quadrant IV. In both of these quadrants, the sine function is negative, so $\sin 2\theta$ must be negative.

101. $\tan(3\theta - 4°) = \dfrac{1}{\cot(5\theta - 8°)}$ Given equation

$\tan(3\theta - 4°) = \tan(5\theta - 8°)$ Reciprocal identity

The second equation above will be true if $3\theta - 4° = 5\theta - 8°$, so solving this equation will give a value (but not the only value) for which the given equation is true.

$$3\theta - 4° = 5\theta - 8°$$
$$4° = 2\theta$$
$$\theta = 2°$$

Chapter 2 Acute Angles and Right Triangles

2.1 Exercises *(pages 51–54)*

77. One point on the line $y = \sqrt{3}x$ is the origin, $(0, 0)$. Let (x, y) be any other point on this line. Then, by the definition of slope, $m = \frac{y-0}{x-0} = \frac{y}{x} = \sqrt{3}$, but also, by the definition of tangent, $\tan\theta = \frac{y}{x}$. Thus, $\tan\theta = \sqrt{3}$. Because $\tan 60° = \sqrt{3}$, the line $y = \sqrt{3}x$ makes a 60° angle with the positive x-axis. (See **Exercise 74.**)

81. Apply the relationships among the lengths of the sides of a 30°–60° right triangle first to the triangle on the left to find the values of x and y, and then to the triangle on the right to find the values of z and w. In a 30°–60° right triangle, the side opposite the 30° angle is $\frac{1}{2}$ the length of the hypotenuse. The longer leg is $\sqrt{3}$ times the shorter leg.

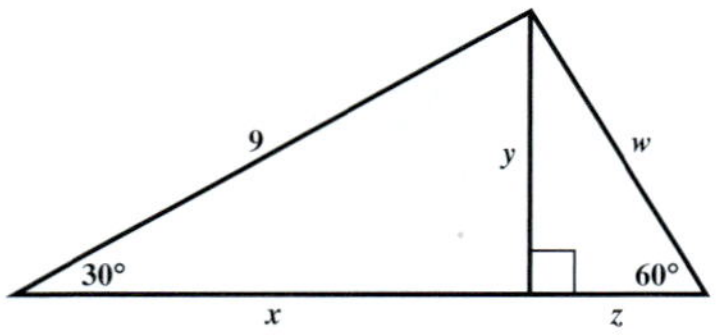

Thus,

$$y = \frac{1}{2}(9) = \frac{9}{2} \quad\text{and}\quad x = y\sqrt{3} = \frac{9\sqrt{3}}{2}.$$

Since $y = z\sqrt{3}$,

$$z = \frac{y}{\sqrt{3}} = \frac{\frac{9}{2}}{\sqrt{3}} = \frac{9}{2\sqrt{3}} \cdot \frac{\sqrt{3}}{\sqrt{3}} = \frac{9\sqrt{3}}{6} = \frac{3\sqrt{3}}{2},$$

and $$w = 2z = 2\left(\frac{3\sqrt{3}}{2}\right) = 3\sqrt{3}.$$

2.2 Exercises *(pages 58–61)*

29. To find the reference angle for $-300°$, sketch this angle in standard position.

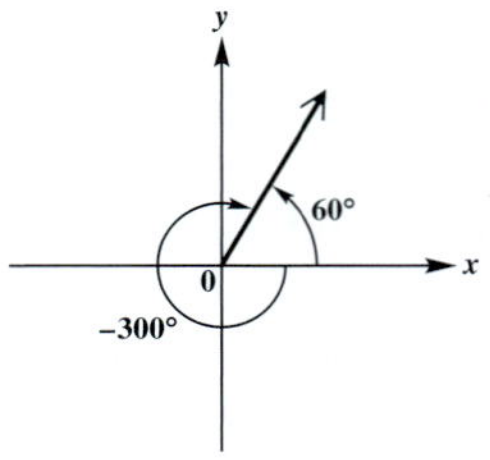

The reference angle for $-300°$ is

$$-300° + 360° = 60°.$$

Because $-300°$ is in quadrant I, the values of all its trigonometric functions are positive, so these values will be identical to the trigonometric function values for 60°. (See the Function Values of Special Angles table that follows **Example 5** in **Section 2.1.**)

$$\sin(-300°) = \frac{\sqrt{3}}{2} \qquad \csc(-300°) = \frac{2\sqrt{3}}{3}$$
$$\cos(-300°) = \frac{1}{2} \qquad \sec(-300°) = 2$$
$$\tan(-300°) = \sqrt{3} \qquad \cot(-300°) = \frac{\sqrt{3}}{3}$$

75. The reference angle for 115° is 65°. Since 115° is in quadrant II, the sine is positive. The function $\sin\theta$ decreases on the interval $(90°, 180°)$ from 1 to 0. Therefore, $\sin 115°$ is closest to 0.9.

2.3 Exercises *(pages 63–67)*

47. $\sin 10° + \sin 10° \stackrel{?}{=} \sin 20°$

Using a calculator, we get

$$\sin 10° + \sin 10° \approx 0.34729636$$

and $$\sin 20° \approx 0.34202014.$$

Thus, the statement is false.

67. For parts (a) and (b), $\theta = 3°$, $g = 32.2$, and $f = 0.14$.

(a) Use the fact that 45 mph = 66 ft per sec.

$$R = \frac{V^2}{g(f + \tan\theta)} = \frac{66^2}{32.2(0.14 + \tan 3°)} \approx 703 \text{ ft}$$

(b) Use the fact that 70 mph $= \frac{70(5280)}{3600}$ ft per sec = 102.67 ft per sec.

$$R = \frac{V^2}{g(f + \tan\theta)} = \frac{102.67^2}{32.2(0.14 + \tan 3°)} \approx 1701 \text{ ft}$$

67. (c) Intuitively, increasing θ would make it easier to negotiate the curve at a higher speed much as is done at a race track. Mathematically, a larger value of θ (acute) will lead to a larger value for $\tan\theta$. If $\tan\theta$ increases, then the ratio determining R will *decrease*. Thus, the radius can be smaller and the curve sharper if θ is increased.

$$R = \frac{V^2}{g(f + \tan\theta)}$$

$$= \frac{66^2}{32.2(0.14 + \tan 4^\circ)}$$

$$\approx 644 \text{ ft}$$

$$R = \frac{V^2}{g(f + \tan\theta)}$$

$$= \frac{102.67^2}{32.2(0.14 + \tan 4^\circ)}$$

$$\approx 1559 \text{ ft}$$

As predicted, both values are less.

2.4 Exercises *(pages 72–76)*

23. Solve the right triangle with $B = 73.0^\circ$, $b = 128$ in., and $C = 90^\circ$.

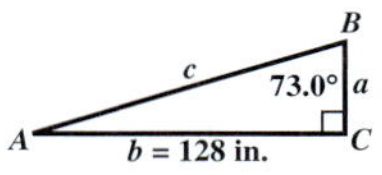

$$A = 90^\circ - 73.0^\circ = 17.0^\circ$$

$$\tan 73.0^\circ = \frac{128}{a} \qquad \tan B = \frac{b}{a}$$

$$a = \frac{128}{\tan 73.0^\circ} \approx 39.1 \text{ in.} \qquad \text{Three significant digits}$$

$$\sin 73.0^\circ = \frac{128}{c} \qquad \sin B = \frac{b}{c}$$

$$c = \frac{128}{\sin 73.0^\circ} \approx 134 \text{ in.} \qquad \text{Three significant digits}$$

43. Let x represent the horizontal distance between the two buildings and y represent the height of the portion of the building across the street that is higher than the window.

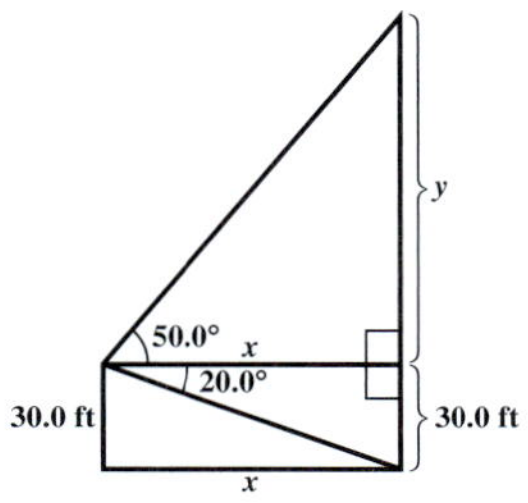

$$\tan 20.0^\circ = \frac{30.0}{x} \qquad \text{Tangent ratio}$$

$$x = \frac{30.0}{\tan 20.0^\circ} \approx 82.4 \qquad \text{Solve for } x.$$

$$\tan 50.0^\circ = \frac{y}{x} \qquad \text{Tangent ratio}$$

$$y = x \tan 50.0^\circ = \left(\frac{30.0}{\tan 20.0^\circ}\right) \tan 50.0^\circ \approx 98.2 \qquad \text{Solve for } y.$$

$$\text{height} = y + 30.0 = \left(\frac{30.0}{\tan 20.0^\circ}\right) \tan 50.0^\circ + 30.0 \approx 128 \qquad \text{Three significant digits}$$

The height of the building across the street is about 128 ft.

49. Let h represent the height of the tower.

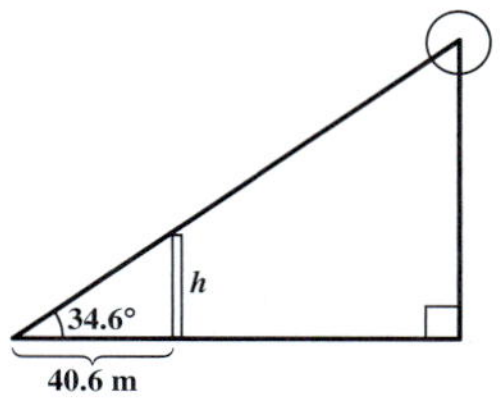

$$\tan 34.6^\circ = \frac{h}{40.6} \qquad \text{Tangent ratio}$$

$$h = 40.6 \tan 34.6^\circ \approx 28.0 \qquad \text{Three significant digits}$$

The height of the tower is about 28.0 m.

2.5 Exercises *(pages 81–85)*

21. Let $x =$ the distance between the two ships. The angle between the bearings of the ships is

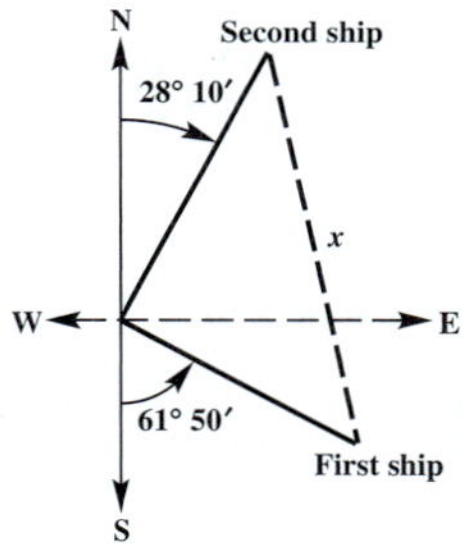

$$180^\circ - (28^\circ\,10' + 61^\circ\,50') = 90^\circ.$$

The triangle formed is a right triangle.

Distance traveled at 24.0 mph:

$$(4 \text{ hr})(24.0 \text{ mph}) = 96 \text{ mi}$$

Distance traveled at 28.0 mph:

$$(4 \text{ hr})(28.0 \text{ mph}) = 112 \text{ mi}$$

Applying the Pythagorean theorem gives the following.

$$x^2 = 96^2 + 112^2$$

$$x^2 = 21{,}760$$

$$x \approx 148$$

The ships are 148 mi apart.

33. Let $x =$ the distance from the closer point on the ground to the base of height h of the pyramid.

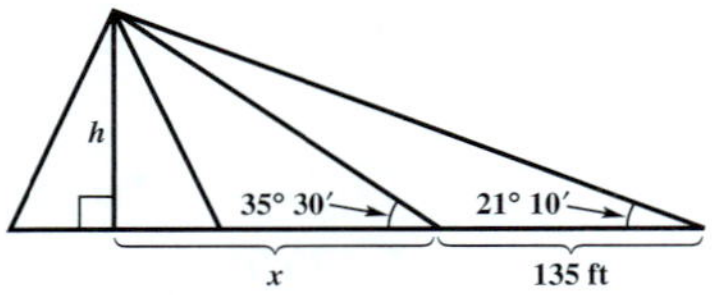

In the larger right triangle,

$$\tan 21° \, 10' = \frac{h}{135 + x}$$

$$h = (135 + x) \tan 21° \, 10'.$$

In the smaller right triangle,

$$\tan 35° \, 30' = \frac{h}{x}$$

$$h = x \tan 35° \, 30'.$$

Substitute for h in this equation, and solve for x.

$$(135 + x) \tan 21° \, 10' = x \tan 35° \, 30'$$
Substitute $(135 + x)\tan 21°10'$ for h.

$$135 \tan 21° \, 10' + x \tan 21° \, 10' = x \tan 35° \, 30'$$
Distributive property

$$135 \tan 21° \, 10' = x \tan 35° \, 30' - x \tan 21° \, 10'$$
Write the x-terms on one side.

$$135 \tan 21° \, 10' = x(\tan 35° \, 30' - \tan 21° \, 10')$$
Factor out x.

$$\frac{135 \tan 21° \, 10'}{\tan 35° \, 30' - \tan 21° \, 10'} = x$$
Divide by the coefficient of x.

Then substitute for x in the equation for the smaller triangle.

$$h = \left(\frac{135 \tan 21° \, 10'}{\tan 35° \, 30' - \tan 21° \, 10'}\right) \tan 35° \, 30' \approx 114$$

The height of the pyramid is about 114 ft.

Chapter 3 Radian Measure and the Unit Circle

3.1 Exercises *(pages 98–100)*

89. (a) In 24 hr, the hour hand will rotate twice around the clock. Since one complete rotation measures 2π radians, the two rotations will measure

$$2(2\pi) = 4\pi \text{ radians.}$$

(b) In 4 hr, the hour hand will rotate $\frac{4}{12} = \frac{1}{3}$ of the way around the clock, which will measure

$$\tfrac{1}{3}(2\pi) = \tfrac{2\pi}{3} \text{ radians.}$$

3.2 Exercises *(pages 103–109)*

35. For the large gear and pedal,

$$s = r\theta = 4.72\pi. \quad 180° = \pi \text{ radians}$$

Thus, the chain moves 4.72π in. Find the angle through which the small gear rotates.

$$\theta = \frac{s}{r} = \frac{4.72\pi}{1.38} \approx 3.42\pi$$

The angle θ for the wheel and for the small gear are the same, so for the wheel,

$$s = r\theta = 13.6(3.42\pi) \approx 146 \text{ in.}$$

The bicycle will move about 146 in.

67. (a)

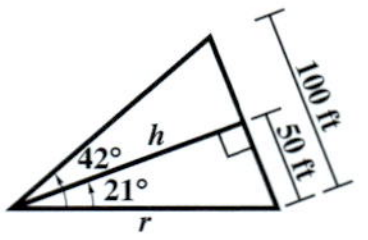

The triangle formed by the sides of the central angle and the chord is isosceles. Therefore, the bisector of the central angle is also the perpendicular bisector of the chord and divides the larger triangle into two congruent right triangles.

$$\sin 21° = \frac{50}{r}$$

$$r = \frac{50}{\sin 21°} \approx 140 \text{ ft}$$

The radius of the curve is about 140 ft.

(b) $r = \dfrac{50}{\sin 21°}$; $\theta = 42°$

$$42° = 42\left(\frac{\pi}{180} \text{ radian}\right) = \frac{7\pi}{30} \text{ radian}$$

$$s = r\theta = \frac{50}{\sin 21°} \cdot \frac{7\pi}{30} = \frac{35\pi}{3 \sin 21°} \approx 102 \text{ ft}$$

The length of the arc determined by the 100-ft chord is about 102 ft.

(c) The portion of the circle bounded by the arc and the 100-ft chord is the shaded region in the figure below.

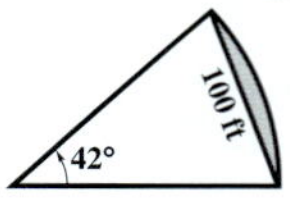

The area of the portion of the circle can be found by subtracting the area of the triangle from the area of the sector. From the figure in part (a),

$$\tan 21° = \frac{50}{h}, \quad \text{so} \quad h = \frac{50}{\tan 21°}.$$

$$\mathcal{A}_{\text{sector}} = \frac{1}{2} r^2 \theta$$

$$= \frac{1}{2}\left(\frac{50}{\sin 21°}\right)^2 \left(\frac{7\pi}{30}\right) \quad \text{From part (b), } 42° = \tfrac{7\pi}{30}.$$

$$\approx 7135 \text{ ft}^2$$

$$\mathcal{A}_{\text{triangle}} = \frac{1}{2} bh = \frac{1}{2}(100)\left(\frac{50}{\tan 21°}\right)$$

$$\approx 6513 \text{ ft}^2$$

$$\mathcal{A}_{\text{portion}} = \mathcal{A}_{\text{sector}} - \mathcal{A}_{\text{triangle}}$$

$$\approx 7135 \text{ ft}^2 - 6513 \text{ ft}^2$$

$$= 622 \text{ ft}^2$$

The area of the portion is about 622 ft^2.

69. Use the Pythagorean theorem to find the hypotenuse of the right triangle, which is also the radius of the sector of the circle.

$$r^2 = 30^2 + 40^2 = 900 + 1600 = 2500$$

$$r = \sqrt{2500} = 50$$

$$\mathcal{A}_{\text{triangle}} = \frac{1}{2}bh = \frac{1}{2}(30)(40)$$

$$= 600 \text{ yd}^2$$

$$\mathcal{A}_{\text{sector}} = \frac{1}{2}r^2\theta$$

$$= \frac{1}{2}(50)^2 \cdot \frac{\pi}{3} \qquad 60° = \frac{\pi}{3}$$

$$= \frac{1250\pi}{3} \text{ yd}^2$$

$$\text{Total area} = \mathcal{A}_{\text{triangle}} + \mathcal{A}_{\text{sector}}$$

$$= 600 \text{ yd}^2 + \frac{1250\pi}{3} \text{ yd}^2$$

$$\approx 1900 \text{ yd}^2$$

The area of the lot is about 1900 yd^2.

3.3 Exercises *(pages 117–120)*

45. cos 2

$\frac{\pi}{2} \approx 1.57$ and $\pi \approx 3.14$, so $\frac{\pi}{2} < 2 < \pi$. Thus, an angle of 2 radians is in quadrant II. (The figure for **Exercises 35–44** also shows that 2 radians is in quadrant II.) Because values of the cosine function are negative in quadrant II, cos 2 is negative.

63. $\left[\pi, \frac{3\pi}{2}\right]$; $\tan s = \sqrt{3}$

Recall that $\tan \frac{\pi}{3} = \sqrt{3}$ and that in quadrant III, tan s is positive.

$$\tan\left(\pi + \frac{\pi}{3}\right) = \tan \frac{4\pi}{3} = \sqrt{3}$$

Thus, $s = \frac{4\pi}{3}$.

3.4 Exercises *(pages 123–126)*

27. The hour hand of a clock moves through an angle of 2π radians (one complete revolution) in 12 hr. Find ω as follows.

$$\omega = \frac{\theta}{t} = \frac{2\pi}{12} = \frac{\pi}{6} \text{ radian per hr}$$

37. At 215 revolutions per min, the bicycle tire is moving $215(2\pi) = 430\pi$ radians per min. This is the angular velocity ω. Find v as follows.

$$v = r\omega = 13(430\pi) = 5590\pi \text{ in. per min}$$

Convert this velocity to miles per hour.

$$v = \frac{5590\pi \text{ in.}}{1 \text{ min}} \cdot \frac{60 \text{ min}}{1 \text{ hr}} \cdot \frac{1 \text{ ft}}{12 \text{ in.}} \cdot \frac{1 \text{ mi}}{5280 \text{ ft}} \approx 16.6 \text{ mph}$$

Chapter 4 Graphs of the Circular Functions

4.1 Exercises *(pages 143–148)*

55. $E = 5 \cos 120\pi t$

(a) The amplitude is $|5| = 5$, and the period is $\frac{2\pi}{120\pi} = \frac{1}{60}$.

(b) Since the period is $\frac{1}{60}$, one cycle is completed in $\frac{1}{60}$ sec. Therefore, in 1 sec, 60 cycles are completed.

(c) For $t = 0$, $E = 5 \cos 120\pi(0) = 5 \cos 0 = 5$.
For $t = 0.03$, $E = 5 \cos 120\pi(0.03) \approx 1.545$.
For $t = 0.06$, $E = 5 \cos 120\pi(0.06) \approx -4.045$.
For $t = 0.09$, $E \approx -4.045$.
For $t = 0.12$, $E \approx 1.545$.

(d)

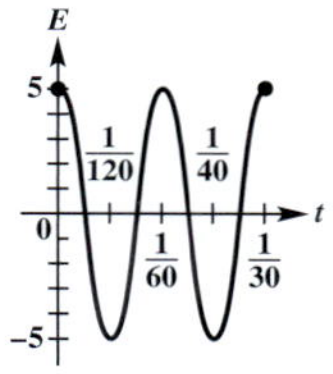

$E = 5 \cos 120\pi t$

4.2 Exercises *(pages 155–158)*

55. $y = \frac{1}{2} + \sin\left[2\left(x + \frac{\pi}{4}\right)\right]$

This equation has the form $y = c + a \sin[b(x - d)]$ with $c = \frac{1}{2}$, $a = 1$, $b = 2$, and $d = -\frac{\pi}{4}$. Start with the graph of $y = \sin x$ and modify it to take into account the amplitude, period, and translations required to obtain the desired graph.

Amplitude: $|a| = 1$

Period: $\frac{2\pi}{b} = \frac{2\pi}{2} = \pi$

Vertical translation: $\frac{1}{2}$ unit up

Phase shift (horizontal translation): $\frac{\pi}{4}$ unit to the left

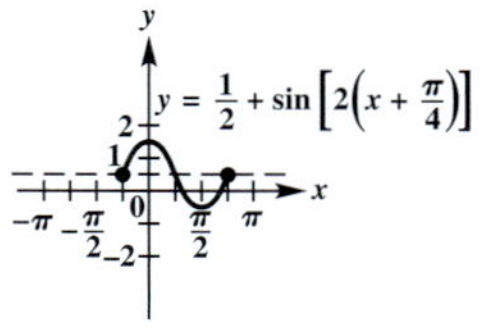

4.3 Exercises *(pages 166–168)*

29. $y = -1 + \frac{1}{2}\cot(2x - 3\pi)$

$$y = -1 + \frac{1}{2}\cot\left[2\left(x - \frac{3\pi}{2}\right)\right] \qquad \text{Rewrite } 2x - 3\pi \text{ as } 2\left(x - \frac{3\pi}{2}\right).$$

Period: $\frac{\pi}{b} = \frac{\pi}{2}$

Vertical translation: 1 unit down

Phase shift (horizontal translation): $\frac{3\pi}{2}$ units to the right

Because the function is to be graphed over a two-period interval, locate three adjacent vertical asymptotes. Because asymptotes of the graph of $y = \cot x$ occur at multiples of π, the following equations can be solved to locate asymptotes.

$$2\left(x - \frac{3\pi}{2}\right) = -2\pi, \quad 2\left(x - \frac{3\pi}{2}\right) = -\pi, \quad \text{and}$$

$$2\left(x - \frac{3\pi}{2}\right) = 0$$

Solve each of these equations.

$$2\left(x - \frac{3\pi}{2}\right) = -2\pi$$

$$x - \frac{3\pi}{2} = -\pi \qquad \text{Divide by 2.}$$

$$x = -\pi + \frac{3\pi}{2} \qquad \text{Add } \tfrac{3\pi}{2}.$$

$$x = \frac{\pi}{2}$$

$$2\left(x - \frac{3\pi}{2}\right) = -\pi$$

$$x - \frac{3\pi}{2} = -\frac{\pi}{2}$$

$$x = -\frac{\pi}{2} + \frac{3\pi}{2}$$

$$x = \frac{2\pi}{2}, \quad \text{or} \quad \pi$$

$$2\left(x - \frac{3\pi}{2}\right) = 0$$

$$x - \frac{3\pi}{2} = 0$$

$$x = \frac{3\pi}{2}$$

Divide the interval $\left(\frac{\pi}{2}, \pi\right)$ into four equal parts to obtain the following key x-values.

first-quarter value: $\frac{5\pi}{8}$; middle value: $\frac{3\pi}{4}$;

third-quarter value: $\frac{7\pi}{8}$

Evaluating the given function at these three key x-values gives the following points.

$$\left(\frac{5\pi}{8}, -\frac{1}{2}\right), \quad \left(\frac{3\pi}{4}, -1\right), \quad \left(\frac{7\pi}{8}, -\frac{3}{2}\right)$$

Connect these points with a smooth curve and continue the graph to approach the asymptotes $x = \frac{\pi}{2}$ and $x = \pi$ to complete one period of the graph. Sketch an identical curve between the asymptotes $x = \pi$ and $x = \frac{3\pi}{2}$ to complete a second period of the graph.

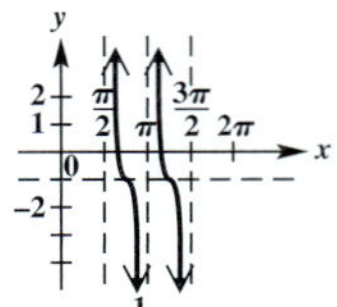

$y = -1 + \frac{1}{2}\cot(2x - 3\pi)$

45. $\tan(-x) = \dfrac{\sin(-x)}{\cos(-x)}$ Quotient identity

$= \dfrac{-\sin x}{\cos x}$ Negative–angle identities

$= -\dfrac{\sin x}{\cos x}$ $\quad \frac{-a}{b} = -\frac{a}{b}$

$= -\tan x$ Quotient identity

4.4 Exercises *(pages 174–176)*

31. $\sec(-x) = \dfrac{1}{\cos(-x)}$ Reciprocal identity

$= \dfrac{1}{\cos x}$ Negative–angle identity

$= \sec x$ Reciprocal identity

4.5 Exercises *(pages 179–180)*

19. (a) We will use a model of the form $s(t) = a \cos \omega t$ with $a = -3$. Since

$$s(0) = -3\cos(\omega \cdot 0) = -3\cos 0 = -3 \cdot 1 = -3,$$

using a cosine function rather than a sine function will avoid the need for a phase shift.

The frequency of $\frac{6}{\pi}$ cycles per sec is the reciprocal of the period.

$$\frac{6}{\pi} = \frac{\omega}{2\pi} \qquad \text{Frequency} = \tfrac{1}{\text{period}}$$

$$6 \cdot 2 = \omega \qquad \text{Multiply by } 2\pi.$$

$$\omega = 12 \qquad \text{Multiply and rewrite.}$$

Therefore, a model for the position of the weight at time t seconds is

$$s(t) = -3\cos 12t.$$

(b) Period $= \dfrac{1}{\frac{6}{\pi}} = 1 \div \dfrac{6}{\pi} = 1 \cdot \dfrac{\pi}{6} = \dfrac{\pi}{6}$ sec

Chapter 5 Trigonometric Identities

5.1 Exercises *(pages 193–196)*

35. $\cot\theta = \frac{4}{3}$, $\sin\theta > 0$

Because $\cot\theta > 0$ and $\sin\theta > 0$, θ is in quadrant I, so all the function values are positive.

$\tan\theta = \frac{1}{\cot\theta} = \frac{1}{\frac{4}{3}} = \frac{3}{4}$ Reciprocal identity

$\sec^2\theta = \tan^2\theta + 1$ Pythagorean identity

$= \left(\frac{3}{4}\right)^2 + 1 = \frac{9}{16} + \frac{16}{16} = \frac{25}{16}$

$\sec\theta = \sqrt{\frac{25}{16}} = \frac{5}{4}$ $\sec\theta > 0$

$\cos\theta = \frac{1}{\sec\theta} = \frac{1}{\frac{5}{4}} = \frac{4}{5}$ Reciprocal identity

$\sin^2\theta = 1 - \cos^2\theta$ Alternative form of Pythagorean identity

$= 1 - \left(\frac{4}{5}\right)^2 = \frac{9}{25}$

$\sin\theta = \sqrt{\frac{9}{25}} = \frac{3}{5}$ $\sin\theta > 0$

$\csc\theta = \frac{1}{\sin\theta} = \frac{1}{\frac{3}{5}} = \frac{5}{3}$ Reciprocal identity

Thus, $\sin\theta = \frac{3}{5}$, $\cos\theta = \frac{4}{5}$, $\tan\theta = \frac{3}{4}$, $\sec\theta = \frac{5}{4}$, and $\csc\theta = \frac{5}{3}$.

57. $\csc x = \frac{1}{\sin x}$ Reciprocal identity

$= \frac{1}{\pm\sqrt{1 - \cos^2 x}}$ Alternative form of Pythagorean identity

$= \frac{\pm 1}{\sqrt{1 - \cos^2 x}}$ Redistribute signs.

$= \frac{\pm 1}{\sqrt{1 - \cos^2 x}} \cdot \frac{\sqrt{1 - \cos^2 x}}{\sqrt{1 - \cos^2 x}}$ Rationalize the denominator.

$\csc x = \frac{\pm\sqrt{1 - \cos^2 x}}{1 - \cos^2 x}$ Multiply.

73. $\sec\theta - \cos\theta = \frac{1}{\cos\theta} - \cos\theta$

$= \frac{1}{\cos\theta} - \frac{\cos^2\theta}{\cos\theta}$ Use a common denominator.

$= \frac{1 - \cos^2\theta}{\cos\theta}$ Subtract fractions.

$= \frac{\sin^2\theta}{\cos\theta}$ $1 - \cos^2\theta = \sin^2\theta$

$= \frac{\sin\theta}{\cos\theta} \cdot \sin\theta$ $\sin^2\theta = \sin\theta \cdot \sin\theta$

$= \tan\theta \sin\theta$ $\frac{\sin\theta}{\cos\theta} = \tan\theta$

85. Since $\cos x = \frac{1}{5} > 0$, x is in quadrant I or quadrant IV.

$\sin x = \pm\sqrt{1 - \cos^2 x} = \pm\sqrt{1 - \left(\frac{1}{5}\right)^2}$

$= \pm\sqrt{\frac{24}{25}} = \pm\frac{2\sqrt{6}}{5}$

$\tan x = \frac{\sin x}{\cos x} = \frac{\pm\frac{2\sqrt{6}}{5}}{\frac{1}{5}} = \pm 2\sqrt{6}$

$\sec x = \frac{1}{\cos x} = \frac{1}{\frac{1}{5}} = 5$

Quadrant I:

$\frac{\sec x - \tan x}{\sin x} = \frac{5 - 2\sqrt{6}}{\frac{2\sqrt{6}}{5}} = \frac{5(5 - 2\sqrt{6})}{2\sqrt{6}}$

$= \frac{25 - 10\sqrt{6}}{2\sqrt{6}} \cdot \frac{\sqrt{6}}{\sqrt{6}}$

$= \frac{25\sqrt{6} - 60}{12}$

Quadrant IV:

$\frac{\sec x - \tan x}{\sin x} = \frac{5 - (-2\sqrt{6})}{-\frac{2\sqrt{6}}{5}} = \frac{5(5 + 2\sqrt{6})}{-2\sqrt{6}}$

$= \frac{25 + 10\sqrt{6}}{-2\sqrt{6}} \cdot \frac{-\sqrt{6}}{-\sqrt{6}}$

$= \frac{-25\sqrt{6} - 60}{12}$

5.2 Exercises *(pages 202–204)*

11. $\frac{1}{1+\cos x} - \frac{1}{1-\cos x} = \frac{1(1-\cos x) - 1(1+\cos x)}{(1+\cos x)(1-\cos x)}$

$= \frac{1-\cos x - 1 - \cos x}{1-\cos^2 x}$

$= \frac{-2\cos x}{\sin^2 x}$

$= -\frac{2\cos x}{\sin^2 x}$

$= -2\left(\frac{\cos x}{\sin x}\right)\left(\frac{1}{\sin x}\right)$

$= -2\cot x \csc x$

15. $(\sin x + 1)^2 - (\sin x - 1)^2$

$= [(\sin x + 1) + (\sin x - 1)][(\sin x + 1) - (\sin x - 1)]$ Factor the difference of squares.

$= [2\sin x][\sin x + 1 - \sin x + 1]$ Simplify.

$= [2\sin x][2]$ Simplify again.

$= 4\sin x$ Multiply.

59. Verify that $\frac{\tan^2 t - 1}{\sec^2 t} = \frac{\tan t - \cot t}{\tan t + \cot t}$ is an identity.

Work with the right hand side.

$\frac{\tan t - \cot t}{\tan t + \cot t} = \frac{\tan t - \frac{1}{\tan t}}{\tan t + \frac{1}{\tan t}}$ $\cot t = \frac{1}{\tan t}$

$= \frac{\tan t}{\tan t}\left(\frac{\tan t - \frac{1}{\tan t}}{\tan t + \frac{1}{\tan t}}\right)$

Multiply numerator and denominator of the complex fraction by the LCD, $\tan t$.

$= \frac{\tan^2 t - 1}{\tan^2 t + 1}$ Distributive property

$= \frac{\tan^2 t - 1}{\sec^2 t}$ $\tan^2 t + 1 = \sec^2 t$

87. Show that $\sin(\csc t) = 1$ is not an identity.

We need find only one value for which the statement is false. Let $t = 2$. Use a calculator to find that $\sin(\csc 2) \approx 0.891094$, which is not equal to 1. Thus, $\sin(\csc t) = 1$ is not true for *all* real numbers t, so it is not an identity.

5.3 Exercises *(pages 212–215)*

39. $\sec\theta = \csc\left(\frac{\theta}{2} + 20^\circ\right)$

By a cofunction identity, $\sec\theta = \csc(90^\circ - \theta)$.

$\csc\left(\frac{\theta}{2} + 20^\circ\right) = \csc(90^\circ - \theta)$ Substitute.

$\frac{\theta}{2} + 20^\circ = 90^\circ - \theta$ Set the angle measures equal.

$\frac{3\theta}{2} = 70^\circ$ Add θ and subtract 20°.

$\theta = \frac{2}{3}(70^\circ) = \frac{140^\circ}{3}$ Multiply by $\frac{2}{3}$.

61. *True* or *false*: $\cos\frac{\pi}{3} = \cos\frac{\pi}{12}\cos\frac{\pi}{4} - \sin\frac{\pi}{12}\sin\frac{\pi}{4}$.

Note that $\frac{\pi}{3} = \frac{4\pi}{12} = \frac{\pi}{12} + \frac{3\pi}{12} = \frac{\pi}{12} + \frac{\pi}{4}$.

$\cos\frac{\pi}{3} = \cos\left(\frac{\pi}{12} + \frac{\pi}{4}\right)$ Substitute.

$= \cos\frac{\pi}{12}\cos\frac{\pi}{4} - \sin\frac{\pi}{12}\sin\frac{\pi}{4}$

Cosine sum identity

The given statement is true.

5.4 Exercises *(pages 220–224)*

47. $\cos s = -\frac{8}{17}$ and $\cos t = -\frac{3}{5}$, s and t in quadrant III

In order to substitute into sum and difference identities, we need to find the values of $\sin s$ and $\sin t$, and also the values of $\tan s$ and $\tan t$. Because s and t are both in quadrant III, the values of $\sin s$ and $\sin t$ will be negative, and $\tan s$ and $\tan t$ will be positive.

$\sin s = -\sqrt{1-\cos^2 s} = -\sqrt{1-\left(-\frac{8}{17}\right)^2}$

$= -\sqrt{\frac{225}{289}} = -\frac{15}{17}$

$\sin t = -\sqrt{1-\cos^2 t} = -\sqrt{1-\left(-\frac{3}{5}\right)^2}$

$= -\sqrt{\frac{16}{25}} = -\frac{4}{5}$

$\tan s = \frac{\sin s}{\cos s} = \frac{-\frac{15}{17}}{-\frac{8}{17}} = \frac{15}{8}$

$\tan t = \frac{\sin t}{\cos t} = \frac{-\frac{4}{5}}{-\frac{3}{5}} = \frac{4}{3}$

(a) $\sin(s+t) = \sin s\cos t + \cos s\sin t$

$= \left(-\frac{15}{17}\right)\left(-\frac{3}{5}\right) + \left(-\frac{8}{17}\right)\left(-\frac{4}{5}\right)$

$= \frac{45}{85} + \frac{32}{85}$

$= \frac{77}{85}$

(b) $\tan(s+t) = \dfrac{\tan s + \tan t}{1 - \tan s \tan t} = \dfrac{\frac{15}{8} + \frac{4}{3}}{1 - \left(\frac{15}{8}\right)\left(\frac{4}{3}\right)}$

$= \dfrac{\frac{45}{24} + \frac{32}{24}}{1 - \frac{60}{24}} = \dfrac{\frac{77}{24}}{-\frac{36}{24}} = -\dfrac{77}{36}$

(c) From parts (a) and (b), $\sin(s+t) > 0$ and $\tan(s+t) < 0$. The only quadrant in which values of sine are positive and values of tangent are negative is quadrant II. Thus, $s+t$ is in quadrant II.

55. $\tan \dfrac{11\pi}{12} = \tan\left(\dfrac{3\pi}{4} + \dfrac{\pi}{6}\right)$ — $\frac{3\pi}{4} = \frac{9\pi}{12}$; $\frac{\pi}{6} = \frac{2\pi}{12}$

$= \dfrac{\tan \frac{3\pi}{4} + \tan \frac{\pi}{6}}{1 - \tan \frac{3\pi}{4} \tan \frac{\pi}{6}}$ — Tangent sum identity

$= \dfrac{-1 + \frac{\sqrt{3}}{3}}{1 - (-1)\left(\frac{\sqrt{3}}{3}\right)}$ — $\tan \frac{3\pi}{4} = -1$ and $\tan \frac{\pi}{6} = \frac{\sqrt{3}}{3}$

$= \dfrac{-1 + \frac{\sqrt{3}}{3}}{1 + \frac{\sqrt{3}}{3}}$ — Simplify.

$= \dfrac{-1 + \frac{\sqrt{3}}{3}}{1 + \frac{\sqrt{3}}{3}} \cdot \dfrac{3}{3}$ — Multiply numerator and denominator by 3.

$= \dfrac{-3 + \sqrt{3}}{3 + \sqrt{3}}$ — Distributive property

$= \dfrac{-3 + \sqrt{3}}{3 + \sqrt{3}} \cdot \dfrac{3 - \sqrt{3}}{3 - \sqrt{3}}$ — Rationalize the denominator.

$= \dfrac{-9 + 6\sqrt{3} - 3}{9 - 3}$ — FOIL

$= \dfrac{-12 + 6\sqrt{3}}{6}$ — Subtract.

$= \dfrac{6\left(-2 + \sqrt{3}\right)}{6}$ — Factor the numerator.

$= -2 + \sqrt{3}$ — Lowest terms

67. Verify that $\dfrac{\sin(x-y)}{\sin(x+y)} = \dfrac{\tan x - \tan y}{\tan x + \tan y}$ is an identity.

Work with the left hand side.

$\dfrac{\sin(x-y)}{\sin(x+y)} = \dfrac{\sin x \cos y - \cos x \sin y}{\sin x \cos y + \cos x \sin y}$ — Sine sum and difference identities

$= \dfrac{\frac{\sin x \cos y}{\cos x \cos y} - \frac{\cos x \sin y}{\cos x \cos y}}{\frac{\sin x \cos y}{\cos x \cos y} + \frac{\cos x \sin y}{\cos x \cos y}}$ — Divide numerator and denominator by $\cos x \cos y$.

$= \dfrac{\frac{\sin x}{\cos x} \cdot 1 - 1 \cdot \frac{\sin y}{\cos y}}{\frac{\sin x}{\cos x} \cdot 1 + 1 \cdot \frac{\sin y}{\cos y}}$ — Divide.

$= \dfrac{\tan x - \tan y}{\tan x + \tan y}$ — Tangent quotient identity

5.5 Exercises *(pages 230–232)*

25. Verify that $\sin 4x = 4 \sin x \cos x \cos 2x$ is an identity.
Work with the left hand side.

$\sin 4x = \sin 2(2x)$ — Factor: $4 = 2 \cdot 2$.

$= 2 \sin 2x \cos 2x$ — Sine double-angle identity

$= 2(2 \sin x \cos x)\cos 2x$ — Sine double-angle identity

$= 4 \sin x \cos x \cos 2x$ — Multiply.

45. $\dfrac{1}{4} - \dfrac{1}{2}\sin^2 47.1° = \dfrac{1}{4}(1 - 2\sin^2 47.1°)$ — Factor out $\frac{1}{4}$.

$= \dfrac{1}{4}\cos 2(47.1°)$ — $\cos 2A = 1 - 2\sin^2 A$

$= \dfrac{1}{4}\cos 94.2°$

51. $\tan 3x = \tan(2x + x)$

$= \dfrac{\tan 2x + \tan x}{1 - \tan 2x \tan x}$ — Tangent sum identity

$= \dfrac{\frac{2\tan x}{1 - \tan^2 x} + \tan x}{1 - \frac{2\tan x}{1 - \tan^2 x} \cdot \tan x}$ — Tangent double-angle identity

$= \dfrac{\frac{2\tan x + (1 - \tan^2 x)\tan x}{1 - \tan^2 x}}{\frac{1 - \tan^2 x - 2\tan^2 x}{1 - \tan^2 x}}$ — Add and subtract using the common denominator.

$= \dfrac{2\tan x + \tan x - \tan^3 x}{1 - \tan^2 x - 2\tan^2 x}$ — Multiply numerator and denominator by $1 - \tan^2 x$.

$\tan 3x = \dfrac{3\tan x - \tan^3 x}{1 - 3\tan^2 x}$ — Combine like terms.

5.6 Exercises *(pages 235–239)*

21. Find $\tan \frac{\theta}{2}$, given $\sin \theta = \frac{3}{5}$, with $90° < \theta < 180°$.

To find $\tan \frac{\theta}{2}$, we need the values of $\sin \theta$ and $\cos \theta$. We know $\sin \theta = \frac{3}{5}$.

$$\cos \theta = \pm\sqrt{1 - \sin^2 \theta} \qquad \text{Fundamental identity}$$

$$= \pm\sqrt{1 - \left(\frac{3}{5}\right)^2} \qquad \text{Substitute.}$$

$$= \pm\sqrt{\frac{16}{25}} \qquad \text{Simplify.}$$

$$\cos \theta = -\frac{4}{5} \qquad \theta \text{ is in quadrant II.}$$

Thus,

$$\tan \frac{\theta}{2} = \frac{\sin \theta}{1 + \cos \theta} \qquad \text{Half-angle identity}$$

$$= \frac{\frac{3}{5}}{1 - \frac{4}{5}} \qquad \text{Substitute.}$$

$$= 3. \qquad \text{Simplify.}$$

45. Verify that $\sec^2 \frac{x}{2} = \frac{2}{1 + \cos x}$ is an identity.

Work with the left hand side.

$$\sec^2 \frac{x}{2} = \frac{1}{\cos^2 \frac{x}{2}} \qquad \text{Reciprocal identity}$$

$$= \frac{1}{\left(\pm\sqrt{\frac{1 + \cos x}{2}}\right)^2} \qquad \text{Cosine half-angle identity}$$

$$= \frac{1}{\frac{1 + \cos x}{2}} \qquad \text{Apply the exponent.}$$

$$= \frac{2}{1 + \cos x} \qquad \text{Divide.}$$

Chapter 6 Inverse Circular Functions and Trigonometric Equations

6.1 Exercises *(pages 257–261)*

87. $\sin\left(2 \cos^{-1} \frac{1}{5}\right)$

Let $\theta = \cos^{-1} \frac{1}{5}$, so $\cos \theta = \frac{1}{5}$. The inverse cosine function yields values only in quadrants I and II, and since $\frac{1}{5}$ is positive, θ is in quadrant I. Sketch θ in quadrant I, and label the sides of a right triangle. By the Pythagorean theorem, the length of the side opposite θ will be

$$\sqrt{5^2 - 1^2} = \sqrt{24} = 2\sqrt{6}.$$

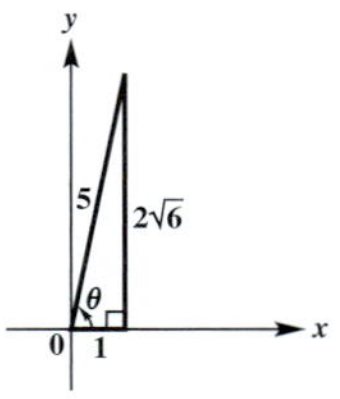

From the figure, $\sin \theta = \frac{2\sqrt{6}}{5}$.

$$\sin\left(2 \cos^{-1} \frac{1}{5}\right) = \sin 2\theta$$

$$= 2 \sin \theta \cos \theta \qquad \text{Sine double-angle identity}$$

$$= 2\left(\frac{2\sqrt{6}}{5}\right)\left(\frac{1}{5}\right)$$

$$= \frac{4\sqrt{6}}{25}$$

93. $\sin\left(\sin^{-1} \frac{1}{2} + \tan^{-1}(-3)\right)$

Let $\sin^{-1} \frac{1}{2} = A$ and $\tan^{-1}(-3) = B$. Then $\sin A = \frac{1}{2}$ and $\tan B = -3$. Sketch angle A in quadrant I and angle B in quadrant IV, and use the Pythagorean theorem to find the unknown side in each triangle.

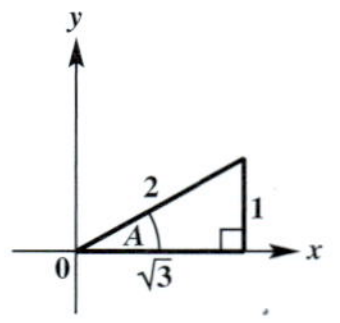

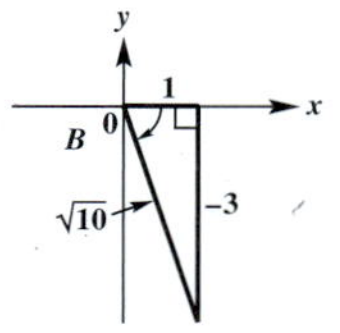

$$\sin\left(\sin^{-1} \frac{1}{2} + \tan^{-1}(-3)\right) = \sin(A + B)$$

$$= \sin A \cos B + \cos A \sin B \qquad \text{Sine sum identity}$$

$$= \frac{1}{2} \cdot \frac{1}{\sqrt{10}} + \frac{\sqrt{3}}{2} \cdot \frac{-3}{\sqrt{10}}$$

$$= \frac{1 - 3\sqrt{3}}{2\sqrt{10}}, \text{ or } \frac{\sqrt{10} - 3\sqrt{30}}{20}$$

6.2 Exercises *(pages 266–269)*

15. $\tan^2 x + 3 = 0$, so $\tan^2 x = -3$.

The square of a real number cannot be negative, so this equation has no solution. Solution set: $\emptyset$

25.
$$2\sin\theta - 1 = \csc\theta \quad \text{Original equation}$$
$$2\sin\theta - 1 = \frac{1}{\sin\theta} \quad \text{Reciprocal identity}$$
$$2\sin^2\theta - \sin\theta = 1 \quad \text{Multiply by } \sin\theta.$$
$$2\sin^2\theta - \sin\theta - 1 = 0 \quad \text{Subtract 1.}$$
$$(2\sin\theta + 1)(\sin\theta - 1) = 0 \quad \text{Factor.}$$
$$2\sin\theta + 1 = 0 \quad \text{or} \quad \sin\theta - 1 = 0 \quad \text{Zero-factor property}$$
$$\sin\theta = -\frac{1}{2} \quad \text{or} \quad \sin\theta = 1$$

Over the interval $[0°, 360°)$, the equation $\sin\theta = -\frac{1}{2}$ has two solutions, the angles in quadrants III and IV that have reference angle 30°. These are 210° and 330°. In the same interval, the only angle θ for which $\sin\theta = 1$ is 90°. All three of these check.

Solution set: $\{90°, 210°, 330°\}$

57.
$$\frac{2\tan\theta}{3 - \tan^2\theta} = 1 \quad \text{Original equation}$$
$$2\tan\theta = 3 - \tan^2\theta \quad \text{Multiply by } 3 - \tan^2\theta.$$
$$\tan^2\theta + 2\tan\theta - 3 = 0 \quad \text{Write in standard quadratic form.}$$
$$(\tan\theta - 1)(\tan\theta + 3) = 0 \quad \text{Factor.}$$
$$\tan\theta - 1 = 0 \quad \text{or} \quad \tan\theta + 3 = 0 \quad \text{Zero-factor property}$$
$$\tan\theta = 1 \quad \text{or} \quad \tan\theta = -3$$

Over the interval $[0°, 360°)$, the equation $\tan\theta = 1$ has two solutions, 45° and 225°. Over the same interval, the equation $\tan\theta = -3$ has two solutions that are approximately $-71.6° + 180° = 108.4°$ and $-71.6° + 360° = 288.4°$. All of these check.

The period of the tangent function is 180°, so the solution set is $\{45° + 180°n, 108.4° + 180°n$, where n is any integer$\}$.

6.3 Exercises *(pages 273–275)*

23. $\cos 2x + \cos x = 0$

We choose the identity for $\cos 2x$ that involves only the cosine function.

$$\cos 2x + \cos x = 0 \quad \text{Original equation}$$
$$2\cos^2 x - 1 + \cos x = 0 \quad \text{Cosine double-angle identity}$$
$$2\cos^2 x + \cos x - 1 = 0 \quad \text{Standard quadratic form}$$
$$(2\cos x - 1)(\cos x + 1) = 0 \quad \text{Factor.}$$
$$2\cos x - 1 = 0 \quad \text{or} \quad \cos x + 1 = 0 \quad \text{Zero-factor property}$$
$$2\cos x = 1$$
$$\cos x = \frac{1}{2} \quad \text{or} \quad \cos x = -1 \quad \text{Solve for } \cos x.$$

Over the interval $[0, 2\pi)$, the equation $\cos x = \frac{1}{2}$ has two solutions, $\frac{\pi}{3}$ and $\frac{5\pi}{3}$. Over the same interval, the equation $\cos x = -1$ has only one solution, π.

Solution set: $\left\{\frac{\pi}{3}, \pi, \frac{5\pi}{3}\right\}$

31.
$$2\sin\theta = 2\cos 2\theta \quad \text{Original equation}$$
$$\sin\theta = \cos 2\theta \quad \text{Divide by 2.}$$
$$\sin\theta = 1 - 2\sin^2\theta \quad \text{Cosine double-angle identity}$$
$$2\sin^2\theta + \sin\theta - 1 = 0 \quad \text{Standard quadratic form}$$
$$(2\sin\theta - 1)(\sin\theta + 1) = 0 \quad \text{Factor.}$$
$$2\sin\theta - 1 = 0 \quad \text{or} \quad \sin\theta + 1 = 0 \quad \text{Zero-factor property}$$
$$\sin\theta = \frac{1}{2} \quad \text{or} \quad \sin\theta = -1 \quad \text{Solve for } \sin\theta.$$

Over the interval $[0°, 360°)$, the equation $\sin\theta = \frac{1}{2}$ has two solutions, 30° and 150°. Over the same interval, the equation $\sin\theta = -1$ has one solution, 270°.

The period of the sine function is 360°, so the solution set is $\{30° + 360°n, 150° + 360°n, 270° + 360°n$, where n is any integer$\}$.

6.4 Exercises *(pages 279–282)*

15. $y = \cos(x + 3)$, for x in $[-3, \pi - 3]$ Original equation
$$x + 3 = \arccos y \quad \text{Definition of arccos}$$
$$x = -3 + \arccos y \quad \text{Subtract 3.}$$

37.
$$\arccos x + 2\arcsin\frac{\sqrt{3}}{2} = \pi \quad \text{Original equation}$$
$$\arccos x = \pi - 2\arcsin\frac{\sqrt{3}}{2} \quad \text{Isolate } \arccos x.$$
$$\arccos x = \pi - 2\left(\frac{\pi}{3}\right) \quad \arcsin\frac{\sqrt{3}}{2} = \frac{\pi}{3}$$
$$\arccos x = \pi - \frac{2\pi}{3} \quad \text{Multiply.}$$
$$\arccos x = \frac{\pi}{3} \quad \text{Subtract.}$$
$$x = \cos\frac{\pi}{3} \quad \text{Rewrite.}$$
$$x = \frac{1}{2} \quad \text{Evaluate.}$$

Solution set: $\left\{\frac{1}{2}\right\}$

41. $\cos^{-1} x + \tan^{-1} x = \frac{\pi}{2}$ — Original equation

$\cos^{-1} x = \frac{\pi}{2} - \tan^{-1} x$ — Subtract $\tan^{-1} x$.

$x = \cos\left(\frac{\pi}{2} - \tan^{-1} x\right)$

Definition of $\cos^{-1} x$

$x = \cos \frac{\pi}{2} \cdot \cos(\tan^{-1} x)$

$+ \sin \frac{\pi}{2} \cdot \sin(\tan^{-1} x)$

Cosine difference identity

$x = 0 \cdot \cos(\tan^{-1} x) + 1 \cdot \sin(\tan^{-1} x)$

$\cos \frac{\pi}{2} = 0$ and $\sin \frac{\pi}{2} = 1$

$x = \sin(\tan^{-1} x)$

Let $u = \tan^{-1} x$, so $\tan u = x$.

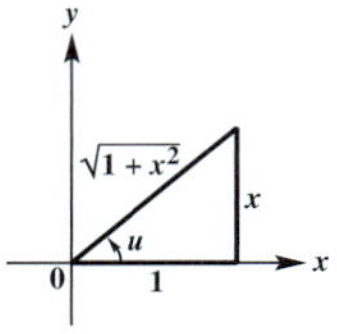

From the triangle, we find that $\sin u = \frac{x}{\sqrt{1 + x^2}}$, so the equation $x = \sin(\tan^{-1} x)$ becomes

$$x = \frac{x}{\sqrt{1 + x^2}}.$$

Solve this equation.

$x = \frac{x}{\sqrt{1 + x^2}}$

$x\sqrt{1 + x^2} = x$ — Multiply by $\sqrt{1 + x^2}$.

$x\sqrt{1 + x^2} - x = 0$ — Subtract x.

$x\left(\sqrt{1 + x^2} - 1\right) = 0$ — Factor.

$x = 0$ or $\sqrt{1 + x^2} - 1 = 0$ — Zero-factor property

$\sqrt{1 + x^2} = 1$ — Isolate the radical.

$1 + x^2 = 1$ — Square each side.

$x^2 = 0$ — Subtract 1.

$x = 0$ — Take square roots.

Solution set: $\{0\}$

Chapter 7 Applications of Trigonometry and Vectors

7.1 Exercises *(pages 295–299)*

33. We cannot find θ directly because the length of the side opposite angle θ is not given. Redraw the triangle shown in the figure, and label the third angle as α.

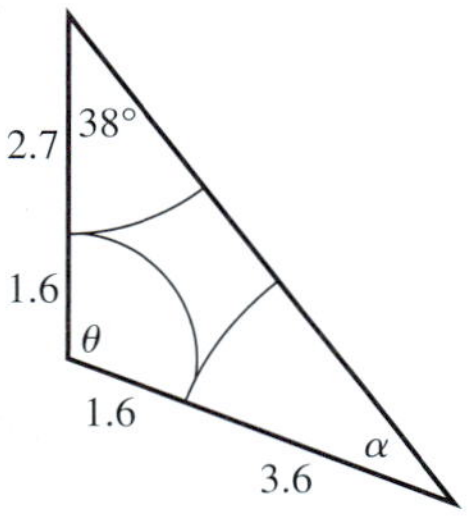

$\frac{\sin \alpha}{1.6 + 2.7} = \frac{\sin 38°}{1.6 + 3.6}$ — Alternative form of the law of sines

$\frac{\sin \alpha}{4.3} = \frac{\sin 38°}{5.2}$ — Add in the denominators.

$\sin \alpha = \frac{4.3 \sin 38°}{5.2} \approx 0.50910468$

$\alpha \approx 31°$ — Use the inverse sine function.

Then $\theta \approx 180° - 38° - 31°$

$\theta \approx 111°$.

41. To find the area of the triangle, use $\mathcal{A} = \frac{1}{2}bh$, with $b = 1$ and $h = \sqrt{2}$.

$$\mathcal{A} = \frac{1}{2}(1)\left(\sqrt{2}\right) = \frac{\sqrt{2}}{2}$$

Now use $\mathcal{A} = \frac{1}{2}ab \sin C$, with $a = 2$, $b = 1$, and $C = 45°$.

$$\mathcal{A} = \frac{1}{2}(2)(1) \sin 45° = \sin 45° = \frac{\sqrt{2}}{2}$$

Both formulas show that the area is $\frac{\sqrt{2}}{2}$ sq unit.

7.2 Exercises *(pages 304–306)*

11. $\frac{\sin B}{b} = \frac{\sin A}{a}$ — Alternative form of the law of sines

$\frac{\sin B}{2} = \frac{\sin 60°}{\sqrt{6}}$ — Substitute values from the figure.

$\sin B = \frac{2 \sin 60°}{\sqrt{6}}$ — Multiply by 2.

$\sin B = \frac{2 \cdot \frac{\sqrt{3}}{2}}{\sqrt{6}}$ — $\sin 60° = \frac{\sqrt{3}}{2}$

$\sin B = \frac{\sqrt{3}}{\sqrt{6}} = \sqrt{\frac{1}{2}} = \frac{\sqrt{2}}{2}$ — Simplify and rationalize.

$B = 45°$ — Use the inverse sine function.

There is another angle between 0° and 180° whose sine is $\frac{\sqrt{2}}{2}$: $180° - 45° = 135°$. However, this is too large because $A = 60°$ and $60° + 135° = 195°$. Since $195° > 180°$, there is only one solution, $B = 45°$.

19. $A = 142.13°$, $b = 5.432$ ft, $a = 7.297$ ft

$\frac{\sin B}{b} = \frac{\sin A}{a}$ — Alternative form of the law of sines

$\sin B = \frac{b \sin A}{a}$ — Multiply by b.

$\sin B = \frac{5.432 \sin 142.13°}{7.297}$ — Substitute given values.

$\sin B \approx 0.45697580$ — Simplify.

$B \approx 27.19°$ — Use the inverse sine function.

Because angle A is obtuse, angle B must be acute, so this is the only possible value for B and there is one triangle with the given measurements.

$C = 180° - A - B$ Angle sum formula, solved for C

$C \approx 180° - 142.13° - 27.19°$

$C \approx 10.68°$

Thus, $B \approx 27.19°$ and $C \approx 10.68°$.

7.3 Exercises *(pages 313–319)*

21. $C = 45.6°$, $b = 8.94$ m, $a = 7.23$ m

First find c.

$c^2 = a^2 + b^2 - 2ab\cos C$ Law of cosines

$c^2 = 7.23^2 + 8.94^2 - 2(7.23)(8.94)\cos 45.6°$ Substitute given values.

$c^2 \approx 41.7493$ Use a calculator.

$c \approx 6.46$ Square root property

Find A next since angle A is smaller than angle B (because $a < b$), and thus angle A must be acute.

$\frac{\sin A}{a} = \frac{\sin C}{c}$ Alternative form of the law of sines

$\sin A = \frac{a \sin C}{c}$ Multiply by a.

$\sin A = \frac{7.23 \sin 45.6°}{6.46}$ Substitute.

$\sin A \approx 0.79963428$ Simplify.

$A \approx 53.1°$ Use the inverse sine function.

Finally, find B.

$$B = 180° - C - A$$
$$B \approx 180° - 45.6° - 53.1°$$
$$B \approx 81.3°$$

Thus, $c \approx 6.46$ m, $A \approx 53.1°$, and $B \approx 81.3°$.

43. Find AC, or b, in this figure.

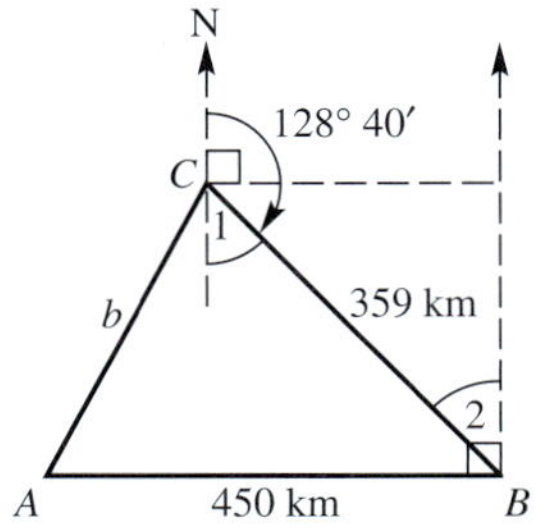

Angle 1 = 180° − 128° 40′ = 51° 20′

Angles 1 and 2 are alternate interior angles formed when two parallel lines (the north lines) are cut by a transversal, line BC, so angle 2 = angle 1 = 51° 20′.

angle ABC = 90° − angle 2 = 90° − 51° 20′ = 38° 40′ Complementary angles

$b^2 = a^2 + c^2 - 2ac\cos B$ Law of cosines

$b^2 = 359^2 + 450^2 - 2(359)(450)\cos 38° 40'$ Substitute values from the figure.

$b^2 \approx 79{,}106$ Use a calculator.

$b \approx 281$ Square root property

C is about 281 km from A.

7.4 Exercises *(pages 328–331)*

19. Use the figure to find the components of **u** and **v**: $\mathbf{u} = \langle -8, 8 \rangle$ and $\mathbf{v} = \langle 4, 8 \rangle$.

(a) $\mathbf{u} + \mathbf{v} = \langle -8, 8 \rangle + \langle 4, 8 \rangle = \langle -8 + 4, 8 + 8 \rangle = \langle -4, 16 \rangle$

(b) $\mathbf{u} - \mathbf{v} = \langle -8, 8 \rangle - \langle 4, 8 \rangle = \langle -8 - 4, 8 - 8 \rangle = \langle -12, 0 \rangle$

(c) $-\mathbf{u} = -\langle -8, 8 \rangle = \langle 8, -8 \rangle$

47. $\mathbf{v} = \langle a, b \rangle = \langle 5\cos(-35°), 5\sin(-35°) \rangle = \langle 4.0958, -2.8679 \rangle$

81. First write the given vectors in component form.

$$3\mathbf{i} + 4\mathbf{j} = \langle 3, 4 \rangle; \quad \mathbf{j} = \langle 0, 1 \rangle$$

$$\cos\theta = \frac{\langle 3, 4 \rangle \cdot \langle 0, 1 \rangle}{|\langle 3, 4 \rangle|\,|\langle 0, 1 \rangle|}$$

$$\cos\theta = \frac{3(0) + 4(1)}{\sqrt{9 + 16} \cdot \sqrt{0 + 1}} = \frac{4}{5} = 0.8$$

$$\theta = \cos^{-1} 0.8 \approx 36.87°$$

7.5 Exercises *(pages 335–338)*

5. Use the parallelogram rule. In the figure, **x** represents the second force and **v** is the resultant.

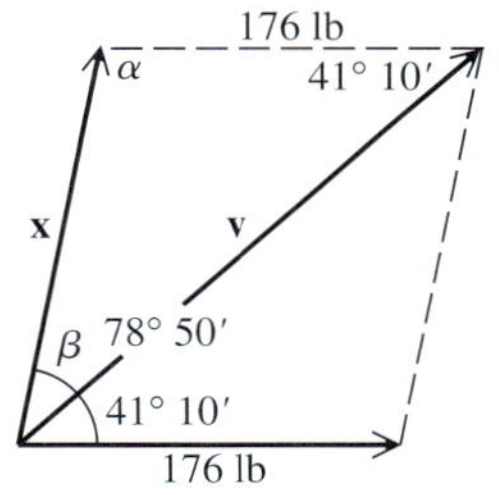

$$\alpha = 180° - 78° 50' = 101° 10'$$

$$\beta = 78° 50' - 41° 10' = 37° 40'$$

$\frac{|\mathbf{x}|}{\sin 41° 10'} = \frac{176}{\sin 37° 40'}$ Law of sines

$$|\mathbf{x}| = \frac{176 \sin 41° 10'}{\sin 37° 40'} \approx 190$$

$\frac{|\mathbf{v}|}{\sin 101° 10'} = \frac{176}{\sin 37° 40'}$ Law of sines

$$|\mathbf{v}| = \frac{176 \sin 101° 10'}{\sin 37° 40'} \approx 283$$

Thus, the magnitude of the second force is about 190 lb, and the magnitude of the resultant is about 283 lb.

27. Let **v** represent the airspeed vector.

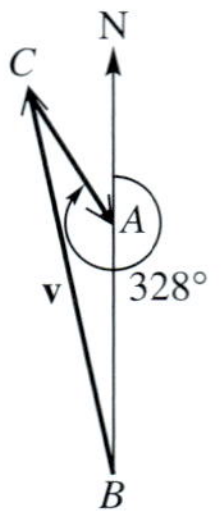

The ground speed is $\frac{400 \text{ mi}}{2.5 \text{ hr}} = 160$ mph.

angle $BAC = 328° - 180° = 148°$

$|\mathbf{v}|^2 = 11^2 + 160^2 - 2(11)(160) \cos 148°$ Law of cosines

$|\mathbf{v}|^2 \approx 28{,}706$

$|\mathbf{v}| \approx 169.4$

The airspeed must be approximately 170 mph.

$\frac{\sin B}{11} = \frac{\sin 148°}{169.4}$ Law of sines

$\sin B = \frac{11 \sin 148°}{169.4} \approx 0.03441034$

$B \approx 2°$

The bearing must be approximately $360° - 2° = 358°$.

Chapter 8 Complex Numbers, Polar Equations, and Parametric Equations

8.1 Exercises *(pages 356–359)*

61. $-i\sqrt{2} - 2 - (6 - 4i\sqrt{2}) - (5 - i\sqrt{2})$

$= (-2 - 6 - 5) + [-\sqrt{2} - (-4\sqrt{2}) - (-\sqrt{2})]i$

$= -13 + 4i\sqrt{2}$ Combine real parts and combine imaginary parts.

79. $(2 + i)(2 - i)(4 + 3i)$

$= [(2 + i)(2 - i)](4 + 3i)$ Associative property

$= (2^2 - i^2)(4 + 3i)$ Product of the sum and difference of two terms

$= [4 - (-1)](4 + 3i)$ $i^2 = -1$

$= 5(4 + 3i)$ Subtract.

$= 20 + 15i$ Distributive property

107. $\left(\frac{\sqrt{2}}{2} + \frac{\sqrt{2}}{2}i\right)^2$

$= \left(\frac{\sqrt{2}}{2}\right)^2 + 2 \cdot \frac{\sqrt{2}}{2} \cdot \frac{\sqrt{2}}{2}i + \left(\frac{\sqrt{2}}{2}i\right)^2$ Square of a binomial

$= \frac{2}{4} + 2 \cdot \frac{2}{4}i + \frac{2}{4}i^2$ Apply exponents and multiply.

$= \frac{1}{2} + i + \frac{1}{2}i^2$

$= \frac{1}{2} + i + \frac{1}{2}(-1)$ $i^2 = -1$

$= \frac{1}{2} + i - \frac{1}{2}$ Multiply.

$= i$ Combine real parts.

Thus, $\frac{\sqrt{2}}{2} + \frac{\sqrt{2}}{2}i$ is a square root of i.

8.2 Exercises *(pages 364–366)*

31. $3 \text{ cis } 150°$

$= 3(\cos 150° + i \sin 150°)$

$= 3\left(-\frac{\sqrt{3}}{2} + i \cdot \frac{1}{2}\right)$ $\cos 150° = -\frac{\sqrt{3}}{2}$; $\sin 150° = \frac{1}{2}$

$= -\frac{3\sqrt{3}}{2} + \frac{3}{2}i$ Rectangular form

43. $-5 - 5i$

Sketch the graph of $-5 - 5i$ in the complex plane.

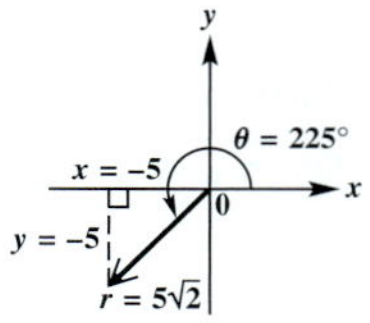

Since $x = -5$ and $y = -5$,

$r = \sqrt{x^2 + y^2} = \sqrt{(-5)^2 + (-5)^2} = \sqrt{50} = 5\sqrt{2}$

and $\tan \theta = \frac{y}{x} = \frac{-5}{-5} = 1.$

Since $\tan \theta = 1$, the reference angle for θ is 45°. The graph shows that θ is in quadrant III, so

$$\theta = 180° + 45° = 225°.$$

Use these results.

$$-5 - 5i = 5\sqrt{2}(\cos 225° + i \sin 225°)$$

8.3 Exercises *(pages 369–371)*

5. $[4(\cos 60° + i \sin 60°)][6(\cos 330° + i \sin 330°)]$

$= 4 \cdot 6[\cos(60° + 330°) + i \sin(60° + 330°)]$ Product theorem

$= 24(\cos 390° + i \sin 390°)$ Multiply and add.

$= 24(\cos 30° + i \sin 30°)$ 390° and 30° are coterminal angles.

$= 24\left(\frac{\sqrt{3}}{2} + i \cdot \frac{1}{2}\right)$ $\cos 30° = \frac{\sqrt{3}}{2}$; $\sin 30° = \frac{1}{2}$

$= 12\sqrt{3} + 12i$ Rectangular form

21. $\frac{-i}{1 + i}$

Numerator: $-i = 0 - 1i$

$r = \sqrt{0^2 + (-1)^2} = 1$

$\theta = 270°$ since $\cos 270° = 0$ and $\sin 270° = -1$. Thus $-i = 1 \text{ cis } 270°$.

Denominator: $1 + i = 1 + 1i$

$$r = \sqrt{1^2 + 1^2} = \sqrt{2}$$

$$\tan \theta = \frac{y}{x} = \frac{1}{1} = 1$$

Since x and y are both positive, θ is in quadrant I, and $\theta = \tan^{-1} 1 = 45°$. Thus, $1 + i = \sqrt{2} \text{ cis } 45°$.

$\frac{-i}{1+i}$

$= \frac{1 \text{ cis } 270°}{\sqrt{2} \text{ cis } 45°}$ Substitute.

$= \frac{1}{\sqrt{2}} \text{cis}(270° - 45°)$ Quotient theorem

$= \frac{\sqrt{2}}{2} \text{ cis } 225°$ Rationalize and subtract.

$= \frac{\sqrt{2}}{2}(\cos 225° + i \sin 225°)$ Equivalent form

$= \frac{\sqrt{2}}{2}\left(-\frac{\sqrt{2}}{2} - i \cdot \frac{\sqrt{2}}{2}\right)$ $\cos 225° = -\frac{\sqrt{2}}{2}$; $\sin 225° = -\frac{\sqrt{2}}{2}$

$= -\frac{1}{2} - \frac{1}{2}i$ Rectangular form

8.4 Exercises *(pages 376–378)*

11. $(-2 - 2i)^5$

First write $-2 - 2i$ in trigonometric form.

$$r = \sqrt{(-2)^2 + (-2)^2} = \sqrt{8} = 2\sqrt{2}$$

$$\tan \theta = \frac{y}{x} = \frac{-2}{-2} = 1$$

Because x and y are both negative, θ is in quadrant III. Thus $\theta = 225°$.

$$-2 - 2i = 2\sqrt{2}(\cos 225° + i \sin 225°)$$

$$(-2 - 2i)^5 = \left[2\sqrt{2}(\cos 225° + i \sin 225°)\right]^5$$

$= \left(2\sqrt{2}\right)^5\left[\cos(5 \cdot 225°) + i \sin(5 \cdot 225°)\right]$ De Moivre's theorem

$= 32 \cdot 4\sqrt{2}(\cos 1125° + i \sin 1125°)$

$= 128\sqrt{2}(\cos 1125° + i \sin 1125°)$

$= 128\sqrt{2}(\cos 45° + i \sin 45°)$ 1125° and 45° are coterminal.

$= 128\sqrt{2}\left(\frac{\sqrt{2}}{2} + i \cdot \frac{\sqrt{2}}{2}\right)$ $\cos 45° = \frac{\sqrt{2}}{2}$; $\sin 45° = \frac{\sqrt{2}}{2}$

$= 128 + 128i$ Rectangular form

41. $x^3 - \left(4 + 4i\sqrt{3}\right) = 0$

$$x^3 = 4 + 4i\sqrt{3}$$

$$r = \sqrt{4^2 + \left(4\sqrt{3}\right)^2} = \sqrt{16 + 48} = \sqrt{64} = 8$$

$$\tan \theta = \frac{4\sqrt{3}}{4} = \sqrt{3}$$

θ is in quadrant I, so $\theta = 60°$.

$$x^3 = 4 + 4i\sqrt{3}$$

$$x^3 = 8\left(\frac{1}{2} + i\frac{\sqrt{3}}{2}\right)$$

$$r^3(\cos 3\alpha + i \sin 3\alpha) = 8(\cos 60° + i \sin 60°)$$

$r^3 = 8$, so $r = 2$.

$\alpha = \frac{60°}{3} + \frac{360° \cdot k}{3}$, k any integer *n*th root theorem

$\alpha = 20° + 120° \cdot k$, k any integer

If $k = 0$, then $\alpha = 20° + 0° = 20°$.

If $k = 1$, then $\alpha = 20° + 120° = 140°$.

If $k = 2$, then $\alpha = 20° + 240° = 260°$.

Solution set: $\{2(\cos 20° + i \sin 20°)$, $2(\cos 140° + i \sin 140°)$, $2(\cos 260° + i \sin 260°)\}$

8.5 Exercises *(pages 387–391)*

57. $r = 2 \sin \theta$

$r^2 = 2r \sin \theta$ Multiply by r.

$x^2 + y^2 = 2y$ $r^2 = x^2 + y^2$, $r \sin \theta = y$

$x^2 + y^2 - 2y = 0$ Subtract $2y$.

$x^2 + y^2 - 2y + 1 = 1$ Add 1 to complete the square on y.

$x^2 + (y - 1)^2 = 1$ Factor the perfect square trinomial.

The graph is a circle with center $(0, 1)$ and radius 1.

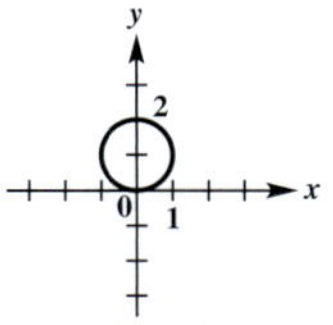

$r = 2 \sin \theta$
$x^2 + (y - 1)^2 = 1$

63. $r = 2 \sec \theta$

$r = \frac{2}{\cos \theta}$ Reciprocal identity

$r \cos \theta = 2$ Multiply by $\cos \theta$.

$x = 2$ $r \cos \theta = x$

The graph is the vertical line through $(2, 0)$.

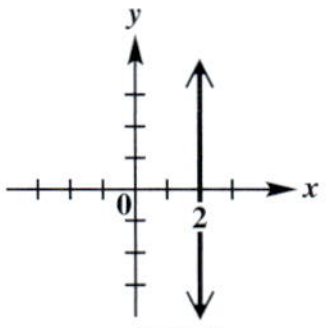

$r = 2 \sec \theta$
$x = 2$

8.6 Exercises *(pages 397–400)*

9. $x = t^3 + 1,\ y = t^3 - 1,\quad$ for t in $(-\infty, \infty)$

(a)

t	x	y
-2	-7	-9
-1	0	-2
0	1	-1
1	2	0
2	9	7
3	28	26

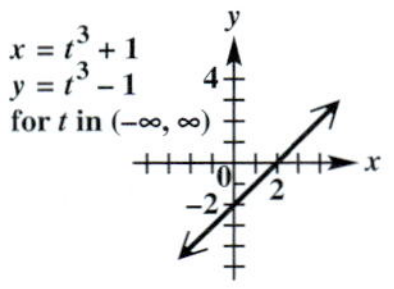

(b)

$$\begin{aligned} x &= t^3 + 1 \\ y &= t^3 - 1 \\ \hline x - y &= 2 \end{aligned}$$ Subtract equations to eliminate t.

$y = x - 2$ Solve for y.

The rectangular equation is $y = x - 2$, for x in $(-\infty, \infty)$. The graph is a line with slope 1 and y-intercept -2.

13. $x = 3\tan t,\ y = 2\sec t,\quad$ for t in $\left(-\frac{\pi}{2}, \frac{\pi}{2}\right)$

(a)

t	x	y
$-\frac{\pi}{3}$	$-3\sqrt{3} \approx -5.2$	4
$-\frac{\pi}{6}$	$-\sqrt{3} \approx -1.7$	$\frac{4\sqrt{3}}{3} \approx 2.3$
0	0	2
$\frac{\pi}{6}$	$\sqrt{3} \approx 1.7$	$\frac{4\sqrt{3}}{3} \approx 2.3$
$\frac{\pi}{3}$	$3\sqrt{3} \approx 5.2$	4

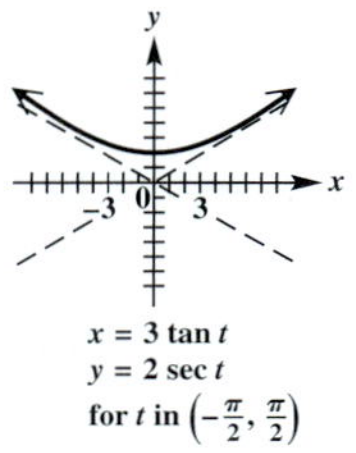

(b) $x = 3\tan t$, so $\frac{x}{3} = \tan t$.

$y = 2\sec t$, so $\frac{y}{2} = \sec t$.

$1 + \tan^2 t = \sec^2 t$ Pythagorean identity

$1 + \left(\frac{x}{3}\right)^2 = \left(\frac{y}{2}\right)^2$ Substitute expressions for $\tan t$ and $\sec t$.

$1 + \frac{x^2}{9} = \frac{y^2}{4}$ Apply the exponents.

$y^2 = 4\left(1 + \frac{x^2}{9}\right)$ Multiply by 4. Rewrite.

$y = 2\sqrt{1 + \frac{x^2}{9}}$ Use the positive square root because $y > 0$ in the given interval for t.

The rectangular equation is $y = 2\sqrt{1 + \frac{x^2}{9}}$, for x in $(-\infty, \infty)$. The graph is the upper half of a hyperbola.

Answers to Selected Exercises

To The Student

In this section we provide the answers that we think most students will obtain when they work the exercises using the methods explained in the text. If your answer does not look exactly like the one given here, it is not necessarily wrong. In many cases there are equivalent forms of the answer. For example, if the answer section shows $\frac{3}{4}$ and your answer is 0.75, you have obtained the correct answer but written it in a different (yet equivalent) form. Unless the directions specify otherwise, 0.75 is just as valid an answer as $\frac{3}{4}$. (In answers with radicals, we give rationalized denominators when appropriate.) In general, if your answer does not agree with the one given in the text, see whether it can be transformed into the other form. If it can, then it is equivalent to the correct answer. If you still have doubts, talk with your instructor.

If you need further help with trigonometry, you may want to obtain a copy of the *Student's Solution Manual* that goes with this book. Your college bookstore either has this manual or can order it for you.

Chapter 1 Trigonometric Functions

1.1 Exercises *(pages 7–10)*

1. (a) 60° **(b)** 150° **3. (a)** 45° **(b)** 135° **5. (a)** 36° **(b)** 126° **7. (a)** 89° **(b)** 179° **9. (a)** 75° 40′ **(b)** 165° 40′ **11. (a)** 69° 49′ 30″ **(b)** 159° 49′ 30″ **13.** 70°; 110° **15.** 30°; 60° **17.** 40°; 140° **19.** 107°; 73° **21.** 69°; 21° **23.** 45° **25.** 150° **27.** 7° 30′ **29.** 130° **31.** $(90 - x)°$ **33.** $(x - 360)°$ **35.** 83° 59′ **37.** 179° 19′ **39.** 23° 49′ **41.** 38° 32′ **43.** 60° 34′ **45.** 30° 27′ **47.** 17° 01′ 49″ **49.** 35.5° **51.** 112.25° **53.** −60.2° **55.** 20.9° **57.** 91.598° **59.** 274.316° **61.** 39° 15′ 00″ **63.** 126° 45′ 36″ **65.** −18° 30′ 54″ **67.** 31° 25′ 47″ **69.** 89° 54′ 01″ **71.** 178° 35′ 58″ **73.** 392° **75.** 386° 30′ **77.** 320° **79.** 235° **81.** 1° **83.** 359° **85.** 179° **87.** 130° **89.** 240° **91.** 120°

In Exercises 93 and 95, answers may vary.

93. 450°, 810°; −270°, −630° **95.** 360°, 720°; −360°, −720° **97.** $30° + n \cdot 360°$ **99.** $135° + n \cdot 360°$ **101.** $-90° + n \cdot 360°$ **103.** $0° + n \cdot 360°$, or $n \cdot 360°$

Angles other than those given are possible in Exercises 107–117.

107.

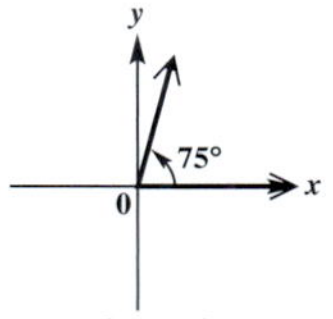

435°; −285°; quadrant I

109.

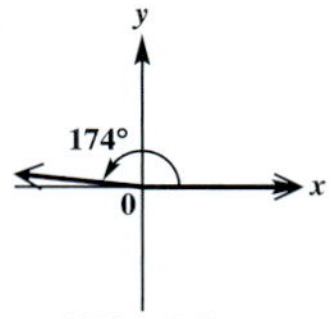

534°; −186°; quadrant II

111.

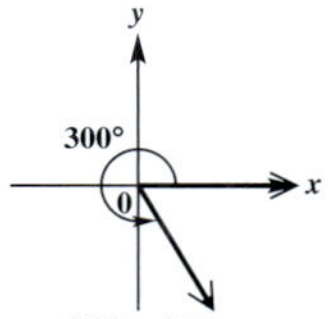

660°; −60°; quadrant IV

113.

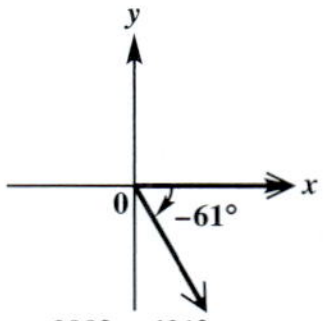

299°; −421°; quadrant IV

115.

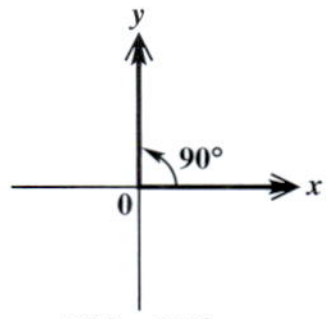

450°; −270°; no quadrant

117.

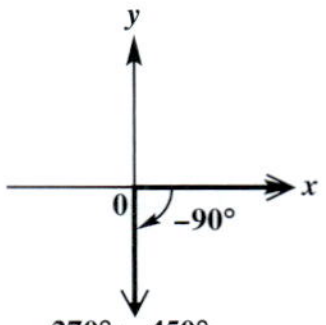

270°; −450°; no quadrant

119. $3\sqrt{2}$

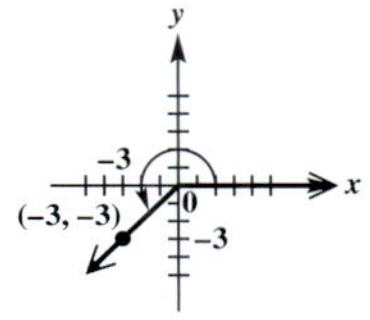

121. $\sqrt{34}$

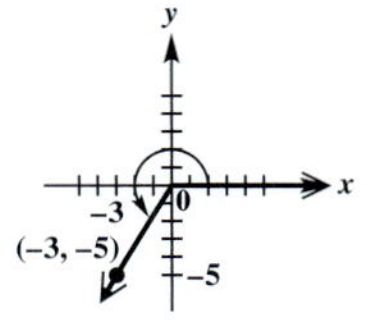

123. 2

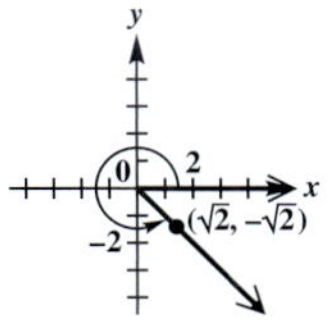

125. 2

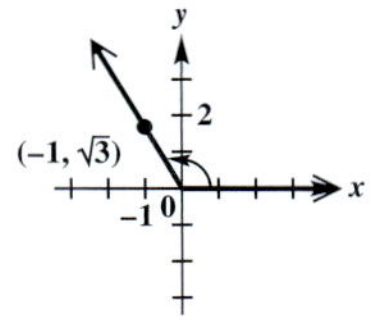

127. 4

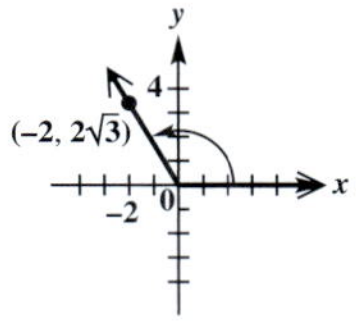

129. 4

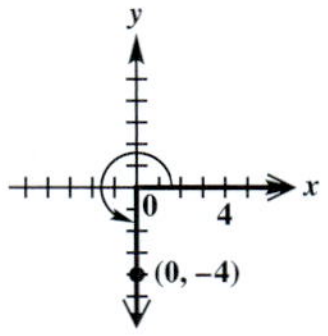

131. $\frac{3}{4}$ **133.** 1800° **135.** 12.5 rotations per hr **137.** 4 sec

1.2 Exercises *(pages 15–20)*

1. Answers are given in numerical order: 55°; 65°; 60°; 65°; 60°; 120°; 60°; 60°; 55°; 55° **3.** 51°; 51° **5.** 50°; 60°; 70° **7.** 60°; 60°; 60° **9.** 45°; 75°; 120° **11.** 49°; 49° **13.** 48°; 132° **15.** 91° **17.** 2° 29′ **19.** 25.4° **21.** 22° 29′ 34″ **25.** right; scalene

27. acute; equilateral **29.** right; scalene **31.** right; isosceles **33.** obtuse; scalene **35.** acute; isosceles **41.** A and P; B and Q; C and R; AC and PR; BC and QR; AB and PQ **43.** A and C; E and D; ABE and CBD; EB and DB; AB and CB; AE and CD **45.** $Q = 42°$; $B = R = 48°$ **47.** $B = 106°$; $A = M = 44°$ **49.** $X = M = 52°$ **51.** $a = 20$; $b = 15$ **53.** $a = 6$; $b = 7.5$ **55.** $x = 6$ **57.** 30 m **59.** 500 m; 700 m **61.** 112.5 ft **63.** $x = 110$ **65.** $c \approx 111.1$ **67.** **(a)** approximately 236,000 mi **(b)** no **69.** **(a)** approximately 2900 mi **(b)** no **71.** **(a)** approximately $\frac{1}{4}$ **(b)** approximately 30 arc degrees

Chapter 1 Quiz *(page 21)*

[1.1] **1.** **(a)** 71° **(b)** 161° **2.** 65°; 115° **3.** 26°; 64° [1.2] **4.** 20°; 24°; 136° **5.** 130°; 50° [1.1] **6.** **(a)** 77.2025° **(b)** 22° 01′ 30″ **7.** **(a)** 50° **(b)** 300° **(c)** 170° **(d)** 417° **8.** 1800° [1.2] **9.** 10 ft **10.** **(a)** $x = 12$; $y = 10$ **(b)** $x = 5$

1.3 Exercises *(pages 26–28)*

In Exercises 1–19 and 45–55, we give, in order, sine, cosine, tangent, cotangent, secant, and cosecant.

1.
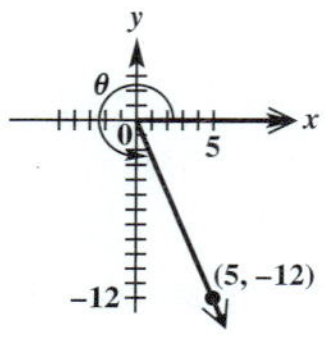

$-\frac{12}{13}$; $\frac{5}{13}$; $-\frac{12}{5}$; $-\frac{5}{12}$; $\frac{13}{5}$; $-\frac{13}{12}$

3.
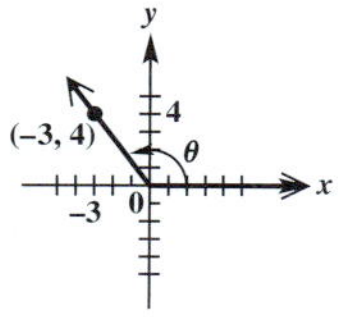

$\frac{4}{5}$; $-\frac{3}{5}$; $-\frac{4}{3}$; $-\frac{3}{4}$; $-\frac{5}{3}$; $\frac{5}{4}$

5.
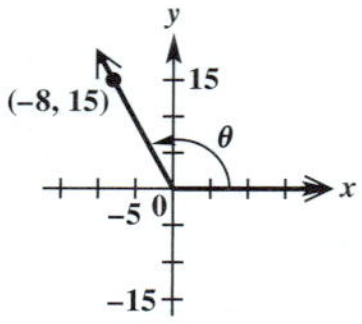

$\frac{15}{17}$; $-\frac{8}{17}$; $-\frac{15}{8}$; $-\frac{8}{15}$; $-\frac{17}{8}$; $\frac{17}{15}$

7.
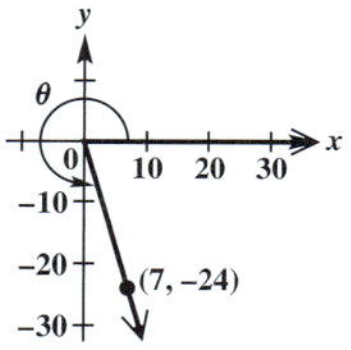

$-\frac{24}{25}$; $\frac{7}{25}$; $-\frac{24}{7}$; $-\frac{7}{24}$; $\frac{25}{7}$; $-\frac{25}{24}$

9.
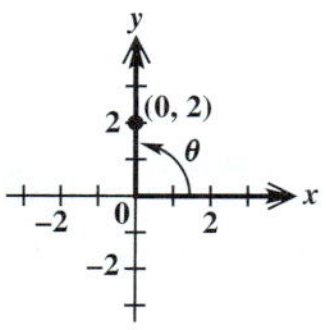

1; 0; undefined; 0; undefined; 1

11.
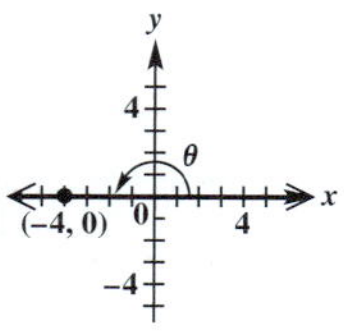

0; −1; 0; undefined; −1; undefined

13.
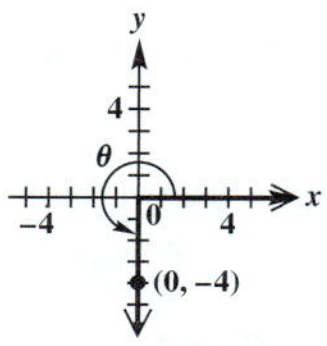

−1; 0; undefined; 0; undefined; −1

15.
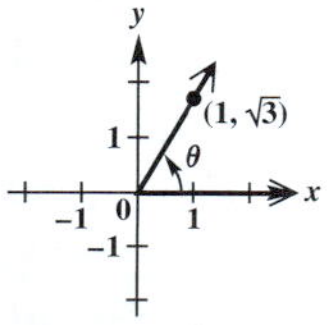

$\frac{\sqrt{3}}{2}$; $\frac{1}{2}$; $\sqrt{3}$; $\frac{\sqrt{3}}{3}$; 2; $\frac{2\sqrt{3}}{3}$

17.
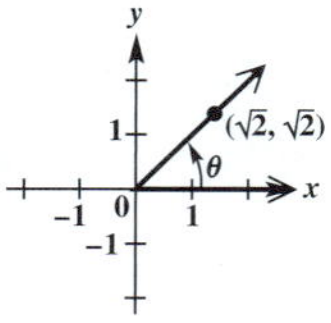

$\frac{\sqrt{2}}{2}$; $\frac{\sqrt{2}}{2}$; 1; 1; $\sqrt{2}$; $\sqrt{2}$

19.
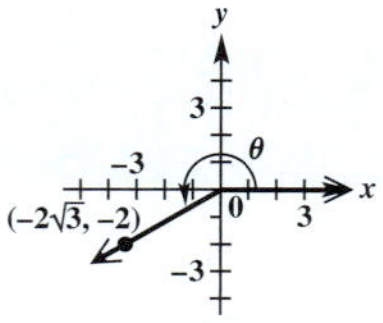

$-\frac{1}{2}$; $-\frac{\sqrt{3}}{2}$; $\frac{\sqrt{3}}{3}$; $\sqrt{3}$; $-\frac{2\sqrt{3}}{3}$; −2

23. 0 **25.** negative **27.** negative **29.** positive **31.** positive **33.** negative **35.** positive **37.** negative **39.** positive **41.** positive **43.** positive

45.
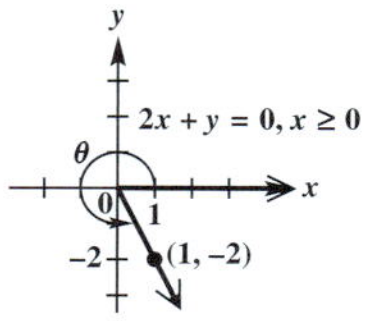

$-\frac{2\sqrt{5}}{5}$; $\frac{\sqrt{5}}{5}$; −2; $-\frac{1}{2}$; $\sqrt{5}$; $-\frac{\sqrt{5}}{2}$

47.
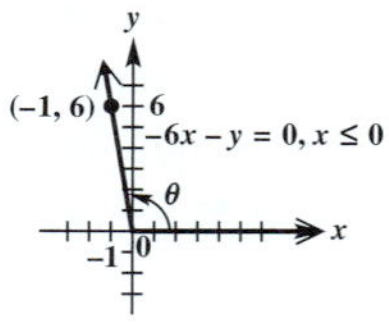

$\frac{6\sqrt{37}}{37}$; $-\frac{\sqrt{37}}{37}$; −6; $-\frac{1}{6}$; $-\sqrt{37}$; $\frac{\sqrt{37}}{6}$

49.
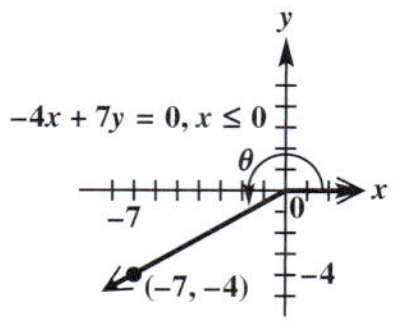

$-\frac{4\sqrt{65}}{65}$; $-\frac{7\sqrt{65}}{65}$; $\frac{4}{7}$; $\frac{7}{4}$; $-\frac{\sqrt{65}}{7}$; $-\frac{\sqrt{65}}{4}$

51.
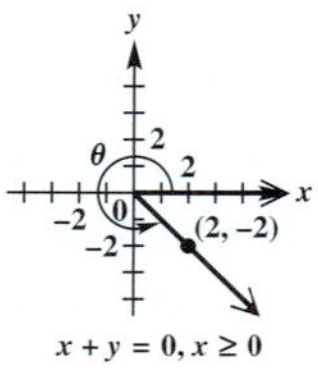

$-\frac{\sqrt{2}}{2}$; $\frac{\sqrt{2}}{2}$; −1; −1; $\sqrt{2}$; $-\sqrt{2}$

53.
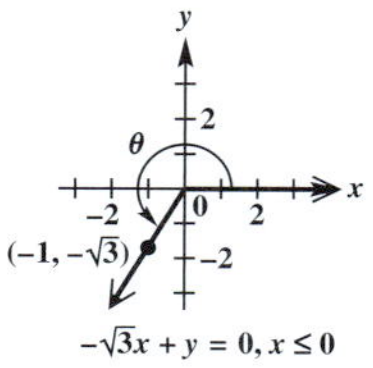

$-\frac{\sqrt{3}}{2}$; $-\frac{1}{2}$; $\sqrt{3}$; $\frac{\sqrt{3}}{3}$; −2; $-\frac{2\sqrt{3}}{3}$

55.
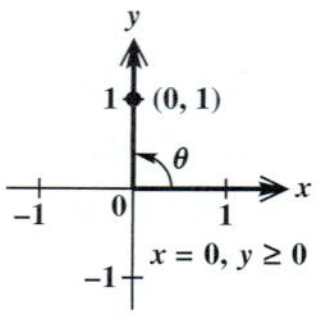

1; 0; undefined; 0; undefined; 1

57. 0 **59.** 0 **61.** −1 **63.** 1 **65.** undefined **67.** −1 **69.** 0 **71.** undefined **73.** 1 **75.** −1 **77.** 0 **79.** −3 **81.** −3 **83.** 5 **85.** 1 **87.** 0 **89.** 0 **91.** 1 **93.** 0 **95.** 0 **97.** −1 **99.** 0 **101.** undefined **103.** They are equal. **105.** They are negatives of each other. **107.** about 0.940; about 0.342 **109.** 35° **111.** decrease; increase

1.4 Exercises *(pages 36–38)*

1. $\frac{3}{2}$ **3.** $-\frac{7}{3}$ **5.** $\frac{1}{5}$ **7.** $-\frac{2}{5}$ **9.** $\frac{\sqrt{2}}{2}$ **11.** −0.4 **13.** 0.70069071 **17.** Because $\cot 90° = 0$, $\frac{1}{\cot 90°}$ and consequently $\tan 90°$ are undefined. **19.** All are positive. **21.** Tangent and cotangent are positive. All others are negative. **23.** Sine and cosecant are positive. All others are negative. **25.** Cosine and secant are positive. All others are negative. **27.** Sine and cosecant are positive. All others are negative. **29.** All are positive. **31.** I, II

33. I **35.** II **37.** I **39.** III **41.** III, IV **45.** impossible **47.** possible **49.** possible **51.** impossible **53.** possible **55.** possible **57.** possible **59.** impossible **61.** $-\frac{4}{5}$ **63.** $-\frac{\sqrt{5}}{2}$ **65.** $-\frac{\sqrt{3}}{3}$ **67.** 3.44701905

In Exercises 69–79, we give, in order, sine, cosine, tangent, cotangent, secant, and cosecant.

69. $\frac{15}{17}$; $-\frac{8}{17}$; $-\frac{15}{8}$; $-\frac{8}{15}$; $-\frac{17}{8}$; $\frac{17}{15}$

71. $\frac{\sqrt{5}}{7}$; $\frac{2\sqrt{11}}{7}$; $\frac{\sqrt{55}}{22}$; $\frac{2\sqrt{55}}{5}$; $\frac{7\sqrt{11}}{22}$; $\frac{7\sqrt{5}}{5}$

73. $\frac{8\sqrt{67}}{67}$; $\frac{\sqrt{201}}{67}$; $\frac{8\sqrt{3}}{3}$; $\frac{\sqrt{3}}{8}$; $\frac{\sqrt{201}}{3}$; $\frac{\sqrt{67}}{8}$

75. $\frac{\sqrt{2}}{6}$; $-\frac{\sqrt{34}}{6}$; $-\frac{\sqrt{17}}{17}$; $-\sqrt{17}$; $-\frac{3\sqrt{34}}{17}$; $3\sqrt{2}$

77. $\frac{\sqrt{15}}{4}$; $-\frac{1}{4}$; $-\sqrt{15}$; $-\frac{\sqrt{15}}{15}$; -4; $\frac{4\sqrt{15}}{15}$

79. 0.164215; −0.986425; −0.166475; −6.00691; −1.01376; 6.08958 **83.** This statement is false. For example, $\sin 180° + \cos 180° = 0 + (-1) = -1 \neq 1$. **85.** negative **87.** positive **89.** negative **91.** negative **93.** positive **95.** negative **97.** positive **99.** negative **101.** 2° **103.** 3° **105.** Quadrant II is the only quadrant in which the cosine is negative and the sine is positive.

Chapter 1 Review Exercises *(pages 40–43)*

1. complement: 55°; supplement: 145° **3.** 186° **5.** $x = 30$; $y = 30$ **7.** 9360° **9.** 119.134° **11.** 275° 06′ 02″ **13.** 40°; 60°; 80° **15.** 0.25 km **17.** $N = 12°$; $R = 82°$; $M = 86°$ **19.** $p = 7$; $q = 7$ **21.** $k = 14$ **23.** 12 ft

In Exercises 25–31, we give, in order, sine, cosine, tangent, cotangent, secant, and cosecant.

25. $-\frac{\sqrt{3}}{2}$; $\frac{1}{2}$; $-\sqrt{3}$; $-\frac{\sqrt{3}}{3}$; 2; $-\frac{2\sqrt{3}}{3}$

27. $-\frac{4}{5}$; $\frac{3}{5}$; $-\frac{4}{3}$; $-\frac{3}{4}$; $\frac{5}{3}$; $-\frac{5}{4}$

29. $\frac{15}{17}$; $-\frac{8}{17}$; $-\frac{15}{8}$; $-\frac{8}{15}$; $-\frac{17}{8}$; $\frac{17}{15}$

31. $-\frac{1}{2}$; $\frac{\sqrt{3}}{2}$; $-\frac{\sqrt{3}}{3}$; $-\sqrt{3}$; $\frac{2\sqrt{3}}{3}$; -2

33. tangent and secant

35.

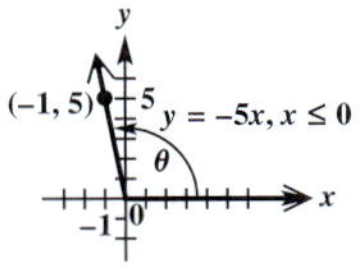

37. 0; −1; 0; undefined; −1; undefined **39.** **(a)** impossible **(b)** possible **(c)** impossible

In Exercises 41–45, we give, in order, sine, cosine, tangent, cotangent, secant, and cosecant.

41. $-\frac{\sqrt{39}}{8}$; $-\frac{5}{8}$; $\frac{\sqrt{39}}{5}$; $\frac{5\sqrt{39}}{39}$; $-\frac{8}{5}$; $-\frac{8\sqrt{39}}{39}$

43. $\frac{2\sqrt{5}}{5}$; $-\frac{\sqrt{5}}{5}$; -2; $-\frac{1}{2}$; $-\sqrt{5}$; $\frac{\sqrt{5}}{2}$

45. $-\frac{3}{5}$; $\frac{4}{5}$; $-\frac{3}{4}$; $-\frac{4}{3}$; $\frac{5}{4}$; $-\frac{5}{3}$

47. 40 yd **49.** approximately 9500 ft

Chapter 1 Test *(pages 43–44)*

[1.1] **1.** **(a)** 23° **(b)** 113° **2.** 145°; 35° **3.** 20°; 70° [1.2] **4.** 130°; 130° **5.** 110°; 110° **6.** 20°; 30°; 130° **7.** 60°; 40°; 100° [1.1] **8.** 74.31° **9.** 45° 12′ 09″ **10.** **(a)** 30° **(b)** 280° **(c)** 90° **11.** 2700° [1.2] **12.** $10\frac{2}{3}$ ft, or 10 ft, 8 in. **13.** $x = 8$; $y = 6$

[1.3] **14.**

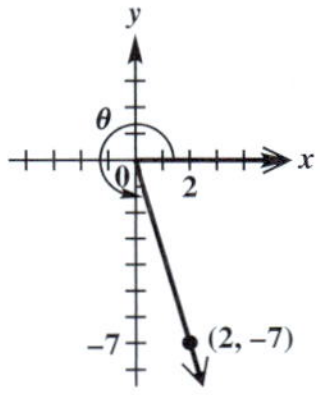

$\sin\theta = -\frac{7\sqrt{53}}{53}$; $\cos\theta = \frac{2\sqrt{53}}{53}$; $\tan\theta = -\frac{7}{2}$; $\cot\theta = -\frac{2}{7}$; $\sec\theta = \frac{\sqrt{53}}{2}$; $\csc\theta = -\frac{\sqrt{53}}{7}$

15.

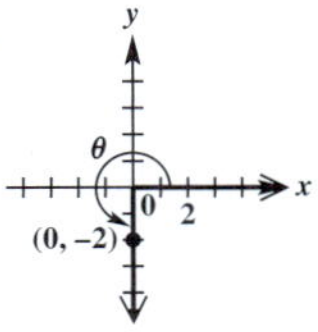

$\sin\theta = -1$; $\cos\theta = 0$; $\tan\theta$ is undefined; $\cot\theta = 0$; $\sec\theta$ is undefined; $\csc\theta = -1$

16.

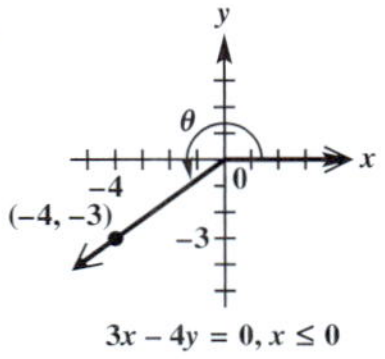

$\sin\theta = -\frac{3}{5}$; $\cos\theta = -\frac{4}{5}$; $\tan\theta = \frac{3}{4}$; $\cot\theta = \frac{4}{3}$; $\sec\theta = -\frac{5}{4}$; $\csc\theta = -\frac{5}{3}$

17. row 1: 1, 0, undefined, 0, undefined, 1; row 2: 0, 1, 0, undefined, 1, undefined; row 3: −1, 0, undefined, 0, undefined, −1 **18.** cosecant and cotangent [1.4] **19.** **(a)** I **(b)** III, IV **(c)** III **20.** **(a)** impossible **(b)** possible **(c)** possible **21.** $\sec\theta = -\frac{12}{7}$

22. $\cos\theta = -\frac{2\sqrt{10}}{7}$; $\tan\theta = -\frac{3\sqrt{10}}{20}$; $\cot\theta = -\frac{2\sqrt{10}}{3}$; $\sec\theta = -\frac{7\sqrt{10}}{20}$; $\csc\theta = \frac{7}{3}$

Chapter 2 Acute Angles and Right Triangles

2.1 Exercises *(pages 51–54)*

In Exercises 1 and 3, we give, in order, sine, cosine, and tangent.

1. $\frac{21}{29}$; $\frac{20}{29}$; $\frac{21}{20}$ **3.** $\frac{n}{p}$; $\frac{m}{p}$; $\frac{n}{m}$ **5.** C **7.** B **9.** E

In Exercises 11–19, we give, in order, the unknown side, sine, cosine, tangent, cotangent, secant, and cosecant.

11. $c = 13$; $\frac{12}{13}$; $\frac{5}{13}$; $\frac{12}{5}$; $\frac{5}{12}$; $\frac{13}{5}$; $\frac{13}{12}$

13. $b = \sqrt{13}$; $\frac{\sqrt{13}}{7}$; $\frac{6}{7}$; $\frac{\sqrt{13}}{6}$; $\frac{6\sqrt{13}}{13}$; $\frac{7}{6}$; $\frac{7\sqrt{13}}{13}$

15. $b = \sqrt{91}$; $\frac{\sqrt{91}}{10}$; $\frac{3}{10}$; $\frac{\sqrt{91}}{3}$; $\frac{3\sqrt{91}}{91}$; $\frac{10}{3}$; $\frac{10\sqrt{91}}{91}$

17. $b = \sqrt{3}$; $\frac{\sqrt{3}}{2}$; $\frac{1}{2}$; $\sqrt{3}$; $\frac{\sqrt{3}}{3}$; 2; $\frac{2\sqrt{3}}{3}$

19. $a = \sqrt{21}$; $\frac{2}{5}$; $\frac{\sqrt{21}}{5}$; $\frac{2\sqrt{21}}{21}$; $\frac{\sqrt{21}}{2}$; $\frac{5\sqrt{21}}{21}$; $\frac{5}{2}$

21. sin 60° **23.** sec 30° **25.** csc 51° **27.** cos 51.3°

29. $\csc(75° - \theta)$ **31.** 40° **33.** 20° **35.** 12° **37.** 35° **39.** 18° **41.** true **43.** false **45.** true **47.** true

49. $\frac{\sqrt{3}}{3}$ **51.** $\frac{1}{2}$ **53.** $\frac{2\sqrt{3}}{3}$ **55.** $\sqrt{2}$ **57.** $\frac{\sqrt{2}}{2}$ **59.** 1

61. $\frac{\sqrt{3}}{2}$ **63.** $\sqrt{3}$

65.

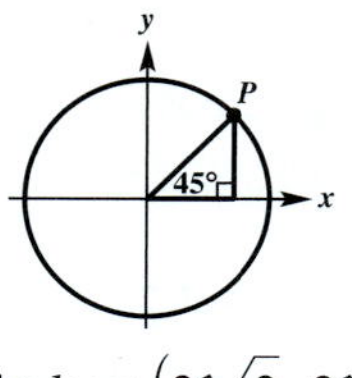

66.

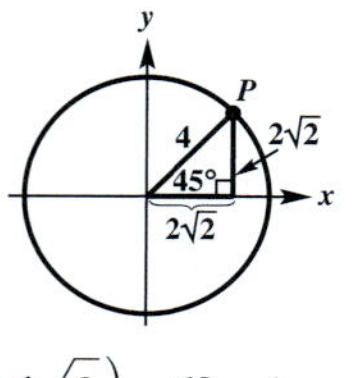

67. the legs; $(2\sqrt{2}, 2\sqrt{2})$ **68.** $(1, \sqrt{3})$ **69.** $\sin x$; $\tan x$

71. 60° **73.** $\left(\frac{\sqrt{2}}{2}, \frac{\sqrt{2}}{2}\right)$; 45° **75.** $y = \frac{\sqrt{3}}{3}x$ **77.** 60°

79. **(a)** 60° **(b)** k **(c)** $k\sqrt{3}$ **(d)** 2; $\sqrt{3}$; 30°; 60°

81. $x = \frac{9\sqrt{3}}{2}$; $y = \frac{9}{2}$; $z = \frac{3\sqrt{3}}{2}$; $w = 3\sqrt{3}$

83. $p = 15$; $r = 15\sqrt{2}$; $q = 5\sqrt{6}$; $t = 10\sqrt{6}$

85. $\mathcal{A} = \frac{s^2}{2}$

2.2 Exercises *(pages 58–61)*

1. C **3.** A **5.** D **11.** $\frac{\sqrt{3}}{3}$; $\sqrt{3}$ **13.** $\frac{\sqrt{3}}{2}$; $\frac{\sqrt{3}}{3}$; $\frac{2\sqrt{3}}{3}$

15. -1; -1 **17.** $-\frac{\sqrt{3}}{2}$; $-\frac{2\sqrt{3}}{3}$

In Exercises 19–35, we give, in order, sine, cosine, tangent, cotangent, secant, and cosecant.

19. $-\frac{\sqrt{3}}{2}$; $\frac{1}{2}$; $-\sqrt{3}$; $-\frac{\sqrt{3}}{3}$; 2; $-\frac{2\sqrt{3}}{3}$

21. $\frac{\sqrt{2}}{2}$; $\frac{\sqrt{2}}{2}$; 1; 1; $\sqrt{2}$; $\sqrt{2}$

23. $\frac{\sqrt{3}}{2}$; $-\frac{1}{2}$; $-\sqrt{3}$; $-\frac{\sqrt{3}}{3}$; -2; $\frac{2\sqrt{3}}{3}$

25. $-\frac{1}{2}$; $-\frac{\sqrt{3}}{2}$; $\frac{\sqrt{3}}{3}$; $\sqrt{3}$; $-\frac{2\sqrt{3}}{3}$; -2

27. $-\frac{\sqrt{2}}{2}$; $-\frac{\sqrt{2}}{2}$; 1; 1; $-\sqrt{2}$; $-\sqrt{2}$

29. $\frac{\sqrt{3}}{2}$; $\frac{1}{2}$; $\sqrt{3}$; $\frac{\sqrt{3}}{3}$; 2; $\frac{2\sqrt{3}}{3}$

31. $-\frac{1}{2}$; $-\frac{\sqrt{3}}{2}$; $\frac{\sqrt{3}}{3}$; $\sqrt{3}$; $-\frac{2\sqrt{3}}{3}$; -2

33. $\frac{1}{2}$; $-\frac{\sqrt{3}}{2}$; $-\frac{\sqrt{3}}{3}$; $-\sqrt{3}$; $-\frac{2\sqrt{3}}{3}$; 2

35. $-\frac{\sqrt{3}}{2}$; $\frac{1}{2}$; $-\sqrt{3}$; $-\frac{\sqrt{3}}{3}$; 2; $-\frac{2\sqrt{3}}{3}$ **37.** $-\frac{\sqrt{2}}{2}$

39. $-\frac{\sqrt{3}}{2}$ **41.** $-\sqrt{2}$ **43.** -1 **45.** 1 **47.** $\frac{23}{4}$

49. $\frac{7}{2}$ **51.** $-\frac{29}{12}$ **53.** false; $0 \neq \frac{\sqrt{3}+1}{2}$

55. false; $\frac{1}{2} \neq \sqrt{3}$ **57.** true **59.** false; $0 \neq \sqrt{2}$

61. $(-3\sqrt{3}, 3)$ **63.** yes **65.** positive **67.** positive

69. negative **75.** 0.9 **77.** 45°; 225° **79.** 30°; 150° **81.** 120°; 300° **83.** 45°; 315° **85.** 210°; 330° **87.** 30°; 210° **89.** 225°; 315°

2.3 Exercises *(pages 63–67)*

1. sin; 1 **3.** reciprocal; reciprocal

In Exercises 5–21, the number of decimal places may vary depending on the calculator used.

5. 0.62524266 **7.** 1.0273488 **9.** 15.055723 **11.** 0.74080460 **13.** 1.4830142

15. $\tan 23.4° \approx 0.43273864$ **17.** $\cot 77° \approx 0.23086819$

19. $\tan 4.72° \approx 0.08256640$ **21.** $\cos 51° \approx 0.62932039$

23. 55.845496° **25.** 16.166641° **27.** 38.491580° **29.** 68.673241° **31.** 45.526434° **33.** 12.227282°

37. 56° **39.** 1 **41.** 1 **43.** 0 **45.** A: 68.94 mph; B: 65.78 mph; **47.** false **49.** true **51.** false **53.** false **55.** true **57.** true **59.** 70 lb **61.** $-2.9°$ **63.** 2500 lb **65.** A 2200-lb car on a 2° uphill grade has greater grade resistance. **67.** **(a)** 703 ft **(b)** 1701 ft **(c)** R would decrease; 644 ft, 1559 ft

69. **(a)** 2×10^8 m per sec **(b)** 2×10^8 m per sec **71.** 48.7° **73.** **(a)** approximately 155 ft **(b)** approximately 194 ft

Chapter 2 Quiz *(pages 67–68)*

[2.1] **1.** $\sin A = \frac{3}{5}$; $\cos A = \frac{4}{5}$; $\tan A = \frac{3}{4}$; $\cot A = \frac{4}{3}$; $\sec A = \frac{5}{4}$; $\csc A = \frac{5}{3}$

2.

θ	$\sin\theta$	$\cos\theta$	$\tan\theta$	$\cot\theta$	$\sec\theta$	$\csc\theta$
30°	$\frac{1}{2}$	$\frac{\sqrt{3}}{2}$	$\frac{\sqrt{3}}{3}$	$\sqrt{3}$	$\frac{2\sqrt{3}}{3}$	2
45°	$\frac{\sqrt{2}}{2}$	$\frac{\sqrt{2}}{2}$	1	1	$\sqrt{2}$	$\sqrt{2}$
60°	$\frac{\sqrt{3}}{2}$	$\frac{1}{2}$	$\sqrt{3}$	$\frac{\sqrt{3}}{3}$	2	$\frac{2\sqrt{3}}{3}$

3. $w = 18$; $x = 18\sqrt{3}$; $y = 18$; $z = 18\sqrt{2}$

4. $\mathcal{A} = 3x^2 \sin\theta$

[2.2] **In Exercises 5–7, we give, in order, sine, cosine, tangent, cotangent, secant, and cosecant.**

5. $\frac{\sqrt{2}}{2}$; $-\frac{\sqrt{2}}{2}$; -1; -1; $-\sqrt{2}$; $\sqrt{2}$

6. $-\frac{1}{2}$; $-\frac{\sqrt{3}}{2}$; $\frac{\sqrt{3}}{3}$; $\sqrt{3}$; $-\frac{2\sqrt{3}}{3}$; -2

7. $-\frac{\sqrt{3}}{2}$; $\frac{1}{2}$; $-\sqrt{3}$; $-\frac{\sqrt{3}}{3}$; 2; $-\frac{2\sqrt{3}}{3}$

8. 60°; 120° **9.** 135°; 225° [2.3] **10.** 0.67301251 **11.** -1.1817633 **12.** 69.497888° **13.** 24.777233°

[2.1–2.3] **14.** false **15.** true

2.4 Exercises *(pages 72–76)*

1. 22,894.5 to 22,895.5 **3.** 8958.5 to 8959.5 **7.** 0.05

Note to student: While most of the measures resulting from solving triangles in this chapter are approximations, for convenience we use = rather than ≈.

9. $B = 53°40'$; $a = 571$ m; $b = 777$ m

11. $M = 38.8°$; $n = 154$ m; $p = 198$ m

13. $A = 47.9108°$; $c = 84.816$ cm; $a = 62.942$ cm
15. $A = 37° 40'$; $B = 52° 20'$; $c = 20.5$ ft
21. $B = 62.0°$; $a = 8.17$ ft; $b = 15.4$ ft
23. $A = 17.0°$; $a = 39.1$ in.; $c = 134$ in.
25. $B = 29.0°$; $a = 70.7$ cm; $c = 80.9$ cm
27. $A = 36°$; $B = 54°$; $b = 18$ m
29. $c = 85.9$ yd; $A = 62° 50'$; $B = 27° 10'$
31. $b = 42.3$ cm; $A = 24° 10'$; $B = 65° 50'$
33. $B = 36° 36'$; $a = 310.8$ ft; $b = 230.8$ ft
35. $A = 50° 51'$; $a = 0.4832$ m; $b = 0.3934$ m
41. 9.35 m **43.** 128 ft **45.** 26.92 in. **47.** 22°
49. 28.0 m **51.** 13.3 ft **53.** 37° 35′ **55.** 42.18°
57. **(a)** 29,000 ft **(b)** shorter

2.5 Exercises *(pages 81–85)*

1. It should be shown as an angle measured clockwise from due north. **3.** A sketch is important to show the relationships among the given data and the unknowns.
5. 270°; N 90° W, or S 90° W **7.** 0°; N 0° E, or N 0° W **9.** 315°; N 45° W **11.** 135°; S 45° E
13. $y = \frac{\sqrt{3}}{3}x,\ x \le 0$ **15.** 220 mi **17.** 47 nautical mi
19. 2203 ft **21.** 148 mi **23.** 430 mi **25.** 140 mi
27. $x = \frac{b}{a - c}$ **29.** $y = (\tan 35°)(x - 25)$
31. 433 ft **33.** 114 ft **35.** 5.18 m
37. **(a)** $d = \frac{b}{2}\left(\cot \frac{\alpha}{2} + \cot \frac{\beta}{2}\right)$ **(b)** 345.4 cm **39.** 10.8 ft
41. **(a)** 320 ft **(b)** $R\left(1 - \cos \frac{\theta}{2}\right)$
43. **(a)** 23 ft **(b)** 48 ft **(c)** As the speed limit increases, more land needs to be cleared inside the curve.

Chapter 2 Review Exercises *(pages 88–91)*

In Exercises 1, 13, and 15, we give, in order, sine, cosine, tangent, cotangent, secant, and cosecant.

1. $\frac{60}{61}$; $\frac{11}{61}$; $\frac{60}{11}$; $\frac{11}{60}$; $\frac{61}{11}$; $\frac{61}{60}$ **3.** 10° **5.** 7° **7.** true **9.** true
13. $-\frac{\sqrt{3}}{2}$; $\frac{1}{2}$; $-\sqrt{3}$; $-\frac{\sqrt{3}}{3}$; 2; $-\frac{2\sqrt{3}}{3}$
15. $-\frac{1}{2}$; $\frac{\sqrt{3}}{2}$; $-\frac{\sqrt{3}}{3}$; $-\sqrt{3}$; $\frac{2\sqrt{3}}{3}$; -2 **17.** 120°; 240°
19. 150°; 210° **21.** $3 - \frac{2\sqrt{3}}{3}$ **23.** $\frac{7}{2}$ **25.** -1.3563417
27. 1.0210339 **29.** 0.20834446 **31.** 55.673870°
33. 12.733938° **35.** 63.008286° **37.** 47.1°; 132.9°
39. false; $1.4088321 \neq 1$ **41.** true **45.** III
47. II **49.** $B = 31° 30'$; $a = 638$; $b = 391$
51. $B = 50.28°$; $a = 32.38$ m; $c = 50.66$ m **53.** 137 ft
55. 73.7 ft **57.** 18.75 cm **59.** 1200 m **61.** 140 mi
65. **(a)** 716 mi **(b)** 1104 mi

Chapter 2 Test *(page 92)*

[2.1] **1.** $\sin A = \frac{12}{13}$; $\cos A = \frac{5}{13}$; $\tan A = \frac{12}{5}$; $\cot A = \frac{5}{12}$; $\sec A = \frac{13}{5}$; $\csc A = \frac{13}{12}$
2. $x = 4$; $y = 4\sqrt{3}$; $z = 4\sqrt{2}$; $w = 8$ **3.** 15°
[2.1, 2.2] **4.** **(a)** true **(b)** false; For $0° \le \theta \le 90°$, as the angle increases, $\cos \theta$ decreases. **(c)** true

In Exercises 5–7, we give, in order, sine, cosine, tangent, cotangent, secant, and cosecant.

[2.2] **5.** $-\frac{\sqrt{3}}{2}$; $-\frac{1}{2}$; $\sqrt{3}$; $\frac{\sqrt{3}}{3}$; -2; $-\frac{2\sqrt{3}}{3}$
6. $-\frac{\sqrt{2}}{2}$; $-\frac{\sqrt{2}}{2}$; 1; 1; $-\sqrt{2}$; $-\sqrt{2}$
7. -1; 0; undefined; 0; undefined; -1 **8.** 135°; 225°
9. 240°; 300° **10.** 45°; 225° [2.3] **11.** Take the reciprocal of $\tan \theta$ to get $\cot \theta = 0.59600119$.
12. **(a)** 0.97939940 **(b)** -1.9056082 **(c)** 1.9362132
13. 16.166641° [2.4] **14.** $B = 31° 30'$; $c = 877$; $b = 458$ **15.** 67.1°, or 67° 10′ **16.** 15.5 ft **17.** 8800 ft
[2.5] **18.** 72 nautical mi **19.** 92 km **20.** 448 m

Chapter 3 Radian Measure and the Unit Circle

3.1 Exercises *(pages 98–100)*

1. 1 **3.** 3 **5.** -3 **7.** $\frac{\pi}{3}$ **9.** $\frac{\pi}{2}$ **11.** $\frac{5\pi}{6}$ **13.** $-\frac{5\pi}{3}$
15. $\frac{5\pi}{2}$ **17.** 10π **19.** 0 **21.** -5π **29.** 60°
31. 315° **33.** 330° **35.** $-30°$ **37.** 126° **39.** $-48°$
41. 153° **43.** $-900°$ **45.** 0.68 **47.** 0.742 **49.** 2.43
51. 1.122 **53.** 0.9847 **55.** -0.832391 **57.** 114° 35′
59. 99° 42′ **61.** 19° 35′ **63.** $-287° 06'$ **65.** In the expression “sin 30,” 30 means 30 radians; $\sin 30° = \frac{1}{2}$, while $\sin 30 \approx -0.9880$. **67.** $\frac{\sqrt{3}}{2}$ **69.** 1 **71.** $\frac{2\sqrt{3}}{3}$ **73.** 1
75. $-\sqrt{3}$ **77.** $\frac{1}{2}$ **79.** -1 **81.** $-\frac{\sqrt{3}}{2}$ **83.** $\frac{1}{2}$
85. $\sqrt{3}$ **87.** We begin the answers with the blank next to 30°, and then proceed counterclockwise from there: $\frac{\pi}{6}$; 45; $\frac{\pi}{3}$; 120; 135; $\frac{5\pi}{6}$; π; $\frac{7\pi}{6}$; $\frac{5\pi}{4}$; 240; 300; $\frac{7\pi}{4}$; $\frac{11\pi}{6}$.
89. **(a)** 4π **(b)** $\frac{2\pi}{3}$ **91.** **(a)** 5π **(b)** $\frac{8\pi}{3}$ **93.** 24π

3.2 Exercises *(pages 103–109)*

1. 2π **3.** 20π **5.** 6 **7.** 1 **9.** 2 **11.** 25.8 cm
13. 3.61 ft **15.** 5.05 m **17.** 55.3 in. **19.** The length is doubled. **21.** 3500 km **23.** 5900 km **25.** 44° N
27. 156° **29.** 38.5° **31.** 18.7 cm **33.** **(a)** 11.6 in. **(b)** 37° 05′ **35.** 146 in. **37.** 3π in. **39.** 27π in.
41. 0.20 km **43.** 6π **45.** 72π **47.** 60° **49.** 1.5
51. 1116.1 m^2 **53.** 706.9 ft^2 **55.** 114.0 cm^2
57. 1885.0 mi^2 **59.** 3.6 **61.** approximately 8060 yd^2
63. 20 in. **65.** **(a)** $13\frac{1}{3}°$; $\frac{2\pi}{27}$ **(b)** 478 ft **(c)** 17.7 ft **(d)** approximately 672 ft^2 **67.** **(a)** 140 ft **(b)** 102 ft **(c)** 622 ft^2 **69.** 1900 yd^2 **71.** radius: 3950 mi; circumference: 24,800 mi **73.** The area is quadrupled.
75. $V = \frac{r^2\theta h}{2}$ (θ in radians) **77.** $r = \frac{L}{\theta}$ **78.** $h = r\cos\frac{\theta}{2}$
79. $d = r\left(1 - \cos\frac{\theta}{2}\right)$ **80.** $d = \frac{L}{\theta}\left(1 - \cos\frac{\theta}{2}\right)$

3.3 Exercises *(pages 117–120)*

1. (a) 1 **(b)** 0 **(c)** undefined **3. (a)** 0 **(b)** 1 **(c)** 0 **5. (a)** 0 **(b)** -1 **(c)** 0 **7.** $-\frac{1}{2}$ **9.** -1 **11.** -2 **13.** $-\frac{1}{2}$ **15.** $\frac{\sqrt{2}}{2}$ **17.** $\frac{\sqrt{3}}{2}$ **19.** $\frac{2\sqrt{3}}{3}$ **21.** $-\frac{\sqrt{3}}{3}$ **23.** 0.5736 **25.** 0.4068 **27.** 1.2065 **29.** 14.3338 **31.** -1.0460 **33.** -3.8665 **35.** 0.7 **37.** 0.9 **39.** -0.6 **41.** 2.3 or 4.0 **43.** 0.8 or 2.4 **45.** negative **47.** negative **49.** positive **51.** $\sin\theta = \frac{\sqrt{2}}{2}$; $\cos\theta = \frac{\sqrt{2}}{2}$; $\tan\theta = 1$; $\cot\theta = 1$; $\sec\theta = \sqrt{2}$; $\csc\theta = \sqrt{2}$ **53.** $\sin\theta = -\frac{12}{13}$; $\cos\theta = \frac{5}{13}$; $\tan\theta = -\frac{12}{5}$; $\cot\theta = -\frac{5}{12}$; $\sec\theta = \frac{13}{5}$; $\csc\theta = -\frac{13}{12}$ **55.** 0.2095 **57.** 1.4426 **59.** 0.3887 **61.** $\frac{5\pi}{6}$ **63.** $\frac{4\pi}{3}$ **65.** $\frac{7\pi}{4}$ **67.** $\frac{4\pi}{3}, \frac{5\pi}{3}$ **69.** $\frac{\pi}{4}, \frac{3\pi}{4}, \frac{5\pi}{4}, \frac{7\pi}{4}$ **71.** $-\frac{11\pi}{6}, -\frac{7\pi}{6}, -\frac{5\pi}{6}, -\frac{\pi}{6}, \frac{\pi}{6}, \frac{5\pi}{6}$ **73.** $(-0.8011, 0.5985)$ **75.** $(0.4385, -0.8987)$ **77.** I **79.** II **81.** 0.9846 **83. (a)** 32.4° **85. (a)** 30° **(b)** 60° **(c)** 75° **(d)** 86° **(e)** 86° **(f)** 60° **87. (a)** $\frac{1}{2}$ **(b)** $\frac{\sqrt{3}}{2}$ **(c)** $\sqrt{3}$ **(d)** 2 **(e)** $\frac{2\sqrt{3}}{3}$ **(f)** $\frac{\sqrt{3}}{3}$

Chapter 3 Quiz *(page 120)*

[3.1] **1.** $\frac{5\pi}{4}$ **2.** $-\frac{11\pi}{6}$ **3.** 300° **4.** $-210°$ [3.2] **5.** 1.5 **6.** 67,500 in.^{2} [3.3] **7.** $\frac{\sqrt{2}}{2}$ **8.** $-\frac{1}{2}$ **9.** 0 **10.** $\frac{2\pi}{3}$

3.4 Exercises *(pages 123–126)*

1. 2π sec **3. (a)** $\frac{\pi}{2}$ radians **(b)** 10π cm **(c)** $\frac{5\pi}{3}$ cm per sec **5.** 2π radians **7.** $\frac{3\pi}{32}$ radian per sec **9.** $\frac{6}{5}$ min **11.** 0.1803 radian per sec **13.** 10.77 radians **15.** 8π m per sec **17.** $\frac{9}{5}$ radians per sec **19.** 1.834 radians per sec **21.** 18π cm **23.** 12 sec **25.** $\frac{3\pi}{32}$ radian per sec **27.** $\frac{\pi}{6}$ radian per hr **29.** $\frac{\pi}{30}$ radian per min **31.** $\frac{7\pi}{30}$ cm per min **33.** 168π m per min **35.** 1500π m per min **37.** 16.6 mph **39. (a)** $\frac{2\pi}{365}$ radian **(b)** $\frac{\pi}{4380}$ radian per hr **(c)** about 67,000 mph **41. (a)** 3.1 cm per sec **(b)** 0.24 radian per sec **43.** 3.73 cm **45.** 523.6 radians per sec

Chapter 3 Review Exercises *(pages 128–131)*

1. A central angle of a circle that intercepts an arc of length 2 times the radius of the circle has a measure of 2 radians. **3.** Three of many possible answers are $1 + 2\pi$, $1 + 4\pi$, and $1 + 6\pi$. **5.** $\frac{\pi}{4}$ **7.** $\frac{35\pi}{36}$ **9.** $\frac{40\pi}{9}$ **11.** 225° **13.** 480° **15.** $-110°$ **17.** π in. **19.** 12π in. **21.** 35.8 cm **23.** 49.06° **25.** 273 m^2 **27.** 4500 km **29.** $\frac{3}{4}$; 1.5 sq units **31. (a)** $\frac{\pi}{3}$ radians **(b)** 2π in. **33.** $\sqrt{3}$ **35.** $-\frac{1}{2}$ **37.** 2 **39.** tan 1 **41.** sin 2 **43.** 0.8660 **45.** 0.9703 **47.** 1.9513 **49.** 0.3898 **51.** 0.5148 **53.** 1.1054 **55.** $\frac{\pi}{4}$ **57.** $\frac{7\pi}{6}$ **59.** $\frac{15}{32}$ sec **61.** $\frac{\pi}{20}$ radian per sec **63.** 1260π cm per sec **65.** 5 in.

Chapter 3 Test *(pages 131–132)*

[3.1] **1.** $\frac{2\pi}{3}$ **2.** $-\frac{\pi}{4}$ **3.** 0.09 **4.** 135° **5.** $-210°$ **6.** 229.18° [3.2] **7. (a)** $\frac{4}{3}$ **(b)** 15,000 cm^2 **8.** 2 radians [3.3] **9.** $\frac{\sqrt{2}}{2}$ **10.** $-\frac{\sqrt{3}}{2}$ **11.** undefined **12.** -2 **13.** 0 **14.** 0 **15.** $\sin\frac{7\pi}{6} = -\frac{1}{2}$; $\cos\frac{7\pi}{6} = -\frac{\sqrt{3}}{2}$; $\tan\frac{7\pi}{6} = \frac{\sqrt{3}}{3}$; $\csc\frac{7\pi}{6} = -2$; $\sec\frac{7\pi}{6} = -\frac{2\sqrt{3}}{3}$; $\cot\frac{7\pi}{6} = \sqrt{3}$ **16.** sine and cosine: $(-\infty, \infty)$; tangent and secant: $\{s \mid s \neq (2n+1)\frac{\pi}{2}$, where n is any integer$\}$; cotangent and cosecant: $\{s \mid s \neq n\pi$, where n is any integer$\}$ **17. (a)** 0.9716 **(b)** $\frac{\pi}{3}$ [3.4] **18. (a)** $\frac{2\pi}{3}$ radians **(b)** 40π cm **(c)** 5π cm per sec **19.** approximately 8.127 mi per sec **20. (a)** 75 ft **(b)** $\frac{\pi}{45}$ radian per sec

Chapter 4 Graphs of the Circular Functions

4.1 Exercises *(pages 143–148)*

1. G **3.** E **5.** B **7.** F **9.** D **11.** C

13. 2

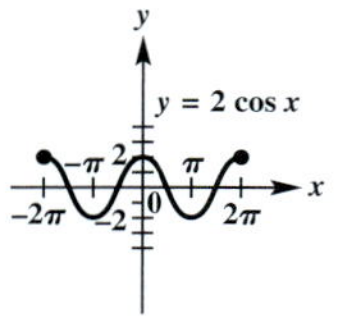

15. $\frac{2}{3}$

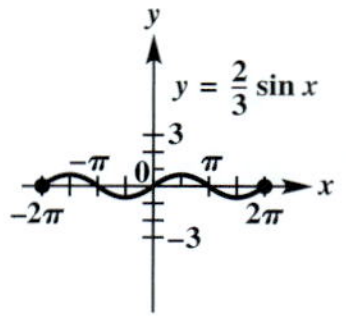

17. 1

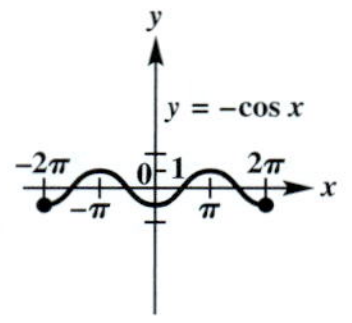

19. 2

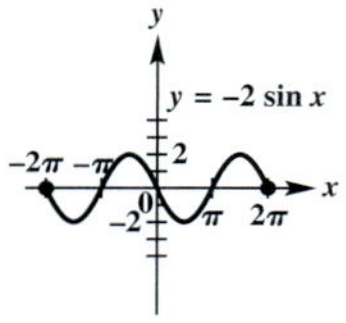

21. 1

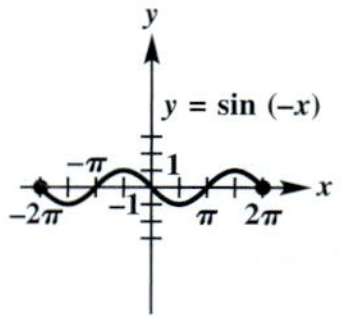

23. 4π; 1

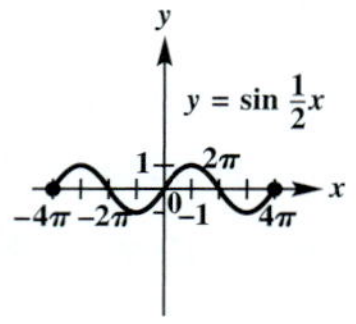

25. $\frac{8\pi}{3}$; 1

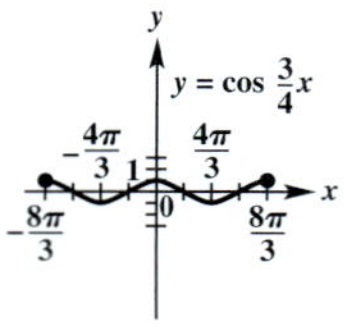

27. $\frac{2\pi}{3}$; 1

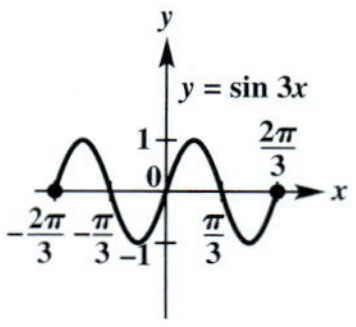

29. 8π; 2

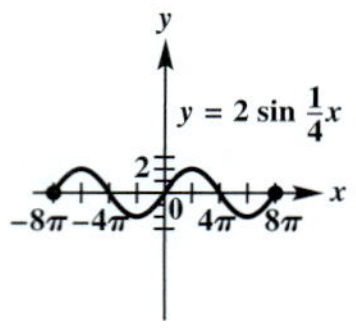

31. $\frac{2\pi}{3}$; 2

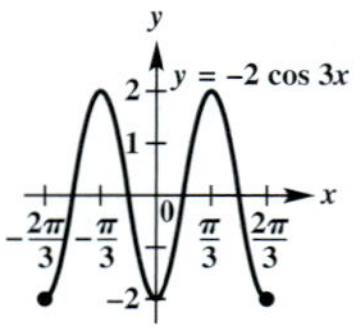

33. 2; 1

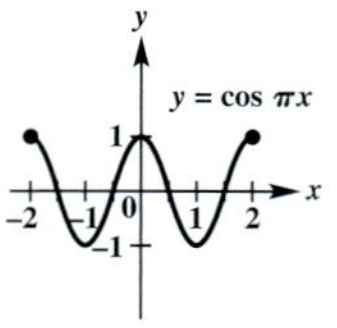

35. 1; 2

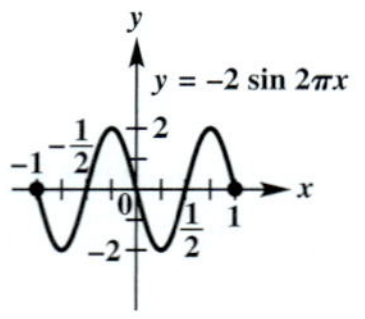

37. 4; $\frac{1}{2}$

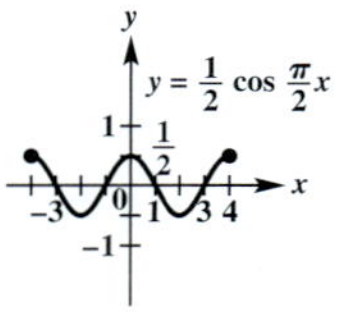

39. 2; π

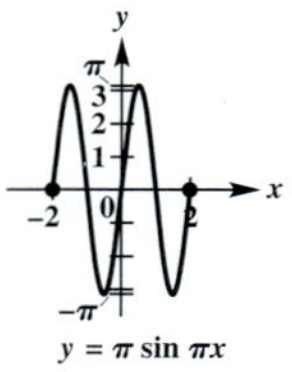

41. $y = 2\cos 2x$ **43.** $y = -3\cos\frac{1}{2}x$ **45.** $y = 3\sin 4x$ **47. (a)** 80°F; 50°F **(b)** 15 **(c)** about 35,000 yr **(d)** downward **49.** 24 hr **51.** approximately 6:00 P.M.; approximately 0.2 ft **53.** approximately 3:18 A.M.; approximately 2.4 ft **55. (a)** 5; $\frac{1}{60}$ **(b)** 60 **(c)** 5; 1.545; −4.045; −4.045; 1.545 **(d)**

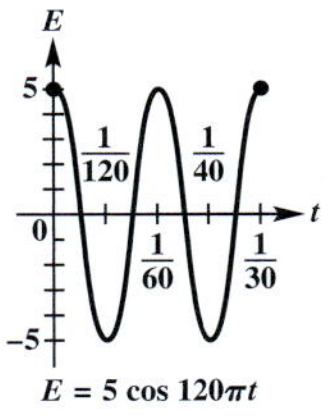

57. (a) $L(x) = 0.022x^2 + 0.55x + 316 + 3.5\sin 2\pi x$

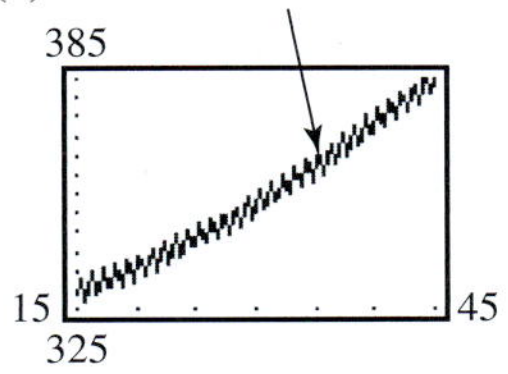

(b) maxima: $x = \frac{1}{4}, \frac{5}{4}, \frac{9}{4}, \ldots$; minima: $x = \frac{3}{4}, \frac{7}{4}, \frac{11}{4}, \ldots$ **59. (a)** 31°F **(b)** 38°F **(c)** 57°F **(d)** 58°F **(e)** 37°F **(f)** 16°F **61.** 1; 240°, or $\frac{4\pi}{3}$ **65.** X = −0.4161468, Y = 0.90929743; X is cos 2 and Y is sin 2. **66.** X = 2, Y = 0.90929743; sin 2 = 0.90929743 **67.** X = 2, Y = −0.4161468; cos 2 = −0.4161468

4.2 Exercises *(pages 155–158)*

1. D **3.** H **5.** B **7.** I **9.** C **11.** A **15.** B **17.** C **19.** right **21.** $y = -1 + \sin x$ **23.** $y = \cos\left(x - \frac{\pi}{3}\right)$ **25.** 2; 2π; none; π to the left **27.** $\frac{1}{4}$; 4π; none; π to the left **29.** 3; 4; none; $\frac{1}{2}$ to the right **31.** 1; $\frac{2\pi}{3}$; up 2; $\frac{\pi}{15}$ to the right

33.

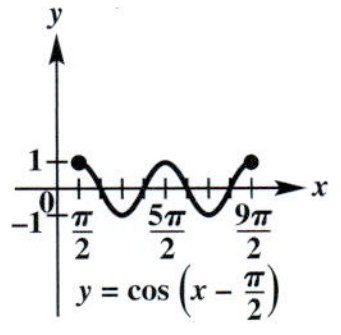

35.

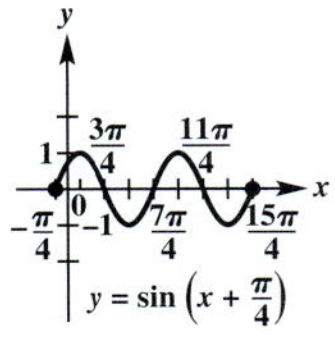

37.

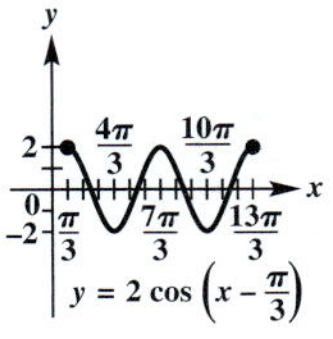

39.

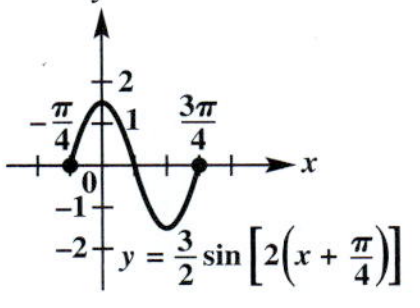

41.

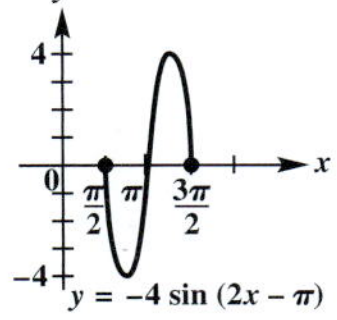

43.

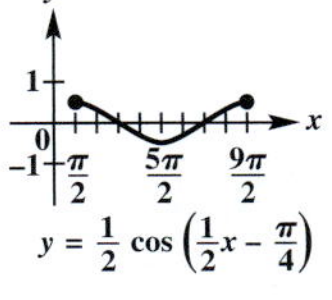

45.

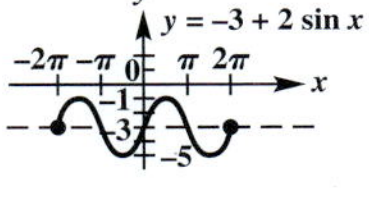

47.

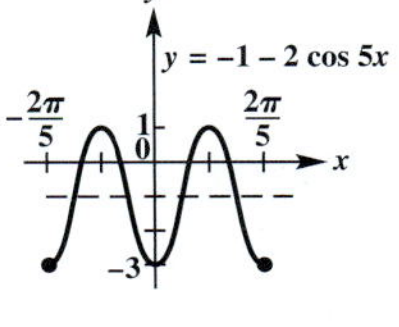

49.

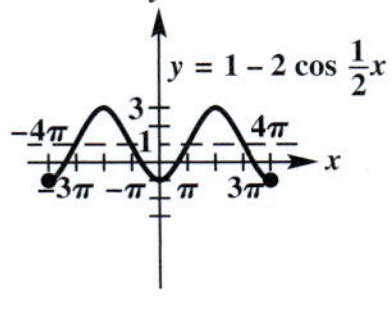

51.

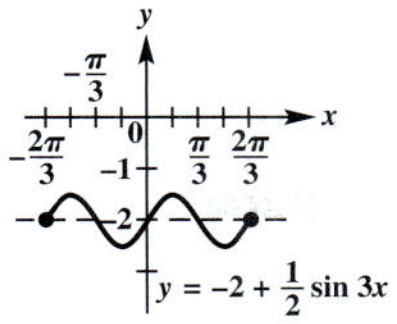

53.

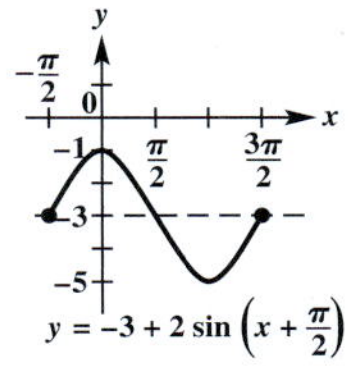

55.

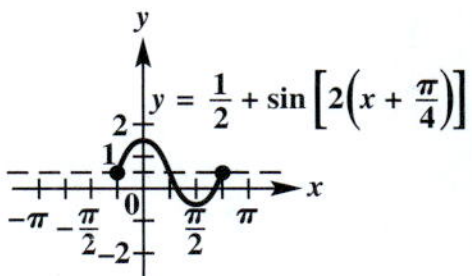

57. (a) yes

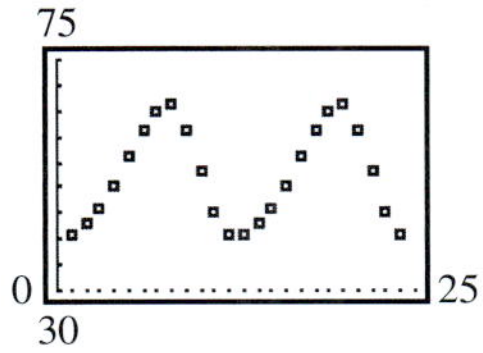

(b) It represents the average yearly temperature.

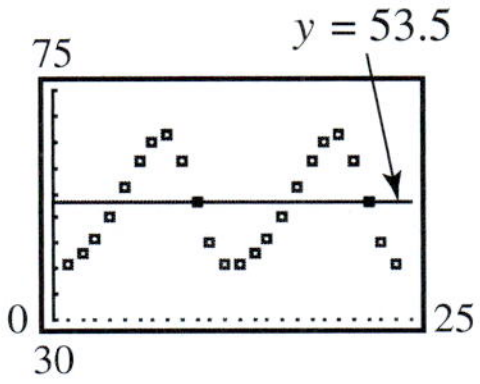

(c) 12.5; 12; 4.5 **(d)** $f(x) = 12.5\sin\left[\frac{\pi}{6}(x - 4.5)\right] + 53.5$

(e) The function gives a good model for the given data.

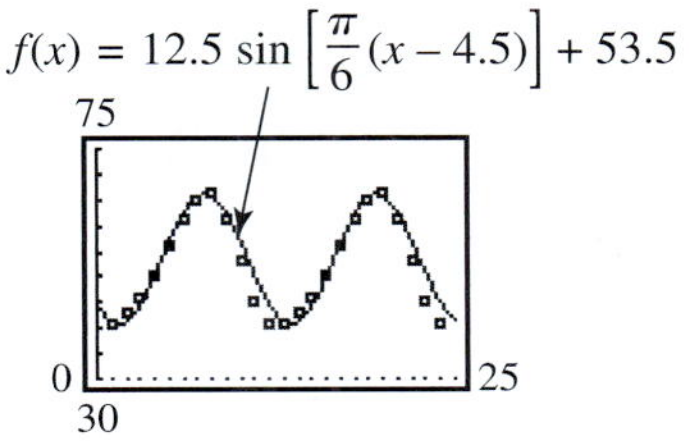

(f)

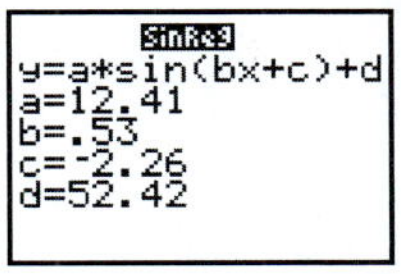

TI-83/84 Plus fixed to the nearest hundredth

59.

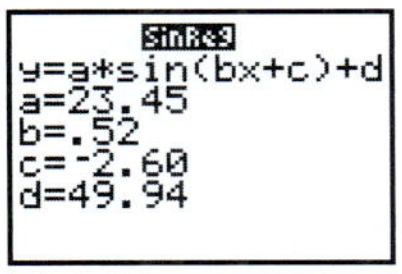

TI-83/84 Plus fixed to the nearest hundredth

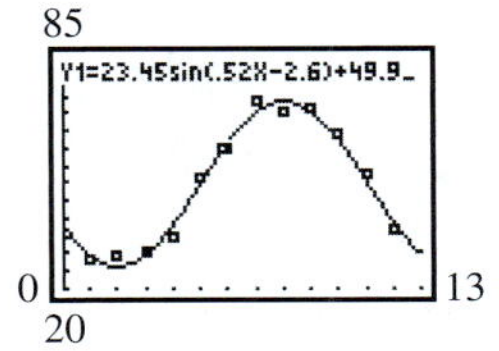

Chapter 4 Quiz *(page 159)*

[4.1, 4.2] **1.** 4; π; 3 up; $\frac{\pi}{4}$ to the left

[4.1] **2.** 2π; 4

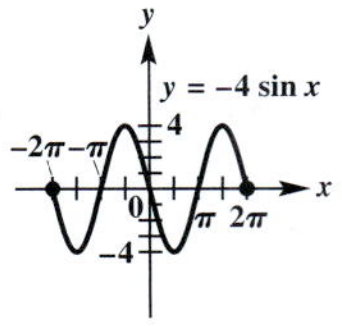

3. π; $\frac{1}{2}$

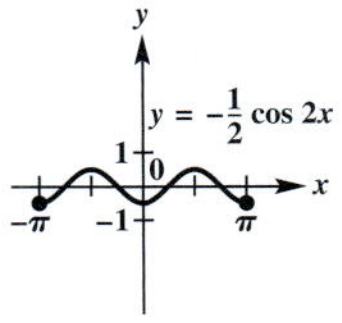

4. 2; 3

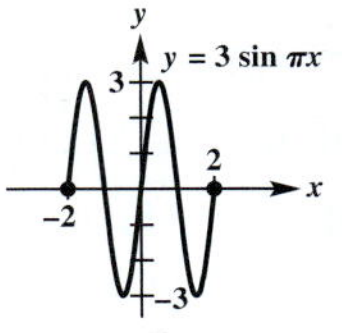

[4.2] **5.** 2π; 2

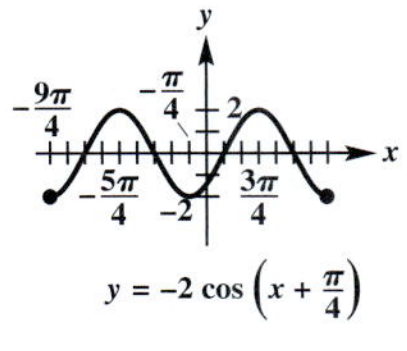

6. π; 1

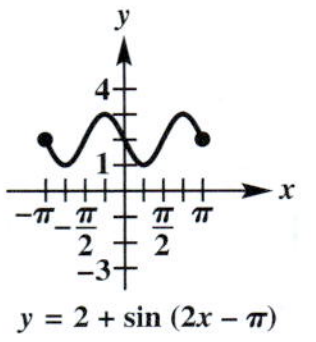

7. 2π; $\frac{1}{2}$

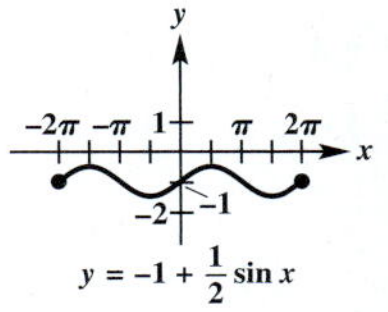

[4.1] **8.** $y = 2\sin x$ **9.** $y = \cos 2x$ **10.** $y = -\sin x$

[4.1, 4.2] **11.** 73°F **12.** 60°F; 84°F

4.3 Exercises (pages 166–168)

1. C **3.** B **5.** F

7.
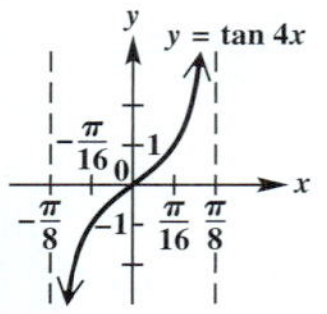

9.
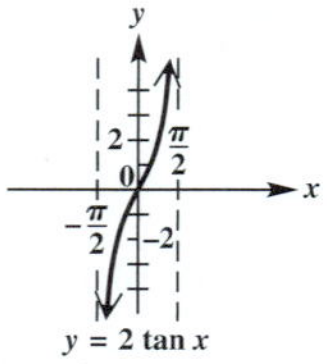

11.
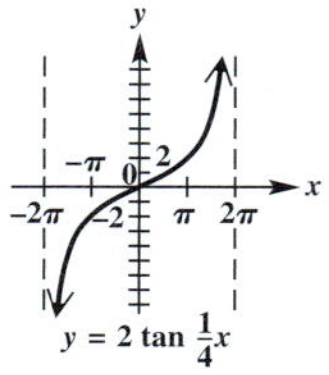

13.
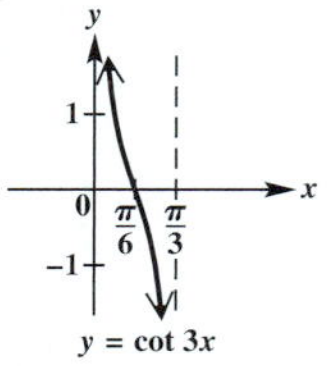

15.
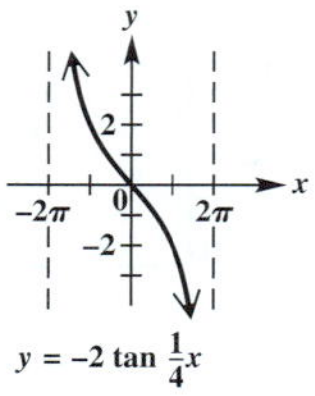

17.
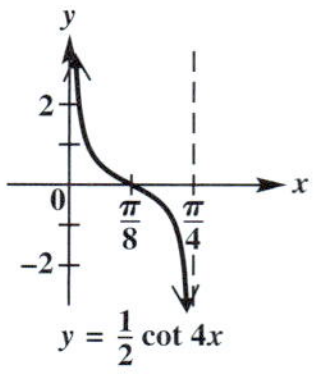

19.
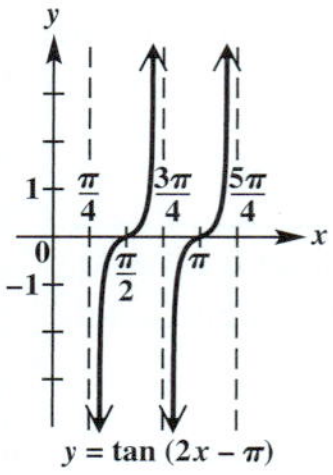

21.
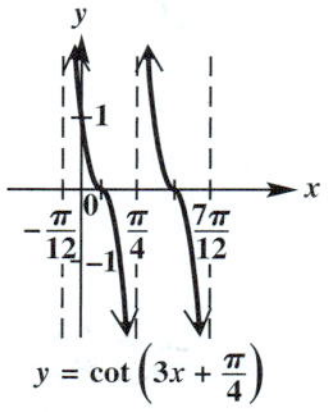

23.
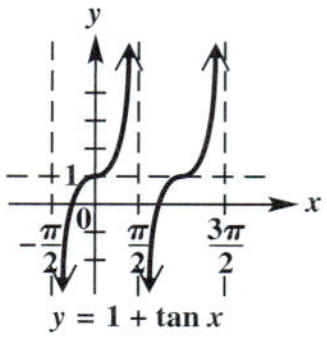

25.
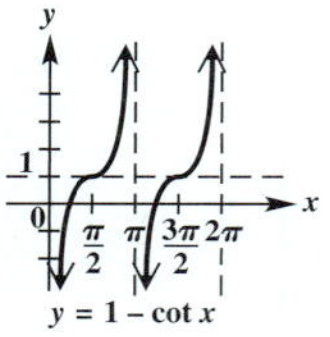

27.
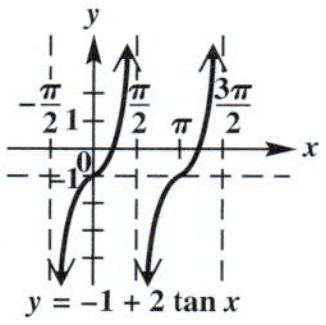

29.
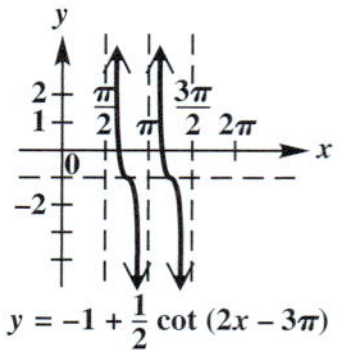

31.
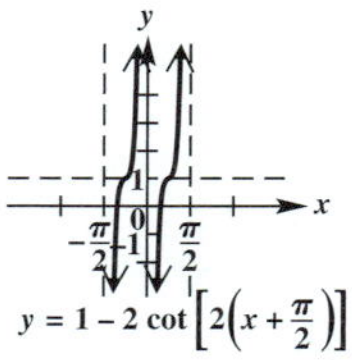

33. $y = -2 \tan x$ **35.** $y = \cot 3x$ **37.** $y = 1 + \tan \frac{1}{2}x$ **39.** true **41.** false; $\tan(-x) = -\tan x$ for all x in the domain. **43.** four

47. **(a)** 0 m **(b)** −2.9 m **(c)** −12.3 m **(d)** 12.3 m **(e)** It leads to $\tan \frac{\pi}{2}$, which is undefined. **49.** π **50.** $\frac{5\pi}{4}$ **51.** $x = \frac{5\pi}{4} + n\pi$ **52.** approximately 0.3217505544 **53.** approximately 3.463343208 **54.** $\{x \mid x = 0.3217505544 + n\pi\}$

4.4 Exercises (pages 174–176)

1. B **3.** D

5.
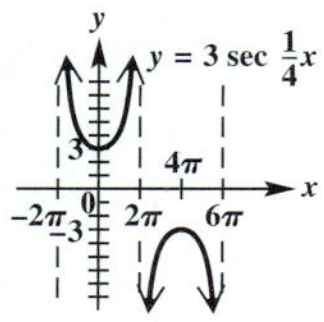

7.
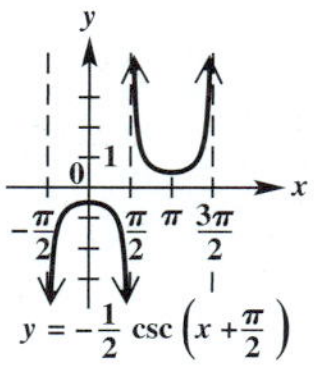

9.
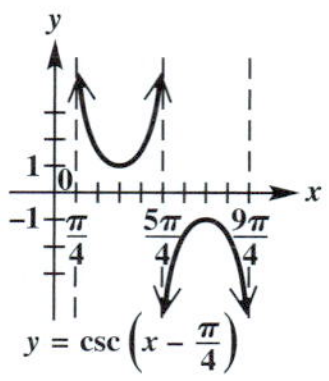

11.
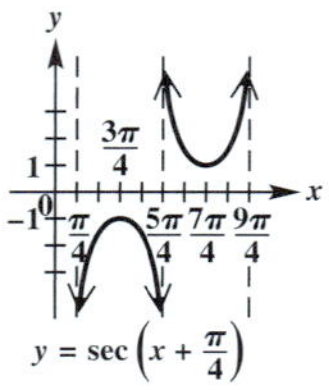

13.
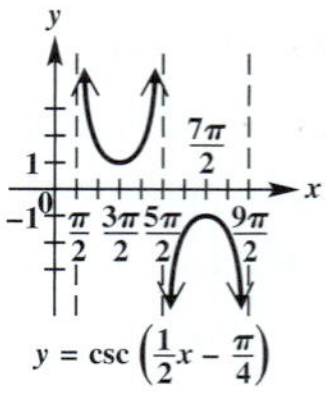

15.
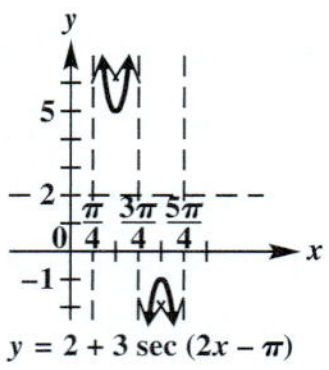

17.
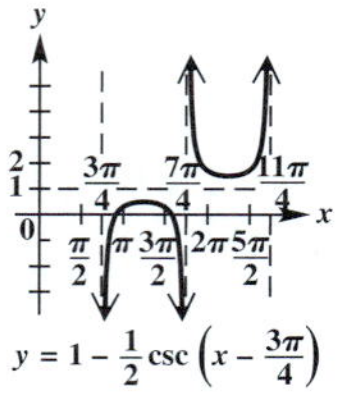

19. $y = \sec 4x$ **21.** $y = -2 + \csc x$ **23.** $y = -1 - \sec x$ **25.** true **27.** true **29.** none **33.** **(a)** 4 m **(b)** 6.3 m **(c)** 63.7 m

35. The display is for $Y_1 + Y_2$ at $X = \frac{\pi}{6}$.

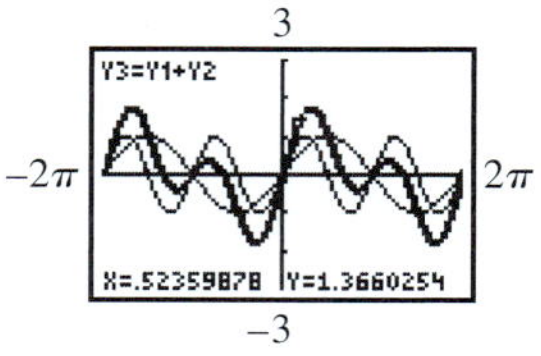

Summary Exercises on Graphing Circular Functions (page 176)

1.
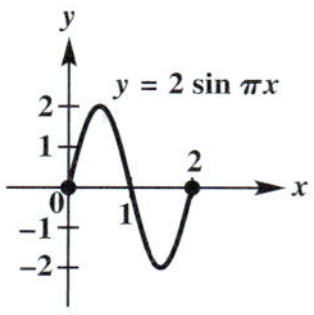

2.
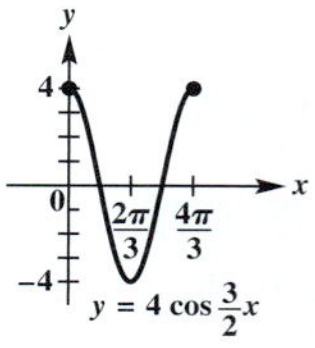

3.
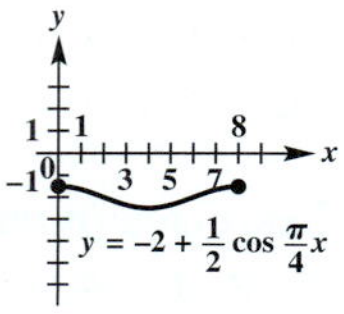

4.
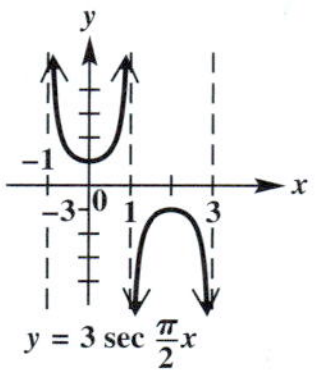

5.
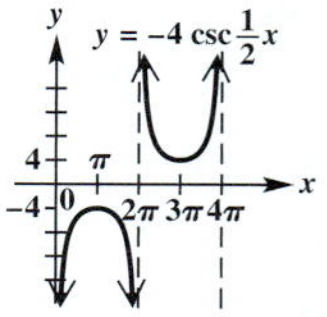

6.
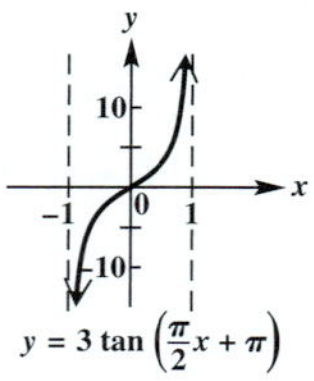

7.
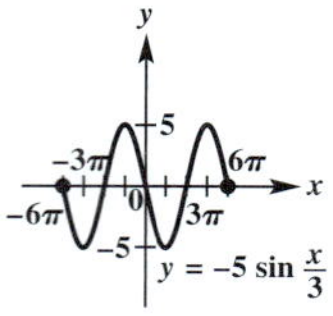

8.
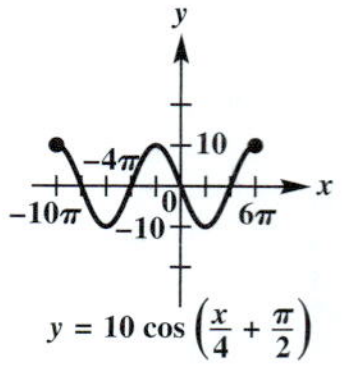

9.
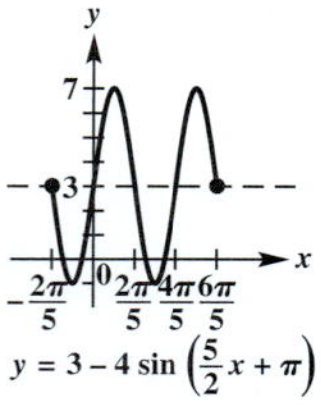

10.
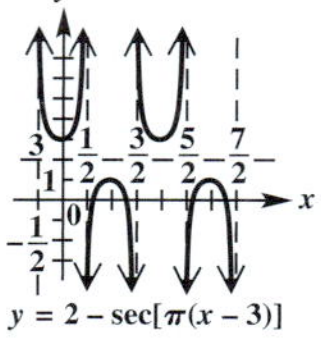

4.5 Exercises (pages 179–180)

1. **(a)** $s(t) = 2 \cos 4\pi t$ **(b)** $s(1) = 2$; The weight is moving neither upward nor downward. At $t = 1$, the motion of the weight is changing from up to down.

3. **(a)** $s(t) = -3 \cos 2.5\pi t$ **(b)** $s(1) = 0$; upward

5. $s(t) = 0.21 \cos 55\pi t$ **7.** $s(t) = 0.14 \cos 110\pi t$

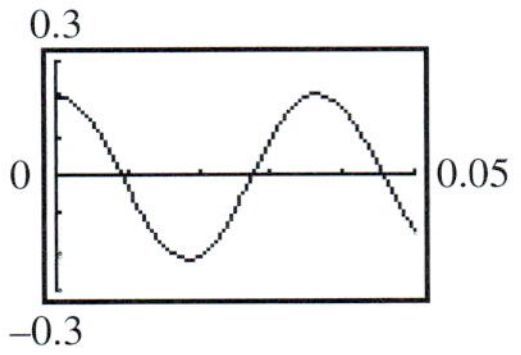

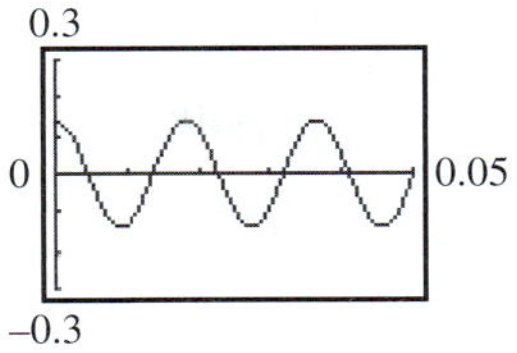

9. **(a)** $s(t) = -4 \cos \frac{2\pi}{3} t$ **(b)** 3.46 units **(c)** $\frac{1}{3}$ oscillation per sec **11.** **(a)** $s(t) = 2 \sin 2t$; amplitude: 2; period: π; frequency: $\frac{1}{\pi}$ rotation per sec **(b)** $s(t) = 2 \sin 4t$; amplitude: 2; period: $\frac{\pi}{2}$; frequency: $\frac{2}{\pi}$ rotation per sec **13.** period: $\frac{\pi}{4}$; frequency: $\frac{4}{\pi}$ oscillations per sec **15.** $\frac{1}{\pi^2}$ **17.** **(a)** 5 in. **(b)** 2 cycles per sec; $\frac{1}{2}$ sec **(c)** after $\frac{1}{4}$ sec **(d)** approximately 4; After 1.3 sec, the weight is about 4 in. above the equilibrium position.

19. **(a)** $s(t) = -3 \cos 12t$ **(b)** $\frac{\pi}{6}$ sec

21. 0; π; They are the same.

Chapter 4 Review Exercises *(pages 183–186)*

1. B **3.** sine, cosine, tangent, cotangent **5.** 2; 2π; none; none **7.** $\frac{1}{2}$; $\frac{2\pi}{3}$; none; none **9.** 2; 8π; 1 up; none **11.** 3; 2π; none; $\frac{\pi}{2}$ to the left **13.** not applicable; π; none; $\frac{\pi}{8}$ to the right **15.** not applicable; $\frac{\pi}{3}$; none; $\frac{\pi}{9}$ to the right **17.** tangent **19.** cosine **21.** cotangent

25.

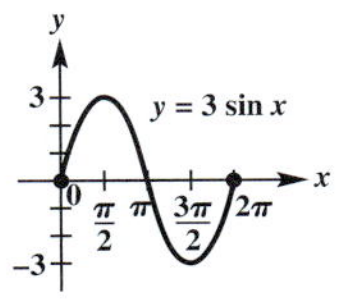

27.

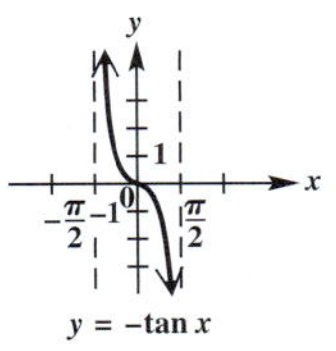

29.

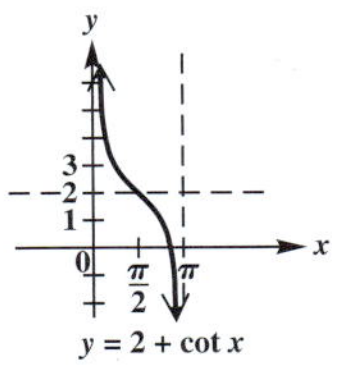

31.

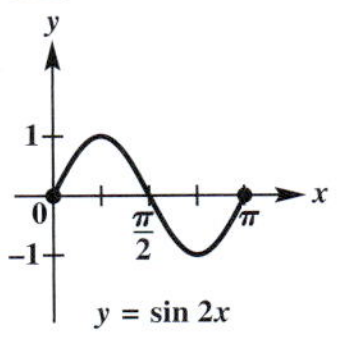

33.

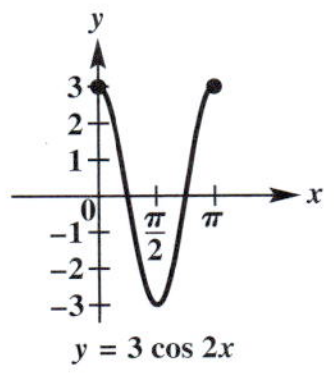

35.

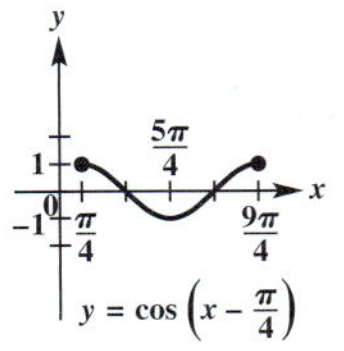

37.

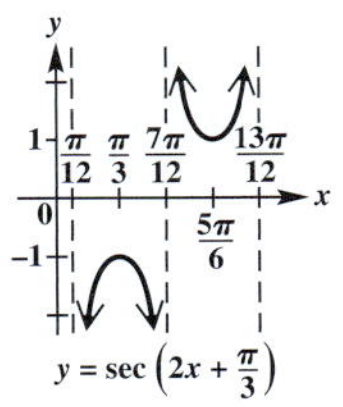

39.

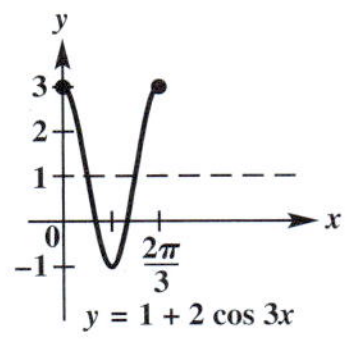

41.

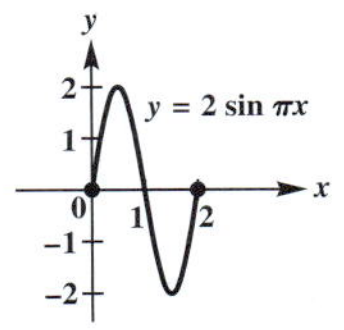

43. $[-2, 2]$ **45.** $y = 1 - \sin x$ **47.** $y = 2 \tan \frac{1}{2}x$

49. **(b)**

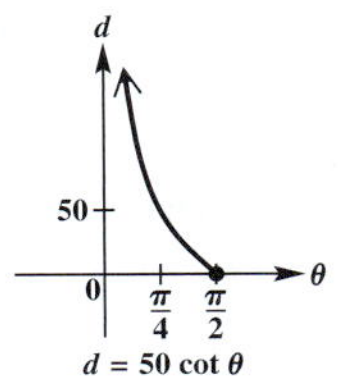

51. **(a)** 30°F **(b)** 60°F **(c)** 75°F **(d)** 86°F **(e)** 86°F **(f)** 60°F

53. **(a)** 100 **(b)** 258 **(c)** 122 **(d)** 296

55. amplitude: 4; period: 2; frequency: $\frac{1}{2}$ cycle per sec

57. The frequency is the number of cycles in one unit of time; -4; 0; $-2\sqrt{2}$

Chapter 4 Test *(pages 187–188)*

[4.1–4.4] **1.** **(a)** $y = \sec x$ **(b)** $y = \sin x$ **(c)** $y = \cos x$ **(d)** $y = \tan x$ **(e)** $y = \csc x$ **(f)** $y = \cot x$

2. **(a)** $y = 1 + \cos \frac{1}{2}x$ **(b)** $y = -\frac{1}{2} \cot x$

[4.1, 4.3, 4.4] **3.** **(a)** $(-\infty, \infty)$ **(b)** $[-1, 1]$ **(c)** $\frac{\pi}{2}$ **(d)** $(-\infty, -1] \cup [1, \infty)$ [4.2] **4.** **(a)** π **(b)** 6 **(c)** $[-3, 9]$ **(d)** -3 **(e)** $\frac{\pi}{4}$ to the left (that is, $-\frac{\pi}{4}$)

5.

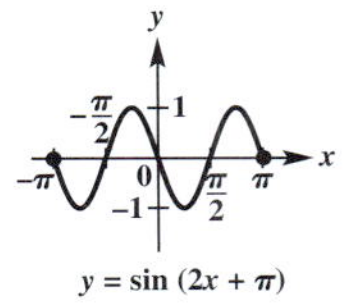

[4.1] **6.**

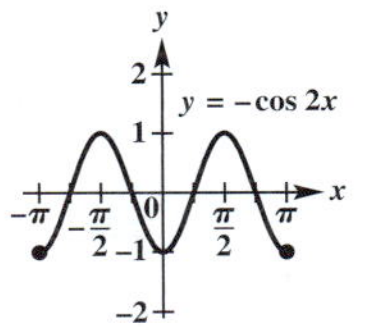

[4.2] **7.**

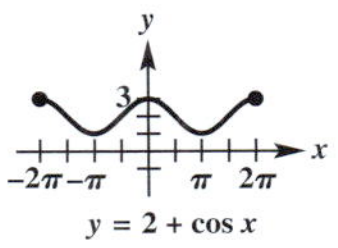

8.

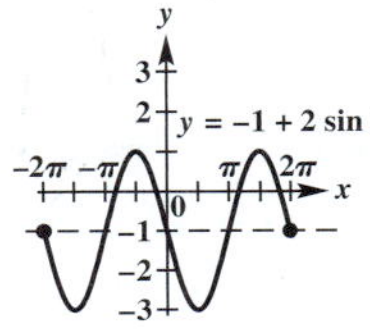

[4.3] **9.**

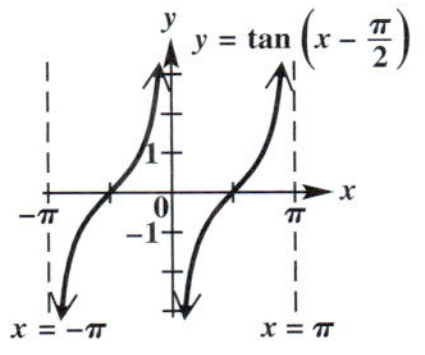

10.

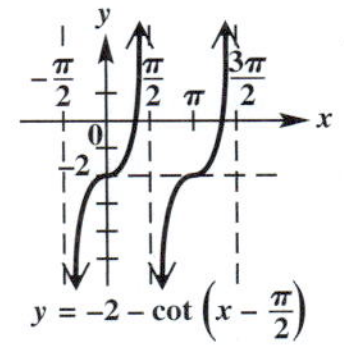

[4.4] **11.**

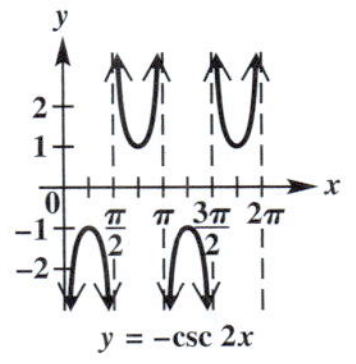

12.

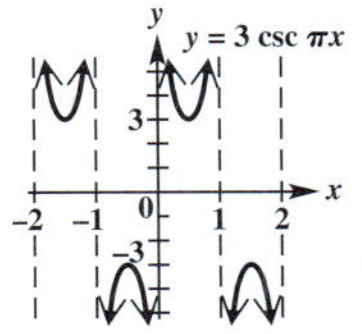

[4.1, 4.2] **13.** **(a)**

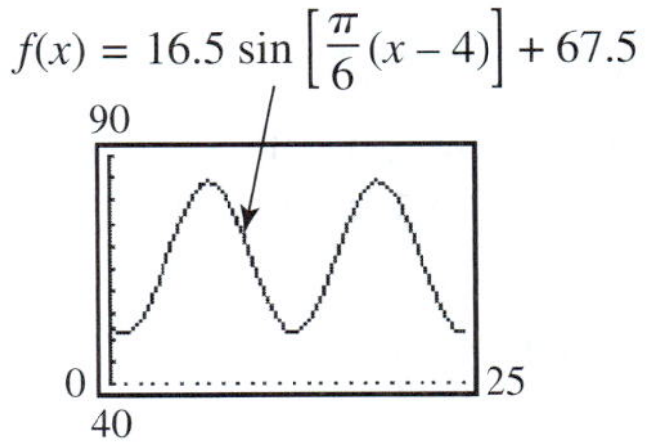

(b) 16.5; 12; 4 to the right; 67.5 up

(c) approximately 53°F **(d)** 51°F in January; 84°F in July

(e) approximately 67.5°F; This is the vertical translation.

[4.5] **14.** **(a)** 4 in. **(b)** after $\frac{1}{8}$ sec

(c) 4 cycles per sec; $\frac{1}{4}$ sec

Chapter 5 Trigonometric Identities

5.1 Exercises *(pages 193–196)*

1. -2.6 **3.** 0.625 **5.** $\frac{2}{3}$ **7.** $\frac{\sqrt{7}}{4}$ **9.** $-\frac{5\sqrt{26}}{26}$ **11.** $-\frac{2\sqrt{5}}{5}$ **13.** $-\frac{\sqrt{15}}{5}$ **15.** $-\frac{\sqrt{105}}{11}$ **17.** $-\frac{4}{9}$ **21.** $-\sin x$ **22.** odd **23.** $\cos x$ **24.** even **25.** $-\tan x$ **26.** odd **27.** $f(-x) = f(x)$ **29.** $f(-x) = -f(x)$ **31.** $\cos\theta = -\frac{\sqrt{5}}{3}$; $\tan\theta = -\frac{2\sqrt{5}}{5}$; $\cot\theta = -\frac{\sqrt{5}}{2}$; $\sec\theta = -\frac{3\sqrt{5}}{5}$; $\csc\theta = \frac{3}{2}$ **33.** $\sin\theta = -\frac{\sqrt{17}}{17}$; $\cos\theta = \frac{4\sqrt{17}}{17}$; $\cot\theta = -4$; $\sec\theta = \frac{\sqrt{17}}{4}$; $\csc\theta = -\sqrt{17}$ **35.** $\sin\theta = \frac{3}{5}$; $\cos\theta = \frac{4}{5}$; $\tan\theta = \frac{3}{4}$; $\sec\theta = \frac{5}{4}$; $\csc\theta = \frac{5}{3}$ **37.** $\sin\theta = -\frac{\sqrt{7}}{4}$; $\cos\theta = \frac{3}{4}$; $\tan\theta = -\frac{\sqrt{7}}{3}$; $\cot\theta = -\frac{3\sqrt{7}}{7}$; $\csc\theta = -\frac{4\sqrt{7}}{7}$ **39.** B **41.** E **43.** A **45.** A **47.** D **51.** $\sin\theta = \frac{\pm\sqrt{2x+1}}{x+1}$ **53.** $\sin x = \pm\sqrt{1-\cos^2 x}$ **55.** $\tan x = \pm\sqrt{\sec^2 x - 1}$ **57.** $\csc x = \frac{\pm\sqrt{1-\cos^2 x}}{1-\cos^2 x}$

In Exercises 59–83, there may be more than one possible answer.

59. $\cos\theta$ **61.** 1 **63.** $\cot\theta$ **65.** $\cos^2\theta$ **67.** $\sec\theta - \cos\theta$ **69.** $-\cot\theta + 1$ **71.** $\sin^2\theta\cos^2\theta$ **73.** $\tan\theta\sin\theta$ **75.** $\cot\theta - \tan\theta$ **77.** $\cos^2\theta$ **79.** $\tan^2\theta$ **81.** $\sec^2\theta$ **83.** $-\sec\theta$ **85.** $\frac{25\sqrt{6}-60}{12}$; $\frac{-25\sqrt{6}-60}{12}$ **87.** $y = -\sin(2x)$ **88.** It is the negative of $y = \sin(2x)$. **89.** $y = \cos(4x)$ **90.** It is the same function. **91. (a)** $y = -\sin(4x)$ **(b)** $y = \cos(2x)$ **(c)** $y = 5\sin(3x)$ **93.** identity **95.** not an identity

5.2 Exercises *(pages 202–204)*

1. $\csc\theta\sec\theta$ **3.** $1 + \sec x$ **5.** 1 **7.** $1 - 2\sin\alpha\cos\alpha$ **9.** $2 + 2\sin t$ **11.** $-2\cot x\csc x$ **13.** $(\sin\theta + 1)(\sin\theta - 1)$ **15.** $4\sin x$ **17.** $(2\sin x + 1)(\sin x + 1)$ **19.** $(\cos^2 x + 1)^2$ **21.** $(\sin x - \cos x)(1 + \sin x\cos x)$ **23.** $\sin\theta$ **25.** 1 **27.** $\tan^2\beta$ **29.** $\tan^2 x$ **31.** $\sec^2 x$ **33.** $\cos^2 x$ **79.** $(\sec\theta + \tan\theta)(1 - \sin\theta) = \cos\theta$ **81.** $\frac{\cos\theta + 1}{\sin\theta + \tan\theta} = \cot\theta$ **83.** identity **85.** not an identity **91.** It is true when $\sin x \leq 0$. **93. (a)** $I = k(1 - \sin^2\theta)$ **(b)** For $\theta = 2\pi n$ and all integers n, $\cos^2\theta = 1$, its maximum value, and I attains a maximum value of k. **95. (a)** The sum of L and C equals 3.

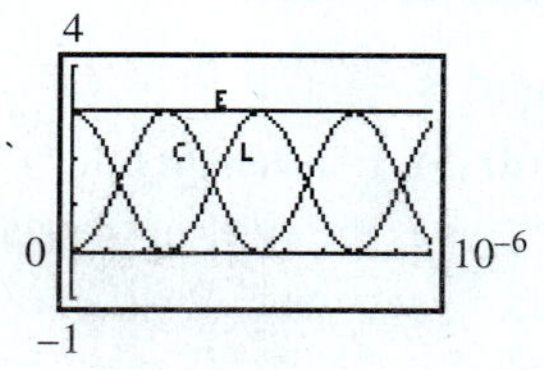

(b) Let $Y_1 = L(t)$, $Y_2 = C(t)$, and $Y_3 = E(t)$. $Y_3 = 3$ for all inputs.

X	Y2	Y3
0	0	3
1E-7	.95646	3
2E-7	2.6061	3
3E-7	2.8451	3
4E-7	1.3688	3
5E-7	.05974	3
6E-7	.58747	3

Y3=Y1+Y2

(c) $E(t) = 3$

5.3 Exercises *(pages 212–215)*

1. F **3.** E **5.** E **7.** $\frac{\sqrt{6}-\sqrt{2}}{4}$ **9.** $\frac{\sqrt{2}-\sqrt{6}}{4}$ **11.** $\frac{\sqrt{2}-\sqrt{6}}{4}$ **13.** $\frac{\sqrt{6}+\sqrt{2}}{4}$ **15.** 0 **17.** The calculator gives a value of 0 for the expression. **19.** $\cot 3°$ **21.** $\sin\frac{5\pi}{12}$ **23.** $\sec 75°\,36'$ **25.** $\cos\left(-\frac{\pi}{8}\right)$ **27.** $\csc(-56°\,42')$ **29.** $\tan(-86.9814°)$ **31.** tan **33.** cos **35.** csc

For Exercises 37–41, other answers are possible. We give the most obvious one.

37. 15° **39.** $\frac{140°}{3}$ **41.** 20° **43.** $\cos\theta$ **45.** $-\cos\theta$ **47.** $\cos\theta$ **49.** $-\cos\theta$ **51.** $\frac{16}{65}$; $-\frac{56}{65}$ **53.** $\frac{4-6\sqrt{6}}{25}$; $\frac{4+6\sqrt{6}}{25}$ **55.** $\frac{2\sqrt{638}-\sqrt{30}}{56}$; $\frac{2\sqrt{638}+\sqrt{30}}{56}$ **57.** true **59.** false **61.** true **63.** true **65.** false **72.** $\frac{-\sqrt{6}-\sqrt{2}}{4}$ **73.** $\frac{-\sqrt{6}-\sqrt{2}}{4}$ **74. (a)** $\frac{\sqrt{2}-\sqrt{6}}{4}$ **(b)** $\frac{-\sqrt{6}-\sqrt{2}}{4}$ **75. (a)** 3 **(b)** 163 and -163; no **77.** $\cos(90° + \theta) = -\sin\theta$ **78.** $\cos(270° - \theta) = -\sin\theta$ **79.** $\cos(180° + \theta) = -\cos\theta$ **80.** $\cos(270° + \theta) = \sin\theta$ **81.** $\sin(180° + \theta) = -\sin\theta$ **82.** $\tan(270° - \theta) = \cot\theta$

5.4 Exercises *(pages 220–224)*

1. C **3.** E **5.** B **9.** $\frac{\sqrt{6}+\sqrt{2}}{4}$ **11.** $2 - \sqrt{3}$ **13.** $\frac{\sqrt{6}+\sqrt{2}}{4}$ **15.** $\frac{-\sqrt{6}-\sqrt{2}}{4}$ **17.** $-2 - \sqrt{3}$ **19.** $\frac{\sqrt{2}}{2}$ **21.** -1 **23.** 0 **25.** 1 **27.** $\frac{\sqrt{3}\cos\theta - \sin\theta}{2}$ **29.** $\frac{\cos\theta - \sqrt{3}\sin\theta}{2}$ **31.** $\frac{\sqrt{2}(\sin x - \cos x)}{2}$ **33.** $\frac{\sqrt{3}\tan\theta + 1}{\sqrt{3} - \tan\theta}$ **35.** $\frac{\sqrt{2}(\cos x + \sin x)}{2}$ **37.** $-\cos\theta$ **39.** $-\tan x$ **41.** $-\tan x$ **45. (a)** $\frac{63}{65}$ **(b)** $\frac{63}{16}$ **(c)** I **47. (a)** $\frac{77}{85}$ **(b)** $-\frac{77}{36}$ **(c)** II **49. (a)** $\frac{4\sqrt{2}+\sqrt{5}}{9}$ **(b)** $\frac{-\sqrt{5}-\sqrt{2}}{2}$ **(c)** II **51.** $\frac{\sqrt{6}-\sqrt{2}}{4}$ **53.** $-2 + \sqrt{3}$ **55.** $-2 + \sqrt{3}$ **57.** $\sin\left(\frac{\pi}{2} + \theta\right) = \cos\theta$ **59.** $\tan\left(\frac{\pi}{2} + \theta\right) = -\cot\theta$ **71.** $180° - \beta$ **72.** $\theta = \beta - \alpha$ **73.** $\tan\theta = \frac{\tan\beta - \tan\alpha}{1 + \tan\beta\tan\alpha}$ **75.** 18.4° **76.** 80.8° **77. (a)** 425 lb **(c)** 0° **79.** $-20\cos\frac{\pi t}{4}$ **81.** $y' = y\cos R - z\sin R$

Chapter 5 Quiz *(page 224)*

[5.1] **1.** $\cos\theta = \frac{24}{25}$; $\tan\theta = -\frac{7}{24}$; $\cot\theta = -\frac{24}{7}$; $\sec\theta = \frac{25}{24}$; $\csc\theta = -\frac{25}{7}$ **2.** $\frac{\cos^2 x + 1}{\sin^2 x}$ [5.4] **3.** $\frac{-\sqrt{6}-\sqrt{2}}{4}$ [5.3] **4.** $-\cos\theta$ [5.3, 5.4] **5. (a)** $-\frac{16}{65}$ **(b)** $-\frac{63}{65}$ **(c)** III [5.4] **6.** $\frac{-1 + \tan x}{1 + \tan x}$

5.5 Exercises *(pages 230–232)*

1. C **3.** B **5.** F **7.** $\cos 2\theta = \frac{17}{25}$; $\sin 2\theta = -\frac{4\sqrt{21}}{25}$
9. $\cos 2x = -\frac{3}{5}$; $\sin 2x = \frac{4}{5}$ **11.** $\cos 2\theta = \frac{39}{49}$;
$\sin 2\theta = -\frac{4\sqrt{55}}{49}$ **13.** $\cos\theta = \frac{2\sqrt{5}}{5}$; $\sin\theta = \frac{\sqrt{5}}{5}$
15. $\cos\theta = -\frac{\sqrt{42}}{12}$; $\sin\theta = \frac{\sqrt{102}}{12}$ **37.** $\frac{\sqrt{3}}{2}$ **39.** $\frac{\sqrt{3}}{2}$
41. $-\frac{\sqrt{2}}{2}$ **43.** $\frac{1}{2}\tan 102°$ **45.** $\frac{1}{4}\cos 94.2°$ **47.** $-\cos\frac{4\pi}{5}$
49. $\sin 4x = 4\sin x\cos^3 x - 4\sin^3 x\cos x$
51. $\tan 3x = \frac{3\tan x - \tan^3 x}{1 - 3\tan^2 x}$ **53.** $\cos^4 x - \sin^4 x = \cos 2x$
55. $\frac{2\tan x}{2 - \sec^2 x} = \tan 2x$ **57.** $\sin 160° - \sin 44°$
59. $\sin\frac{\pi}{2} - \sin\frac{\pi}{6}$ **61.** $3\cos x - 3\cos 9x$
63. $-2\sin 3x\sin x$ **65.** $-2\sin 11.5°\cos 36.5°$
67. $2\cos 6x\cos 2x$ **69.** $a = -885.6$; $c = 885.6$; $\omega = 240\pi$

5.6 Exercises *(pages 235–239)*

1. − **3.** + **5.** C **7.** D **9.** F **11.** $\frac{\sqrt{2+\sqrt{2}}}{2}$
13. $2 - \sqrt{3}$ **15.** $-\frac{\sqrt{2+\sqrt{3}}}{2}$ **19.** $\frac{\sqrt{10}}{4}$ **21.** 3
23. $\frac{\sqrt{50 - 10\sqrt{5}}}{10}$ **25.** $-\sqrt{7}$ **27.** $\frac{\sqrt{5}}{5}$ **29.** $-\frac{\sqrt{42}}{12}$
31. 0.127 **33.** $\sin 20°$ **35.** $\tan 73.5°$ **37.** $\tan 29.87°$
39. $\cos 9x$ **41.** $\tan 4\theta$ **43.** $\cos\frac{x}{8}$ **55.** $\frac{\sin x}{1 + \cos x} = \tan\frac{x}{2}$
57. $\frac{\tan\frac{x}{2} + \cot\frac{x}{2}}{\cot\frac{x}{2} - \tan\frac{x}{2}} = \sec x$ **59.** 106° **61.** 2
63. **(a)** $\cos\frac{\theta}{2} = \frac{R-b}{R}$ **(b)** $\tan\frac{\theta}{4} = \frac{b}{50}$ **65.** They are both radii of the circle. **66.** It is the supplement of a 30° angle. **67.** Their sum is 180° − 150° = 30°, and they are equal.
68. $2 + \sqrt{3}$ **70.** $\frac{\sqrt{6}+\sqrt{2}}{4}$ **71.** $\frac{\sqrt{6}-\sqrt{2}}{4}$ **72.** $2 - \sqrt{3}$
73. $\frac{\sqrt{10 + 2\sqrt{5}}}{4}$ **75.** $\frac{\left(\sqrt{10 + 2\sqrt{5}}\right)\left(\sqrt{5}+1\right)}{4}$ **77.** $1 + \sqrt{5}$
79. $\frac{\sqrt{10 + 2\sqrt{5}}}{4}$ **81.** $\frac{\left(\sqrt{10 + 2\sqrt{5}}\right)\left(-5 + 3\sqrt{5}\right)}{20}$
83. $1 + \sqrt{5}$

Chapter 5 Review Exercises *(pages 242–244)*

1. B **3.** C **5.** D **7.** 1 **9.** $\frac{1}{\cos^2\theta}$ **11.** $-\frac{\cos\theta}{\sin\theta}$
13. $\sin x = -\frac{4}{5}$; $\tan x = -\frac{4}{3}$; $\cot(-x) = \frac{3}{4}$
15. $\sin 165° = \frac{\sqrt{6}-\sqrt{2}}{4}$; $\cos 165° = \frac{-\sqrt{6}-\sqrt{2}}{4}$;
$\tan 165° = -2 + \sqrt{3}$; $\csc 165° = \sqrt{6} + \sqrt{2}$;
$\sec 165° = -\sqrt{6} + \sqrt{2}$; $\cot 165° = -2 - \sqrt{3}$
17. I **19.** H **21.** G **23.** J **25.** F **27.** $\frac{117}{125}$; $\frac{4}{5}$; $-\frac{117}{44}$; II
29. $\frac{2 + 3\sqrt{7}}{10}$; $\frac{2\sqrt{3} + \sqrt{21}}{10}$; $\frac{-25\sqrt{3} - 8\sqrt{21}}{9}$; II
31. $\frac{4 - 9\sqrt{11}}{50}$; $\frac{12\sqrt{11} - 3}{50}$; $\frac{\sqrt{11} - 16}{21}$; IV **33.** $\sin\theta = \frac{\sqrt{14}}{4}$;
$\cos\theta = \frac{\sqrt{2}}{4}$ **35.** $\sin 2x = \frac{3}{5}$; $\cos 2x = -\frac{4}{5}$ **37.** $\frac{1}{2}$
39. $\frac{\sqrt{5}-1}{2}$ **41.** 0.5 **43.** $-\frac{\sin 2x + \sin x}{\cos 2x - \cos x} = \cot\frac{x}{2}$
45. $\frac{\sin x}{1 - \cos x} = \cot\frac{x}{2}$ **47.** $\frac{2(\sin x - \sin^3 x)}{\cos x} = \sin 2x$
71. **(a)** $D = \frac{v^2\sin 2\theta}{32}$ **(b)** approximately 35 ft

Chapter 5 Test *(page 244)*

[5.1] **1.** $\sin\theta = -\frac{7}{25}$; $\tan\theta = -\frac{7}{24}$; $\cot\theta = -\frac{24}{7}$;
$\sec\theta = \frac{25}{24}$; $\csc\theta = -\frac{25}{7}$ **2.** $\cos\theta$ **3.** −1
[5.3] **4.** $\frac{\sqrt{6}-\sqrt{2}}{4}$ [5.3, 5.4] **5.** **(a)** $-\sin x$ **(b)** $\tan x$
[5.6] **6.** $-\frac{\sqrt{2-\sqrt{2}}}{2}$ **7.** $\cot\frac{1}{2}x - \cot x = \csc x$
[5.3, 5.4] **8.** **(a)** $\frac{33}{65}$ **(b)** $-\frac{56}{65}$ **(c)** $\frac{63}{16}$ **(d)** II
[5.5, 5.6] **9.** **(a)** $-\frac{7}{25}$ **(b)** $-\frac{24}{25}$ **(c)** $\frac{24}{7}$ **(d)** $\frac{\sqrt{5}}{5}$ **(e)** 2
[5.3] **15.** **(a)** $V = 163\cos\left(\frac{\pi}{2} - \omega t\right)$ **(b)** 163 volts; $\frac{1}{240}$ sec

Chapter 6 Inverse Circular Functions and Trigonometric Equations

6.1 Exercises *(pages 257–261)*

1. one-to-one **3.** $\cos y$ **5.** π **7.** **(a)** $[-1, 1]$
(b) $\left[-\frac{\pi}{2}, \frac{\pi}{2}\right]$ **(c)** increasing **(d)** −2 is not in the domain.
9. **(a)** $(-\infty, \infty)$ **(b)** $\left(-\frac{\pi}{2}, \frac{\pi}{2}\right)$ **(c)** increasing **(d)** no
11. $\cos^{-1}\frac{1}{a}$ **13.** 0 **15.** π **17.** $\frac{\pi}{4}$ **19.** 0 **21.** $-\frac{\pi}{3}$
23. $\frac{5\pi}{6}$ **25.** $\sin^{-1}\sqrt{3}$ does not exist. **27.** $\frac{3\pi}{4}$ **29.** $-\frac{\pi}{6}$
31. $\frac{\pi}{6}$ **33.** 0 **35.** $\csc^{-1}\frac{\sqrt{2}}{2}$ does not exist. **37.** −45°
39. −60° **41.** 120° **43.** 120° **45.** −30° **47.** $\sin^{-1} 2$ does not exist. **49.** −7.6713835° **51.** 113.500970°
53. 30.987961° **55.** 121.267893° **57.** −82.678329°
59. 1.1900238 **61.** 1.9033723 **63.** 0.83798122
65. 2.3154725 **67.** 2.4605221
69.

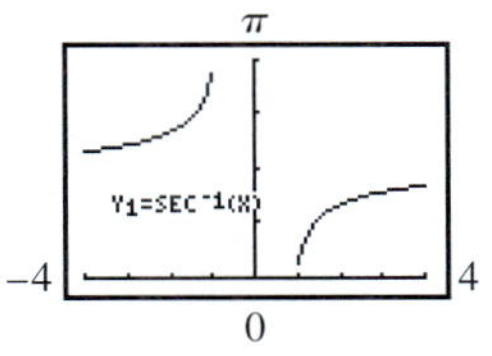

71.

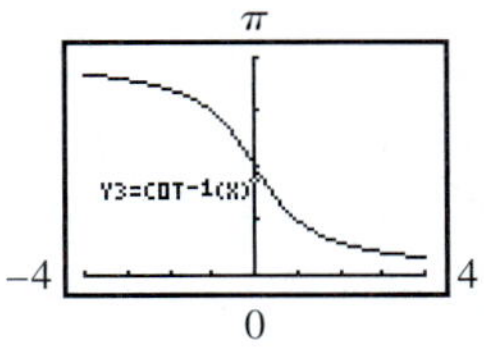

73.

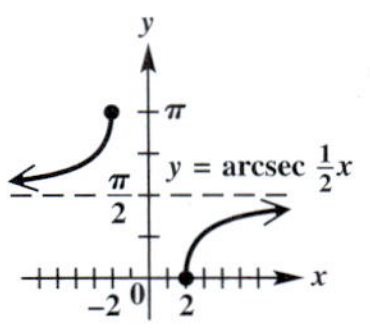

75. 1.003 is not in the domain of $y = \sin^{-1} x$. **76.** In both cases, the result is x. In each case, the graph is a straight line bisecting quadrants I and III (i.e., the line $y = x$).

77. It is the graph of $y = x$.

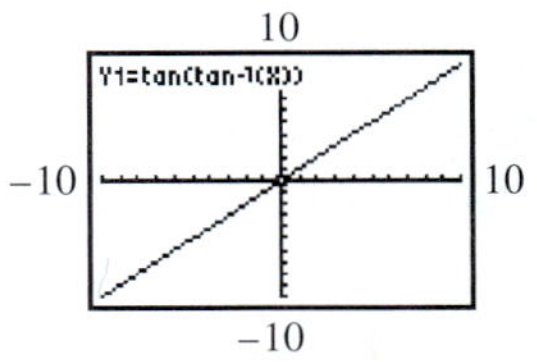

78. It does not agree because the range of the inverse tangent function is $\left(-\frac{\pi}{2}, \frac{\pi}{2}\right)$, not $(-\infty, \infty)$, as was the case in Exercise 77.

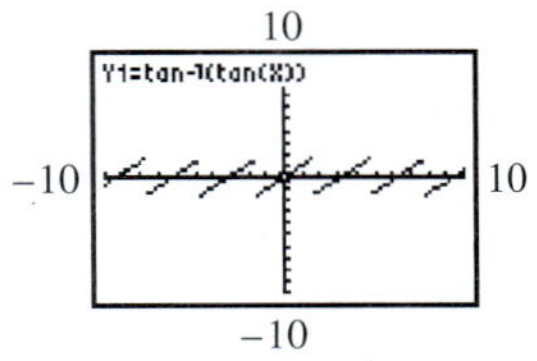

79. $\frac{\sqrt{7}}{3}$ **81.** $\frac{\sqrt{5}}{5}$ **83.** $\frac{120}{169}$ **85.** $-\frac{7}{25}$ **87.** $\frac{4\sqrt{6}}{25}$ **89.** 2

91. $\frac{63}{65}$ **93.** $\frac{\sqrt{10}-3\sqrt{30}}{20}$ **95.** 0.894427191

97. 0.1234399811 **99.** $\sqrt{1-u^2}$ **101.** $\sqrt{1-u^2}$

103. $\frac{4\sqrt{u^2-4}}{u^2}$ **105.** $\frac{u\sqrt{2}}{2}$ **107.** $\frac{2\sqrt{4-u^2}}{4-u^2}$ **109.** 41°

111. (a) 18° **(b)** 18° **(c)** 15°

(e) 1.4142151 m (Note: Due to the computational routine, there may be a discrepancy in the last few decimal places.)

(f) $\sqrt{2}$

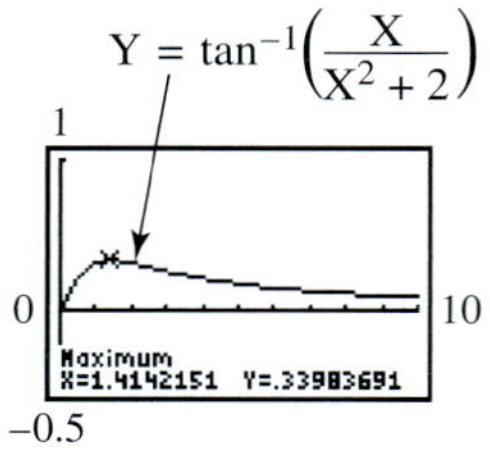

Radian mode

113. 44.7%

6.2 Exercises *(pages 266–269)*

1. Solve the linear equation for cot x. **3.** Solve the quadratic equation for sec x by factoring. **5.** Solve the quadratic equation for sin x using the quadratic formula. **7.** Use an identity to rewrite as an equation with one trigonometric function. **11.** $\{\frac{3\pi}{4}, \frac{7\pi}{4}\}$ **13.** $\{\frac{\pi}{6}, \frac{5\pi}{6}\}$

15. ∅ **17.** $\{\frac{\pi}{4}, \frac{2\pi}{3}, \frac{5\pi}{4}, \frac{5\pi}{3}\}$ **19.** $\{\pi\}$ **21.** $\{\frac{7\pi}{6}, \frac{3\pi}{2}, \frac{11\pi}{6}\}$

23. {30°, 210°, 240°, 300°} **25.** {90°, 210°, 330°} **27.** {45°, 135°, 225°, 315°} **29.** {45°, 225°} **31.** {0°, 30°, 150°, 180°} **33.** {0°, 45°, 135°, 180°, 225°, 315°} **35.** {53.6°, 126.4°, 187.9°, 352.1°} **37.** {149.6°, 329.6°, 106.3°, 286.3°} **39.** ∅ **41.** {57.7°, 159.2°} **43.** {180° + 360°n, where n is any integer} **45.** $\{\frac{\pi}{3} + 2n\pi, \frac{2\pi}{3} + 2n\pi$, where n is any integer$\}$ **47.** {19.5° + 360°n, 160.5° + 360°n, 210° + 360°n, 330° + 360°n, where n is any integer} **49.** $\{\frac{\pi}{3} + 2n\pi, \pi + 2n\pi, \frac{5\pi}{3} + 2n\pi$, where n is any integer$\}$ **51.** {180°n, where n is any integer} **53.** $\{0.8751 + 2n\pi, 2.2665 + 2n\pi, 3.5908 + 2n\pi, 5.8340 + 2n\pi$, where n is any integer$\}$ **55.** {33.6° + 360°n, 326.4° + 360°n, where n is any integer} **57.** {45° + 180°n, 108.4° + 180°n, where n is any integer} **59.** {0.6806, 1.4159} **61. (a)** 0.00164 and 0.00355 **(b)** [0.00164, 0.00355] **(c)** outward

63. (a) $\frac{1}{4}$ sec **(b)** $\frac{1}{6}$ sec **(c)** 0.21 sec **65. (a)** One such value is $\frac{\pi}{3}$. **(b)** One such value is $\frac{\pi}{4}$.

6.3 Exercises *(pages 273–275)*

1. $\{\frac{\pi}{3}, \pi, \frac{4\pi}{3}\}$ **3.** {60°, 210°, 240°, 310°}

7. $\{\frac{\pi}{12}, \frac{11\pi}{12}, \frac{13\pi}{12}, \frac{23\pi}{12}\}$ **9.** {90°, 210°, 330°}

11. $\{\frac{\pi}{18}, \frac{7\pi}{18}, \frac{13\pi}{18}, \frac{19\pi}{18}, \frac{25\pi}{18}, \frac{31\pi}{18}\}$

13. {67.5°, 112.5°, 247.5°, 292.5°} **15.** $\{\frac{\pi}{2}, \frac{3\pi}{2}\}$

17. $\{0, \frac{\pi}{3}, \pi, \frac{5\pi}{3}\}$ **19.** ∅ **21.** {180°} **23.** $\{\frac{\pi}{3}, \pi, \frac{5\pi}{3}\}$

25. $\{\frac{\pi}{12} + \frac{2n\pi}{3}, \frac{\pi}{4} + \frac{2n\pi}{3}$, where n is any integer$\}$

27. {720°n, where n is any integer}

29. $\{\frac{2\pi}{3} + 4n\pi, \frac{4\pi}{3} + 4n\pi$, where n is any integer$\}$

31. {30° + 360°n, 150° + 360°n, 270° + 360°n, where n is any integer} **33.** $\{n\pi, \frac{\pi}{6} + 2n\pi, \frac{5\pi}{6} + 2n\pi$, where n is any integer$\}$ **35.** $\{1.3181 + 2n\pi, 4.9651 + 2n\pi$, where n is any integer$\}$ **37.** {11.8° + 180°n, 78.2° + 180°n, where n is any integer} **39.** {30° + 180°n, 90° + 180°n, 150° + 180°n, where n is any integer} **41.** $\{\frac{\pi}{2}\}$

43. {0.4636, 3.6052} **45.** {1.2802}

47. (a) For $x = t$,

$$P(t) = 0.003 \sin 220\pi t + \frac{0.003}{3} \sin 660\pi t + \frac{0.003}{5} \sin 1100\pi t + \frac{0.003}{7} \sin 1540\pi t$$

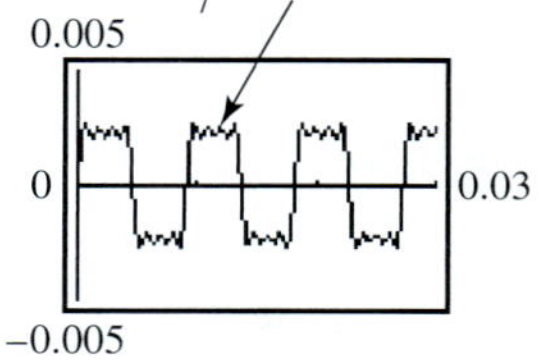

(b) The graph is periodic, and the wave has "jagged square" tops and bottoms. **(c)** This will occur when t is in one of these intervals: (0.0045, 0.0091), (0.0136, 0.0182), (0.0227, 0.0273).

49. (a) For $x = t$,

$$P(t) = \frac{1}{2}\sin[2\pi(220)t] + \frac{1}{3}\sin[2\pi(330)t] + \frac{1}{4}\sin[2\pi(440)t]$$

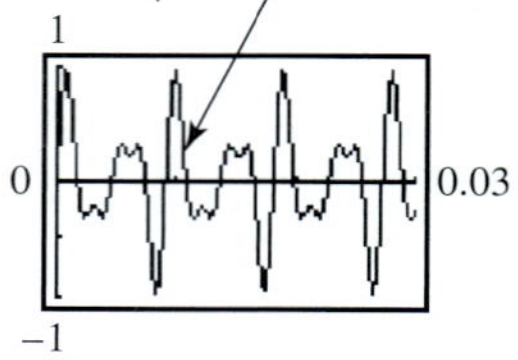

(b) 0.0007576, 0.009847, 0.01894, 0.02803 **(c)** 110 Hz

(d) For $x = t$,

$$P(t) = \sin[2\pi(110)t] + \frac{1}{2}\sin[2\pi(220)t] + \frac{1}{3}\sin[2\pi(330)t] + \frac{1}{4}\sin[2\pi(440)t]$$

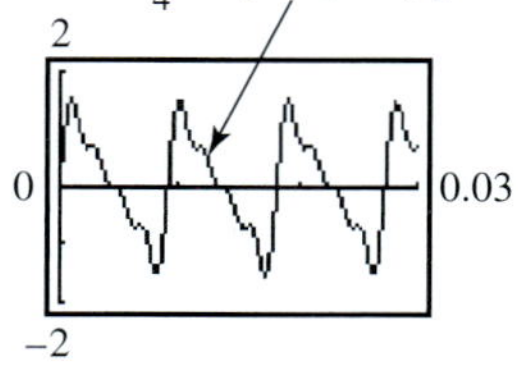

51. (a) when $x = 7$ (during July) **(b)** when $x = 2.3$ (during February) and when $x = 11.7$ (during November)

53. 0.001 sec

55. 0.004 sec

Chapter 6 Quiz *(page 276)*

[6.1] **1.** $[-1, 1]$; $[0, \pi]$

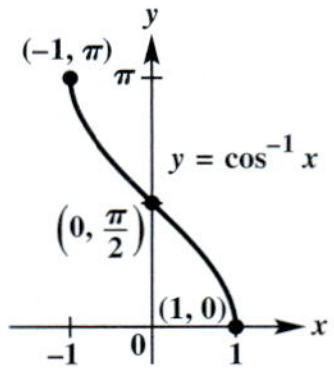

2. (a) $-\frac{\pi}{4}$ **(b)** $\frac{\pi}{3}$ **(c)** $\frac{5\pi}{6}$ **3. (a)** $22.568922°$ **(b)** $137.431085°$ **4. (a)** $\frac{5\sqrt{41}}{41}$ **(b)** $\frac{\sqrt{3}}{2}$

[6.2] **5.** $\{60°, 120°\}$ **6.** $\{60°, 180°, 300°\}$

7. $\{0.6089, 1.3424, 3.7505, 4.4840\}$

[6.3] **8.** $\left\{\frac{\pi}{6}, \frac{2\pi}{3}, \frac{7\pi}{6}, \frac{5\pi}{3}\right\}$ **9.** $\left\{\frac{5\pi}{3} + 4n\pi, \frac{7\pi}{3} + 4n\pi,\right.$ where n is any integer$\left.\right\}$ [6.2] **10. (a)** 0 sec **(b)** 0.20 sec

6.4 Exercises *(pages 279–282)*

1. C **3.** C **5.** $x = \text{arccos}\,\frac{y}{5}$ **7.** $x = \frac{1}{3}\,\text{arccot}\,2y$

9. $x = \frac{1}{2}\arctan\frac{y}{3}$ **11.** $x = 4\,\text{arccos}\,\frac{y}{6}$

13. $x = \frac{1}{5}\,\text{arccos}\left(-\frac{y}{2}\right)$ **15.** $x = -3 + \text{arccos}\,y$

17. $x = \arcsin(y + 2)$ **19.** $x = \arcsin\left(\frac{y + 4}{2}\right)$

21. $x = \frac{1}{2}\sec^{-1}\left(\frac{y - \sqrt{2}}{3}\right)$ **25.** $\left\{-\frac{\sqrt{2}}{2}\right\}$ **27.** $\{-2\sqrt{2}\}$

29. $\{\pi - 3\}$ **31.** $\left\{\frac{3}{5}\right\}$ **33.** $\left\{\frac{4}{5}\right\}$ **35.** $\{0\}$ **37.** $\left\{\frac{1}{2}\right\}$

39. $\left\{-\frac{1}{2}\right\}$ **41.** $\{0\}$

43. Y = arcsin X – arccos X – $\frac{\pi}{6}$

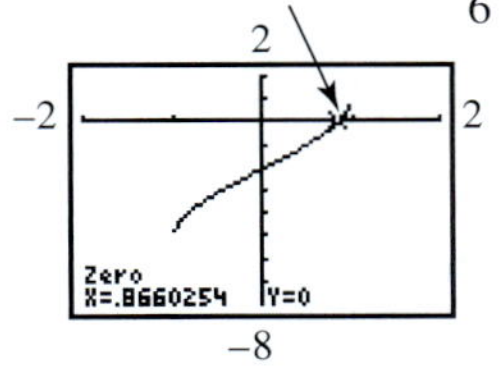

45. $\{4.4622\}$ **47. (a)** $A \approx 0.00506$, $\phi \approx 0.484$; $P = 0.00506\sin(440\pi t + 0.484)$

(b) The two graphs are the same.

For $x = t$,
$P(t) = 0.00506\sin(440\pi t + 0.484)$
$P_1(t) + P_2(t) = 0.0012\sin(440\pi t + 0.052) + 0.004\sin(440\pi t + 0.61)$

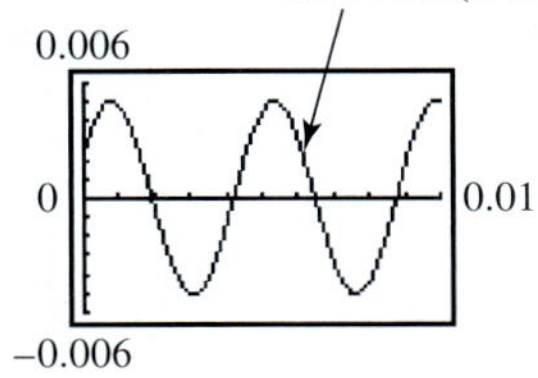

49. (a) $\tan\alpha = \frac{x}{z}$; $\tan\beta = \frac{x + y}{z}$ **(b)** $\frac{x}{\tan\alpha} = \frac{x + y}{\tan\beta}$

(c) $\alpha = \arctan\left(\frac{x\tan\beta}{x + y}\right)$ **(d)** $\beta = \arctan\left(\frac{(x + y)\tan\alpha}{x}\right)$

51. (a) $t = \frac{1}{2\pi f}\arcsin\frac{E}{E_{max}}$ **(b)** 0.00068 sec

53. (a) $t = \frac{3}{4\pi}\arcsin 3y$ **(b)** 0.27 sec

Chapter 6 Review Exercises *(pages 284–287)*

1.

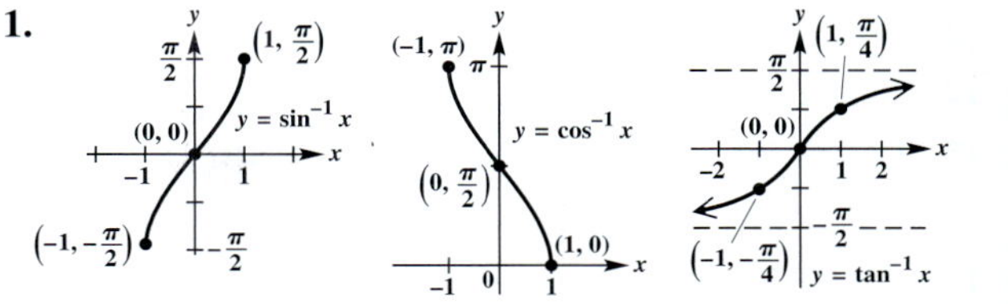

$[-1, 1]$; $\left[-\frac{\pi}{2}, \frac{\pi}{2}\right]$ $[-1, 1]$; $[0, \pi]$ $(-\infty, \infty)$; $\left(-\frac{\pi}{2}, \frac{\pi}{2}\right)$

3. false; $\arcsin\left(-\frac{1}{2}\right) = -\frac{\pi}{6}$, not $\frac{11\pi}{6}$. **5.** $\frac{\pi}{4}$ **7.** $-\frac{\pi}{3}$

9. $\frac{3\pi}{4}$ **11.** $\frac{2\pi}{3}$ **13.** $\frac{3\pi}{4}$ **15.** $-60°$ **17.** $60.68°$

19. $36.49°$ **21.** $73.26°$ **23.** -1 **25.** $\frac{3\pi}{4}$ **27.** $\frac{\pi}{4}$ **29.** $\frac{\sqrt{7}}{4}$

31. $\frac{\sqrt{3}}{2}$ **33.** $\frac{294 + 125\sqrt{6}}{92}$ **35.** $\frac{1}{u}$ **37.** $\{0.4636, 3.6052\}$

39. $\left\{\frac{\pi}{4}, \frac{3\pi}{4}, \frac{5\pi}{4}, \frac{7\pi}{4}\right\}$ **41.** $\left\{\frac{\pi}{8}, \frac{3\pi}{8}, \frac{5\pi}{8}, \frac{7\pi}{8}, \frac{9\pi}{8}, \frac{11\pi}{8}, \frac{13\pi}{8}, \frac{15\pi}{8}\right\}$

43. $\left\{\frac{\pi}{3} + 2n\pi, \pi + 2n\pi, \frac{5\pi}{3} + 2n\pi,\right.$ where n is any integer$\left.\right\}$

45. $\{270°\}$ **47.** $\{45°, 90°, 225°, 270°\}$

49. $\{70.5°, 180°, 289.5°\}$ **51.** $\{300° + 720°n, 420° + 720°n$, where n is any integer$\}$ **53.** $\{180° + 360°n$, where n is any integer$\}$ **55.** $\emptyset$ **57.** $\left\{-\frac{1}{2}\right\}$

59. $x = \arcsin 2y$ **61.** $x = \left(\frac{1}{3}\arctan 2y\right) - \frac{2}{3}$

63. (b) 8.6602567 ft; There may be a discrepancy in the final digits.

Y = arctan$\left(\frac{15}{X}\right)$ – arctan$\left(\frac{5}{X}\right)$

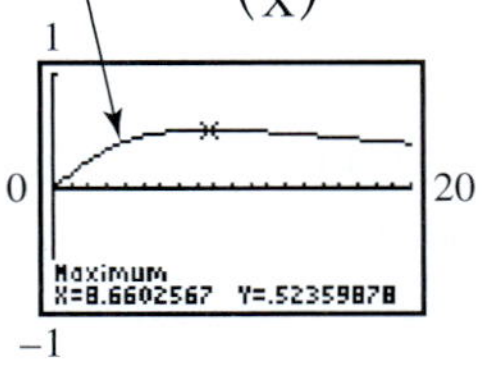

65. No light will emerge from the water.

67.

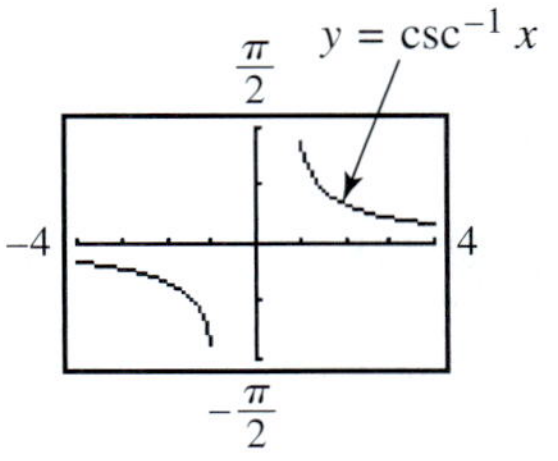

Radian mode

Chapter 6 Test *(pages 287–288)*

[6.1] **1.** $[-1, 1]$; $\left[-\frac{\pi}{2}, \frac{\pi}{2}\right]$

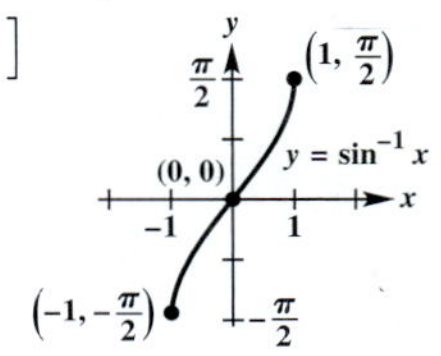

2. (a) $\frac{2\pi}{3}$ **(b)** $-\frac{\pi}{3}$ **(c)** 0 **(d)** $\frac{2\pi}{3}$ **3. (a)** $30°$ **(b)** $-45°$ **(c)** $135°$ **(d)** $-60°$ **4. (a)** $42.54°$ **(b)** $22.72°$ **(c)** $125.47°$ **5. (a)** $\frac{\sqrt{5}}{3}$ **(b)** $\frac{4\sqrt{2}}{9}$ **8.** $\frac{u\sqrt{1 - u^2}}{1 - u^2}$

[6.2, 6.3] **9.** $\{30°, 330°\}$ **10.** $\{90°, 270°\}$

11. $\{18.4°, 135°, 198.4°, 315°\}$ **12.** $\left\{0, \frac{2\pi}{3}, \frac{4\pi}{3}\right\}$
13. $\left\{\frac{\pi}{12}, \frac{7\pi}{12}, \frac{3\pi}{4}, \frac{5\pi}{4}, \frac{17\pi}{12}, \frac{23\pi}{12}\right\}$
14. $\{0.3649, 1.2059, 3.5065, 4.3475\}$
15. $\{90° + 180°n$, where n is any integer$\}$
16. $\left\{\frac{2\pi}{3} + 4n\pi, \frac{4\pi}{3} + 4n\pi\right.$, where n is any integer$\left.\right\}$
17. $\left\{\frac{\pi}{2} + 2n\pi\right.$, where n is any integer$\left.\right\}$
[6.4] **18. (a)** $x = \frac{1}{3}\arccos y$ **(b)** $x = \text{arccot}\left(\frac{y-4}{3}\right)$
19. (a) $\left\{\frac{4}{5}\right\}$ **(b)** $\left\{\frac{\sqrt{3}}{3}\right\}$ **20.** $\frac{5}{6}$ sec, $\frac{11}{6}$ sec, $\frac{17}{6}$ sec

Chapter 7 Applications of Trigonometry and Vectors

Note to student: Although most of the measures resulting from solving triangles in this chapter are approximations, for convenience we use = rather than ≈ in the answers.

7.1 Exercises *(pages 295–299)*

1. C **3.** $\sqrt{3}$ **5.** $C = 95°$, $b = 13$ m, $a = 11$ m **7.** $B = 37.3°$, $a = 38.5$ ft, $b = 51.0$ ft **9.** $C = 57.36°$, $b = 11.13$ ft, $c = 11.55$ ft **11.** $B = 18.5°$, $a = 239$ yd, $c = 230$ yd **13.** $A = 56°\,00'$, $AB = 361$ ft, $BC = 308$ ft **15.** $B = 110.0°$, $a = 27.01$ m, $c = 21.36$ m **17.** $A = 34.72°$, $a = 3326$ ft, $c = 5704$ ft **19.** $C = 97°\,34'$, $b = 283.2$ m, $c = 415.2$ m **25.** 118 m **27.** 17.8 km **29.** first location: 5.1 mi; second location: 7.2 mi **31.** 0.49 mi **33.** 111° **35.** The distance is about 419,000 km, which compares favorably to the actual value. **37.** approximately 6600 ft **39.** $\frac{\sqrt{3}}{2}$ sq unit **41.** $\frac{\sqrt{2}}{2}$ sq unit **43.** 46.4 m² **45.** 356 cm² **47.** 722.9 in.² **49.** 65.94 cm² **51.** 100 m² **53.** $a = \sin A$, $b = \sin B$, $c = \sin C$ **55.** $x = \frac{d \sin\alpha \sin\beta}{\sin(\beta - \alpha)}$

7.2 Exercises *(pages 304–306)*

1. A **3. (a)** $4 < L < 5$ **(b)** $L = 4$ or $L > 5$ **(c)** $L < 4$ **5.** 1 **7.** 2 **9.** 0 **11.** 45° **13.** $B_1 = 49.1°$, $C_1 = 101.2°$, $B_2 = 130.9°$, $C_2 = 19.4°$ **15.** $B = 26°\,30'$, $A = 112°\,10'$ **17.** no such triangle **19.** $B = 27.19°$, $C = 10.68°$ **21.** $B = 20.6°$, $C = 116.9°$, $c = 20.6$ ft **23.** no such triangle **25.** $B_1 = 49°\,20'$, $C_1 = 92°\,00'$, $c_1 = 15.5$ m; $B_2 = 130°\,40'$, $C_2 = 10°\,40'$, $c_2 = 2.88$ m **27.** $B = 37.77°$, $C = 45.43°$, $c = 4.174$ ft **29.** $A_1 = 53.23°$, $C_1 = 87.09°$, $c_1 = 37.16$ m; $A_2 = 126.77°$, $C_2 = 13.55°$, $c_2 = 8.719$ m **31.** 1; 90°; a right triangle **35.** 664 m **37.** 218 ft **42.** $\mathcal{A} = 1.12257R^2$ **43. (a)** 8.77 in.² **(b)** 5.32 in.² **44.** red

7.3 Exercises *(pages 313–319)*

1. (a) SAS **(b)** law of cosines **3. (a)** SSA **(b)** law of sines **5. (a)** ASA **(b)** law of sines **7. (a)** SSS **(b)** law of cosines **9.** 5 **11.** 120° **13.** $a = 7.0$, $B = 37.6°$, $C = 21.4°$ **15.** $A = 73.7°$, $B = 53.1°$, $C = 53.1°$ (The angles do not sum to 180° due to rounding.) **17.** $b = 88.2$, $A = 56.7°$, $C = 68.3°$ **19.** $a = 2.60$ yd, $B = 45.1°$, $C = 93.5°$ **21.** $c = 6.46$ m, $A = 53.1°$, $B = 81.3°$ **23.** $A = 82°$, $B = 37°$, $C = 61°$ **25.** $C = 102°\,10'$, $B = 35°\,50'$, $A = 42°\,00'$ **27.** $C = 84°\,30'$, $B = 44°\,40'$, $A = 50°\,50'$ **29.** $a = 156$ cm, $B = 64°\,50'$, $C = 34°\,30'$ **31.** $b = 9.53$ in., $A = 64.6°$, $C = 40.6°$ **33.** $a = 15.7$ m, $B = 21.6°$, $C = 45.6°$ **35.** $A = 30°$, $B = 56°$, $C = 94°$ **37.** The value of $\cos\theta$ will be greater than 1. Your calculator will give you an error message (or a nonreal complex number) when using the inverse cosine function. **39.** 257 m **41.** 163.5° **43.** 281 km **45.** 438.14 ft **47.** 10.8 mi **49.** 40° **51.** 26° and 36° **53.** second base: 66.8 ft; first and third bases: 63.7 ft **55.** 39.2 km **57.** 47.5 ft **59.** 5500 m **61.** 16.26° **63.** $24\sqrt{3}$ sq units **65.** 78 m² **67.** 12,600 cm² **69.** 3650 ft² **71.** Area and perimeter are both 36. **73.** 390,000 mi² **75. (a)** 87.8° and 92.2° both appear possible. **(b)** 92.2° **(c)** With the law of cosines we are required to find the inverse cosine of a negative number. Therefore, we know that angle C is greater than 90°.

77.

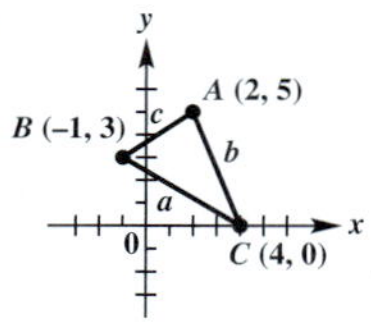

$a = \sqrt{34}$, $b = \sqrt{29}$, $c = \sqrt{13}$

78. 9.5 sq units **79.** 9.5 sq units **80.** 9.5 sq units

Chapter 7 Quiz *(pages 319–320)*

[7.1] **1.** 131° [7.3] **2.** 201 m **3.** 48.0° [7.1] **4.** 15.75 sq units [7.3] **5.** 189 km² [7.2] **6.** 41.6°, 138.4° [7.1] **7.** $a = 648$, $b = 456$, $C = 28°$ **8.** 3.6 mi [7.3] **9.** 25.24983 mi **10.** 3921 m

7.4 Exercises *(pages 328–331)*

1. **m** and **p**; **n** and **r** **3.** **m** and **p** equal 2**t**, or **t** equals $\frac{1}{2}$**m** and $\frac{1}{2}$**p**. Also **m** = 1**p** and **n** = 1**r**.

5.

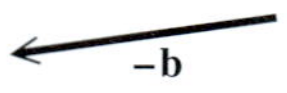

7.

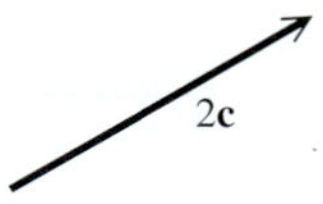

9.

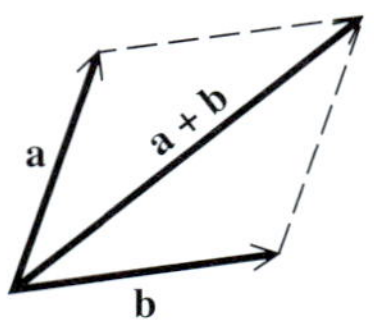

11.

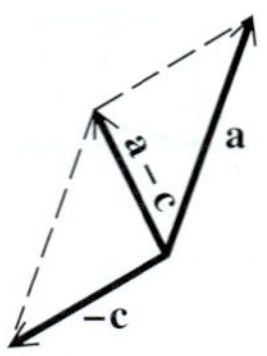

13.

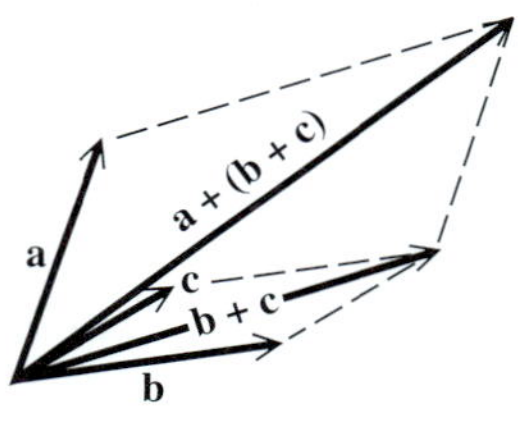

15.

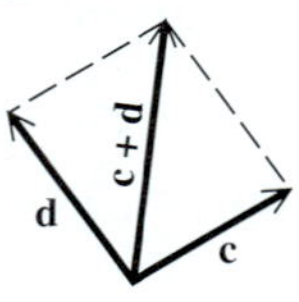

17. Yes, it appears that vector addition is associative (and this is true, in general). **19. (a)** $\langle -4, 16 \rangle$ **(b)** $\langle -12, 0 \rangle$ **(c)** $\langle 8, -8 \rangle$ **21. (a)** $\langle 8, 0 \rangle$ **(b)** $\langle 0, 16 \rangle$ **(c)** $\langle -4, -8 \rangle$ **23. (a)** $\langle 0, 12 \rangle$ **(b)** $\langle -16, -4 \rangle$ **(c)** $\langle 8, -4 \rangle$ **25. (a)** $4\mathbf{i}$ **(b)** $7\mathbf{i} + 3\mathbf{j}$ **(c)** $-5\mathbf{i} + \mathbf{j}$ **27. (a)** $\langle -2, 4 \rangle$ **(b)** $\langle 7, 4 \rangle$ **(c)** $\langle 6, -6 \rangle$

29.

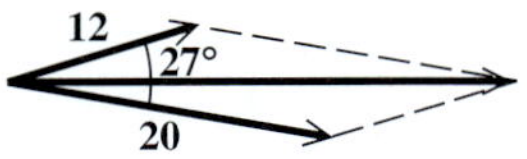

31.

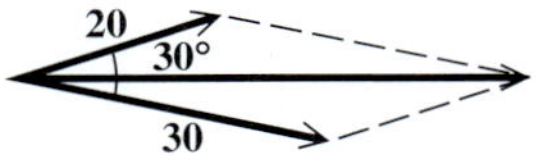

33. 17; 331.9° **35.** 8; 120° **37.** 47, 17 **39.** 38.8, 28.0 **41.** 123, 155 **43.** $\left\langle \frac{5\sqrt{3}}{2}, \frac{5}{2} \right\rangle$ **45.** $\langle -3.0642, 2.5712 \rangle$ **47.** $\langle 4.0958, -2.8679 \rangle$ **49.** 530 newtons **51.** 88.2 lb **53.** 94.2 lb **55.** 24.4 lb **57.** $\langle a + c, b + d \rangle$ **59.** $\langle -6, 2 \rangle$ **61.** $\langle 8, -20 \rangle$ **63.** $\langle -30, -3 \rangle$ **65.** $\langle 8, -7 \rangle$ **67.** $-5\mathbf{i} + 8\mathbf{j}$ **69.** $2\mathbf{i}$, or $2\mathbf{i} + 0\mathbf{j}$ **71.** 7 **73.** 0 **75.** 20 **77.** 135° **79.** 90° **81.** 36.87° **83.** −6 **85.** −24 **87.** orthogonal **89.** not orthogonal **91.** not orthogonal

In Exercises 93–97, answers may vary due to rounding.
93. magnitude: 9.5208; direction angle: 119.0647° **94.** $\langle -4.1042, 11.2763 \rangle$ **95.** $\langle -0.5209, -2.9544 \rangle$ **96.** $\langle -4.6252, 8.3219 \rangle$ **97.** magnitude: 9.5208; direction angle: 119.0647° **98.** They are the same. Preference of method is an individual choice.

7.5 Exercises *(pages 335–338)*

1. 2640 lb at an angle of 167.2° with the 1480-lb force **3.** 93.9° **5.** 190 lb and 283 lb, respectively **7.** 18° **9.** 2.4 tons **11.** 17.5° **13.** 226 lb **15.** 13.5 mi; 50.4° **17.** 39.2 km **19.** current: 3.5 mph; motorboat: 19.7 mph **21.** bearing: 237°; ground speed: 470 mph **23.** ground speed: 161 mph; airspeed: 156 mph **25.** bearing: 74°; ground speed: 202 mph **27.** bearing: 358°; airspeed: 170 mph **29.** ground speed: 230 km per hr; bearing: 167° **31. (a)** $|\mathbf{R}| = \sqrt{5} \approx 2.2$, $|\mathbf{A}| = \sqrt{1.25} \approx 1.1$; About 2.2 in. of rain fell. The area of the opening of the rain gauge is about 1.1 in.^{2}. **(b)** $V = 1.5$; The volume of rain was 1.5 in.^{3}.

Summary Exercises on Applications of Trigonometry and Vectors *(pages 338–339)*

1. 29 ft; 38 ft **2.** 38.3 cm **3.** 5856 m **4.** 15.8 ft per sec; 71.6° **5.** 42 lb **6.** 7200 ft **7. (a)** 10 mph **(b)** $3\mathbf{v} = 18\mathbf{i} + 24\mathbf{j}$; This represents a 30-mph wind in the direction of **v**. **(c)** **u** represents a southeast wind of $\sqrt{128} \approx 11.3$ mph. **8.** 380 mph; 64° **9.** It cannot exist. **10.** Other angles can be 36° 10′, 115° 40′, third side 40.5, or other angles can be 143° 50′, 8° 00′, third side 6.25. (Lengths are in yards.)

Chapter 7 Review Exercises *(pages 343–346)*

1. 63.7 m **3.** 41.7° **5.** 54° 20′ or 125° 40′ **9. (a)** $b = 5$, $b \geq 10$ **(b)** $5 < b < 10$ **(c)** $b < 5$ **11.** 19.87°, or 19° 52′ **13.** 55.5 m **15.** 19 cm **17.** $B = 17.3°$, $C = 137.5°$, $c = 11.0$ yd **19.** $c = 18.7$ cm, $A = 91° 40'$, $B = 45° 50'$ **21.** 153,600 m^2 **23.** 0.234 km^2 **25.** 58.6 ft **27.** 13 m **29.** 53.2 ft **31.** 115 km **33.** 25 sq units **35.**

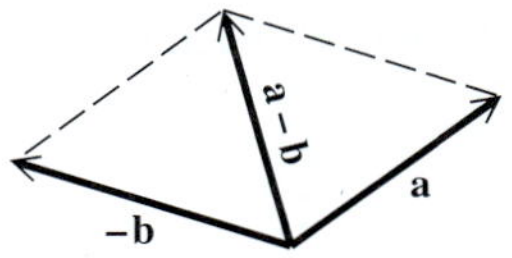

37. 207 lb **39.** 869; 418 **41.** 15; 126.9° **43. (a)** **i** **(b)** $4\mathbf{i} - 2\mathbf{j}$ **(c)** $11\mathbf{i} - 7\mathbf{j}$ **45.** 90°; orthogonal **47.** 29 lb **49.** bearing: 306°; ground speed: 524 mph **51.** 34 lb **53.** Both expressions equal $\frac{1 + \sqrt{3}}{2}$. **55.** Both expressions equal $-2 + \sqrt{3}$.

Chapter 7 Test *(pages 347–348)*

[7.1] **1.** 137.5° [7.3] **2.** 179 km **3.** 49.0° **4.** 168 sq units [7.1] **5.** 18 sq units [7.2] **6. (a)** $b > 10$ **(b)** none **(c)** $b \leq 10$ [7.1–7.3] **7.** $a = 40$ m, $B = 41°$, $C = 79°$ **8.** $B_1 = 58° 30'$, $A_1 = 83° 00'$, $a_1 = 1250$ in.; $B_2 = 121° 30'$, $A_2 = 20° 00'$, $a_2 = 431$ in. [7.4] **9.** $|\mathbf{v}| = 10$; $\theta = 126.9°$ **10.**

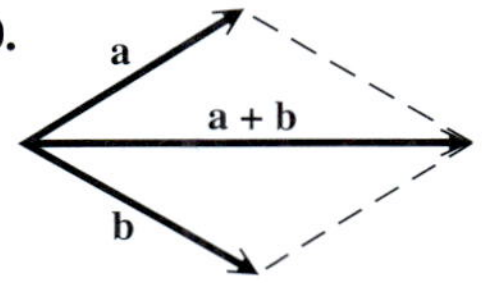

11. (a) $\langle 1, -3 \rangle$ **(b)** $\langle -6, 18 \rangle$ **(c)** −20 **(d)** $\sqrt{10}$ **12.** 41.8° **13.** Show that $\mathbf{u} \cdot \mathbf{v} = 0$. [7.1] **14.** 2.7 mi [7.4] **15.** $\langle -346, 451 \rangle$ [7.5] **16.** 1.91 mi [7.1] **17.** 14 m [7.5] **18.** 30 lb **19.** bearing: 357°; airspeed: 220 mph **20.** 18.7°

Chapter 8 Complex Numbers, Polar Equations, and Parametric Equations

8.1 Exercises *(pages 356–359)*

1. true **3.** true **5.** false; *Every* real number is a complex number. **7.** real, complex **9.** pure imaginary, nonreal complex, complex **11.** nonreal complex, complex **13.** real, complex **15.** pure imaginary, nonreal complex, complex **17.** $5i$ **19.** $i\sqrt{10}$ **21.** $12i\sqrt{2}$ **23.** $-3i\sqrt{2}$ **25.** $\{\pm 4i\}$ **27.** $\{\pm 2i\sqrt{3}\}$ **29.** $\left\{-\frac{2}{3} \pm \frac{\sqrt{2}}{3}i\right\}$ **31.** $\{3 \pm i\sqrt{5}\}$ **33.** $\left\{\frac{1}{2} \pm \frac{\sqrt{6}}{2}i\right\}$ **35.** $\left\{-\frac{1}{2} \pm \frac{\sqrt{3}}{2}i\right\}$ **37.** −13 **39.** $-2\sqrt{6}$ **41.** $\sqrt{3}$ **43.** $i\sqrt{3}$ **45.** $\frac{1}{2}$ **47.** −2 **49.** $-3 - i\sqrt{6}$ **51.** $2 + 2i\sqrt{2}$ **53.** $-\frac{1}{8} + \frac{\sqrt{2}}{8}i$ **55.** $12 - i$ **57.** 2 **59.** 0 **61.** $-13 + 4i\sqrt{2}$ **63.** $8 - i$ **65.** $-14 + 2i$

67. $5 - 12i$ **69.** 10 **71.** 13 **73.** 7 **75.** $25i$
77. $12 + 9i$ **79.** $20 + 15i$ **81.** $2 - 2i$ **83.** $\frac{3}{5} - \frac{4}{5}i$
85. $-1 - 2i$ **87.** $5i$ **89.** $8i$ **91.** $-\frac{2}{3}i$ **93.** i **95.** -1
97. $-i$ **99.** 1 **101.** $-i$ **103.** $-i$ **111.** $E = 30 + 60i$
113. $Z = \frac{233}{37} + \frac{119}{37}i$ **115.** $110 + 32i$

8.2 Exercises *(pages 364–366)*

1. length (magnitude)

3.

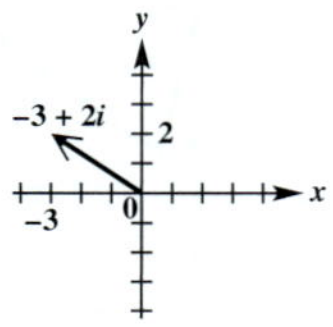

5.

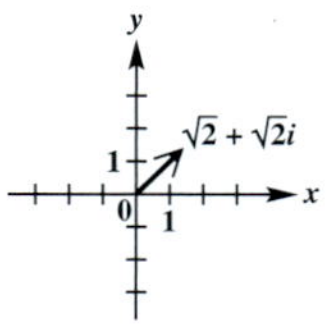

7.

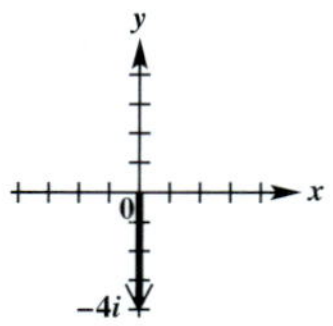

9.

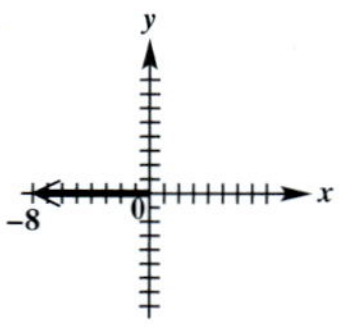

11. $1 - 4i$

13. $3 - i$

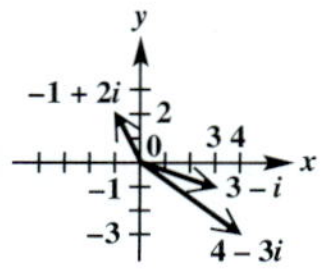

15. $-3i$

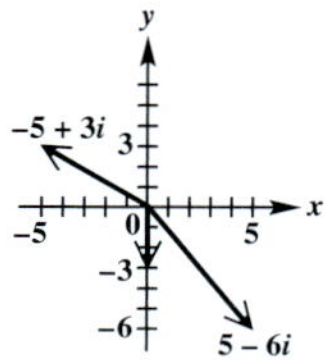

17. $-3 + 3i$ **19.** $-6 - 8i$ **21.** $7 + 9i$ **23.** $\frac{7}{6} + \frac{7}{6}i$
25. $\sqrt{2} + i\sqrt{2}$ **27.** $10i$ **29.** $-2 - 2i\sqrt{3}$
31. $-\frac{3\sqrt{3}}{2} + \frac{3}{2}i$ **33.** $\frac{5}{2} - \frac{5\sqrt{3}}{2}i$ **35.** $-1 - i$
37. $2\sqrt{3} - 2i$ **39.** $6(\cos 240° + i \sin 240°)$
41. $2(\cos 330° + i \sin 330°)$
43. $5\sqrt{2}(\cos 225° + i \sin 225°)$
45. $2\sqrt{2}(\cos 45° + i \sin 45°)$ **47.** $5(\cos 90° + i \sin 90°)$
49. $4(\cos 180° + i \sin 180°)$
51. $\sqrt{13}(\cos 56.31° + i \sin 56.31°)$
53. $-1.0261 - 2.8191i$ **55.** $12(\cos 90° + i \sin 90°)$
57. $\sqrt{34}(\cos 59.04° + i \sin 59.04°)$ **59.** It is the circle of radius 1 centered at the origin. **61.** It is the vertical line $x = 1$. **63.** yes **67.** B **69.** A

8.3 Exercises *(pages 369–371)*

1. multiply; add **3.** $-3\sqrt{3} + 3i$ **5.** $12\sqrt{3} + 12i$ **7.** 4
9. $-3i$ **11.** $-\frac{15\sqrt{2}}{2} + \frac{15\sqrt{2}}{2}i$ **13.** $\sqrt{3} + i$ **15.** -2
17. $-\frac{1}{6} - \frac{\sqrt{3}}{6}i$ **19.** $2\sqrt{3} - 2i$ **21.** $-\frac{1}{2} - \frac{1}{2}i$
23. $\sqrt{3} + i$ **25.** $0.6537 + 7.4715i$
27. $30.8580 + 18.5414i$ **29.** $1.9563 + 0.4158i$
31. $-3.7588 - 1.3681i$ **33.** 2 **34.** $w = \sqrt{2}$ cis 135°; $z = \sqrt{2}$ cis 225° **35.** 2 cis 0° **36.** 2; It is the same.
37. $-i$ **38.** cis$(-90°)$ **39.** $-i$; It is the same.
43. $1.18 - 0.14i$ **45.** approximately $27.43 + 11.50i$

8.4 Exercises *(pages 376–378)*

1. $27i$ **3.** 1 **5.** $\frac{27}{2} - \frac{27\sqrt{3}}{2}i$ **7.** $-16\sqrt{3} + 16i$
9. $4096i$ **11.** $128 + 128i$

13. **(a)** $\cos 0° + i \sin 0°$, $\cos 120° + i \sin 120°$, $\cos 240° + i \sin 240°$

(b)

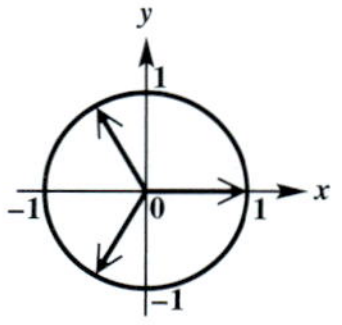

15. **(a)** 2 cis 20°, 2 cis 140°, 2 cis 260°

(b)

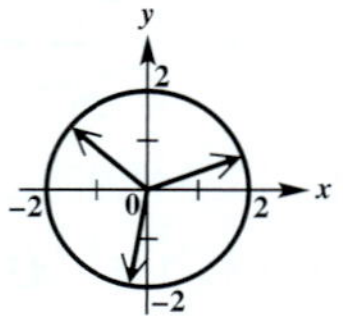

17. **(a)** $2(\cos 90° + i \sin 90°)$, $2(\cos 210° + i \sin 210°)$, $2(\cos 330° + i \sin 330°)$

(b)

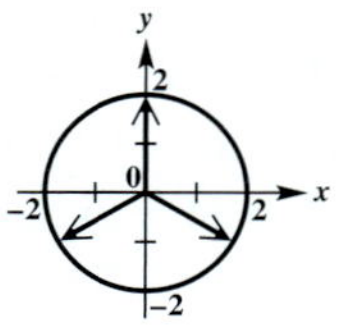

19. **(a)** $4(\cos 60° + i \sin 60°)$, $4(\cos 180° + i \sin 180°)$, $4(\cos 300° + i \sin 300°)$

(b)

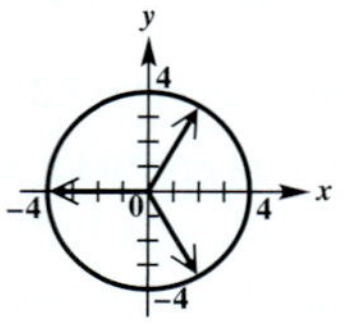

21. **(a)** $\sqrt[3]{2}(\cos 20° + i \sin 20°)$, $\sqrt[3]{2}(\cos 140° + i \sin 140°)$, $\sqrt[3]{2}(\cos 260° + i \sin 260°)$

(b)

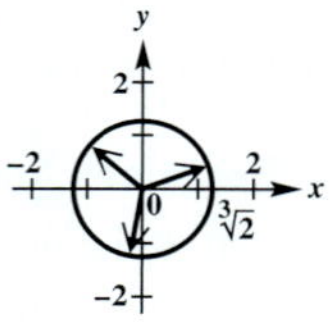

23. **(a)** $\sqrt[3]{4}(\cos 50° + i \sin 50°)$, $\sqrt[3]{4}(\cos 170° + i \sin 170°)$, $\sqrt[3]{4}(\cos 290° + i \sin 290°)$

(b)

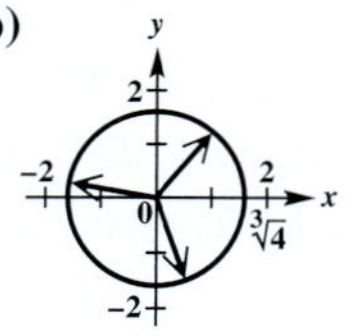

25. $\cos 0° + i \sin 0°$, $\cos 180° + i \sin 180°$

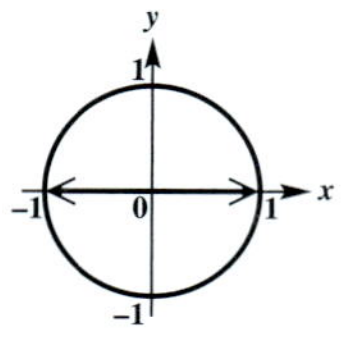

27. $\cos 0° + i \sin 0°$, $\cos 60° + i \sin 60°$, $\cos 120° + i \sin 120°$, $\cos 180° + i \sin 180°$, $\cos 240° + i \sin 240°$, $\cos 300° + i \sin 300°$

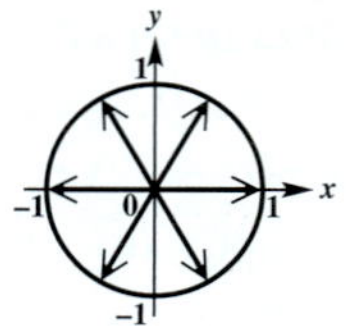

29. $\cos 30° + i \sin 30°$, $\cos 150° + i \sin 150°$, $\cos 270° + i \sin 270°$

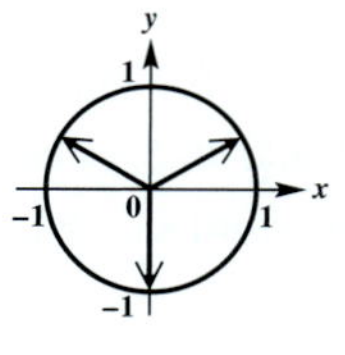

31. {cos 0° + *i* sin 0°, cos 120° + *i* sin 120°, cos 240° + *i* sin 240°} **33.** {cos 90° + *i* sin 90°, cos 210° + *i* sin 210°, cos 330° + *i* sin 330°} **35.** {2(cos 0° + *i* sin 0°), 2(cos 120° + *i* sin 120°), 2(cos 240° + *i* sin 240°)} **37.** {cos 45° + *i* sin 45°, cos 135° + *i* sin 135°, cos 225° + *i* sin 225°, cos 315° + *i* sin 315°} **39.** {cos 22.5° + *i* sin 22.5°, cos 112.5° + *i* sin 112.5°, cos 202.5° + *i* sin 202.5°, cos 292.5° + *i* sin 292.5°} **41.** {2(cos 20° + *i* sin 20°), 2(cos 140° + *i* sin 140°), 2(cos 260° + *i* sin 260°)}
43. $1, -\frac{1}{2} + \frac{\sqrt{3}}{2}i, -\frac{1}{2} - \frac{\sqrt{3}}{2}i$ **45.** $\cos 2\theta + i \sin 2\theta$
46. $(\cos^2 \theta - \sin^2 \theta) + i(2 \cos \theta \sin \theta) = \cos 2\theta + i \sin 2\theta$
47. $\cos 2\theta = \cos^2 \theta - \sin^2 \theta$ **48.** $\sin 2\theta = 2 \sin \theta \cos \theta$
49. **(a)** yes **(b)** no **(c)** yes **51.** 1, 0.30901699 + 0.95105652*i*, −0.809017 + 0.58778525*i*, −0.809017 − 0.5877853*i*, 0.30901699 − 0.9510565*i*
53. $-4, 2 - 2i\sqrt{3}$ **55.** {−1.8174 + 0.5503*i*, 1.8174 − 0.5503*i*} **57.** {0.8771 + 0.9492*i*, −0.6317 + 1.1275*i*, −1.2675 − 0.2524*i*, −0.1516 − 1.2835*i*, 1.1738 − 0.5408*i*} **59.** false

Chapter 8 Quiz *(page 379)*

[8.1] **1.** **(a)** $-6\sqrt{2}$ **(b)** $\frac{1}{3}i$
[8.1, 8.2] **2.** **(a)** $-1 + 6i$ **(b)** $7 + 4i$ **(c)** $-17 - 17i$ **(d)** $-\frac{7}{17} - \frac{23}{17}i$

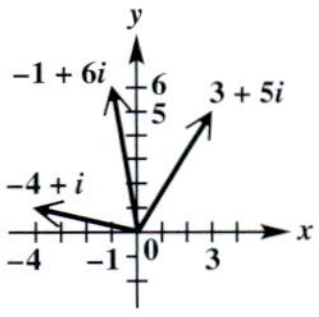

3. **(a)** $-2 - 2i$ **(b)** *i*, or 0 + *i* [8.1] **4.** $\left\{\frac{1}{6} \pm \frac{\sqrt{47}}{6}i\right\}$
[8.2] **5.** **(a)** 4(cos 270° + *i* sin 270°)
(b) 2(cos 300° + *i* sin 300°)
(c) $\sqrt{10}$ (cos 198.4° + *i* sin 198.4°)
6. **(a)** $2 + 2i\sqrt{3}$ **(b)** $-3.2139 + 3.8302i$
(c) −7*i*, or 0 − 7*i* **(d)** 2, or 2 + 0*i*
[8.3, 8.4] **7.** **(a)** 36(cos 130° + *i* sin 130°) **(b)** $2\sqrt{3} + 2i$
(c) $-\frac{27\sqrt{3}}{2} + \frac{27}{2}i$ **(d)** $-864 - 864i\sqrt{3}$
[8.4] **8.** 2(cos 45° + *i* sin 45°), 2(cos 135° + *i* sin 135°), 2(cos 225° + *i* sin 225°), 2(cos 315° + *i* sin 315°); $\sqrt{2} + i\sqrt{2}, -\sqrt{2} + i\sqrt{2}, -\sqrt{2} - i\sqrt{2}, \sqrt{2} - i\sqrt{2}$

8.5 Exercises *(pages 387–391)*

1. **(a)** II **(b)** I **(c)** IV **(d)** III

Graphs for Exercises 3(a), 5(a), 7(a), 9(a), 11(a), 13(a)

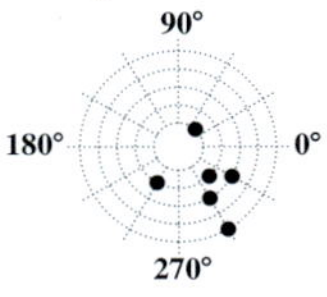

Answers may vary in Exercises 3(b)–13(b).

3. **(b)** (1, 405°), (−1, 225°) **(c)** $\left(\frac{\sqrt{2}}{2}, \frac{\sqrt{2}}{2}\right)$
5. **(b)** (−2, 495°), (2, 315°) **(c)** $\left(\sqrt{2}, -\sqrt{2}\right)$
7. **(b)** (5, 300°), (−5, 120°) **(c)** $\left(\frac{5}{2}, -\frac{5\sqrt{3}}{2}\right)$
9. **(b)** (−3, 150°), (3, −30°) **(c)** $\left(\frac{3\sqrt{3}}{2}, -\frac{3}{2}\right)$
11. **(b)** $\left(3, \frac{11\pi}{3}\right), \left(-3, \frac{2\pi}{3}\right)$ **(c)** $\left(\frac{3}{2}, -\frac{3\sqrt{3}}{2}\right)$
13. **(b)** $\left(-2, \frac{7\pi}{3}\right), \left(2, \frac{4\pi}{3}\right)$ **(c)** $\left(-1, -\sqrt{3}\right)$

Graphs for Exercises 15(a), 17(a), 19(a), 21(a), 23(a), 25(a)

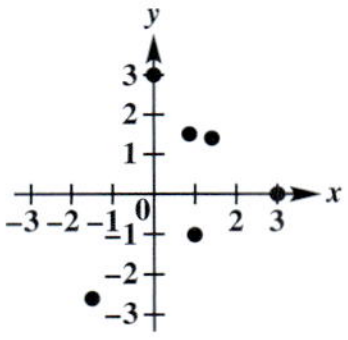

Answers may vary in Exercises 15(b)–25(b).

15. **(b)** $(\sqrt{2}, 315°), (-\sqrt{2}, 135°)$ **17.** **(b)** (3, 90°), (−3, 270°) **19.** **(b)** (2, 45°), (−2, 225°)
21. **(b)** $(\sqrt{3}, 60°), (-\sqrt{3}, 240°)$ **23.** **(b)** (3, 0°), (−3, 180°) **25.** **(b)** (3, 240°), (−3, 60°)

27. $r = \frac{4}{\cos \theta - \sin \theta}$

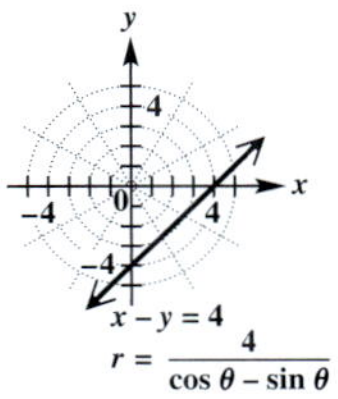

29. $r = 4$ or $r = -4$

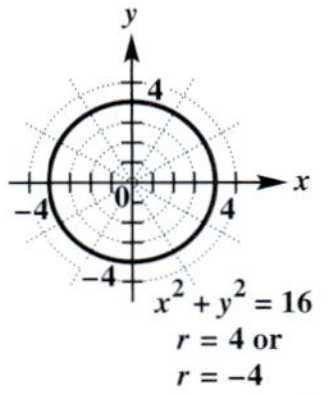

31. $r = \frac{5}{2 \cos \theta + \sin \theta}$

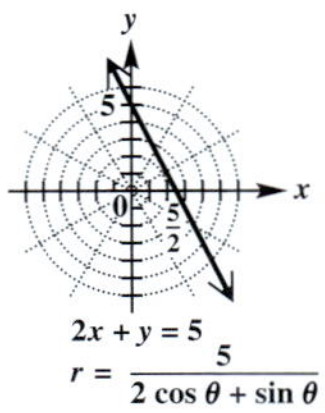

33. $r \sin \theta = k$
34. $r = \frac{k}{\sin \theta}$
35. $r = k \csc \theta$

36. $y = 3$

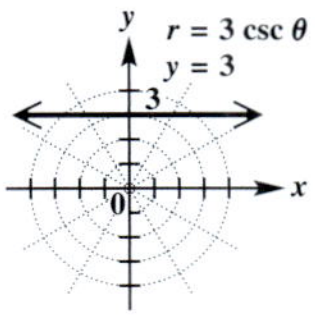

37. $r \cos \theta = k$
38. $r = \frac{k}{\cos \theta}$
39. $r = k \sec \theta$

40. $x = 3$

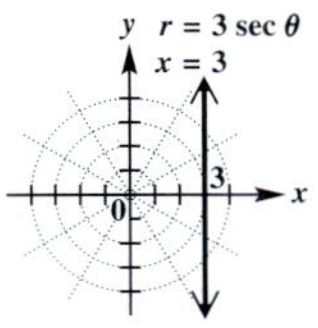

41. C **43.** A **45.** cardioid

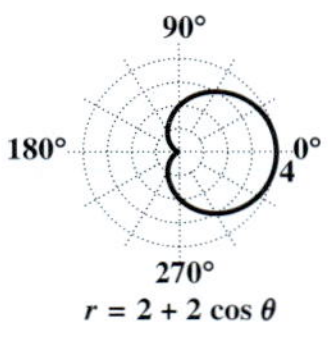

47. limaçon

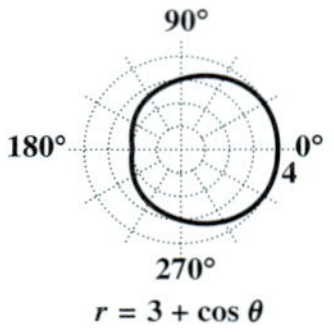

49. four-leaved rose

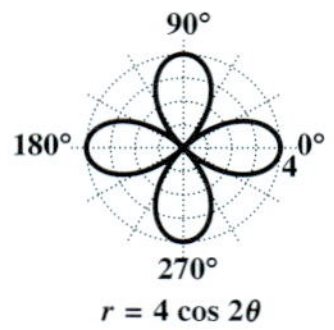

51. lemniscate

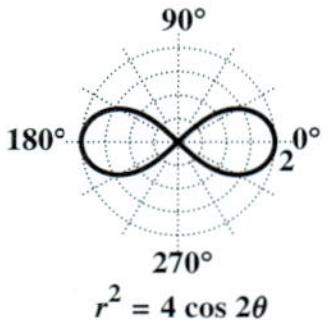

53. cardioid

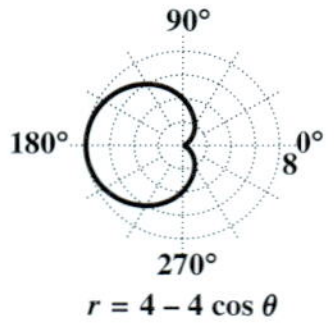

55.

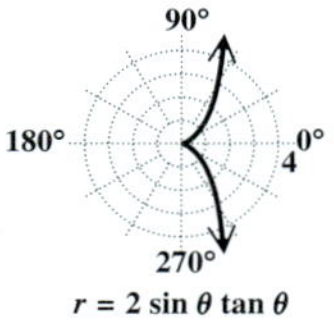

57. $x^2 + (y - 1)^2 = 1$

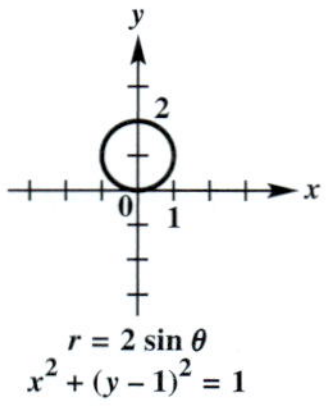

59. $y^2 = 4(x + 1)$

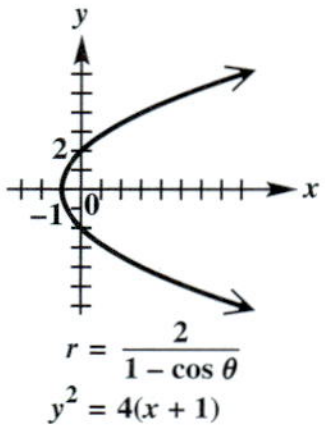

61. $(x + 1)^2 + (y + 1)^2 = 2$

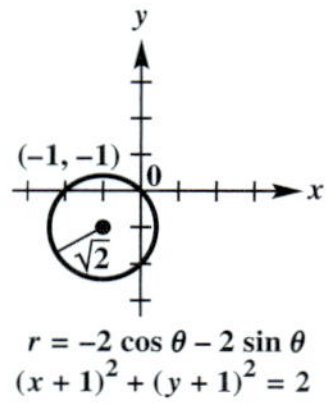

63. $x = 2$

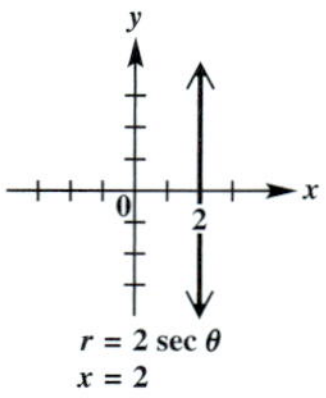

65. $x + y = 2$

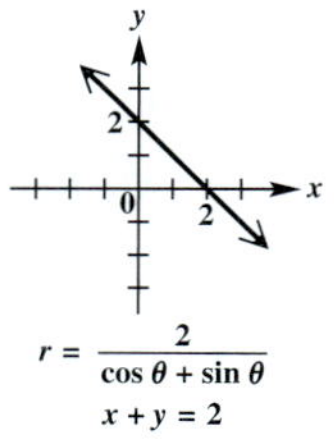

67.

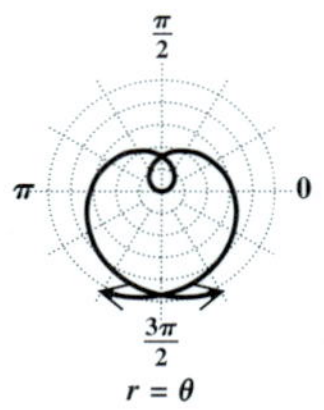

69. $r = \frac{2}{2\cos\theta + \sin\theta}$

71. **(a)** $(r, -\theta)$ **(b)** $(r, \pi - \theta)$ or $(-r, -\theta)$ **(c)** $(r, \pi + \theta)$ or $(-r, \theta)$

73. $r = \theta, 0 \le \theta \le 4\pi$

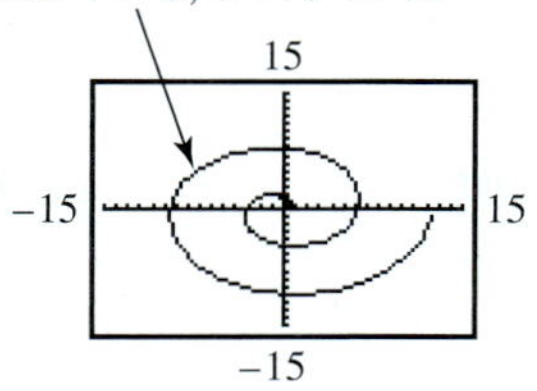

75. $r = 1.5\theta, -4\pi \le \theta \le 4\pi$

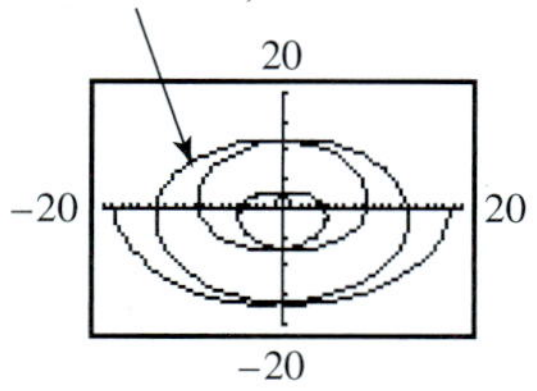

77. $\left(2, \frac{\pi}{6}\right), \left(2, \frac{5\pi}{6}\right), (0, 0)$ **79.** $\left(\frac{4 + \sqrt{2}}{2}, \frac{\pi}{4}\right), \left(\frac{4 - \sqrt{2}}{2}, \frac{5\pi}{4}\right)$

81. **(a)**

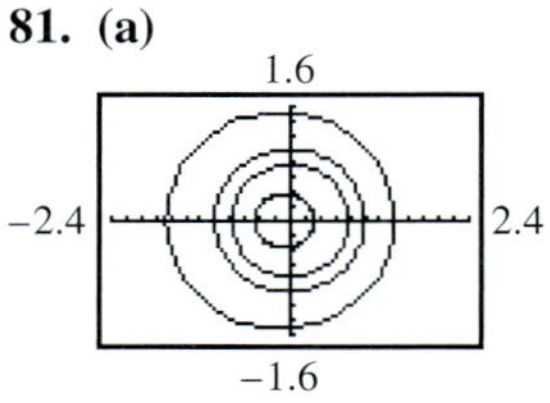

(b)

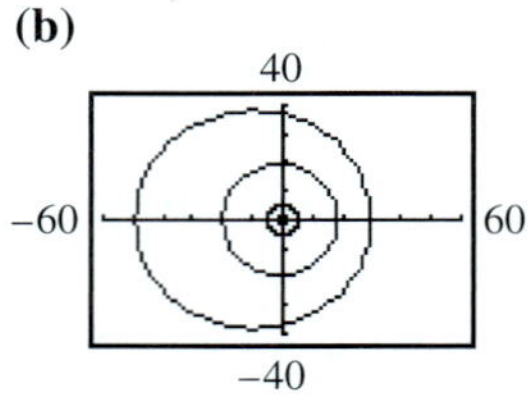

Earth is closest to the sun.

(c) no

8.6 Exercises *(pages 397–400)*

1. C **3.** A

5. **(a)**

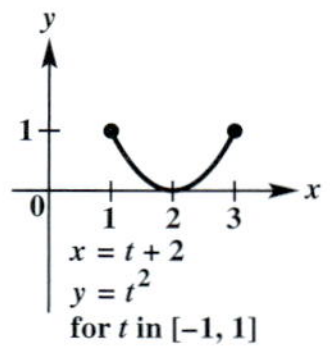

(b) $y = x^2 - 4x + 4$, for x in $[1, 3]$

7. **(a)**

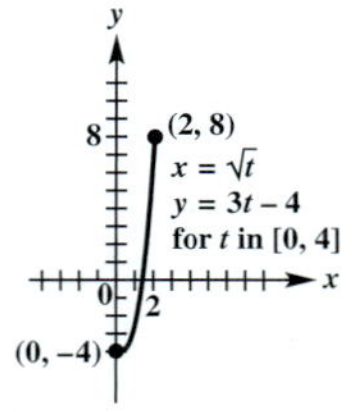

(b) $y = 3x^2 - 4$, for x in $[0, 2]$

9. **(a)**

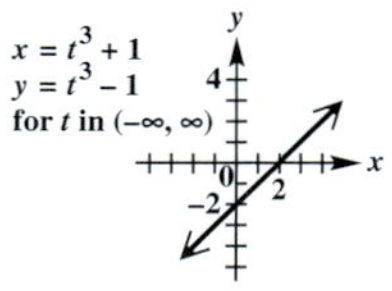

(b) $y = x - 2$, for x in $(-\infty, \infty)$

11. **(a)**

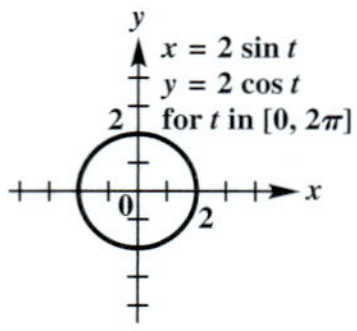

(b) $x^2 + y^2 = 4$, for x in $[-2, 2]$

13. **(a)**

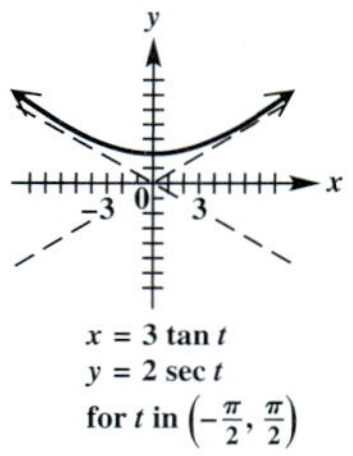

(b) $y = 2\sqrt{1 + \frac{x^2}{9}}$, for x in $(-\infty, \infty)$

15. **(a)**

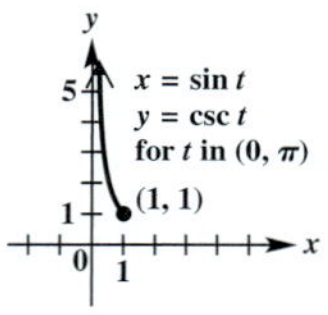

(b) $y = \frac{1}{x}$, for x in $(0, 1]$

17. **(a)**

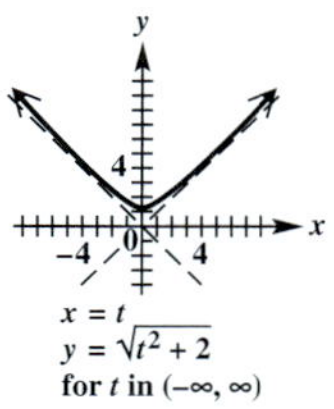

(b) $y = \sqrt{x^2 + 2}$, for x in $(-\infty, \infty)$

19. **(a)**

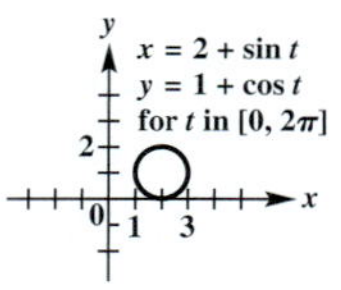

(b) $(x - 2)^2 + (y - 1)^2 = 1$, for x in $[1, 3]$

21. **(a)**

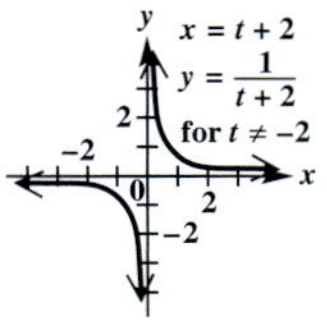

(b) $y = \frac{1}{x}$, for x in $(-\infty, 0) \cup (0, \infty)$

23. **(a)**

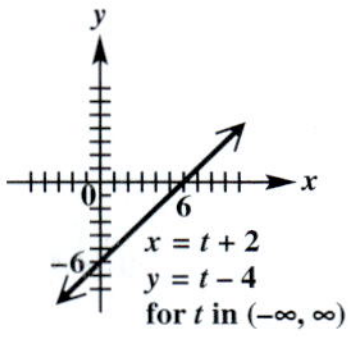

(b) $y = x - 6$, for x in $(-\infty, \infty)$

25.

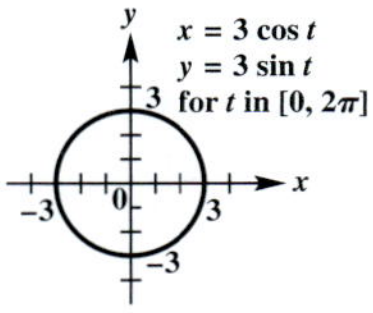

$x^2 + y^2 = 9$

27.

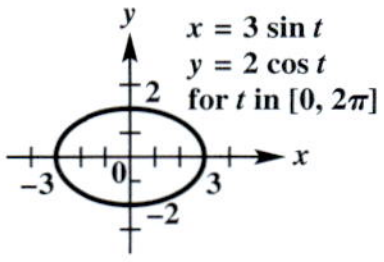

$\frac{x^2}{9} + \frac{y^2}{4} = 1$

Answers may vary for Exercises 29 and 31.

29. $x = t$, $y = (t + 3)^2 - 1$, for t in $(-\infty, \infty)$; $x = t - 3$, $y = t^2 - 1$, for t in $(-\infty, \infty)$ **31.** $x = t$, $y = t^2 - 2t + 3$, for t in $(-\infty, \infty)$; $x = t + 1$, $y = t^2 + 2$, for t in $(-\infty, \infty)$

33.

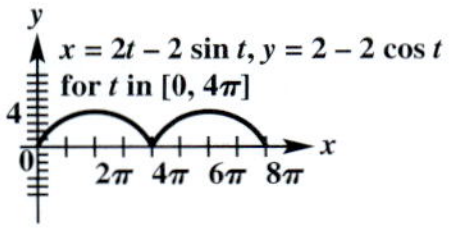

35.

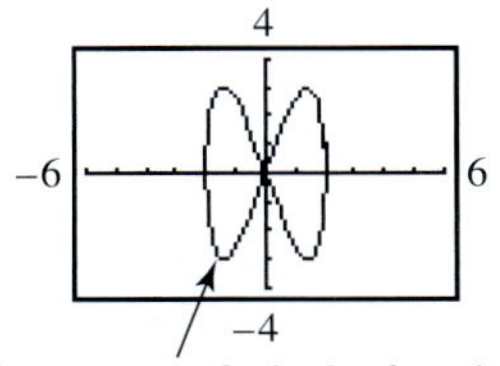

$x = 2 \cos t$, $y = 3 \sin 2t$, for t in $[0, 6.5]$

37.

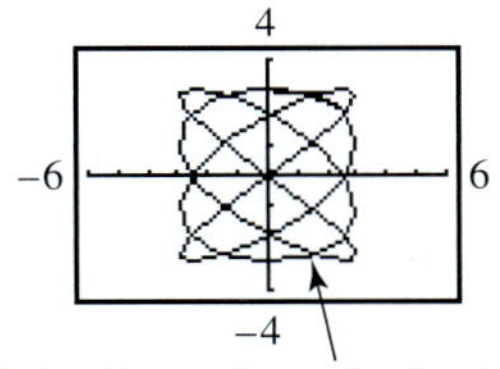

$x = 3 \sin 4t$, $y = 3 \cos 3t$, for t in $[0, 6.5]$

39. **(a)** $x = 24t$, $y = -16t^2 + 24\sqrt{3}t$
(b) $y = -\frac{1}{36}x^2 + \sqrt{3}x$ **(c)** 2.6 sec; 62 ft

41. **(a)** $x = (88 \cos 20°)t$, $y = 2 - 16t^2 + (88 \sin 20°)t$
(b) $y = 2 - \frac{x^2}{484 \cos^2 20°} + (\tan 20°)x$ **(c)** 1.9 sec; 161 ft

43. **(a)** $y = -\frac{1}{256}x^2 + \sqrt{3}x + 8$; parabolic path
(b) approximately 7 sec; approximately 448 ft

45. **(a)** $x = 32t$, $y = 32\sqrt{3}t - 16t^2 + 3$
(b) about 112.6 ft **(c)** 51 ft maximum height; The ball had traveled horizontally about 55.4 ft. **(d)** yes

47. **(a)**
$x = 56.56530965t$
$y = -16t^2 + 67.41191099t$

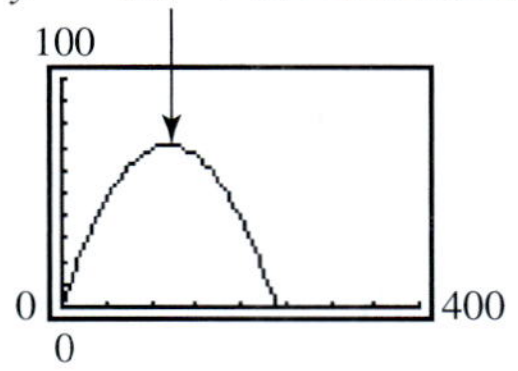

(b) 50.0°
(c) $x = (88 \cos 50.0°)t$, $y = -16t^2 + (88 \sin 50.0°)t$

49. Many answers are possible; for example, $y = a(t - h)^2 + k$, $x = t$ and $y = at^2 + k$, $x = t + h$.

51. Many answers are possible; for example, $x = a \sin t$, $y = b \cos t$ and $x = t$, $y^2 = b^2\left(1 - \frac{t^2}{a^2}\right)$.

55. The graph is translated c units to the right.

Chapter 8 Review Exercises *(pages 404–406)*

1. $3i$ **3.** $\{\pm 9i\}$ **5.** $-2 - 3i$ **7.** $5 + 4i$ **9.** $29 + 37i$
11. $-32 + 24i$ **13.** $-2 - 2i$ **15.** $2 - 5i$ **17.** $-\frac{3}{26} + \frac{11}{26}i$
19. i **21.** $-30i$ **23.** $-\frac{1}{8} + \frac{\sqrt{3}}{8}i$ **25.** $8i$ **27.** $-\frac{1}{2} - \frac{\sqrt{3}}{2}i$

29.

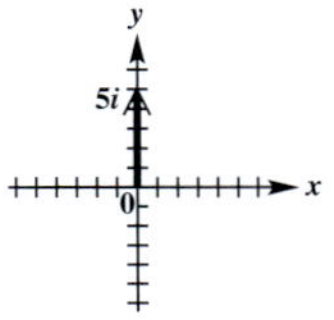

31.

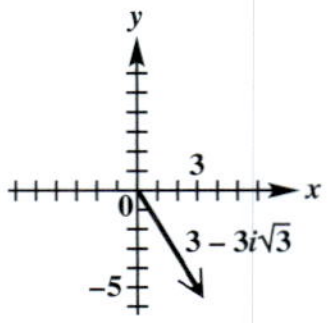

33. $2\sqrt{2}(\cos 135° + i \sin 135°)$ **35.** $-\sqrt{2} - i\sqrt{2}$
37. $\sqrt{2}(\cos 315° + i \sin 315°)$
39. $4(\cos 270° + i \sin 270°)$ **41.** It is the line $y = -x$.
43. $\sqrt[6]{2}(\cos 105° + i \sin 105°)$, $\sqrt[6]{2}(\cos 225° + i \sin 225°)$, $\sqrt[6]{2}(\cos 345° + i \sin 345°)$ **45.** none
47. $\{2(\cos 45° + i \sin 45°), 2(\cos 135° + i \sin 135°), 2(\cos 225° + i \sin 225°), 2(\cos 315° + i \sin 315°)\}$
49. $\{\cos 135° + i \sin 135°, \cos 315° + i \sin 315°\}$
51. $(2, 120°)$

53. circle

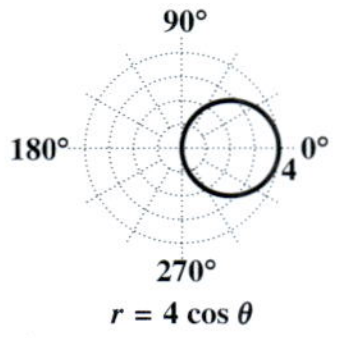

55. eight-leaved rose

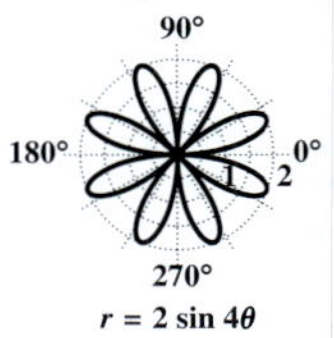

57. $y^2 = -6\left(x - \frac{3}{2}\right)$, or $y^2 + 6x - 9 = 0$
59. $x^2 + y^2 = 4$
61. $r = \tan \theta \sec \theta$, or $r = \frac{\tan \theta}{\cos \theta}$ **63.** B **65.** C
67. $r = 2 \sec \theta$, or $r = \frac{2}{\cos \theta}$ **69.** $r = \frac{4}{\cos \theta + 2 \sin \theta}$

71.

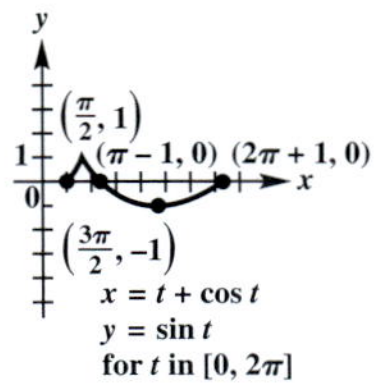

73. $y = \sqrt{x^2 + 1}$, for x in $[0, \infty)$
75. $y = 3\sqrt{1 + \frac{x^2}{25}}$, for x in $(-\infty, \infty)$
77. $y^2 = -\frac{1}{2}(x - 1)$, or $2y^2 + x - 1 = 0$, for x in $[-1, 1]$

79. **(a)** $x = (118 \cos 27°)t$, $y = 3.2 - 16t^2 + (118 \sin 27°)t$ **(b)** $y = 3.2 - \frac{4x^2}{3481 \cos^2 27°} + (\tan 27°)x$ **(c)** 3.4 sec; 358 ft

Chapter 8 Test *(pages 406–407)*

[8.1] **1.** **(a)** $-4\sqrt{3}$ **(b)** $\frac{1}{2}i$ **(c)** $\frac{1}{3}$

[8.1, 8.2] **2.** **(a)** $7 - 3i$ **(b)** $-3 - 5i$ **(c)** $14 - 18i$ **(d)** $\frac{3}{13} - \frac{11}{13}i$

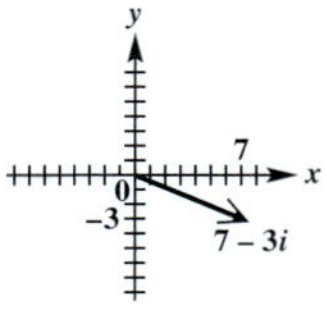

3. **(a)** $-i$ **(b)** $2i$ [8.1] **4.** $\left\{\frac{1}{4} \pm \frac{\sqrt{31}}{4}i\right\}$

[8.2] **5.** **(a)** $3(\cos 90° + i \sin 90°)$ **(b)** $\sqrt{5}$ cis 63.43° **(c)** $2(\cos 240° + i \sin 240°)$ **6.** **(a)** $\frac{3\sqrt{3}}{2} + \frac{3}{2}i$ **(b)** $3.06 + 2.57i$ **(c)** $3i$

[8.3, 8.4] **7.** **(a)** $16(\cos 50° + i \sin 50°)$ **(b)** $2\sqrt{3} + 2i$ **(c)** $4\sqrt{3} + 4i$ [8.4] **8.** 2 cis 67.5°, 2 cis 157.5°, 2 cis 247.5°, 2 cis 337.5° [8.5] **9.** Answers may vary. **(a)** (5, 90°), (5, −270°) **(b)** $(2\sqrt{2}, 225°)$, $(2\sqrt{2}, -135°)$

10. **(a)** $\left(\frac{3\sqrt{2}}{2}, -\frac{3\sqrt{2}}{2}\right)$ **(b)** (0, −4)

11. cardioid

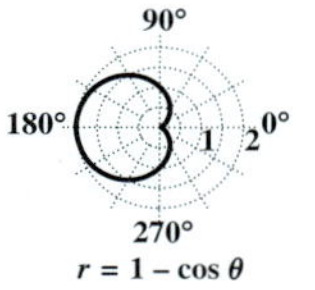

12. three-leaved rose

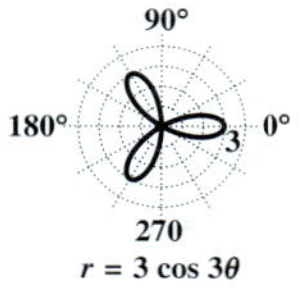

13. **(a)** $x - 2y = -4$

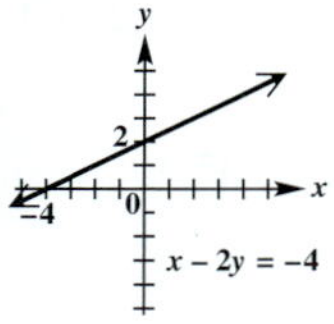

(b) $x^2 + y^2 = 36$

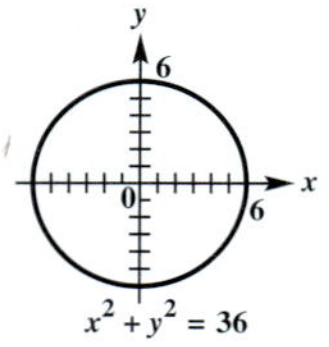

[8.6] **14.**

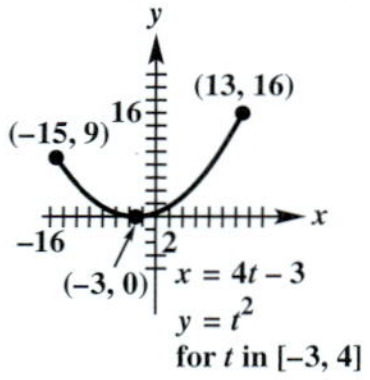

15.

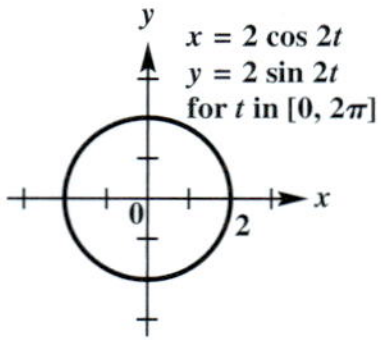

[8.2] **16.** $z^2 - 1 = -1 - 2i$; $r = \sqrt{5}$ and $\sqrt{5} > 2$

Appendices

Appendix A Exercises *(pages 416–418)*

1. true **3.** false **5.** B **7.** {−4} **9.** {1} **11.** $\left\{-\frac{2}{7}\right\}$ **13.** $\left\{-\frac{7}{8}\right\}$ **15.** {−1} **17.** {10} **19.** {75} **21.** {0} **23.** identity; {all real numbers} **25.** conditional equation; {0} **27.** contradiction; ∅ **29.** D; $\left\{\frac{1}{3}, 7\right\}$ **31.** A, B, C, D **33.** {2, 3} **35.** $\left\{-\frac{2}{5}, 1\right\}$ **37.** $\left\{-\frac{3}{4}, 1\right\}$ **39.** {±10} **41.** $\left\{\frac{1}{2}\right\}$ **43.** $\left\{-\frac{3}{5}\right\}$ **45.** {±4} **47.** $\left\{\pm 3\sqrt{3}\right\}$ **49.** $\left\{\frac{1 \pm 2\sqrt{3}}{3}\right\}$ **51.** {1, 3} **53.** $\left\{-\frac{7}{2}, 4\right\}$ **55.** $\left\{1 \pm \sqrt{3}\right\}$ **57.** $\left\{-\frac{5}{2}, 2\right\}$ **59.** $\left\{\frac{2 \pm \sqrt{10}}{2}\right\}$ **61.** $\left\{\frac{-1 \pm \sqrt{97}}{4}\right\}$ **63.** $\left\{\frac{-2 \pm \sqrt{10}}{2}\right\}$ **65.** $\left\{\frac{-3 \pm \sqrt{41}}{8}\right\}$ **67.** F **69.** A **71.** I **73.** B **75.** E **79.** [−4, ∞) **81.** [−1, ∞) **83.** (−∞, ∞) **85.** (−∞, 4) **87.** $\left[-\frac{11}{5}, \infty\right)$ **89.** $\left(-\infty, \frac{48}{7}\right]$ **91.** (−5, 3) **93.** [3, 6] **95.** (4, 6) **97.** [−9, 9] **99.** (−16, 19]

Appendix B Exercises *(pages 424–425)*

1., 3., 5., 7.

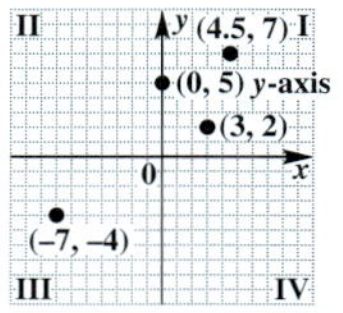

9. **(a)** $\sqrt{34}$ **(b)** $\left(\frac{11}{2}, \frac{7}{2}\right)$ **11.** **(a)** $3\sqrt{41}$ **(b)** $\left(0, \frac{5}{2}\right)$ **13.** **(a)** $\sqrt{133}$ **(b)** $\left(2\sqrt{2}, \frac{3\sqrt{5}}{2}\right)$

15. 25.35%; This is very close to the actual figure of 25.2%.

Other ordered pairs are possible in Exercises 17–27.

17. **(a)**

x	y
0	−2
4	0
2	−1

(b)

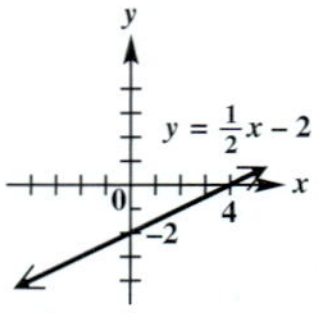

19. **(a)**

x	y
0	$\frac{5}{3}$
$\frac{5}{2}$	0
4	−1

(b)

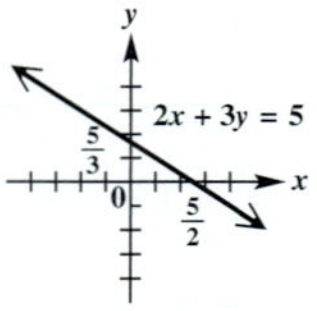

21. **(a)**

x	y
0	0
1	1
−2	4

(b)

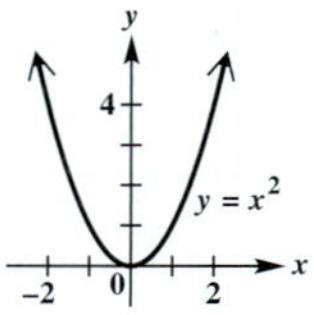

23. **(a)**

x	y
3	0
4	1
7	2

(b)

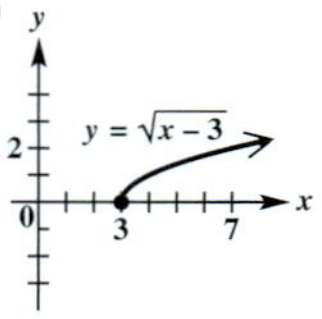

25. **(a)**

x	y
4	2
−2	4
0	2

(b)

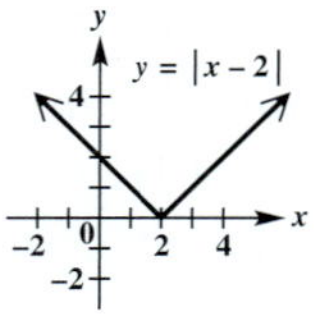

27. **(a)**

x	y
0	0
−1	−1
2	8

(b)

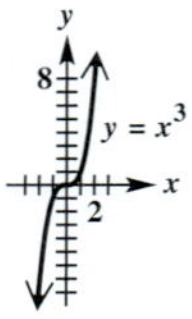

29. **(a)** $x^2 + y^2 = 36$ **(b)**

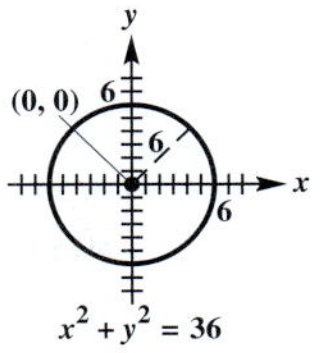

31. **(a)** $(x-2)^2 + y^2 = 36$ **(b)**

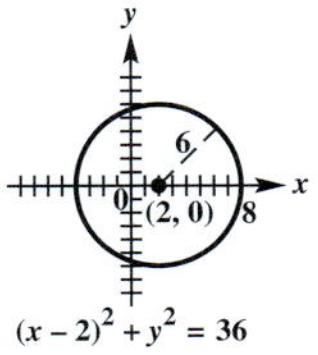

33. **(a)** $x^2 + (y-4)^2 = 16$ **(b)**

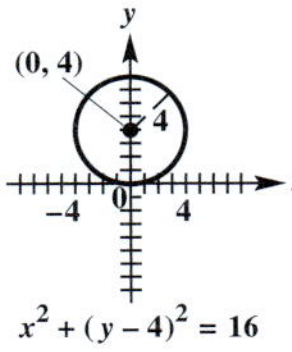

35. **(a)** $(x+2)^2 + (y-5)^2 = 16$ **(b)**

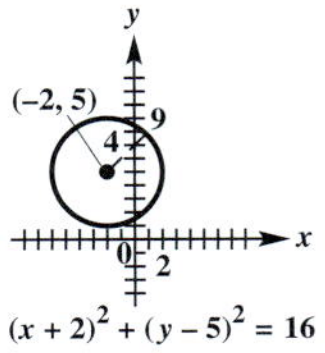

37. **(a)** $(x-5)^2 + (y+4)^2 = 49$ **(b)**

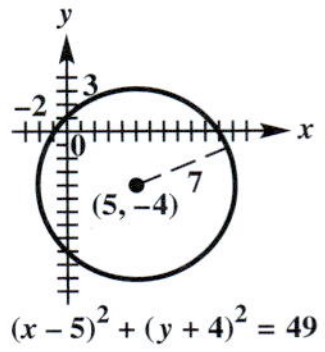

39. **(a)** $\left(x - \sqrt{2}\right)^2 + \left(y - \sqrt{2}\right)^2 = 2$

(b)

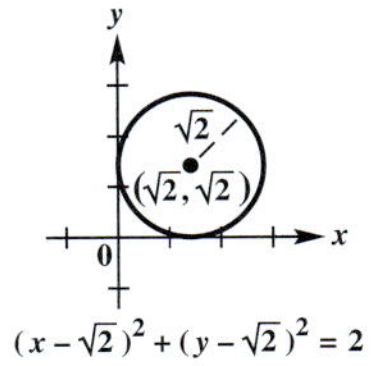

41. $(x-3)^2 + (y-1)^2 = 4$ **43.** $(x+2)^2 + (y-2)^2 = 4$

Appendix C Exercises *(pages 434–436)*

1. function **3.** not a function **5.** function **7.** function **9.** not a function; domain: $\{0, 1, 2\}$; range: $\{-4, -1, 0, 1, 4\}$ **11.** function; domain: $\{2, 3, 5, 11, 17\}$; range: $\{1, 7, 20\}$ **13.** function; domain: $\{0, -1, -2\}$; range: $\{0, 1, 2\}$ **15.** function; domain: $(-\infty, \infty)$; range: $(-\infty, \infty)$ **17.** not a function; domain: $[3, \infty)$; range: $(-\infty, \infty)$ **19.** function; domain: $(-\infty, \infty)$; range: $(-\infty, \infty)$ **21.** function; domain: $(-\infty, \infty)$; range: $[0, \infty)$ **23.** not a function; domain: $[0, \infty)$; range: $(-\infty, \infty)$ **25.** function; domain: $(-\infty, \infty)$; range: $(-\infty, \infty)$ **27.** function; domain: $[0, \infty)$; range: $[0, \infty)$ **29.** function; domain: $(-\infty, 0) \cup (0, \infty)$; range: $(-\infty, 0) \cup (0, \infty)$ **31.** function; domain: $\left[-\frac{1}{4}, \infty\right)$; range: $[0, \infty)$ **33.** function; domain: $(-\infty, 3) \cup (3, \infty)$; range: $(-\infty, 0) \cup (0, \infty)$ **35.** 4 **37.** -11 **39.** 3 **41.** $\frac{11}{4}$ **43.** $-3p + 4$ **45.** $3x + 4$ **47.** $-3x - 2$ **49.** $-6m + 13$ **51.** **(a)** 2 **(b)** 3 **53.** **(a)** 15 **(b)** 10 **55.** **(a)** 3 **(b)** -3 **57.** **(a)** 0 **(b)** 4 **(c)** 2 **(d)** 4 **59.** **(a)** -3 **(b)** -2 **(c)** 0 **(d)** 2 **61.** **(a)** $[4, \infty)$ **(b)** $(-\infty, -1]$ **(c)** $[-1, 4]$ **63.** **(a)** $(-\infty, 4]$ **(b)** $[4, \infty)$ **(c)** none **65.** **(a)** none **(b)** $(-\infty, -2]$; $[3, \infty)$ **(c)** $(-2, 3)$

Appendix D Exercises *(pages 444–448)*

1. **(a)** B **(b)** D **(c)** E **(d)** A **(e)** C **3.** **(a)** B **(b)** A **(c)** G **(d)** C **(e)** F **(f)** D **(g)** H **(h)** E **(i)** I **5.** **(a)** F **(b)** C **(c)** H **(d)** D **(e)** G **(f)** A **(g)** E **(h)** I **(i)** B

7.

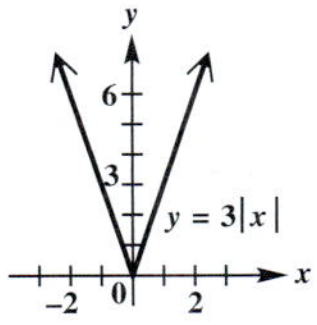

9.

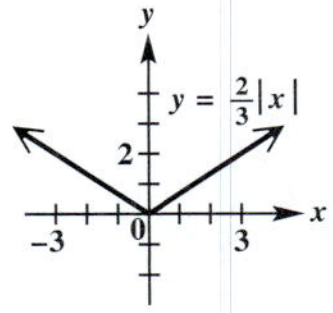

11.

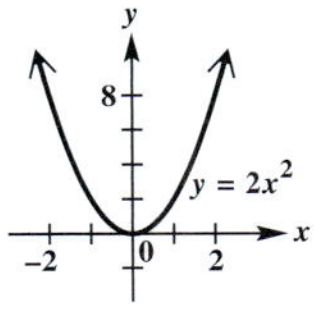

13.

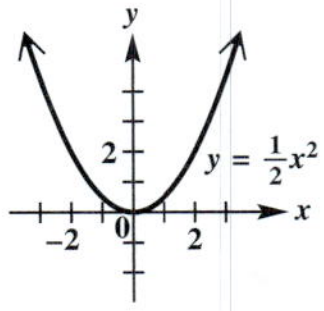

15.

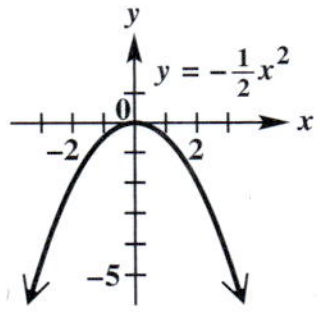

17.

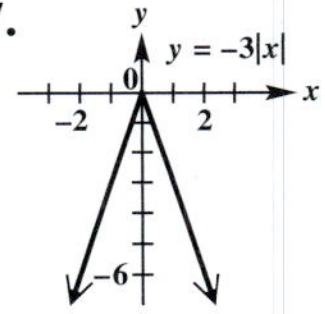

19.

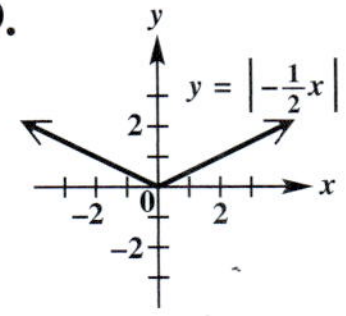

21.

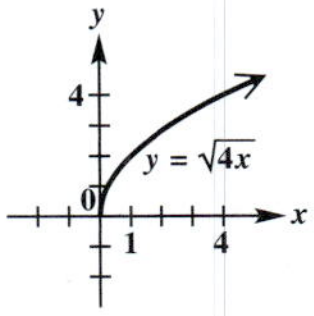

23.

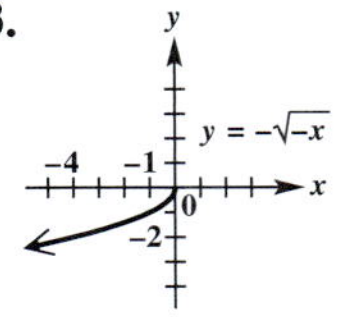

25. **(a)** $(4, 12)$ **(b)** $(8, 16)$ **27.** **(a)** $(2, 12)$ **(b)** $(32, 12)$

29.

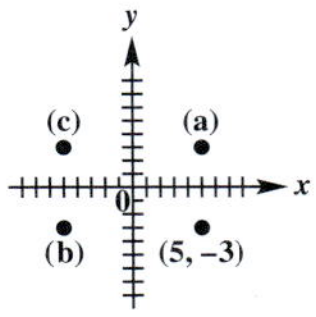

31.

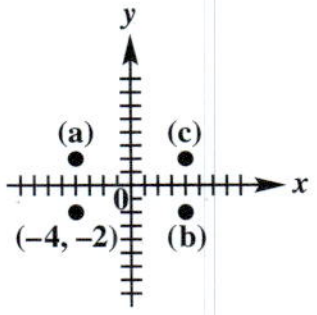

33. y-axis **35.** x-axis, y-axis, origin **37.** origin
39. none of these

41.
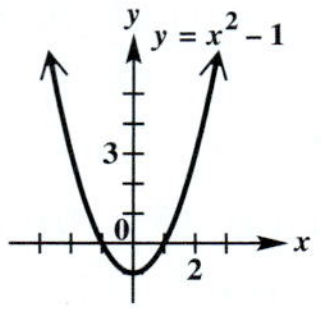

43.
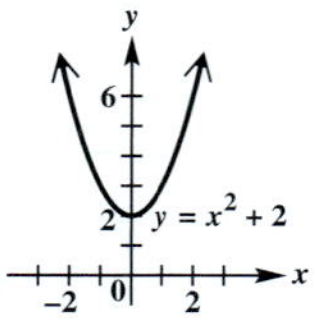

45.
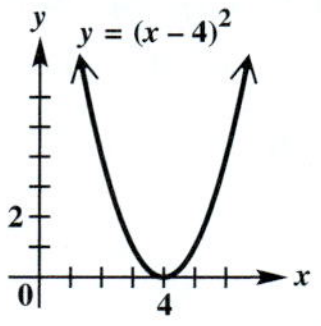

47.
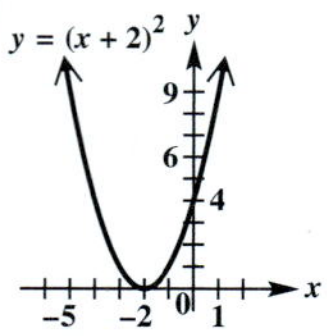

49.
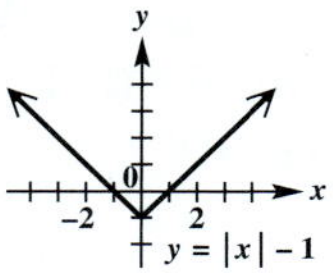

51.
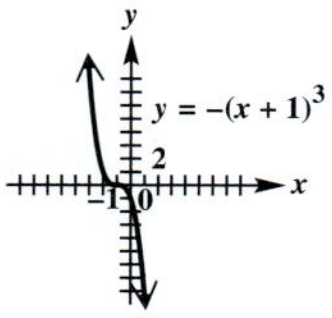

53.
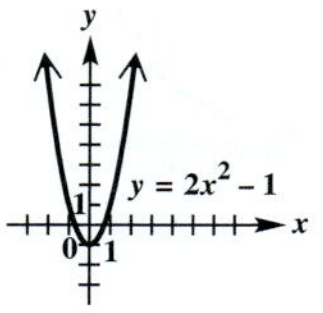

55.
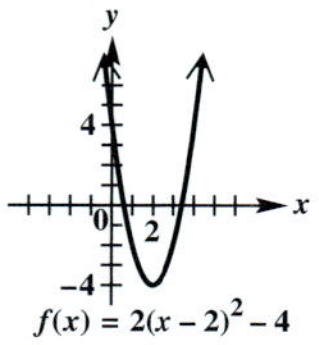

57.
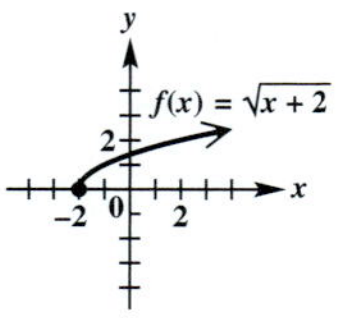

59.
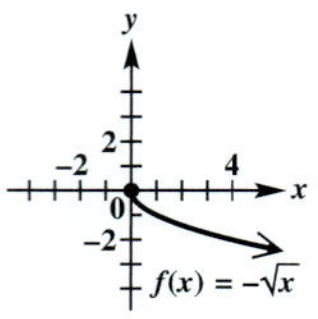

61.
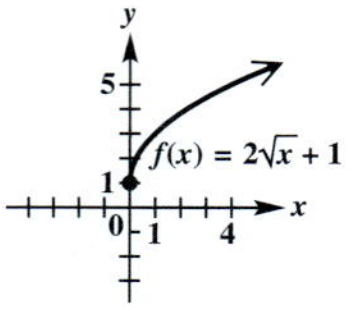

63.
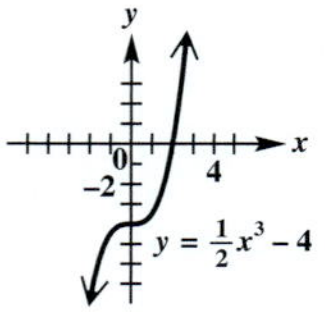

65. It is the graph of $f(x) = |x|$ translated 1 unit to the left, reflected across the x-axis, and translated 3 units up. The equation is $y = -|x + 1| + 3$. **67.** It is the graph of $g(x) = \sqrt{x}$ translated 1 unit to the right and translated 3 units down. The equation is $y = \sqrt{x - 1} - 3$.
69. It is the graph of $g(x) = \sqrt{x}$ translated 4 units to the left, stretched vertically by a factor of 2, and translated 4 units down. The equation is $y = 2\sqrt{x + 4} - 4$.
71. $f(-3) = -6$ **73.** $f(9) = 6$ **75.** $f(-3) = -6$

Photo Credits

page 1: EdBockStock/Shutterstock **page 13:** Pinggr/Shutterstock **page 20:** (sun) Silver Tiger/Shutterstock; (redwoods) Harry Spurling/Corbis **page 45:** Carroteater/Shutterstock **page 64:** VladKol/Shutterstock **page 67:** Manfredxy/Shutterstock **page 71:** AP Images **page 75:** Holbox/Shutterstock **page 85:** Engine Images/Fotolia **page 90:** Goodshoot/Thinkstock **page 93:** Triff/Shutterstock **page 100:** Joseph Kim/Shutterstock **page 103:** Comstock/Thinkstock **page 121:** Pixland/Thinkstock **page 125:** J. Bell, Cornell University/M. Wolff, Space Science Center/NASA **page 130:** Trinacria Photo/Shutterstock **page 133:** Christopher Meder Photography/Shutterstock **page 146:** TextArt/Fotolia **page 153:** Natalia Bratslavsky/Shutterstock **page 158:** Lisa F. Young/Shutterstock **page 159:** Rob Hainer/Shutterstock **page 179:** Alexander Sakhatovsky/Shutterstock **page 186:** Photos/Thinkstock **page 189:** Photos/Thinkstock **page 201:** Paul Matthew Photography/Shutterstock **page 214:** Lifesize/Thinkstock **page 223:** Iconogenic/iStockphoto **page 232:** Digital Vision/Thinkstock **page 238:** Comstock/Thinkstock **page 245:** Blazej Lyjak/Shutterstock **page 257:** Engine Images/Fotolia **page 272:** Ingrid Balabanova/Shutterstock **page 274:** Alenavlad/Shutterstock **page 275:** Lijuan Guo/Shutterstock **page 288:** Pete Saloutos/Shutterstock **page 289:** Kadmy/Fotolia **page 298:** Courtesy of Courtney Bolluyt **page 306:** Shutterstock **page 336:** (pyramids) Sculpies/Shutterstock; (motorboat) Mypokcik/Shutterstock **page 349:** Mishan/Shutterstock **page 363:** Ross Anderson **page 373:** North Wind Picture Archives/Alamy **page 377:** (both fractals) Ross Anderson **page 391:** NatUlrich/Shutterstock **page 406:** Xymm/Fotolia **page 427:** Courtesy of John Hornsby

Index of Applications

Geology

Geometry

Medical

Music

Physics

Sports and Entertainment

Travel

Weather

Index

G

H

I

J

L

M

N

T

U

V

W

X

Y

Z